HISTORY OF AERONAUTICS

HISTORY OF AERONAUTICS

A Selected List of References
to Material in
The New York Public Library

Compiled by WILLIAM B. GAMBLE
Chief of the Science and Technology Division

THE NEW YORK PUBLIC LIBRARY
ASTOR, LENOX AND TILDEN FOUNDATIONS
&
ARNO PRESS INC.
A PUBLISHING AND LIBRARY SERVICE OF THE NEW YORK TIMES
New York

Originally published 1938 by
The New York Public Library

Reprint edition 1971

LC 72-137699
NYPL ISBN 0-87104-503-6
ARNO ISBN 0-405-01747-2

REPRINTED FROM THE
BULLETIN OF THE NEW YORK PUBLIC LIBRARY
OF JANUARY 1936-SEPTEMBER 1937

Manufactured in the United States of America

TABLE OF CONTENTS

Bibliography	1
Traditional and Literary Accounts	3
General Works	7
Lighter-than-air Craft	22
Balloons	22
Airships	43
Heavier-than-air Craft	67
Airplanes and Ornithopters	67
Seaplanes	106
Gliders and Gliding	108
Parachutes	113
Helicopters and Autogiros	115
Kites	119
Engines	122
Propellers	128
Airplane Flights	132
Military Aeronautics	159
World War	172
Civil Aeronautics	189
Air Mail	197
Polar Expeditions	200
Associations and Conferences	205
Jurisprudence	207
Aeronautics and Science	210
Photography and Surveying	232
Animal Flight	237
Women in Aeronautics	242
Miscellaneous	246
Index of Authors	255
Index of Subjects	273
Corrections (1971)	325

HISTORY OF AERONAUTICS

A Selected List of References to Material in The New York Public Library

BIBLIOGRAPHY

1. The **Aeronautic** library. (National Aeronautic Association review, Washington, D. C., 1925, v. 3, p. 127, 159; 1926, v. 4, p. 31, 47, 63, 112, 142.) **VDS**
Extensive list of books, 1685–1793.

2. **Aktiebolaget C. E. Fritzes kungliga hofbokhandel.** Katalog över flyglitteratur 1910 med anledning av Internationella luftfarts utställningen Göteborg 1923. Utgiven av A.-B. C. E. Fritzes kungl. hofbokhandel. Stockholm: C. E. Fritze, 1923. 148, xxvi p. illus. 8°. **VDT**
Extensive classified list covering the entire field of aeronautics from 1910.

3. **Alcanter de Brahm.** La collection Nadar. illus. (L'Aérophile, Paris, 1923, année 31, p. 238–243.) **3–VDS**
Account of the collections of prints, manuscripts, etc., of Félix Tournachon, familiarly known as Nadar. See also entry no. 9.

4. **Baer, Joseph, & Company,** Frankfurt am Main. Flug und Luftschiffahrt in Wort und Bild. Meteorologie. Lagerkatalog 567 zur Internationalen Luftschiffahrt-Ausstellung in Frankfurt am Main herausgegeben im Juli 1909. Frankfurt am Main [1909]. 23 p. illus. 12°. **VDT p.v.2, no.10**
Lists 296 titles of eighteenth and nineteenth century works and illustrations.

5. **Bertùccioli,** Americo. À vol d'avion. Pages de littérature française de l'aviation, avec préface de Henry Bordeaux. Milano: Fratelli Treves, 1926. 269 p. 12°. **VDW**
Extracts from the works of numerous French authors on the development of aerial navigation and on its heroic aspects during the World War.

6. **Boffito,** Giuseppe. Biblioteca aeronautica italiana illustrata. Precede uno studio sull' aeronautica nella letteratura, nell' arte e nel folklore. Firenze: Leo S. Olschki, 1929. cxv, 544 p. 4°. **† VDT**
An extensive bibliography of early and modern Italian aeronautics, with introductory text on aeronautics and aviation in literature, art, and folklore. Reproductions of old illustrations. For a list of Boffito's other writings see p. 74–76.

7. —— Note retrospettive documentaire d'aerotecnica. (L'Aerotecnica, Roma, 1930, v. 10, p. 29–37, 163–170, 287–292, 486–493, 746–748, 850–858; 1931, v. 11, p. 1482–1485.) **VDS**
Supplements the author's *Biblioteca aeronautica italiana.* See entry no. 6.

8. **Brockett,** Paul. Bibliography of aeronautics. Washington: The Smithsonian Institution, 1910. xiv, 940 p. 8°. (Smithsonian miscellaneous collections. v. 55. Publication no. 1920.) **VDT**
Comprises 13,487 author, subject, and title entries, covering the entire field of aeronautics previous to July, 1909. Continued by United States. — National Advisory Committee for Aeronautics, Bibliography of aeronautics (see entry no. 34).

9. **Catalogue** des publications et des manuscrits composant la collection Nadar à la Bibliothèque [Nationale, Paris]. (Paris. — Bibliothèque Historique de la Ville. Bulletin de la Bibliothèque et des travaux historiques, Paris, 1913, tome 6, p. 1–124.) **DO**
Catalogue of the remarkable collection of Félix Tournachon, familiarly known as Nadar.

10. **Darmon,** J. E. Dictionnaire des estampes & livres illustrés sur les ballons & machines volantes des débuts vers 1880, avec leurs prix... Montpellier: H. Barral, 1929. 129 p. illus. 4°. **VDW**

11. **Díaz Arquer,** Graciano, and P. Vindel. Historia bibliográfica de la aeronáutica en España, Portugal, paises hispano-americanos y Filipina desde los orígenes hasta 1900... con un prólogo de Emilio Herrera. Madrid: P. Vindel, 1930. xvii, 63 p. illus. ports. f°. **† VDT**
Reproductions of early Spanish illustrations and title-pages, with descriptive texts.

12. The **Engineering** index. Years 1884–1933. Chicago, 1892; New York, 1896–1934. 8° and 4°. **Room 118**
Changes of title: Descriptive index of current engineering literature, 1884/1891; The Engineering index, 1892–1905; The Engineering index annual, 1906–1918; The Engineering index, 1919–1933.

13. **Fales,** Elisha Noel. Two hundred books on aeronautics. (Aviation, New York, 1917, v. 2, p. 306–308.) **VDS**
Classified list, mostly of contemporary material.

14. **Galloupe,** Francis Ellis. An index to engineering periodicals, comprising engineering; railroads; science; manufactures and trade. Boston, 1888–93. 2 v. 12° and 8°.
v. 1, 1883–1887; v. 2, 1888–1892. **Room 118**

15. **Hagen,** Freiherr vom. Nachweis aëronautischer Litteratur seit 1823. (Zeitschrift für Luftschiffahrt, Berlin, 1882, Jahrg. 1, p. 220–223, 252–255, 280–283, 315–316; 1883, Jahrg. 2, p. 180–183, 340–343; 1884, Jahrg. 3, p. 52–54, 345–347.) **3–VDS**
Indexes books, periodicals, and patents.

Bibliography, continued

16. **Hirschauer,** Auguste Édouard. Les collections de l'aéronautique. illus. (La technique moderne, Paris, 1922, tome 14, p. 12–14.) **VA**

Illustrated description of historical collection installed in the Parc de Chalais-Meudon, Paris.

17. The **Industrial** arts index. Subject index to a selected list of engineering, trade, and business periodicals, books and pamphlets. Annual volumes nos. 1–23 (1913–1933). White Plains, N. Y., 1914–16; New York, 1916–34. 4°. **Room 118**

v. 7, 9, 11, 13, 15, 17, 19 are each a two-years' cumulation.

18. **Jotti,** da Badia, polesine. Bibliofilia. Il giornalismo italiano d'aeronautica. illus. (Aeronautica, Milano, 1927, anno 1, p. 687–697.) **3–VDS**

History of Italian aeronautical periodicals, with reproductions of cover and title-page designs.

19. **Klinckowstroem,** Carl, Graf von. Luftfahrten in der Literatur. illus. (Zeitschrift für Bücherfreunde, Leipzig, 1911/12, N. F. Jahrg. 3, p. 250–264.) **†* GAA**

Brief descriptions of some early German aeronautical works from the fifteenth to the eighteenth century inclusive, with reproductions of pictures. An extension of Minor's list (see entry no. 25).

20. **Landauer,** Bella Clara, and H. B. Weiss. Some early American children's books on aeronautics. illus. (American book collector, Metuchen, N. J., 1932, v. 2, p. 269–274, 334–337, 376.) *** GAA**

21. **Liebmann,** Louis, and Gustav Wahl. Katalog der historischen Abteilung der ersten Internationalen Luftschiffahrts-Ausstellung zu Frankfurt a. M. 1909. Frankfurt a. M.: Wüsten & Co., 1912. xxi, 513 p. illus. plates. 4°. **VDY**

Contains 1,554 titles of books, pictures, maps, and portraits, early and modern, in various libraries of Germany.

22. **Maggs Brothers.** A descriptive catalogue of books and engravings illustrating the evolution of the airship and the aeroplane ... London: Maggs Bros., 1920–30. 3 v. plates. 8°. **VDT**

Catalogue of early and modern works. Reader will find much of interest in the annotations and illustrations.

23. **Maire,** Albert. Aérostation et aviation. Catalogue de la Bibliothèque de l'Université de Paris. (Revue des bibliothèques, Paris, 1910, année 20, p. 233–287.) *** HA**

Covers book titles and some magazine and patent citations, eighteenth to twentieth century. Also published as a separate, *VDT* (*Maire*).

24. **Marsh,** William Lockwood. In search of treasure. Thrills and humours of twenty years spent in collecting an aeronautical library. illus. (Airways, London, 1928, v. 4, p. 367–368.) **VDS**

25. **Minor,** Jacob. Die Luftfahrten in der deutschen Literatur. (Zeitschrift für Bücherfreunde, Leipzig, 1909–10, N. F. Jahrg. 1, Hälfte 1, p. 64–73.) **†* GAA**

For an extension of this list see Klinckowstroem (entry no. 19).

26. **National Advisory Committee for Aeronautics.** Bibliography of publications. (In: United States. — National Advisory Committee for Aeronautics. Bibliography of aeronautics, 1909–30. Washington, 1921–32.) **VDT**

See volumes for 1909/16, p. 937–939; 1917/19, p. 321–326; 1920/21, p. 289–295; 1922, p. 102–114; 1923, p. 91–98; 1924, p. 74–79; 1925, p. 117–126; 1926, p. 94–105; 1927, p. 118–129; 1928, p. 133–144; 1929, p. 151–163; 1930, p. 161–174.

27. The **New York Public Library.** Card index to biographies of aviators and aeronauts, compiled from books, periodicals, and newspapers, by the staff of the Science and Technology Division. **Room 121**

Contains (1933) some 4,000 references.

28. —— List of works in The New York Public Library relating to aeronautics. [New York, 1908.] 16 p. 4°. **VDT**

Reprint from *Bulletin* of The New York Public Library, 1908, v. 12, p. 628–643.

In addition to a list of books in the Library, 1908, there are many references to periodical literature.

29. **Rappaport,** C. E. Aérostation (1670–1890). [Livres et gravures.] Catalogue 22 de la Librairie C. E. Rappaport. Rome: C. E. Rappaport [1912]. 40 p. illus. 12°. **VDY p.v.7**

145 titles of early works.

30. **Royal Meteorological Society.** Index[es] to the Quarterly journal of the Royal Meteorological Society. Vols. viii–xxvi (1882–1900). Vols. xxvii–li (1901–1925). London: Edward Stanton, 1901–28. 2 v. 8°. **PRA**

The Library has the *Quarterly journal,* new series, v. 1 – date (1872 – date), v. 1–2 incomplete. See the indexes under the headings Balloon, Aeronautics, Air, etc.

31. **Squier,** George Owen. Bibliography of aeronautics. (American Society of Mechanical Engineers, Journal, New York, 1908, v. 30, p. 1624–1639.) **VFA**

32. **Tissandier,** Gaston. Bibliographie aéronautique. Catalogue de livres d'histoire, de science, de voyages et de fantaisie, traitant de la navigation aérienne ou des aérostats. Paris: H. Launette et Cie., 1887. 62 p. 4°. **VDT**

Books antedating the discovery of balloons in 1783, history, voyages, periodical publications, almanacs, dramas, poetry, tales, music, and foreign works.

33. **United States.** — Library of Congress. Reports of the Librarian of Congress for the fiscal years 1929/30 – 1933/34. 5 v. 8°. *** SAB**

Contain the reports of the Chief of the Aeronautics Division, Dr. Albert F. Zahm, the first of which contains an interesting account of the Library's important collection and the work of organizing the department.

Bibliography, continued

34. **United States.** — National Advisory Committee for Aeronautics. Bibliography of aeronautics, 1909–1931. Washington: Gov. Prtg. Off., 1921–35. 12 v. 8°. **VDT**

Each volume of this comprehensive compilation comprises references to periodicals, society publications, and books, from 1909, arranged in one alphabet of authors and subjects. It is preceded by Paul Brockett's *Bibliography of aeronautics,* Washington, 1910 (see entry no. 7).

35. **Venturini,** Galileo. ...Da Icaro a Montgolfier... Roma: A. Macioce e Pisani, 1928. 2 v. in 1. plates, port. 4°. **VDW**

Extensive history down to the time of the Montgolfiers, with a bibliography and footnote references. Part 2 contains an aeronautical poem by Bernardo Zamagna (1735–1820).

36. **Zahm,** Albert Francis. The Division of Aeronautics of the Library of Congress. (Mechanical engineering, New York, 1930, v. 52, p. 845.) **VFA**

Account of the various collections of aeronautical works.

TRADITIONAL AND LITERARY ACCOUNTS

37. **Aeligius.** La figurazione delle ali nell' antichità. illus. (Aeronautica, Milano, 1930, anno 4, p. 31–32.) **3–VDS**

Representations of winged animals, mainly Egyptian.

38. Die **Älteste** Darstellung eines fliegenden Menschen. illus. (Deutsche Zeitschrift für Luftschiffahrt, Berlin, 1911, Jahrg. 15, No. 1, p. 24, 28.) **VDS**

Representation of Babylonian brick dating from about 2000 B. C.

39. **Ahmad Zakī.** L'aviation chez les Musulmans: précédé du discours prononcé à la cérémonie d'inauguration du monument Mouillard. Le Caire: Imprimerie Les pyramides, 1912. v, 19 p. 8°. **VDY**

Claims that Ibn Firnas in the ninth century A.D. successfully operated a heavier-than-air machine.

39a. —— L'aviation chez les Arabes. illus. (Institut égyptien, Bulletin, Alexandrie, 1911, série 5, tome 5, p. 92–101.) *** EY**

References to al Djawhari and Ibn Firnas.
Abridged in *Revue aérienne,* Paris, 1912, année 5, p. 93–95, *3–VDS.*

40. **Almanacchi** e aerotecnici del '700. (Rivista aeronautica, Roma, 1931, anno 7, Dec., p. 625–628.) **VDS**

Excerpts from the writings of Jean Jacques Rousseau and of the poet Giovanni Pascoli.

41. **Bertùccioli,** Americo, and B. PINCHETTI. Ali nel cielo. Antologia dell' aviazione. Roma: Libreria del Littorio [1929]. vi, 464 p. 12°. **VDY**

Extracts from the poetry and prose of Italian writers (mainly), covering aeronautics from early times.

42. **Bosis,** Lauro de. Icaro...with a translation from the Italian by Ruth Draper, and a preface by Gilbert Murray. London: Oxford University Press, 1933. x, 201 p. 8°. **NNR**

The Icarus myth done in the manner of a Greek tragedy by an Italian poet who met his death by dropping anti-Mussolini literature over the city of Rome. Library also has a copy of the Italian edition, Milan, 1930, *NNR.*

43. **Bouchot,** Jean. Le vol aérien at la légende. (Mercure de France, Paris, 1912, tome 98, p. 556–567.) *** DM**

Reviews Scandinavian legends.

44. **Cambridge,** Richard Owen. An aerial race between a Briton and a German, together with brief references to a submarine from a mock-heroic poem printed in 1751. London: Printed for H. C. Levis at the Chiswick Press, 1918. 15 p. 4°. **NCC (Cambridge)**

45. —— The scribleriad, an heroic poem in six books. (In: R. O. Cambridge, Works, London, 1803, p. 71–224.) **† NCF**

For account of air race between a Briton and a German, see p. 170–175. For reprint see previous entry.

46. **Cordier,** Henri. Un précurseur dans l'aviation. (Revue hebdomadaire, Paris, 1918, année 27, tome 3, p. 329–332.) *** DM**

The Marquis de Bacqueville's wings and his attempt to fly over the Seine (eighteenth century).

47. **Crispolti,** C. Early gliding experiments. (Aeronautical journal, London, 1909, v. 13, p. 91.) **VDS (Royal)**

Brief account of the experiments of Giovanni Battista Dante, nicknamed Dedalus, translated from Crispolti's *Perugia augusta* (1648).

48. **Cyrano de Bergerac,** Savinien. Les oeuvres diverses de Monsieur de Cyrano Bergerac, enrichi de figures en taille douce. Amsterdam: Daniel Pain, 1699. 2 v. illus. 16°. **NKE**

Includes accounts of his imaginary voyages to the sun and to the moon, describing the various apparatus. Library also has a copy of his *Voyages to the moon and sun,* translated by Richard Aldington, London: Routledge, [1923], 329 p. 12°. *NKT.*

49. **Decremps,** Henri. La magie blanche dévoilée; ou, Explication des tours surprenans qui font depuis peu l'admiration de la capitale & de la province... Paris: F. J. Desoer, 1789. 2 v. illus. 12°. **MZC**

See tome 2, also called supplement, p. 101–112, and frontispiece, for fantastical love story in which three montgolfières in human form are used to frighten African savages.

50. **Douglas,** Norman. A pioneer of aviation. (North American review, New York, 1913, v. 198, p. 100–107.) *** DA**

Interesting account of the famous flying monk of the seventeenth century, St. Joseph of Copertino. He was canonized in 1753.

Traditional and Literary Accounts, continued

51. Dunbar, William. The poems of William Dunbar, edited by John Small. Edinburgh and London: William Blackwood and Sons, 1893. 3 v. 8°. (Scottish Text Society. Publications. nos. 2, 4, 16, 21, 29.) **NDP**

See v. 2, p. 139–143 for the legendary flights of Damian, the abbott of Tungland. Comments in v. 1, p. ccxv.

52. [**Early** French poems on aeronautics.] (Almanach des muses, Paris, 1784–86.) **NKH**

See 1784, p. 25–27, Le globe ascendant, by Gudin de la Brenellerie; p. 87–89, Aux navigateurs aérostatiques, by Comte Raiecki; p. 177–179, Vers à M. Faujas de Saint-Fond sur le premier voyage fait dans les airs, by Gudin de la Brenellerie; p. 194, Vers sur l'expérience de Charles & Robert, by the Comte de Ségur; 1785, p. 51–55, La navigation aérienne, from the Italian of the Abbé Monti; 1786, p. 23–24, Vers adressés à M. de Montgolfier, by de Choisy; p. 189, Vers faits pour la pyramide projettée par le Corps municipal de Calais à la gloire de MM. Blanchard & Jeffries, by De la Place.

53. Farina, Emanuele. Dante and his inferno flight. port. (Aeronautica, Milano, 1927, anno 1, p. 479–480.) **† 3–VDS**

The author's use of the English language is rather amusing.

54. Feldhaus, Franz Maria. Altmeister des Segelfluges; eine Studie... Berlin-Lichterfeld: A. Schultz [1927]. 32 p. illus. port. 8°. **VDY p.v.42, no.4**

A few classical and mediaeval references. Picture of a clay cylinder 2200 B.C. showing flight of Etana to Heaven. The second part of the pamphlet is a biography of Otto Lilienthal.

55. —— Die Darstellung einer Luftfahrt von 1320. (Illustrierte aëronautische Mitteilungen, Berlin, 1909, Jahrg. 13, p. 134–135.) **VDS**

Fanciful air flight of Alexander the Great as depicted in a French manuscript.

56. —— Die Technik der Antike und des Mittelalters. Berlin: Akademische Verlagsgesellschaft [cop. 1931]. 442 p. 4°. **V**

References to ancient and mediaeval facts, myths and traditions, with pictures. Babylonian carving (p. 95); Archytas (p. 129); Daedalus (p. 160); the magician Simon (p. 169); Faust (p. 170); fire balloons (p. 211–212, 247–248, 342, 344, 353–354); Roger Bacon (p. 300); Abbas Ben Firnas (p. 236–237); flight of the unknown Saracen (p. 238); Oliver of Malmesbury (p. 256); Alexander the Great (p. 271); Leonardo da Vinci (p. 390–424). See Tafel 11, opp. p. 320, for reproduction of Walter von Millimete's manuscript (1326) showing hot-air balloon.

57. Fliegende Zauberer in der deutschen Volkssage. (Zeitschrift für Luftschiffahrt, Berlin, 1887, Jahrg. 6, p. 208–213, 241–244.) **3–VDS**

Excerpts from Georg Rudolf Widman's *D. Johannes Faustus,* Hamburg, 1599, in which Faust appears as a magic airman.

58. Flying in the air. illus. (Mechanic's magazine, London, 1823, v. 1, p. 9–11.) **3–VA**

Accounts of Elmer, the Malmesbury monk; the Turkish flier of 1147 A.D.; flight of French monks in Siam; John Baptist Dante, of Perugia; Murray and Degan.

59. Flying machines of 20,000 years ago. (U. S. air services, Washington, D. C., 1933, v. 18, Nov., p. 19.) **VDS**

Legend of Hindu airplane.

60. Fuller, Moyra. The flying man of Florence. illus. (Air, London, 1929, v. 2, p. 75.) **VDS**

Picture and brief description of Giotto's carving on the campanile, Florence, Italy.

61. G., H. Einiges zur Geschichte des Flugwesens. (Deutsche Luftfahrer-Zeitschrift, Berlin, 1915, Jahrg. 19, p. 30, 75–76.) **VDS**

Brief references to mythology and early writings.

62. Geoffrey of Monmouth. The Historia regum Britanniæ of Geoffrey of Monmouth; with contributions to the study of its place in early British history, by Acton Griscom, together with a literal translation of the Welsh manuscript no. LXI of Jesus College, Oxford by Robert Ellis Jones. London: Longmans, Green and Co., 1929. xiii, 672 p. illus. 8°. **CD**

See p. 261–262 for brief reference to King Bladud in Latin and English texts.

63. Greenhill, G. The legend of Alexander and his flying machine. illus. (Nature, London, 1918, v. 101, p. 25–26.) **OA**

Comments on the Roman carving of the twelfth century in the choir of Bâle cathedral.

64. Hagen, Friedrich Heinrich von der. Nordische Heldenromance; übersetzt durch F. H. von der Hagen. Breslau, 1814. 5 v. 16°. **NIDC**

See v. 1, chapter 30, p. 122–127, for Norse version of the Daedalus myth, in which the hero Wieland makes himself a flying shirt of feathers. For reprint of most of the chapter, see *Zeitschrift für Luftschiffahrt,* Berlin, 1893, Jahrg. 12, p. 82–83, *3–VDS.*

65. Hennig, Richard. Beiträge zur Frühgeschichte der Aeronautik. (Verein deutscher Ingenieure, Beiträge zur Geschichte der Technik und Industrie, Berlin, 1918, Bd. 8, p. 100–116.) **VA (Verein)**

References to ancient and mediaeval writings, with comments.

66. —— Zur Vorgeschichte der Luftfahrt. (Verein deutscher Ingenieure, Beiträge zur Geschichte der Technik und Industrie, Berlin, 1928, Bd. 18, p. 87–94.) **VA (Verein)**

Mythology of aeronautics. Aeronautics in mediaeval times. 56 references.

67. Higgins, John. How King Bladud, taking on him to fly, fell upon the Temple of Apollo, and brake his necke, the yeare before Christ, 844. (In: Mirror for magistrates, edited by Joseph Haslewood, London, 1815, v. 1, p. 106–122.) **NCL (Mirror)**

This edition of the *Mirror* is reprinted from that of 1587, collated with those of 1575 and 1610.

68. Jacobi, Max. Humoresken aus der Frühgeschichte der Luftschiffahrt. (Illustrierte aëronautische Mitteilungen, Strassburg, 1903, Jahrg. 7, p. 65–68.) **VDS (Deutsche)**

Brief accounts of some of the early dreams and experiences, notably of Gusmão, Kircher, Gasparus Schott, Besnier, Burattini, Lana, von Guericke, Joseph Galien, Borelli, Georg Paschius, Peter Wilkins, and Cyrano de Bergerac.

Traditional and Literary Accounts, continued

69. **Jacobius,** Helene. Luftschiff und Pegasus. Der Widerhall der Erfindung des Luftballons in der zeitgenössischen Literatur. Halle a. S.: Max Niemeyer, 1909. vii, 130 p. 8°. **VDT**

Selections of early prose and poetry relating to aeronautics in France, Italy, England, Switzerland, and Germany.

70. **Johnson,** Samuel. The history of Rasselas, Prince of Abissinia, a tale. Edited by R. W. Chapman. Oxford: Clarendon Press, 1927. xxi, 221 p. 8°. **NCV**

Chapter 6 of this famous tale, first published in 1759, is a fanciful dissertation on the art of flying. The Library has numerous editions.

71. **Jones,** Ernest Lester. Flight in literature. [Washington, 1925?] 22 p. illus. 8°. **NAC p.v.142, no.7**

72. **Kebra Nagast.** The Queen of Sheba & her only son, Menyelek; being the history of the departure of God & His Ark of the Covenant from Jerusalem to Ethiopia, and the establishment of the religion of the Hebrews & the Solomonic lines of kings in that country. A complete translation...with an introduction by Sir E. A. Wallis Budge. London: The Medici Society, Ltd., 1922. xc, 241 p. front., plates. 8°. *** OEG**

Translation of a sixth-century manuscript, containing (p. 36–37) a reference to "a vessel wherein one could traverse the air (or winds), which Solomon had made by the wisdom that God had given unto him." For a discussion of this reference see *Science*, London, 1923, v. 58, p. 179–180, *OA*.

73. **Klinckowstroem,** Carl, Graf von. Luftfahrten in der Literatur. illus. (Zeitschrift für Bücherfreunde, Leipzig, 1911, N. F. Jahrg. 3, p. 250–264.) *** GAA**

74. **Knolles,** Richard. The Turkish history from the original of the nation, to the growth of the Ottoman empire: with the lives and conquests of their princes and emperors... London: Awnsham Churchill, 1687. 3 v. 6. ed. f°. **† GIB**

See v. 1, p. 26 for account of attempted flight about 1147 A.D. in Constantinople before the Turkish sultan Clisasthlan and the Greek emperor Emanuel.

75. **Krüger,** Herbert. Ein Flugversuch in Byzanz. (Illustrierte aeronautische Mitteilungen, Berlin, 1908, Jahrg. 12, p. 343.) **VDS**

Account of the Saracen's flight in Constantinople (twelfth century).

76. **Laufer,** Berthold. The prehistory of aviation... Chicago, 1928. 96 p. illus. 8°. (Field Museum of Natural History. Publication 253. Anthropological series, v. 18, no. 1.) **QOA (Field)**

The romance of flying in ancient China. Kites as precursors of aeroplanes. The dawn of airships in ancient India. From Babylon and Persia to the Greeks and Arabs. The air mail of ancient times (i. e. carrier pigeons). Library has a second copy, *VDT*.

77. **Lawton,** Harold Walter. Bishop Godwin's *Man in the moone*. (Review of English studies, London, 1931, v. 7, p. 23–55.) **RNA**

First edition of Godwin's tale was published in 1638. The above article describes the work, quoting liberally, and also compares it with a similar work by Cyrano de Bergerac.

78. **Lettre** écrite par un gentil-homme polonais de la ville de Warsovie le 22 décembre 1647 sur une merveilleuse proposition de voler en l'air faite au roy de Pologne. (L'Aérophile, Paris, 1900, année 8, p. 79–80.) **3–VDS**

Informs the king of Poland concerning a marvellous flying machine invented by an Arab.

79. **Levis,** Howard Coppuck. The British king who tried to fly. Extracts from old chronicles and histories relating to Bladud, the ninth king of Britain, together with several portraits. London: Chiswick Press, 1919. viii, 123 p. illus. 4°. **C**

See p. 64–67 for poem (1587) entitled "How King Bladud, taking on him to fly, fell upon the Temple of Apollo, and brake his necke, The yeare before Christ, 844."

80. **Lucian.** The works of Lucian of Samosata. Complete with exceptions specified in the preface. Translated by H. W. Fowler and F. G. Fowler. Oxford: Clarendon Press, 1905. 4 v. 12°. **NSPG**

See v. 3, p. 126 for the amusing account of Icaromenippus, an aerial expedition. The Library has other editions in the original Greek and other translations.

81. **Macgill,** Ronald. Aerial prophecies of the past. The advent of the modern aeroplane is claimed to have been foretold in the Book of Revelations. (Airways, London, 1928–29, v. 5, p. 176, 184.) **VDS**

82. **Marolles,** Michel de. Tableaux du temple des muses tirez du cabinet de feu Mr. Fayereau...avec les descriptions, remarques & annotations composées par Mre Michel de Marolles. A Paris: Chez A. de Sommaville, 1655. 477 p. plates. f°. **3–† MEM (Bloemart)**

Sixty copper engravings by Cornelis Bloemart after paintings by Diepenbeeck. See p. 267–274, with accompanying plate no. 34, for the Icarus legend and quotations from various classical authors. Compare with entry no. 104.

83. **Massip.** Une victime de l'aviation au onzième siècle. (Académie des sciences, inscriptions, & belles-lettres de Toulouse, Mémoires, Toulouse, 1910, série 10, tome 10, p. 199–217.) *** EN**

Discussion of the traditional flight of a Benedictine monk of the abbey of Malmesbury, who was called Egelmer, Elmer, or Oliver.

84. **Mazet,** Horace S. Demi-flights of aerial alchemists. illus. (Sportsman pilot, Albany, N. Y., 1931, v. 5, March, p. 16–17, 45.) **VDS**

Ancient and mediaeval accounts of flying.

85. **Mewes,** Rudolf. Ueber den künstlichen und natürlichen Segelflug. illus. (Zeitschrift für Luftschiffahrt, Berlin, 1887, Jahrg. 6, p. 257–269.) **3–VDS**

Egyptian and Greek conceptions as shown in carvings. Leonardo da Vinci, Dante of Perugia, Degen, and von Helmholtz.

Traditional and Literary Accounts, continued

86. **Moedebeck,** Hermann W. L. Alte Darstellungen fliegender Menschen. illus. (Zeitschrift für Luftschiffahrt, Berlin, 1887, Jahrg. 6, p. 24–25.) **3–VDS**

Reproduction of Egyptian and Javanese figures of flying gods.

87. **Morris,** Ralph. Flying and no failure; or, Aerial transit more than a century ago... Totham: Charles Clark, 1848. 12 l. 8°. **NDH**

Excerpts from the author's *Narrative of the life and astonishing adventures of John Daniel,* 1751. See entry no. 88.

88. —— The life and astonishing adventures of John Daniel. [By Ralph Morris.] London: Robert Holden & Co., Ltd., 1926. xv, 276 p. illus. 8°. **NCT**

Reprint of a fanciful tale first published in 1751. A flying machine figures prominently in the plot.

89. **Munchausen.** The travels and surprising adventures of Baron Munchausen, illustrated with 23 curious engravings, from the Baron's own designs, and five woodcuts by G. Cruikshank. London: William Tegg, 1869. 268 p. illus. 12°. **MEM C 95 MT**

See p. 123–129 for author's flight on the back of an eagle; p. 155–159, for his balloon-borne carriage; p. 260–262, for his success in raising the sunken ship Royal George by means of a balloon. All of these exploits are illustrated.

Among other editions the Library has that of London, 1865, illustrated by Gustave Doré, 4° († *NGV*); and that of London, 1878, illustrated by A. Bichard, 4° († *NGV*).

90. **Nagel,** Robert. Der erste Flieger. (Deutsche Luftfahrer-Zeitschrift, Berlin, 1918, Jahrg. 22, No. 17/18, p. 24.) **VDS**

References to ancient and mediaeval literature.

91. **Narasimm,** V. N. Aviation in ancient Hindustan. (Living age, Boston, 1914, series 7, v. 64, p. 161–166.) *** DA**

Comments on Hindu legendary aviation.

92. **Ovid** und Luftschiffahrt. (Illustrierte aëronautische Mitteilungen, Berlin, 1909, Jahrg. 13, p. 1148–1150.) **VDS**

93. **Paltock,** Robert. The life and adventures of Peter Wilkins, a Cornish man. Relating particularly, his shipwreck near the South Pole; his wonderful passage through a subterraneous cavern into a new kind of world; his there meeting with a gawry, or flying woman...his extraordinary conveyance to the country of glums and gawrys, or men and women that fly...[By Robert Paltock.] Berwick: Printed for W. Phorson, and B. Law, London, 1784. 2 v. 12°. *** KF**

First edition was published in 1751. The Library has several later editions.

94. **Panzer,** Friedrich. Das Flugproblem in Mythus, Sage und Dichtung. (Frankfurt a M. — Internationale Luftschiffahrts-Ausstellung. Denkschrift, Berlin, 1910, v. 1, p. 118–134.) **† VDW**

95. **Patrick,** Mason Mathews. Prophecies of flight. (Aero digest, New York, 1925, v. 7, p. 521–522, 566–567.) **VDS**

Quotations from early romancers.

96. **Poe,** Edgar Allan. The balloon-hoax. (In his: Works, London, 1900, v. 3, p. 239–255.) **NBG**

Fictitious account of Monck Mason's transatlantic flight in a dirigible. Originally appeared in the *New York Evening Sun,* April 14, 1844, of which the Library has a facsimile copy, with editorial comments by Clarence S. Brigham, * *KF* (*1932 Brigham*).

97. —— The unparalleled adventure of one Hans Pfaall. (In his: Works, London, 1900, v. 3, p. 175–238.) **NBG**

Fanciful voyage to the moon in a balloon using a gas whose density is 34.7 less than that of hydrogen.

98. **Rey,** Robert. Aviation et diableries. illus. (La Revue aérienne, Paris, 1913, année 6, p. 228–231.) **3–VDS**

Some of the miraculous stories concerning persons transported through the air.

99. —— Quelques légendes aériennes. illus. (La Revue aérienne, Paris, 1912, année 5, p. 481–483, 631–633.) **3–VDS**

Discusses the legend of Icarus, with reproductions of sculptures.

100. **Rousseau,** Jean Jacques. Le nouveau Dédale; ouvrage inédit. [Edited, with introduction by Pierre Paul Plan.] (Mercure de France, Paris, 1910, v. 87, p. 577–597.) *** DM**

Reprint of pamphlet attributed to Rousseau, copied from a manuscript of 1742. He discusses the possibilities of human flight.

101. **Schumann,** Herbert. Luftfahrwesen in einem Zukunftsroman aus dem Jahre 1819. illus. (Deutsche Luftfahrer-Zeitschrift, Berlin, 1919, Jahrg. 25, No. 11/12, p. 17.) **VDS**

Abstract of romance by Julius von Voss (1819), with picture of an airship drawn by eagles.

102. **Surcouf,** Édouard. L'aérostation au Cambodge six cents ans avant Montgolfier. (L'Aérophile, Paris, 1894, année 2, p. 200–205.) **3–VDS**

Translation of Cambodian text relative to the use of balloons six centuries before the Montgolfiers.

103. **Taylor,** John. All the works of John Taylor, the water poet, being 63 in number; collected into one volume...with sundry new additions, corrected, revised, and newly imprinted. London: James Boler, 1630. var. paged. 4°. **Reserve**

See A memorial of all the English monarchs, p. 271, for portrait of King Bladud, with a verse and a very brief biography.

104. The **Temple** of the muses; or, The principal histories of fabulous antiquity, represented in sixty sculptures; designed and engraved by Bernard Picart le Romain and other celebrated masters. With explications and remarks... Amsterdam: Z. Chatelain, 1733. 142 p. plates. f°. **†† MEBL (Picart)**

Work is somewhat similar to entry no. 82, with new texts and some changes in the illustrations. See p. 92–93 and plate no. 36 for the Icarus myth.

105. **Teutenberg,** A. Goethe und die Luftschiffahrt. (Illustrierte aëronautische Mitteilungen, Berlin, 1908, Jahrg. 12, p. 524–528.) **VDS (Deutsche)**

Extracts from Goethe's writings.

Traditional and Literary Accounts, continued

106. **Thorpe,** W. H. Dr. Johnson on aeronautics. (Engineering, London, 1917, v. 104, p. 695.) **VDA**

Refers briefly to Dr. Johnson's knowledge of aeronautics as shown in his *Rasselas, prince of Abyssinia,* published in 1759. Also in *Scientific American supplement,* New York, 1918, v. 85, p. 99, *VA.*

107. **Verne,** Jules. De la terre à la lune; trajet direct en 97 heures 20 minutes. Paris: J. Hetzet et Cie. [1886.] 169 p. 4°. **† NKV**

Celebrated tale involving a cannon 900 feet long as the propulsive agent of his car. The Library has several editions of the English translation, *NKV.*

108. **Wilhelm,** Balthasar. Schweikart und Mohr, zwei schwäbische Flieger aus alter Zeit. (Illustrierte aëronautische Mitteilungen, Berlin, 1909, Jahrg. 13, p. 441–445.) **VDS**

Account of the miller Schweikart (about 1750) and "the flying monk" Mohr (1575–1625).

109. **William of Malmesbury.** Chronicle of the kings of England from the earliest period to the reign of King Stephen, with notes and illustrations, by G. A. Giles. London: Henry G. Bohn, 1847. xvi, 544 p. 12°. **Stuart 7229**

For brief account of the flight of Elmer the monk see p. 251–252. William of Malmesbury's great history was written originally in Latin during the twelfth century A.D.

110. **Wolberuffener** und Vielbeschreyeter Aero Nauta oder Lufft-Schiffer / das ist neu auffgerichteter Helden-Schatz / gleichsam Rüst- und Zeug Kammer der Lufft / darinnen gar artige und vernünfftige / seltzame jedoch wahrhafftige / beynebens lächerliche jedoch ernste Theologische / Physikalische / Juristische / Medicinische Abhandlungen über allerley Lufft-Fahrten und Lufft-Schifferey ältester / neuerer und jetziger Zeit... Bibliopolae, sub signo Martis et Aprilis. [Leipzig, 1916.] 55, 15 p. illus. 8°. **VDT**

Reprint of a work published about 1750 which deals with the legends of Biblical times, of the Greeks and the Romans, the story of Doctor Faustus and other mediaeval allusions, the writings of John Wilkins, Roger Bacon, and the airships of Lana and Gusmão. Section 2 tells of the exploits of Christopher Wagner in the West Indies and in China.

111. **Wolff,** Maurice. Les romans de l'aviation. (Revue politique et littéraire, Paris, 1912, série 5, année 50, semestre 2, p. 151–153, 180–183.) *** DM**

Refers to some of the imaginative writers — Jules Verne, Cyrano de Bergerac, Bishop Wilkins, Reinser, Voltaire, Fontenelle, Caveau, Rousseau, Robids, Paul Adam, Kistemackers, Danrit, d'Annunzio, Richepin, Mandelstam, Victor Hugo, and Corday.

112. —— Les vieux romans de l'aviation. Un voyage en Utopie ou le monde transformé par l'aviation en 1788. illus. (La Revue aérienne, Paris, 1913, année 6, p. 181–184.) **3–VDS**

Account of book by Reinser entitled *La république universelle, ou l'humanité ailée sous l'empire de la raison,* 1788. A flying machine is illustrated and described.

113–114. **Worcester** (2. marquis), Edward Somerset. The century of inventions of the marquis of Worcester. From the original ms., with historical and explanatory notes and a biographical memoir by Charles F. Partington. London: J. Murray, 1825. lxxxiv, 138 p. nar. 16°. **V**

See p. 69 for oft-quoted passage: "How to make a man to fly; which I have tried with a little boy of ten years old, in a barn, from one end to the other, on a hay-mow." Worcester (1601–1667), however, does not tell how he accomplished the flight. The editor has a note on prime movers for aeronautical purposes.

GENERAL WORKS

115. **Adams,** Darwin J. A brief history of aerostation and aviation. illus. 24 p. (Sportsman pilot, New York, 1930, v. 3, March, section 2, supplement.) **VDS**

Has well-executed balloon and airplane pictures done in color.

116. **Aerial** navigation. illus. (Scientific American, New York, 1869, new series, v. 21, p. 309–310, 325–326, 346–347, 356, 373.) **VA**

General history, with some reference to American developments. Rufus Porter's airship. Kaufmann's flying machine.

117. **Aero Club of America.** A scientific statement of the progress of aëronautical science up to the present time. New York: Doubleday, Page & Company, 1907. xli, 259 p. illus. 12°. **VDW**

Twenty-three chapters, each by an authority.

118. The **Aero Club of America's** exhibit of aeronautical apparatus. illus. (Scientific American, New York, 1906, v. 94, p. 93–94.)

Illustrations of historic interest. **VA**

119. **Aeronautica,** or The history of aviation and aerostation, told in contemporary autograph letters, broadsides, drawings, engravings, manuscripts, newspapers, paintings, posters, press notices, etc. Dating from the year 1557 to 1880. [London, 191–?] 17 p., 12 pl. 4°. **VDT**

Reprinted from the catalogue of J. Pearson & Co.

120. **Aeronautical** expenditures of foreign governments. charts. (Aviation, New York, 1929, v. 27, p. 715–716.) **VDS**

European budgets in 1925–1929.

121. **Aeronautics** in 1908. A forecast of progress. (Scientific American supplement, New York, 1908, v. 65, p. 74.) **VA**

122. **Aeronautics** in 1921. illus. (Engineer, London, 1922, v. 133, p. 18–21.) **VA**

British developments in civil aviation, airplanes, and airships.

123. **Agnoletti,** V. L'areoplano: e la conquista dell' aria. Milano: Società editoriale milanese [1909]. 63 p. 16°. **VDY**

Brief history of aeronautics.

General Works, continued

124. **Aircraft** progress in 1914. illus. (Flight, London, 1914, v. 6, p. 1223–1226.) **VDS**

125. **Alace.** Breve cronistoria aerostatica. (Aeronautica, Milano, 1927, anno 1, p. 265–266.) **3–VDS**

Brief chronology of Italian aeronautics.

126. **Alexander-Katz,** Bruno. Die deutschen Patente über Flugapparate. Vollständige Sammlung deutscher Patente vom Jahre 1879 bis Ende Juni 1911. Systematisch bearbeitet. Berlin: M. Krayn, 1912. iv, 723 p. illus. 8°. *** V**

Classified illustrated abstracts.

127. **American Society of Mechanical Engineers.** Technological progress in 1931. A review of the progress reports of the A. S. M. E. professional divisions... (Mechanical engineering, New York, 1931, v. 53, p. 885–925.) **VFA**

See p. 886–889 for aeronautics.

128. **Archibald,** Douglas. The story of the earth's atmosphere. New York: D. Appleton and Company, 1897. 194 p. illus. 16°. **PRC**

Chapter 13 deals with efforts of man to fly and with the development of kites.

129. **Armengaud,** J., the younger. Discours. (L'Aéronaute, Paris, 1902, année 35, p. 44–57.) **3–VDS**

A history of aeronautics in France as exemplified by the work of prominent members of the Société française de navigation aérienne.

130. **Arnesen,** Odd, and E. O. Sem-Jacobsen. ...Til veirs på norske vinger; av flyvningens historie i Norge. Oslo: Gyldendal norsk forlag, 1930. 229 p. illus. ports. 8°. **VDY**

History of aerial navigation in Norway.

131. The **Arrival** of the airship. illus. (World's work, London, 1908, v. 12, p. 597–610.) *** DA**

How the Wrights learned to fly. Dangers of dirigibles. Capt. Baldwin's airship. Possibilities of the monoplane. Helicopters.

132. **Aus** Russland. (Zeitschrift für Luftschiffahrt, Berlin, 1884, Jahrg. 3, p. 215–220, 244–250.) **3–VDS**

Contemporary news from Russia as reported in the St. Petersburg "Herold."

133. **Aviation** development in Soviet Russia. illus. map. (Aero digest, New York, 1934, v. 24, April, p. 20–24.) **VDS**

Air transportation, non-transport activities, manufacture, stratosphere flights, and military aviation.

134. **Aviation** in 1924. (Flight, London, 1925, v. 17, p. 9–10.) **VDS**

Record covers both heavier-than-air and lighter-than-air craft.

135. **Aviation's** greatest year. illus. (Popular aviation, Chicago, 1928, v. 2, Jan., p. 7–9.) **VDS**

Flights and developments of 1927.

136. **Bacon,** Gertrude. Balloons, airships, and flying machines. New York: Dodd, Mead & Co., 1905. vii, 9–24 p. illus. 12°. (Practical science series.) **VDW**

"Popular" type of history.

137. **Baden-Powell,** Baden Fletcher Smyth. Aerial navigation. (Society of Arts, Journal, London, 1907, v. 55, p. 596–605.) **VDA**

With discussion.

Prognostications. General principles of the airplane. Discussion of some of the recent machines.

138. —— The problem of aerial navigation. (Nineteenth century and after, London, 1908, v. 64, p. 777–785.) *** DA**

A reply to Simon Newcomb. See entry no. 318.

139. —— Recent aeronautical progress, and deductions to be drawn therefrom, regarding the future of aerial navigation. (Aeronautical journal, London, 1903, v. 7, p. 3–8.) **VDS**

Also in *Scientific American supplement,* New York, 1903, v. 56, p. 23310–23312, *VA.*

For a list of writings by and about Baden-Powell see Paul Brockett, *Bibliography of aeronautics,* Washington, 1910, p. 81–84, *VDT.*

140. **Banet-Rivet,** P. L'aviation. (Revue des deux mondes, Paris, 1909, période 5, v. 49, p. 98–130.) *** DM**

Comparison of French and German airships. Discussion of airplane principles, the work of the Wright brothers, and possibilities of the airplane in war.

141. —— Early experiments in air flight. (Popular science monthly, New York, 1900, v. 56, p. 603–607.) *** DA**

Brief history of aviation and aeronautics down to the time of the Montgolfiers.

142. —— La navigation aérienne. illus. (Bulletin technique, Paris, 1899, année 1, no. 3, p. 193–211, no. 6, p. 185–199.) **VDA**

General history, with particular attention to the flying machine of Ader.

143. —— Les nouveaux progrès de la navigation aérienne. (Revue des deux mondes, Paris, 1909, période 5, v. 53, p. 135–159.) *** DM**

Mountain flights of Spelterini; cross-channel flights of Blériot and Latham; and brief account of the *Zeppelin II.*

144. **Bannerman-Phillips,** H. Progress in aeronautics. (United service magazine, London, 1909–20, v. 38–61.) *** DA**

This series, comprising over one hundred installments, is an excellent running account, especially of the war period. See author index in each volume.

145. **Bek,** B. Roustam. Aerial Russia; the romance of the giant aeroplane... London: J. Lane, 1916. vii–xvi, 154 p. illus. ports. 12°. **VDY**

Accounts of Chessborough Mackenzie's and Captain Kostovich's prototype of the Zeppelin. Biography of I. I. Sikorsky. Russia's air service during the World War. Reviewed in *Aviation,* New York, 1916, v. 1, p. 251, *VDS.*

146. **Berget,** Alphonse. Ballons, dirigeables et aéroplanes. Paris: Librairie universelle, 1906. iii, 276 p. 12°. **VDY**

A popular work showing state of aeronautics (1906) with a brief history.

General Works, continued

147. —— The conquest of the air; aeronautics, aviation; history; theory; practice. New York: G. P. Putnam's Sons, 1911. xx, 249 p. illus. new ed. 8°. **VDT**

Part 1, chapter 5: History and description of the principal airships. Part 2, chapter 6: Early days of aviation.

The Library has a copy of the French edition, Paris: Hachette et Cie, 1909, entitled *La route de l'air, VDY.*

148. Berry, William Hibbert. Aircraft in war and commerce, with introduction by Col. Lord Montagu of Beaulieu... Illustrated by drawings by Geoffrey Watson, and photographs. London: E. J. Burrow & Co., Ltd. [1918.] illus. 8°. **VDY**

Airplane development since the Wright Brothers. Advent of the Zeppelin. War-time developments. History and development of the engine. Comparison of the sea plane and the airplane. Air raids during the war.

149. Besançon, Georges. L'année aéronautique à l'Aéro-Club de France. Années 1904–1931. (L'Aérophile, Paris, 1905–32, années 13–42.) **3–VDS**

Aeronautical developments in France for the period covered. Some of the entries are incorporated in the proceedings of the Aero-club of France, also published in *L'Aérophile.* See index under author or Aéro-club

For a list of Besançon's writings on balloons, airships and biography, published principally in the periodical *L'Aérophile,* see Paul Brockett, *Bibliography of aeronautics,* Washington, 1910, p. 121–125, *VDT.*

150. Bescherelle, Louis Nicolas, aîné. L'instruction popularisée par l'illustration. Histoire des ballons et des locomotives aériennes. Paris [1852]. 32 p. illus. 4°. *** C p.v.876**

Small cyclopedia of pioneers, and of early machines and events.

151. Béthuys, G. Chronique de l'Exposition [Paris, 1889]. illus. (Revue de l'aéronautique, Paris, 1889, année 2, p. 80–92, 122–125; 1890, année 3, p. 70–75; 1891, année 4, p. 27–31.) **3–† VDS**

Detailed account of aeronautical exhibits has much of historical interest. See année 2, p. 83 for chronology of hydrogen manufacture.

152. [**Bibliography** of A. von Parseval.] (In: Paul Brockett, Bibliography of aeronautics, Washington, 1910, p. 654–655, *VDT*; continued in United States. — National Advisory Committee for Aeronautics, Bibliography of aeronautics, 1909–31, Washington, 1921–35, v. for 1909/16, p. 1006–1008; 1917/19, p. 345; 1920/21, p. 319; 1922, p. 121; 1923, p. 105; 1924, p. 84; 1927, p. 137; 1928, p. 154; 1931, p. 218. *VDT.*)

Citations to Von Parseval's writings and work on balloons, airships, and airplanes.

[**153. Bibliography** of the Alps and aeronautics.] (United States. — National Advisory Committee for Aeronautics, Bibliography of aeronautics, 1909–31. Washington, 1921–35.) **VDT**

See volume for 1909/16, p. 54; 1920/21, p. 23; 1924, p. 6; 1926, p. 7; 1928, p. 12, for references.

154. [**Bibliography** of Carl Dienstbach.] (In: Paul Brockett, Bibliography of aeronautics, Washington, 1910, p. 243–246.) **VDT**

References to Dienstbach's numerous articles in German and American periodicals.

155. [**Bibliography** of Georges Espitallier.] (In: Paul Brockett, Bibliography of aeronautics, Washington, 1910, p. 284–288.) **VDT**

List of Espitallier's numerous periodical articles on various phases of aeronautics.

156. [**Bibliography** of Victor Silberer.] (In: Paul Brockett, Bibliography of aeronautics, Washington, 1910, p. 770–776.) **VDT**

References to Silberer's numerous writings covering various fields of aeronautics.

157. Big men of the movement. ports. (Aircraft, New York, 1910, v. 1, p. 22–23, 64–65, 108–109, 140–141, 184–185, 250–251.) **VDS**

Biographies and portraits of Hudson Maxim, Cortlandt F. Bishop, Samuel P. Langley, A. Lawrence Rotch, Thomas S. Baldwin, Octave Chanute, James Gordon Bennett, Henry Deutsch, Colgate Hoyt, J. C. McCoy, A. Leo Stevens, Count von Zeppelin, A. B. Lambert, Israel Ludlow, Charles J. Edwards, Felix Tournachon (Nadar), William R. Hearst, Hermann Moedebeck, Robert Esnault-Pelterie, Louis Blériot, Orville Wright, Wilbur Wright, Henry Farman, Glenn H. Curtiss, Clifford B. Harmon, Santos-Dumont, Alexander G. Bell, Hugo C. Gibson, Gabriel Voisin, Horace B. Wild, Gage E. Tarbell, Clement Ader, Sir Hiram Maxim, Charles K. Hamilton, Charles S. Rolls, and Alfred R. Shrigley.

158. Blanchet, Georges. Georges Besançon. port. (L'Aérophile, Paris, 1934, année 42, p. 321–323.) **3–VDS**

Besançon (1886–1934), for many years associated with the Aéro-club de France, was prominent in the development of French aeronautics both as a writer and an experimenter.

159. —— Henry Coxwell. Percy Sinclair Pilcher. (L'Aérophile, Paris, 1900, année 8, p. 5–7.) **3–VDS**

Brief biographies of these two English aeronautical pioneers.

160. Boffito, Giuseppe. L'aeronautica nelle città italiane. Salerno. (Rivista aeronautica, Roma, 1928, anno 4, Oct., p. 181–183.) **VDS**

Stanislao Abate's rocket propulsion (1877); mechanical studies of Vincenzo Fruscione (1875).

161. —— L'aeronautica nelle città italiane. Torino. (Rivista aeronautica, Roma, 1929, anno 5, Aug., p. 421–425, Sept., p. 645–648, Oct., p. 187–189; Dec., p. 603–606; 1930, anno 6, Jan., p. 199–201.) **VDS**

Early balloon ascensions in Turin. Airships of Antonio Comaschi (1841), Vittorio Angius (1854–1857), G. Ossari (1848), Luigi Melano Calcina (1827), and Vincenzo Lanzillo (1875). Flying machine of Carlo Emanuele Boglioni.

162. —— L'aeronautica nelle città italiane. Udine. (Rivista aeronautica, Roma, 1928, anno 4, Aug., p. 447–450.) **VDS**

Balloon studies of Fr. Stella and A. M. Cortenovis (1784). Flying machine of Leonardo Andervolti (1836). Flight theories of Lodovico Lestani (1869) and Virgilio Moretti (1904).

163. —— L'aeronautica nelle città italiane. Venezia. illus. (Rivista aeronautica, Roma, 1928, anno 4, March, p. 605–607; April, p. 198–201; June, p. 613–617; July, p. 209–212.) **VDS**

(1) Parachute of Fausto Veranzio; (2) Tito Livio Burattini (1615–82) construttore d'una macchini volante; (3) La relazione dell' ambasciatore veneto à Parigi sulle prime ascensioni aerostatiche (1873); (4) Pesaro: Il razzo come motore dell' aerostato. Ancora del bombardemento aereo di Venezia nel 1849.

General Works, continued

164. **Bowen,** Robert Sidney. Trends of the industry during 1930. (Aviation, New York, 1931, v. 30, p. 15–17.) **VDS**

165. Les **Brevets** relatifs à l'aéronautique délivrés pendant les dernières années. (L'Aéronaute, Paris, 1876, année 9, p. 280–285; 1877, année 10, p. 60–62, 92–97, 123–126, 152–154, 176–179, 236–238, 263–266, 356–360; 1878, année 11, p. 33–34, 69–70, 104–106, 178, 249–250, 278, 360–362.) **3–VDS**

French patents, 1869–1874.

166. **Brewer,** Griffith, and P. Y. Alexander. Aëronautics: an abridgment of aëronautical specifications filed at the Patent Office from A. D. 1815 to A. D. 1891. London: Taylor and Francis, 1893. vi, 160 p. 8°. **VDY**

Abstracts of British patents, with name and subject indexes.

167. **Brittain,** Sir Harry Ernest. Wings of speed, with an introduction by the Most Hon. the Marquess of Londonderry. London: Hutchinson & Co., Ltd. [1934.] 288 p. illus. 8°. **VDYC**

Brief history of aeronautics. Author's experiences at Croydon. History of the Imperial Airways and the establishment of world routes. Famous flights and fliers. Dangers of desert and mountain flying. Pioneer work of the Royal Air Force. Aerial photography and surveying. Recent technical advances.

168. **Bröckelmann,** Karl, editor. Wir Luftschiffer: die Entwicklung der modernen Luftschifftechnik in Einzeldarstellungen, unter Mitarbeit erster Fachleute... Berlin: Ullstein & Co., 1909. 433 p. map. illus. 8°. **VDW**

Much of historical interest by various authors.

169. **Brown,** Cecil Leonard Morley. The conquest of the air; an historical survey... London: Oxford University Press, 1927. 126 p. illus. ports. 12°. (The world's manuals.) **VDY**

Reviewed in *Aeroplane,* London, 1927, v. 33, p. 382, *VDS.*

170. **Bruel,** François Louis. Histoire aéronautique par les monuments peints, sculptés, dessinés et gravés des origines à 1830. Deux cents reproductions en noir et en couleur; texte par F. L. Bruel. Paris: A. Marty, 1909. 93 p., 151 facsims., 10 ports. f°. **†* KF**

De luxe collection of early plates, caricatures, and manuscripts, with brief descriptive texts.

171. **Bryan,** George Hartley. Progress in aviation. illus. (Nature, London, 1908, v. 78, p. 668–672.) **OA**

Based on notable literature of 1907 and 1908.

172. **Caslant,** Charles Eugène Éloi. Passé et avenir de la navigation aérienne, l'hélicoplane futur. Paris: R. Chapelot et Cie., 1911. iii, 369 p. illus. 8°. **VDT**

Includes well-diagramed histories of the airship and the airplane.

173. **Chanute,** Octave. Aerial navigation. (Popular science monthly, New York, 1904, v. 64, p. 385–393.) *** DA**

Reviews "the present status of the question, the prospects of its solution and the probable uses of the hoped-for airships."

174. —— Aerial navigation. (Smithsonian Institution, Annual report for 1902–1903, Washington, 1904, p. 173–181.) *** EA**

Brief account of development of airships and flying machines.

Also in *Scientific American supplement,* New York, 1904, v. 57, p. 23598–23600, *VA.*

175. —— Aerial navigation. Balloons and flying machines from an engineering standpoint. illus. (Cassier's magazine, New York, 1901, v. 20, p. 111–123.) **VDA**

Review of progress of airships and airplanes.

176. —— Progress in aerial navigation. illus. (Engineering magazine, New York, 1891, v. 2, p. 1–13.) **VDA**

Krebs-Renard airship *La France.* Flying machine experiments of Hureau de Villeneuve, Hargrave, Gustave Trouvé, A. Goupil, Ader, Mouillard, and H. S. Maxim.

177. **Chronology** of some important aeronautical events [for the years 1928 to 1933]. (Aero digest, New York, 1928, v. 13, p. 1144, 1146, 1148; 1929, v. 15, Dec., p. 76, 78, 80; 1930, v. 17, Dec., p. 80, 84, 162; 1931, v. 19, Dec., p. 52, 54, 77; 1932, v. 21, Dec., p. 22–23; 1934, v. 24, Jan., p. 15–16.) **VDS**

178. **Cleveland,** Reginald M. A year of aeronautic progress. (Air law review, New York, 1933, v. 4, p. 1–16.) **3–VDS**

Reviews the year 1932, dealing with passenger and freight transport, military aircraft, engines, private flying, notable flights, safety, and lighter-than-air craft.

179. [**Clippings** from various newspapers, 1858–1880, relating to aërial navigation, astronomy, etc., including articles upon the sun, moon and planets: mounted in album. n. p., 1857–1880.] 127 p. f°. **† VDW**

180. **Cochrane,** Charles H. Recent progress in aerial navigation. illus. (Popular science monthly, New York, 1901, v. 58, p. 616–624.) *** DA**

Airships of Danilevsky, Frederick R. Merritt, Theodore Lieband, Carl E. Myers. Flying machines of Jerome B. Blanchard, Daniel C. Funcheon, Arthur Stentzel, and Beecher Moore.

181. **Colwell,** J. H. The origin and development of aeronautics. (Patent Office Society, Journal, Washington, 1920, v. 3, p. 12–35.) *** V**

Comprehensive but brief review from earliest times to the present by an examiner of the United States Patent Office.

182. **Cressy,** Edward. Discoveries and inventions of the twentieth century. London: George Routledge & Sons, Ltd., 1930. xxi, 476 p. illus. 3. ed. rev. & enl. 8°. **V**

See chapter 17: The conquest of the air.

General Works, continued

183. Crocco, Gaetano Arturo. Il momento aeronautico. illus. (Rivista de artiglieria e genio, Roma, 1906, annata 23, v. 2, p. 371–393.) **VWW**

A review of contemporary developments in various fields of aeronautics.

184. Crosara, Leonardo. . . . Cronologia aeronautica. . . Roma: Alfieri & Lacroix [191–?]. 2 v. illus. 8°. **VDY**

Chronological history, 390 B. C. to 1909 A. D. For review of v. 1, see *Aviation,* Highland, N. Y., 1922, v. 12, p. 635, *VDS.*

185. Crowther, James Gerald. Aviation. (In: The frustration of science, London, 1935, p. 30–41.) **OAP**

"Aviation has been frustrated by its submission to military needs, by possessors of real estate, by nationalistic governments; and its influence on human culture has been frustrated because it has been developed for the use of soldiers and sportsmen, and not for the service of the working civilian."

186. Curry, John F. An outline of aeronautical development. *Aviation's* birthday the zero hour of aeronautics in the United States. (Aviation, Highland, N. Y., 1926, v. 21, p. 174–175.) **VDS**

Comments on poor showing compared with other nations for 10-year period 1916–1926.

187. David, Evan John. Aircraft; its development in war and peace and its commercial future. . . New York: C. Scribner's Sons, 1919. xiv, 379 p. illus. 8°. **VDT**

"Emphasis has been laid on the changes in the design or construction of aeroplanes and dirigibles, which show the evolution of flight and aircraft from early experiments with balloons and gliders to the transatlantic flights of the NC-4, the Vickers 'Vimy' Bomber, and the R-34."

188. Davies, Alec. Air history in postage stamps. illus. (Air and airways, London, 1933, v. 10, p. 70.) **VDS**

Brief account of a set of Latvian stamps.

189. Developments and trends in mechanical engineering. A brief review of science, mechanical technology, and professional development in 1932. illus. (Mechanical engineering, New York, 1932, v. 54, p. 845–858.) **VFA**

For aeronautics see p. 848–849, covering the autogiro, tailless machines, speed, airplane control at low speeds, propellers, lubricants, photography, and motors.

190. Doane, Robert R. Rate of growth of the aircraft industry. charts. (Aviation, New York, 1930, v. 28, p. 755–757.) **VDS**

Statistical study from 1909.

191. Dobias, Frank. The picture book of flying. New York: The Macmillan Company, 1928. 64 p. illus. 4°. **VDY**

Historical picture book in color with brief text.

192. Dollfus, Charles. Histoire de l'aéronautique; texte et documentation de Charles Dollfus & Henri Bouché. Paris: L'Illustration, 1932. xxiii, 569 p. illus. plates (part col'd). f°. **† VDY**

Copiously illustrated, partly in color. "The book is an invaluable work of reference and is worthy of the reputation of the authors." — *Royal Aeronautical Society, Journal, London, 1933, v. 37, p. 205–206, VDS.*

193. Donalies, Hans. Im Fluge durch die Jahrhunderte. illus. (In: Karl Bröckelmann, Wir Luftschiffer, Berlin, 1909, p. 194–238.) **VDW**

Concise, well-illustrated historical account of heavier- and lighter-than-air craft.

194. Draper, O. Griswold. New England's share in the history of aeronautics. (Airscout, New York, 1910, v. 1, no. 2, p. 23.) **3–VDS**

195. Encyclopaedia Britannica. Editions 1, 3, 5–14, 1771–1929. *** AL**

Under the heading *Aerostation,* aeronautics was introduced as a subject in the 3d edition, 1797, with 13 pages of text and some interesting illustrations. This article was revised by Sir John Leslie in the 7th edition (1842) and again in the 9th edition (1875) by James Glaisher, at which time J. Bell Pettigrew introduced the additional heading of Flight and flying machines. Succeeding editions show more or less revision; but not until the 14th edition were any radical changes in text or arrangement made. The present headings are Aviation, aerial law, aerial navigation, aerodromes, aero engines, aeroplane, balloon, balloons in war, flight, and parachute.

196. Die **Eroberung** der Luft; ein Handbuch der Luftschiffahrt und Flugtechnik. Nach den neuesten Erfindungen und Erfahrungen gemeinverständlich dargestellt für alt und jung, mit einem Geleitwort des Grafen Zeppelin. . . Stuttgart: Union deutsche Verlagsgesellschaft [1920]. viii, 402 p. illus. 3. ed., rev. 8°. **VDY**

Well-illustrated chapters by various authors covering early and modern phases. Numerous portraits.

197. Estimated cost of the air service of the United States for six years, including the army, navy, post office, and the National Advisory Committee for Aeronautics, fiscal years 1920–1925 inclusive. (Aviation, Highland, N. Y., 1926, v. 20, p. 108.) **VDS**

198. Les **Étapes** de la navigation aérienne. illus. (La Nature, Paris, 1909, année 37, semestre 2, p. 177–192.) **OA**

Brief history of aeronautics with numerous interesting illustrations.

199. The **Evolution** of the great war-ships of the air: balloons, non-dirigible and dirigible, and aeroplanes — from the seventeenth century to the twentieth century. illus. (Illustrated London news, London, 1909, v. 135, July 3, supplement, p. ii–iii.) *** DA**

Development shown by double-page illustration. Also in *Fly,* Philadelphia, 1909, v. 2, Nov., p. 12–13, *3-† VDS.*

200. Fairey, C. R. Growth of aviation. diagrs. (Royal Aeronautical Society, Journal, London, 1931, v. 35, p. 4–28.) **VDS**

With discussion.

Progress in flight records, 1905–1930; organizations; civil aviation; subsidies; and exports of aircraft and engines.

Abridged in *Flight,* London, 1930, v. 22, p. 1147–1151, 1171–1174, *VDS; Aeroplane,* London, 1930, v. 39, p. 856, 858, 860, 862, *VDS.*

201. Fechet, James Edmond. Flying. New York: The Century Co., 1933. ix, 138 p. port. 12°. (A century of progress series.) **VDY**

A concise history of aeronautics in the first four chapters.

General Works, continued

202. **Feldhaus,** Franz Maria. Luftfahrten einst und jetzt... Berlin: H. Paetel Verlag, 1923. 125 p. illus. 2. rev. ed. 12°. (Sammlung belehrender Unterhaltungsschriften. Bd. 28.) **VDT**

Especially good account of early aeronautics, particularly before the invention of the balloon.

203. —— Ruhmsblätter der Technik von den Urerfindungen bis zur Gegenwart. Leipzig: Fr. Brandstetter, 1924. 2 v. illus. 8°. **V**

See v. 2, p. 1–69 for general history from ancient times.

204. **Ferris,** Richard. How it flies; or, The conquest of the air; the story of man's endeavor to fly and of the inventions by which he has succeeded. Illustrated by over one hundred and fifty half-tones and line drawings, showing the stages of development from the earliest balloon to the latest monoplane and biplane. New York: T. Nelson and Sons [1910]. 11–475 p. 12°. **VDT**

Includes brief biographies of prominent aeronauts, and notable aviation records, 1903–1910.

205. **Fijnje van Salverda,** J. G. W. Aërial navigation, translated from the Dutch by George E. Waring, Jr., with notes concerning some recent developments in the art. New York: D. Appleton & Co., 1894. vi, 209 p. illus. 12°. **VDY**

Author, a distinguished Dutch engineer, reviews the subject from the Montgolfiers down to 1892. He gives a resumé of the ideas of Langley and of John P. Holland.

206. The **First** aeronautical exhibition, June 25 – August 5, 1868. illus. (Aeronautical journal, London, 1910, v. 14, p. 123–124.) **VDS (Royal)**

Brief account of exhibition at Crystal Palace, London, with an interesting engraving. See also *La Nature,* Paris, 1920, année 48, semestre 2, p. 127–128, *OA.*

207. The **Flight** of man. (All the year round, London, 1894, series 3, v. 12, p. 318–323.) *** DA**

Gusmão's airship. The Abbot of Tungland. Professor Ritchell. Sir Hiram Maxim.

208. **Fonvielle,** Wilfrid de. Histoire de la navigation aérienne. Paris: Hachette & Cie., 1907. 270 p. illus. 4°. **VDY**

Illustrated account of the development of balloons and airships.

For numerous citations to the writings of de Fonvielle see Paul Brockett, *Bibliography of aeronautics,* Washington, 1910, p. 324–336, *VDT.*

209. **Foreign** aeronautical activities. (Aviation, New York, 1930, v. 28, p. 600–605.) **VDS**

Expenditures for civil and military aviation, and statistics of operation of civil air lines, extending over a number of years, in various countries of the world are shown by charts.

210. **Fraser,** Chelsea Curtis. Heroes of the air. New York: Thomas Y. Crowell Company [cop. 1927]. xvi, 487 p. illus. ports. 12°. **VDY**

Stories of famous airship and airplane flights (including polar flights), preceded by a brief history of aeronautics.

211. —— The story of aircraft. New York: Thomas Y. Crowell Company [cop. 1933]. ix, 510 p. illus. 8°. **VDY**

Comprehensive general history from the earliest time, with a chronology from 1250 A. D.

212. **French,** Joseph Lewis, editor. The big aviation book for boys. Introduction by Commander Richard E. Byrd. Springfield, Mass.: McLoughlin Bros., Inc. [cop. 1929.] 284 p. illus. 8°. **VDY**

Includes chapters by various authors on history of airships, military balloons, airplanes (especially of the Wright brothers), Rickenbacker and the Flying circus, Lindbergh, the Wilkins Arctic expeditions, the *Bremen's* flight to America, and the Byrd Antarctic expedition.

213. —— Conquerors of the sky... Introduction by Amelia Earhart. Springfield, Mass.: McLoughlin Bros., Inc. [cop. 1932.] 305 p. illus. 8°. **VDY**

Among this miscellany of brief articles are an account of America's first flying sportswoman, Bessica Raiche; Count Zeppelin's visit to America during the Civil War; brief account of Gen. Benjamin D. Foulois; Walter Wellman; and the Post-Gatty flight around the world.

214. **Freyer,** Wilhelm. Das Problem des menschlichen Fluges. (Velhagen & Klasings Monatshefte, Bielefeld, 1902, Jahrg. 16, Bd. 2, p. 625–642.) *** DF**

Brief general history with technical comments.

215. **Frion,** O. Brevets relatifs à l'aéronautique. (L'Aéronaute, Paris, 1869, année 2, p. 167–172, 177–183; 1870, année 3, p. 70–75, 85–96; 1871, année 4, p. 27–29, 105–112.) **3–VDS**

Abstracts of French aeronautical patents of 1867 and 1868.

216. **Fullerton,** J. D. Modern aerial navigation. illus. (Royal United Service Institution, Journal, London, 1892, v. 36, p. 719–749.) **VWA**

With discussion by George Phillips, J. Stringfellow, Baden-Powell, and others.

History of airships and airplanes to date.

217. —— Recent progress in aerial navigation. (Royal United Service Institution, Journal, London, 1907, v. 51, p. 11–54.) **VWA**

With discussion.

A fifteen-year review of airships, airplanes, and motors.

218. **Giacomelli,** R. An historical survey of Italian aeronautics. illus. (Royal Aeronautical Society, Journal, London, 1929, v. 33, p. 947–964.) **VDS**

Brief history from the time of Leonardo da Vinci to the present.

219. **Golding,** Harry, editor. The wonder book of aircraft. London: Ward, Lock & Co., Ltd. [1929.] 256 p. illus. 8°. **VDY**

Copiously illustrated book of the "popular" type, with historical chapters and a chronological record of famous flights.

220. **Goldstrom,** John. A narrative history of aviation. New York: The Macmillan Company, 1930. xii, 319 p. illus. ports. 8°. **VDT**

Contains list of official world records and a classified bibliography.

General Works, continued

221. **Golikere,** R. K. Through wonderlands of the universe. Bombay: D. B. Taraporevala Sons & Co., 1933. xvii, 400 p. 8°. **OMY**

References to notable flights and ascensions.

222. **Grahame-White,** Claude. Flying; an epitome and a forecast... London: Chatto and Windus, 1930. xi, 263 p. illus. ports. 8°. **VDY**

Well-illustrated early history of balloons and airplanes. First engine-driven flights. Cross-channel flights. Aeronautics in the Great War. Civil aviation. Post-war flights. Development of airplane design. Reviewed in *Aeroplane*, London, 1930, v. 39, p. 1218, 1220, *VDS*.

223. **Grahame-White,** Claude, and Harry Harper. Heroes of the air. A book for boys. Illustrated in colour by Cyrus Cuneo and with numerous portraits. London: Henry Frowde, 1912. 272 p. illus. ports. 8°. **VDY**

Careers of Otto Lilienthal, the Wright brothers, Santos-Dumont, Henri Farman, Louis Blériot, Hubert Latham, Louis Paulhan, Claude Grahame-White, and others.

224. **Grand-Carteret,** John, and L. Delteil. La conquête de l'air par l'image (1495–1909). Ascensions célèbres — inventions et projets — portraits... Paris: Librairie des annales, 1909. xiv, 175 p. illus. f°. † **VDT**

A splendid collection of old and modern illustrations, caricatures, and portraits.

225. **Great Britain.** — Patent Office. Abridgments of specifications. Class 4 (aeronautics). London, 1855–date. * **V**

After 1925 this classification becomes Group 33.

226. **Grey,** Charles Grey. 1913. [By Charles G. Grey.] (Aeroplane, London, 1914, v. 6, p. 3–4, 6, 8, 10, 12.) **VDS**

Month by month review of events.

227. —— 1923. (Aeroplane, London, 1924, v. 26, p. 2, 4, 22, 24.) **VDS**

A review of events and developments in British aviation for the year.

228. —— On British progress in 1929. illus. (Aeroplane, London, 1930, v. 38, p. 1–2, 4, 6, 8, 10, 12, 14, 16, 37–38, 40, 42, 44, 46, 48–50, 52, 77–78, 80, 82, 84, 86, 88, 109–110, 112, 114, 116, 118, 120.) **VDS**

229. —— On Canada's part in aviation. illus. (Aeroplane, London, 1927, v. 32, p. 513–515, 518.) **VDS**

230. —— On 1915. (Aeroplane, London, 1915, v. 9, p. 809–810, 814, 816, 818, 820, 822, 824, 826; 1916, v. 10, p. 6, 8, 10, 12, 14, 16.) **VDS**

Month by month account of significant events.

231. —— On 1920. (Aeroplane, London, 1921, v. 20, p. 3–4.) **VDS**

Brief comments on business slump during the year.

232. —— On 1921. (Aeroplane, London, 1922, v. 22, p. 1–2.) **VDS**

Brief review of British events for the year.

233. —— On 1922. (Aeroplane, London, 1923, v. 24, p. 1–2, 4, 21–22, 41–42.) **VDS**

Developments and events of British aviation for the year.

234. —— On 1928. (Aeroplane, London, 1928, v. 35, p. 997–998, 1000, 1002.) **VDS**

Deals mainly with developments in the various British airplane factories.

235. —— On 1932. (Aeroplane, London, 1933, v. 44, p. 2, 4, 6, 8, 10, 45–48.) **VDS**

236. —— On the philosophy of delay. (Aeroplane, London, 1931, v. 41, p. 453–454, 456, 458, 460.) **VDS**

A discussion of the slow progress of aeronautics.

For a list of Grey's writings on aeronautics see United States. — National Advisory Committee for Aeronautics, *Bibliography of aeronautics*, 1916–31, Washington, 1921–35, v. for 1909/16, p. 586; 1917/19, p. 199–200; 1920/21, p. 180; 1922, p. 61; 1923, p. 54; 1924, p. 41–42; 1925, p. 73; 1926, p. 57; 1927, p. 70; 1928, p. 80–82; 1929, p. 94; 1930, p. 94–95; 1931, p. 114, *VDT*.

237. **Günther,** S. Die Vorgeschichte der Luftschiffahrt. (Zeitschrift für Luftschiffahrt, Berlin, 1892, Jahrg. 11, p. 93–99.)

History before 1800. **3–VDS**

238. **Guggenheim,** Harry Frank. The seven skies. New York: G. P. Putnam's Sons, 1930. 216 p. illus. 8°. **VDY**

See chapter 1: Aviation's first one hundred years; and chapter 2: The significance of the Lindbergh flight. Reviewed in *Aviation*, New York, 1930, v. 29, p. 253, *VDS*; *Aeroplane*, London, 1930, v. 39, p. 604, *VDS*.

239. **Gulf Refining Company.** Aviation atlas ... 2. ed. Pittsburgh, Penn.: Gulf Refining Company, 1931. 20 p. illus. ports., maps. f°. † **VDY**

Photographs of pioneers, prominent fliers, famous aircraft, chronology from early times, historical flight map, transport routes, and statistics of development. Library has also copy of the 3. ed., Pittsburgh, Penn., 1933.

240. **Hagen,** Freiherr vom. Vorläufer der Gebrüder Montgolfier. (Zeitschrift für Luftschiffahrt, Berlin, 1883, Jahrg. 2, p. 78–85.) **3–VDS**

Brief accounts of Magnus Pegelius, Gaspar Schott, Joseph Galien, Franz Lana, Professor Lohmeier, Tiberius Cavallo, and Gusmão.

For a list of vom Hagen's writings, many of which are of historical interest, see Paul Brockett, *Bibliography of aeronautics*, Washington, 1910, p. 389–391, *VDT*.

241. **Harney,** Laura B. The skycraft book, with a foreword by Roland H. Spaulding. Boston: D. C. Heath & Co. [cop. 1932.] vi, 338 p. illus. 12°. **VDY**

Has much of historical interest for boys and girls.

242. **Harper,** Harry. The evolution of the flying machine; balloon: airship: aeroplane ... London: Hutchinson & Co., Ltd. [1930.] 288 p. illus. 8°. **VDY**

Comprehensive, with good illustrations. Chronology 400 B.C. to 1929 A.D. Reviewed in *Flight*, London, 1930, v. 22, p. 361, *VDS*.

243. **Hartwig,** Georg Ludwig. The aerial world: a popular account of the phenomena and life of the atmosphere. London: Longmans, Green, and Co., 1877. xviii, 556 p. illus. 8°. **PRB**

See chapters 32–37 for history of aeronautics.

General Works, continued

244. **Hearne,** R. P. The second Paris aeronautical salon. An analysis of the types: — monoplanes, biplanes, dirigibles, and engines. illus. (Scientific American supplement, New York, 1909, v. 68, p. 280–283, 294–295, 318–319.) **VA**

Pictures include the car of the Blanchard balloon, and a view of the old balloon *Volta,* used for scientific purposes in 1870.

245. **Hildebrandt,** Alfred Louis Heinrich. Airships past and present, together with chapters on the use of balloons in connection with meteorology, photography and the carrier pigeon... Translated by W. H. Story. New York: D. Van Nostrand Company, 1908. xvi, 364 p. illus. 8°. **3–VDW**

Well-illustrated accounts of development of balloons, airships, airplanes, parachutes, kites, military ballooning, and scientific ascensions.

246. —— Flugmaschinen und Lenkballons. illus. (Glasers Annalen für Gewerbe und Bauwesen, Berlin, 1908, Bd. 62, p. 220–227, 239–247.) **3–† VDA**

Illustrated history of aeronautics.

247. —— Die Luftschiffahrt nach ihrer geschichtlichen und gegenwärtigen Entwicklung... München: R. Oldenbourg, 1907. v, 426 p. illus. 8°. **VDW**

Includes also history of kites, parachutes, military aeronautics, aerial scientific investigations, and aerial photography.

248. **Historical** design section. illus. (Aircraft year book, New York, 1921, p. 172–175; 1922, p. 121–159; 1923, p. 211–266; 1924, p. 259–288; 1925, p. 217–254; 1926, p. 219–268; 1927, p. 247–300; 1928, p. 330–411; 1929, p. 248–355; 1930, p. 322–469; 1931, p. 300–454; 1932, p. 321–414.) **VDS**

A series of drawings of airplanes, airships, and engines from time of the Wright brothers to the present. Title varies.

249. **Historique** de l'aéronautique. (L'année aéronautique 1925/1926, Paris, 1926, année 7, p. 285–307.) **VDS**

Covers significant dates for free balloon flights, 1783–1919; airships, 1852–1919; of aviation, 1903–1919; speed, altitude, duration, and distance records, 1906–1926; and notable voyages, 1919–1926.

250. **History** of air navigation. (Congressional digest, Washington, D. C., 1925, v. 4, p. 219–221, 249.) **† SEA**

Brief outline of heavier-than-air and lighter-than-air craft, with official world records as of April 1, 1925.

251. **History** and progress of aeronautics. Air Ministry and Royal Aeronautical exhibits at Olympia. illus. (Flight, London, 1929, v. 21, p. 821–825.) **VDS**

Exhibition of scale models shows development from the flying boat of Francesco Lana (1670) to the present time.

252. **Hodgins,** Eric, and F. A. Magoun. Sky high; the story of aviation... Boston: Little, Brown and Company, 1929. xx, 337 p. ports. illus. 8°. **VDY**

Well-illustrated account from legendary attempts down to the autogiro.

253. **Hodgson,** J. E. The history of aeronautics in Great Britain from the earliest times to the latter half of the nineteenth century, with 150 illustrations from contemporary sources, chronology... London: H. Milford, 1924. xxii, 436 p. illus. 4°. **VDT**

Also includes an excellent history of early aeronautics in other countries. Bibliographies. Has a good index of names and subjects.

254. —— Notes on Sir George Cayley as a pioneer of aeronautics. port. (Newcomen Society, Transactions. London, 1924, v. 3, 1922/23, p. 69–89.) **VA**

With discussion. For portrait see plate 15 at back of volume.
Cayley (1773–1857) designed airships and experimented with mechanical flight.
See also *Aviation,* Highland, N. Y., 1923, v. 15, p. 744–747, *VDS.*

255. **Hoernes,** Hermann. Buch des Fluges. Unter Mitwirkung von K. K. Rat von Angeli, Baron Berlepsch, Prof. Berson [und anderes], hrsg. von H. Hoernes. Buchschmuck von Tanna Hoernes. Wien: G. Szelinski, 1911. 3 v. illus. 4°. **VDW**

Contains chapters on the evolution of airships, models, gliders, military aeronautics, biographies of pioneers in various countries, and records of fatalities. See index in each volume.

256. **Hof,** J. J. De geschiedenis der luchtscheepvaart. Gent: A. Hoste, 1911. 182 p. illus. 12°. (Willems-Fonds. Uitgaven. no. 152.) **VDY**

Comprehensive illustrated account of aeronautical history from early times.

257. **Holland,** Rupert Sargent. Historic airships...illustrated by Manning de V. Lee. Philadelphia: Macrae Smith Company [cop. 1928]. 343 p. illus. 8°. **VDT**

The high spots in the history of airplanes, balloons, and airships.

258. **Hunsaker,** Jerome Clarke. The story of transportation by air. (Aviation, Highland, N. Y., 1923, v. 14, p. 215–218.) **VDS**

History of aeronautics from 1783.
For a list of Hunsaker's writings see United States. — National Advisory Committee for Aeronautics, *Bibliography of aeronautics,* 1909–31, Washington, 1921–35, volumes for 1909/16, p. 657–658; 1917/19, p. 225–226; 1920/21, p. 205–206; 1923, p. 61; 1924, p. 48; 1925, p. 81, *VDT.*

259. **Hureau de Villeneuve,** Abel. Les progrès accomplis en navigation aérienne [1875]. (L'Aéronaute, Paris, 1876, année 9, p. 16–24.) **3–VDS**

260. —— Le progrès de la navigation aérienne. (L'Aéronaute, Paris, 1875, année 8, p. 16–28.) **3–VDS**

Events of 1873 and 1874.

261. —— Le tableau d'aviation de M. Dieuaide. illus. (L'Aéronaute, Paris, 1880, année 13, p. 241–245.) **3–VDS**

Shows some of the pictures from the author's Dieuaide's collection: — The mechanical birds of Gérard, Degan, Esterno, and Lebris; Henson's airplane; and de Letur's flying machine.
For citations to Hureau de Villeneuve's articles on the various phases of aeronautics see Paul Brockett, *Bibliography of aeronautics,* Washington, 1910, p. 859–861, *VDT.*

General Works, continued

262. **Huth,** Fritz. Luftfahrzeugbau. Konstruktion von Luftschiffen und Flugmaschinen. Berlin: M. Krayn, 1909. 10–240 p. illus. 8°. **VDY**

Descriptions of contemporary airplanes and airships. Has chronological lists of prominent craft, 1842–1897, with brief characterization of each.

263. **Hylander,** Clarence John. Cruisers of the air. The story of lighter-than-air craft from the days of Roger Bacon to the making of the ZRS-4. Illustrated with many photographs and with drawings and diagrams by the author. New York: The Macmillan Company, 1931. xxiv, 308 p. illus. 12°. **VDW**

Reviewed in *Aviation*, New York, 1931, v. 30, p. 553, *VDS*.

264. **Important** events in the history of heavier-than-air flying machines. (Aircraft year book, New York, 1919, p. 310–314; 1920, p. 248–258; 1921, p. 107–138; 1922, p. 232–243; 1923, p. 376–386; 1924, p. 221–242; 1925, p. 168–196; 1926, p. 117–136; 1927, p. 175–204; 1928, p. 245–253; 1929, p. 155–187; 1930, p. 475–480; 1931, p. 455–460; 1932, p. 485–515.) **VDS**

Chronologies. After the first installment lighter-than-air craft are included. Title varies.

265. **Jackson,** George Gibbard. The world's aeroplanes and airships... London: S. Low, Marston & Co., Ltd. [1930.] xii, 244 p. illus. 8°. **VDY**

"Popular" work with much of historic interest, including notable flights.

266. **Janssen,** Pierre Jules César. Les progrès de l'aéronautique. (Revue scientifique, Paris, 1900, série 4, tome 14, p. 385–389.) **OA**

Reviews ten years' developments. English translation in Smithsonian Institution, *Annual report for 1899–1900*, Washington, 1901, p. 187–193.

267. **Jauneaud,** Marcel. ...L'évolution de l'aéronautique. Paris: E. Flammarion, 1923. 284 p. illus. 12°. (Bibliothèque de philosophie scientifique.) **VDT**

Concise history from beginnings down to and through the World War.

268. **Johns,** William Earl. Some milestones in aviation. London: J. Hamilton, Ltd. [1935.] 232 p. illus. 12°. **VDY**

Covers balloons and airplanes.

269. **Jones,** Ernest Lester. Believe it or not of aeronautics. (Aero digest, New York, 1930, v. 17, Sept., p. 48, 188, 190, 192, 194.) **VDS**

Extensive chronology of aeronautics from Roger Bacon to 1914.

270. **Jotti,** da Badia, polesine. Documenti poco noti di storia aeronautica. illus. (Aeronautica, Milano, 1928, anno 2, p. 951–956.) **3–VDS**

Copies of interesting old prints of airships and balloons.

271. **Jünger,** Ernst, editor. Luftfahrt ist not! [Leipzig: Vaterländischer Buchvertrieb T. Rudolph, 1928.] 399 p. illus. ports. 4°. **VDY**

Various aspects of German aeronautics by experts, including much of historical interest, especially the work of Otto Lilienthal, military aeronautics, and oversea flights.

272. **Klemin,** Alexander. Aeronautics. illus. (Mechanical engineering, New York, 1930, v. 52, p. 485–490.) **VFA**

Brief comprehensive account of development of practical aeronautics, military and commercial.

273. —— Progress in aeronautics [for 1929]: aerodynamics. (Mechanical engineering, New York, 1930, v. 52, p. 5–6.) **VFA**

274. —— Recent aircraft engineering research. illus. (Mechanical engineering, New York, 1931, v. 53, p. 594–598.) **VFA**

Developments in slots, spins, dynamometers, load distribution and stress analysis, strength of thin-walled cylinders, two-cycle direct-injection engine, reduction of fire hazard, landings with locked propeller, passenger comfort, airship investigations, the new American seaplane tank, and the new wind tunnel at Langley Field.

275. —— With the men who fly. illus. (Scientific American, New York, 1923, v. 129, p. 226–227, 294, 326–327; 1924, v. 130, p. 26–27, 72.) **VA**

Coast-to-coast flights and their lessons, engine development, night-flying, gliding experiments, the Pulitzer race, the night mail, and the *Zeppelin ZR-1* (*Shenandoah*).

276. **Krauss,** J. S. Aerial navigation: how far is it practicable? London: Simpkin, Marshall & Co., 1901. 23 p. 8°. **VDC p.v.10, no.2**

Author's outlook is pessimistic for both airships and airplanes.

277. **Krogh,** Chr. von. In die Lüfte empor! Entwicklung und Technik der Luftschiffahrt. Charlottenburg: M. Teschner [1908]. 110 p. illus. 8°. **VDT**

Brief history of aeronautics dealing principally with airships.

278. **La Landelle,** Guillaume Joseph Gabriel de. Dans les airs, histoire élémentaire de l'aéronautique. Paris: R. Hatton, 1884. 288 p. 12°. **VDY**

Reviewed in *L'Aéronaute*, Paris, 1884, année 17, p. 14–17, *3–VDS*.

279. The **Late** Lieut.-Colonel Moedebeck. port. (Aeronautics, London, 1910, v. 3, p. 35.) **VDS**

Hermann W. L. Moedebeck was prominent in the development of aeronautics in Germany; also a prolific writer on the subject.

280. **La Vaulx,** Henri, comte de. ...Les vainqueurs de l'air; histoire de l'aéronautique: ses débuts sportifs, son application militaire, sa réalisation commerciale. Préface autographe de M. le maréchal Foch. [Paris:] Hachette [cop. 1922]. 246 p. illus. ports. f°. **VDT**

Part 1 deals with significant flights and events prior to 1914. Part 2 has to do chiefly with the World War. Well illustrated.

General Works, continued

281. La Vaulx, Henri, comte de, and others. L'aéronautique; des origines à 1922. Paris: H. Floury, 1922. 46 p. illus. ports. f°. † VDT

A folio volume de luxe of reproductions of documents and pictures, with descriptive texts.

282. Ledeboer, John Henry. Charles C[yril] Turner. By L. port. (Aeronautics, London, 1911, v. 4, p. 69.) VDS

Portrait and brief biography of prolific English writer on aeronautics.

283. Leechman, Douglas. Some early flights. (Flying, London, 1902, no. 4, Sept., p. 186–189.) VDS

Chronology, 400 B. C. to 1815.

284. Liebmann, Louis. Die historische Entwicklung der Luftfahrt. illus. (Zeitschrift für Flugtechnik und Motorluftschiffahrt, Berlin, 1928, Jahrg. 19, p. 441–449.) VDY

Brief general history.

285. Linke, Franz. Moderne Luftschiffahrt. Berlin: A. Schall, 1903. 296 p. illus. 8°. VDW

Principally concerned with modern developments, especially in Germany, in balloons and airships, with a brief section on airplanes. Historical introduction.

286. Lobo d'Avila Lima, J. Contribution des Portugais au développement de la navigation aérienne. (International Congress of Aerial Navigation, Fourth, Rome, 1927, Rapports, Memorials, Rome, 1928, v. 2, p. 507–517.) VDS

287. London. — Science Museum. Catalogue of the collections in the Science Museum, South Kensington. With descriptive and historical notes and illustrations. Aeronautics. London: H. M. Stationery Off., 1922. 71 p. plates. 8°. VBA p.v.78, no.6

Also a separately-paged supplement, London, 1924. 57 p. 8°. *VDY p.v.33, no.4.* This includes the war period.

288. —— Handbook of the collections illustrating aeronautics...a brief outline of the history and development of mechanical flight ...and a catalogue of the exhibits; compiled by M. J. B. Davy. London: H. M. Stationery Off., 1929–34. 3 v. plates. 8°. VDY

Part 1 covers heavier-than-air craft; part 2, lighter-than-air craft; and part 3, propulsion. Explanatory texts and bibliographies.

289. Lougheed, Victor. Vehicles of the air. A popular exposition of modern aeronautics, with working drawings. Chicago: The Reilly and Britton Co. [1909.] 7–479 p. illus. 8°. VDY

290. M., E. Un pioniere dell' aeronautica: Enrico Forlanini. port. (Rivista aeronautica, Roma, 1930, anno 6, Dec., p. 589–593.) VDS

Forlanini (1848–1930) was a prominent Italian experimenter with airships, rocket and steam propulsion, and the helicopter.

291. Männer der Luftfahrt. ports. (Illustrierte Flug-Welt, Leipzig, 1920, Jahrg. 2, p. 197–199, 230–231, 254–255, 278–279, 302–303, 326–327, 350–351, 374–375, 398–399, 424–425, 442–443, 460–461, 498–499, 538–539, 578–579; Illustrierte Flug-Woche, Leipzig, 1920, Jahrg. 3, p. 2–3, 40–41, 140–141.) VDS

Biographies and portraits: August Euler, E. Rumpler, A. von Parseval, Graf Zeppelin, Kurt Hermann, Franz Schneider, Hugo Junkers, Ernst Udet, Walter Mackenthun, Franz Z. Diemer, Hans C. von Schlick, Adolf Miethe, Hugo Hergesell, Gustav Basser, Alexander Baumann, Richard Knoller, Ernst von Hoeppner, and D. Siegert.

292. Magoun, Frederick Alexander, and Eric Hodgins. A history of aircraft... New York and London: Whittlesey House, McGraw-Hill Book Company, Inc., 1931. xx, 495 p. illus. 8°. VDT

Notable, comprehensive, illustrated work with numerous foot-note references and an extensive chronology, 400 B. C. – 1931 A. D., and a complete index.
Reviewed in *Aviation,* New York, 1931, v. 30, p. 712, *VDS.*

292a. —— The first twenty centuries. illus. (In their: A history of aircraft, New York, 1931, p. 3–19.) VDT

The period of mythology. Early apocrypha. Roger Bacon. Francisco di Mendoza. Gaspar Schott. Lauretus Laurus. John Wilkins. The Marquis of Worcester. Francis Willoughby. Speculations and publications of Borelli. Francesco de Lana. The Friar de Gusman. Attempts of Allard. Supposed success of Besnier. Joseph Addison's objections. Failure of de Bacqueville.

293. Marchis, Lucien René André Edmond. L'épopée aérienne. Préface de M. le commandant P. Renard. Paris: H. Dunod & E. Pinat, 1910. 4–104 p. illus. obl. 8°. † VDY

Pictorial history with brief texts. Numerous portraits.

294. —— Leçons sur la navigation aérienne. (Ballons sphériques — Aérostation militaire — Aérostation scientifique — Aéronautique maritime — Ballons dirigeables.) Paris: Vve. C. Dunod [1906]. 14, 704, 105 p. illus. 4°. (Université de Bordeaux. Faculté des sciences. Année 1903–1904.) 3–VDW

This technical work, dealing with lighter-than-air craft, has much of historic interest, notably on military aeronautics. In the appendix are: a detailed record of ascensions during the siege of Paris; a record of accidents, 1804–1902; and notable ascensions, 1784–1901. The work is in facsimile manuscript.

295. Maria, A. de. Navigazione aerea (aviazione); con 103 incisioni ed appendice: Biblioteca aeronautica. Milano: U. Hoepli, 1909. viii–xvi, 338 p. 16°. VDY

History of the determination of K. Leonardo da Vinci. History of ornithopters, helicopters, airscrews, and airplanes. Possibilities of military aviation.

296. Marquis, Raoul. L'aéronautique et l'aviation en 1909. [Par Raoul Marquis.] Paris: H. Desforges, 1910. 137 p. illus. 8°. VDT

297. —— Les ballons dirigeables et la navigation aérienne, par Henry de Graffigny [pseud.]... Paris: J. B. Baillière et fils, 1902. 380 p. illus. new ed. enl. 12°. (Bibliothèque des connaissances utiles.) VDY

Comprehensive history of balloons, airships, and airplanes.

298. —— La navigation aérienne et les ballons dirigeables, par Henri de Graffigny [pseud.]. Paris: J. B. Baillière et fils, 1888.

General Works, continued

2 parts in 1 v. illus. 8°. (Bibliothèque scientifique contemporaine.) **VDY**

History of balloons; military aeronautics, with list of balloons in Franco-Prussian war; development of airships; use of electricity in propulsion; development of airplanes.

299. —— La navigation aérienne en 1900, par H. de Graffigny, pseud. (La Nature, Paris, 1901, année 29, semestre 1, p. 103–106.) **OA**

Brief accounts of Zeppelin, Santos-Dumont, and Roze airships.

300. Marsh, William Lockwood. Aeronautical prints and drawings, with text by Lt.-Col. W. Lockwood Marsh...and Foreword by Major-General Sir Frederick H. Sykes... London: H. and T. Smith, Ltd., 1924. 36 p. illus. f°. **† VDW**

A collection de luxe of early plates.

301. —— The story of flight, a chronological record of the history of aviation. (Airways, London, 1927, v. 3, p. 261, 263, 300, 344, 376, 416; 1927/28, v. 4, p. 34, 70, 186, 224, 263, 298, 337, 367, 408, 446; 1928/29, v. 5, p. 35, 66, 106, 147, 220, 256, 291, 324, 360, 396, 424, 481; 1929/30, v. 6, p. 30, 68, 106, 140, 180, 215, 252, 341, 422, 461; 1930/31, v. 7, p. 495, 568, 605, 639; 1931/32, v. 8, p. 36, 275, 338, 377, 408; 1932/33, v. 9, p. 70, 111, 183, 252, 278, 361, 394, 431; 1933/34, v. 10, p. 40, 69.) **VDS**

An excellent history of ballooning and aviation from 1783, including deeds of the World War. Later numbers of the periodical have the title *Air and airways.*

302. Marshall, Alfred William, and H. Greenly. Flying machines: past, present, and future. A popular account of flying machines, dirigible balloons and aeroplanes. London: Percival Marshall & Co. [1907.] vi, 128 p. illus. 12°. **VDY**

303. Martin, Rudolf Emil. Das Zeitalter der Motorluftschiffahrt. Leipzig: T. Thomas, 1907. viii, 101 p. 8°. **VDY p.v.5, no.4**

Discussion of industrial, social, military, and political aspects of aeronautics.

304. The **Mastery** of the skies. How the Scientific American has been privileged to chronicle the development of airship and airplane. illus. (Scientific American, New York, 1920, v. 123, p. 338–339, 362–364.) **VA**

305. Maxim, Sir Hiram Stevens. Progress in aerial navigation. (Fortnightly review, London, 1892, new series, v. 52, p. 438–449.) *** DA**

Giffard. Dupuy de Lôme. The Tissandier brothers. Renard and Krebs. Langley. Engines.

306. Milestones. The Vickers machines. illus. (Flight, London, 1919, v. 11, p. 760–769, 1121–1130.) **VDS**

Development of Vickers airplanes and airships.

307. Miller, Francis Trevelyan. The world in the air; the story of flying in pictures... First historic collection of official prints and photographs from government archives and private collections recording five thousand years of man's struggles to conquer the air ... New York and London: G. P. Putnam's Sons, 1930. 2 v. illus. ports. 4°. **VDY**

308. Mingos, Howard. The birth of an industry, printed for a few of the pioneers. New York: [Printed by W. B. Conkey Company,] 1930. 95 p. ports. 8°. **TAK**

"The purpose here is to...narrate, possibly for the first time publicly, the incidents which have helped mightily in placing the [American] aeronautical industry upon its present firm and substantial foundation."

309. Mitchell, William. America trails in world aeronautics. (Popular aviation, Chicago, 1929, v. 5, Dec., p. 14–16, 46, 48, 50, 52.) **VDS**

Tells how various nations have approached and solved problems in military and commercial aeronautics.

310. Moedebeck, Hermann W. L. The development of aerial navigation in Germany. (Aeronautical journal, London, 1902, v. 6, p. 24–28.) **VDS (Royal)**

Ballooning in the eighteenth century. Development of meteorology. Formation of societies.

For lists of Moedebeck's writings on the various phases of aeronautics, see Paul Brockett, *Bibliography of aeronautics,* Washington, 1910, p. 588–597, *VDT;* also United States. — National Advisory Committee for Aeronautics, *Bibliography of aeronautics,* 1909–31, Washington, 1921–35, volume for 1909/16, p. 912–914, *VDT.*

311. Montagu (2. baron), John Walter Edward Douglas-Scott-Montagu. A short history of balloons and flying machines. Edited by Lord Montagu. With a chapter by B. Baden-Powell. London: "The Car Illustrated," 1907. 123 p. illus. 12°. **VDW**

Brief, with chronological arrangement. Has several early verses on aeronautics.

312. Moore, Willis Luther. [Important achievements.] (American magazine of aeronautics, New York, 1907, v. 1, Nov., p. 19–25.) **VDS (Aeronautics)**

Reviews developments in airships, airplanes, and meteorological aeronautics.

313. Mort de M. Henry Deutsch de la Meurthe. port. (Aérophile, Paris, 1919, année 27, p. 348–350.) **3–VDS**

Obituary of a prominent French patron of aeronautics.

314. Mowbray, Jay Henry. Conquest of the air by airships and other flying machines; being a comprehensive history of aerial navigation, with graphic accounts of thrilling flights, daring dashes, and desperate courage, hairbreadth escapes... [New York, 1910.] viii, 432 p. illus. 8°. **VDT**

315. Myers, Carl E. Aerial navigation: the progressive development of airships. illus. (Sibley journal of engineering, Ithaca, N. Y., 1902, v. 16, p. 353–368.) **VDA**

Brief history of balloons and airships, with account of the author's own experiments, notably his "sky-cycle" and the exploding of balloons to produce rain.

General Works, continued

316. **Nayler,** J. L., and E. Ower. Aviation of to-day; its history and development... With a chapter on aircraft engines by W. J. Stern...and a foreword by Lt.-Col. Mervyn O'Gorman... London: F. Warne & Co., Ltd. [1930.] xix, 491 p. maps, illus. 16°. **VDY**

Development traced along technical lines, but adapted to the requirements of the non-expert. Also considers outstanding historical events. Reviewed in *Aeroplane,* London, 1930, v. 39, p. 1220, *VDS.*

317. **New** European aeroplanes and airships. illus. (Scientific American, New York, 1908, v. 98, p. 40–41.) **VA**

The Gastambide-Mangin, Blériot, and Antoinette airplanes, the De Marcay airship, and the Dufaux 120 H. P. motor.

318. **Newcomb,** Simon. The problem of aerial navigation. (Nineteenth century and after, London, 1908, v. 64, p. 430–442.) *** DA**

Discusses possible practical uses of aeronautics. For reply by Baden-Powell see entry no. 138.

319. **Nimfuehr,** Raimund. Leitfaden der Luftschiffahrt und Flugtechnik in gemeinverständlicher Darstellung und mit besonderer Berücksichtigung der historischen Entwicklung. Wien: A. Hartleben, 1909. xiv, 444 p. illus. 8°. **VDW**

Has chapters on historical developments in various phases of aeronautics, with numerous illustrations.

320. —— Die Luftschiffahrt. Ihre wissenschaftlichen Grundlagen und technische Entwicklung. Leipzig: B. G. Teubner, 1909. 152 p. 12°. (Aus Natur und Geisteswelt. Bd. 300.) **VDW**

Part 2 is a brief account of developments down to the time of the Wright brothers.

321. **1914** — a retrospect. (Aeroplane, London, 1915, v. 8, p. 3–4, 6, 8, 10, 23–24, 26, 28, 30.) **VDS**

In two installments, the second of which is entitled Five months of war.

322. **1927.** (Flight, London, 1927, v. 19, p. 875–876.) **VDS**

Great flights, technical and commercial developments of the year.

323. **1932** passes into history. Closing months of a lean year show promise for the future. illus. (Aviation, New York, 1933, v. 32, p. 1–5.) **VDS**

American developments in transport, design and construction, research, autogiros, airships, military and naval aeronautics, airports, gliders, and some of the record flights.

324. **North,** H. C. Airships 2,000 years old! Steam engines, motion pictures, and many other "modern" inventions were used by ancient priests to mystify. illus. (Popular science monthly, New York, 1927, v. 111, Sept., p. 41–42, 133.) *** DA**

325. **Obermayer,** A. von. Ueber Luftschiffahrt. (Repertorium für Experimental-Physik, München, 1878, Bd. 14, p. 305–335.) **PAA**

Details of historic balloons and airships, with a chronological bibliography of aeronautics, 1627 to 1876 A. D.

326. **Official** world and American air records. (Aviation, Highland, N. Y., 1927, v. 23, p. 947–949, 952.) **VDS**

Covers all kinds of aircraft as of Aug. 1, 1927.

327. The **Old** year. The turning point was passed in 1933 and all forces are moving upward. illus. (Aviation, New York, 1934, v. 33, p. 1–4.) **VDS**

History for the year 1933 in transport, research, design, naval aviation, Air Corps activities, gliding and soaring, flights, competition, and records.

328. **Orcy,** Ladislas d'. The passing of a great aeronautical pioneer. Thomas Scott Baldwin is dead. (Aviation, Highland, N. Y., 1923, v. 14, p. 584–585.) **VDS**

Baldwin (1854–1923) was an American pioneer in both lighter- and heavier-than-air craft. Built the first U. S. Army airship.

329. **P.,** E. L. La conquista del dominio dei ciele. (Rivista aeronautica, Roma, 1928, anno 4, Dec., p. 603–619.) **VDS**

Brief comprehensive history of aeronautics.

330. **Payne,** Enoch George, and H. R. Barrows. The story of aviation. New York: American Viewpoint Society, 1930. 232 p. illus. 12°. **VDY**

Concise work for children including some appropriate poems.

331. **Piazza,** Lino. Storia aeronautica d'Italia. Prefazione di Italo Balbo. Milano: Istituto editoriale nazionale [1934]. 246 p. 12°. **VDT**

332. **Platte,** August. Die Aëronautik im Jahre 1896. (Oesterreichischer Ingenieur- und Architekten-Verein, Zeitschrift, Wien, 1897, Jahrg. 49, p. 187–189.) **VDA**

Year's developments in heavier-than-air and lighter-than-air craft.

333. **Popper,** Josef. Ueber die Fortschritte und die Aussichten im Gebiete der Luftschiffahrt. (Zeitschrift für Luftschiffahrt, Berlin, 1888, Jahrg. 7, p. 2–11, 33–40, 72–78, 109–114, 136–148, 178–185, 203–213, 234–242, 264–272, 297–306, 326–335, 363–387.) **3–VDS**

Critical, theoretical and technical examination of aeronautical progress during the decade, covering both lighter- and heavier-than-air craft.

334. **Protheroe,** Ernest. A book about aircraft. London: J. A. Sharpe [1923]. 128 p. front. illus. 12°. **VDT**

Concise "popular" account from ancient times. Chapter on aircraft in war. Log of the *R–34.*

335. **Rameau,** Louis. Revue de l'aérostation en 1880. (L'Aéronaute, Paris, 1880, année 13, p. 283–288.) **3–VDS**

336. **Recent** airship and aeroplane experiments in Europe. illus. (Scientific American, New York, 1906, v. 95, p. 470–472.) **VA**

Accounts of the *Patrie,* Deutsch, and Zeppelin airships; and of the Santos-Dumont airplane.

General Works, continued

337. **Renard,** Paul. L'aéronautique. Paris: E. Flammarion, 1909. 368 p. 12°. (Bibliothèque de philosophie scientifique.) **VDW**

The science, practice, and history of aeronautics set forth by a well-known French author.

338. —— Ce qui constitue la superiorité d'un navire aérien. (Revue des deux mondes, Paris, 1909, période 5, v. 54, p. 181–199.) *** DM**

Discussion of the relative advantages of airships and airplanes.

339. —— Un demi-siècle de navigation aérienne. (Technique aéronautique, Paris, 1922, nouv. série, année 13, p. 194–198, 243–248, 277–281, 307–312, 393–401.) **3–VDS**

Traces the subject, especially in France, 1872 to 1922.

340. —— Notice biographique du lieutenant-colonel Georges Espitallier. port. (Technique aéronautique, Paris, 1923, nouv. série, année 14, p. 522–527.) **3–VDS**

Espitallier (1849–1923) was a prolific writer who did much to popularize aeronautics in France.

341. **Rentschler,** F. B. The evolution of flight. illus. (U. S. air service, Washington, 1922, v. 7, Dec., p. 11–15.) **VDS**

Compares automobile and aeronautical progress through nineteen years.

342. **Répertoire** descriptif des brevets d'invention délivrés en France. Aéronautique. Paris: Librairie aéronautique [1911]. 146 p. illus. 8°. **VDY**

Illustrated abridgments of French aeronautical patents of 1910.

343. **Rodnykh,** Aleksandr A. Istoriya vozdukhoplavaniya i letaniya v Rossii. St. Petersburg: "Gramotnost," 1911–12. 2 v. illus. 8°. *** QH**

Well-illustrated account of early aeronautics in Russia.

344. —— Kurzer Abriss der Geschichte der russischen Luftschiffahrt. (Oesterreichische Flug-Zeitschrift, Wien, 1912, Jahrg. 6, p. 209–210.) **VDS**

345. **Rojas,** Francisco de Paula. Globos esféricos libres provistos de cámara de aire y sus ventajas para viajes de larga duración. illus. (Memorial de ingenieros del ejército, Madrid, 1905, época 4, tomo 22, p. 206–215, 225–234, 257, 290.) **VWA**

Study on the use of ballonets of air in the interior of balloons.

346. **Rotch,** Abbott Lawrence. The conquest of the air, or the advent of aërial navigation. Edited and revised by Alexander Klemin. New York: Moffat, Yard and Company, 1918. 200 p. illus. [2. ed.] 12°. **VDY**

A general history of aeronautics with some account of the distinguished author's meteorological investigations. Library has a copy of the 1. ed., New York: Moffatt, Yard and Company, 1909, 192 p. 12°, *VDW*.

347. **Saint-Fégor,** L. de. Le royaume de l'air. Paris: Société d'édition et de publications [1909]. 318 p. illus. 4°. **VDT**

Comprehensive history of aeronautics with numerous early illustrations. Chapter on military aeronautics.

348. **Sayers,** W. H. Lessons of 1924. (Institution of Aeronautical Engineers, Proceedings, London, 1925, no. 15, p. 27–34.) **VDS**

Comments upon notable events of the year. Abstracted in *Aeroplane*, London, 1925, v. 28, p. 462, 464, 466, *VDS*.

349. **Sibbald,** Andrew T. Aëropoedia. (Dublin review, London, 1890, v. 106 [series 3, v. 23], p. 95–110.) *** DA**

Brief history of aeronautics with accounts of notable ascensions.

350. **Sinclair,** Mrs. J. Herbert. Summary of human flight. illus. (Aircraft, New York, 1910, v. 1, p. 3–4, 43, 91–92, 133, 245.) **VDS**

Brief general history.

351. **Smith,** Laurence Yard. The romance of aircraft. New York: Frederick A. Stokes Company [cop. 1919]. viii, 264 p. illus. 8°. **VDT**

Comprehensive "popular" work. Reviewed in *Aeroplane*, London, 1920, v. 18, p. 704, *VDS*.

352. **Soreau,** Rodolphe, and A. Volmerange. L'aéronautique, de 1880 à 1930. illus. (Le génie civil, Paris, Nov., 1930, special number, p. 16–26.) **† V**

353. **Stade,** H. Hermann W. L. Moedebeck. (Deutsche Zeitschrift für Luftschiffahrt, Berlin, 1910, Jahrg. 14, Heft 10, p. 25–26.) **VDS**

Biography of prominent German writer (1857–1910) on aeronautics.

354. **Stevens,** Walter Le Conte. The history of aeronautics. illus. (Scientific American supplement, New York, 1880, v. 29, p. 11792–11794, 11806–11810.) **VA**

355. —— Recent progress in aerial navigation. illus. (Popular science monthly, New York, 1885, v. 27, p. 296–307.) *** DA**

History of the balloon and the airship.

356. The **Story** of 1931. illus. (Aviation, New York, 1932, v. 31, p. 1–6.) **VDS**

Brief accounts of developments during the year by C. W. Howard, W. A. Moffett, C. Fayette Taylor, and the editor.

357. **Suain,** L. G. La conquête de l'air. (Revue de Belgique, Bruxelles, 1910, série 2, v. 58, p. 5–31, 155–166, 218–229; v. 59, p. 73–92, 261–285.) *** DM**

Unusually complete chronological account from legendary events to the present.

General Works, continued

358. Sumner, Percy James Hammond. Aircraft progress and development. London: Crosby Lockwood & Son, Ltd., 1935. xiv, 295 p. illus. 8°. **VDY**

Although this work deals mainly with modern technical development of the airplane and of its engine, there is a good account of the early history of aeronautics in general. Has tabular data covering airplane records for speed, altitude, duration, and for long-distance flights, 1918 to date.

359. Supf, Peter. Das Buch der deutschen Fluggeschichte; Vorzeit, Wendezeit, Werdezeit. Berlin-Grunewald: H. Klemm A. G. [1935.] 515 p. illus. 4°. **VDT**

Comprehensive history of German aeronautics from mediaeval times to the present. Bibliography, p. 502–508.

360. Suplee, Henry Harrison. The development of aerial navigation. illus. (Cassier's magazine, New York, 1909, v. 35, p. 647–677.) **VDA**

Reviews modern developments and gives an account of some motors in use in 1909.

361. Teros, A. P. The long-sought flying machine. illus. (English illustrated magazine, London, 1898, v. 19, p. 3–11.) *** DA**

General history of early nineteenth-century flying machines and airships.

362. Theuermeister, Robert. Vom Luftballon zum Zeppelin. Ein Stück Menschenarbeit den Kindern erzählt. Leipzig: Ernst Wunderlich, 1931. 204 p. illus. ports. 8°. **VDW**

Children's history of balloons and airships.

363. Thomson, Jay Earle. Aviation stories. New York: Longmans, Green and Co., 1929. vi, 250 p. illus. 12°. **VDY**

Concise history of aeronautics for children.

364. Thurston, Robert Henry. The problem of air navigation. (Forum, New York, 1890, v. 8, p. 542–554.) *** DA**

Brief history of aeronautics with comments on possibilities.

365. —— The status of aeronautics in 1884. (Science, Cambridge, Mass., 1885, v. 5, p. 295.) **OA**

366. Tinker, Clifford A. Is America first in the air? In spectacular records — yes! In regulation and government control aid — no! (Aeronautical digest, New York, 1924, v. 4, p. 28–29, 62.) **VDS**

367. Tissandier, Gaston. La locomotion aérienne avant les Montgolfier. illus. (Revue scientifique, Paris, 1885, année 2, semestre 1, p. 769–776.) **OA**

For citations to the works and writings of Gaston Tissandier and his brother Albert, see Paul Brockett, *Bibliography of aeronautics,* Washington, 1910, p. 822–829, *VDT.*

368. Todd, Frederick. The man in the air. Personal experiences of Messrs. Wright, Farman, Curtiss, Baldwin, and others, and explanations of the different kinds of aircraft — Unsolved problems of the air. illus. (World's work, New York, 1908, v. 16, p. 10802–10809.) *** DA**

369. Turgan, L. La aviación; reseña histórica documentada y elementos de aviación, traducción y prólogo de José Muñoz Escámez... Cubierta de Geo Dorival. Paris: L. Michaud [1909]. 271 p. illus. ports. 8°. **VDY**

Comprehensive history of aeronautics from earliest times. A translation of the author's *Histoire de l'aviation.*

370. Turner, Charles Cyril. Aircraft of today; a popular account of the conquest of the air. Philadelphia: J. B. Lippincott Co., 1917. 315 p. illus. 12°. **VDY**

Includes the history of balloons, airships, and airplanes. Table of long-distance balloon voyages, p. 49; of early airships, p. 62–63; of airplanes, p. 81; of aircraft in war, p. 215–217, 266–268.

371. —— The old flying days... London: S. Low, Marston & Co., Ltd. [1927.] x, 374 p. tables, plates. 8°. **VDY**

These reminiscences of an English writer and aviator throw interesting sidelights on the history of aviation in England down to and through the World war. See especially: Early days at Brooklands, p. 26–45; Casualties [1908–1912], p. 224–238; Early exhibitions, p. 316–336; Big races, p. 109–122; Government and flying, p. 293–315; British flying meetings, p. 100–108; Some of the pioneers, p. 71–99; Prizes, p. 123–133; Periodicals, p. 344–346; The Royal Aero Club, p. 337–343. The book has a good index.

Reviewed in *Air,* London, 1927, v. 1, Dec., p. 53, *VDS; Aeroplane,* London, 1927, v. 33, p. 580, *VDS.*

372. —— Romance of aeronautics; an interesting account of the growth & achievements of all kinds of aerial craft. Philadelphia: J. B. Lippincott Company, 1912. 9–313 p. plates. 12°. **VDY**

Good comprehensive account. Includes a chapter: Mechanics who have become famous.

373. Turnor, Christopher Hatton. Astra castra. Experiments and adventures in the atmosphere. London: Chapman and Hall, 1865. xxiii, 530 p., 34 pl., 5 ports. f°. **VDT**

Notable work with many classical, mediaeval, and early modern citations, anecdotes, portraits, and other illustrations, with a chronology 625 B. C. to 1865 A. D. The reader is advised to consult its unusually good *index,* p. 519–525.

374. Tziolkovski, K. Uspekhi vozdukhoplavaniya v 19 veke. St. Petersburg: Tip. P. P. Soikina, 1901. 10 p. 8°. **VDY p.v.4, no.9**

Progress of aeronautics during the nineteenth century.

375. Die **Ueberlegenheit** Frankreichs. (Zeitschrift für Luftschiffahrt, Berlin, 1887, Jahrg. 6, p. 367–370.) **3–VDS**

Caustic remarks on French claims to superiority in aeronautical progress.

376. United States. — Census Bureau. Fifteenth census of the United States. Manufactures. v. 2, Reports by industries. Washington: Gov. Prtg. Off., 1933. **† SDG**

See p. 1191–1193 for statistics showing development of the aircraft industry in the United States, 1914–1929.

General Works, continued

377. **Valentine,** E. Seton, and F. L. Tomlinson. Travels in space, a history of aerial navigation... With an introduction by Sir Hiram Maxim. London: Hurst & Blackett, 1902. xvi, 328 p. 8°. **VDT**

Interesting pictures of early balloons and airships. Chronology of accidents.

378. **Van Muffling,** Adrian. The antiquity of flight. Historical research shows that ancient and medieval inventors and scientists understood present-day use of aircraft. illus. (Aero digest, New York, 1924, v. 5, p. 330–338.) **VDS**

Brief general history to 1793.

379. —— Human flight throughout the ages. (American Academy of Political and Social Science, Annals, Philadelphia, 1927, v. 131, May, p. 1–6.) **SA**

Brief accounts of ancient stories; the middle ages; Leonardo da Vinci; Francisco Lana; Bartholomy de Gusmão; the eighteenth century.

380. **Vivian,** Evelyn Charles H. A history of aeronautics; with a section on progress in aeroplane design, by Lieut.-Col. W. Lockwood Marsh... London: W. Collins Sons & Co., Ltd. [cop. 1921.] x, 521 p. illus. plates, ports. 8°. **VDT**

Covers airplanes, airships, and engines. Good bibliography, p. 504–511. For synopsis see *Aeroplane,* London, 1921, v. 20, p. 437–440, *VDS.*

381. **Vivien,** F. Louis. Modèles d'appareils d'aviation, de l'antiquité à nos jours. Paris: Librairie des sciences aéronautiques, F. L. Vivien, 1910. 36 p. illus. obl. 8°. **VDY**

A series of sketches, old cuts, and photographs showing development, principally of heavier-than-air machines from earliest times.

382. **Vorreiter,** Ansbert. Der gegenwärtige Stand der Motorluftschiffahrt. illus. (Dinglers polytechnisches Journal, Berlin, 1909, Bd. 324, p. 134–137, 149–153, 170–173, 232–237, 249–251, 265–268, 282–285, 617–619, 628–631, 648–651, 724–729.) **3–VA**

Descriptions and illustrations of contemporary airplanes, helicopters, and airships.

383. **Walker,** Frederick. Aerial navigation; a practical handbook on the construction of dirigible balloons, aerostats, aeroplanes, and airships. London: Crosby Lockwood and Son, 1910. 164 p. illus. 2. rev. ed. 12°. **VDY**

Descriptions of contemporary craft and engines. Library has also a copy of the 1. ed., London, 1902, 151 p., *VDY.*

384. **Warner,** Edward Pearson. The onward march of aircraft. (Yale review, Concord, 1925, v. 14, p. 239–256.) *** DA**

Discusses broadly the progress of aeronautics as shown in increase of speed, altitudes gained, aircraft in war, long-distance flights, improved construction, engines, gliders, airships, and the mail service.

385. **Watkins,** John Elfreth. Bird-men of centuries past. illus. (Air-scout, New York, 1910, v. 1, no. 2, p. 17, 46–47.) **3–† VDS**

386. —— The modern Icarus. illus. (Scientific American, New York, 1909, v. 101, p. 243–245.) **VA**

Airplane history, with portraits of some of the pioneers. Includes a list of notable balloon and airplane fatalities.

387. **Wellner,** Georg. Ueber die Frage der Luftschiffahrt. illus. (Oesterreichischer Ingenieur- und Architekten-Verein, Zeitschrift, Wien, 1902, Jahrg. 54, p. 337–343.) **VDA**

A review of airships and flying machines. Author's preference is for the helicopter type.

388. —— Die Zukunft der Luftschiffahrt. (Oesterreichischer Ingenieur- und Architekten-Verein, Zeitschrift, Wien, 1907, Jahrg. 59, p. 311–317.) **VDA**

Modern developments in airships and airplanes. Table shows details of airships and their motors, 1852–1906.

389. **Wenham,** Francis Herbert. Some remarks on the present state of aeronautical science. (Engineering, London, 1867, v. 3, p. 419.) **VDA**

Paper read before the Aeronautical Society of Great Britain. "Scarcely two writers agree upon the precise conditions upon which flight is performed."

390. **Whittaker,** W. E. de B. "1912." (Aeroplane, London, 1913, v. 4, p. 8, 10–12.) **VDS**

Month by month record of events.

391. **Who's** who in American aeronautics. New York: The Gardner Moffat Co., 1922–28. Years 1922, 1925, 1928. 3 v. 8° and f°.

Brief biographies. **† VDY**

392. **Williams,** Archibald. Conquering the air; the romance of the development and use of aircraft. Rev. and enl. by Marion Barton Crowell. New York: T. Nelson and Sons, 1930. vii, 366 p. illus. 8°. **VDY**

393. **Winter,** Lumen, and Glenn Degner. Minute epics of flight. New York: Grosset & Dunlap [cop. 1933]. 160 p. illus. 4°. **VDY**

Covers aeronautical history from ancient times, each page carrying a brief story and a picture.

394. **Woodhouse,** Henry. History of the conquest of the air. illus. (Flying, New York, 1916, v. 5, p. 229–242.) **VDS**

A well-illustrated history from early times, "prepared especially to supply members of the National Educational Association of the United States...with a textbook for use in school rooms" as a basis for prize essays.

395. The **World** we live in. (Blackwood's magazine, London, 1836, v. 40, p. 609–626.) *** DA**

See p. 624–626 for a well-written prophecy concerning aeronautics in peace and war. These pages are reprinted in *Mechanic's magazine,* London, 1836, v. 26, p. 190–191, *3–VA.*

396. **Zahm,** Albert Francis. Aërial navigation. (Franklin Institute, Journal, Philadelphia, 1894, v. 138, p. 265–287, 347–356.) **VA**

Hermite's test balloon (1893) rose to 52,494 feet. The airship *La France.* An excellent early history of gliding, soaring, and other heavier-than-air craft.

LIGHTER-THAN-AIR CRAFT

Balloons

397. Abercron, Hugo Wilhelm von. 500 Fahrten im Freiballon. Erlebnisse und Erfahrungen. Berlin: R. C. Schmidt & Co., 1929. 154 p. illus. 4°. **VDW**

Experiences of 29 years with a free balloon by the author, a celebrated German aeronaut.

398. Account of an aerial voyage made in a balloon on Saturday the 30th of April 1831, by T. Forster, M. B., F. L. S. &C. (Philosophical magazine, London, 1831, v. 10, p. 67–69.) **OA**

399. Account of an aerostatic experiment performed at Dijon. (London magazine, London, 1784, new series, v. 2, p. 449–452.) *** DA**

Gas was made from potatoes by distillation.

400. An **Account** of the aerostatical ball which has lately been made to ascend up into the air at Paris, and the principles on which it is constructed; together with a short history of the discoveries that have led to them. illus. (London magazine, London, 1783, new series, v. 1, p. 260–264.) *** DA**

401. Account of Mr. Blanchard. port. (European magazine, London, 1785, v. 7, p. 401–402.) *** DA**

Includes a full-page engraving of Blanchard.

402. Adams, George. On air balloons. (In his: Lectures on natural and experimental philosophy, London, 1794, v. 3, p. 442–462.) **PAD**

The author discusses the principles of balloons after relating, with considerable detail, the principal events to date.

403. Aerial travelling. illus. (European magazine, London, 1783, v. 4, p. 272.) *** DA**

Two engravings depicting the animals that ascended in the Montgolfier balloon.

404. Aerial voyage across the Atlantic. (Mechanics' magazine, London, 1843, v. 39, p. 155–156.) **3–VA**

Jocular remarks concerning "Honest Tom Steele's" proposal to cross the Atlantic ocean in a balloon.

405. The **Aeronef.** (Chambers' journal, London, 1864, v. 41 [series 4, no. 1], p. 37–39.)
Description of Nadar's *Geant.* *** DA**

406. Aerostatics. From the foreign journals, May 23, 1784. (London magazine, London, 1784, new series, v. 3, p. 305–313.) *** DA**

Accounts of ascensions by Blanchard, the Robert brothers, and Lunardi, and of an attempt to raise a fire-balloon in Edinburgh.

407. Aerostatics in France. (Cornhill magazine, London, 1873, v. 27, p. 336–344.) *** DA**

Cyrano de Bergerac. The Montgolfiers. Pilâtre de Rozier. Blanchard and Jefferies.

408. Aerostation. (American journal of science, New Haven, 1827, v. 12, p. 166–168.) **OA**

Account of ascent by Eugene Robertson, New York City, Oct. 16, 1826, to 21,000 feet.

409. Aerostation. Inquiry into the uses and capabilities of the balloon-projected voyage across the Atlantic, &c. (Polytechnic journal, London, 1840, v. 2, p. 70–77, 134–142, 267–279.) *** DA**

Deals largely with the ideas and experiences of Charles Green.

410. Aerostation and aeronautics. (Engineer, London, 1887, v. 64, p. 196.) **VA**

Proposal of M. Jovis to cross the Atlantic ocean in a balloon. Renard expects to double the speed of his airship. The author is pessimistic concerning the future.

411. Aimé, Emmanuel. Edward Spelterini. port. (L'Aérophile, Paris, 1896, année 4, p. 97–102.) **3–VDS**

Brief account of popular Swiss aeronaut.

412. —— Frédéric L'Host. port. (L'Aérophile, Paris, 1896, année 4, p. 33–40.) **3–VDS**

Biography of famous French aeronaut.

413. —— M. Georges Besançon. port. (Aérophile, Paris, 1894, année 2, p. 217–222.) **3–VDS**

Portrait and brief biography of French balloonist and director of the periodical *L'Aérophile.*

414. —— La thermosphère. Ballon gonflé de vapeur d'eau pour voyage. illus. (L'Aérophile, Paris, 1893, année 1, p. 93–108, 117–135.) **3–VDS**

Suggests the use of steam as a balloon gas.

415. Air balloons. (Mechanic's magazine, London, 1826, v. 6, p. 285–287.) **3–VA**

Account of ascents of the Duc de Chartres (1784) and of M. Testu (1786).

416. Ancelle, Paul. Une ascension de Xavier de Maistre. illus. (L'Aérophile, Paris, 1900, année 8, p. 9–11.) **3–VDS**

Ascension in 1784.

417. [The **Anchored** balloon of M. Henry Giffard.] (Chemical news, London, 1867, v. 16, p. 158.) **PKA**

Describes construction and inflation of famous captive balloon.

418. Andrée, Salomon August. Beobachtung während einer Ballonfahrt am 14. Juli 1894. Ein Versuch, den Ballon bei einer Schleppfahrt zu steuren. illus. (Zeitschrift für Luftschiffahrt, Berlin, 1895, Jahrg. 14, p. 113–124.) **3–VDS**

Use of drag-anchor for steering.

419. Another monster balloon. (Scientific American, New York, 1860, new series, v. 3, p. 273.) **VA**

Balloon of Carlos C. Coe, of Rome, N. Y., stated to be 290 feet in height, and 118 feet in diameter.

Lighter-Than-Air Craft — Balloons, continued

420. Archdeacon, Ernest. La mort des aéronautes L'Hoste et Mangot. ports. (L'Aérophile, Paris, 1894, année 2, p. 1–11.) **3–VDS**

Account of fatal attempt to cross the English Channel in 1887.

421. —— Voyage de "Arago." illus. (L'Aérophile, Paris, 1894, année 2, p. 41–51.) **3–VDS**

Fatal ascension of L'Hoste and Mangot.

422. Archibald, Douglas. The captive kite-balloon. illus. (Nature, London, 1887, v. 36, p. 278.) **OA**

Illustrated account of the author's invention.

423. Arlandes, Marquis d'. Lettre de M. le marquis d'Arlandes, à M. Faujas de St. Fond, sur le voyage aérien du 21 novembre [1783]. (L'Esprit des journaux, Paris, 1784, année 13, tome 1, Jan., p. 383–391.) ***DM**

The author accompanied Pilâtre de Rozier on the ascension of the above date.

424. Army team wins national balloon race. Lieut. R. S. Olmstead, winner. Lieut. J. B. Lawrence, U.S.N., second. H. E. Honeywell, third. (Aviation, Highland, N. Y., 1923, v. 15, p. 121–123.) **VDS**

Race started from Indianapolis, Ind., July 4, 1923.

425. Arnold, Stuart Amos. Copy of a letter from Mr. Arnold, of the Royal George rotunda to William Robertson... (New London magazine, London, 1785, v. 1, p. 371–373.) ***DA**

Relates Arnold's unfortunate ascension, 1785.

426. Ascension de 24 heures sans escale. illus. (L'Aérophile, Paris, 1897, année 5, p. 186–197.) **3–VDS**

Log of the balloon *Leipzig* on non-stop flight in 1897.

427. August Riedinger. port. (Wiener Luftschiffer-Zeitung, Wien, 1904, Jahrg. 3, p. 169–171.) **VDS**

Biography and portrait of prominent German balloon manufacturer.

428. Babinet, Léon. Notice sur Pilâtre de Rozier. (Académie impériale de Metz, Mémoires, Metz, 1864/65, série 2, année 13, p. 161–228.) ***EN**

Lengthy account of de Rozier's ascensions.

429. —— Rapport sur mémoire de M. le docteur Van Hecke, ayant pour titre: Nouveau système de locomotion aérienne. (Académie des sciences, Comptes rendus, Paris, 1847, tome 24, p. 152–153.) ***EO**

Method of raising balloon by means of vertical screw.

430. Bacon, Gertrude. Memories of land and sky... London: Methuen & Co., Ltd. [1928.] viii, 244 p. ports. illus. 12°. **AN (Bacon)**

These recollections of half a century by the daughter of the famous balloonist, John M. Bacon, throw sidelights on the development of the balloon, the airship, and the airplane. Reviewed in *Air,* London, 1928, v. 1, July, p. 53–54, *VDS; Aeroplane,* London, 1928, v. 34, p. 814, *VDS.*

431. —— The record of an aeronaut; being the life of John M. Bacon. London: J. Long, 1907. 357 p. plates, ports. 8°. **VDW**

Biography of British balloonist and scientist, by his daughter.

432. Bacon, John Mackenzie. Balloon ascents in thunderstorms. (Aeronautical journal, London, 1902, v. 6, p. 62–65.) **VDS**

Experiences of the author and of several earlier aeronauts.

433. —— By land and sky. London: Osbister & Co., 1901. viii, 9–276 p. illus. 8°. **VDW**

Author's balloon experiences. Reviewed in *Aeronautical journal,* London, 1901, v. 5, p. 31–32, *VDS.*

434. —— The dominion of the air; the story of aërial navigation. London: Cassell and Co., Ltd., 1902. viii, 348 p. plates. 12°. **VDT**

Principally a history of ascents down to 1900.
For list of writings by and about Bacon see Paul Brockett, *Bibliography of aeronautics,* Washington, 1910, p. 80–81, *VDT.*

435. Baden-Powell, Baden Fletcher Smyth. Ballooning as a sport. London: W. Blackwood and Sons, 1907. xx, 135 p. illus. 12°. **VDW**

Author's experiences and a brief history of the sport.

436. Baeder, F. de, and G. DUBOUCHET. Dictionnaire illustré de la navigation aérienne. Paris: Librairie aéronautique [1913]. liv, 138 p. 12°. **VDY**

Contains a brief history of aeronautics and an account of the use of the balloon in decorative art.

437. Balichou. Voyage aérien de M. Blanchard de Douvres à Calais. Extrait d'une lettre datée de Douvres du 7 janvier 1785. (L'Esprit des journaux, Paris, 1785, année 14, tome 2, Feb., p. 331–334.) ***DM**

437a. Balloon ascension. illus. (Gleason's pictorial drawing-room companion, Boston, 1851, v. 1, p. 248.) ***DA**

Ascension of John Wise, at Philadelphia, in 1851.

438. The **Balloon** ascent from the top of the Righi. (Aeronautical journal, London, 1900, v. 4, p. 162.) **VDS**

Ascent of Captain Spelterini.

439. Balloon intelligence. (New London magazine, London, 1785, v. 1, p. 46–49, 103–105, 158-160, 212–214.) ***DA**

Contemporary accounts of ascensions.

440. A **Balloon** railway. illus. (Scientific American supplement, New York, 1892, v. 34, p. 13851.) **VA**

Invention of Lucien Payn of Paris. Balloon acts as motive force on an inclined railway.

441. A **Balloon** trip over the Alps. (Scientific American supplement, New York, 1903, v. 56, p. 23252–23253.) **VA**

Voyage of Spelterini, 1903.

Lighter-Than-Air Craft — Balloons, continued

442. **Balloon-making** in France. (American artisan, New York, 1870, new series, v. 11, p. 355.) 3–† VA

Methods at the Godard factory during the siege of Paris.

442a. **Balloons.** (The Mirror, London, 1824, v. 3, p. 391–393.) * DA

Brief early history.

443. **Balloons** and ballooning. illus. (Frank Leslie's popular monthly, New York, 1877, v. 3, Jan., p. 33–48.) * DA

Illustrated history. Library has a second copy in *VDW p.v.1, no.6.*

443a. **Balloons** and railways. (The Mirror, London, 1836, v. 28, p. 346–348.) * DA

"Might it [the balloon] not pour conflagration on sleeping cities, bring sudden invasion, shed poison in all our streams, fling infection in all our fields, and exhaust us in perpetual vigilance."

444. **Barnes,** Henry. On the supposed advantage of auxiliary balloons in aerial navigation. (Mechanic's magazine, London, 1849, v. 50, p. 327–328.) 3–VA

Criticism of George B. Gales' invention.

445. **Barr,** Guy. Experiments relating to the electrification of balloon fabrics. London: H. M. Stationery Off., 1926. 10 p. 8°. (Great Britain. — Aeronautical Research Committee. Reports and memoranda. no. 1017.) VDW

446. **Barré,** L. L'exposition aéronautique et les ballons. (Revue scientifique, Paris, 1883, tome 31 [série 3, tome 5], p. 811–813.) OA

Account of Paris exhibition of 1883.

447. **Beaufoy,** Henry. Journal kept by Henry Beaufoy, Esq., during an aerial voyage with Mr. James Sadler, Sen. from Hackney, Middlesex, to East Thorpe, Essex, five miles from Colchester, August 29, 1811. (Annals of philosophy, London, 1814, v. 4, p. 282–291.) 3–OA

448. **Belgium** again wins Gordon Bennett race. port. (Aviation, Highland, N. Y., 1923, v. 15, p. 438–439.) VDS

"Lieutenant E. Demuyter repeats victory of last year in most tragic balloon race ever held."

449. **Berg,** Alfred. Die historische Abteilung der Internationalen Luftschiffahrt-Ausstellung zu Frankfurt a. M. illus. (Deutsche Zeitschrift für Luftschiffahrt, Berlin, 1909, Jahrg. 13, p. 745–753.) VDS

Reproductions of old ballooning plates shown at the exhibition.

450. **Besançon,** Georges. Henri Lachambre. port. (L'Aérophile, Paris, 1904, année 12, p. 18–19.) 3–VDS

Portrait and brief biography of well-known French balloon manufacturer.

451. —— M. Anatole Brissonnet. M. Pierre de Balaschoff. ports. (L'Aérophile, Paris, 1900, année 8, p. 54–57.) 3–VDS

Brissonnet was a prominent builder of balloons and is remembered for the large number of toy-balloons which he introduced. Balaschoff was a Russian patron of aeronautics.

452. [**Bibliographies** of the Gordon Bennett contests.] (In: Paul Brockett, Bibliography of aeronautics, Washington, 1910, p. 367–369; continued in United States. — National Advisory Committee for Aeronautics, Bibliography of aeronautics, 1909–31, Washington, 1921–35, v. for 1909/16, p. 556–560; 1920/21, p. 171–173; 1922, p. 59; 1923, p. 51; 1924, p. 39; 1925, p. 71; 1926, p. 54; 1927, p. 68; 1928, p. 77; 1929, p. 91; 1930, p. 90.) VDT

Includes balloon and airplane contests.

453. [**Bibliography** of] balloons. (United States. — National Advisory Committee for Aeronautics. Bibliography of aeronautics, 1909–31. Washington, 1921–35.) VDT

See volume for 1909/16, p. 115–117; 1917/19, p. 43; 1920/21, p. 40; 1922, p. 14; 1923, p. 12; 1924, p. 8; 1925, p. 17; 1926, p. 11–12; 1927, p. 14; 1928, p. 20; 1929, p. 27; 1930, p. 22; 1931, p. 29.

454. [**Bibliography** of captive balloons.] (In: Paul Brockett, Bibliography of aeronautics, Washington, 1910, p. 93.) VDT

List of writings on the subject.

455. [**Bibliography** of Eric Stuart Bruce.] (In: Paul Brockett, Bibliography of aeronautics, Washington, 1910, p. 158–159.) VDT

List of Bruce's writings, principally on balloons.

456. [**Bibliography** of Gabriel Yon.] (In: Paul Brockett, Bibliography of aeronautics, Washington, 1910, p. 906–907.) VDT

List of writings by and about Yon, chiefly concerning captive balloons.

457. [**Bibliography** of gas.] (In: Paul Brockett, Bibliography of aeronautics, Washington, 1910, p. 352.) VDT

Numerous references to the use of gas for balloons.

458. [**Bibliography** of H. Gross.] (In: Paul Brockett, Bibliography of aeronautics, Washington, 1910, p. 379.) VDT

References to Gross's writings on the development of balloons and airships.

459. [**Bibliography** of helium.] (United States. — National Advisory Committee for Aeronautics. Bibliography of aeronautics, 1909–31. Washington, 1921–35.) VDT

See volume for 1909/16, p. 629; 1917/19, p. 215–216; 1920/21, p. 195–196; 1922, p. 66–67; 1923, p. 58; 1924, p. 45–46; 1925, p. 77; 1926, p. 62; 1928, p. 88; 1929, p. 100–101; 1930, p. 103–104; 1931, p. 124.

460. [**Bibliography** of hydrogen.] (United States. — National Advisory Committee for Aeronautics, Bibliography of aeronautics, 1909–31. Washington, 1921–35.) VDT

See volume for 1909/16, p. 663–665; 1917/19, p. 226; 1920/21, p. 207; 1922, p. 71; 1923, p. 61; 1924, p. 48; 1925, p. 81; 1926, p. 66; 1927, p. 83; 1928, p. 93; 1929, p. 105.

461. [**Bibliography** of kite-balloons.] (United States. — National Advisory Committee for Aeronautics, Bibliography of aeronautics, 1909–31. Washington, 1921–35.) VDT

See volume for 1909/16, p. 718; 1917/19, p. 243; 1920/21, p. 226–227; 1926, p. 73.

Lighter-Than-Air Craft — Balloons, continued

462. [**Bibliography** of the Montgolfier brothers.] (In: Paul Brockett, Bibliography of aeronautics, Washington, 1910, p. 600–601.) **VDT**

List of citations to the Montgolfier brothers and their balloons.

463. [**Bibliography** of pilot balloons.] (United States. — National Advisory Committee for Aeronautics, Bibliography of aeronautics, 1909–31. Washington, 1921–35.) **VDT**

See volume for 1909/16, p. 1043–1044; 1917/19, p. 354; 1920/21, p. 325; 1925, p. 137; 1926, p. 114; 1928, p. 157.

464. **Bienaimé,** Maurice. A 10,081 mètres d'altitude. Nouveau record français d'altitude. ports. (L'Aérophile, Paris, 1913, année 21, p. 281–283.) **3–VDS**

Ascension in the balloon *Icare* in 1913 by Jacques Schneider, Maurice Bienaimé, and Albert Senouque.

465. **Black,** Joseph. Dr. Black on aerostation. (Glasgow mechanics' magazine, Glasgow, 1825, v. 2, p. 185–186.) **3–VA**

His letter to James Lind, 1784, in which he discusses balloons.

466. **Blanchard,** Jean Pierre. Détail des observations de M. Blanchard dans son voyage aérien à Rouen, le 23 mai, 1784. (L'Esprit des journaux, Paris, 1784, année 13, tome 7, July, p. 372–376.) *** DM**

467. —— Détail des observations...pendant son troisième voyage aérien, le 18 juillet 1784. (L'Esprit des journaux, Paris, 1784, année 13, tome 9, Sept., p. 337–345.) *** DM**

468. —— An exact narrative of M. Blanchard's observations during his third aerial voyage, on the 18th of July, 1784. Extracted from a pamphlet translated from the French of M. Blanchard. (European magazine, London, 1784, v. 6, p. 243–246.) *** DA**

469. —— Journal and certificates on the fourth voyage of Mr. Blanchard, who ascended from the Royal Military Academy, at Chelsea, the 16th of October, 1784... London: Printed for the author, by Baker and Galabin, 1784. 29 p. sq. 8°. *** C p.v.1083, no.10**

470. —— Journal of my forty-fifth ascension, being the first performed in America, on the ninth of January, 1793... Philadelphia: Charles Cist, 1793. 27 p. front. 12°. *** KD (1793)**

Reprinted in *Magazine of history,* Tarrytown, N. Y., 1918, extra number 64, p. 260–282, *IAG.*

471. —— Nouveau voyage aérostatique de M. Blanchard. (L'Esprit des journaux, Paris, 1785, année 14, tome 7, July, p. 318–320.) *** DM**

Account of ascension, London, 1785, with Mlle. Simonet.

472. —— Suite des expériences du Sr. Blanchard en Angleterre. (L'Esprit des journaux, Paris, 1785, année 14, tome 9, Sept., p. 310–321.) *** DM**

Letter from London.

473. —— Voyage aérostatique de M. Blanchard. (L'Esprit des journaux, Paris, 1784, année 13, tome 4, April, p. 360–364.) *** DM**

Ascension of March 2, 1784.

For list of writings by and about Blanchard see Paul Brockett, *Bibliography of aeronautics,* Washington, 1910, p. 132–133, *VDT.*

474. **Boffito,** Giuseppe. L'aeronauta Francesco Zambeccari. port. (L'Ala d'Italia, Milano, 1926, anno 5, no. 1, p. 11.) **VDS**

Zambeccari lived from 1756 to 1812. Description of his balloon.

475. —— L'aeronautica nelle città italiane: Firenze. illus. (Rivista aeronautica, Roma, 1928, anno 4, Feb., p. 393–396; 1931, anno 7, July, p. 161–163; Sept., p. 587–591.) **VDS**

Early history of ballooning in Florence. The first article of the series was written by R. Giacomelli; see entry no. 598.

476. —— L'aeronautica nelle città italiane: Genova. illus. (Rivista aeronautica, Roma, 1930, anno 6, Feb., p. 401–402; March, p. 601–604; May, p. 387–389.) **VDS**

Balloon ascension in Genoa (1784). Madame Blanchard's flight, Milan-Genoa (1811). J. Bruno's attempt to traverse the Mediterranean sea in a balloon.

477. —— L'aeronautica nelle città italiane: Mantova. (Rivista aeronautica, Roma, 1931, anno 7, June, p. 584–587.) **VDS**

History of aeronautics in Mantua.

478. —— L'aeronautica nelle città italiane: Palermo. (Rivista aeronautica, Roma, 1930, anno 6, Nov., p. 389–394.) **VDS**

Includes a letter by Vincenzo Lunardi describing ascension at Palermo in 1790.

479. —— L'aeronautica nelle città italiane: Napoli. (Rivista aeronautica, Roma, 1930, anno 6, June, p. 605–608; Aug., p. 421–423; Oct., p. 183–187.) **VDS**

Early suggestions for directing balloons by V. Lamberte, G. D. Bruno, Vincenzo Curzio, and Marco Antonio Costa. Lunardi's ascensions. Other ascensions above Naples.

480. —— L'aeronautica nelle città italiane: Roma. (Rivista aeronautica, Roma, 1931, anno 7, Dec., p. 628–630; 1932, anno 8, March, p. 583–585; April, p. 213–218; June, p. 669–677; July, p. 219–220; Oct., p. 183–188.) **VDS**

Aeronautics in Rome from ancient times to the last years of the eighteenth century.

481. —— La prima ascensione o esperienza aerostatica veneziana. illus. (L'Ala d'Italia, Milano, 1925, anno 4, p. 462–463.) **VDS**

Ascension by Zambeccari, 1784.

482. **Boiler-plate** balloon building. (Practical mechanic's journal, London, 1868, new series, v. 4, p. 161–163, 193–195.) **VA**

Discusses the possibility of making a balloon shell of iron.

Lighter-Than-Air Craft — Balloons, continued

483. Bouvet. Du moyen de produire à volonté, à bord des aérostats, un excédent de force ascensionnelle pour opèrer des montées et des descentes partielles. illus. (Académie des sciences, Comptes rendus, Paris, 1870, tome 71, p. 881–885.) *** EO**

Continuation of entry no. 484.

484. —— Sur un procédé de réchauffement méthodique du gaz d'un aérostat, par la combustion d'une partie de ce gaz lui-même, pour compenser les pertes de force ascensionnelle. (Académie des sciences, Comptes rendus, Paris, 1870, tome 71, p. 841–845.) *** EO**

Combined hot-air and gas balloon. See entry no. 483 for continuation.

485. Boyé, Pierre. Les premières expériences aérostatiques faites en Lorraine (1783–1788). Paris: Berger-Levrault & Cie., 1909. 48 p. illus. 8°. **VDT p.v.2, no.11**

Has an old cut of a hydrogen apparatus. Also in Société d'archéologie lorraine et du Musée historique lorrain, *Mémoires,* Nancy, 1909, tome 59 [série 4, v.9], p. 5–48, *ETI.*

486. Broszus, J. E. Bemerkungen zu dem Referat: Ueber Ballons mit Segelflächen. illus. (Zeitschrift für Luftschiffahrt, Berlin, 1883, Jahrg. 2, p. 274–282.) **3–VDS**

Discussion of Wellner's ideas for wing-regulated balloons. See entry no. 843.

487. Browne, C. Orde. The Woolwich balloon ascent. illus. (Popular science review, London, 1874, v. 13, p. 404–410.) **3–OA**

Engraving shows two propellers, one for horizontal and the other for vertical motion.

488. Bruce, Eric Henry Stuart. The balloon work of the late Mr. Henry Coxwell. port. (Aeronautical journal, London, 1899, v. 3, p. 118–120.) **VDS (Royal)**

Coxwell (1819–1900) acted as pilot in the British Association ascents of James Glaisher.

489. Burnaby, Frederick Gustavus. Possibilities of ballooning. (Fortnightly review, London, 1884, new series, v. 35, p. 668–674.) *** DA**

"So far as the solution of the problem how to navigate the air is concerned, we believe that balloons have done more harm than good."

490. —— A ride across the Channel, and other adventures in the air. London: Sampson Low, 1882. 128 p. map. 16°. **VDW**

Balloon voyage, March 23, 1882.

490a. Burrell, George Arthur. Beginnings of the helium development. The authentic account of the developments that resulted in the discovery of the helium supply at Petrolia, Texas. (Chemical and metallurgical engineering, New York, 1923, v. 29, p. 1013–1014.) **VOA**

Efforts during the World War to introduce helium for balloon purposes.

491. Burrell, Thomas. Balloons on old French fabrics. illus. (Antiques, Boston, 1927, v. 12, p. 306–309.) **MAA**

492. Burton, Walter E. The gas made famous by airships. illus. (Popular aviation, Chicago, 1932, v. 10, p. 83–84, 124.) **VDS**

Story of helium.

493. Capazza's lenticular balloon. illus. (Scientific American supplement, New York, 1885, v. 19, p. 7831–7832.) **VA**

494. Captain [Carl W.] Petersen's navigable double conical balloon. (Scientific American supplement, New York, 1885, v. 19, p. 7864.) **VA**

495. The **Captive** balloon. (Engineer, London, 1869, v. 27, p. 374.) **VA**

Exhibition of the Giffard balloon in London.

496. Captive balloon of 1878. illus. (Scientific American, New York, 1878, v. 39, p. 130–132.) **VA**

Full-page illustrations show Giffard's balloon, gas factory, and winding machinery.

497. Caron, Félix. Le monument des frères Montgolfier. illus. (L'Aéronaute, Paris, 1883, année 16, p. 163–174.) **3–VDS**

Monument erected at Annonay, France. Article includes a portion of address by Dupuy de Lome, tracing the history of aeronautics.

498. Castro, Giovanni de. Cento anni fa. (Archivio storico lombardo, Milano, 1881, anno 8, p. 267–314.) **BWS**

References to Luigi Cagnola, Paolo Andreani, and other Italian pioneer balloonists, with some contemporary poetry.

499. Cavallo, Tiberius. The history and practice of aerostation... London: Printed for the author and sold by C. Dilly [etc.], 1785. viii, 326 p. illus. 8°. **VDY**

Library has also a copy of French edition, Paris, 1786, *VDT.*

Detailed account of Montgolfier's invention and of the flights during the years 1783–1785. Author's celebrated experiments on soap bubbles and the lifting power of hot air.

500. —— On aerostation. (In his: Elements of natural or experimental philosophy, London, 1803, v. 4, p. 316–356.) **3–PAE**

Account of the author's balloon experiments precedes a history of balloons.

501. Cavendish, Henry. Three papers, containing experiments on factitious air. illus. (Royal Society of London, Philosophical transactions for 1766, London, 1767, v. 56, p. p. 141–184.) *** EC**

See p. 144–159 for author's account of his famous experiments with hydrogen.

502. Un **Centenaire.** (L'Aéronaute, Paris, 1904, année 37, p. 83–87.) **3–VDS**

Interesting account of the relations between Garnerin and the Emperor Napoleon.

503. A **Century** of ballooning. illus. (Scientific American supplement, New York, 1884, v. 17, p. 6834–6836.) **VA**

Early portraits and illustrations with text.

504. Chandler, Charles de Forest. James Comly McCoy: pioneer aeronaut. port. (U. S. air services, Washington, 1934, v. 19, June, p. 27–28.) **VDS**

Brief account of early American balloonist.

Lighter-Than-Air Craft — Balloons, continued

505. **Charles,** Jacques Alexandre César. Récit des deux premiers voyages aériens. illus. (L'Aérophile, Paris, 1923, année 31, p. 134–137.) **3–VDS**

Dr. Charles and M. Robert were the first to ascend in a gas-filled balloon, 1783.

506. **Chumakov,** D. Osnovy k razresheniyu zadachi vozdukhplavaniya. (Chast 1–2). Askhabad: Tip. "Ashkabab," 1901. 148 p. 8°. **VDY p.v.1, no.4**

Principles for the solution of aeronautical problems, mainly of balloon forms.

507. **Cléry,** A. André Schelcher. port. (L'Aérophile, Paris, 1910, année 18, p. 49.) **3–VDS**

Brief biography and portrait of French balloonist and pioneer in aerial photography.

508. —— Maurice Bienaimé. port. (L'Aérophile, Paris, 1911, année 19, p. 521.) **3–VDS**

Portrait and brief account of noted French balloonist.

509. **Copie** du procès-verbal, dressé au château de la Muette après l'expérience de la machine aérostatique de M. de Montgolfier. (L'Esprit des journaux, Paris, 1784, année 13, tome 1, Jan., p. 381–383.) *** DM**

Report on ascension of Pilâtre de Rozier and the Marquis d'Arlandes, Nov. 21, 1783.

510. **Copper** balloon. (Mechanic's magazine, London, 1844, v. 40, p. 208, 272.) **3–VA**

Invention of Marey Monge.

510a. **Costello,** Dudley. Balloons in their infancy. (The Mirror, London, 1845, new series, v. 2, p. 53–54.) *** DA**

A miracle story of 1786, involving the descent of Blanchard's balloon on a church.

511. **Count** de la Vaulx balloon "Méditerranéen." illus. (Scientific American supplement, New York, 1901, v. 52, p. 21648.) **VA**

Account of disastrous trip on the Mediterranean sea.

512. **Count** de la Vaulx's balloon trip across the Mediterranean. illus. (Scientific American, New York, 1901, v. 85, p. 265–266.) **VA**

Disastrous voyage began Oct. 13, 1901.

513. La **Coupe** aéronautique Gordon-Bennett 1906. illus. map. (L'Aérophile, Paris, 1906, année 14, p. 255–276.) **3–VDS**

Detailed account of famous balloon competition.

514. **Coxwell,** Henry. My life and balloon experiences. Series 2. London: W. H. Allen, 1889. vi, 283 p. illus. 12°. **VDYC**

Experiences during the '50s and '60s of the nineteenth century. Library has also a copy of the first edition, London, 1887, which includes a photograph of the author.

For references by and about Coxwell see Paul Brockett, *Bibliography of aeronautics,* Washington, 1910, p. 217, *VDT.*

515. **Crestadoro,** Andrea. Air locomotion dispensing with gas and ballast. An explanatory account of the experimental model of a metallic balloon exhibited by A. Crestadoro ...at the Crystal Palace, June, 1868. London: Spottiswoode & Co., 1868. 15 p. 12°. **VDC p.v.19, no.3**

Experimental balloon using sheet copper or sheet brass.

516. **Cumberland,** G. On the origin of air balloons. (Royal Institution of Great Britain, Quarterly journal of science, literature, and art, London, 1828, p. 29–34.) *** EC**

A review of Lana's *Prodromo,* 1670.

See also Franklin Institute, *Journal,* Philadelphia, 1829, new series, v. 3, p. 20–24, *VA.*

517. **Darce,** Léon. Un délesteur automatique pour ballons libres. illus. (La Nature, Paris, 1903, année 31, semestre 1, p. 33–34.) **OA**

Automatic ballast regulator proposed by Léo Dex and de Castillon de Saint Victor, for balloon voyage over the Sahara desert. Translation in *Scientific American supplement,* New York, 1903, v. 55, p. 22808, *VA.*

518. **Daring** feats of acrobats and aeronauts. (Scientific American, New York, 1859, new series, v. 1, p. 39.) **VA**

Describes balloon voyage of Wise, La Mountain, Gager, and Hyde from St. Louis, Mo., to Jefferson county, N. Y.

519. **Deburaux,** Édouard Léopold Joseph. Essais préliminaires à la navigation aérienne au long cours. illus. (La Revue technique, Paris, 1900, tome 21, p. 193–196, 218–222.) **3–† VA**

Signed with the author's pseud., Léo Dex.

520. —— Projet d'exécution d'aérostats non montés capables de traverser le Sahara. Ballon de 1,000 mètres cubes, ballon, agrès et apparaux. (La Revue technique, Paris, 1902, tome 23, p. 209–219.) **3–† VA**

Signed with the author's pseud., Léo Dex.

For references to Deburaux's writings on long-distance balloon flights see Paul Brockett, *Bibliography of aeronautics,* Washington, 1910, p. 241–242, under heading of Dex, Léo, *VDT.*

521. **Deburaux,** Édouard Léopold Joseph, and Maurice Dibos. Voyages aériens au long cours; les aérostats et la traversée de l'Afrique australe. Annexe de l'ouvrage: Les aérostats et l'exploration du continent africain. Paris: L. Baudoin, 1894. 24 p. map. 4°. **VDY p.v.9, no.9**

A study of the winds of southern Africa and the possibility of free ballooning in the region.

522. **Decazes,** vicomte. Conditions d'utilisation de l'hélice à axe vertical pour remplacer le lest et la soupape dans les voyages aériens au long cours. illus. (L'Aérophile, Paris, 1896, année 4, p. 149–151, 203–207.) **3–VDS**

523. **Degérando.** Éloge de M. Conté. (Société d'encouragement pour l'industrie nationale, Bulletin, Paris, 1806, année 4, p. 241–255.) **3–VA**

Nicholas Jacques Conté (1755–1805), a distinguished French artist and inventor, was active during the French revolution in the manufacture of balloons and in the large-scale production of hydrogen.

524. **Degouy,** Marcel. L'équilibre des ballons captifs. diagrs. (Revue de l'aéronautique, Paris, 1889, année 2, p. 23–31, 41–71.) **3–† VDS**

Lighter-Than-Air Craft — Balloons, continued

525. **Demuyter,** Ernest. La navigation aérienne et les randonnées victorieuses du "Belgica." Conquête de la coupe Gordon-Bennett. Bruxelles: Expansion belge vers le Levant [1925]. 135 p. illus. maps, ports. 8°. **VDY p.v.39, no.11**

Account of the author's numerous flights in free balloons, culminating in the winning of the Gordon-Bennett cup in 1924.

526. **Deparcieux,** Antoine. Dissertation sur les globes aérostatiques. Paris: l'Auteur, 1783. 38 p. illus. 12°. **VDT p.v.1, no.2**

Author's ideas on balloon design.

527. **Deussen,** Ernst. Ueber einen Flugapparat aus dem Jahre 1878. illus. (Archiv für die Geschichte der Naturwissenschaften und der Technik, Leipzig, 1912–13, Bd. 4, p. 162–164.) **OA**

Describes Gebhard Ackermann's balloon and wing combination for which a German patent was granted.

528. **Dibos,** Maurice, and others. Exposé détaillé de la possibilité des voyages aériens au long cours à travers le Sahara central et le Soudan français [par M. Dibos, Léo Dex, et H.]. (La revue technique, Paris, 1898, tome 19, p. 201–206.) **3–† VA**

529. **Dime,** Eric Adolphus. America's first aeronaut. port. illus. (Air travel, New York, 1918, v. 1, p. 214–217, 239.) **VDS**

Biography of Charles F. Durant (1805–1873). He made twelve balloon ascensions, beginning in 1830.

530. **Documents** inédits sur la mort de Pilâtre de Rozier. illus. (L'Aérophile, Paris, 1916, année 24, p. 264–266.) **VDS**

Copy of a letter (1825) describing de Rozier's balloon in which he attempted to cross the English Channel, with an account of investigation of causes which led to its destruction.

531. **Dollfus,** Charles. Étude historique sur les projets de traversée de l'Atlantique en ballon. illus. (L'Aéronautique, Paris, 1920, année 1, p. 380–386.) **VDS**

Illustrated history beginning with project of Charles Green. Bibliography.

532. —— Henry de la Vaulx. port. (L'Aéronautique, Paris, 1930, année 12, p. 205–206.) **VDS**

La Vaulx (1870–1930) was a well-known balloonist, airship pioneer, and writer.

533. —— Il y cent ans. (L'Aérophile, Paris, 1914, année 22, p. 18–19, 234–236; 1915, année 23, p. 150–152; 1916, année 24, p. 199–200; 1917, année 25, p. 354–355; 1918, année 26, p. 359–362; 1919, année 27, p. 313–317; 1920, année 28, p. 369–370.) **3–VDS**

History of ballooning, 1813–1820.

534. **Domarus,** M. Die Luftreise Blanchards von Frankfurt nach Weilburg am 3. Oktober 1785. (Verein für nassauische Altertumskunde und Geschichtsforschung, Annalen, Wiesbaden, 1907, Bd. 36, p. 87–108.) **† EKL**

535. **Donaldson,** Washington Harrison. History of Donaldson's balloon ascensions... compiled and arranged by M. L. Amick, M.D... Illustrated from original drawings of Donaldson. Cincinnati: Cincinnati News Co., 1875. 200 p. illus. port. 8°. **VDW**

Donaldson (1840–1875) made 139 ascensions, 1871–1875, when he lost his life in Lake Michigan.

536. **Dubosc,** André. Le physicien Charles, aéronaute. (Société industrielle de Rouen, Bulletin, Rouen, 1923, année 51, p. 582–600.) **VA**

Charles was the first aeronaut to use a balloon inflated with hydrogen (1783).

537. **Dubrunfaut.** Observations sur la mort de l'aéronaute Mosment en 1806. (L'Aéronaute, Paris, 1878, année 11, p. 147–151, 219–224.) **3–VDS**

538. **Dupuis-Delcourt,** Jules François. Nouveau manuel complet d'aérostation, ou guide pour servir à l'histoire des ballons. Paris: Roret, 1850. vi, 292 p. illus. 24°. **VDW**

Includes some early aeronautical poetry and a chronological list of ascensions, 1783–1848. For German opinion of the book see article by vom Hagen in *Zeitschrift für Luftschiffahrt,* Berlin, 1885, Jahrg. 4, p. 52–57, *3–VDS.*

For a list of Dupuis-Delcourt's writings, 1824–1857, see Paul Brockett, *Bibliography of aeronautics,* Washington, 1910, p. 262–263, *VDT.*

539. **Dupuy de Lome,** Stanislas Charles Henri Laurent. Montgolfier. (Revue scientifique, Paris, 1883, série 3, tome 6 [tome 32], p. 392–402.) **OA**

Discourse at the unveiling of the Montgolfier statue.

540. —— The Montgolfier brothers. (Scientific American supplement, New York, 1883, v. 16, p. 6587–6588.) **VA**

541. **Durant,** Charles Ferson. The aeronaut to the people! Distributed on his ninth grand ascension. Observatory Garden (Federal Hill,) Baltimore, October 14th, 1833. [Poem.] [Baltimore, 1833.] broadside. 4°. **† VDW**

With this are bound two large mounted posters announcing ascensions by Durant in 1830.

542. —— The balloon ascent. (New York journal of commerce, New York, June 5, 1833, n. p.) *** A**

Account of author's ascension from Castle Garden, New York City, 1833. Also in *Magazine of history,* Poughkeepsie, N. Y., 1916, v. 23, p. 225–228, *IAA.*

543. **Duriez.** Extract of a letter from M. Duriez, dated Boulogne, June 25 [1785]. (New London magazine, London, 1785, v. 1, p. 46–47.) *** DA**

Testimony of eye-witness to the death of Pilâtre de Rozier.

544. **E.,** H. [Opinion of balloons.] (London magazine, London, 1784, new series, v. 2, p. 301–302.) *** DA**

Criticizes French enthusiasm and says: "I have very little hopes that the means of directing the aerostat will be discovered; and have doubts whether it would contribute to the extension of natural knowledge even if they were."

545. **Early** aeronautics. illus. (Harper's new monthly magazine, New York, 1869, v. 39, p. 145–157.) *** DA**

Brief history of early ballooning.

Lighter-Than-Air Craft — Balloons, continued

546. **Einiges** über die ersten Berliner Luftschiffahrts-Versuche. (Zeitschrift für Luftschiffahrt, Berlin, 1885, Jahrg. 4, p. 289–294, 329–335, 369–376.) **VDS**

Account of early balloon experiments in Berlin, beginning in 1783.

547. L'**Électrolyse** industrielle de l'eau. Procédé Latchinoff, d'Arsonval, et Ducretet. (Revue de l'aéronautique, Paris, 1891, année 4, p. 63–71.) **3–† VDS**

547a. **Elworthy,** Reginald Thomas. The history of helium and its production. Discovery of helium, occurrences and properties. Work of Professor McLennan in England and Canada. (Canadian chemistry and metallurgy, Toronto, 1921, v. 5, p. 7–9.) **VOA**

Brief history from 1785.

548. **Erdmann,** Hugo. Über die Verwendung flüssigen Wasserstoffs in der Luftschiffahrt. illus. (Frankfurt a M. — Internationale Luftschiffahrts-Ausstellung, Denkschrift, Berlin, 1910, v. 1, p. 30–45.) **† VDW**

Describes the preparation of liquid hydrogen and suggests how it may be used in filling balloons.

549. **Espitallier,** Georges. Aérostiers et aviateurs. Nombreuses illustrations de Gil Baer. Paris: Société française d'imprimerie et de librairie [1913?]. 384 p. illus. maps. f°. **† VDY**

Careers of Coutelle and Conté during the latter eighteenth century. Some of Conté's pictures. How the first balloons were made. The siege of Paris. Meusnier, Giffard, Dupuy de Lome, Tissandier, and Renard.

550. —— La ballon "Svenske." illus. (La Nature, Paris, 1904, année 32, semestre 1, p. 83.) **OA**

Illustrated description of Eric Unge's balloon.

551. —— La ballon Unge et les idées du capitaine Voyer. (La Nature, Paris, 1902, année 30, semestre 2, p. 218–219.) **OA**

Claims that the ideas of the Swedish Captain Unge are based upon those of Voyer.

552. —— L'emploi des gaz comprimés pour l'aérostation et la lumière oxhydrique. illus. (Revue de l'aéronautique, Paris, 1890, année 3, p. 2–10.) **3–† VDS**

Apparatus manufactured in Birmingham, England.

553. —— Les expériences du Méditerranéen et les ascensions aéro-maritimes. illus. (La Nature, Paris, 1902, année 30, semestre 2, p. 374–378.) **OA**

Voyage of de la Vaulx and Hervé. Six illustrations.

554. —— La fabrication rapide de l'hydrogène. illus. (Revue aéronautique, Paris, 1888, année 1, p. 51–59, 85–104.) **3–† VDS**

Reviews the processes of Coutelle, Guyton de Morveau, Giffard, Dupuy de Lome, Renard, Tissandier, and others.

555. —— Les grands aérostats de baudruche. illus. (Revue de l'aéronautique, Paris, 1890, année 3, p. 33–39.) **3–† VDS**

Brief history of the use of goldbeater's skin for balloon coverings.

556. **Expérience** aérostatique par MM. Robert. (L'Esprit des journaux, Paris, 1784, année 13, tome 9, Sept., p. 332–333.) *** DM**

Ascension of the Robert brothers at St. Cloud, July 15, 1784.

557. **Expérience** fait en grand, à Versailles, le 19 septembre, par M. de Montgolfier, avec sa machine aérostatique. (L'Esprit des journaux, Paris, 1783, année 12, tome 11, Nov., p. 329–334.) *** DM**

558. **Expérience** singulier faite sur le gas inflammable. (L'Esprit des journaux, Paris, 1783, année 12, tome 9, Sept., p. 348–349.) *** DM**

Brief account of the ascent of Montgolfier's first balloon as reported in the *Journal de Paris.*

559. **Experiences** in the national balloon race. Interesting accounts by Major Westover, Ralph Upson, Lieutenant Reed and Commander Norfleet. illus. port. (Aviation, Highland, N. Y., 1922, v. 13, p. 746–749.) **VDS**

Race started from Milwaukee, Wis., May 31, 1922.

560. **Extrait** du journal & des procès-verbaux du quatrième voyage aérien de M. Blanchard en Angleterre, au mois d'octobre dernier. (L'Esprit des journaux, Paris, 1785, année 14, tome 2, p. 322–331.) *** DM**

561. **Farman,** Maurice. 3,000 kilomètres en ballon. illus. maps. (L'Aérophile, Paris, 1901, année 9, p. 109–119.) **3–VDS**

Graphic account of the author's flights.

562. The **Fate** of the first aeronaut. (Harper's magazine, New York, 1855, v. 10, p. 826–829.) *** DA**

Account of Pilâtre de Rozier.

563. **Faujas de Saint-Fond,** Barthélemy. Description des expériences de la machine aérostatique de MM. de Montgolfier et de celles auxquelles cette découverte a donné lieu. Paris: Chez Cuchet, 1783. xl, 299 p. illus. 12°. **VDT p.v.1, no.1**

Library has also 2. ed., Paris, 1784, including a supplementary second volume of 366 pages, *VDT.* In volume 1 of each edition, p. xiii–xxiv is a lengthy quotation from Joseph Galien's brochure *L'art de naviger dans les airs,* Avignon, 1775.

564. —— Nouvelles expériences avec la machine de Montgolfier. (L'Esprit des journaux, Paris, 1783, année 12, tome 12, Dec., p. 336–339.) *** DM**

Brief account of first ascensions of Pilâtre de Rozier in captive Montgolfier balloon.

565. —— Travels in England, Scotland, and the Hebrides...translated from the French ... London: Printed for James Ridgway, 1799. 2 v. 12°. **CBD**

Account of author's visit to Priestley in Birmingham, in 1784, when he was shown the latter's process of preparing hydrogen.

566. **Feldhaus,** Franz Maria. Der Warmluftballon, eine deutsche Erfindung des Mittelalters. illus. (Illustrierte aëronautische Mitteilungen, Strassburg, 1906, Jahrg. 10, p. 113–116; 1907, Jahrg. 11, p. 53–54.) **VDS (Deutsche)**

Dragon hot-air balloons seem to date from the time of the Emperor Trajan.

Lighter-Than-Air Craft — Balloons, continued

567. Figuier, Guillaume Louis. Les aérostats. Paris: Jouvet & Cie., 1882. 268 p. illus. 12°. (Bibliothèque instructive.) **VDW**

Well-illustrated account of early ballooning and of the first dirigibles.
Library has also 2. ed., Paris, 1887, 291 p., *VDT (Room 120).*

568. —— Les merveilles de la science; ou, Description populaire des inventions modernes... Paris: Furne, Jouvet et Cie. [1867–69.] 4 v. 4°, and Supplément, Paris [1888–90]. 4 v. 4°. **3-† V**

See v. 2, p. 423–626; supplement, v. 1, p. 618–732, under title Les Aérostats, for comprehensive and well-illustrated account of early balloons, airships and parachutes. Numerous portraits.

569. The **First** aerial voyage in England. (Chambers's journal, London, 1857, v. 8, p. 60–62.) *** DA**

Account of Lunardi's adventure.

570. The **First** aeronaut. (Harper's magazine, New York, 1854, v. 8, p. 521–523.) *** DA**

Account of ascensions of Pilâtre de Rozier. Also in *Leisure hour,* London, 1854, v. 3, p. 30–32, ** DA.*

571. The **First** international balloon race. illus. (Scientific American, New York, 1906, v. 95, p. 300–301.) **VA**

Gordon Bennett cup won by Lieut. Frank P. Lahm of the United States Army.

572. The **First** long voyage in a balloon. illus. (English illustrated magazine, London, 1897, v. 18, p. 107–112.) *** DA**

Voyage of Charles Green, Robert Holland, and Monck Mason in the *Nassau,* 1836.

573. Fonvielle, Wilfrid de. La catastrophe du ballon l'Arago. ports. (L'Aéronaute, Paris, 1887, année 20, p. 223–226.) **3-VDS**

Fatal ascension of F. Lhoste and J. Mangot.

574. —— Édouard Surcouf. port. (L'Aérophile, Paris, 1897, année 5, p. 145–148.) **3-VDS**

Portrait and brief biography of French captive balloon constructor.

575. —— Gabriel Yon. port. (Aérophile, Paris, 1894, année 2, p. 63–66.) **3-VDS**

Portrait and brief account of distinguished French balloonist.

576. —— Jacques Faure. port. (L'Aérophile, Paris, 1900, année 8, p. 29–31.) **3-VDS**

Portrait and biography of French builder of small man-carrying balloons.

577. —— Manuel pratique de l'aéronaute. Paris: B. Tignol [1894]. iv, 246 p. illus. 12°. (Bibliothèque des actualités industrielles. no. 59.) **VDW**

Manual of practical ballooning.

578. —— M. Auguste Riedinger. port. (L'Aérophile, Paris, 1900, année 8, p. 73–75.) **3-VDS**

Portrait and account of work of the inventor of the military kite balloon.

579. —— M. Ernest Archdeacon. port. (L'Aérophile, Paris, 1899, année 7, p. 97–99.) **3-VDS**

Portrait and account of work of prominent automobile and balloon pioneer of France.

580. —— M. Maurice Farman. port. (L'Aérophile, Paris, 1899, année 7, p. 109–111.) **3-VDS**

Portrait and brief biography of French writer and aeronaut.

581. —— M. Victor Silberer. port. (L'Aérophile, Paris, 1900, année 8, p. 61–62.) **3-VDS**

Portrait and brief account of prominent Austrian aeronaut and writer.

581a. Forrest, Joanna. A lady's account of a nocturnal balloon ascent. (The Mirror, London, 1841, v. 38, p. 189–190.) *** DA**

Author's ascension with Green, 1841.

582. Franklin, Benjamin. Benjamin Franklin on balloons; a letter from Passy, France, January sixteenth, MDCCLXXXIV... St. Louis: Privately printed by W. K. Bixby, 1922. 7 l. facsims. sq. 4°. *** KP (Bixby)**

Facsimile reprint.

583. —— The ingenious Dr. Franklin. Selected scientific letters of Benjamin Franklin, edited by Nathan G. Goodman. Philadelphia: University of Pennsylvania Press, 1931. xi, 244 p. illus. port. 8°. **OAP**

First hydrogen balloon, p. 89–93. A hot-air balloon, p. 94–95. First aerial voyage by man, p. 96–98. Second aerial voyage by man, p. 99–102. A prophecy on aerial navigation, p. 103–105.

584. —— The works of Benjamin Franklin; containing several political and historical tracts not included in any former edition, and many letters official and private not hitherto published; with notes and a life of the author. By Jared Sparks. Boston: Hilliard, Gray, and Company, 1836–40. 10 v. 4°. **IAW**

See v. 6, p. 452–454; v. 10, p. 13–14, 32–33, 123–125, 146–147, 186–187, for letters concerning balloons.

585. Franklin investigates aerostatics. illus. (Sportsman pilot, Albany, N. Y., 1932, v. 7, Jan., p. 26, 40–41.) **VDS**

Franklin's comments on the early ascensions in France.

586. Franklin's aeronautical correspondence. (Aeronautical annual, Boston, 1895, no. 1, p. 115–122.) **3-VDS**

Reprints of letters, 1783–86.

587. Fribourg, André. L'aviation et le public au XVIIIe siècle. (Revue de Paris, Paris, 1911, année 18, v. 6, p. 837–858.) *** DM**

What Frenchmen thought of the new art of aeronautics, with excerpts from various writings, especially those of David Bourgeois.

588. From America to England by balloon. (Nature, London, 1873, v. 8, p. 364–366.) **OA**

Includes a letter from Professor Joseph Henry to Professor Wise concerning the latter's proposed balloon journey across the Atlantic ocean.

589. Frossard, Benjamin Sigismond. An account of the last experiment made by M. Montgolfier, with his grand aërostatic machine, launched at Lyons, on Monday, January 19, 1784, carrying the celebrated inventor

Lighter-Than-Air Craft — Balloons, continued

himself, and six other persons. (Universal magazine, London, 1784, v. 74, p. 126–128.) * DA

590. A **Full** account of the grand aerostatic machine launched at Lyons, January 19, 1784. (London magazine, London, 1784, new series, v. 2, p. 203–205.) * DA

591. **Fuller,** Moyra. A balloon ascent from Cremorne Gardens, 1854. illus. (Popular flying, London, 1934, v. 3, p. 246–247, 268.) VDS

Reproduction of a painting on glass. The article describes some of the famous ascents from the place, 1837–74.

592. **G.,** L. Cenni storici sull' aeronautica fino alle recenti ascensioni fatte dal Sig. Green e compagni da Londra e da Parigi, con appendice fino agli ultimi voli e tentativi per la direzione degli aerostati. Firenze: Birindelli, 1838. 175 p. 12°. VDW

Record of early balloon ascensions. See index for extensive list of aeronauts.

593. **Gärtner,** Jean Paul. Des Herrn Jean Paul Gärtner Beschreibung seiner am 13. Aprill 1803 mit Herrn Garnerin und dessen Gattin unternommen Luftreise zu Berlin. illus. (Illustrierte aëronautische Mitteilungen, Strassburg, 1905, Jahrg. 9, p. 237–244.) VDS

594. **Genuine** memoirs of Mr. Blanchard. Embellished with a striking likeness, elegantly engraved by Thornton. port. (New London magazine, London, 1785, v. 1, p. 92.) * DA

595. **George** Washington was air-minded. illus. (Popular aviation, Chicago, 1931, v. 8, June, p. 25, 56.) VDS

Account of Blanchard's balloon ascension, Philadelphia, 1793.

596. **Gerlach,** A. Die Anfänge der Luftschiffahrt. Aus Berichten der Zeitgenossen ... Hamburg: A. Jensen, 1910. 114 p. illus. 12°. VDT

History of balloon ascensions, 1783–84, with reproductions of old cuts.

597. **Gerli,** Agostino. Opuscoli di Agostino Gerli. Parma: Dalla Stamperia Reale, 1785. 99 p. illus. f°. 3–† MAS

Includes a description and excellent engravings of the balloon constructed by the author and his two brothers.

598. **Giacomelli,** R. L'aeronautica nelle città italiane. Firenze. illus. (Rivista aeronautica, Roma, 1927, anno 3, Dec., p. 185–189.) VDS

Early history of ballooning in Florence. For continuation see articles by G. Boffito, entry no. 475.

599. **Giffard,** Henri. Improvements in balloons. illus. (English mechanic, London, 1868, v. 7, p. 375, 379–380.) VA

Illustrations and descriptions of Giffard's captive balloon of 1867.

600. —— Système d'aérostats captifs et fabrication industrielle du gaz hydrogène. (Génie industriel, Paris, 1868, tome 36, p. 150–152.) VA

601. **Giroud de Villette,** Andre. Lettre. (L'Esprit des journaux, Paris, 1783, année 12, tome 12, Dec., p. 339–340.) * DM

Author's experience as companion to de Rozier in captive Montgolfier balloon.

602. **Giulini,** Gino. L'aerostatico nel secolo XVIII. Un primato da rivendicare. illus. (Le Opere e i giorni, Genova, 1928, anno 7, no. 7, p. 44–51.) * DO

Accounts of Paolo Andreani and the brothers Gerli, pioneer Italian balloonists. Interesting plates.

603. **Glaisher,** James. The first balloon voyage in England. illus. (Good words, London, 1863, v. 4, p. 719–725.) * DA

Lunardi's ascension, 1784.

604. **Godard,** Louis. Des records aérostatiques. (L'Aéronaute, Paris, 1904, année 37, p. 106–115, 129–132.) 3–VDS

Chronological list of notable balloon records for duration, height, and distance, 1803–1904.

605. **Goodman,** Henry. "The Enterprise" goes up...illustrated by Stewart C. Hinds. illus. (Air travel, New York, 1917, v. 1, p. 162–164, 183.) VDS

Story of observation balloon used by the French against the Austrians in the battle of Fleurus, 1794. This is stated to be the first use of aircraft in war.

606. **Grand** balloon. (Mechanic's magazine, London, 1825, v. 4, p. 408.) 3–VA

States that Selim Ogal, of Smyrna, has completed one of the largest balloons ever made — 140 feet in diameter.

607. Les **Grandes** ascensions. illus. (L'Aérophile, Paris, 1906, année 14, p. 296–298.) 3–VDS

Balloon flight over the Alps by Usuelli and Crespi in the *Milano*, 1906.

607a. **Great** balloon ascension. illus. (Gleason's pictorial drawing-room companion, Boston, 1851, v. 1, p. 121.) * DA

Ascension of Mlle. Delon, Philadelphia, 1851.

608. **Green,** Charles. On the height attainable by balloons. (Mechanic's magazine, London, 1837, v. 27, p. 79–80.) 3–VA

609. —— Proposition to cross the Atlantic in a balloon. (Mechanic's magazine, London, 1839, v. 32, p. 185–186.) 3–VA

610. **Gregory,** Austin. America's foremost and most famous aeronaut, Captain Thomas Scott Baldwin. port. (Aeronautics, New York, 1908, v. 3, June, p. 36–41.) VDS

611. **Gregory,** George. A dictionary of arts and sciences. London, 1806. 2 v. illus. 4°. 3–VB

See article entitled Air balloon in v. 1, p. 35–39, for good history of the balloon. Same in 2. ed., 1815–16, * *AL;* also in American ed., 1821–22, * *AK.*

612. **Grimaldi,** Gabbriello. Memorie sopra la direzione, utilitatà, ed invenzione dei globi aerostatici recitate nella R. Accademia fiorentina. Firenze: G. Cambiagi, 1788. 88 p. 8°. VDT (Room 120)

Lighter-Than-Air Craft — Balloons, continued

613. **Grosse,** Margarete. Hochgebirgsfahrten. illus. (Deutsche Luftfahrer Zeitschrift, Berlin, 1912, Jahrg. 16, p. 400–402, 417–420, 443–445, 512–515, 534–538, 563–566.) **VDS**

Illustrated history of Alpine balloon ascensions.

614. **Grüel,** C. A. Ueber die Darstellung von Aerostaten aus Collodion. (Annalen der Physik und Chemie, Leipzig, 1848, Bd. 75, p. 333–335.) **PAA**

Proposes to spray balloons with collodion.

615. **Gustave** Hermite. port. (L'Aérophile, Paris, 1914, année 22, p. 407.) **3–VDS**

Obituary of French balloon pioneer, who invented several aeronautical instruments.

616. **H.,** E. Rarefied air balloons. (American journal of science, New Haven, 1827, series 1, v. 12, p. 372–373.) **OA**

How an experimental hot-air balloon was made at Amherst College.

617. **Hackstetter,** K. Die Geschichte der Luftschiffahrt. illus. (In: Die Luftschiffahrt dem heutigen Stande... Stuttgart, 1908, p. 11–49.) **VDY**

Illustrated history of balloons.

618. **Hagen,** Freiherr vom. Ein Blick in die Ballonwerkstatt. (Zeitschrift für Luftschiffahrt, Berlin, 1884, Jahrg. 3, p. 225–234, 258–270.) **3–VDS**

The making of a balloon with frequent reference to previous methods.

619. —— Die Opfer der Luftschiffahrt. (Zeitschrift für Luftschiffahrt, Berlin, 1884, Jahrg. 3, p. 1–14.) **3–VDS**

Record of balloon casualties beginning with Pilâtre de Rozier (1785).

620. —— Ueber Ballonhüllen aus Metall. (Zeitschrift für Luftschiffahrt, Berlin, 1885, Jahrg. 4, p. 149–156, 169–171.) **VDS**

History of various suggestions for metal balloons.

621. —— Ueber Heissluftballons und einige Abarten derselben. (Zeitschrift für Luftschiffahrt, Berlin, 1884, Jahrg. 3, p. 299–305, 327–335, 367–372.) **3–VDS**

Comparative advantages of gas and hot-air balloons. Historical ascensions of the latter. Various types of hot-air balloons and attempts to combine this type with the gas balloon.

622. **Hastings,** George E. Notes on the beginnings of aeronautics in America. (American historical review, New York, 1919, v. 25, p. 68–72.) *** R-IAA**

Early ballooning.

623. **Hay,** James, jr. America's first air journey. illus. (Aero digest, New York, 1930, v. 16, April, p. 77, 242, 244, 246.) **VDS**

Account of Jean Pierre Blanchard's ascent over Philadelphia in 1793.

624. **Hazen,** H. A. Glaisher's highest balloon ascension. (Aeronautical journal, London, 1899, v. 3, p. 13.) **VDS**

Discussion as to height attained.

625. **Heath,** John. On the control of balloons by pressure. illus. (Engineering, London, 1868, v. 5, p. 318.) **VDA**

Abstract of paper read before the Aeronautical Society of Great Britain. Regulates ascent and descent by means of a belt around the balloon.

626. **Hengler,** F. Einiges über Aëronautik. (Dingler's polytechnisches Journal, Stuttgart, 1832, Bd. 47, p. 321–341.) **3–VA**

Discusses faults of the balloon and suggests improvements.

627. **Henry,** Edmond. Étude sur la forme des aérostats. diagrs. (Revue de l'aéronautique théorique et appliquée, Paris, 1893, année 6, p. 20–36.) **3–† VDS**

628. **Herbert,** Charles. Over the Alps in a balloon. illus. port. (Strand magazine, London, 1899, v. 18, p. 175–183.) *** DA**

Edward Spelterini's flight, with photographs.

629. **Hergesell,** Hugo. Die Vertikalbewegungen eines Freiballons. (Illustrierte aëronautische Mittheilungen, Strassburg, 1899, Jahrg. 3, p. 101–112.) **VDS**

630. **Hersey,** Henry B. Experiences in the sky. illus. (Century magazine, New York, 1908, v. 75, p. 643–655.) *** DA**

American and French experiences of the author, an inspector of the United States Weather Bureau.

631. **Hervé,** Henri. Matériel aéronautique. Fascicule VIe: Les ancres de cape. Paris: "Le Yacht," 1900. 216 p. illus. 4°. (Revue de l'aéronautique, tome 15.) **3–† VDS (Revue)**

A detailed history of marine and aerial drag-anchors.

632. —— Historique des engins de ralentissement et des déviateurs aériens. (L'Aéronaute, Paris, 1898, année 31, p. 104–108.) **3–VDS**

History of balloon anchors and deviators.
For list of Hervé's writings on balloons and airships see Paul Brockett, *Bibliography of aeronautics*, Washington, 1910, p. 415–417, *VDT*.

633. **Hildenbrand.** Konstruktion, Leistungen und Schicksale der bisher gebauten Riesen-Fesselballons. (Zeitschrift für Luftschiffahrt, Berlin, 1889, Jahrg. 8, p. 81–88.) **3–VDS**

History of large captive balloons and accidents.

634. **Hirschauer,** Ch. Les premières expériences aérostatiques à Versailles. (19 septembre 1783–23 juin 1784.) illus. (Revue de l'histoire de Versailles et de Seine-et-Oise, Versailles, 1915/16, p. 26–48; 1917/18, p. 56–95.) **DQI**

635. An **History** of the discovery of the aërostatic machine, or air balloon; with a discussion of the philosophical principles on which it is constructed, and a full account of the late wonderful aërial excursions. (Universal magazine, London, 1784, v. 74, p. 18–23.) *** DA**

636. **Hoffman,** Bernard von. Experiences in the Gordon Bennett balloon race. illus. (Aviation, Highland, N. Y., 1921, v. 11, p. 503–504.) **VDS**

Author was pilot of the *City of St. Louis.*

Lighter-Than-Air Craft — Balloons, continued

637. **Hofmann,** A. W. Ballooning. (Franklin Institute, Journal, Philadelphia, 1876, v. 101, p. 123–126, 209–212.) **VA**

Deals mainly with the development of hydrogen in aeronautics.

638. **Hofwerberg,** Johannes. Constructionem geometricam navis aërostaticæ...proponit J. H. Lundae, 1786. 20 p. 8°. **VDC p.v.22, no.7**

Balloon calculations.

639. **Hopkins,** Manley. Balloons and aeronauts. illus. (Once a week, London, 1865, v. 13, p. 386–392, 526–532.) *** DA**

Brief history of balloons and parachutes.

640. **Hot** air ballooning, Weehawken, N. J. illus. (Scientific American, New York, 1891, v. 65, p. 143, 147–148.) **VA**

Interesting pictures of flights and inflation.

641. **Hureau de Villeneuve,** Abel. L'ordonnance concernant les montgolfières. (L'Aéronaute, Paris, 1877, année 10, p. 159–166.)

Facsimile of police regulations of 1784. **3–VDS**

642. **Illustrations** and illustrated clippings from English magazines and newspapers from about 1840 to 1890, relating to balloons and ballooning, mounted and bound on folio sheets. **†† VDW**

643. **Important** events in balloon history. (Aircraft year book, New York, 1919, p. 333–335.) **VDS**

644. **Improvement** in mooring balloons. illus. (Mechanic's magazine, London, 1837, v. 28, p. 40–42, 85, 114, 138.) **3–VA**

Letters by Edward C. Hitchings, William Baddeley, and Frederick Gye, jr.

644a. **Improvements** in aerostation. illus. (Illustrated London news, London, 1846, v. 8, p. 356.) *** DA**

Deviating and compensation apparatus of Charles Green. Double balloon device of MacSweeny.

645. **Inarre,** C. F. d'. Umstaendliche Beschreibung der aerostatischen Maschinen und der bisher damit angestellten Versuchen; nebst einer Anweisung wie solche sowohl im kleinen als grosen zu verfertigen und zu berechnen sind. Frankfurt: H. L. Brönner, 1784. 124 p. illus. 12°. **VDW**

Account of contemporary developments.

646. **Indicator,** pseud. Ballooning. (Mechanic's magazine, London, 1826, v. 6, p. 114–115, 231.) **3–VA**

Discussion of the tendency of balloons to burst at high altitudes.

647. **Institut de France.** — Académie royale des sciences. Rapport fait à l'Académie des sciences, sur la machine aérostatique, de Mrs. de Montgolfier; par Mrs. le Roy, Tillet, Brisson, Cadet, Lavoisier, Bossut, de Condorcet at Desmarest. (In its: Mémoires, 1783, Paris, 1786, p. 5–23.) *** EO**

648. An **Interesting** sketch of the distinguished aeronaut, John La Mountain. (Scientific American, New York, 1870, v. 22, p. 233.) **VA**

"La Mountain did not become an aeronaut for the purpose of the mountebank exhibitor... His higher object was to render aerial navigation of practical use in the great enterprise of modern progress and commerce."

649. **Jackson,** Joseph. The first balloon hoax. illus. (Pennsylvania magazine of history, Philadelphia, 1911, v. 35, p. 51–58.) *** R–IAA**

States that the reputed flight (1783) of David Rittenhouse and Francis Hopkinson never occurred.

650. **Jamieson,** Alexander. A dictionary of mechanical science, arts, manufactures, and miscellaneous knowledge... London: Henry Fisher, Son, and Co., 1827. 1066 p. illus. 4°. **VB**

See section on aerostation, p. 14–16, for principles, inflammable air, gas making, and construction of balloons.

651. **Jamin,** J. Les ballons. (Revue des deux mondes, Paris, 1885, tome 67, p. 113–141.) *** DM**

Early history of balloons and airships with especial notice of James Glaisher.

651a. **Janet.** On aerial travelling. (The Mirror, London, 1824, v. 3, p. 402–403.) *** DA**

Comments on the flight of Mr. and Mrs. Graham.

652. **Janssen,** Pierre Jules César. Voyage aéronautique du Volta, entrepris le 2 décembre 1870, en vertu d'une mission scientifique. (Académie des sciences, Comptes rendus, Paris, 1871, tome 73, p. 546–559.) *** EO**

Library has a second copy in † *VBA, p.v.16, no.3.*

For a portrait and a brief characterization of Janssen's career see *Aérophile,* Paris, 1894, année 2, p. 121–127, *3–VDS.*

653. **Jaubert,** George F. L'hydrolithe et la préparation instantanée de l'hydrogène pour le gonflement des ballons de campagne. (Revue générale de chimie pure et appliquée, Paris, 1907, tome 10, p. 266–268.) **3–PKA**

The greater part of this article is translated in *Scientific American,* New York, 1907, v. 97, p. 78–79, *VA.*

654. **Jeffries,** John. First aerial voyage across the English Channel. ports. (Magazine of American history, New York, 1885, v. 13, p. 66–88.) **IAA**

Diary of Dr. John Jeffries.

655. **Jones,** Bradley. The international balloon race [of 1933] is won by Poland. map. (U. S. air services, Washington, D. C., 1933, v. 18, Oct., p. 22–24.) **VDS**

Balloon was operated by Franciszek Hynek and Zbigniew Burzynsky.

656. —— This stratosphere business. (U. S. air services, Washington, 1934, v. 19, Jan., p. 14–16.) **VDS**

Brief accounts of stratosphere flights, 1931–1933.

Lighter-Than-Air Craft — Balloons, continued

657. Jotti, da Badia, polesine. L'aeronauta Francesco Arban. illus. maps. (Aeronautica, Milano, 1930, anno 4, p. 53–58, 123–129, 196–200, 293–296.) **3–VDS**

Illustrated account of ascents, 1845–1847.

658. —— L'aeronautica, a Gallarate. (Rivista aeronautica, Roma, 1931, anno 7, Feb., p. 401–402.) **VDS**

Ballooning (1784–1785) at Gallarte, northern Italy.

659. —— L'aerostatica nell' Emilia nell' anno 1784. (Rivista aeronautica, Roma, 1931, anno 7, May, p. 387–396.) **VDS**

Early balloon history in the province of Emilia, Italy.

660. —— Ai margini della storia. Documenti sull' italianità di Blanchard. port. (Rivista aeronautica, Roma, 1930, anno 6, p. 629–631.) **VDS**

Discussion of the nativity of the aeronaut Blanchard.

661. —— Documenti sopra le ascensioni aerostatiche eseguite da Francesco Arban... Raccolti, ordinati e notati da Jotti da Badia polesine. Milano: G. Modiano, 1930. xvii, 88 p. facsims., maps, plates, ports. 8°. **VDY**

Stories of Arban's ascensions, 1845–1849.

662. —— Esperienzi aerostatiche di Marsilio Landriani, Giacomo Veneziani e Francesco Maria Regibus. (Rivista aeronautica, Roma, 1931, anno 7, Oct., p. 201–210.) **VDS**

Ascensions, 1783–1784.

663. Kay, John. A series of original portraits, by the late John Kay, miniature painter, with biographical sketches and illustrative anecdotes. Edinburgh: Hugh Paton Carver and Gilder, 1837–42. 2 v. in 4. 8°. **CR (Edinburgh)**

See v. 1, p. 79–85 for portrait and account of Vincenzo Lunardi; p. 86–89 for picture of a group of aeronauts and an account of James Tytler.

664. Klaussmann, Anton Oskar Eugen. Abenteuer der Luft in Ballon und Flugmaschine... Illustrationen...von H. R. Schulze. Breslau: Phönix Verlag [1909]. iv, 236 p. ports. illus. 12°. **VDT**

A record of hazardous adventures and fatalities in balloon, airplane, and airship.

665. Kleinenn, H. Graf Franz Zambeccari, ein Vorgänger Zeppelins im 19. Jahrhundert. (Die Umschau, Frankfurt a. M., 1908, Jahrg. 12, p. 677–679.) **OA**

"Zambeccari [born 1756]...investigated the peculiarities of atmospheric currents and was the first to conceive the idea of controlling the course of a balloon by rising or sinking from an unfavorable to a favorable current in which oars could be used to advantage." Abstracted in *Scientific American supplement,* New York, 1908, v. 66, p. 215, *VA.*

666. Klemin, Alexander. Aids for stratosphere flight. ports. illus. (Scientific American, New York, 1934, v. 151, p. 147–150.) **VA**

Describes equipment for the ascension of William E. Kepner and Albert W. Stevens.

667. Knights of the Channel. (Flight, London, 1909, new series, v. 1, p. 432–433.) **VDS**

Table shows crossings of the English Channel, 1784–1908.

668. Kotzebue, August Friedrich Ferdinand von. Erinnerungen von einer Reise aus Liefland nach Rom und Neapel. Berlin: Heinrich Frölich, 1805. 3 Theilen. 16°. **BTYA**

See Theil 1, p. 125–141 for account of Zambeccari's ascension at Bologna; Theil 3, p. 356–372 for an interview with this aeronaut.

669. Kremser, Victor Felix Karl. Richard Opitz. (Zeitschrift für Luftschiffahrt, Berlin, 1893, Jahrg. 12, p. 80–81.) **3–VDS**

Brief biography of prominent German balloonist and instructor (1855–1892).

670. Kruckman, Arnold. Aeronaut Leo Stevens. illus. port. (Fly, Philadelphia, 1909, v. 1, Jan., p. 12–13.) **3–VDS**

Brief account of American aeronaut and airship builder.

671. Krull, Fritz. Der Hydrolith. (Zeitschrift für angewandte Chemie, Berlin, 1906, Jahrg. 19, Heft 27, p. 1233–1234.) **PKA**

Use of a hydrogen-calcium compound for hydrogen production.

672. Laforet, Auguste. Les ballons; depuis leur invention jusqu'à nos jours. (Revue de Marseille, Marseille, 1871, année 17, p. 109–125, 218–244, 248–269.) **DRN**

673. Lahm, F. S. A pioneer American aeronaut. illus. port. (Flying, New York, 1912–13, v. 1, no. 12, p. 25–26.) **VDS**

Portrait and biography of John Wise (1808–1879).

674. Laporte, Albert. Les naufrages aériens. Paris: T. Lefévre [188–?]. viii, 292 p. illus. 8°. **VDW**

Illustrated history of misadventures with balloons and parachutes.

675. Launay, L. de. La fièvre aérienne au XVIIIe siècle. illus. (La Nature, Paris, 1914, année 42, semestre 1, p. 341–345.) **OA**

Blanchard and Jeffries' crossing of the English Channel. Cuts from eighteenth-century French papers.

676. La Vaulx, Henri, comte de. L'aéronautique maritime. La traversée du "Méditerranéen." (Revue française de l'étranger, Paris, 1902, v. 27, p. 1–24.) **KAA**

677. —— De France en Russie en ballon. illus. map. (L'Aérophile, Paris, 1900, année 8, p. 164–174.) **3–VDS**

The author commanded the balloon *Centaure* in flight from Paris to Kiev.

678. —— L'emploi des ballons à ballonnet d'après la théorie du général Meusnier. illus. (Académie des sciences, Comptes rendus, Paris, 1903, tome 137, p. 749–753.) ***EO**

Illustrated description of a modern ballonnet balloon, is preceded by reference to earlier applications.

679. —— Expérience d'aéronautique sur la Méditerranée. illus. (L'Aérophile, Paris, 1902, année 10, p. 8–14, 39–44, 53–58.) **3–VDS**

680. —— La navigation aérienne. (Société normande de géographie, Bulletin, Rouen, 1904, année 26, p. 31–47.) **KAA**

Describes race between the *Centaure* and the *St. Louis,* from Paris to Kiev, 1900.

For list of references to writings by and about La Vaulx, see Paul Brockett, *Bibliography of aeronautics,* Washington, 1910, p. 500–502, *VDT.*

Lighter-Than-Air Craft — Balloons, continued

681. **La Vaulx,** Henri, comte de, and P. Tissandier. Joseph et Étienne de Montgolfier. Annonay, 1926. 76 p. illus. facsims., maps, ports. f°. † **VDW**

A de luxe biography with reproduction of prints and manuscripts.

682. **Lecocq,** Georges. La céramique et les aérostats. illus. (L'Aéronaute, Paris, 1877, année 10, p. 8–21.) **3–VDS**

Pictures and descriptions of decorated chinaware of the late eighteenth century.

683. **Ledeboer,** John Henry. Death of Nadar. port. (Aeronautics, London, 1910, v. 3, p. 72.) **VDS**

684. **Leher,** Max. Joseph Maximilian Freiherr v. Lütgendorf, der erste deutsche Luftschiffer. illus. (Illustrierte aëronautische Mitteilungen, Strassburg, 1903, Jahrg. 7, p. 281–306.) **VDS**

He made a balloon ascension in 1786.

685. —— "Die Kunst zu fliegen" in historischer Beleuchtung. illus. (Illustrierte aëronautische Mitteilungen, Strassburg, 1904, Jahrg. 8, p. 269–302.) **VDS**

Early balloons and attempts at airships, especially in Germany.

686. —— Lustige und traurige Episoden aus den ersten Jahren der Ballon-Aera (1783–1786). (Illustrierte aëronautische Mitteilungen, Strassburg, 1905, Jahrg. 9, p. 108–115, 304–312, 387–396; 1907, Jahrg. 11, p. 123–128, 210–215, 335–341; 1908, Jahrg. 12, p. 610–615, 681–683, 769–776.) **VDS**

687. **Leloup,** Jules. L'aérostat à dispositif équilibreur atmosphérique. Paris: P. Rosier [1909]. 14 p. 12°. **VDY**

Author proposes to regulate ascension by means of an interior "ballonet."

688. **Lesueur.** Relation des deux premières ascensions du ballon Montgolfier, en 1783, par un témoin oculaire. (Académie des sciences, Comptes rendus, Paris, 1851, tome 32, p. 727–728.) *** EO**

Farmers take the first balloon to be an animal of the air.

689. **Lettre** à M. de *** sur son projet de voyager avec la sphère aérostatique de M. de Montgolfier. Aéropolis: Zéphirolin, n. d. [Paris, 1783.] 16 p. illus. 12°. **VDT p.v.1, no.4**

Bound with: B. Faujas de St. Fond, Description des expériences.

690. **Lhoste,** F., and J. Mangot. Traversée de Cherbourg à Londres. illus. map. (La Nature, Paris, 1886, année 14, semestre 2, p. 183–186.) **OA**

Account of authors' voyage, July 29, 1886. Translation in *Scientific American,* New York, 1886, v. 55, p. 249–250, *VA.*

691. A **Life** preserver for balloonists. illus. (Scientific American, New York, 1909, v. 100, p. 282.) **VA**

Floats are attached to the car.

692. **Linke,** Franz. Moderne Luftschiffahrt. Berlin: A. Schall, 1903. 296 p. illus. 8°. **VDW**

The author, a member of the staff of the Geophysical Institute of Göttingen, tells of modern balloon and airship developments, principally in Germany.

693. The **Long** distance balloon races from St. Louis. illus. (Scientific American, New York, 1907, v. 97, p. 308–309.) **VA**

Gordon Bennett cup contest of 1907.

694. **Loring,** John. An aeronautical classic. The Gordon Bennett race, first held in 1909, is still an annual event of international importance. illus. (Air and airways, London, 1932, v. 9, p. 229–230.) **VDS**

695. **Lunardi,** Vincenzo. An account of the first aërial voyage in England, in a series of letters to his guardian, Chevalier Gherardo Compagni, written under the impressions of the various events that affected the undertaking. London: the author, 1784. 66 p. 2. ed. nar. 8°. **VDT (Room 120)**

696. —— An account of five aerial voyages in Scotland, in a series of letters to his guardian, Gerardo Compagni, written under the impression of the various events that affected the undertaking. London: the author, 1786. 114 p. illus. port. nar. 8°. **VDT (Room 120)**

Bound with his: An account of the first aërial voyage in England. London, 1784. 2. ed. nar. 8°.

697. **McClean,** Sir Francis Kennedy. From Zürich to Bohemia by balloon, being an account of the Gordon-Bennett balloon race, 1909. (Flight, London, 1909, v. 1, p. 763.) **VDS**

Author was the British representative in this race.

697a. **Magoun,** Frederick Alexander, and Eric Hodgins. The balloon's greatness and decline. illus. (In their: A history of aircraft. New York, 1931, p. 60–97.) **VDT**

The beginnings of military balloons. Zambeccari's disasters. Robertson's *Minerva.* Scientific ascents of Gay-Lussac. Great balloon of Nassau. Green's proposal to cross the Atlantic. Flights of Henry Coxwell. Flights of John Wise in America. First crossing of the Alps in a balloon. Flights and photographs of Nadar. Balloons in the American Civil war. Polar flight of Andrée.

697b. —— Historic early flights. illus. (In their: A history of aircraft, New York, 1931, p. 39–59.) **VDT**

Activities of Blanchard. Crossing the English Channel. First ascent in America. Early flights of women. Flights of James Sadler. Attempts of James Tytler. Flights of Vincent Lunardi. Richard Crosbie in Ireland. Attempt of Dr. Potain. Disaster to de Rozier.

697c. —— The invention of the "globes aerostatiques." illus. (In their: A history of aircraft, New York, 1931, p. 20–38.) **VDT**

Galien. Isolation of hydrogen. The brothers Montgolfier and Professor Charles. First public showing. Montgolfier's gas. Ascent from the Champ de Mars. Ascent at Versailles. First ascent of a human being. Ascent of Professor Charles.

698. **Maisoncourt,** marquis de la. Lettre au rédacteur des Affiches, annonces & avis divers, sur la fin malheureuse de MM. Pilâtre de Rozier & Romain. (L'Esprit des journaux, Paris, 1785, année 14, tome 8, Aug., p. 320–325.) *** DM**

Lighter-Than-Air Craft — Balloons, continued

699. **Mance,** H. Osborne. Long-distance ballooning. (Aeronautical journal, London, 1899, v. 3, p. 78–83.) VDS

Has historical data on the use of the trail rope.

700. **Marey-Monge,** Edmond. Sur un ballon à enveloppe en cuivre laminé. (Académie des sciences, Comptes rendus, Paris, 1844, tome 18, p. 419–421.) * EO

Abstract of author's communication on a copper-shelled balloon. Also in *Dingler's polytechnisches Journal,* Stuttgart, 1844, Bd. 92, p. 138–140, *3–VA.*

701. **Marion,** Fulgence. Les ballons et les voyages aériens, ouvrage illustré de 30 vignettes par P. Sellier. Paris: L. Hachette & Cie., 1869. 336 p. 2. ed. 12°. (Bibliothèque des merveilles.) VDW

Well-illustrated history of early ballooning.

702. —— Wonderful balloon ascents; or, The conquest of the skies. A history of balloons and balloon voyages. From the French of F. Marion. New York: Scribner, Armstrong & Co., 1874. xvi, 218 p. illus. 16°. (Illustrated library of wonders.) VDW

703. **Marquis,** Raoul. Albert Tissandier. [By Henry de Graffigny, pseud. of Raoul Marquis.] port. (L'Aérophile, Paris, 1901, année 9, p. 97–99.) 3–VDS

Portrait and brief account of French aeronaut, also noted as an illustrator. He was the brother of the well-known Gaston Tissandier.

704. —— Les ballons dans la littérature. (L'Aérophile, Paris, 1899, année 7, p. 15–18, 32–33.) 3–VDS

The balloon in French fiction.

705. —— L'électricité dans l'aérostation. illus. (La Lumière électrique, Paris, 1892, tome 43, p. 213–217.) VGA

Electric lights for balloons; electric propulsion; communication; reheating of gas.

706. —— Gabriel Mangin. port. (L'Aérophile, Paris, 1902, année 10, p. 49–51.) 3–VDS

Portrait and brief biography of French aeronaut and inventor.

707. —— Récits d'un aéronaute — Histoire de l'aérostation — Fantaisies aérostatiques, par H. de Graffigny [pseud.]... Paris: C. Delagrave, 1886. 2 parts in 1 v. illus. 2. ed. 4°. VDW

Includes some aeronautical tales by the author.

708. —— Traité d'aérostation théorique et pratique...par Henry de Graffigny [pseud.]. Guide complet à l'usage...de toutes les personnes s'intéressant à l'aérostation et à la navigation aérienne... Paris: Baudry et Cie., 1891. vii, 267 p. illus. 12°. VDY

Has a list of the author's ascensions; also of the 65 balloons used during the siege of Paris. Records.

709. **Marsh,** William Lockwood. A French balloon of 1872 whose motive power was eight men turning a crank. (Airways, London, 1927, v. 3, p. 198.) VDS

Story of Dupuy de Lôme's man-propelled airship.

710. **Marvin,** Charles Frederick. Note upon economical shapes for cutting envelopes for balloons. (United States Weather Bureau, Monthly weather review, Washington, 1904, v. 31, p. 314–317.) † PRS

710a. **Mason,** Thomas Monck. Account of the late aeronautical voyage from London to Weilburg. (The Mirror, London, 1837, v. 29, p. 10–13.) * DA

Green's voyage, 1836.

711. —— Aeronautica; or, Sketches illustrative of the theory and practice of aerostation: comprising an enlarged account of the late aerial expedition to Germany. London: F. C. Westley, 1838. 355 p. illus. ports. 8°. VDW

Has portraits of Robert Holland, Monck Mason, and Charles Green.

712. **Mathon de la Cour.** Lettre...sur l'expérience du grand ballon de M. Montgolfier. (L'Esprit des journaux, Paris, 1784, année 13, tome 3, March, p. 312–322.) * DM

Describes tribulations of the ascent of Pilâtre de Rozier, Montgolfier, and five others at Lyons, 1784.

713. **Meek,** Robert. A biographical sketch of the life of James Tytler, for a considerable time a liberal contributor to the Encyclopœdia Britannica. Edinburgh: Denovan, 1805. 71 p. 16°. * C p.v.2243

Brief account with a portrait of Edinburgh balloon enthusiast who failed to accomplish anything of importance.

714. **Mehl,** Adolf. Geschichte des Freiballons. illus. (In his: Der Freiballon in Theorie und Praxis, Stuttgart, 1911, Bd. 1, p. 11–25.) VDW

Concise early history.

715. **Meissel,** E. Ueber die Verwendung eines grossen Luftschiffes zu geographischen Entdeckungsreisen. illus. (Zeitschrift für Luftschiffahrt, Berlin, 1882, Jahrg. 1, p. 10–23.) 3–VDS

Describes a double balloon for the purpose. For comments by Buchholtz see p. 44–46.

716. **Meylan,** E. L'électrolyse et le gonflement des ballons. (La Lumière électrique, Paris, 1889, tome 31, p. 277–278.) † VGA

Description of the Latchinoff process.

717. **Mr.** Green's grand new balloon. (Mechanic's magazine, London, 1836, v. 25, p. 395–396.) 3–VA

Description of craft, 157 feet in circumference.

718. **Mr.** Green's intended aerial voyage across the Atlantic. illus. (Mirror, London, 1840, v. 35, p. 221–223.) * DA

719. **Moedebeck,** Hermann W. L. Blanchard à Francfort-sur-Mein en 1785. (L'Aérophile, Paris, 1897, année 5, p. 159–161.) 3–VDS

720. —— Coxwell. port. (Illustrierte aëronautische Mitteilungen, Strassburg, 1900, Jahrg. 4, p. 53–54.) VDS

Biography of Henry Coxwell (1819–1900), English balloonist and associate of James Glaisher.

Lighter-Than-Air Craft — Balloons, continued

721. **Monaghan,** Frank. With Washington's good wishes. How the Father of his Country fathered the first American flight and saved Blanchard from walking home. illus. (Sportsman pilot, Albany, N. Y., 1932, v. 8, Oct., p. 30–31, 45.) **VDS**

722. **M.** Gaston Tissandier. (Aeronautical journal, London, 1899, v. 3, p. 91.) **VDS (Royal)**

Brief biography of famous French balloonist and author. He and his brother founded the well-known periodical *La Nature.*

723. **Monsieur** Menier's hot-air balloon. illus. (Engineer, London, 1874, v. 38, p. 217.) **VA**

Burns paraffine oil. Experiments made at the Woolwich arsenal. Abridged in *Scientific American,* New York, 1874, v. 31, p. 377, *VA.*

724. **Montgolfiere** lancée à Versailles, le 23 juin 1784. (L'Esprit des journaux, Paris, 1784, année 13, tome 8, August, p. 338–341.) *** DM**

Ascension of Pilâtre de Rozier.

725. **Montgolfieri,** Umbra, pseud. Aeronautics. Comparative safety of Montgolfier and gas balloons. Proposal for an aeronautic club. (Mechanic's magazine, London, 1836, v. 25, p. 441–442.) **3–VA**

726. **Moore,** Richard Bishop. Helium: its history, properties, and commercial development. illus. (Franklin Institute, Journal, Philadelphia, 1921, v. 191, p. 145–197.) **VA**

727. **Morazzoni,** Giuseppe. Un pioniere dell' aeronautica, Vincenzo Lunardi [1759–1806]; documenti inediti e saggio iconografico... Milano: Antiquariato W. Toscanini, 1931. 157 p. illus. ports. 4°. **VDW**

Has illustrations of balloons, portraits, and poetry.

728. **Mortane,** Jacques. L'historique du sphérique. (In: L'Aéronautique, par Roland Garros [and others], Paris: Pierre Lafitte & Cie., 1914, p. 1–21.) **VDW (Aéronautique)**

729. **Müllenhoff,** Karl. Ueber die Wirkung bedeutender Luftverdünnung auf den menschlichen Körper. (Zeitschrift für Luftschiffahrt, Berlin, 1891, Jahrg. 10, p. 3–8.) **3–VDS**

Effects of altitude on early balloonists.

730. **Munro,** Robert. Improved double balloon, with hydrogen and carbonic acid gases. illus. (Mechanic's magazine, London, 1836, v. 25, p. 393–394.) **3–VA**

731. —— Mode of preserving the gas of balloons. Carbonic acid for ballast. (Mechanic's magazine, London, 1836, v. 26, p. 101–102.) **3–VA**

Manufactures carbonic acid gas in transit.

732. **Myers,** Carl E. The construction of balloons. (Scientific American supplement, New York, 1889, v. 28, p. 11596.) **VA**

733. —— Manœuvring of balloons. (International Conference on Aerial Navigation, Chicago, 1893, Proceedings, New York, 1894, p. 346–349.) **VDS**

Experiences of author and his condemnation of parachute descents.

734. —— Natural gas balloon ascensions and change of temperature in gas balloons. (International Conference on Aerial Navigation, Chicago, 1893, Proceedings, New York, 1894, p. 331–333.) **VDS**

Author's experiences.

735. —— Sphere of the gas balloon. ports. (Fly, Philadelphia, 1909, v. 1, April, p. 3–4.) **3–† VDS**

Author does not believe in the "assumed supremacy of the coming flying machine, citing some of the ascensions of himself and of Mrs. Myers."

For references to Myers' writings on balloons see Paul Brockett, *Bibliography of aeronautics,* Washington, 1910, p. 613–614, *VDT.*

736. **N.** [Aerostatical experiments lately made in France and in London.] (London magazine, London, 1783, new series, v. 1, p. 499–502, 566–567.) *** DA**

Explanation of balloon principles.

737. **N.,** K. Die Durchquerung der Sahara mittelst Ballon-Schlepp-Fahrt. illus. (Illustrierte aëronautische Mittheilungen, Strassburg, 1903, Jahrg. 7, p. 36–49.) **VDS**

Explains Captain Deburaux's proposition to cross the Sahara desert by means of a balloon guided by means of a drag anchor.

738. The **Nassau** balloon. (Mechanic's magazine, London, 1840, v. 33, p. 160.) **3–VA**

Sold in bankruptcy auction to Charles Green.

739. The **National** balloon race [of 1926]. Ward T. Van Orman wins Litchfield trophy with balloon flight of over 800 miles. port. (Aviation, Highland, N. Y., 1926, v. 20, p. 708.) **VDS**

740. The **National** elimination balloon race. (Aviation, Highland, N. Y., 1925, v. 18, p. 548.) **VDS**

The race for 1925 was won by Ward T. Van Orman, the winner of the 1924 event.

741. **National Geographic Society.** U. S. Army Corps stratosphere flight of 1934 in the balloon "Explorer." Washington: National Geographic Society, 1935. (Contributed technical papers. Stratosphere series. no. 1.)

742. Das **Neue** System der Luftschiffahrt von Dr. van Hecke. (Dingler's polytechnisches Journal, Stuttgart, 1847, Bd. 103, p. 312.) **3–VA**

Proposes to elevate or depress a balloon by means of a motor.

743. **Neuhäuser,** O. Ueber metallene Ballonhüllen. (Zeitschrift für Luftschiffahrt, Berlin, 1883, Jahrg. 2, p. 282–284.) **3–VDS**

Author suggests the use of copper-plated aluminium.

Lighter-Than-Air Craft — Balloons, continued

743a. The **New** great Montgolfier balloon. illus. (The Mirror, London, 1838, v. 31, p. 338, 358, 359, 373–375.) ***DA**

Description of giant smoke balloon at the Surrey Zoological Gardens, with a history of earlier Montgolfier craft.

744. A **New** sectional balloon. illus. (Scientific American, New York, 1874, new series, v. 31, p. 99.) **VA**

Patent of James Hartness, Detroit, Mich., is composed of parts resembling in form those of an orange.

745. **New** system of aero-locomotion. illus. (Scientific American, New York, 1867, new series, v. 16, p. 197.) **VA**

Invention of J. A. A. Fontaine, of New York, adapts a balloon to an inclined railway.

746. [A **Newly** discovered portrait of Vincent Lunardi.] port. (Aeronautical journal, London, 1910, v. 14, opp. p. 158.) **VDS**

Silhouette by Torond, 1784.

747. [**Newspaper** clippings about early aeronautics and Charles F. Durant's experiences and accomplishments as an aeronaut, silk culturist, naturalist and scientist.] 2 v. **3–VDT** and **3–† VDT**

Two scrapbooks covering the period 1830–1851.

748. **Nicholson,** William. Short account of some of the most remarkable facts and observations in an aerostatic voyage, made from Petersburgh, by Messrs. Robertson and Sacharoff, under the sanction of the Imperial Academy. By W. N. (Journal of natural philosophy, [Nicholson's], London, 1805, v. 11, p. 52–55.) **VDS**

748a. **Niemeyer,** Victor. Wasserfahrten. illus. (In: Bröckelmann, Wir Luftschiffer, Berlin, 1909, p. 393–407.) **VDW**

Notable cross-Channel, Baltic, and Mediterranean balloon flights. Andrée's Polar flight.

749. **Notable** balloon ascent. (Aeronautical journal, London, 1898, v. 2, p. 83–85.) **VDS**

Ascent of Stanley Spencer and Dr. Berson to 27,000 feet, 1898.

750. **Notice** sur l'art aérostatique; par un aéronaute; suivi d'une nomenclature biographique des aéronautes célèbres. Lille, 1845. 15 p. 8°. **VDT p.v.2, no.8**

Includes a list of aeronauts. Brief biographies are given for a few of these.

751. **Nouvel** appareil de M. Gaston Tissandier pour la fabrication du gaz hydrogène. illus. (Société d'encouragement pour l'industrie nationale, Bulletin, Paris, 1884, série 3, tome 11, p. 148–152.) **3–VA**

752. **Nouvelle** expérience faite avec le globe aérostatique. (L'Esprit des journaux, Paris, 1783, année 12, tome 10, Oct., p. 333–344.) ***DM**

Account of the balloon of Dr. Charles and the Robert brothers; also of the career of the Montgolfiers.

753. A **Novel** balloon farm. illus. (Aeronautical world, Glenville, Ohio, 1903, v. 1, p. 220–224.) **VDS**

Account of Carl E. Myers' establishment at Frankfort, N. Y.

754. **Observations** au sujet de l'invention de MM. de Montgolfier. (L'Esprit des journaux, Paris, 1784, année 13, tome 1, Jan., p. 377–380.) ***DM**

A prophecy of what the balloon might do for science and the arts.

755. **Old** stories retold. Balloon ascents. (All the year round, London, 1867, v. 18, p. 447–453.) ***DA**

History of ballooning to 1836.

756. **On** air balloons. (Mechanic's magazine, London, 1826, v. 6, p. 92–95, 102–104.) **3–VA**

List of early flights in warm-air balloons.

757. Les **Origines** du gaz d'éclairage et les ballons. illus. (L'Aérophile, Paris, 1905, année 13, p. 10–19.) **3–VDS**

A discussion as to the inventor of coal gas and as to its first use in the inflation of balloons.

758. **Our** progress in the art of making balloons. (Once a week, London, 1869, new series, v. 4, p. 96–98.) ***DA**

Includes an account of Zambeccari's experiment, London, 1783; also of anonymous author's ascent in Giffard's captive balloon, London, 1869. Also in *Scientific American,* New York, 1869, new series, v. 21, p. 257–258, *VA.*

759. **Over** the water. (Household words, London, 1853, v. 7, p. 483–488.) ***DA**

Cross-Channel flight of Blanchard and Jeffries, 1785.

760. **Pappafava,** Vladimir. Un précurseur, l'aéronaute comte Francesco Zambeccari. Paris: Per Orbem, 1924. 26 p. 8°. **VDY p.v.34, no.7**

761. **Parseval,** August von. Der Drachen-Ballon. illus. (Zeitschrift für Luftschiffahrt, Berlin, 1896, Jahrg. 15, Beilage, p. 1–32.) **3–VDS**

Author's researches on the kite balloon.

762. —— The dragon balloon. illus. (Journal of the United States Artillery, Fort Monroe, Va., 1900, v. 13, p. 312–320; v. 14, p. 310–325.) **VWA**

763. **Perrotin,** Henri. Les premières expériences d'aérostation faites en Lorraine. illus. (L'Aérophile, Paris, 1917, année 25, p. 255–257.) **3–VDS**

Record of events, 1783–1787.

764. **Pilâtre de Rozier,** François. Extrait d'une lettre à M. le Roi, de l'Académie des sciences...sur son dernier voyage aérien. (L'Esprit des journaux, Paris, 1784, année 13, tome 9, p. 345–352.) ***DM**

765. **Platte,** August. Ein Ballon mit Segelfläche. (Zeitschrift für Luftschiffahrt, Berlin, 1884, Jahrg. 3, p. 271–284.) **3–VDS**

Theory of sails for balloons. In a footnote are several references to previous literature.

Lighter-Than-Air Craft — Balloons, continued

766. **Poeschel,** Johannes. Luftreisen. Leipzig: F. W. Grunow, 1908. 237 p. illus. maps. 8°. **VDW**

Author's balloon ascensions, 1904–1907.

767. **Poole,** John. Crotchets in the air; or, An (un)scientific account of a balloon-trip, in a familiar letter to a friend. London: Henry Colburn, 1838. 98 p. 8°. **VDW**

Humorous account of a trip with Charles Green.

768. **Prechtl,** Johann Joseph, Ritter von. Ueber die Mittel, den Luftbällen eine sichere und dauerhafte Konstruktion zu geben, damit sie bequem als Luftschiffe zur Unternehmung grosser Reisen gebraucht werden können. (K. K. polytechnisches Institute in Wien, Jahrbuch, Wien, 1924, Band 5, p. 99–113.) **3–VA (Technische)**

Proposes a thin copper globe with interior ballonets.

769. Le **Premier** voyage aérien du physicien Charles. (Vie des peuples, Paris, 1923, année 4, p. 751–765.) *** DM**

Charles' account of his first balloon ascension, 1783.

770. **Preston,** R. A. D. Winning the national balloon championship. illus. (Aero and hydro, Chicago, 1913, v. 6, p. 369–370.) **3–VDS**

Account of competition starting at Kansas City in 1913 and won by the *Goodyear.*

771. **Priestley,** Joseph. Experiments and observations on different kinds of air...in three volumes; being the former six volumes abridged and methodized, with many additions. Birmingham: Thomas Pearson, 1790. 3 v. 8°. **PLO**

See v. 1, p. 290–297 for his account of making hydrogen for balloons.

772. Le **Prime** ascensioni aeree in Italia. illus. (Almanacco aeronautico, Milano, 1932, anno 3, p. 37–44.) **VDS**

Ascension of Lunardi, 1789.

773. The **Principles,** history, & use of air-balloons; also a prospectus of Messrs. Blanchard & Baker's intended aerial voyage from the city of New York. New York, 1796. 46 p. 12°. *** KD (1796, Principles)**

History of early ballooning, including an account of Blanchard's ascension at Philadelphia in 1793, also a chronological list of his earlier ascents.

774. The **Progress** of practical aeronautics during 1899. illus. (Scientific American, New York, 1900, v. 82, p. 117.) **VA**

Carl E. Myers' experiments with captive balloons.

774a. **R——y,** F. On balloons and parachutes. (The Mirror, London, 1823, v. 2, p. 259–261.) *** DA**

Early history of each.

775. **Rae,** John. Notice of the effect of solar heat in raising a balloon, in a letter to the editor, dated Hamilton, Upper Canada, April 17th, 1837. (American journal of science, New Haven, 1838, v. 33, p. 196–198.) **OA**

Experiments with cylinders of blackened paper.

776. **Ramsay,** Sir William. The life and letters of Joseph Black, M.D., with an introduction dealing with the life and work of Sir William Ramsay, by F. G. Donnan. London: Constable and Company, Ltd., 1918. xix, 148 p. illus. ports. 8°. **AN**

Black (1728–1799) experimented with small hydrogen balloons as early as 1768. See p. 76–84.

777. **Rapport** fait à l'Académie des sciences, sur la machine aérostatique, inventée par MM. de Montgolfier. Paris: Imprimerie de Moutard, 1784. 27 p. 8°. **OAI p.v.14, no.12**

Reprinted in *Journal de physique,* Paris, 1784, tome 24, p. 81–94, *3–VOA.*

778. **Renard,** Charles. Étude sur les aérostats à volume maximum variable. illus. (L'Aéronaute, Paris, 1881, année 14, p. 97–108, 121–129, 146–155, 169–175, 193–204.) **3–VDS**

779. **Robert,** brothers. Lettre...aux auteurs du Journal de Paris, sur leur expérience aérostatique. (L'Esprit des journaux, Paris, 1784, année 13, tome 9, Sept., p. 334–337.) *** DM**

780. **Robertson,** Étienne Gaspard. Mémoires récréatifs scientifiques et anecdotiques du physicien-aéronaute E. G. Robertson... Paris, 1840. 2 v. illus. 8°. **AN**

Author's experiences as a balloonist in Russia, Germany, Sweden, and Austria. Also his relations with the physician Charles. For a picture of balloons used by the French army at Meudon during the Revolution, see opp. p. 19.

781. —— La Minerve, vaisseau aërien destiné aux découvertes, et proposé à toutes les académies de l'Europe. Vienne, 1804, réimprimé à Paris, 1820. 36 p. illus. 2. ed. 8°.

Amusing illustrations. **VDT p.v.2, no.3**

782. **Roch,** Eugène. Bosquejo de los viajes aéreos de Eugenio Robertson en Eurpopa, los Estados Unidos y las Antillas...traducido del frances por D. José María Heredia. Mejico: Galvan, 1835. 84 p. 24°. **VDY**

Account of the ascents of Eugène Robertson (1819–1836), principally in America.

783. **Rolls,** Charles Stewart. The Bath balloon centenary. illus. (Flying, London, 1902, no. 4, Dec., p. 149–155.) **VDS**

Includes an account of Garnerin's ascent in 1802.

784. **Romocki,** S. J. von. Geschichte der Explosivstoffe. Berlin: Robert Oppenheim, 1895–96. 2 v. illus. 8°. **VOG**

See v. 1, p. 161–162 for a brief reference to early use of dragon hot-air balloons.

785. **Rose,** Friedrich. Ueber einen Fallschirmballon. illus. (Zeitschrift für Luftschiffahrt, Berlin, 1893, Jahrg 12, p. 24–25.) **3–VDS**

Description and illustrations of Capazza's combined balloon and parachute.

Lighter-Than-Air Craft — Balloons, continued

786. Rotch, Abbott Lawrence. Benjamin Franklin and the first balloons. (American Antiquarian Society, Proceedings, Worcester, Mass., 1907, new series, v. 18, p. 259–274.) **IAA**

Letters written from France in 1783.

787. —— The highest balloon ascension in America. (Science, New York, 1909, new series, v. 30, p. 302–303.) **OA**

Account of *ballon sonde* which rose to 17,700 meters in 1908 from Pittsfield, Mass.

788. Rouland. Tableau historique des propriétés et des phénomènes de l'air... Paris: Gueffier, 1784. xvi, 636 p. 8°. **PLO**

See p. 311–371 for accounts of the first balloon ascensions in France.

789. Rowland, John T. Sky boats. The "blimp" fills a definite niche in the general scheme of aeronautics; the author makes a splendid case for its versatile accomplishments. illus. (Scientific American, New York, 1934, v. 150, p. 117–119.) **VA**

790. Russian ascent to the stratosphere. (Science, New York, 1933, v. 78, p. 328.) **OA**

791. Sacharow. Bericht...an die kaiserl. Akad. der Wissenschaften zu Petersburg, über die Luftfahrt, welche er zu Folge ihres Auftrags in Begleitung des Physicus Robertson am 30sten Junius 1804 unternommen hat. (Annalen der Physik, Halle, 1805, Bd. 20, p. 107–127.) **PAA**

792. Sailing in the air. History of aeronautics. (Harper's magazine, New York, 1851, v. 2, p. 168–173.) *** DA**

History of ballooning.

793. Saint-Félix, T. Notice biographique sur Eugène Godard, aéronaute se S. M. l'Empereur à la campagne d'Italie. Paris, 1862. 16 p. 12°. **AGI p.v.4, no.18**

794. Saladini, Girolamo. Memoria intorno ai globi aereostatici. Recitata ai 13 brinatore anno IX repubblicano (4 dicembre 1800 v. s.). Bologna, 1800. 16 p. 8°. **VDT p.v.2, no.7**

Calculations for a balloon.

795. Samuel A. King, the oldest living aeronaut. port. (Fly, Philadelphia, 1908, v. 1, Nov., p. 8.) **3–† VDS**

It is stated that King, in 1857, was the first aeronaut successfully to use the drag rope.

796. Sander, Alexander. Die Darstellung der Ballongase. (Zeitschrift für angewandte Chemie, Leipzig, 1911, Jahrg. 24, p. 193–200.) **PKA**

Brief histories of use of hot-air, illuminating gas, and hydrogen. Abstract in translation, *Scientific American supplement*, New York, 1911, v. 71, p. 210–211, *VA*.

797. —— Die erste Anwendung des Steinkohlengases in der Luftschiffahrt. (Journal für Gasbeleuchtung und Wasserversorgung, München, 1913, Jahrg. 56, p. 1262–1264.) **3–† VOA**

Priority is claimed for J. P. Minckelers in the first use of coal gas for balloon purposes (1783).

798. Sartwell, Franklin G. An early Atlantic flight that came to grief. (Literary digest, New York, 1928, v. 98, Sept. 22, p. 64, 66–69.) *** DA**

Attempt of Donaldson and Wise, 1871.

799. Serph, André. Le centenaire du physicien Charles. illus. (L'Aérophile, Paris, 1923, année 31, p. 131–133.) **3–VDS**

M. Serph, grand-nephew of the aeronaut Charles, presents notes on the latter's ascensions during the eighteenth century.

800. Shaw, B. Russell. Ballooning and its history. (National Aeronautic Association review, Washington, 1925, v. 3, p. 34–36, 59–60, 63.) **VDS**

Concise account from earliest times, but largely deals with flight competitions.

801. Silberer, Herbert. Viertausend Kilometer im Ballon. Leipzig: O. Spamer [1903]. 136 p. illus. 8°. **VDW**

Author's balloon ascensions, 1899–1903, with excellent aerial photographs.

802. Silberer, Victor. Gaston Tissandier. (Zeitschrift für Luftschiffahrt, Berlin, 1899, Jahrg. 18, p. 304–306.) **3–VDS**

Biography of French balloonist, airship builder, and editor of the periodical *La Nature*.

803. Sircos, A., and T. PALLIER. ...Histoire des ballons et des ascensions célèbres, avec une préface de Nadar. Dessins de A. Tissandier et des meilleurs artistes. Paris: F. Roy, 1876. 476 p. illus. 4°. **VDW**

Excellent early history with numerous engravings.

804. Small balloons provide new sport. illus. (Scientific American, New York, 1927, v. 137, p. 39.) **VA**

Describes balloon jumping.

805. Snow, Julia D. Sophronia. The "Clayton's ascent" bandbox. illus. (Antiques, Boston, 1928, v. 14, p. 240–241.) **MAA**

Account of Clayton and of a bandbox having a picture of the ascent.

806. Some account of Mr. Arnold, the unsuccessful aerial adventurer. Embellished with a striking likeness, finely engraved by Thornton. (New London magazine, London, 1785, v. 1, p. 300–301.) *** DA**

807. Some account of Mr. Sadler, the first English aerostatist. Embellished with a striking likeness, curiously engraved by Thornton. port. (New London magazine, London, 1785, v. 1, p. 301–302.) *** DA**

808. Some account of Mr. Vincent Lunardi. Embellished with a striking likeness, elegantly engraved by Thornton. port. (New London magazine, London, 1785, v. 1, p. 93.) *** DA**

Lighter-Than-Air Craft — Balloons, continued

809. Some balloon record. (Aeronautical journal, London, 1898, v. 2, p. 86–87.) **VDS**

Table shows high ascents, 1852–1894.

810. Souvenirs rétrospectifs. (L'Aéronaute, Paris, 1903, année 36, p. 229–237.) **3–VDS**

Account of ascension of the Robert brothers, 1784, includes a reprint of a contemporary description.

811. Spangler, Robert W. C. A. Coey. illus. ports. (Fly, Philadelphia, 1908, v. 1, Dec., p. 8–13, 18.) **3–VDS**

Relates the career of American pioneer automobilist as well as balloon expert.

811a. Steering balloons. (The Mirror, London, 1837, v. 29, p. 340–342.) *** DA**

Discussion of several means proposed.

811b. Steering balloons. (The Mirror, London, 1837, v. 29, p. 71–72.) *** DA**

Discussion of Charles Green's ideas.

812. Sterne, Carus. Aërostatische Figuren. Ein Beitrag zur Geschichte der Luftschiffahrt. illus. (Prometheus, Berlin, 1904, Jahrg. 15, p. 5–8, 19–20.) **3–VA**

Historical account of balloons in human and animal forms.

813. Straus, Émile. Tombes d'aéronautes. (L'Aérophile, Paris, 1899, année 7, p. 123–125.) **3–VDS**

Descriptions of tombs in Père-Lachaise cemetery, Paris, of Sivel and Crocé Spinelli, Robertson, and Madame Blanchard.

814. Surcouf, Édouard. Aérostation militaire. Le ballon cerf-volant (Drachen-ballon.) (Aérophile, Paris, 1898, année 5, p. 204–212.)

Describes Parseval's kite balloon. **3–VDS**

815. Théorie des aérostats. (Encyclopédie méthodique. Arts et métiers mécaniques, Paris, 1791, tome 8, whole no. 162, p. 103–111.) *** AP**

816. Tissandier, Gaston. L'ascension à grande hauteur du 15 avril 1875. illus. (L'Aéronaute, Paris, 1875, année 8, p. 167–185.) **3–VDS**

Fatal ascension of Th. Sivel and J. Crocé Spinelli, 1875.

817. —— Les ballons captifs à Paris en 1889. illus. (La Nature, Paris, 1889, année 17, semestre 2, p. 243–246.) **OA**

Yon-Godard and Lachambre exhibits at the Paris exposition, with brief references to earlier balloons of Godard and Giffard.

818. —— Le centenaire de Blanchard et de la traversée de la Manche en ballon avec le Dr. Jeffries. illus. ports. (La Nature, Paris, 1885, année 13, semestre 1, p. 401–404.) **OA**

Also in *Scientific American supplement,* New York, 1885, v. 20, p. 7983–7984, *VA.*

819. —— Exposition internationale de 1878. Projet de construction d'un grand ballon captif à vapeur. illus. (L'Aéronaute, Paris, 1876, année 9, p. 263–271.) **3–VDS**

Project of Henry Giffard.

820. —— Le grand ballon captif à vapeur de M. Henry Giffard... Avec de nombreuses illustrations par A. Tissandier. Paris: G. Masson, 1878. 68 p. illus. new ed. 8°. **VDC p.v.10, no.5**

821. —— Histoire des ballons et des aéronauts célèbres, 1783–1890. Paris, 1887–90. 2 v. ports., plates. 4°. **† VDW**

"...a work essentially original. It is not a compilation from the texts of preceding historians, but is based upon original documents." — *Preface.* Elaborately illustrated with reproductions of old plates, many of them in color.

822. —— Histoire de mes ascensions; récit de vingt-quatre voyages aériens. Paris: M. Dreyfous [1879]. 244 p. port. 2. ed. 12°. **VDY**

Tissandier (1843–1899) was a pioneer in airship construction. In 1884 he and his brother Albert installed a one and one-third H. P. electric motor, weighing 121 pounds and taking current from a bichromate battery.

For an extensive review of the first edition, 1878, see *Nature,* London, 1878, v. 18, p. 639–640, *OA.*

823. —— Industries parisiennes. Une fabrique d'aérostats. illus. (La Nature, Paris, 1883, année 11, semestre 2, p. 280–282.) **OA**

Illustration and description of a toy balloon factory. Also in *Scientific American,* New York, 1883, v. 16, p. 6588, *VA.*

824. —— Nouveaux appareils de M. H. Giffard pour la préparation en grand de l'hydrogène pur. illus. (L'Aéronaute, Paris, 1877, année 10, p. 308–317.) **3–VDS**

825. —— Sur l'ancienne école aérostatique de Meudon. illus. (La Nature, Paris, 1876, année 4, semestre 1, p. 311–314.) **OA**

French national school founded in 1794. Interesting illustrations. Also in *L'Aéronaute,* Paris, 1876, année 9, p. 151–158, *3–VDS.*

826. —— Xavier de Maistre. (L'Aéronaute, Paris, 1876, année 9, p. 95–102.) **3–VDS**

Account of the author of *Voyage autour de ma chambre,* who made a balloon ascension at Chambéry in 1784.

827. Tridon, J. Louis. Nacelle close pour les ascensions au-delà des limites de l'air respirable. illus. (L'Aéronaute, Paris, 1871, année 4, p. 129–136.) **3–VDS**

828. Troisième expérience aérostatique de MM. Robert. (L'Esprit des journaux, Paris, 1784, année 13, tome 11, Nov., p. 340–345.) *** DM**

829. Twenty-four hours in a balloon. (Aeronautical journal, London, 1898, v. 2, p. 45–46.) **VDS**

Voyage of L. Godard, 1897. Estimated distance 1,032 miles.

830. Twenty-four years of ballooning. port. (Flying, London, 1903, no. 6, April, p. 270–274.) **VDS**

Portrait and experiences of Auguste E. Gaudron, French balloonist.

831. Unge, Eric. Luftballons welche längere Zeit die nöthige Tragfähigkeit beibehalten können. illus. (Illustrierte aëronautische Mittheilungen, Strassburg, 1902, Bd. 6, p. 159–165.) **VDS**

Swedish engineer tells how he constructs a balloon for long voyages.

Lighter-Than-Air Craft — Balloons, continued

832. Upshaw, H. C. Helium production. illus. (Popular aviation, Chicago, 1930, v. 6, March, p. 37, 74–76.) **VDS**

Early history of the gas and of its development for aeronautical uses.

833. Upson, Ralph Hazlett. Lessons of the Gordon Bennett balloon race. illus. map. (Aviation, Highland, N. Y., 1921, v. 11, p. 502–503.) **VDS**

Race of 1921 "showed up very clearly the various mistakes and hazards of the game."

834. The **Valveless** balloon voyage. (Aeronautical journal, London, 1900, v. 4, p. 99–101.) **VDS**

Ascent of John M. Bacon.

835. Van Orman, Ward Tunte. How Goodyear III won the national balloon race [of 1924]. (Aviation, Highland, N. Y., 1924, v. 16, p. 618–619.) **VDS**

Author was pilot of the winner.

836. Van Orman wins national balloon race. Sails with aide, W. W. Morton, to beach near Bar Harbor, Me. (Aviation, Highland, N. Y., 1927, v. 22, p. 1286–1287.) **VDS**

Gets Litchfield cup, having won the race for the third time.

836a. The **Vauxhall** balloon. illus. (The Mirror, London, 1836, v. 28, p. 177–181.) *** DA**

Picture and description of Green's balloon.

837. Verhaegen, P. Les premiers ballons en Belgique, 1783–1786. (Société d'archéologie de Bruxelles, Annales, Bruxelles, 1899, tome 13, p. 361–372.) **GBA**

838. A **View** and description of Mr. Lunardi's aerial voyage from the Artillery-Ground, London, to a field near Ware, in Hertfordshire, on Wednesday the 15th of September, 1784. illus. (European magazine, London, 1784, v. 6, p. 241–243.) *** DA**

839. Voyage aérien du premier décembre, par MM. Charles & Robert jeune. (L'Esprit des journaux, Paris, 1784, année 13, tome 1, Jan., p. 391–404.) *** DM**

Includes Charles' account of the ascension, Dec. 1, 1783.

840. Voyer. Das Ballonet von Meusnier. (Illustrierte aëronautische Mitteilungen, Strassburg, 1905, Jahrg. 9, p. 353–361.) **VDS**

Quotations from Meusnier's descriptions of gas-bags to be used inside the balloon.

841. —— L'équilibre de l'aérostat et les ascensions au long cours. (Revue de l'aéronautique, Paris, 1901, tome 8, p. 13–76.) **3–† VDS**

For long-distance balloons the author advises ballonnets, sustaining propellers, and reheating of the gas.

841a. W., I. H. Recent balloon ascent from Hungerford (New) Market. (The Mirror, London, 1833, v. 22, p. 19–20.) *** DA**

Ascent with Graham, 1833.

841b. W., P. T. Recent balloon ascent. (The Mirror, London, 1829, v. 13, p. 418.) *** DA**

Ascent with Green, 1829.

842. Weekes, W. H. Memoranda on aerostation; being suggestions designed to promote the objects of the scientific aeronaut, the improvement of aeronautic machinery, and the general advancement of aeronautics. illus. (Mechanic's magazine, London, 1839, v. 31, p. 289–299.) **3–VA**

Discusses imperfection of the balloon. Describes his method of condensing hydrogen during an ascension.

843. Wellner, Georg. Der lenkbare Segelballon. illus. (Zeitschrift für Luftschiffahrt, Berlin, 1883, Jahrg. 2, p. 161–172.) **3–VDS**

Calculations for balloons with wing control.

843a. Wenz zu Niederlahnstein, Oskar von. General Richepance und Pilâtre de Rozier, zwei Helden aus dem alten Metz. (Gesellschaft für lothringische Geschichte und Altertumskunde, Jahrbuch, Leipzig, 1909, Jahrg. 21, Hälfte 2, p. 332–341.) **ETI**

844. Whymper, Frederick. Adventures above the clouds. A retrospect of a century's work. (Good words, London, 1883, v. 24, p. 425–431, 516–523.) *** DA**

Brief history of balloon ascents.

845. Wicks, Z. W. Helium and its relation to lighter-than-air ships. port. (Aeronautical digest, New York, 1923, v. 3, p. 404–405.) **VDS**

History of its conception.

845a. —— Six years with the Navy in helium production. illus. (American Society of Naval Engineers, Journal, Washington, 1925, v. 37, p. 698–718.) **VXA**

846. Wild, Horace B. My forty years of flying. (Popular science monthly, New York, 1930–31, v. 117, Dec., p. 23–25, 150–153; v. 118, Jan., p. 40–42, 132.) *** DA**

Adventurous career of American balloonist and aviator (born 1871).

847. Windham, William. The Windham papers. The life and correspondence of the Rt. Hon. William Windham, 1750–1810... London: Herbert Jenkins, Ltd., 1913. 2 v. 8°. **AN**

For account of his balloon ascension with James Sadler, see v. 1, p. 75–83.

848. Wise, John. Aerial navigation. (Scientific American, New York, 1870, new series, v. 22, p. 33.) **VA**

Proposes to direct balloon by using propitious air currents.

849. —— A system of aeronautics, comprehending its earliest investigations, and modern practice and art; designed as a history for the common reader, and a guide to the student of the art... Philadelphia: J. A. Speel, 1850. 310 p., 9 pl., 1 port. 8°. **VDT**

Early history of balloons and parachutes, with accounts of the author's flights during the middle years of the nineteenth century.

Lighter-Than-Air Craft — Balloons, continued

850. —— Through the air: a narrative of forty years' experience as an aëronaut. Comprising a history of the various attempts in the art of flying by artificial means from the earliest period down to the present time... Philadelphia: To-day Printing and Publishing Co., 1873. 650 p. illus. ports. 8°. **VDW**

Enlarged edition of the author's *System of aeronautics*, 1850.

851. **Wraxall,** Sir Frederick Charles Lascelles, 3. bart. Five weeks in a balloon. (St. James magazine, London, 1863, v. 8, p. 315–323.) *** DA**

Remarkable trans-African voyage of Dr. Samuel Ferguson in 1862. Doubts were expressed as to the authenticity of this record.

852. **Wynne,** Clarence P. Professor Samuel Archer King. (Flying, New York, 1915, v. 3, p. 358, 376.) **VDS**

King (1828–1914) was a pioneer American aeronaut and balloon maker.

Airships

853. **Adler,** Milton. The early adventures of Count Zeppelin. illus. port. (Popular aviation, Chicago, 1932, v. 10, Jan., p. 21–22, 50, 54.) **VDS**

854. **Admiral** Moffett. (U. S. air services, Washington, 1933, v. 18, May, p. 9.) **VDS**

"Rear Admiral William Adger Moffett, U.S.N., became in 1922 the first Chief of the Bureau of Aeronautics and until his death in the *Akron* disaster last month [April, 1933] carried on the broad policies of airplane and airship development fostered by the Congress and the Navy Department."

855. **Aerial** navigation. (Engineer, London, 1884, v. 58, p. 161.) **VA**

Has a paragraph on the fanciful flight of Gavarni's dirigeable double balloon, Paris to Algiers and return, in 1857.

856. **Aerial** navigation. (Industries, London, 1892, v. 13, p. 224–225.) **3–† VA**

Describes combined airplane and balloon of W. J. Brewer, of New York City.

857. **Aerial** navigation by birds. (Mechanic's magazine, London, 1835, v. 24, p. 200–201.) **3–VA**

"The best would be to train up an eagle as a leader, with a light rein to direct him; if that was impracticable, I should fix a piece of meat at the end of a long, light pole...and working upon a universal joint."

858. **Aerial** navigation in France. (Nature, London, 1872, v. 5, p. 334.) **OA**

Description of Dupuy de Lôme's airship.

859. The **Aerovolt.** New York [: Leggo & Co., 18—?]. 10 p. illus. 4°. **VDW p.v.1, no.9**

Prospectus of the Universal Aerial Navigation Company.

860. **Aimé,** Emmanuel. Alberto Santos-Dumont. port. illus. (L'Aérophile, Paris, 1901, année 9, p. 69–86.) **3–VDS**

Good account of works of celebrated Brazilian who constructed several early airships.

861. —— Le Santos-Dumont no. 5. illus. (L'Aérophile, Paris, 1901, année 9, p. 180–210.) **3–VDS**

Illustrated account.

862. The **Airship** Ezekiel. illus. (Aeronautical world, Glenville, O., 1903, v. 1, p. 250–251.) **VDS**

Reverend B. Cannon's airship is stated to be based upon specifications laid down by the prophet Ezekiel in chapters 1 and 10.

863. **Airship** pioneers. illus. (Flight, London, 1916, v. 8, p. 954–958, 977–980, 998–1001.) **VDS**

Pictures and brief descriptions of craft from the Duc de Chartres, 1784, to Yamada, 1911. On p. 977 are tabular data, 1784–1912.

864. **Airships** for commercial purposes. Flight, London, 1919, v. 11, p. 144–148.) **VDS**

See p. 145 for progress of airplanes and airships during the period 1914–1918.

865. **Alanson** Wood's airship. (Aeronautical world, Glenville, O., 1902, v. 1, p. 102–104.) **VDS**

866. **Alban,** and Vallet. [Le projet d'un cours de direction aéronautique.] (Journal de Paris, 1785, no. 240, Aug. 28, p. 991–993.) *** DM**

Announcement of the authors' dirigible balloon, with an account of an ascent.

867. The **All-metal** airship. (Domestic air news, Washington, D. C., 1929, no. 50, April 15, p. 7–9.) **VDS (U. S. Commerce)**

Suggestions of Roger Bacon (1214–1294), Lana (1670), Dom Dauthey (1783), Dupuis Delcourt (1824), Marey Monge (1843), and A. de Bausset (1888). Account of the Schwarz airships (1893–1897).

868. **American** aeronautics. A flying machine to be built. illus. (Scientific American, New York, 1870, new series, v. 22, p. 48.) **VA**

Description of the *Avitor*, built in San Francisco. For illustration see above magazine, 1869, new series, v. 21, p. 356.

869. **André,** Henri. Les dirigeables. Étude complète de la direction des ballons des tentatives réalisées et des projets nouveaux. Paris: C. Béranger, 1902. iii, 346 p. 8°. **VDY**

See p. 183–337 for history of airships.

870. **Andrews,** Solomon. Aerial navigation and a proposal to form an aerial navigation company. [New York: J. F. Trow,] 1865. 32 p. 8°. **VDY p.v.26, no.4**

871. **Another** discoverer of a mode of propelling and directing balloons. (Mechanic's magazine, London, 1836, v. 26, p. 57.) **3–VA**

Invention of Leonardo Andervolti.

872. **Arco.** Die Beziehungen zwischen dem Luftschiff von Meudon und seinen Vorgängern. (Zeitschrift für Luftschiffahrt, Berlin, 1885, Jahrg. 4, p. 7–13.) **VDS**

Meudon airship compared with those of Haenlein and Giffard.

Lighter-Than-Air Craft — Airships, continued

873. **Armengaud,** J., the younger. Les progrès de la navigation aérienne et les expériences de M. Santos-Dumont. (Société des ingénieurs civils de France, Mémoires, Paris, 1901, année 1901, v. 2, p. 887–923.) **VDA**

A second copy in *VDY p.v.9, no.2.*

874. **Artificial** flight; an aerial ship. illus. (Journal of science, London, 1872, new series, v. 2, p. 457–463.) **OA**

Description of Dupuy de Lôme's airship.

875. The **Ascension** of Count Zeppelin's airship. illus. (Scientific American, New York, 1900, v. 83, p. 88.) **VA**

876. **Aus** Oscar Erbslöhs Leben. port. (Deutsche Zeitschrift für Luftschiffahrt, Berlin, 1910, Jahrg. 14, Heft 15, p. 15–18.) **VDS**

Biography of Oscar Erbslöh, preceded by an account of fatal airship disaster, 1910.

877. **B.,** G. Le "De Bradsky." (L'Aérophile, Paris, 1902, année 10, p. 236–245.) **3–VDS**

878. **B.,** N. Aerial travelling. (Mechanic's magazine, London, 1853, v. 59, p. 167–168.) **3–VA**

A review of a pamphlet by James Nye entitled *Thoughts on aerial travelling and the best means of propelling balloons* (1852), involving rocket propulsion.

879. **Bach,** Carl von. Die Stellungnahme deutscher Ingenieure zu dem Projekt des Grafen von Zeppelin, betreffend den Bau lenkbarer Luftschiffe, im letzten Jahrzehnt des vorigen Jahrhunderts. (Verein deutscher Ingenieure, Zeitschrift, Berlin, 1908, v. 52, p. 1549–1552.) **3–† VDA**

Shows the encouraging attitude of German engineers toward the projects of Count Zeppelin.

880. **Bacon,** John Mackenzie. The development of the airship. (Nineteenth century, London and New York, 1902, v. 52, p. 481–491.) *** DA**

Brief but well-written account of the development of balloons and airships, and of the experiments of Hiram Maxim, Santos-Dumont, Pilcher, and Langley.

881. **Baddeley,** William. The first aerial airship, "The Eagle." illus. (Mechanic's magazine, London, 1835, v. 23, p. 289–291.) **3–VA**

Comments on the craft of the European Aeronautical Society.

882. —— Mechanical flying *v.* navigable balloons. (Mechanics' magazine, London, 1843, v. 38, p. 339.) **3–VA**

States that he is withholding his invention of an airship until he can achieve "some more tangible benefit than the prospect of future immortality."

883. —— Mode discovered of propelling balloons in any direction. (Mechanic's magazine, London, 1836, v. 26, p. 39–40.) **3–VA**

Invention of the author.

884. **Balloon** communication between London and Paris. (Mechanic's magazine, London, 1835, v. 23, p. 272.) **3–VA**

Proposal of the "European Aeronautical Society" to establish a regular service with the *Eagle* airship. Copy of a letter from one Campenas to Napoleon, suggesting balloon attacks on England.

885. A **Balloon** with power. (Builder, London, 1876, v. 34, p. 117–118.) **3–† MQA**

Hänlein's experiments, 1870–1872. Also in *Van Nostrand's eclectic engineering magazine,* New York, 1876, v. 14, p. 354–357, *3–VDA.*

886. **Baltimore,** J. Mayne. A successful airship. (American inventor, Washington, 1903, v. 11, p. 225–226.) *** V**

Flight of Dr. August Greth's *California eagle,* Oct. 18, 1903, at San Francisco. Greth's name is misspelled.

887. **Banet-Rivet,** P. Les dirigeables. (Revue des deux mondes, Paris, 1908, période 5, v. 46, p. 415–446.) *** DM**

Various airship problems, referring to the Lebaudy, Patrie, Ville de Paris, Zeppelin, and Santos-Dumont craft. Possibilities of the airship in war.

888. **Barkeley,** Armand Sylvain. Captain Thomas S. Baldwin. The way he builds airships. illus. port. (Fly, Philadelphia, 1909, v. 1, Feb., p. 8–9.) **3–VDS**

889. **Barry,** Edward A. The Meudon balloon experiments. (Aeronautical Society of Great Britain, 20th annual report, London, 1885, p. 9–29.) **3–VDS**

Describes ascension of the Renard-Krebs airship.

890. **Barton,** F. A. Further notes on the Barton airship. illus. (Aeronautical journal, London, 1902, v. 6, p. 30–38.) **VDS**

Author's early experiments.

891. The **Barton** airship. illus. (Engineer, London, 1902, v. 93, p. 392.) **VA**

892. The **Barton** 150-horsepower airship's forthcoming trial. illus. (Scientific American, New York, 1904, v. 90, p. 44, 46.) **VA**

Designed by F. A. Barton, of England.

893. The **Barton-Rawson** airship. (Engineer, London, 1905, v. 100, p. 96.) **VA**

894. **Bauer-Keiper,** Karl. Prinzipien moderner Luftschiffahrt vor 250 Jahren. (Deutsche Zeitschrift für Luftschiffahrt, Berlin, 1910, Jahrg. 14, Heft 16, p. 11–13.) **VDS**

Airship as imagined by Cyrano de Bergerac (1619–1655).

895. **Bausset,** A. de. Aerial navigation on the vacuum principle. Lowell, Mass.: Citizen Printing Works, 1889. 48 p. 16°. **VDC p.v.19, no.2**

Description of proposed airship provided with a vacuum vessel made with very thin steel plates.

896. The **Beachy** and Knabenshue airship experiments. (Scientific American, New York, 1908, v. 98, p. 443.) **VA**

Experiments in stabilization.

897. The **Beedle** airship. illus. (Scientific American, New York, 1904, v. 90, p. 313.) **VA**

Frame of this English airship is constructed with steel bicycle tubing.

Lighter-Than-Air Craft — Airships, continued

898. **Besançon,** Georges. L'aéronat "Lebaudy." illus. (L'Aérophile, Paris, 1903, année 11, p. 110–118.) 3–VDS
History of the French airship constructed by Henri Julliot.

899. —— L'aéronat "La Ville-de-Paris." illus. (L'Aérophile, Paris, 1904, année 12, p. 12–16.) 3–VDS
Description of French airship designed by Henry Deutsch de la Meurthe.

900. —— Auguste Severo. (L'Aérophile, Paris, 1902, année 10, p. 98–100.) 3–VDS
Biography of the designer of the ill-fated airship *Pax*.

901. —— Le "de Bradsky." illus. (L'Aérophile, Paris, 1902, année 10, p. 236–245.) 3–VDS

902. —— Mort d'Albert Tissandier. (L'Aérophile, Paris, 1906, année 14, p. 196–197.) 3–VDS
Albert Tissandier (1839–1906), prominent French balloonist and illustrator, experimented with an electrically-driven airship (1881–1884) with his brother, Gaston Tissandier.

903. —— L'orthoptère, aérostat dirigeable, adopté par l'armée russe. illus. (L'Aérophile, Paris, 1893, année 1, p. 28–32.) 3–VDS
Airship of Armand Le Compagnon uses lateral sail propellers.

904. **Béthuys,** G. Un projet de ballon dirigeable. (Le Génie civil, Paris, 1889, tome 16, p. 154–156.) VA
Wing-regulated airship designed by Arsène Olivier.

905. —— Sur le ballonnet à compression de l'aérostat du général Meusnier. (Revue de l'aéronautique, Paris, 1888, année 1, p. 135–139.) 3–† VDS

906. [**Bibliographies** of airships.] (In: Paul Brockett, Bibliography of aeronautics, Washington, 1910, p. 252–253; continued in United States. — National Advisory Committee for Aeronautics, Bibliography of aeronautics, 1909–1931, Washington, 1921–35, volumes for 1909/16, p. 46, 359–362; 1917/19, p. 19–22; 1920/21, p. 19–21, 112–113; 1922, p. 8, 40; 1923, p. 7, 34; 1924, p. 4–5, 25; 1925, p. 8–9, 47; 1926, p. 4–5; 1927, p. 6–7; 1928, p. 9–11; 1929, p. 15–17; 1930, p. 10–12; 1931, p. 15–17.) VDT
Some of the earlier references have the heading Dirigibles.

907. [**Bibliography** of Auguste Severo.] (In: Paul Brockett, Bibliography of aeronautics, Washington, 1910, p. 766.) VDT
Citations to writings concerning Severo and his airship.

908. [**Bibliography** of Clément-Bayard airship.] (In: United States. — National Advisory Committee for Aeronautics, Bibliography of aeronautics, 1909–1931, Washington, 1921–35.) VDT
For references to this airship see 1909/16, p. 272–223; 1917/19, p. 90.

909. [**Bibliography** of E. T. Willows and his airships.] (In: United States. — National Advisory Committee for Aeronautics, Bibliography of aeronautics, 1909–1930. Washington, 1921–32.) VDT
See 1909/16, p. 1426; 1917/19, p. 474; 1920/21, p. 429.

910. [**Bibliography** of Gustave Koch's writings on the theory and construction of airships.] (In: Paul Brockett, Bibliography of aeronautics, Washington, 1910, p. 466–468.) VDT

911. [**Bibliography** of Ottokar de Bradsky-Laboun.] (In: Paul Brockett, Bibliography of aeronautics, Washington, 1910, p. 148.) VDT
List of writings by and about Bradsky-Laboun.

912. [**Bibliography** of Pierre and Paul Lebaudy.] (In: Paul Brockett, Bibliography of aeronautics, Washington, 1910, p. 503–505.) VDT

913. [**Bibliography** of Schütte-Lanz airships.] (In: United States. — National Advisory Committee for Aeronautics, Bibliography of aeronautics, 1909–1930. Washington, 1921–32.) VDT
See 1909/16, p. 1200; 1917/19, p. 399; 1920/21, p. 369; 1923, p. 122; 1924, p. 98; 1925, p. 158; 1926, p. 129.

914. [**Bibliography** of Thomas S. Baldwin.] (In: Paul Brockett, Bibliography of aeronautics, Washington, 1910, p. 86.) VDT
List of writings by and about Baldwin, especially on airships and parachutes.

915. **Big** French airship probably lost in storm. Dixmude, ex-Zeppelin, vanishes near North African coast with crew of fifty men. (Aviation, Highland, N. Y., 1924, v. 16, p. 14–15.) VDS

916. **Billwiller,** C. F. Miniatur-Modell eines lenkbaren Luftschiffs im Maassstab 1:200, ausgestellt von C. F. Billwiller. Zürich: Ulrich & Co., 1894. 24 p. illus. 12°. VDY p.v.13, no.5

917. **Blanchet,** Georges. Léonce Girardot. port. (L'Aérophile, Paris, 1902, année 10, p. 73–76.) 3–VDS
Illustrated description of Girardot's airship.

918. —— Le premier aéronaut italien. illus. port. (L'Aérophile, Paris, 1905, année 13, p. 173–175.) 3–VDS
Illustrated account of the airship *Italia* designed by Count Almerico da Schio.

919. The **Boerner** airship. illus. (Scientific American, New York, 1926, v. 135, p. 302–304.) VA
Proposed German semi-rigid craft, 1926.

920. **Boffito,** Giuseppe. L'aeronautica nelle città italiane. Brescia. illus. (Rivista aeronautica, Roma, 1929, anno 5, Jan., p. 197–201; March, p. 641–642; May, p. 428–430.) VDS
Airships (1784) of Carlo Bettoni and Gaspare Turbini. Ascension and death of Agostino Magnoni. Studies of Achille Bertelli and Cosimo Canovetti.

921. —— L'aeronautica nelle città italiane. Vicenza. illus. (Rivista aeronautica, Roma, 1928, anno 4, Nov., p. 395–405.) VDS
Describes airship investigations of Pasquale Cordenons and Count Almerico da Schio.

Lighter-Than-Air Craft — Airships, continued

922. **Bonaparte,** Napoleone Luigi. Intorno alla direzione degli aerostati. Lettera al prof. G. Gazzeri. illus. (Antologia, Firenze, 1828, tomo 30, April, p. 178–183; tomo 31, Sept., p. 193–195.) **NNA**

Interesting plates showing balloon driven by a helice. Also in *VDC p.v.10, no.3.* French abstract in *Bulletin des sciences technologiques,* Paris, 1829, tome 13, p. 325–330, *3-VA.*

923. **Booth,** R. S. Airship development abroad. (Royal Aeronautical Society, Journal, London, 1933, v. 37, p. 366–380.) **VDS**

With discussion.
Contemporary developments in Germany and the United States. Also, without the discussion, in *Flight,* London, 1933, v. 25, p. 52–55, *VDS.*

924. **Bouttieaux,** Victor Paul. Les ballons dirigeables. (Revue scientifique, Paris, 1907, série 5, v. 8, p. 452–458, 481–486.) **OA**

Development of French types.

925. —— La navigation aérienne par ballons dirigeables. Paris: Ch. Delagrave [1909]. 103 p. illus. 8°. **VDY**

Much of historic interest.

926. **Braun's** electro-dynamic air ship. illus. (Scientific American, New York, 1887, v. 56, p. 249.) **VA**

Invention of Dr. Martin Braun.

927. **Braving** ocean winds in the Graf Zeppelin. illus. (Literary digest, New York, 1928, v. 99, Nov. 3, p. 36, 38, 48, 50, 52.) *** DA**

Passengers' experiences during trans-Atlantic flight.

928. **Breithaupt,** Joachim. Mit "Graf Zeppelin" nach Süd- u. Nordamerika; Reisseeindrücke und Fahrterlebnisse. Lahr, Baden: M. Schauenburg, 1930. 152 p. illus. map, port. 8°. **VDY**

929. **Broszus,** J. E. Einige Bemerkungen zu dem neuesten lenkbaren Luftballon des Herrn Maximilian Wolff. (Zeitschrift für Luftschiffahrt, Berlin, 1885, Jahrg. 4, p. 76–81.) **3-VDS**

930. —— Der lenkbare Luftballon mit Dampfmaschine. (Zeitschrift für Luftschiffahrt, Berlin, 1882, Jahrg. 1, p. 161–169.) **3-VDS**

Discusses the adaptability of the steam engine to airships.

931. —— Die untere Ausführungsgrenze der lenkbaren Luftballons. illus. (Zeitschrift für Luftschiffahrt, Berlin, 1885, Jahrg. 4, p. 106–119.) **VDS**

Comparisons of performances of airships of Giffard, Haenlein, and Renard.
For a list of Broszus' writings on airships see Paul Brockett, *Bibliography of aeronautics,* Washington, 1910, p. 156, *VDT.*

932. **Brown,** Alphonse. La conquête de l'air; 40 jours de navigation aérienne. Paris: Glady frères, 1875. 452 p. 12°. **VDW**

Imaginary adventures of the airship *Céleste.*

933. **Brown,** Enos. The Heaton airship failure. (Scientific American, New York, 1905, v. 92, p. 159.) **VA**

Account of accident to the *California messenger,* San Francisco, Feb. 12, 1905.

934. **Bruce,** Eric Henry Stuart. The "Peace" balloon of the late Senhor Augusto Severo. (Aeronautical journal, London, 1902, v. 6, p. 59–60.) **VDS**

Abridged in *Aeronautical world,* Glenville, O., 1902, v. 1, p. 111–112.

935. **Bruignac,** Duroy de. Exposé sommaire de l'état présent de l'aéronautique. illus. (Société des ingénieurs civils de France, Mémoires et Compte rendu, Paris, 1884, année 1884, v. 2, p. 393–414.) **VDA**

Discusses technical features of the airships of Giffard, Dupuy de Lôme, Tissandier, Renard, and Krebs.

936. **Buchholtz,** F. Das Graf Zeppelinische Luftschiff. illus. (Glaser's Annalen für Gewerbe und Bauwesen, Berlin, 1901, Bd. 48, p. 133–138.) **3-† VDA**

Technical discussion of the Zeppelin airship and of its predecessors.

937. —— [Die Lenkbarkeit der Luftschiffe mit Anwendung der Elektrizität betreffend.] (Elektrotechnische Zeitschrift, Berlin, 1884, Jahrg. 5, p. 431–433.) **VGA**

Tabular comparison of the airships of Giffard, Dupuy de Lôme, Haenlein, Tissandier brothers, and Renard-Krebs.

938. —— Die neueren Luftschiffe, ihre Bauart und technischen Einrichtungen. illus. (Glasers Annalen für Gewerbe und Bauwesen, Berlin, 1908, Bd. 63, p. 1–7.) **3-† VDA**

Brief illustrated account of new German and French airships.

939. —— Die neueren Versuche über die Fortbewegung von Luftschiffen und ihre Ergebnisse. illus. (Glaser's Annalen für Gewerbe und Bauwesen, Berlin, 1904, Bd. 54, p. 24–28.) **3-† VDA**

Discussion of the airships of Severo, De Bradsky, Santos-Dumont, and Lebaudy.

940. **Burgess,** Charles Paine. Airships. (Mechanical engineering, New York, 1929, v. 51, p. 12–13.) **VFA**

Progress during 1928.

941. —— Progress in aeronautics [for 1929]. Airships. (Mechanical engineering, New York, 1930, v. 52, p. 11–12.) **VFA**

942. —— Progress in airship design from U.S.S. Shenandoah to U.S.S. Akron. (American Society of Naval Engineers, Journal, Washington, 1931, v. 43, p. 419–425.) **VXA**

943. **Burton,** Walter E. The dirigible *is* practical. illus. (Popular aviation, Chicago, 1931, v. 9, Oct., p. 21, 54, 56.) **VDS**

"The 'Akron' concludes a notable series of experimental dirigibles. It will be of inestimable value in time of war."

944. **Cabot,** Godfrey L. Lessons from the Shenandoah. illus. (Aero digest, New York, 1925, v. 7, p. 540, 570.) **VDS**

Lighter-Than-Air Craft — Airships, continued

945. **California** airship comes to grief. illus. (Aeronautics, New York, 1908, v. 2, June, p. 24–26.) **VDS**

Disaster to large ship of the National Airship Company, at Berkeley, California, 1908.

946. **Carelli,** Jules. Ballon dirigeable pratique. illus. (L'Aérophile, Paris, 1903, année 11, p. 44–45.) **3–VDS**

Description of author's airship.

947. **Caron,** Félix. L'aérostat électrique à hélice de MM. les capitaines Ch. Renard et Krebs. illus. (L'Aéronaute, Paris, 1884, année 17, p. 163–178.) **3–VDS**

948. —— La mort de M. Giffard. (L'Aéronaute, Paris, 1882, année 15, p. 103–112.) **3–VDS**

Account of funeral, with discourses by Hervé Mangon and Gaston Tissandier.

949. **Carra.** Air balloons. Extract from a pamphlet lately published at Paris...containing the art of directing the aerostatic machines at pleasure and of accelerating their motion in the air... illus. (London magazine, London, 1784, new series, v. 2, p. 443–449.) *** DA**

Uses taffeta wings and auxiliary balloon.

950. **Cassé,** Émile. Cinquantenaire de la 1re ascension d'un aérostat dirigeable. (L'Aéronaute, Paris, 1903, année 36, p. 10–16.) **3–VDS**

Anniversary of Giffard's airship, with an account of its ascension.

951. **Castelli,** Gaetano. Il problema della dirigibilità dei palloni. illus. (Rivista di fisica, mathematica e scienze naturali, Pisa, 1909, anno 10, v. 19, p. 350–365.) **OA**

Study of air resistance.

952. **Cayley,** Sir George. Aerostation. Mr. Luntley's pamphlet. illus. (Mechanic's magazine, London, 1851, v. 55, p. 292–293.) **3–VA**

A brief discussion of Luntley's rotary balloon. For Luntley's reply see entry no. 1151.

953. —— On aerial navigation. illus. (Philosophical magazine, London, 1816, v. 47, p. 81–86, 321–329.) **OA**

Discusses means for directing balloons, replying to communications by John Evans, jr. (see entry no. 1027).

954. —— Practical remarks on aerial navigation. (Mechanic's magazine, London, 1837, v. 26, p. 418–428.) **3–VA**

Interesting pictures of author's airship.

955. **Chanute,** Octave. Aerial navigation... A lecture delivered to the students of Sibley College, Cornell University. New York: Railroad and Engineering Journal, 1891. 36 p. illus. 12°. **VDY p.v.32, no.1**

Describes power-driven balloons of Giffard (1852), Dupuy de Lôme (1872), Tissandier (1884), French war balloon (1884–1885). Discusses possibilities of the heavier-than-air machine. Also in Military Service Institution of the United States, *Journal,* New York, 1892, v. 13, p. 1193–1207; 1893, v. 14, p. 154–165, *VWA.*

956. **Charpentier,** Paul. Arbeit und Kampf. Erlebnisse eines deutschen Luftschiff-Ingenieurs. Strassburg: Heitz & Cie., 1930. 179 p. illus. 8°. **VDW**

Account of the career of Gustave Unger, inventor of steel airship bearing his name.

957. **Choinski,** Tim. Aerial navigation and its possibility. illus. (Scientific American supplement, New York, 1881, v. 11, p. 4507–4508.) **VA**

Suggests an airship, using hydrogen and wings, and based upon the principles of bird-flight.

958. **Christening** of the Army airship Roma. Largest semi-rigid in the world is christened with liquid air after stormy trip from Langley Field. illus. (Aviation, Highland, N. Y., 1922, v. 12, p. 15–16.) **VDS**

959. **Cléry,** A. Capitaine Hildebrandt. port. (L'Aérophile, Paris, 1909, année 17, p. 313.) **3–VDS**

Portrait and brief account of prominent German aeronaut, investigator, and writer.

960. **Collins,** Thomas Byard. The airship system of M. Frederick L'Hoste. illus. (Scientific American supplement, New York, 1903, v. 55, p. 22673.) **VA**

961. —— Electricity as a motive power in mechanical flight. illus. (Electrical world, New York, 1903, v. 41, p. 440–442.) **VGA**

Discusses progress of airships and the advantages and disadvantages of electricity.

962. **Cordenons,** Pascal. Navigazione nell' aria. (Il politecnico, Milano, 1878, anno 26, p. 604–621.) **3–† VDA**

Author's suggestions for an airship.

963. —— Le problème de la navigation aérienne. Solution. Vérone: H. F. Münster, 1868. 15 p. illus. 8°. **VDY p.v.6, no.21**

Author's suggestions for an airship.

964. **Count** von Zeppelin's dirigible airship. illus. (Scientific American supplement, New York, 1899, v. 48, p. 19951–19952.) **VA**

965. **Count** von Zeppelin's great work expressed in figures. (Aero Club of America, Bulletin, New York, 1912, v. 1, May, p. 38.)

Statistics and chronology. **VDS (Flying)**

966. **Coxwell,** Henry. On flying and balloon steering. (Contemporary review, London, 1884, v. 46, p. 532–542.) *** DA**

Comments on progress and prospects by the well-known associate of James Glaisher.

967. **Crocco,** Gaetano Arturo. Sur la stabilité des dirigeables. (Académie des sciences, Comptes rendus, Paris, 1904, tome 139, p. 1195–1198.) *** EO**

968. **Dallwitz-Wegner,** Richard von. Der praktische Luftschiffer. Eine geschichtliche und technische Uebersicht über den Stand der Luftschiffahrt und eine Entwicklung ihrer physikalischen und technischen Bedingungen. Rostock i. M.: C. J. E. Volckmann Nachfolger, 1909. x, 109 p. 8°. **VDY**

A study of contemporary balloons and airships, with some historical references.

Lighter-Than-Air Craft — Airships, continued

969. **Danilevski,** Konstantin. Upravlyayemyi letatelnyi snaryad. Kharkov: Tip. I. M. Varskavchika, 1900. ii, 78 p. illus. 4°. **VDY p.v.3, no.6**

Describes airship of the author. Also published in German with the title *Ein lenkbarer Flug-Apparat,* Kharkov: I. M. Varshavchik, 1900, *VDY.*

For references to Danilevski's airship see Paul Brockett, *Bibliography of aeronautics,* Washington, 1910, p. 225–226, *VDT.*

970. The **Danilewsky** dirigible flying machine. illus. (Scientific American, New York, 1898, v. 79, p. 424.) **VA**

Illustrations and brief description of airship.

971. The **Danilewsky** flying machine. illus. (Scientific American, New York, 1900, v. 82, p. 43.) **VA**

972. **Daubanton,** C. Abel. De la meilleure forme d'avant à donner aux ballons dirigeables. illus. (L'Aéronaute, Paris, 1871, année 4, p. 113–119.) **3–VDS**

973. **De ***,** vte. Le dauphin enlevé; ou, L'art de se diriger dans les airs. Londres: T. Spilsbury & fils, 1793. 76 p. illus. nar. 8°. **VDT (Room 120)**

Bound with: V. Lunardi, An account of the first aërial voyage in England, London, 1784. 2. ed. nar. 8°.

974. The **Death** of Melvin Vaniman. (Scientific American, New York, 1912, v. 107, p. 22.) **VA**

Vaniman was an American pioneer in the effort to construct an airship of constant displacement.

975. **De Bradsky** airship disaster. illus. (Scientific American, New York, 1902, v. 87, p. 331.) **VA**

Paul Morin and the inventor, Baron de Bradsky, were killed Oct. 30, 1902.

976. **Defects** of the first British airship. The need of harbors for aerial craft. illus. (Scientific American supplement, New York, 1907, v. 64, p. 379.) **VA**

Comments on the *Nulli secundus.*

977. The **De la Vaulx** airship. illus. (Scientific American supplement, New York, 1907, v. 63, p. 25928.) **VA**

978. **Deminne,** Edm. Solution du problème de la navigation aérienne. Le ballon Deminne. [By Edm. Deminne.] Bruxelles, 1865. 32 p. illus. 8°. **VDY p.v.11, no.9**

Picture and description of an odd looking airship designed by the author.

979. **Denza,** Francesco. I primi cultori italiani dell' aeronautica. Torino: Stamperia dell' Ateneo, 1889. 17 p. 8°. **VDY p.v.6, no.10**

Brief accounts of the work of Lana-Terzi and Serafino Serrati; and of early ballooning in Italy.

979a. **Descripção** burlesca de um imaginario aeróstato e de seus petrechos, satyra ao P[e.] Bartholomeu Lourenço de Gusmão. illus. (Biblioteca da Universidade de Coimbra, Coimbra, 1935, v. 12, p. 138–148.) *** GRS**

Has several early Portuguese satirical poems on Gusmão.

980. A **Description** of two machines, proper to be navigated through the air. Translated from a pamphlet lately published at Paris, by Mons. B——. illus. (London magazine, London, 1784, new series, v. 2, p. 13–16.)

Airship conceptions, with illustrations. *** DA**

981. **Deslandres,** Henri. Méthode permettant de déterminer la vitesse propre des aérostats dirigeables. Application aux expériences de M. Santos-Dumont. (Académie des sciences, Comptes rendus, Paris, 1901, tome 133, p. 993–996.) *** EO**

982. **Destruction** of the airship Akron [no. 1]. Vaniman's career and tragic end. illus. port. (Scientific American, New York, July 13, 1912, v. 107, p. 28–29.) **VA**

983. **Destruction** of the Severo navigable balloon. illus. port. (Flying, London, 1902, no. 3, Jan., p. 101–108.) **VDS**

Illustrated history of the *Pax,* and a portrait of Senor Severo.

984. **Detroit's** all-metal dirigible. illus. (Literary digest, New York, 1929, v. 102, Sept. 7, p. 10.) *** DA**

Account of first all-metal *ZMC–2* of the United States Navy.

985. **Dettmann,** Ludwig J. Chr. Mit dem "Zeppelin" nach Amerika; das Wunder von Himmel und Ozean; mit einem Geleitwort von Dr. Hugo Eckener. Berlin: R. Hobbing [cop. 1929]. 145 p. illus. obl. 8°. **† VDW**

986. A **Device** to navigate the air. illus. (Scientific American, New York, 1890, v. 62, p. 388.) **VA**

Airship patented by Charles E. Bechtel of Udall, Kansas.

987. **Dienstbach,** Carl. The burning of the German military dirigible M III. (Scientific American, New York, 1911, v. 105, p. 312.) **VA**

988. —— Our first naval dirigible. An American-built airship possessing novel features of control and of anchorage. illus. (Scientific American, New York, 1915, v. 113, p. 44.) **VA**

989. **Diners** travel via air route. illus. (Aerial age weekly, New York, 1919, v. 9, p. 691.) **VDS**

Brief account of the landing of U. S. Army airship on the roof of the Statler Hotel, Cleveland, Ohio.

990. The **Dirigible** balloon of de Santos Dumont. illus. (Scientific American, New York, 1898, v. 79, p. 281.) **VA**

Ascension of Sept. 20, 1898.

991. **Dirigible** balloons. illus. (American inventor, Washington, 1901, v. 7, Jan. 1, p. 12.) *** V**

Very brief history of airships. References to the craft of Dr. De Basset, of New York, and of Leo Stevens. Statement that Rufus Porter, in 1833, was the first to apply power for propulsion.

Lighter-Than-Air Craft — Airships, continued

992. **Dodoret,** Thomas. Concerning an explanation of a method to combine and put in opposition the two powers of the levity of the aerostatic globe, and the weight of the car... illus. (London magazine, London, 1784, new series, v. 3, p. 114–118.) * DA

Thinks that by changing the center of gravity he can impart motion to the balloon.

993. —— The explication of a simple method of directing the new aerostatic machines, in paths which are oblique to the motion of the wind. (London magazine, London, 1784, new series, v. 3, p. 52–54.) * DA

Uses sails above and below the balloon.

994. **Dollfus,** Charles. Les précurseurs du ballon dirigeable rigide. illus. (L'Aéronautique, Paris, 1920, année 2, p. 221–226.) VDS

Has interesting pictures of early suggestions, beginning with the aeronautical fish of 1783.

995. **Duerr,** L. Fuenfundzwanzig Jahre Zeppelin-Luftschiffbau... Berlin: V. D. I. Verlag, G.m.b.H., 1925. 83 p. illus. port. f°. † VDY

Well-illustrated account of the development of the various elements of Zeppelin ships. Reviewed in *Aviation,* New York, 1925, v. 18, p. 162, *VDS.*

996. **Du Plessis de Grenédan,** Jean Joseph Anne Marie Julien, comte. Les grands dirigeables dans la paix et dans la guerre. Leur passé, leur avenir, l'expérience du "Dixmude" ... Paris: Plon-Nourrit et Cie. [cop. 1925.] 2 v. illus. 12°. VDW

Contains a brief history of the modern airship, p. 28–46; also accounts of the author's experiences with the *Dixmude.*

997. **Duponchel,** A. L'aérostat élastique automoteur. (Revue scientifique, Paris, 1884, tome 33 [série 3, tome 7], semestre 1, p. 709–717.) OA

Describes the heating and cooling of hydrogen in his "aerial fish."

998. —— Essai d'aéronautique: le poisson aérien. illus. (Revue scientifique, Paris, 1883, tome 32 [série 3, tome 6], p. 236–242.) OA

Descriptions and calculations for a fish-shaped airship.

999. **Dupuy de Lôme,** Stanislas Charles Henri Laurent. Essai de l'aérostat à hélice. (L'Aéronaute, Paris, 1872, année 5, p. 50–60.) 3-VDS

Account of trial trip of author's famous airship.

1000. —— Navigation aérienne. Résumé de la note sur l'aérostat à hélice, remise en décembre 1871, à la Commission d'essai. (Académie des sciences, Comptes rendus, Paris, 1872, tome 74, p. 337–354.) * EO

Description of airship designed during the Franco-Prussian war.

1001. —— Note sur l'aérostat à hélice construit pour le compte de l'État, sur les plans et sous la direction de M. Dupuy de Lôme. 67 p. exclusive of plates. (Académie des sciences, Mémoires, Paris, 1876, tome 40, no. 1.) * EO

1002. —— Projet d'aérostat dirigé, muni d'un propulseur. illus. (Académie des sciences, Comptes rendus, Paris, 1870, tome 71, p. 502–521, 546–550.) * EO

Title varies.

1003. —— Résumé de la note sur l'aérostat à hélice. (L'Aéronaute, Paris, 1872, année 5, p. 37–45.) 3-VDS

Description of airship, construction of which was begun during the siege of Paris.

For writings on Dupuy de Lôme's airships see Paul Brockett, *Bibliography of aeronautics,* Washington, 1910, p. 263–264, *VDT.*

1004. **Duvernoy,** von. Graf Zeppelin und der gegenwärtige Stand der Luftschiffahrt. illus. (Westermanns Monatshefte, Braunschweig, 1909, v. 106, p. 421–437.) * DF

1005. **E.,** M. Une expérience d'aviation en Espagne à la fin du XVIIIe siècle. (Revue des études historiques, Paris, 1913, année 79, p. 688–692.) DA

Experiments of the Comte de Galvez in 1784 with means for propelling and steering a balloon.

1006. An **Early** idea in airships. The Roze aeroplane-airship-helicopter-parachute- boat. (Flight, London, 1909, v. 1, p. 417.) VDS

1007. **Eckener,** Hugo. Die Amerikafahrt des "Graf Zeppelin"; herausgegeben von Rolf Brandt. Berlin: Verlag A. Scherl [cop. 1928]. 114 p. illus. ports., map. 12°. VDW

Cruise of the *Graf Zeppelin,* Germany to America and return in 1924, under the command of the author. Diary of one of the passengers.

1008. —— The first airship flight around the world. illus. (National geographic magazine, Washington, D. C., 1930, v. 57, p. 653–688.)

Flight of the *Graf Zeppelin.* KAA

1009. —— Graf Zeppelin: sein Luftschiff und seine Persönlichkeit. port. (Nord und Süd, Berlin, 1908, Bd. 127, p. 123–134.) * DF

1010. —— Luftschiff und Luftverkehr. Letzte Errungenschaften und Zukunftsperspektiven in uraltem Streben. Stuttgart: Greiner & Pfeiffer, 1904. 36 p. 12°. (Gegenwartsfragen. Heft 11.) VDW

Discusses characteristics and possibilities of airships.

1011. —— Mit Graf Zeppelin 1900–1908. (In: Karl Bröckelmann, Wir Luftschiffer, Berlin, 1909, p. 269–284.) VDW

1012. —— The remodeled "Zeppelin III" airship. Description of the first flights since the changes have been made. illus. (Scientific American supplement, New York, 1908, v. 66, p. 380–381.) VA

For list of Eckener's papers on Zeppelin airships see Paul Brockett, *Bibliography of aeronautics,* Washington, 1910, p. 270, *VDT.*

1013. **Edgeworth,** Richard Lovell. On aerostation. (Philosophical magazine, London, 1816, v. 47, p. 185–187.) OA

Claims to have anticipated John Evans, jr., in the use of inclined planes for guiding balloons. See entry no. 1027.

Lighter-Than-Air Craft — Airships, continued

1014. **Edwards,** R. S. The Stanley airship. illus. (American inventor, Washington, 1903, v. 10, p. 305–306.) * V

Invention of Charles Stanley, of California.

1015. **Ellicott,** Eugene C. The first airship. facsim., port. (Illinois State Historical Society, Journal, Springfield, Ill., 1931, v. 24, p. 105–109.) IAA

Brief account of an attempted flight by Hugh Newell, near Danville, Illinois, in 1841.

1016. **Espitallier,** Georges. L'aviateur de M. Roze. illus. (La Nature, Paris, 1901, année 29, semestre 2, p. 274–278.) OA

Airship uses two elongated balloons, lifting screws, and a parachute.

1017. —— Le ballon dirigeable Severo. illus. (La Nature, Paris, 1902, année 30, semestre 1, p. 387–391.) OA

1018. —— Le ballon Lebaudy. illus. (La Nature, Paris, 1903, année 31, semestre 1, p. 337–339.) OA

Translation in *Scientific American supplement*, New York, 1903, v. 56, p. 23221, *VA*.

1019. —— La campagne de 1904 du ballon Lebaudy. illus. (Le Génie civil, Paris, 1904, tome 45, p. 362–364.) VA

1020. —— La catastrophe du ballon le "Bradsky." illus. (La Nature, Paris, 1902, année 30, semestre 2, p. 321–322.) OA

1021. —— Le dirigeable "La Ville de Paris." illus. (Le Génie civil, Paris, 1903, tome 44, p. 123–124.) VA

Built by Victor Tatin.

1022. —— Les nouvelles expériences du "Lebaudy." illus. (Le Génie civil, Paris, 1905, tome 48, p. 89–92.) VA

1023. —— Les progrès de l'aéronautique. illus. (Le Génie civil, Paris, 1902, tome 40, p. 289–294, 309–313, 329–333.) VA

History of airships, especially those of Santos-Dumont; review of investigations on stability, air resistance, and speed measurement; long-distance balloon voyages, notably that of De la Vaulx across the Mediterranean Sea.

1024. —— Le "Santos-Dumont no. 9." illus. (La Nature, Paris, 1903, année 31, semestre 2, p. 17–18.) OA

1025. —— Ueber die Ursachen des Unfalls des Luftschiffes von Severo. illus. (Illustrierte aëronautische Mittheilungen, Strassburg, 1903, Jahrg. 7, p. 33–36.) VDS

Causes of the accident to the airship *Pax,* 1902.

1026. **Evans,** Edward S. The all-metal dirigible ZMC–2. illus. (Aero digest, New York, 1929, v. 15, Oct., p. 82–83, 261.) VDS

Story of experimental ship built for the United States Navy.

1027. **Evans,** John, jr. Experiments tending to prove the possibility of causing balloons to deviate considerably from the general direction of the current of wind which carries them along in the atmosphere. illus. (Philosophical magazine, London, 1815, v. 46, p. 321–327.) OA

Author's experiments with inclined plane below the balloon. Illustration on plate 6 at end of volume. For reply by Cayley see entry no. 953.

1028. **Evans,** S. Maurice. The earlier British airships. illus. (Air, London, 1930, v. 3, p. 19–21.) VDS

1029. **Fairbairn,** A. R. Evolution of airship sheds. illus. (American architect, New York, 1919, v. 116, p. 345–346.) 3–† MQA

Construction is compared with that of some of the famous English cathedrals.

1030. **Fanciulli,** Jerome S. The first Government airship. illus. port. (Aeronautics, New York, 1908, v. 3, Sept., p. 13–16.) VDS

United States Signal Corps airship built by Thomas Scott Baldwin.

1031. **Farcot,** Eugène. La navigation atmosphérique. Paris: A. Bourdillat & Cie., 1859. x, 12–104 p. plate. 12°. VDW

Suggestions for an airship.

1032. **Faria,** Antonio de, vicomte. Bartholomeu Lourenço de Gusmão (1685–1724); inventeur des aérostats. En commémoration du 202e anniversaire de la première ascension publique qu'il a faite à Lisbonne le 8 août 1709. Lausanne: Imprimeries réunis S. A., 1911. xi, 217 p. illus. ports. 8°. (Académie aéronautique Bartholomeu de Gusmão.) AN

Proceedings of the anniversary, press notices, and the text of Gusmão's patent.

1033. **Feldhaus,** Franz Maria. Leinberger's Dampfluft. illus. (Deutsche Zeitschrift für Luftschiffahrt, Berlin, 1909, Jahrg. 13, p. 1058–1060.) VDS

Illustration and description of steam-driven airship designed 1835–1839.

1034. **Ferro,** Guido Gattegno. L'aeronave dirigible dell' abate Vincenzo Curzio di Napoli. illus. (Rivista aeronautica, Roma, 1932, anno 8, March, p. 581–583.) VDS

Idea for airship, 1805.

1035. The **First** air liner. The "Deutschland" and its ill-fated voyage. illus. (Scientific American supplement, New York, 1910, v. 70, p. 66–67.) VA

Describes accident to *Zeppelin VII,* 1910.

1036. The **First** airship to carry two planes; the "R33" launching test. illus. (Illustrated London news, London, 1926, v. 169, p. 82.) * DA

Excellent illustrations with brief descriptive texts.

1037. The **First** dirigible in America to carry 3 passengers. illus. (Aeronautics, New York, 1908, v. 3, July, p. 21–22.) VDS

Brief account of the Knabenshue craft, 1908.

1038. The **First** two trial trips of Von Zeppelin's airship. illus. (Scientific American supplement, New York, 1900, v. 50, p. 20754–20755.) VA

Lighter-Than-Air Craft — Airships, continued

1039. A **Flight** and no flight. (Mechanic's magazine, London, 1835, v. 23, p. 464.) 3–VA

Tells how the *Eagle* of the European Aeronautical Society got into the hands of the sheriff.

1040. **Flight** of the "Zeppelin I." Recent performances of the famous German airship. (Scientific American supplement, 1909, v. 67, p. 243.) VA

1041. **Flight** of the ZR–1. illus. (Aeronautical digest, New York, 1923, v. 3, p. 232–236, 282–283.) VDS

Test flights of U. S. Navy rigid airship, Sept., 1923.

1042. **Flying** machines in America. (Aeronautical journal, London, 1899, v. 3, p. 13–14.) VDS

Brief description of Professor Charles E. Hite's airship.

1043. **Fonvielle,** Wilfrid de. La direction des ballons. illus. (La Lumière électrique, Paris, 1884, tome 13, p. 500–502.) † VGA

Comments on electrical equipment of the Krebs-Renard airship.

1044. —— Eugène Farcot. port. (L'Aéronaute, Paris, 1896, année 19, p. 75–79.) 3–VDS

Farcot (1830–1896), a watchmaker, designed an airship in 1859, and took a prominent part in aerial operations during the Franco-Prussian war.

1045. **Fonvielle,** Wilfrid de, and Georges Besançon. Notre flotte aérienne. Paris: Gauthier-Villars, 1908. 234 p. illus. 8°. VDY

Principally a history of French airships from the time of General Meusnier.

1046. **Forlanini,** Enrico. Studi sperimentali sulla navigazione aerea. (Il politecnico, Milano, 1877, anno 25, p. 740–748.) 3–† VDA

Airships and helicopters compared. For other references to the work of Forlanini see Giuseppe Boffito, *Biblioteca aeronautica italiana,* Firenze, 1929, p. 175–176, † *VDT.*

1047. **Fortmüller.** Betrachtungen über die Versuche von G. Baumgarten und Dr. Wölfert. (Zeitschrift für Luftschiffahrt, Berlin, 1888, Jahrg. 7, p. 99–109.) 3–VDS

1048. **France** produces its first rigid dirigible. illus. (Aero and hydro, Chicago, 1913, v. 6, p. 87.) 3–VDS

Description and illustrations of the *Spiess* airship.

1049. **Freire de Carvalho,** Francisco. Memoria que tem por objecto revindicar para a nação portugueza a gloria da invenção das machinas aerostaticas. (Academia real das sciencias de Lisboa, Memorias, Lisboa, 1843, serie 2, tomo 1, parte 1, Classe de sciencias moraes e bellas lettras, p. 133–155.) * ES

The airship of Bartholomeu Lourenço de Gusmão as cited in early documents and manuscripts.

1050. The **French** airship catastrophe. The Dixmude now assumed lost. (Flight, London, 1924, v. 16, p. 9.) VDS

This ship was the former German airship *L 72.*

1051. The **French** balloon experiments. (English mechanic, London, 1884, v. 39, p. 574–575.) VA

Conflicting accounts of the Renard-Krebs airship are followed by a brief history of balloons.

1052. **Frenchman's** proposed airship in the Civil war. illus. (U. S. air service, New York, 1921, v. 5, April, p. 36.) VDS

Illustration and brief description of Metiviev's aircraft of 1862.

1053. **Frion,** O. L'appareil Haenlein. illus. (L'Aéronaute, Paris, 1876, année 9, p. 293–305.) 3–VDS

1054. **Fritsche,** Carl B. The metalclad airship. illus. (Mechanical engineering, New York, 1929, v. 51, p. 905–911.) VFA

Construction and building of the United States Navy craft *ZMC–2.*

1055. —— The metalclad airship. illus. (Royal Aeronautical Society, Journal, London, 1931, v. 35, p. 819–883.) VDS

Reviews ten years' research. Abridged in *Flight,* London, 1931, v. 23, p. 461–464, 483–487, *VDS.*

1056. **Fromage,** Lucien. Ce que seront les aérostats quand ils n'auront plus de ballons; suite à mon projet d'aérostat à tube central. Rouen: E. Cagniard, 1889. 19 p. illus. 4°. † VBA p.v.6, no.6

Author's idea of a huge airship dispenses with the balloon.

1057. **Fürstenberg zu Fürstenberg,** A., Graf. Das erste Luftschiff. Teilweise nach Briefen zwischen Bartholomeu de Guzmão und Elisabeth von Braunschweig-Blankenburg. (Die Umschau, Frankfurt a. M., 1908, Jahrg. 12, p. 346–348.) OA

1058. **Fulton,** Garland. Airship development. illus. (Aviation, New York, 1931, v. 30, p. 29–30.) VDS

Progress in the United States during 1930.

1059. **Fyfe,** Herbert C. The newest flying machine. illus. (Strand magazine, London, 1899, v. 17, p. 596–601.) * DA

K. I. Danilevski's wing-steered airship.

1060. **Gaede,** R. Entwürfe zum Baue von lenkbaren Luftschiffen. illus. (Practische Maschinen-Constructeur, Leipzig, 1873, Bd. 6, p. 10–11, 22–24, 38–39.) 3–† VDA

Author, a captain in the Prussian artillery, presents two of his designs.

1061. **Galvez,** comte de. Sur un moyen de donner la direction aux machines äerostatiques. Communicated by Sir Joseph Banks. Read July 1, 1784. illus. (Royal Society of London, Philosophical transactions, 1784, London, 1784, v. 74, part 1, p. 469–470.) * EC

Proposes adjustable sails. See plate 21 at back of volume.

1062. **Ganswindt,** Hermann. Die Lenkbarkeit des aërostatischen Luftschiffes, gemeinfasslich mit ausführlichen Berechnung und Zeichnungen dargestellt. Berlin: Gsellius, 1884. 84 p. illus. 8°. VDY

Project for propelling and steering airships.

Lighter-Than-Air Craft — Airships, continued

1063. Geiger, Fr. Navire aérien par M. Camille Vert. Ingénieur-mécanicien à Paris als Vorgänger des Grafen Zeppelin in der Anwendung des starren Systems. illus. (Deutsche Zeitschrift für Luftschiffahrt, Berlin, 1911, Jahrg. 15, No. 15, p. 21.) **VDS**

Description and illustration of famous "flying fish" airship of the time of Napoleon III.

1064. General Thayer's dirigible balloons. illus. (Scientific American, New York, 1885, v. 53, p. 405.) **VA**

Illustration on p. 399.

1065. Gerville-Réache, Léo. Autour du monde en Zeppelin. Paris: La Nouvelle revue critique [1929]. 223 p. illus. ports. [21. ed.] 12°. (Collection La vie d'aujourd'hui. v. 7.) **VDY**

Author was a passenger of the *Graf Zeppelin* during the round-the-world flight of 1929.

1066. Giffard, Henri. Description du premier aérostat à vapeur. (Académie des sciences, Comptes rendus, Paris, 1870, tome 71, p. 683–687.) *** EO**

Description of the author's steam-driven airship of 1852.

1067. —— Système de navigation aérienne. illus. (Génie industriel, Paris, 1865, tome 29, p. 251–265.) **VA**

Description of and calculations for author's airship of 1855.

For citations to writings by and about Giffard see Paul Brockett, *Bibliography of aeronautics,* Washington, 1910, p. 360, *VDT.*

1068. A **Gigantic** airship disaster. illus. (Scientific American, New York, 1908, v. 98, p. 430–431.) **VA**

Wreck, May 23, 1908, Berkeley, Cal., of ship designed and built by John A. Morrell.

1069. Girard, E., and A. de Rouville. Les ballons dirigeables; théorie — applications. Paris: Berger-Levrault & Cie., 1907. 307 p. illus. 8°. **VDW**

Discusses form, airscrews, motors, and reviews the investigations of Meusnier, Dupuy de Lôme, Giffard, Tissandier, Renard, Zeppelin, Santos-Dumont, and Lebaudy. Also in *Revue du génie militaire,* Paris, 1906–07, v. 32, p. 5–46, 97–138, 231–269, 325–368, 413–452, 555–592; v. 33, p. 5–52, *VWA.*

1070. Goldsmith, Margaret Leland. Zeppelin, a biography. New York: W. Morrow and Company, 1931. 278 p. illus. ports. 8°. **AN**

Bibliography, p. 273. Reviewed in *Aviation engineering,* Washington, N. J., 1931, v. 1931, May, p. 30, *VDS.*

1071. The **Goodyear-Zeppelin** airship deal. (Flight, London, 1923, v. 15, p. 723.) **VDS**

Goodyear Tire and Rubber Company, of Akron, Ohio, acquires manufacturing rights from the Zeppelin Company.

1072. Gouilly, Alexandre. Analyse de l'œuvre de Henri Giffard. illus. (Société des ingénieurs civils de France, Mémoires, Paris, année 1888, v. 2, p. 365–504.) **VDA**

Extensive account of Giffard's life and works, with analyses of his patents.

1073. Goupil, Albert Louis. Arrêt et stabilité d'un ballon dirigeable contre un vent violent. illus. (Société des anciens élèves des écoles nationales d'arts et métiers, Bulletin technologique, Paris, 1908, p. 1041–1077.) **3–VA**

1074–1075. Govi, Gilberto. Sur la date d'un travail inédit de Meusnier relatif à l'équilibre des machines aérostatiques, et sur celle de l'extrait que Monge en a laissé, et que l'Académie des sciences de Paris vient de publier. (R. Accademia delle scienze di Torino, Atti, Torino, 1870–71, v. 6, p. 286–293.) *** ER**

Discussion as to the date of Meusnier's original paper. See entry no. 1180.

1076. Gradenwitz, Alfred. The Parseval dirigeable airship. illus. (Scientific American, New York, 1906, v. 95, p. 341–342.) **VA**

1077. Great Britain. — Courts: Court of Inquiry into the Loss of the R. 101. Report of the R. 101 inquiry. Presented by the Secretary of State for Air to Parliament... March, 1931. London: H. M. Stationery Office, 1931. 129 p. illus. 8°. (Cd. 3825.) **VDY**

Full report of airship disaster as reported by the Board of Assessors under the chairmanship of Sir John Simon.

1078. Gregg, Willis Ray. The trans-Atlantic flight of the British dirigible "R. 34." illus. map. (Monthly weather review, Washington, 1919, v. 47, p. 541–543.) **† PRS**

Study of weather conditions.

1079. Grey, Charles Grey. The death of Count von Zeppelin. (Aeroplane, London, 1917, v. 12, p. 659–660.) **VDS**

1080. —— The Kempston balloon accident. (Aeroplane, London, 1926, v. 31, p. 174.) **VDS**

Account of the death of Ernest T. Willows, England's pioneer airship builder.

1081. —— Lord Thomson. (Aeroplane, London, 1930, v. 39, p. 800.) **VDS**

Brigadier-General Thomson, Air Minister in the MacDonald cabinet, was actively interested in the building of airships.

1082. —— On the coming of age of the British airship. illus. (Aeroplane, London, 1926, v. 30, p. 529–530, 532, 534.) **VDS**

Describes the experiments of Ernest T. Willows, beginning in 1905.

1083. Grilleau, B. de. Les aérostats dirigeables, leur passé, leur présent, leur avenir. Le ballon de Meudon et les progrès les plus récents de l'aéronautique. Paris: E. Dentu, 1884. xv, 20–211 p. illus. 12°. **VDW**

Reviewed in *L'Aéronaute,* Paris, 1884, année 17, p. 203–207, *3–VDS.*

1084. Gross, Hans. Das Aluminium-Luftschiff C. Schwarz. illus. (Zeitschrift für Luftschiffahrt, Berlin, 1897, Jahrg. 16, p. 291–296.) **3–VDS**

1085. —— Die Entwickelung der Motor-Luftschiffahrt im 20. Jahrhundert. Berlin: Otto Salle, 1906. 31 p. illus. 8°. **VDW p.v.2, no.8**

Zeppelin, Parseval, and Lebaudy types.

Lighter-Than-Air Craft — Airships, continued

1086. **Guarini,** Emile. The "Lebaudy II." illus. (Scientific American, New York, 1904, v. 91, p. 216, 218.) **VA**

1087. **Gustav** Koch's System zur Herstellung von lenkbaren Luftschiffen. (Zeitschrift für Luftschiffahrt, Berlin, 1883, Jahrg. 2, p. 1–12.) **3–VDS**

1088. **Haenlein,** Paul. Haenlein's lenkbares Luftschiff. illus. (Zeitschrift für Luftschiffahrt, Berlin, 1882, Jahrg. 1, p. 46–57, 79–89.) **3–VDS**

Author's illustrated description of his famous airship.

1089. —— Ueber Wolff's Projekt eines lenkbaren Luftschiffes. (Zeitschrift für Luftschiffahrt, Berlin, 1882, Jahrg. 1, p. 188–189.)

A criticism of Wolff's airship. **3–VDS**

For references to Haenlein's writings see Paul Brockett, *Bibliography of aeronautics,* Washington, 1910, p. 389, *VDT.*

1090. **Haenlein's** lenkbares Luftschiff. illus. (Der practische Maschinen-Constructeur, Leipzig, 1874, Jahrg. 7, p. 3–6, 20–22, 38–40, 353–356, 371.) **3–† VDA**

For illustrations see atlas volume for 1874, plates 5 and 89.

1091. **Hagen,** Freiherr vom. Historisches und Kritisches über lenkbare Luftballons. (Zeitschrift für Luftschiffahrt, Berlin, 1883, Jahrg. 2, p. 353–368.) **3–VDS**

Good history of airships.

1092. **Hanawalt,** Wilbur R. Around the world by airship. illus. port. (Aero digest, New York, 1929, v. 15, Oct., p. 51, 261–263.) **VDS**

Account of voyage of the *Graf Zeppelin,* under the command of Hugo Eckener.

1093. **Harper,** Harry. Santos Dumont — the invincible. (Airways, London, 1927, v. 3, p. 258.) **VDS**

1094. **Hearne,** R. P. Airships in peace and war: being the second edition of Aerial warfare, with seven new chapters...an introduction by Sir H. Maxim... London: J. Lane, 1910. vii–xlviii, 324 p. illus. port. 8°. **VDY**

Contains a brief history of dirigibles for war purposes.

1095. **Heilig,** Sterling. The dirigible balloon of M. Santos-Dumont. illus. (Century magazine, New York, 1901, v. 63 [new series, v. 41], p. 66–75.) *** DA**

1096. —— The over-sea experiments of Santos-Dumont. (McClure's magazine, New York, 1902, v. 19, p. 195–206.) *** DA**

1097. **Heinen,** Anton. Leviathans of the sky. Safe and sane rigid airship enterprises are profitable. illus. port. (Aero digest, New York, 1924, v. 4, p. 284–287, 304.) **VDS**

Lessons learned by the author while captain of the German airship *Bodensee.*

1098. **Helmar,** N. Is the air navigable? illus. (Scientific American supplement, New York, 1891, v. 31, p. 12543.) **VA**

Brief description of author's proposed airship.

1099. **Henderson,** John. Claim to the discovery of a means of directing balloons. (Mechanic's magazine, London, 1844, v. 40, p. 19.) **3–VA**

Author claims the distinction and tries to form a joint-stock company.

1100. **Henri** Mager's airship. (Aeronautical world, Glenville, O., 1903, v. 1, p. 133–134.) **VDS**

Airship of Henri Mager and Charles Cadet uses containers for regulating the amount of gas in the balloon.

1101. **Henrion,** Francesco. Fondamenti teorico-pratici delle' arte aereonautica nuovamenta proposti...dedicati a...Giovanni Luigi di Durfort... Firenze: Stamperia di P. Allegrini, 1789. 138 p. illus. 8°. **VDT (Room 120)**

Theory of directing balloons and project.

1102. **Hervé,** Henri. Sur la stabilisation de route des ballons dirigeables. illus. (Académie des sciences, Comptes rendus, Paris, 1904, tome 139, p. 37–39.) *** EO**

1103. L'**Histoire** des "dirigeables" Santos-Dumont. 100 photogravures; texte de MM. Santos-Dumont, H. Deutsch...[and others.] Un document historique publié par La vie au grand air. Paris: P. Lafitte & Cie. [1901.] 16 p. illus. f°. **† VDY**

1104. **Hoernes,** Hermann. Das Zeppelin'sche Ballonproblem. illus. (Oesterreichischer Ingenieur- und Architekten-Verein, Zeitschrift, Wien, 1901, Jahrg. 53, p. 197–202, 217–222.) **VDA**

A study of Zeppelin construction and flights, with comments.

1105. **Hunsaker,** Jerome Clarke. Airship engineering progress in the United States. illus. (Aviation, New York, 1919, v. 7, p. 72–76, 123–128.) **VDS**

Progress of research and construction from 1916.

1106. —— The present state of airships in Europe. illus. (American Society of Naval Engineers, Journal, Washington, 1914, v. 26, p. 773–812.) **VXA**

1107. **Hunt,** Edmund. Plan of an aerial locomotive or governable balloon. illus. (Mechanic's magazine, London, 1848, v. 49, p. 169–171.) **3–VA**

For criticisms see above magazine, p. 257–258.

1108. **Hureau de Villeneuve,** Abel. Ernest Pétin. illus. port. (L'Aéronaute, Paris, 1878, année 11, p. 255–260.) **3–VDS**

Account of Pétin's airships, 1850–1851.

1109. **Hutchinson,** W. N. Aërial navigation. (United service magazine, London, 1882, part 2, p. 255–279; part 3, p. 158–167.) *** DA**

Details and illustrations of author's design for an airship.

Lighter-Than-Air Craft — Airships, continued

1110. **Hydravion,** pseud. Bartholomeu Lourenço de Gusmão. illus. (Het Vliegveld, Amsterdam, 1928, jaarg. 12, p. 263–266.) **3–† VDS**

1111. **Im** Fluge über den Ozean. Graf Zeppelin und sein Lebenswerk. Vorwort von Dr. Hugo Eckener. Berlin: Lichtbild-Bühne, 1924. 39 p. illus. 8°. **VDY p.v.33, no.3**

Brief account of the discouragements and accomplishments of Count Zeppelin.

1112. An **Improved** air ship. illus. (Scientific American, New York, 1889, v. 60, p. 358.) **VA**

Description and illustration of craft patented by Herman A. J. Rieckert, New York City.

1113. An **Improved** airship. illus. (Scientific American, New York, 1892, v. 67, p. 150.) **VA**

Patented by B. F. Barnes, of Ohio.

1114. **Improved** flying machine. illus. (Scientific American, New York, 1863, new series, v. 9, p. 129.) **VA**

Patent of Arthur Kinsella, Washington territory, using air jet for propulsion. The engine works by compressed hydrogen.

1115. The **Improved** Parseval airship. illus. (Scientific American supplement, New York, 1908, v. 66, p. 204.) **VA**

1116. The **Inquiry** into the loss of the R 101. (Aeroplane, London, 1930, v. 39, p. 1022, 1026–1027, 1078–1082, 1254, 1256, 1450, 1452, 1454.) **VDS**

1117. **Institut de France.** — Académie des sciences, Paris. Rapport sur un mémoire de M. Torres, concernant un avant-projet de ballon dirigeable, présenté à l'Académie dans la séance du 26 mai 1902. (Académie des sciences, Comptes rendus, Paris, 1902, tome 135, p. 141–146.) *** EO**

Report on airship with an interior keel.

1118. **Is** the dirigible outstripping the airplane? (Scientific American, New York, 1919, v. 120, p. 366–367.) **VA**

1119. **Jaray,** P. Studien zur Entwicklung der Luftfahrzeuge unter besonderer Berücksichtigung der Zeppelin-Luftschiffe. illus. (Zeitschrift für Flugtechnik und Motorluftschiffahrt, München, 1920, Jahrg. 11, p. 153–159, 173–178.) **VDS**

Technical account of the development of airships, especially of the Zeppelin type. English translation in *Aviation,* Highland, N. Y., 1921, v. 10, p. 424–426, 456–460, *VDS.*

1120. **Jotti,** da Badia, polesine. Fantasie — Dirigibilistiche. (Aeronautica, Milano, 1928, anno 2, p. 1037–1043.) **3–VDS**

Pictures and descriptions of oddly constructed early airships.

1121. —— Pionieri che scompaiono Louis Capazza. illus. (Aeronautica, Milano, 1929, anno 3, p. 98–101.) **3–VDS**

Career of pioneer airship constructor.

1122. **Julliot,** H. Le dirigeable Lebaudy. illus. (Société des ingénieurs civils de France, Mémoires, Paris, 1905, année 1905, tome 1, p. 754–801.) **VDA**

Illustrated description and a record of flights.

1123. **Keller,** H. R. U.S.S. Langley, our first aircraft carrier. (American Society of Naval Engineers, Journal, Washington, 1923, v. 35, p. 500–525.) **VXA**

"The conversion of the *Jupiter,* which was our first electric-drive ship, into the *Langley* was authorized by act of Congress dated July 11, 1919."

1124. **Klinckowstroem,** Carl Ludwig Friedrich Otto, Graf von. Beitrag zur Gusmão-Bibliographie. (Archiv für die Geschichte der Naturwissenschaften und der Technik, Leipzig, 1910–12, Bd. 3, p. 214–223.) **OA**

Examination of early references to the airship of the Brazilian Bartholomeu Lourenço de Gusmão (1685–1724).

1125. —— Die Gusmão-Flugblätter von 1709. illus. (Zeitschrift für Bücherfreunde, Leipzig, 1911, N. F., Jahrg. 3, p. 36–41.) **†* GAA**

Facsimile title-pages and descriptions of the pamphlets of Bartholomeu Lourenço de Gusmão, 1709.

1126. **Kollmann,** Franz. Das Zeppelinluftschiff; seine Entwicklung, Tätigkeit und Leistungen. Berlin: M. Krayn, 1924. 89 p. illus. 8°. **VDW**

History of the various Zeppelin types, with good illustrations.

1127. **Kratzenstein,** Christian Gottlieb. L'art de naviguer dans l'air. Copenhaven & Leipzig: Faber & Nitschke, 1784. 99 p. illus. 12°. **VDW**

Discusses propulsion and steering of balloons.

1128. **Kretschmar,** Julius. Versuch einer Entwickelung der Gesetze nach denen die Bewegung von Flugapparaten in atmosphärischer Luft stattfindet. (Zeitschrift für Luftschiffahrt, Berlin, 1882, Jahrg. 1, p. 139–145, 169–177, 199–204, 245–252, 270–274.) **3–VDS**

Aerodynamic calculations for the proper construction of airships.

1129. **Lagrange,** L. Adolph Clément. port. (L'Aérophile, Paris, 1908, année 16, p. 181.) **3–VDS**

Portrait and brief biography of French manufacturer who, with Louis Capazza, built the famous airship *Bayard-Clément.*

1130. **Lake,** John. Aerial navigation. illus. (Mechanic's magazine, London, 1837, v. 26, p. 393–395.) **3–VA**

Proposes to drive an airship with air jets.

1131. **Lambrecht,** Wolfgang. Transoceanic airship service. illus. (Scientific American, New York, 1933, v. 149, p. 20–22.) **VA**

Story of the *Graf Zeppelin,* beginning operations in 1928.

Lighter-Than-Air Craft — Airships, continued

1132. **Lana Terzi,** Francesco. The aerial ship. London: Royal Aeronautical Society of Great Britain, 1910. vi, 27 p. 12° **VDY (Royal)**

Fifth and sixth chapters of Lana's famous *Prodromo,* published in 1670, translated into English.

1133. —— Magisterium naturae, et artis. Opus physico-mathematicum... Brixiae: per Io: Mariam Ricciardum, 1684–92. 3 v. plates. f°. **†† PAD**

See Book 6, p. 291–294 for further discussion of airship proposed in the author's celebrated *Prodromo.* See entry no. 1400.

1134. —— Prodromo; ouero, Saggio di alcune inuentioni nuoue premesso all' Arte maestra, opera che prepara il p. Francesco Lana Bresciano della Compagnia di Giesv. Per mostrare li piu reconditi principij della naturale filosofia, riconosciuti con accurata teorica nelle piu segnalate inuentioni, ed isperienze sin'hora ritrouate da gli scrittori di questa materia & altre nuoue dell' autore medesimo. Dedicato alla Sacra Maesta Cesarea del imperatore Leopoldo I. Brescia, Per li Rizzardi, MDCLXX. 4 p.l., 252 p., 20 pl. f°. **†* KB 1670**

One of the famous books in aeronautical history. Lana (1631–1687) formulated a theory of aerostatics and proposed to raise a wooden car by suspending it to four metal globes from which the air had been evacuated.

1135. **Lansdowne,** Zachary. With the Shenandoah. illus. (Saturday evening post, Philadelphia, 1925, v. 198, Oct. 24, p. 7, 214, 217–218.) *** DA**

Discussion of the ship's possibilities written shortly before the disaster.

1136. The **Latest** aerial humbug. illus. (Engineering news, New York, 1891, v. 25, p. 182–183.) **VDA**

Description of the Pennington airship, promoted by the Mt. Carmel Aerial Navigation Company, of Chicago.

1137. **Laussedat.** Sur les tentatives effectuées à diverses époques pour la direction des aérostats. (L'Aéronaute, Paris, 1884, année 17, p. 207–210.) **3–VDS**

1138. The **Lebaudy** airship. illus. (Scientific American, New York, 1903, v. 89, p. 411.) **VA**

1139. **Lehmann,** Ernst A., and HOWARD MINGOS. The Zeppelins. The development of the airship, with the story of the Zeppelin air raids in the World war. New York: J. H. Sears & Co. [cop. 1927.] xii, 329 p. illus. 8°. **BTZY**

"The narrative is personal, largely the experiences of a naval architect [Capt. Lehmann], who joined Count Zeppelin's staff early in 1913." Has an account of post-war development. Follows closely the authors' series of articles in *World to-day,* London, 1927, v. 50, p. 571–587; 1927–28, v. 51, p. 21–35; 186–199, 264–279, 367–377, 488–499, * *DA.*

Reviewed in *Air,* London, 1929, v. 2, p. 90–91, *VDS; Aeroplane,* London, 1928, v. 34, p. 878, *VDS.*

1140. **Leo** Stevens und sein selbstfahrender Ballon. port. illus. (Illustrierte aëronautische Mitteilungen, Strassburg, 1903, Jahrg. 7, p. 69–75.) **VDS**

Account of Stevens' airship.

1141. **Lericolais,** Eugène. Le premier conquérant de l'air. Bartholomeu de Gusmão. (La Revue, Paris, 1910, série 6, v. 88, p. 642–653.) *** DM**

1142. **Létonné.** Sur un projet de machine aéronautique, redigé par le général Meusnier... (Académie des sciences, Comptes rendus, Paris, 1886, tome 103, p. 237–240.) *** EO**

Account of Meusnier's writings and of his famous atlas of aeronautical drawings (1784 to 1789). Also in *L'Aéronaute,* Paris, 1886, année 19, p. 183–187, *3–VDS.*

1143. The **Lighter-than-air** situation. Official inquiry into the Roma accident... illus. (Aviation, Highland, N. Y., 1922, v. 12, p. 280–282.) **VDS**

1144. **Linke,** Franz. Die Luftschiffahrt von Montgolfier bis Graf Zeppelin...mit einem Beitrage über Militärluftschiffahrt von D. A. Hildebrandt... Berlin: Alfred Schall [1909]. 378 p. illus. port. 8°. **VDT**

1145. **Liversedge,** A. J. Possibilities of the large airship. (Fortnightly review, New York, 1915, v. 104 [new series, v. 98], p. 1079–1092.) *** DA**

1146. **Loessl,** F., Ritter von. Das Zeppelin-Luftschiff als neuester Versuch zur Lenkbarmachung von Gasballons. (Oesterreichischer Ingenieur- und Architekten-Verein, Zeitschrift, Wien, 1900, Jahrg. 52, p. 543–547.) **VDA**

1147. **Loss** of the Campbell air ship. illus. (Scientific American, New York, 1889, v. 61, p. 47, 54.) **VA**

Designed by Peter C. Campbell and piloted by E. D. Hogan. Lost at sea, July 16, 1889.

1148. The **Loss** of the French airship "La Patrie." (Scientific American, New York, 1907, v. 97, p. 438.) **VA**

1149. The **Loss** of H. M. airship R 101. Death of Lord Thomson and 47 others. illus. (Flight, London, 1930, v. 22, p. 1107–1113.) **VDS**

Details of disaster, Oct. 5, 1930.

1150. Die **Luftschiffahrt** dem heutigen Stande der Wissenschaft entsprechend dargestellt, von K. Hackstetter, S. Hartmann... und Graf F. von Zeppelin [jr.]. Stuttgart: Franckh [1908]. 157 p. 8°. **VDY**

A series of contributions on contemporary German airships, with a history of balloons.

1151. **Luntley,** John. The rotary balloon. (Mechanic's magazine, London, 1851, v. 55, p. 328.) **3–VA**

A reply to criticism of Sir George Cayley. See entry no. 952.

Lighter-Than-Air Craft — Airships, continued

1152. Lyle, Eugene P. Circling the Eiffel tower in an air-ship. M. Santos-Dumont and his experiments in air navigation. The features which make possible the success of his new flying-machine. illus. (Everybody's magazine, New York, 1901, v. 5, p. 515–532.) *** DA**

Also in *Smithsonian report for 1901,* Washington, 1902, p. 575–592, ** EA.* A second copy of the latter in *VDC p.v.10, no.9.*

1153. Lyman, Lauren D. Eckener, publicist, navigator. illus. port. (Aero digest, New York, 1930, v. 17, Sept., p. 46–47, 182, 184.) **VDS**

Biography and portrait of famous Zeppelin captain.

1154. McCluer, C. E. Rufus Porter and his "flying ship." illus. (Scientific American, New York, 1910, v. 102, p. 30.) **VA**

Porter (1792–1884) organized an aerial navigation company in 1852 for the purpose of exploiting airships.

1155. Maceroni, Francis. Impracticability of aerial navigation — Montgolfier's preferable to gas balloons. (Mechanics' magazine, London, 1836, v. 25, p. 408–410.) **3–VA**

Sees no promise in an elongated fish-like balloon, because it cannot be kept in an horizontal position, and because no means can be found for propelling it.

1156. Mackinnon, A. Aerial projects — Royal Condor Company. (Mechanic's magazine, London, 1829, v. 12, p. 87–88.) **3–VA**

Proposes to harness condors to aerial carriages.

1157. Mackintosh, T. S. Observations on the practicability of guiding and directing air-balloons or aerial ships. (Mechanic's magazine, London, 1835, v. 24, p. 55–60.) **3–VA**

Deprecating the use of oars or other mechanical means, the author suggests the use of large birds.

1158. McLean, William P. The flight of the airship Toledo... Photographs by H. G. Gager. [New York:] A. Fuller Co., cop. 1905. 4 l. illus. 8°. **VDW**

Story of Roy Knabenshue's airship flight over New York City in 1905.

1159. MacSweeny, Joseph. An essay on aerial navigation, pointing out modes of directing balloons. Cork: George Purcell & Co., 1844. vi, 122 p. illus. 2. ed. 12°. **VDW**

First edition was published in 1824. Has historical introduction.

1159a. Magoun, Frederick Alexander, and Eric Hodgins. Cayley and those who followed. illus. port. (In their: A history of aircraft, New York, 1931, p. 114–130.) **VDT**

Cayley's experiments. Cayley's study of hulls. Cayley's ideas of power. Cayley's second airship design. Cayley's other proposals. Lennox's eagles. Design of Monck Mason. Jullien's model. Pétin's multiple balloons. Bell's locomotive balloon. The first successful dirigible (Giffard's). Nye's rocket balloon. Contributions of Luntley. Lassie's gimlet. The metallic balloon of Richard Boyman. Contributions of Dupuy de Lôme.

1159b. —— Count Zeppelin and the rigid airship. illus. (In their: A history of aircraft, New York, 1931, p. 157–186.) **VDT**

The beginnings of the Zeppelins. Early experiments. First commercial airship routes. English pre-war lighter-than-air experiments. The "Mayfly" disaster. Zeppelins in the World war. The *Dixmude* disaster. British rigid airships. First lighter-than-air trans-Atlantic flight. The *R38* or *ZR2* disaster. The airship *Shenandoah.* Fatal disaster to the *Shenandoah.* Post-war Zeppelin construction. The airship in Polar exploration. The *Italia* disaster. The *Graf Zeppelin.* Recent British rigid airships. The tragedy of the *R-101.* Present airship practice.

1159c. —— The creation of the practical airship. illus. (In their: A history of aircraft, New York, 1931, p. 131–156.) **VDT**

Experiments of Haenlein. The Tissandier dirigible. Success of Renard and Krebs. The first rigid airship. Alberto Santos-Dumont. The brothers Lebaudy. The *Clément-Bayards.* Dirigibles "pour le sport." Early semi-rigids. Baldwin in the United States. Wellman's *America.* Vaniman disaster. Blimps. The *Roma* disaster.

1159d. —— Early attempts at navigation. illus. (In their: A history of aircraft, New York, 1931, p. 101–113.) **VDT**

Experiments of Blanchard, and of Montgolfier. Work of General Meusnier. Attempt at flight with Meusnier's ballonet. Miolan and Janinet. The "necklace" of Marey Monge. Vincent Lunardi. Inclined planes of Bourgeois. Joseph Binns. The promises of Uncles. Scott's fish-shaped balloon. Construction of a fish-shaped balloon (Pauly's). Experiments of Pauly and Egg. Early Russian cylindrical balloon. Ovoid balloon of Guillie. Proposals of Joseph MacSweeney. Balloon à chevaux. The faint dawn of sanity.

1160. Maitland, Edward Maitland. The log of H.M.A. R34 journey to America and back...with a letter from Rudyard Kipling. London: Hodder and Stoughton, Ltd. [1921.] xii, 168 p. illus. 2. ed. 8°. **VDYC**

Journey was made from and to England in July, 1919.

1161. Mammoth aluminum air-ships. illus. (Aeronautical world, Glenville, O., 1903, v. 1, p. 218–219.) **VDS**

Brief account of Schwartz's craft and its fatal trial.

1162. Mansfield, Charles Blachford. Aerial navigation...edited by his brother Robert Blachford Mansfield, with a preface by J. M. Ludlow. London: Macmillan & Co., 1877. xxiii, 513 p. illus. 12°. **VDW**

Mansfield completed the manuscript in 1851. The work is a study of the conditions under which an airship should be constructed. Bibliography.

1163. Marey-Monge, Edmond. Études sur l'aérostation. Paris: Bachelier, 1847. 353 p. plates. 8°. **VDW**

A discussion of the conditions requisite for a successful airship and the author's suggestions for meeting the problem. Also has notes on Meusnier's airship.

1164. Marsh, William Lockwood. The evolution of rigid airship design. (Air annual of the British Empire, London, 1930, v. 2, p. 95–104.) **VDC**

States that Dr. Schwartz, an Austrian, experimented with a rigid (in the modern sense) ship in 1897 and 1898.

1165. —— A French balloon of 1872 whose motive power was eight men turning a crank. (Airways, London, 1927, v. 3, p. 198.) **VDS**

Account of the Dupuy de Lôme craft.

1166. —— Twenty-one years of airship progress. illus. (Flight, London, 1930, v. 22, p. 84–88.) **VDS**

Lighter-Than-Air Craft — Airships, continued

1167. **Marsh** and Ranwell's proposed method of aerial navigation. illus. (Mechanic's magazine, London, 1840, v. 33, p. 353–354.) **3–VA**

Airship comprises 21 small balloons and two wing propellers.

1168. **Martini,** Tito. Per la storia della conquista dell' aria; l'areonave del P. Francesco Lana. (Rivista di fisica, matematica e scienze naturali, Firenze, 1912, v. 25, p. 101–109.) **OA**

Excerpts from Lana de Terzi's famous work (1670) *Prodromo,* in which he proposes an aerial ship.

1169. **Masfrand,** A. de. Le dirigeable militaire "Patrie." illus. (L'Aérophile, Paris, 1906, année 14, p. 299–306.) **3–VDS**

1170. —— Les expériences du "Lebaudy." illus. (L'Aérophile, Paris, 1905, année 13, p. 151–159.) **3–VDS**

History of pioneer French airship.

1171. —— Les premiers exploits du Clément-Bayard. illus. map. (L'Aérophile, Paris, 1908, année 16, p. 449–451.) **3–VDS**

Early flights of celebrated French airship.

1172. **Massau,** J. Discours sur les ballons dirigeables, suivi d'une note sur l'aviation. Mons: Duquesne-Masquillier & fils, 1904. 26 p. 4°. **VDY p.v.8, no.11**

Brief histories of some of the famous French airships.

1173. **Masterman,** Edward Alexander Dimsdale. The evolution of mooring and handling devices for airships. illus. (Air annual of the British Empire, London, 1930, v. 2, p. 105–117.) **VDS**

1174. **Material** on Dr. Solomon Andrews' flying ship "Aeron" and the Aerial Navigation Company. [v. p., n. d.] 10 l. f°. **† VDW**

Scrap-book made by Arthur T. Hill, 1890.

1175. **Maxson,** Herburt W. First semi-rigid airship to be designed and built in the United States. illus. (U. S. air services, Washington, 1925, v. 10, May, p. 37–38.) **VDS**

The *R.-31* built for the United States Army Air Service, 1925.

1176. **Meltzl,** David. Ueber die Lenkung der Aerostaten. (Dingler's polytechnisches Journal, Stuttgart, 1824, Bd. 14, p. 63–80.) **3–VA**

Author proposes, with the help of sails, to use upper air currents blowing in different directions. For comments see *Bulletin des sciences technologiques,* Paris, 1826, tome 5, p. 34–35, *3–VA.*

1177. **Mesa y Ramos,** José. Un proyecto de globo dirigible. (Revista contemporánea, Madrid, 1902, v. 125, p. 391–399.) **NPA**

Describes design of Leonardo Torres for a metal-framed airship.

1178. A **Metal** balloon. (Engineer, London, 1883, v. 55, p. 110.) **VA**

Aluminium airship of Count Buonaccorsi di Pistoja.

1179. **Metalclad** airship development. illus. (Society of Automotive Engineers, Journal, New York, 1926, v. 18, p. 94–96.) **TOL**

1180. **Meusnier,** Jean Baptiste Marie. Mémoire sur l'équilibre des machines aérostatiques... (Journal de physique, Paris, 1784, tome 25, p. 39–69.) **3–OA**

Library has reprint of above in *OAI p.v.14, no.13.* For same, as abridged by Marey-Monge see Académie des sciences, *Comptes rendus,* Paris, 1870, tome 71, p. 569–577, ** EO.*

1181. **Micciollo,** Alfred. Aéronef dirigeable. [By Alfred Micciollo.] n. t.-p. [Le Puy: Imp. R. Marchessou & Co., 1905.] 42 p. 8°. **VDY**

Author describes his proposed airship.

1182. **Milla,** Karl. Ladislaus R. v. Dutczynski. (Zeitschrift für Luftschiffahrt, Berlin, 1894, Jahrg. 13, p. 214.) **3–VDS**

Brief biography of Austrian military airship pioneer.

1183. **Mingos,** Howard. The story of the Zeppelins. Some of their feats in war and peace. illus. (World's work, Garden City, N. Y., 1928, v. 57, Nov., p. 37–46.) *** DA**

1184. **Mr.** Monck Mason's aerial machine. (Mechanic's magazine, London, 1843, v. 39, p. 467–472.) **3–VA**

Description of model airship exhibited in London.

1185. **Mitchell,** Harley W. Is the dirigible utterly worthless? illus. (Popular aviation, Chicago, 1932, v. 11, p. 144–147.) **VDS**

Author answers in the affirmative.

1186. **Moedebeck,** Hermann W. L. Bartholomeu Lourenço de Gusmão, der Erfinder des Luftballons. illus. (Zeitschrift für Luftschiffahrt und Physik der Atmosphäre, Berlin, 1893, Jahrg. 12, p. 1–10.) **3–VDS**

1187. —— Betrachtungen über das lenkbare Luftschiff und Bericht über den Versuch mit dem Aluminium-Luftschiff in Berlin. illus. (Illustrierte aëronautische Mittheilungen, Strassburg, 1898, Jahrg. 2, p. 18–20.) **VDS**

Brief description of the Schwarz airship and of its motor.

1188. —— Fortschritte in der Luftschiffahrt, insbesondere im Luftschiffbau. illus. (Verein deutscher Ingenieure, Zeitschrift, Berlin, 1908, Bd. 52, p. 901–906.) **VDA**

Principally a history of airships, beginning with Haenlein, 1872.

1189. —— Das Geheimnis beim Luftschiff "La France" von Renard-Krebs. illus. (Illustrierte aëronautische Mitteilungen, Berlin, 1908, Jahrg. 12, p. 282–284.) **VDS**

Account of airship constructed by Charles Renard and Captain Krebs in 1884.

1190. —— The German aluminium balloon. illus. (Aeronautical journal, London, 1898, v. 2, Jan., p. 11–15.) **VDS**

Account of David Schwarz's airship and its disaster.

1191. —— Die lenkbaren Luftschiffe unter besonderer Berücksichtigung ihrer Motoren. illus. (Glaser's Annalen für Gewerbe und

Lighter-Than-Air Craft — Airships, continued

Bauwesen, Berlin, 1885, Bd. 17, p. 12–19, 43–44.) 3–† VDA

With discussion.
History of modern airships with special reference to their motors.

1192. —— Die Luftschiffahrt; ihre Vergangenheit und ihre Zukunft, insbesondere das Luftschiff im Verkehr und im Kriege. Strassburg: K. J. Trübner, 1906. vi, 137 p. illus. 8°. **VDT p.v.2, no.2**

1193. —— Paul Haenlein. port. (Illustrierte aëronautische Mitteilungen, Strassburg, 1905, Jahrg. 9, p. 65–70.) **VDS**

Account of airships built by German pioneer in 1865 and 1872.

1194. —— Rückblick auf die Fortschritte im Luftschiffbau im Jahre 1904. illus. (Prometheus, Berlin, 1905, Jahrg. 16, p. 417–422, 442–445.) **3–VA**

Airship developments during 1904, principally concerning the Lebaudy craft and the Hervé kite balloon.

1195. —— The termination of the trials of Count von Zeppelin's airship. (Scientific American supplement, New York, 1901, v. 51, p. 21138–21140.) **VA**

1196. **Moffett,** William Adger. Rigid airship development and the U.S.S. Akron. illus. (Aeronautic review, Washington, 1932, v. 10, Jan., p. 6–12.) **VDS**

1197. **Moigno,** F. Steering of balloons. (Practical magazine, London, 1874, v. 4, p. 239.) **3–† VA**

Account of Henry Giffard's work in 1863.

1198. **M.** Santos-Dumont wins the Deutsch prize. illus. (Scientific American supplement, New York, 1901, v. 52, p. 21631–21632.) **VA**

1199. **Mort** du colonel Charles Renard. (L'Aérophile, Paris, 1905, année 13, p. 80–81.) **3–VDS**

Renard (1847–1905) is stated to have been the first to demonstrate the practicability of the airship.

1200. **Mortane,** Jacques. L'historique des dirigeables. (In: L'Aéronautique, par Roland Garros [and others], Paris: Pierre Lafitte & Cie., 1914, p. 57–83.) **VDW**

History of airships through 1912, with brief accounts of notable flights.

1201. **Moses,** Stanford E. Life on the Shenandoah. illus. (Saturday evening post, Philadelphia, 1924, v. 197, Dec. 20, p. 4–5, 51–52.) *** DA**

1202. **Müllenhoff,** Karl. Dr. Wölfert. (Zeitschrift für Luftschiffahrt, Berlin, 1897, Jahrg. 16, p. 239–243.) **3–VDS**

Account of pioneer German designer of airships.

1203. **Müller-Breslau,** Heinrich Franz Bernard. Zur Geschichte des Zeppelin-Luftschiffes. illus. (Verein zur Beförderung des Gewerbfleisses, Verhandlungen, Berlin, 1914, Jahrg. 93, p. 35–52.) **3–VA (Gewerbefleiss)**

1204. **Myers,** Carl E. Airship experiments. illus. (Scientific American, New York, 1899, v. 80, p. 41–42.) **VA**

Author's man-driven balloon.

1205. —— Progress in airships. Forms of gas bags. (Scientific American supplement, New York, 1907, v. 63, p. 26007.) **VA**

Author's experiments in balloon shapes.

1206. **N.,** K. Die Katastrophe des Ballons Bradsky. illus. ports. (Illustrierte aëronautische Mittheilungen, Strassburg, 1903, Jahrg. 7, p. 1–7.) **VDS**

1207. —— Neues zur Katastrophe Bradsky. illus. (Illustrierte aëronautische Mittheilungen, 1903, Strassburg, Jahrg. 7, p. 141–162.) **VDS**

Illustrated description of the Bradsky airship.

1208. The **Navigable** balloon at the Antwerp exhibition. illus. (Engineer, London, 1894, v. 78, p. 198–199.) **VDS**

Description and illustrations of airship operated by electricity from the ground.

1209. **Navigable** trains of airships. illus. (Scientific American, New York, 1885, v. 52, p. 99.) **VA**

Sectional airship proposed by Capt. Carl W. Petersen.

1210. A **New** airship. illus. (Scientific American, New York, 1892, v. 67, p. 262.) **VA**

Invention of S. B. Battey of New York City. Uses explosive pellets for driving power.

1211. The **New** air ship of M. de Santos-Dumont. illus. (Scientific American, New York, 1900, v. 83, p. 7.) **VA**

1212. **New** airship under construction for the British War Office. illus. (Scientific American, New York, 1902, v. 86, p. 308.) **VA**

Description of the Barton airship.

1213. A **New** dirigible balloon. illus. (Scientific American, New York, 1896, v. 75, p. 271.)

Illustrated description of Wölfert's airship. **VA**

1213a. The **New** flying ship. illus. (Gleason's pictorial drawing-room companion, Boston, 1851, v. 1, p. 256.) *** DA**

Robjohn's airship, Hoboken, N. J.

1214. **Nippoldt.** Ueber Hänlein's lenkbares Luftschiff. (Dingler's polytechnisches Journal, Augsburg, 1875, Bd. 217, p. 507–511.) **3–VA**

1215. **Norway,** N. S. R 100 Canadian flight, 1930. (Royal Aeronautical Society, Journal, London, 1931, v. 35, p. 401–414.) **VDS**

England to Montreal and return, 1930.

1216. **Novel** experiment in aerostation. (Mechanic's magazine, London, 1840, v. 32, p. 480.) **3–VA**

Model demonstration of airship by Charles Green at the Polytechnic Institution, London.

1217. A **Novel** form of aerial vessel. illus. (Scientific American, New York, 1887, v. 56, p. 1.) **VA**

Invention of Moses S. Cole.

Lighter-Than-Air Craft — Airships, continued

1218. **Oakes,** E. Mechanical flight. illus. (Scientific American supplement, New York, 1892, v. 34, p. 13818.) **VA**

Author's airship, a combination of balloon and airplane, is driven by fan wheels.

1219. **Omri.** On aerostation. (Mechanic's magazine, London, 1836, v. 25, p. 307–308, 394–395.) **3–VA**

Discusses proper shape of balloon for purposes of direction.

1220. **On** aerial navigation. (Philosophical magazine, London, 1823, v. 61, p. 6–8.) **OA**

Discusses difficulties of steering a balloon.

1221. **On** dirigible balloons. (Engineering, London, 1909, v. 87, p. 312–313.) **VDA**

Comments on the danger of airships, citing notable accidents. Characteristics of the rigid and non-rigid types; table gives chief particulars of eleven important craft.

1222. **Opinions** of thirty-two scientific gentlemen, on the invention of Muzio Muzzi, in English and Italian. New York: Herald Book & Job Printing Off., 1845. 14 p. 8°. **VDC p.v.10, no.6**

Method of steering a balloon of lenticular form.

1223. **Orcy,** Ladislas d'. Development of the American airship industry. (Aviation, New York, 1919, v. 6, p. 226–227.) **VDS**

Statistics for the period of the World war.

1224. —— D'Orcy's airship manual, an international register of airships with a compendium of the airship's elementary mechanics. New York: The Century Co. [cop. 1917.] 232 p. illus. obl. 16°. **VDY**

1225. **Our** first rigid airship the "Shenandoah." The design, construction and successful trials of the great airship built by the United States Navy. illus. (Scientific American, New York, 1924, v. 130, p. 82–83.) **VA**

1226. "An **Outsider,**" pseud. The motor car in excelsis. illus. (Engineer, London, 1897, v. 83, p. 174–175.) **VA**

Illustrated account of Pennington's airship, a combination of helicopter and balloon.

1227. **P.,** F. X. [Concerning a method of directing the air balloon.] (New London magazine, London, 1785, v. 1, p. 93–95.) *** DA**

1228. **Parseval,** August von. Graf Zeppelin und die deutsche Luftfahrt... Berlin-Grunewald: H. Klemm A. G. [1925.] vi, 147 p. illus. ports. sq. 4°. **VDY**

Well-illustrated account of development of Zeppelin and Parseval ships, and of the dirigible in the World war.

1229. —— Motorballon und Flugmaschine. Wiesbaden: J. F. Bergmann, 1908. 39 p. illus. 8°. **VDY p.v.5, no.8**

Descriptions of contemporary airships and airplanes.

1230. —— Die Segelrad-Flugmaschine des Professors Wellner. (Zeitschrift für Luftschiffahrt, Berlin, 1894, Jahrg. 13, p. 240–245.) **3–VDS**

For Wellner's reply see above magazine, 1894, Jahrg. 13, p. 298–299.

1231. **Partridge,** J. S. Description of the pneudromon, or aero-locomotive vehicle designed by Mr. J. S. Partridge — communicated by the inventor. illus. (Mechanics' magazine, London, 1843, v. 38, p. 395–406.) **3–VA**

Steam-driven airship with lifting propellers.

1232–1233. The **Passing** of a pioneer. Death of Santos-Dumont. illus. (Flight, London, 1932, v. 24, p. 703.) **VDS**

Brief account of his work.

1234. **Patterson,** T. T. Million goldbeaters skins in Navy's new dirigibles. (U. S. air service, Washington, 1922, v. 7, Aug., p. 13–14.) **VDS**

1235. **Pearson,** William. Aerostation. (Mechanic's magazine, London, 1835, v. 23, p. 389–390.) **3–VA**

Comments on the *Eagle* airship, designed to travel between London and Paris.

1236. The **Pennington** airship. illus. (Scientific American, New York, 1891, v. 64, p. 150.) **VA**

Invention of E. J. Pennington of Mt. Carmel, Ill.

1237. **Perkins,** Frank C. Modern air ships at the St. Louis world's fair. illus. (American inventor, New York, 1904, v. 12, p. 465–466.) *** V**

Mainly an account of flights of Captain T. S. Baldwin's *The California arrow* at the Louisiana Purchase Exposition.

1238. The **Peterson** airship. illus. (American inventor, New York, 1905, v. 13, p. 2.) *** V**

Elongated balloon equipped with sails — the invention of P. H. Peterson, of San Francisco.

1239. **Picca,** Paolo. I prodromi dell' aeronautica. Dal P. F. Lana e B. L. Gusmão a G. G. Rousseau. (Nuova antologia, Roma, 1910, serie 5, v. 150, p. 309–321.) **NNA**

1240. **Pierre,** J. La navigation aérienne en 1870. (Revue du Berry et du Centre, Châteauroux, 1910, année 39, p. 287–295.) **DQL**

Airship of Charles Boutet.

1241. **Pitter,** Joseph. Archimedean balloon. illus. (Mechanic's magazine, London, 1847, v. 46, p. 531–536; 1848, v. 48, p. 553–556.) **3–VA**

Author's elongated airship driven by four large paddle-wheels.

1242. —— Is aerial navigation practicable? (Mechanic's magazine, London, 1856, v. 65, p. 541–542, 589–590; 1857, v. 66, p. 251–252, 469–470, 542, 589–590; 1857, v. 67, p. 15–16.) **3–VA**

Discusses the possibility of aerial navigation and again places faith in his Archimedean airship of 1847–48.

1243. **Platte,** August. Die Aëronautik im Jahre 1897. (Oesterreichischer Ingenieur- und Architekten-Verein, Zeitschrift, Wien, 1898, Jahrg. 50, p. 82–83.) **VDA**

Account of aluminium airship of C. Schwarz; also airship of Koch.

Lighter-Than-Air Craft — Airships, continued

1244. **Plumb,** C. M. The aereon, or flying-ship invented by Solomon Andrews. [By C. M. Plumb.] New York: Aerial Navigation Company, 1866. 15 p. 8°. *** C p.v.1714, no.3**

1245. **Pochhammer,** Bruno, editor. Logbuch ZR III, mit authentischen Beiträgen von folgenden Mitgliedern der Besatzung während der Atlantikfahrt: A. Wittemann... G. Belser... W. Speck... W. Siegle... L. Marz... L. Knorr... M. Pruss... Freiburg i. Br.: T. Fisher, 1924. 94 p. illus. maps. 12°. **VDY p.v.34, no.2**

History of German flight in 1924, Germany to America.

1246. **Pole,** William. Aërial navigation. (Nature, London, 1886, v. 33, p. 444–445.) **OA**

Believes an efficient airship possible.

1247. —— Further data on aerial navigation. (Institution of Civil Engineers, Proceedings, London, 1884/85, v. 81, p. 233–240.) **VDA**

Use of electric power for airships as exemplified by experiences of the Tissandier brothers and Renard and Krebs, with comparisons by the author.

1248. —— A study of the problem of aerial navigation, as affected by recent mechanical improvements. illus. (Institution of Civil Engineers, Proceedings, London, 1881/82, v. 67, p. 369–394.) **VDA**

Discusses airships of Giffard and Dupuy de Lôme. Also in *Van Nostrand's engineering magazine,* New York, 1882, v. 27, p. 1–15, *3–VDA.*

1249. **Primbault,** H. La navigation aérienne. (Revue générale, Bruxelles, 1908, v. 88, p. 802–824.) *** DM**

Passing of the balloon and the advent of the airship as exemplified by the *Ville de Paris, Patrie,* and the Zeppelins.

1250. **Proctor,** Richard Anthony. French balloon experiments. (Knowledge, London, 1884, v. 6, p. 380.) **OA**

"Until it is shown to be possible to urge a monstrous balloon through the air at a rate considerably exceeding that of a stiff breeze, nothing is to be hoped for from balloon-supported propellers."

1251. **Projet** de ballon dirigeable de M. Torres. illus. (Le Génie civil, Paris, 1902, tome 41, p. 224–225.) **VA**

1252. **Quelques** vues sur les machines aérostatiques. illus. (Magazin für das Neueste aus der Physik und Naturgeschichte, Gotha, 1785, Bd. 3, Stück 1, p. 73–97.) **3–OA**

Illustrations at back of Stück 1. The article describes in detail a proposed airship (Dalberg's perhaps).

1253. The **R 33** emulates the Shenandoah. The reconstructed British airship R–33 is blown from her mast and returns safely to her hangar. illus. (Aviation, Highland, N. Y., 1925, v. 18, p. 464–465.) **VDS**

1254. The **"R 38"** disaster. (Flight, London, 1921, v. 13, p. 589–592.) **VDS**

Account of accident at Hull, England, August 24, 1921.

1255. **R 100** flies to Montreal. charts. (Flight, London, 1930, v. 22, p. 891–894.) **VDS**

1256. **R 100** starts for Montreal. illus. (Flight, London, 1930, v. 22, p. 872–873.) **VDS**

Cardington, England, to Montreal, 1930. Brief biographies of the officers.

1257. **R 101.** Funeral of victims. illus. (Flight, London, 1930, v. 22, p. 1137–1141.) **VDS**

Impressive pictures of the state funeral.

1258. **R 101** Simon inquiry report. Disaster due to loss of gas. (Flight, London, 1931, v. 23, p. 290–291.) **VDS**

See also *Aeroplane,* London, 1931, v. 40, p. 589–596, *VDS.*

1259. **Rameau,** Louis. Dupuy de Lôme. port. (L'Aéronaute, Paris, 1885, année 18, p. 43–45.) **3–VDS**

Account of Dupuy de Lôme's aeronautical work.

1260. **Ranwell,** William. Suggestions on aerial locomotion. (Mechanic's magazine, London, 1836, v. 26, p. 168.) **3–VA**

Suggests the use of eagles for drawing the balloon.

1261. **Rathbun,** John B. Disaster visits the Akron. illus. (Popular aviation, Chicago, 1933, v. 12, p. 353–354, 396–397.) **VDS**

A discussion of major dirigible accidents, with table, 1912–1933.

1262. The **Record** flight of "Zeppelin II." illus. (Scientific American supplement, New York, 1909, v. 68, p. 13–14.) **VA**

1263. **Reis,** P. Neue Ideen über Luftschiffahrt. (Dingler's polytechnisches Journal, Stuttgart, 1859, Bd. 151, p. 13–16.) **3–VA**

Author's suggestion for directing airship, using batteries of air jets.

1264. **Renard,** Charles. L'aérostat dirigeable de Chalais-Meudon. Expériences de 1885. illus. map. (La Nature, Paris, 1886, année 14, semestre 1, p. 86–92.) **OA**

Author's report on the ascensions of the airship *La France.* Includes introduction and conclusions by Gaston Tissandier. Translation in *Scientific American supplement,* New York, 1886, v. 21, p. 8492–8493, *VA.* Abstract in *Nature,* London, 1886, v. 33, p. 421–423, *OA.*

1265. —— Ballons dirigeables. Stabilité longitudinale. illus. (Académie des sciences, Comptes rendus, Paris, 1904, tome 139, p. 183–185.) *** EO**

1266. —— La navigation aérienne. illus. (Revue de l'aéronautique, Paris, 1888, année 1, p. 13–30, 42–50.) **3–† VDS**

Discussion of airship difficulties precedes a brief historical outline.

1267. —— Les piles légères (piles chlorochromiques) du ballon dirigeable "La France." Paris: G. Masson, 1890. 35 p. illus. diagrs. 8°. **3–† VDS**

Use of chromium batteries on the airship *La France.* Bound with *Revue de l'aéronautique,* Paris, 1890, année 3.

Lighter-Than-Air Craft — Airships, continued

1268. —— Sur l'empennage des carènes des ballons dirigeables. illus. (Académie des sciences, Comptes rendus, Paris, 1904, tome 138, p. 1576–1578.) * **EO**

A discussion of fins for airships.

1269. —— Sur les nouvelles expériences exécutées en 1885 au moyen du ballon dirigeable "La France." map, charts. (Académie des sciences, Comptes rendus, Paris, 1885, tome 101, p. 1111–1118.) * **EO**

See also *L'Aéronaute,* Paris, 1886, année 19, p. 5–12, *3–VDS.*

1270. —— Sur la vitesse critique des ballons dirigeables. illus. (Académie des sciences, Comptes rendus, Paris, 1904, tome 138, p. 1405–1408.) * **EO**

1271. **Renard,** Charles, and A. Krebs. Sur un aérostat dirigeable. (Académie des sciences, Comptes rendus, Paris, 1884, tome 99, p. 316–319.) * **EO**

1272. **Renard,** Paul. Les aérostats dirigeables. illus. (Revue générale des sciences, Paris, 1908, année 19, p. 426–442, 479–494.) **OA**

Discussion of airship principles and accounts of different projects from the time of Meusnier.

1273. —— La sécurité dans les ballons dirigeables. illus. (Société d'encouragement pour l'industrie nationale, Bulletin, Paris, 1903, tome 104, semestre 1, p. 196–223.) **3–VA**

Referring to several airship disasters the author makes suggestions for future prevention.

1274. —— Sur la mesure indirecte de la vitesse propre des navires aériens. diagrs. (Académie des sciences, Comptes rendus, Paris, 1904, v. 139, p. 353–356.) * **EO**

1275. **Renard** and Krebs' electric balloon. illus. (Scientific American supplement, New York, 1884, v. 18, p. 7294–7295.) **VA**

1276. **Report** on accident to the airship Roma. Official War Department report fails to determine with absolute accuracy the causes of the accident. illus. (Aviation, Highland, N. Y., 1922, v. 13, p. 148–152.) **VDS**

Disaster to United States airship, with 34 lives lost, Feb. 21, 1922.

1277. The **Return** voyage of R 34. illus. ports., map. (Flight, London, 1919, v. 11, p. 944–951.) **VDS**

1278. **Riat,** Georges. Les ballons dirigeables et la caricature en 1784. (La Contemporaine, Paris, 1901, v. 5, p. 99–107.) * **DM**

1279. **Richmond,** V. C. A review of the present position with regard to airship research and experiment. illus. diagrs. (Royal Aeronautical Society, Journal, London, 1926, v. 30, p. 547–586.) **VDS**

With discussion.

1280. **Rivera y Sempere,** Manuel. Navegación aerea. Memoria descriptiva de un sistema. Recopilación corregida y aumentada de varios escritos publicados. Madrid: Imp. de E. Arias, 1907. 123 p. illus. 8°. **VDW**

Airship designed by the author.

1281. **Robert** Lebaudy. port. (L'Aérophile, Paris, 1901, année 9, p. 127.) **3–VDS**

Portrait of Lebaudy.

1282. **Roeser.** Aus der Geschichte des deutschen Starrluftschiffes. illus. (Zeitschrift für Flugtechnik und Motorluftschiffahrt, München, 1925, Jahrg. 16, p. 251–272.) **VDS**

Extensive history of the German rigid airships, with table of dimensions, characteristics, etc. Includes a bibliography.

1283. **Rogers,** Bogart. The dirigible is *not* practical. illus. (Popular aviation, Chicago, 1931, v. 9, Oct., p. 20, 52–53.) **VDS**

"The record of past dirigibles is not an enviable one. I can see very little in the future for the big gas bags."

1284. **Roma** destroyed with great loss of life. illus. (Aviation, Highland, N. Y., 1922, v. 12, p. 251.) **VDS**

United States Army airship destroyed Feb. 21, 1922.

1285. **Rosendahl,** Charles Emery. Up ship! New York: Dodd, Mead & Co., 1931. xiv, 311 p. illus. 8°. **VDW**

Deals mainly with the evolution of the American airship, although there are two chapters on the flights of the *Graf Zeppelin.* Account of disaster to the *Shenandoah,* of which the author was commander.

1286–1287. **Rothschild,** Zeta. Count Zeppelin comes across. How the genius of the dirigible joined the Yankee ranks and took ride in a balloon. illus. ports. (Sportsman pilot, Albany, N. Y., 1931, v. 6, Sept., p. 22–23, 52, 54.) **VDS**

1288. **Roze,** Louis. L'aviateur Roze. illus. port. (L'Aérophile, Paris, 1901, année 9, p. 146–151; 1902, année 10, p. 65–71.) **3–VDS**

French airship builder gives a brief autobiographical account of his work.

1289. The **Roze** dirigible airship. illus. (Scientific American, New York, 1900, v. 83, p. 392.) **VA**

1290. **Rummel,** Walter, Freiherr von. Graf Zeppelin. Bielefeld: Velhagen & Klasing [1914]. 34 p. port. 4°. (Velhagen & Klasings Volksbücher. Nr. 119. Volksbücher der Geschichte.) **AN p.v.177, no.13**

1291. **Rumpler,** Edmund. Das Parsevalsche Motorluftschiff Modell 1908. illus. (Verein deutscher Ingenieure, Zeitschrift, Berlin, 1909, Bd. 53, p. 896–901.) **3–VDA**

1292. **Saager,** Adolf. Zeppelin, der Mensch — der Kämpfer — der Sieger; bunte Bilder von gestern und heute... Stuttgart: R. Lutz [1915]. 262 p. port. 7. ed. 12°. (Anekdoten-Bibliothek. Bd. 19.) **AN**

Lighter-Than-Air Craft — Airships, continued

1293. **Sabatier,** J. État actuel de la question de l'hélium et de son emploi dans l'aérostation. (Aéronautique, Paris, 1919, année 1, p. 67–70.) **VDS**

1294. **Sailed** five miles in airship. (American inventor, New York, 1904, v. 12, p. 244.) *** V**

Flight of Greth's airship, 1904.

1295. **Sampaio,** Carlos. The "Peace" balloon of the late Senhor Augusto Severo. port. (Aeronautical journal, London, 1902, v. 6, p. 56–59.) **VDS**

Abridged in *Aeronautical world,* Glenville, O., 1902, v. 1, p. 110–111, *VDS.*

1296. **Sandford,** W. T. Some recent developments in the design of rigid airships. (Royal Aeronautical Society, Journal, London, 1927, v. 31, p. 1029–1036.) **VDS**

1297. **Santos-Dumont,** Alberto. My airships; the story of my life. London: G. Richards, 1904. viii, 318 p. port. illus. 8°. **VDYC**

For lists of writings concerning Santos-Dumont see Paul Brockett, *Bibliography of aeronautics,* Washington, 1910, p. 746–749, *VDT;* also United States. — National Advisory Committee on Aeronautics, *Bibliography of aeronautics,* 1909–1931, Washington, 1921–35, vols. for 1909/16, p. 1184–1185; 1917/19, p. 396; 1922, p. 139; 1926, p. 127; 1927, p. 153, *VDT.*

1298. **Santos-Dumont** dead; pioneer aviator was 59. Developer of motored lighter-than-air craft passes in remote Brazil state. port. (New York Herald Tribune, New York, July 25, 1932, p. 13.) *** A**

1299. The **Santos-Dumont** navigable balloon. illus. (Engineer, London, 1901, v. 92, p. 89–91.) **VA**

History of various experiments.

1300. **Santos-Dumont** wins the Deutsch prize. illus. (Scientific American, New York, 1901, v. 85, p. 314.) **VA**

1301. **Santos-Dumont's** combined aeroplane and airship. illus. (Scientific American, New York, 1907, v. 97, p. 12–13.) **VA**

1302. **Santos-Dumont's** "Demoiselle." illus. (Airways, London, 1929, v. 6, p. 100.) **VDS**

Description and illustration of early plane in which, by a flight of five miles, Santos-Dumont won a wager of 1000 francs.

1303. **Santos-Dumont's** Versuche und Erfolge mit einem Luftschiff. illus. (Prometheus, Berlin, 1901, Jahrg. 13, p. 262–267, 279–284.) **3–VA**

Histories of Santos-Dumont's airships.

1304. **Sazerac de Forge,** Léonide. Le ballon dirigeable. (Correspondant, Paris, 1907, année 79, v. 226 [nouv. série, v. 190], p. 975–1002.) *** DM**

History of the airship.

1305. —— La conquête de l'air. illus. (Revue d'artillerie, Paris, 1907, v. 70, p. 178–199, 256–278, 281–317.) **VWA**

Construction and flights of the Lebaudy and Patrie airships.

1306. —— La conquête de l'air. Le ballon dirigeable. Paris: Berger-Levrault & Cie., 1910. xii, 811 p. 2. ed. rev. 8°. **VDW**

Comprehensive account of the development of the airship in different countries. Library has also a copy of the first edition, Paris: Berger-Levrault & Cie., 1907, 378 p. illus., 8°, which contains a chapter on heavier-than-air craft, *VDW.*

1307. **Schiavone,** Mario. Il principo della dirigibilità orizzintale degli aerostati ed il binaerostato. Potenza: Garramone e Marchesiello, 1898. 48 p. 8°. **VDC p.v.28, no.9**

A study of the principles governing the design of airships.

1308. **Schildhauer,** Edward. Development of the lighter-than-air airship. illus. (Boston Society of Civil Engineers, Journal, Boston, 1922, v. 9, p. 269–283.) **VDA**

Deals principally with developments in Germany.

1309. **Schiller,** Hans von. Log of the ZR–3. Famous Zeppelin navigator tells brilliant story of the great land and sea flight. illus. port. (Aero digest, New York, 1924, v. 5, p. 266–268, 300.) **VDS**

Trans-Atlantic flight of craft subsequently named *Los Angeles.*

1310. **Schio,** Almerico, conte da. Di un progresso nell' aeronautica. illus. (Nuova antologia, Roma, 1911, serie 5, v. 155, p. 257–275.) **NNA**

History of the airship *Italia,* designed and built by the author, and subsequently reconstructed several times.

1311. —— I primi sperimenti dell' aeronave "Italia" a Schio, 17 giugno-4 luglio 1905. illus. (Reale istituto veneto, Atti, Venezia, 1905, v. 64, parte 1, p. 149–154.) *** ER**

1312. **Schneider,** von. The de Bradsky airship. A short account of its development, and of the causes that led to the accident on October 13th, 1902. illus. (Aeronautical journal, London, 1903, v. 7, p. 66–72.) **VDS**

1313. The **Science** of aerostation. illus. (Scientific American, New York, 1863, new series, v. 8, p. 193.) **VA**

Invention of Thomas L. Shaw has a hand-operated propeller and cylinders for condensed gas.

1314. **Scott,** G. H. The development of airship mooring. (Royal Aeronautical Society, Journal, London, 1926, v. 30, p. 459–481.) **VDS**

With appendices and discussion. Appendix 1 covers history.

1315. **Scrutator Mechanicus,** pseud. The London and Paris aeronautical scheme. (Mechanic's magazine, London, 1835, v. 23, p. 372–374.) **3–VA**

Contains a French notice of the project of the European Aeronautical Society.

1316. The **Severo** airship. illus. (Scientific American, New York, 1902, v. 86, p. 224.) **VA**

1317. The **Severo** airship catastrophe. illus. (Scientific American, New York, 1902, v. 86, p. 395–396.) **VA**

Lighter-Than-Air Craft — Airships, continued

1318. **The Shenandoah** court findings. (Aviation, Highland, N. Y., 1926, v. 20, p. 44–46.) **VDS**

Full text of facts and opinions concerning disaster of Sept. 3, 1925.

1319. The **Shenandoah** investigation. (Aviation, Highland, N. Y., 1925, v. 19, p. 437.) **VDS**

Commander Lansdowne's protest against the flight (1925) and some official correspondence relating thereto.

1320. The **Shenandoah** verdict. (Literary digest, New York, 1926, v. 88, Jan. 16, p. 11–12.) *** DA**

1321. **Silberer,** Victor. Der Tod Severo's. illus. (Wiener Luftschiffer-Zeitung, Wien, 1902, Jahrg. 1, p. 79–83.) **VDS**

Account of disaster to the *Pax* airship and of its constructor, Augusto Severo.

1322. **Smith,** James Walter. A metal balloon. illus. (Strand magazine, London, 1898, v. 15, p. 285–288.) *** DA**

Tragic end of Schwarz's aluminium airship that rose to 820 feet in Germany (1897).

1323. **Soreau,** Rodolphe. Le problème de la direction des ballons. illus. (Société des ingénieurs civils de France, Mémoires, Paris, 1893, année 1893, v. 1, p. 223–301.) **VDA**

Early experiments. The airship *La France*. A study of various resistances. Motors. Library has a reprint in *VDY p.v.9, no.8.*

1324. **Spanner,** Edward Frank. About airships. London: E. F. Spanner, 1929. 206 p. illus. 8°. **VDW**

Vigorous comments on the uselessness of the dirigible, with reference to *R 34* and the *Shenandoah.*

1325. —— Gentlemen prefer aeroplanes. Complementary to This airship business. London: E. F. Spanner, 1928. xxi, 457 p. illus. 8°. **VDW**

Contains a bibliography of airships, especially of some British patents.

1326. —— This airship business. London: Williams and Norgate, Ltd., 1927. xv, 435 p. illus. 8°. **VDW**

Comments on Great Britain's airship policy, with a bibliography, including patents. Critical review in *Flight,* London, 1928, v. 20, p. 51–53, *VDS.* Briefly reviewed in *Aeroplane,* London, 1928, v. 34, p. 368, *VDS.*

1327. —— The tragedy of the "R 101." London: E. F. Spanner [1931]. 2 v. 8°. **VDW**

Discussion of disaster to British airship in France, 1930.

1328. **Sprigg,** Christopher St. John. The airship; its design, history, operation and future. London: S. Low, Marston & Co., Ltd. [1931.] viii, 248 p. illus. 8°. **VDW**

See chapters 6 to 10 inclusive for history. Reviewed in *Aeroplane,* London, 1931, v. 40, p. 558, *VDS.*

1329. **Sprigg,** Theodore Stanhope. British airships of the past. Pioneer naval and military airships that led the way to the modern leviathans of Germany and the U. S. A. illus. (Air and airways, London, 1932, v. 9, p. 241–243.) **VDS**

1330. Das **Starre** System. David Schwarz, Carl Berg, Graf Zeppelin. Ein historischer Rückblick. port. (Deutsche Zeitschrift für Luftschiffahrt, Berlin, 1909, Jahrg. 13, p. 742–744.) **VDS**

Brief accounts of Schwarz and Berg, pioneers in the design of rigid airships.

1331. **Steam** flying ship. illus. (Scientific American, New York, 1872, new series, v. 26, p. 319–320.) **VA**

Proposes to use steam as lifting power.

1332. **Steele,** G. W. ZR–3 makes historic journey. Narrative of record trans-Atlantic flight, with glamorous details, from log of American passenger. illus. ports. (U. S. air services, Washington, 1924, v. 9, Nov., p. 13–17.) **VDS**

Voyage of Zeppelin airship which subsequently was renamed *Los Angeles.*

1333. **Stevens'** airship. illus. (Scientific American, New York, 1902, v. 87, p. 223–224.) **VA**

Invention of Leo Stevens.

1334. **Stewart,** Andrew. Production of helium for use in airships. (United States. — Mines Bureau, Bulletin 178c, Washington, 1919, p. 75–87.) **VHCA (U. S.)**

History of research and production in the United States. Also in *Aviation,* New York, 1919, v. 7, p. 32–35, *VDS.*

1335. **Strait,** H. On aerial navigation. (American journal of science, New Haven, 1834, v. 25, p. 15–26.) **OA**

Author describes his balloon equipped with wings.

1336. **Sturm,** Johann Christoph. Collegium experimentale, sive curiosum, in quo primaria seculi superioris inventa & experimenta physico-mathematica... Norimbergæ: Sumtibus W. M. Endteri, 1685–1701. 2 parts in 1 v. illus. 4°. **PAD**

See part 1, p. 56–66 for a discussion of Lana's airship and a full-page picture.

1337. **Success** — and disaster. The story of commercial airship development. (Aviation year book, London, 1931, p. 55–64.) **VDS**

Brief account of developments in Germany, Great Britain, Spain, and the United States. Has a list of airship disasters, 1909–1930.

1338. **Surcouf,** Édouard. L'aéronautique maritime. illus. (Société des ingénieurs civils de France, Mémoires, Paris, 1902, année 1902, v. 1, p. 87–155.) **VDA**

A study of the control of balloons and airships at sea. Chronologies on p. 91–93, 121–125.

1339. **Talbot,** Frederick Arthur Ambrose. The Spencer airship for 1903. illus. (Scientific American, New York, 1903, v. 89, p. 169–170.) **VA**

1340. The **Tale** of Zeppelin disasters. (The New York times current history, New York, 1918, v. 8, p. 309–310.) **BAA (N. Y. Times)**

Brief records of most of the Zeppelin airships from the *L–1* to *L–57.*

Lighter-Than-Air Craft — Airships, continued

1341. Technical aspects of the loss of the U. S. S. *Shenandoah.* illus. (American Society of Naval Engineers, Journal, Washington, 1926, v. 38, p. 487–694.) **VXA**

Lengthy abstract of the proceedings of the court of inquiry.

1342. Technical development, balloons and airships, 1914–1919. (Aircraft yearbook, New York, 1920, p. 123–138.) **VDS**

1343. Teffé, baron de. [Les aérostats dirigeables.] (L'Aéronaute, Paris, 1890, année 23, p. 5–11.) **3–VDS**

Discusses the airship of Ribeiro de Souza of Brazil, 1880.

1344. Thayer, Russell. The dirigible balloon with gyroscope control. illus. (Franklin Institute, Journal, Philadelphia, 1909, v. 168, p. 19–39.) **VA**

Experiments of the author.

1345. —— The navigation of the air. illus. (Engineers' Club of Philadelphia, Proceedings, Philadelphia, 1885, v. 4, p. 365–375.) **3–† VDA**

Author's idea for an airship using jet propulsion.

1346. Tissandier, Gaston. L'aérostat dirigeable électrique de MM. Albert et Gaston Tissandier. illus. (La Nature, Paris, 1883, année 11, semestre 2, p. 326–330, 343–346.) **OA**

Translation in *Science,* Cambridge, Mass., 1884, v. 3, p. 152–155, 196–199, *OA;* and *Scientific American supplement,* New York, 1883, v. 16, p. 6631–6634, *VA.*

1347. —— L'aérostat dirigeable électrique de MM. Ch. Renard et Krebs. illus. (La Nature, Paris, 1884, année 12, semestre 2, p. 193–195, 241–245, 374–375.) **OA**

Account of flights.

1348. —— L'aérostat électrique à hélice de MM. Tissandier frères. illus. (L'Aéronaute, Paris, 1883, année 16, p. 223–232; 1884, année 17, p. 23–31.) **3–VDS**

1349. —— Applications de l'électricité à la navigation aérienne. L'aérostat électrique à hélice de MM. Albert et Gaston Tissandier. illus. (Société d'encouragement pour l'industrie nationale, Bulletin, Paris, 1884, série 3, tome 11, p. 140–148.) **3–VA**

1350. —— Les ballons dirigeables. Application de l'électricité à la navigation aérienne ... Paris: Gauthier-Villars, 1885. xii, 108 p. illus. 12°. **VDY**

Illustrated account of developments by the author.

1351. —— Le problème de la direction des aérostats; application de l'électricité à la navigation aérienne. Conférence faite à la Sorbonne le samedi 3 mars, 1883, et au groupe de l'Association des anciens élèves de l'École centrale de Paris, le 8 mars. Paris: Génie civile, 1883. 32 p. illus. 8°. **VDY p.v.8, no.8**

Good early history of airships. Describes Tissandier brothers' airship and its electrical equipment.
Also in *Le Génie civil,* Paris, 1883, tome 3, p. 225–229, 250–253, *VA.*

1352. —— Le projet d'aérostat dirigeable du général Meusnier. illus. (Revue de l'aéronautique, Paris, 1888, année 1, p. 130–135.) **3–† VDS**

1353. —— Sur la deuxième expérience de l'aérostat électrique à hélice de MM. Tissandier frères. (Académie des sciences, Comptes rendus, Paris, 1884, tome 99, p. 530–532.) *** EO**

Also in *L'Aéronaute,* Paris, 1884, année 17, p. 203–207, *3–VDS.*

1354. To fly or not to fly. illus. (American inventor, Washington, 1902, v. 9, July 1, p. 116.) *** V**

Facetious remarks on Alanson Wood's substitute for the "Wings of a dove."

1355. Torres, L. Un avant-projet de ballon dirigeable. illus. (L'Aérophile, Paris, 1902, année 10, p. 212–215.) **3–VDS**

Proposes an interior keel for airship.

1356. —— Sur la stabilité longitudinale des ballons dirigeables. illus. (Académie des sciences, Comptes rendus, Paris, 1905, tome 140, p. 1019–1021.) *** EO**

1357. Total wreck of the Severo airship. illus. (Scientific American, New York, 1902, v. 86, p. 362–363.) **VA**

The airship *Pax* was wrecked May 12, 1902.

1358. The **Transatlantic** dirigible. illus. (Scientific American, New York, 1919, v. 121, p. 58–59, 66, 68, 70, 72.) **VA**

Account of the *R 34* and its flight.

1359. The **Transatlantic** voyage of R 34. port. map. (Flight, London, 1919, v. 11, p. 906–912.) **VDS**

Log of the flight.

1360. Triaca, Albert C. The dirigibles. illus. (Fly, Philadelphia, 1909, v. 1, March, p. 8–9; April, p. 16–17; May, p. 11.) **3–† VDS**

Brief illustrated history of airships.

1361. Tronnier, Adolph. Karl Theodor von Dalberg, der Erfinder des starren Luftschiffes. illus. port. (Deutsche Luftfahrer-Zeitschrift, 1913, Jahrg. 17, p. 439–441.) **VDS**

Dalberg (1744–1793) proposed an airship in 1784.

1362. Tschudi, Georg Julius Friedrich von. Aus 34 Jahren Luftfahrt; persönliche Erinnerungen... Berlin: R. Hobbing [cop. 1928]. 186 p. illus. ports. 8°. **VDY**

Tschudi (born 1862) was prominent in the development of German airships.

1363. Turner, E. E. The Atlantic cruise of H.M. airship R–34. illus. (Society of Engineers, Transactions, London, 1922, p. 233–265.) **VDA**

1364. A **"Twister"** wrecks the Shenandoah. Broken in three sections but 28 of crew land safely. illus. (Aviation, Highland, N. Y., 1925, v. 19, p. 310–314.) **VDS**

Disaster of Sept. 3, 1925, over the Ohio valley. Biographies of some of the lost officers.

Lighter-Than-Air Craft — Airships, continued

1365. **Two** airplanes successfully launched from a dirigible. illus. (Scientific American, New York, 1927, v. 136, p. 112.) **VA**

Launched from the British airship *R 33.*

1366. **Tziolkovski,** K. Aerostat metallickeski upravlyayemyĭ. Izdaniye Chertkova. Moscow: tip. M. G. Volchaninova, 1892. 83 p. 8°. **VDY p.v.1, no.3**

Details of author's design for a metal airship.

1367. The **United States** and British government airships. illus. (Scientific American, New York, 1908, v. 99, p. 125.) **VA**

Illustrations and brief descriptions of the ship built by Captain Thomas A. Baldwin, and of the *Nulli Secundus,* by Colonel Cody.

1368. **Upson,** Ralph Hazlett. Airships and their commercial possibilities. illus. (Society of Automotive Engineers, Journal, New York, 1919, v. 5, p. 349–352, 375.) **TOL**

1369. —— Metalclad rigid airship development. illus. (Society of Automotive Engineers, Journal, New York, 1926, v. 18, p. 117–131.) **TOL**

History of early ideas; details of the United States *MC-2;* and a bibliography. Abstract in *Scientific American,* New York, 1926, v. 134, p. 352–354, *VA.*

1370. **Vélez,** Miguel. Relatorio descriptivo do projecto de viação aérea equilibrada denominada Tram-Vélez. Rio de Janeiro: Imprensa Mont' alverne, 1891. 19 p. illus. 8°. **VDY p.v.6, no.3**

Pictures and illustrations of airship proposed by author.

1371. **Vissering,** Harry. Zeppelin; the story of a great achievement. [By Harry Vissering.] [Chicago: Printed by Wells and Company, cop. 1922.] 59 numb. l. illus. ports., map. 8°. **VDW**

1372. **Voemel,** Alexander. Graf Ferdinand von Zeppelin; ein Mann der Tat... Konstanz (Baden): C. Hirsch [cop. 1929]. 173 p. illus. ports. 5. ed., rev. and enl. 8°. **AN**

1373. —— Graf Zeppelin, der Eroberer der Lufte. Ein Vorbild für das deutsche Heer und Volk. Konstanz: J. Blanke, 1914. 47 p. illus. ports. 8°. **AN**

1374. **Vorreiter,** Ansbert. Die deutschen und französischen Motorballons. illus. (Mitteleuropäischer Motorwagen-Verein, Zeitschrift, Berlin, 1908, Jahrg. 7, p. 562–566, 585–588.) **TOL (Automobil)**

Comparison of French airships with those of Parseval and Zeppelin.

1375. —— The German military "Zeppelin I." Detailed description of the remodelled third airship of Count von Zeppelin. illus. (Scientific American supplement, New York, 1909, v. 67, p. 381–383.) **VA**

1376. **Voyer.** Le ballon dirigeable "Patrie." (Revue du génie militaire, Paris, 1908, v. 35, p. 105–146.) **VWA**

1377. —— Les cones sphériques et leur application à la construction des ballons allonges. illus. (Revue de l'aéronautique, Paris, 1894, année 7, p. 1–51.) **3–† VDS**

Geometry of end forms for airships.

1378. —— Les expériences du ballon "Lebaudy." illus. (Revue du génie militaire, Paris, 1907, v. 33, p. 97–124.) **VWA**

1379. —— General Meusnier und die lenkbaren Ballons. illus. (Illustrierte aëronautische Mitteilungen, Strassburg, 1905, Jahrg. 9, p. 373–387.) **VDS**

Illustrated account of Meusnier's design for airship of 1784.

1380. **Wade,** Walter Lewis. Early efforts in airships. illus. (Aeroplane, London, 1916, v. 11, p. 448–450.) **VDS**

1381. —— A short history of the airship. illus. (Aeroplane, London, 1916, v. 11, p. 806, 808, 810, 822, 824, 826, 828, 830, 832, 834, 836, 840, 842.) **VDS**

Continues entry no. 1380.

1382. **Wager-Smith,** Curtis. An artist's impressions of Beachey. illus. (Fly, Philadelphia, 1908, v. 1, Dec., p. 14.) **3–† VDS**

Account of prominent American airship and airplane performer.

1383. —— The world's youngest aeronaut. illus. port. (Fly, Philadelphia, 1909, v. 1, Feb., p. 11.) **3–VDS**

Brief account of 16-year-old airship constructor, Cromwell Dixon.

1384. **Walker,** John Bernard. The tragedy of the "Shenandoah." The wrecking of this superb and well-tried ship by a furious thunderstorm. illus. (Scientific American, New York, 1925, v. 133, p. 301.) **VDS**

1385. **Walton,** Francis D. The L. Z. 127 comes to Lakehurst. illus. port. (Aero digest, New York, 1928, v. 13, p. 894–895, 1022.) **VDS**

Flight of the *Graf Zeppelin,* under command of Hugo Eckener, across the Atlantic ocean, 1928.

1386. **Ward,** William Edward. The Spencer airship and its thirty-mile trip. illus. (Scientific American, New York, 1902, v. 87, p. 259–260.) **VA**

Invention of the Spencer brothers of London, England.

1387. **Warner,** Edward Pearson. What of the airship? illus. (American review of reviews, New York, 1928, v. 78, p. 614–620.) *** DA**

Post-war airship history, especially noting the *Graf Zeppelin.*

1388. **Wead,** Frank Wilbur. Lighter than air. An interesting review of lighter than air achievements, climaxed recently by the successful flight of the R 101. illus. (Popular aeronautics, Chicago, 1930, v. 6, Feb., p. 32–34, 54, 56, 58.) **VDS**

Brief history of airships. This number of *Popular aviation* is published under the title of *Aeronautics.*

Lighter-Than-Air Craft — Airships, continued

1389. Weldon. Aerial navigation. (Engineer, London, 1863, v. 16, p. 156–157.) **VA**

Author deprecates the spherical balloon as a commercial impossibility and argues for the elongated airship.

1390. Wellner, Georg. Die lenkbaren Ballons. illus. (Illustrierte aëronautische Mittheilungen, Strassburg, 1903, Jahrg. 7, p. 228–233.) **VDS**

Has a table giving details of fifteen airships from that of Giffard (1852) to that of Deutsch de la Meurthe (1903).

1391. —— Die Segelradflugmaschine. illus. (Zeitschrift für Luftschiffahrt, Berlin, 1894, Jahrg. 13, p. 86–96.) **3–VDS**

Combination airship and airplane suggested by the author.

1392. Whale, George. British airships, past, present and future. London: J. Lane, 1919. 244 p. illus. 12°. **VDW**

1393. What caused the Zeppelin L 2 disaster? illus. (Aero and hydro, Chicago, 1913, v. 7, p. 69–70.) **3–VDS**

Account of disaster October 17, 1913.

1394. Whitcomb, Russell. An old time flying machine in 1878. (Fly, Philadelphia, 1912, v. 4, Aug., p. 14–15.) **3–† VDS**

Brief account of airship built by Professor Ritchel, of Maine.

1395. Whitman, Roger Bradbury. An airship of '63; a forgotten page of aeronautical history. illus. (U. S. air service, New York, 1920, v. 3, May, p. 29–32.) **VDS**

Author claims that Dr. Solomon Andrews of New Jersey was the first man to use a cylindrical balloon.

1396. Why the Shenandoah failed. illus. (Aviation, Highland, N. Y., 1925, v. 19, p. 346–349.) **VDS**

Ship was designed by J. C. Hunsaker, C. P. Burgess, and S. Truscott, who have prepared a paper showing inherent weakness of airships. The above article quotes extensively from the paper.

1397. A **Wild** night on the Shenandoah. Blown from her mooring-mast, the ship weathers the storm and is brought back to port. illus. (Scientific American, New York, 1924, v. 130, p. 158, 207.) **VA**

1398. Wilhelm, Balthasar. An der Wiege der Luftschiffahrt. Francesco Lana, S.J., der Erfinder des Luftschiffes und Barthol. Lourenço de Gusmão, der erste Luftschiffer. illus. (Frankfurter zeitgamässe Broschüren, Hamm, 1909, Bd. 28, p. 137–198.) **EAA**

Extensive accounts of Lana and Gusmão.

1399. —— Die Anfänge der Luftfahrt Lana-Gusmão. Zur Erinnerung an den 200. Gedenktag des ersten Ballonaufstieges (8. Aug. 1709–8. Aug. 1909). Hamm i. W.: Breer & Thiemann, 1909. viii, 204 p. illus. 12°. **VDT**

1400. —— Neues über Francesco Lana, S. J., den Erfinder des Luftschiffes. port. (Illustrierte aëronautische Mitteilungen, Berlin, 1909, Jahrg. 13, p. 399–402.) **VDS**

A digest of that part of Lana's *Magisterium naturae* which supplements his earlier work the *Prodromo*. See entry no. 1133.

1401. —— Vor 200 Jahren. illus. (Deutsche Zeitschrift für Luftschiffahrt, Berlin, 1909, Jahrg. 13, p. 736–751.) **VDS**

Brief account of Gusmão and references to him in early literature.

1402. Wilkins, Sir George Hubert. Some aspects of the Graf Zeppelin flight around the world. illus. ports. (Aviation, New York, 1929, v. 27, p. 596–600.) **VDS**

1403. Willows, E. T. The airship: past, present and future. illus. (Aeronautics, London, 1913, v. 6, p. 410–412, 450–451; 1914, v. 7, p. 139–141.) **VDS**

1404–1405. The **Willows** dirigible airship. illus. (Scientific American supplement, New York, 1910, v. 70, p. 364–365.) **VA**

1406. Wittemann, A. Mit dem Luftschiff über den Atlantischen Ozean. Die Amerikafahrt des Z. R. III... Die Geschichte des gesamten Zeppelinbaues. Wiesbaden: Amsel-Verlag, 1925. 119 p. port, illus. 12°. **VDY**

Brief history of Zeppelin development and an account of the voyage of the *Z. R. III* in 1924.

1407. Woelfert's new aerial ship. illus. (Scientific American supplement, New York, 1883, v. 16, p. 6377.) **VA**

1408. Wolff, Maximilian. Projekt eines lenkbaren Luftschiffes. illus. (Zeitschrift für Luftschiffahrt, Berlin, 1882, Jahrg. 1, p. 6–9.) **3–VDS**

1409. The **World's** largest hangar. illus. (Literary digest, New York, 1929, v. 101, April 20, p. 25.) *** DA**

Built at Akron, Ohio, for the construction of two airships for the United States Navy.

1410. The **Wreck** of the "Zeppelin IV." Why the great craft was destroyed. illus. (Scientific American supplement, New York, 1908, v. 66, p. 149.) **VA**

1411. Wurzer, Ferdinand. Wer gab die erste Idee zu einer ärostatischen Maschine an? (Archiv für die gesammte Naturlehre, Nürnberg, 1829, Bd. 17, p. 366–372.) **3–OA**

Explains the ideas of Philip Lohmeier, who published a thesis in 1676, following closely the scheme laid down in Lana's *Prodromo* of 1670.

1412. Z., O. Zur Geschichte des Fluges. (Deutsche Luftfahrer-Zeitschrift, Berlin, 1918, Jahrg. 22, no. 9/10, p. 23.) **VDS**

Brief account of Heinrich Zeise, who, about 1850, investigated bird flight and designed an airship.

1413. Zabel, Eugen. Deutsche Luftfahrt. Rückblicke und Ausblicke. Berlin: Gustav Braunbeck [cop. 1918]. 136 p. illus. 8°. **VDT**

Comments on early writings, especially those of Goethe. History of the Schütte-Lanz firm of airship builders. Heavier-than-air investigations of Arnold Böcklin.

Lighter-Than-Air Craft — Airships, continued

1414. **Zeppelin,** Ferdinand Adolf August Heinrich, Graf von. Dirigeable airships. (Scientific American supplement, New York, 1906, v. 62, p. 25886–25888.) **VA**

Lesson learned from his experiences.

1415. —— Die Entwicklung des Baues meiner Luftschiffe. illus. (Schiffbautechnische Gesellschaft, Jahrbuch, 1915, Berlin, 1915, Bd. 17, p. 178–191.) **† VXA**

Translation in *Scientific American supplement,* New York, 1916, v. 81, p. 77–79, *VA*.

1416. —— Entwürfe für lenkbare Luftfahrzeuge. (Verein deutscher Ingenieure, Zeitschrift, Berlin, 1896, Bd. 40, p. 408–414.)

With discussion. **3–† VDA**
A record of difficulties overcome.

1417. —— Erfahrungen beim Bau von Luftschiffen; Vortrag gehalten auf der 49. Hauptversammlung des Vereines deutscher Ingenieure zu Dresden am 29. Juni 1908. Berlin: J. Springer, 1908. 23 p. 8°. **VDW p.v.2, no.2**

1418. —— Die Eroberung der Luft. Ein Vortrag. Stuttgart: Deutsche Verlags Anstalt, 1908. 31 p. 8°. **VDY p.v.5, no.2**

For references to Zeppelin and his various airships see Paul Brockett, *Bibliography of aeronautics,* Washington, 1910, p. 911–914, *VDT;* also United States. — National Advisory Committee for Aeronautics, *Bibliography of aeronautics,* 1909–1931, Washington, 1921–1935, volumes for 1909/16, p. 1476–1489; 1917/19, p. 491–494; 1920/21, p. 446–447; 1922, p. 167; 1923, p. 144; 1924, p. 114; 1925, p. 188; 1926, p. 154; 1927, p. 184; 1928, p. 214; 1929, p. 241–242; 1930, p. 261; 1931, p. 311–312, *VDT.*

1419. The **Zeppelin** airships. (Engineering, London, 1908, v. 86, p. 130–132.) **VDA**

Details of construction, based on paper by Count Zeppelin before the Verein deutscher Ingenieure.

1420. **Zeppelin-Denkmal** für das deutsche Volk; aus Anlass des fünfundzwanzigjährigen Jubiläums des ersten Luftschiff-Aufstiegs des Grafen Zeppelin... Stuttgart: Germania-Verlag [1925]. xii, 385 p. illus. facsims., tables, ports. f°. **† AN**

1421. **Z R 1,** the first American-built rigid airship. illus. (Flight, London, 1923, v. 15, p. 721–723.) **VDS**

1422. **Z R 3** reaches Lakehurst from Germany. Last Zeppelin covers 5000 miles in 81 hours across Atlantic. illus. (Aviation, Highland, N. Y., 1924, v. 17, p. 1190–1193.) **VDS**

HEAVIER-THAN-AIR CRAFT

Airplanes and Ornithopters

1423. **Abbot,** Charles Greeley. The relations between the Smithsonian Institution and the Wright brothers. Washington, 1928. 27 p. 8°. (Smithsonian Institution, Miscellaneous collections, v. 81, no. 5. Publication no. 2977.) *** EA**

"An attempt...to clarify an unfortunate controversy, and to correct errors where errors have been made, in order to do justice alike to three great pioneers of human flight — Wilbur and Orville Wright, and Samuel Pierpont Langley — as well as to the Smithsonian Institution."

1424. **Adams,** Heinrich. Flug, von Heinrich Adams. Unser Flieger, von Wilbur und Orville Wright. Leipzig: C. F. Amelangs Verlag, 1909. viii, 144 p. illus. 8°. **VDY**

Work of the airplane pioneers, including the Wright brothers, with a chapter contributed by the latter.

1425. **Ader,** Clément Agnès. L'aéroplane "Éole." illus. (Revue de l'aéronautique, Paris, 1893, année 6, p. 69–99.) **3–† VDS**

Illustrated description of the author's airplane.

1426. —— Sur des appareils d'aviation. (Académie des sciences, Comptes rendus, Paris, 1898, tome 126, p. 1553–1555.) *** EO**

For references to Ader's experiments and writings see United States. — National Advisory Committee for Aeronautics, *Bibliography of aeronautics,* 1909–1931, Washington, 1921–33, volumes for 1909/16, p. 10; 1917/19, p. 4; 1924, p. 1; 1925, p. 2; 1927, p. 2; 1928, p. 4; 1930, p. 4, *VDT.*

1427. The **"Ader"** avion. illus. (Scientific American supplement, New York, 1906, v. 62, p. 25837–25838.) **VA**

1428. **Ader's** flying machine. illus. (Aeronautical journal, London, 1898, v. 2, Oct., p. 73–75.) **VDS (Royal)**

Ader succeeded in getting a steam-driven airplane off the ground for a few seconds. The fuel was alcohol.

1429. **Advance** of American aviation in 1911 reviewed. illus. (Aero, St. Louis, 1911–12, v. 3, p. 255–256, 279.) **3–VDS**

1430. **Aerial** locomotion. (Engineering, London, 1906, v. 82, p. 792–793.) **VDA**

Good general review of airplane experiments from the time of Henson.

1431. **Aerial** navigation. illus. (Scientific American, New York, 1883, v. 48, p. 87.) **VA**

Illustration and brief description of Baranowski's steam-driven flying machine model.

1432. **Aerial** navigation. illus. (Scientific American, New York, 1895, v. 73, p. 53.) **VA**

Patent for combined helicopter and airplane patented by Estanislao Caballero de los Olivos, of New York City.

1433. **Aerial** steam machine designed by Mr. Joseph M. Kaufmann, engineer, Glasgow. illus. (Engineering, London, 1868, v. 5, p. 198–199.) **VDA**

Model flying machine with flapping wings.

1434. **Aero-club de France.** Aeroplane accidents. Report of the aviation committee of the Aero-club de France. (Aeronautics, London, 1911, v. 4, p. 85–86.) **VDS**

Analysis of accidents for the years 1909–1910.

Heavier-Than-Air Craft — Airplanes, cont'd

1435. **"Aero-amateur,"** pseud. Flying; the why and wherefore. London: "The Aero" [1909]. 96 p. illus. 12°. **VDY**

Has drawings of eleven contemporary airplanes.

1436. **Aeronautics** [in 1908]. (Scientific American, New York, 1909, v. 100, p. 6.) **VA**

"The historian of the future will point to the year 1908 as that in which the problem of mechanical flight was first fully mastered."

1437. **Aeronautics** in 1933. illus. (Engineer, London, 1934, v. 157, p. 7–8, 46–48.) **VA**

Deals mainly with the progress made by British airplane manufacturers.

1438. **Aeroplane** frames. illus. (Engineering, London, 1909, v. 88, p. 611.) **VDA**

Describes current practice (1909).

1439. **Aeroplanes** and motors at the first Paris aeronautical salon. illus. (Scientific American, New York, 1909, v. 100, p. 81–82.) **VA**

1440. **Agen,** Myer. Aeronautics progresses by leaps and bounds. French pilots and inventors come to the fore in improvements on heavier-than-air machines. illus. (Aeronautical digest, New York, 1923, v. 2, p. 429–430, 457.) **VDS**

Brief reference to the automatic airplane of Marçay and to the tailless airplane of René Arnoux.

1441. **Air** speeds step up. A review of the manner in which speeds have gained but not "overnight" as many suppose. (U. S. air services, Washington, D. C., 1933, v. 18, Sept., p. 18–19.) **VDS**

1442. The **Aircraft** carrier Saratoga launched. Largest and fastest ship of its kind to be ready next year. (Aviation, Highland, N. Y., 1925, v. 18, p. 434–435.) **VDS**

1443. **Airplane** fuel from the air. (Literary digest, New York, 1929, v. 102, Sept. 7, p. 20.) *** DA**

Professor Rindine, of Russia, proposes to use hydrogen, which he believes can be obtained in the higher atmosphere.

1444. **Airplane** production and export 1909–1928. (Domestic air news, Washington, D. C., 1929, no. 50, April 15, p. 4.) **VDS (U. S. Commerce)**

1445. **Airplanes** and pilots. Concentration on development in anticipation of better markets. diagrs. (Aviation, New York, 1934, v. 33, p. 79–83.) **VDS**

Statistics of airplane production shown for several years by means of colored charts.

1446. **Airplanes** that fly themselves. illus. (Aviation, New York, 1933, v. 32, p. 213–215.) **VDS**

History of automatic pilots.

1447. The **Airplane's** twentieth anniversary. illus. (Aeronautical digest, New York, 1924, v. 4, p. 97.) **VDS**

Celebration at Dayton, Ohio. Picture of the Wright brothers' shop.

1448. The **Albany-New York** aeroplane flight. How Curtiss competed for the Scientific American trophy and the New York World prize. illus. (Scientific American, New York, 1910, v. 102, p. 480–481.) **VA**

Illustration on p. 473.

1449. **Alberto** Santos-Dumont and his new mission. illus. port. (Flying, New York, 1915, v. 4, p. 731–734, 748.) **VDS**

Biography and portrait.

1450. **Alexander-Katz,** Bruno. Eine Flugmaschine aus dem Jahre 1842. illus. (Deutsche Luftfahrer-Zeitschrift, Berlin, 1913, Jahrg. 17, p. 264–265.) **VDS**

Description and illustration of Henson's patent.

1451. **Alford,** Leon Pratt. The Wright aeroplane — its construction. It has flown for over an hour in a 10-mile breeze, has attained a height of 300 feet, and a speed of over 40 miles per hour. illus. (Scientific American supplement, New York, 1908, v. 66, p. 280–282.) **VA**

1452. **Allow** us to present — Frank H. Russell. port. (Aero digest, New York, 1925, v. 7, p. 662.) **VDS**

Russell (born 1878) was associated with the Wright brothers and with Glenn H. Curtiss as manufacturer and salesman.

1453. **Allow** us to present — Sikorsky. port. (Aero digest, New York, 1925, v. 7, p. 477.) **VDS**

Brief account of famous Russian designer, with a caricature portrait.

1454. **Amans,** P. C. Rapport sur l'exposition aéronautique de Vienne en 1888. illus. (L'Aéronaute, Paris, 1888, année 21, p. 183–190, 209–214, 231–239; 1889, année 22, p. 8–17, 25–36.) **3–VDS**

Includes descriptions of exhibits, notably the flying machines of William Kress and Ernst von Wechmar, and Von Loessl's air resistance measuring apparatus.

1455. **American** airplanes at home and abroad. illus. (Aviation, New York, 1933, v. 32, p. 133–160.) **VDS**

Has histories of thirty-two American manufacturing companies.

1456. **American** export trade in aeronautical equipment. charts. (Aviation, New York, 1929, v. 27, p. 710–711.) **VDS**

Development shown by charts and statistics.

1457. **American** exports. charts. (Aviation, New York, 1930, v. 28, p. 594–596.) **VDS**

Charts and statistics covering the period 1922–1929.

1458. **Anthony** Herman Gerard Fokker. illus. (Aero digest, New York, 1926, v. 9, p. 377, 391.) **VDS**

1459. **Apitz.** Bewertung des deutschen Wright-Patentes. (Zeitschrift für Flugtechnik, München, 1910, Jahrg. 1, p. 39–41, 109–111, 135–136.) **VDS**

Discussion of the merits of the original German patents of the Wright brothers.

Heavier-Than-Air Craft — Airplanes, cont'd

1460. Armengaud, J., the younger. Le problème de l'aviation et sa solution par l'aéroplane. (Société d'encouragement pour l'industrie nationale, Bulletin, Paris, 1907, année 106, July, p. 847–873.) 3–VA

Excellent history of the airplane.

1461. —— Le problème de l'aviation, sa solution par l'aéroplane; conférence faite le 16 février 1908 au Conservatoire national des arts et métiers. Paris: C. Delagrave [1908]. viii, 102 p. illus. 3. ed. rev. 8° VDY p.v.23, no.1

1462. The **Armstrong-Whitworth** machines. illus. (Flight, London, 1919, v. 11, p. 434–438.) VDS

Development of well-known British airplane.

1463. Artingstall, F. D. [A flying machine.] illus. (Engineering, London, 1868, v. 5, p. 383.) VDA

Describes mechanism of flapping wings.

1464. The **Attainment** of high speeds. (Aviation, Highland, N. Y., 1925, v. 19, p. 633.) VDS

Discusses human limit to speed, and cites importance of engine and propeller development.

1465. Automatic stabilizing system of the Wright brothers. An interesting control devised by the American inventors. illus. (Scientific American supplement, New York, 1911, v. 71, p. 20–21.) VA

1466. The **Aveline** automatic pilot. illus. (Engineer, London, 1921, v. 131, p. 184–186.) VA

1467. Der **Aveline-Fluglagenregler.** illus. (Illustrierte Flug-Woche, Leipzig, 1921, Jahrg. 3, p. 348–349.) VDS

1468. Aviatik "milestones." illus. (Flight, London, 1919, v. 11, p. 1226–1230.) VDS

Development of famous German airplane.

1469. Aviation loses a man of vision. illus. port. (Popular aviation, Chicago, 1928, v. 2, May, p. 29.) VDS

Brief account of Rodman Wanamaker's contributions to aeronautics.

1470. Aviation pioneer honoured. Knighthood conferred upon A. V. Roe. port. (Flight, London, 1929, v. 21, p. 183.) VDS

1471. Aviation in Russia, 1917 to 1921. (Aeroplane, London, 1921, v. 21, p. 552.) VDS

1472. L'**Aviation** triomphant, par MM. d'Estournelles de Constant, C. Bouchard, E. Lavisse [and others]. Paris: Librairie aéronautique, 1910. 4–381 p. illus. 12°. VDT

Among the various papers are accounts of Blériot's cross-Channel flight, the Rheims meeting, and aeronautics in Germany.

1473. Aviation's barometers of business. diagrs. (Aircraft yearbook, New York, 1932, p. 11–22.) VDS

Statistical diagrams show development, 1925–1931.

1474. Aviation's twenty-five birthday candles. illus. (Literary digest, New York, 1929, v. 100, Jan. 5, p. 47, 50, 52–56.) * DA

Deals mainly with the careers of the Wright brothers, with a brief account of Lincoln Beachey.

1475. Ayres, W. O. Aerial traveling. (Scientific American, New York, 1885, v. 52, p. 369.) VA

Description of author's man-propelled glider.

1476. B., L. A notable birthday. port. (Aeroplane, London, 1933, v. 44, p. 12.) VDS

Portrait and biography of Hugo Ursinus, founder of periodical *Flugsport,* and a pioneer in the building of bombing airplanes.

1477. Baciocchi, Alighiero. Flying Spaniards. Enormous progress in aviation made by Spain in last two years. illus. (Aero digest, New York, 1925, v. 7, p. 478–479, 510.) VDS

1478. Bairstow, Leonard. The work of S. P. Langley. (Aeronautical journal, London, 1922, v. 26, p. 420–434.) VDS

With discussion.

For review see *Aeroplane,* London, 1922, v. 23, p. 307, *VDS.*

1479. Baldwin, W. H. A mechanical machine. illus. (Aeronautical world, Glenville, O., 1903, v. 1, p. 265–270.) VDS

Flying machine with revolving sails.

1480. Barrenbrugge, Charles C. The story of the famous "Mystery 'S'." illus. (Popular aviation, Chicago, 1932, v. 11, p. 99, 125–126.) VDS

Racing plane of Herbert Rawdon and Walter E. Burnham.

1481. Bates, Lindon Wallace. The first admiral of the air. illus. (Air travel, New York, 1917, v. 1, p. 120–125.) VDS

"William Samuel Henson, an English inventor, in 1843, made public his designs for a steam-driven airplane. Structurally it recognized practically every principle, except that of warping the wings, embodied in the successful Wright machines of more than sixty years later."

1482. Bauer. Jakob Degen's Flugmaschine im Jahre 1808. (Archiv für die Geschichte der Naturwissenschaften und der Technik, Leipzig, 1909–10, Bd. 2, p. 156–158.) OA

Description of flying machine invented by a Viennese watchmaker.

1483. Beatty, W. D., and A. D. de R. Martin. Aeroplanes. A short account of some of the present day aeroplanes and motors for aeronautical purposes. illus. (Royal Engineers journal, Chatham, 1909, v. 9, p. 245–260.) VDA

1484. Beck, James Montgomery. Wilbur Wright's last flight[?]. ports. (U. S. air services, Washington, 1934, v. 19, Nov., p. 22–23.) VDS

Heavier-Than-Air Craft — Airplanes, cont'd

1485. Bell, Alexander Graham. Aerial locomotion, with a few notes of progress in the construction of an aërodrome. illus. (National geographic magazine, Washington, 1907, v. 18, p. 1–34.) **KAA**

Bibliography. Reprint in *VDY p.v.5, no.11.*

1486. —— Experiments in mechanical flight. (Nature, London, 1896, v. 54, p. 80.) **OA**

Brief account of author's experiences at flights of Langley's aerodromes.

1487. —— Langley's contributions to aeronautics. (Nature, London, 1910, v. 83, p. 263–264.) **OA**

Address delivered at the presentation of the Langley medal to the Wright brothers, Feb. 1, 1910.

For references to writings by and about Bell see Paul Brockett, *Bibliography of aeronautics,* Washington, 1910, p. 108–110, *VDT;* also United States. — National Advisory Committee for Aeronautics, *Bibliography of aeronautics,* 1909–1931, Washington, 1921–35, volumes for 1909/16, p. 136–137; 1917/19, p. 51; 1920/21, p. 45; 1930, p. 25, *VDT.*

1488. Bergonzi, P. C. Il volo a vela nel passato nel presente e nell' avvenire. illus. (Aeronautica, Milano, 1929, anno 3, p. 139–147, 217–221.) **3–VDS**

Early history of gliding and of the airplane.

1489. The **Berliner** aeroplane. illus. (Scientific American, New York, 1903, v. 89, p. 216.)

Invention of Emile Berliner. **VA**

1490. Bertin, E. L. Sur la giration des aéroplanes. (Académie des sciences, Comptes rendus, Paris, 1908, v. 147, semestre 2, p. 895–902.) *** EO**

1491. Besançon, Georges. Capitaine Ferber. illus. port. (L'Aérophile, Paris, 1905, année 13, p. 25–26.) **3–VDS**

Portrait and brief account of the airplane experiments of well-known French pioneer.

1492. —— Robert Esnault-Pelterie. port. (L'Aérophile, Paris, 1907, année 15, p. 329.) **3–VDS**

Portrait and brief biography of prominent French airplane and engine designer.

1493. [**Bibliographies** of Henri and Maurice Farman.] (In: Paul Brockett, Bibliography of aeronautics, Washington, 1910, p. 298–301; continued in United States. — National Advisory Committee for Aeronautics, Bibliography of aeronautics, 1909–1931, Washington, 1921–35, volumes for 1909/16, p. 458–465; 1917/19, p. 157; 1920/21, p. 143; 1922, p. 49; 1923, p. 42; 1924, p. 32; 1925, p. 59; 1926, p. 45; 1927, p. 56; 1928, p. 65–66; 1929, p. 76; 1930, p. 75; 1931, p. 90.) **VDT**

1494. [**Bibliographies** of ornithopters.] (In: Paul Brockett, Bibliography of aeronautics, Washington, 1910, p. 646; continued in United States. — National Advisory Committee for Aeronautics, Bibliography of aeronautics, 1909–1931, Washington, 1921–35, volumes for 1909/16, p. 980–982; 1917/19, p. 338; 1920/21, p. 310; 1923, p. 103; 1925, p. 132; 1926, p. 109; 1927, p. 134; 1928, p. 150; 1930, p. 182.) **VDT**

1495. [**Bibliographies** of wings.] (In: Paul Brockett, Bibliography of aeronautics, Washington, 1910, p. 895; continued in United States. — National Advisory Committee for Aeronautics, Bibliography of aeronautics, 1909–1931, Washington, 1921–35, volumes for 1909/16, p. 1432–1436; 1917/19, p. 477; 1920/21, p. 433–438; 1922, p. 163–164; 1923, p. 140–141; 1924, p. 111; 1925, p. 183–185; 1926, p. 150–151; 1927, p. 180; 1928, p. 208–209; 1929, p. 235–237; 1930, p. 255–257; 1931, p. 305–307.) **VDT**

1496. [**Bibliography** of Albert Bazin.] (In: Paul Brockett, Bibliography of aeronautics, Washington, 1910, p. 105.) **VDT**

List of his articles in *L'Aérophile* and *L'Aéronaute.*

1497. [**Bibliography** of Charles Stewart Rolls.] (In: United States. — National Advisory Committee for Aeronautics, Bibliography of aeronautics, 1909–1930, Washington, 1921–35.) **VDT**

See volume for 1909/16, p. 1147.

1498. [**Bibliography** of Ernest Archdeacon.] (In: Paul Brockett, Bibliography of aeronautics, Washington, 1910, p. 52–54.) **VDT**

List of some of Archdeacon's writings.

1499. [**Bibliography** of Glen L. Martin.] (In: United States. — National Advisory Committe for Aeronautics, Bibliography of aeronautics, 1909–1931, Washington, 1921–35.) **VDT**

See volume for 1909/16, p. 844; 1917/19, p. 286; 1920/21, p. 267–268; 1922, p. 93; 1923, p. 83; 1924, p. 66; 1926, p. 86; 1927, p. 109; 1929, p. 139; 1930, p. 147; 1931, p. 177.

1500. [**Bibliography** of Hubert Latham.] (In: United States. — National Advisory Committee for Aeronautics, Bibliography of aeronautics, 1909–1931, Washington, 1921–35.)

See volume for 1909/16, p. 762–764. **VDT**

1501. [**Bibliography** of Hudson Maxim.] (In: United States. — National Advisory Committee for Aeronautics, Bibliography of aeronautics, 1909–1931, Washington, 1921–35.) **VDT**

See volume for 1909/16, p. 854; 1925, p. 107; 1926, p. 86.

1502. [**Bibliography** of Igo Etrich.] (In: United States. — National Advisory Committee for Aeronautics, Bibliography of aeronautics, 1909–1931, Washington, 1921–35.) **VDT**

See volumes for 1909/16, p. 440–442; 1917/19, p. 150.

1503. [**Bibliography** of Igor Sikorsky.] (In: United States. — National Advisory Committee for Aeronautics, Bibliography of aeronautics, 1909–1931, Washington, 1921–35.) **VDT**

See volumes for 1909/16, p. 1228–1229; 1917/19, p. 407; 1925, p. 161; 1926, p. 132–133; 1927, p. 158; 1928, p. 181–182; 1929, p. 203–204; 1930, p. 220; 1931, p. 259.

Heavier-Than-Air Craft — Airplanes, cont'd

1504. [**Bibliography** of John Stringfellow.] (In: United States. — National Advisory Committee for Aeronautics, Bibliography of aeronautics, 1909–1931, Washington, 1921–35.) **VDT**

See volumes for 1909/16, p. 1296; 1917/19, p. 426; 1925, p. 167.

1505. [**Bibliography** of landing gear.] (In: United States. — National Advisory Committee for Aeronautics, Bibliography of aeronautics, 1909–1931, Washington, 1921–35.) **VDT**

See volumes for 1909/16, p. 755–757; 1917/19, p. 253; 1920/21, p. 235; 1924, p. 56; 1925, p. 92; 1926, p. 76; 1927, p. 94; 1929, p. 122; 1930, p. 131; 1931, p. 155.

1506. [**Bibliography** of Louis Blériot.] (In: Paul Brockett, Bibliography of aeronautics, Washington, 1910, p. 136–137; continued in United States. — National Advisory Committee for Aeronautics, Bibliography of aeronautics, 1909–1931, Washington, 1921–35, volume for 1909/16, p. 166–176.) **VDT**

1507. [**Bibliography** of Louis Bréguet.] (In: United States. — National Advisory Committee for Aeronautics, Bibliography of aeronautics, 1909–1931, Washington, 1921–35.) **VDT**

For references to Bréguet and his airplanes see volumes for 1909/16, p. 203–207; 1917/19, p. 67; 1920/21, p. 59; 1922, p. 21–22; 1923, p. 20; 1924, p. 14; 1925, p. 27; 1926, p. 20; 1927, p. 23; 1928, p. 31; 1929, p. 39; 1930, p. 34; 1931, p. 41–42.

1508. [**Bibliography** of Louis Paulhan and his airplanes.] (In: United States. — National Advisory Committee for Aeronautics, Bibliography of aeronautics, 1909–1931, Washington, 1921–35.) **VDT**

See volume for 1909/16, p. 1013–1016.

1509. [**Bibliography** of Max Michael Munk.] (In: United States. — National Advisory Committee for Aeronautics, Bibliography of aeronautics, 1909–1931, Washington, 1921–35.) **VDT**

See volumes for 1917/19, p. 318–319; 1920/21, p. 287–288; 1922, p. 101; 1923, p. 90; 1924, p. 72–73; 1925, p. 116–117; 1926, p. 93–94; 1927, p. 116; 1929, p. 150; 1930, p. 159–160; 1931; p. 189.

1510. [**Bibliography** of models.] (In: United States. — National Advisory Committee for Aeronautics, Bibliography of aeronautics, 1909–1931, Washington, 1921–35.) **VDT**

See volumes for 1909/16, p. 908–912; 1917/19, p. 313–314; 1920/21, p. 281–282; 1922, p. 99; 1923, p. 88; 1924, p. 71; 1925, p. 114; 1927, p. 114; 1928, p. 129–130; 1929, p. 147; 1930, p. 157; 1931, p. 185.

1511. [**Bibliography** of the Nieuport airplane.] (In: United States. — National Advisory Committee for Aeronautics, Bibliography of aeronautics, 1909–1931, Washington, 1921–35.) **VDT**

See volumes for 1909/16, p. 952–955; 1917/19, p. 331–332; 1920/21, p. 300–301; 1924, p. 81; 1925, p. 128; 1927, p. 132; 1929, p. 168.

1512. [**Bibliography** of speed.] (In: United States. — National Advisory Committee for Aeronautics, Bibliography of aeronautics, 1909–1931, Washington, 1921–35.) **VDT**

See volumes for 1909/16; p. 1255–1257; 1917/19, p. 414; 1920/21, p. 384–385; 1922, p. 147–148; 1923, p. 127; 1924, p. 102; 1925, p. 164; 1926, p. 135; 1927, p. 161; 1928, p. 185; 1929, p. 207–208; 1930, p. 226; 1931, p. 264.

1513. [**Bibliography** of speed indicators.] (In: United States. — National Advisory Committee for Aeronautics, Bibliography of aeronautics, 1909–1931, Washington, 1921–35.) **VDT**

See volumes for 1909/16, p. 1258; 1917/19, p. 414; 1920/21, p. 385; 1922, p. 148; 1923, p. 127; 1925, p. 164; 1926, p. 135; 1927, p. 161; 1931, p. 264.

1514. [**Bibliography** of streamlines.] (In: United States. — National Advisory Committee for Aeronautics, Bibliography of aeronautics, 1909–1931, Washington, 1921–35.) **VDT**

See volumes for 1909/16, p. 1295; 1917/19, p. 425; 1920/21, p. 392–393; 1924, p. 103; 1925, p. 167; 1926, p. 137; 1927, p. 165; 1929, p. 212; 1930, p. 231; 1931, p. 271.

1515. [**Bibliography** of T. O. M. Sopwith and of the Sopwith airplane.] (In: United States. — National Advisory Committee for Aeronautics, Bibliography of aeronautics, 1909–1931, Washington, 1921–35.) **VDT**

See volumes for 1909/16, p. 1250–1251; 1917/19, p. 411–412; 1920/21, p. 381; 1924, p. 101.

1516. [**Bibliography** of Thomas Moy.] (In: Paul Brockett, Bibliography of aeronautics, Washington, 1910, p. 609.) **VDT**

Citations to Moy's papers on mechanical flight, 1868–1904.

1517. [**Bibliography** of the Voisin airplane.] (In: United States. — National Advisory Committee for Aeronautics, Bibliography of aeronautics, 1909–1931, Washington, 1909–35.) **VDT**

See volumes for 1909/16, p. 1388–1390; 1917/19, p. 462; 1920/21, p. 420.

1518. [**Bibliography** of W. Sterling Burgess.] (In: United States. — National Advisory Committee for Aeronautics, Bibliography of aeronautics, 1909–1931, Washington, 1921–35.) **VDT**

See volumes for 1909/16, p. 226–230; 1917/19, p. 73 for references to Burgess's airplanes.

1519. [**Bibliography** of warping.] (In: United States. — National Advisory Committee for Aeronautics, Bibliography of aeronautics, 1909–1931, Washington, 1921–35.) **VDT**

See volumes for 1909/16, p. 1405–1406; 1920/21, p. 423; 1928, p. 202.

1520. **Bingham,** Hiram. The future of aviation. port. (Aeronautic review, Washington, 1930, v. 8, Jan., p. 25–26.) **VDS**

Extracts from article by Rear Admiral G. W. Melville, 1901, in which aviation is regarded as a chimerical hope. See entry no. 1961.

Heavier-Than-Air Craft — Airplanes, cont'd

1521. **Blackford,** William W. A new form of flying machine. illus. (Scientific American, New York, 1875, new series, v. 32, p. 100.) **VA**
Airplane with screw propellers.

1522. **Blake,** A. E. Twenty years of British aviation. illus. (Air, London, 1928, v. 1, June, p. 25–27, 40.) **VDS**
Relates the career of A. V. Roe.

1523. **Blanchet,** Georges. Louis Bréguet. port. (L'Aérophile, Paris, 1922, année 30, p. 33–34.) **3–VDS**
Portrait and brief biography of prominent French airplane constructor.

1524. The **Blériot** aeroplane. illus. (Scientific American, New York, 1907, v. 96, p. 388.)
A new design. **VA**

1525. The **Blériot** short-span monoplane — the Channel flyer. illus. (Flight, London, 1909, new series, v. 1, p. 453–456.) **VDS**
Illustrated description of machine used by Blériot in crossing the Channel in 1909.

1526. The **Blue** book of aviation. A biographical history of American aviation. Being a record of the life achievements of the pioneers of mechanical flight, World war flyers, noted men and women pilots, aeronautical executives and engineers of the present day, edited by Roland W. Hoagland. 1932 edition. Los Angeles: The Hoagland Company, 1932. 291 p. illus. f°. **† VDY**
Has also a chronology of American aviation dating from 1861; also a brief history of the American operations during the World war, as well as a list of aces of various countries.

1527. **Bock.** Fortschritte im deutschen Flugmaschinenbau 1911–1912. illus. (Deutsche Luftfahrer-Zeitschrift, Berlin, 1913, Jahrg. 17, p. 14–15, 33–35.) **VDS**

1528. The **Boeing** "Monomail," new type transport plane. illus. (U. S. air services, Washington, 1930, v. 15, Aug., p. 47.) **VDS**

1529. **Boffito,** Giuseppe. L'aeronautica nelle città italiane. Todi. (Rivista aeronautica, Roma, 1931, anno 7, Feb., p. 403–406.) **VDS**
Life and works of Don Abdon Menicali (1827–1886), Italian experimenter with flying machines.

1530. **Boffito,** Giuseppe, and Adolfo Mabellini. Un Francescano aviatore d'un secolo fa. illus. (Aeronautica, Milano, 1927, anno 1, p. 150–153.) **3–VDS**
Flying machine of Nicolò Betti, 1810.

1531. **Boltzmann,** Artur. Ueber moderne Flugtechnik. illus. (Oesterreichischer Ingenieur- und Architekten-Verein, Zeitschrift, Wien, 1909, Jahrg. 61, p. 89–92, 105–108, 121–125.) **VDA**
Discusses theory of airships and airplanes, with reference to contemporary construction.

1532. **Bonbright,** John Miller. A new American aeroplane. illus. (Scientific American, New York, 1907, v. 97, p. 357.) **VA**
Invention of J. W. Roshon.

1533. **Bonfort,** Helene. Wellner's sail-wheel flying machine. illus. (Popular science monthly, New York, 1895, v. 46, p. 627–633.) *** DA**

1534. **Bostwick,** A. E. [Use of the word aeroplane.] (Nature, London, 1925, v. 115, p. 650–651.) **VDS**
See also entry no. 5574.

1535. **Botts,** R. H. The Botts flying machine. illus. (Aeronautical world, Glenville, O., 1903, v. 1, p. 135–136.) **VDS**
Airplane carries a small gas bag and a parachute.

1536. **Bowen,** Robert Sidney. 1928 airplane and engine production. illus. (Aviation, New York, 1929, v. 26, p. 718–721.) **VDS**
Statistics.

1537. **Bower,** F. F. The rotaflier. illus. (Aeronautical world, Glenville, O., 1903, v. 1, p. 247–250.) **VDS**
The construction of this strange craft involves an inclined top-like rotating chamber to which revolving wings are attached.

1538. **Brearey,** Frederick W. Artificial flight. (Journal of science, London, 1879, v. 16 [new series, v. 1], p. 473–478.) **OA**
Author's experiments with heavier-than-air models. Also in *Scientific American supplement,* New York, 1879, v. 8, p. 3036, *VA.*

1539. —— Flight and its imitation. (Popular science review, London, 1879, v. 18 [new series, v. 3], p. 395–402.) **3–OA**
Problems to be solved.

1540. —— Flighty aspirations. (Journal of science, London, 1880, series 3, v. 2, p. 381–391.) **OA**
Author discusses heavier-than-air flight, suggesting a kite-like structure with movable wings. Also in *Scientific American supplement,* New York, 1880, v. 10, p. 3825–3826, *VA.*

1541. —— Flying machines. illus. (Popular science review, London, 1869, v. 8, p. 1–11.) **3–OA**
Discussion of the work of de Lucy, Henson, Stringfellow, Wenham, Charles Spencer, and William Gibson.
For list of writings by and about Brearey see Paul Brockett, *Bibliography of aeronautics,* Washington, 1910, p. 149–150, *VDT.*

1542. **Breary's** aeronautical machine. illus. (English mechanic, London, 1884, v. 40, p. 172.) **VA**
Also in *Scientific American supplement,* New York, 1884, v. 18, p. 7453–7454, *VA.*

1543. **Brett,** R. Dallas. History of British aviation, 1908–1914. London: John Hamilton, Ltd. [1933.] xxiv, 388 p. illus. 8°. **VDY**
"The War dealt practical aviation a blow which knocked it senseless for eleven years, and it was not until 1925 that consciousness returned."

1544. **Brewer,** Griffith. Langley flier tests. Pertinent questions as to their efficacy and results. (New York Times, New York, June 22, 1914, p. 8.) *** A**
Editorial comment in *New York Times,* June 23, 1914, p. 10, ** A.* Also discussed in *Scientific American,* New York, 1914, v. 111, p. 2, *VA.*

Heavier-Than-Air Craft — Airplanes, cont'd

1545. —— The life and work of Wilbur Wright. ports. illus. (Aeronautical journal, London, 1916, v. 20, p. 68–135.) **VDS**

With several appendices. Facsimiles of two youthful magazines published by the Wright boys. Numerous pictures of the North Carolina experiments.

1546. —— A new Langley controversy. Griffith Brewer reopens discussion at scientific meeting in London. illus. port. (Aviation, Highland, N. Y., 1921, v. 11, p. 532–535.) **VDS**

Abstract of lecture before the Royal Aeronautical Society, with replies by Charles M. Manly and Dr. Charles D. Walcott. Brief abstract also in *Flight*, London, 1921, v. 13, p. 703–704, *VDS*.

1547. —— Wilbur Wright. (Flight, London, 1912, v. 4, p. 609.) **VDS**

Appreciation written at the time of Wright's death.

1548. —— Wilbur Wright. port. (Aeronautical journal, London, 1912, v. 16, p. 148–153.) **VDS (Royal)**

Author's experiences with Wilbur Wright in France; the Wright patents and litigation.

1549. —— With the Wrights in America. illus. (Flight, London, 1910, v. 2, p. 706–708.) **VDS**

Interesting account of author's visit to the Wright brothers' flying field at Dayton, Ohio.

1550. Brief history of the aeroplane movement. illus. (Aero manual, London, 1910, 2. ed., p. 41–79.) **VDT**

Good, concise illustrated account from time of Leonardo da Vinci.

1551. Brief history of the Wright achievements. (Airway age, New York, 1928, v. 9, Dec., p. 37–38.) **VDC**

Arranged chronologically.

1552. Brillouin, Louis Marcel. Stabilité des aéroplanes. Surface métacentrique. illus. (Revue de mécanique, Paris, 1909, tome 24, semestre 1, p. 109–126.) **3–VFA**

1553. The **"Bristol"** machines. illus. (Flight, London, 1919, v. 11, p. 100–105.) **VDS**

Development of well-known British military airplane.

1554. The **British** aircraft industry. Brief biographical and historic sketch. (Flight, London, 1924, v. 16, p. 302–304.) **VDS**

Historical outlines of famous aircraft and engine firms.

1555. The **British** aircraft industry. illus. (Flight, London, 1930, v. 22, p. 1265–1369.) **VDS**

This extensive compilation deals with the products of numerous British airplane and engine makers, with occasional bits of historical interest.

1556. Broadwell, L. W. Anton Weczera's flying machine. (Aeronautical journal, London, 1900, v. 4, p. 150–151.) **VDS**

Describes model designed by Hungarian architect.

1557. Brower, Edward M. History of aviation, with interesting records. [Chicago: Emil Forslund Printing Co., 1932.] x, 267 p. illus. 12°. **VDT**

Chiefly concerned with notable airplane flights.

1558. Brown, Harold A. On Wright's trials at Ft. Myers. illus. (Aeronautics, New York, 1909, v. 5, p. 44–45.) **VDS**

1559. Brown, William G. The Guggenheim safety competition tests and the Curtiss "Tanager." illus. (Aviation, New York, 1930, v. 28, p. 236–241.) **VDS**

Describes instruments and methods of testing.

1560. Bruce, Eric Henry Stuart. Mechanical flight. (Royal Society of Arts, Journal, London, 1908–09, v. 57, p. 49–60.) **VA**

With discussion.
An excellent account of developments from the time of F. H. Wenham, 1866, to the present. Abstracted in *Flight*, London, 1909, v. 1, p. 97–98, 108–109, *VDS*.

1561. Brunet y Viadera, Gaspar. Curso de aviación; historia retrospectiva de la navegación aérea por lo más pesado que el aire. Técnica de la aviación... Barcelona: Feliu y Susanna, 1910. 512 p. illus. 8°. **VDW**

Part one covers history of heavier-than-air developments from early times, with fifty-three illustrations.

1562. Bryan, George Hartley. History and progress of aerial locomotion. (Royal Institution of Great Britain, Proceedings, 1899/1901, London, 1902, v. 16, p. 487–492.) *** EC**

Results of gliding and airplane experiments of some of the modern experimenters. Abstracted in *Nature*, London, 1901, v. 63, p. 526–527, *OA*.

1563. —— The problem of the flying machine. (Cornhill magazine, London, 1907, new series, v. 22, p. 605–619.) *** DA**

1564. —— Stability in aviation; an introduction to dynamical stability as applied to the motions of aeroplanes. London: Macmillan and Co., Ltd., 1911. x, 192 p. illus. 8°. **VDY**

For review of theories of Bryan-Williams, Ferber, Lanchester, Brillouin, Reissner, Crocco, Soreau, and Lecornu, see p. 171–177.

1565. The **Buhl-Verville** Aircraft Corporation formed. port. (Aviation, Highland, N. Y., 1925, v. 18, p. 324.) **VDS**

Portrait and biography of Alfred V. Verville, American builder of fast airplanes.

1566. Burke, Newton. Airplanes hurled to bits in test. Terrific crashes against wall show faults. illus. (Popular science monthly, New York, 1925, v. 106, April, p. 31.) *** DA**

Pilotless test planes at Wright Field, Dayton, Ohio.

1567. Butman, Carl H. Langley's flying machine. port. illus. (Aviation and aeronautical engineering, New York, 1917, v. 3, p. 528–531.) **VDS**

"Samuel Pierpont Langley, the inventor of the flying machine, has been long in coming into his own, but at last his name has come to stand for the pioneer in aviation."

1568. Caldwell, Cy. [Giuseppe Mario Bellanca.] port. (Aero digest, New York, 1933, v. 22, May, p. 20–21.) **VDS**

Biography of prominent airplane designer.

Heavier-Than-Air Craft — Airplanes, cont'd

1569. **Camm,** Frederick James. Some early English model aeroplane experiments. (Aeronautics, London, 1917, v. 12, p. 16–18.) **VDS**

Work of Sir George Cayley, George Pocock, W. H. Phillips, W. S. Henson, John Stringfellow, F. H. Wenham, Charles Spencer, Percy S. Pilcher, Sir Hiram Maxim, Horatio Phillips, and Thomas Moy.

1570. **Capt.** Cody's British army aeroplane. illus. (Scientific American, New York, 1909, v. 101, p. 198, 200.) **VA**

1571. **Capt.** Ferber's aeroplane experiments. illus. (Scientific American, New York, 1907, v. 96, p. 315.) **VA**

1572. **Caron,** Félix. La vérité sur l'appareil de Vincent de Groof dit L'Homme Volant. illus. (L'Aéronaute, Paris, 1874, année 7, p. 231–245.) **3–VDS**

Describes de Groof's flying machine (1874) and his fatal fall.

1573. **Carpenter,** F. G. Langley's flying machine. (Aeronautical journal, London, 1897, v. 1, Jan., p. 5–9.) **VDS (Royal)**

1574. **Carr,** G. W. Evolution of metal construction. illus. (U. S. air services, Washington, 1932, v. 17, Aug., p. 34–36.) **VDS**

1575. The **Case** of the Junkers Company. Junkers Aircraft Company of Germany in difficulties. A crisis in the German aircraft industry. (Aviation, Highland, N. Y., 1926, v. 21, p. 57–58.) **VDS**

1576. **Casson,** Herbert Newton. At last we can fly. The story of the Wright brothers who...have made flying practicable. illus. (American magazine, New York, 1907, p. 616–624.) *** DA**

1577. **Castañeda,** Vicente. El primer libro impreso sobre aviación en español. (Revista de archivos, bibliotecas y museos, Madrid, 1915, época 3, tomo 33, p. 350–360.) **BXA**

Discusses book by Antonio de Fuente la Peña, entitled *El Ente dilucidado,* 1677.

1578. **Celebration** commemorating the thirtieth anniversary of the Wright brothers at Kitty Hawk, North Carolina, and the dedication of the aviation section of the Museum of the Franklin Institute, December sixteenth and seventeenth, 1933. (Franklin Institute, Journal, Philadelphia, 1934, v. 217, p. 237–255.) **VA**

1579. **Cesare,** Oscar. The first man to fly. port. (U. S. air services, Washington, 1927, v. 12, Sept., p. 23–25.) **VDS**

Interview with Wilbur Wright. Portrait by the author.

1580. **Chanute,** Octave. The evolution of the "two-surface" flying machine. port. (Aeronautics, New York, 1907, v. 1, Sept., p. 9–10; Oct., p. 28–29.) **VDS**

The bi-plane type built by F. H. Wenham, 1866.

1581. —— First steps in aviation and memorable flights. (Aeronautics, New York, 1909, v. 4, p. 24.) **VDS**

Chronological list from Oliver of Malmesbury (11th century) to July 22, 1908.

1582. —— Langley, sa vie et son œuvre. port. (L'Aérophile, Paris, 1906, année 14, p. 95–96.) **3–VDS**

1583. —— Pending European experiments in flying. illus. (American aeronaut, St. Louis, 1907, v. 1, Oct., p. 13–16.) **3–VDS**

Brief accounts of the contemporary work of Santos-Dumont, Vuia, Delagrange, Blériot, Kapferer, Ferber, Esnault-Pelterie, De la Vaulx, Barlatier Blanc, the Zens brothers, Archdeacon, and others.

1584. —— The present status of aerial navigation. illus. (Engineering magazine, New York, 1896, v. 11, p. 47–58.) **VDA**

Brief review of the work of Langley, H. S. Maxim, Hargrave, Otto Lilienthal, Herring, and Pilcher.

1585. —— Progress in flying machines. New York: American engineer and railroad journal, 1894. iv, 308 p. illus. 8°. **VDY**

Extensive, copiously illustrated account from early times, with a chapter on screws for propelling and lifting.

1586. —— Recent progress in aviation. illus. (Western Society of Engineers, Journal, Chicago, 1910, v. 15, p. 111–147.) **VDA**

Illustrated accounts of work of Langley, Maxim, the Wright brothers, Santos-Dumont, Delagrange, Farman, Blériot, Ferber, Latham, and others. Includes an extensive chronology of memorable flights, with data. Also in *Scientific American supplement,* New York, 1910, v. 70, p. 56–58, 72–74, 88–90, 106–107, *VA.*

1587. —— Twenty-four uses for flying machines. (Fly, Philadelphia, 1909, v. 2, Nov., p. 18.) **3–† VDS**

For references by and about Chanute see Paul Brockett, *Bibliography of aeronautics,* Washington, 1910, p. 183–187, *VDT;* also United States. — National Advisory Committee for Aeronautics, *Bibliography of aeronautics,* 1909–1931, Washington, 1921–35, volume for 1909/16, p. 253–254, *VDT.*

1588. **Characteristics** of the leading American aeroplanes, hydro-aeroplanes, and flying boats, 1917–1918. illus. (Aerial age weekly, New York, 1918, v. 7, p. 36–40.) **VDS**

Tabular data with illustrations.

1589. **Charnley,** Mitchell Vaughn. The boy's life of the Wright brothers... New York: Harper & Brothers [cop. 1928]. ix, 291 p. illus. 12°. **AN**

1590. **Charnock,** Richard Everett. Adventures of an aeronaut in search of a patron. (Mechanics' magazine, London, 1843, v. 38, p. 340–341.) **3–VA**

Tribulations of the author in trying to interest his own and foreign governments in his flying machine.

1591. **Chatley,** Herbert. Aeroplane problems. Difficulties in construction. (Scientific American supplement, New York, 1909, v. 67, p. 326–327, 346–347.) **VA**

A classification and study of early airplanes.

Heavier-Than-Air Craft — Airplanes, cont'd

1592. —— Principles of rocket propulsion. (Royal Aeronautical Society, Journal, London, 1933, v. 37, p. 723–728.) **VDS**

Bibliography of twenty-four titles covers early literature.

1593. —— The problem of flight: a textbook of aerial engineering. London: C. Griffin & Co., Ltd., 1907. x, 119 p. illus. 8°. **VDW**

The library also has a copy of the 2. ed., London, 1910.

For list of Chatley's writings see Paul Brockett, *Bibliography of aeronautics*, Washington, 1910, p. 188, *VDT*.

1594. The **Chauvière** aeroplane. illus. (Engineering, London, 1909, v. 88, p. 662.) **VDA**

1595. **Chung,** De Senn. Aviation in China. illus. (Aviation, Highland, N. Y., 1921, v. 11, p. 511–512.) **VDS**

History begins in 1910.

1596. **Clark,** Virginius Evans. The historical development of the American airplane. (International Civil Aeronautics Conference, Proceedings, 1928, Washington, 1929, p. 85–94.) **VDS**

Also deals briefly with pioneering work in other countries. Also in *Mechanical engineering*, New York, 1929, v. 51, p. 103–108, *VFA*.

1597. **Claudy,** Carl Harry. The army's first airplane. illus. (Aero digest, New York, 1927, v. 11, p. 637–640.) **VDS**

Story of Orville Wright's flight over Fort Myer in 1908.

1598. —— The Langley flying machine. illus. (American inventor, Washington, 1903, v. 11, p. 208–209.) *** V**

Brief illustrated account of the machine and its disaster.

1599. —— Two American conquerors of the air. illus. (St. Nicholas, New York, 1909, v. 36, part 2, p. 785–787.) **Room 78**

Brief account of the Wright brothers' plane, with portraits.

1600. —— With the Wright brothers at Fort Myer. illus. (World today, Chicago, 1909, v. 17, p. 929–936.) *** DA (Hearst)**

1601. **Claxton,** William J. The mastery of the air... London: Blackie & Son, Ltd. [1930.] vii, 264 p. illus. [4. ed.] 12°. (The enterprise library.) **VDY**

Deals "mainly with the real achievements of pioneers who have helped to make aviation what it is to-day."

1602. **Cléry,** A. Louis Bréguet. port. (L'Aérophile, Paris, 1909, année 17, p. 145.) **3–VDS**

Brief account of well-known French pioneer in helicopters and airplanes.

1603. The **Coanda** aeroplane. illus. (Engineer, London, 1910, v. 110, p. 461.) **VA**

1604. **Cody,** Samuel Franklin. Experiences with the "power kite." illus. (Aeronautical journal, London, 1909, v. 13, p. 15–19.) **VDS**

Paper read before the Aeronautical Society of Great Britain describes author's motor-driven airplane.

For references to Cody and his work see United States. — National Advisory Committee for Aeronautics, *Bibliography of aeronautics*, 1909–1931, Washington, 1921–35, volume for 1909/16, p. 281–284, *VDT*.

1605. **Colas,** François. Le brevet de l'aéroplane de M. Hiram Maxim. illus. (L'Aéronaute, Paris, 1894, année 27, p. 243–254.) **3–VDS**

1606. **Collins,** Archie Frederick. A new Russian flying machine. illus. (American inventor, New York, 1906, v. 15, p. 87–88.) *** V**

Account of A. W. H. Warshavsky's model that "closely reproduces the effects obtained by the complex movements of the swiftest soaring and flying birds."

1607. **Col.** Thurman H. Bane. port. (U. S. air services, Washington, 1932, v. 17, April, p. 23.) **VDS**

An outstanding figure in the development of aviation in the United States.

1608. **Congreve,** Sir William. Aerial carriage. An inquiry into the possibility of man raising himself into the air by the application of his own power. [By W. C.] illus. (Mechanic's magazine, London, 1828, v. 9, p. 130–134, 186–187.) **3–VA**

1609. The **Construction** of the Wright aeroplane. illus. (Scientific American, New York, 1908, v. 99, p. 208–210.) **VA**

1610. **Crandell,** Roy. Bell and his boys. illus. (Sportsman pilot, Albany, N. Y., 1932, v. 7, April, p. 32–35, 50, 52.) **VDS**

Alexander Graham Bell's experiences with Glenn Curtiss and Capt. Tom Baldwin.

1611. **Curtis,** Frank L. A review of aviation accomplishments. The president of the Guggenheim Fund surveys three years of progress. An interview with Harry F. Guggenheim. port. (Aviation, New York, 1929, v. 27, p. 1010–1011.) **VDS**

1612. **Curtiss,** Glenn Hammond. Aeroplanes of Aerial Experimental Association. Details of four American-built machines... illus. (American machinist, New York, 1909, v. 32, p. 999–1009.) **VFA**

Machines were the *Red wing*, the *White wing*, the *June bug*, and the *Silver dart*.

1613. —— The Wright patents decision. (Aeronautics, London, 1914, v. 7, p. 186.) **VDS**

Comments on court's decision, 1914.

For numerous references to Curtiss and his work see United States. — National Advisory Committee for Aeronautics, *Bibliography of aeronautics*, 1909–1931, Washington, 1921–35, volumes for 1909/16, p. 311–323; 1917/19, p. 107–108; 1920/21, p. 101–102; 1922, p. 34; 1923, p. 31; 1924, p. 22; 1925, p. 43; 1926, p. 33; 1927, p. 40; 1928, p. 49; 1929, p. 60; 1930, p. 54, *VDT*.

1614. **Dacey,** Norman F. A biographical sketch — Igor Ivan Sikorsky. port. (Popular aviation, Chicago, 1930, v. 6, March, p. 31–32.) **VDS**

This number of *Popular aviation* is published under the title of *Aeronautics*.

Heavier-Than-Air Craft — Airplanes, cont'd

1615. **Dallwitz-Wegner,** Richard von. Der praktische Flugschiffer; eine Anleitung zur Konstruktion von Gleitfliegern, Schraubenfliegern und Schaufelfliegern, ihrer Tragdecken... Rostock: C. J. E. Volckmann Nachfolger, 1909. 78 p. 8°. **VDY**

Describes contemporary airplanes. Library has also 2. ed., Rostock, 1910, *VDY*.

1616. **Dam,** H. J. W. The Maxim airship. An interview with the inventor. illus. (McClure's magazine, New York, 1894, v. 2, p. 148–158.) ***DA**

Also in *Scientific American*, New York, 1894, v. 70, p. 233–235, *VA*.

1617. **Dandrieux.** [Un appareil destiné à reproduire le vol des mouches...par la force humaine.] illus. (L'Aéronaute, Paris, 1880, année 13, p. 198–202.) **3–VDS**

Description of ornithopter.

1618–1619. **Daniel Guggenheim Fund for the Promotion of Aeronautics.** The Daniel Guggenheim international safe aircraft competition. Final report. January 31, 1930. New York [1930]. 147 p. illus. tables. 4°. **VDY**

Safety tests of airplanes in competition held in 1929.

1620. **Davy,** M. J. B. ...Henson and Stringfellow, their work in aeronautics; the history of a stage in the development of mechanical flight. London: His Majesty's Stationery Office, 1931. 114 p. illus. ports. 8°. **VDT**

"Henson [1805–1888] was responsible for the 'first aeroplane project' in 1842, and Stringfellow [1799–1883] carried it to a point of limited success when he demonstrated in 1848 that a model could fly under its own power." Illustrations of models now in the Science Museum, London.

1621. **Dayton Wright Airplane Company.** The early history of the airplane. Dayton, O.: Dayton-Wright Airplane Co. [1908?] 24 p. f°. **†* C p.v.1742, no.9**

Contents: Wright, O. and W., The Wright brothers' aeroplane. Wright, O., How we made the first flight. Wright, W., Some aeronautical experiments.

1622. **Death** of Eustace Short. port. (Flight, London, 1932, v. 24, p. 332.) **VDS**

Short, the second of three brothers, famous as British manufacturers of airplanes, began his career in 1897 as a balloonist.

1623. **Death** of Gaston Caudron. (Aeronautics, London, 1915, v. 9, p. 417–418.) **VDS**

Brief obituary of well-known French airplane constructor.

1624. The **De Havilland,** or "Airco" machines. illus. (Flight, London, 1919, v. 11, p. 36–45.) **VDS**

History of development of famous British airplane.

1625. **Denham,** Athel F. New Vickers "Vanguard" largest passenger carrying plane in world. illus. (Automotive industries, Philadelphia, 1926, v. 54, p. 408.) **VDS**

1626. **Development** of airplane ambulances. illus. (International aeronautics, New York, 1923–24, v. 1, p. 257–261.) **VDS**

Brief history with pictures of some of the earlier equipment.

1627. **Devèze.** Mémoire sur le vol ou la navigation aérienne sans ballons, au moyen d'un appareil mis en mouvement par une machine à vapeur. illus. (Les mondes, Paris, 1868, tome 16, p. 621–638.) **3–OA**

Gives calculations for his flying machine using flapping wings. For discussion see above periodical, 1868, tome 17, p. 149–157.

1628. **Didiée,** Joseph. Aviation. Travaux et inventions de deux Vauclusiens, M. Louis Constantin & le Dr. Joseph Cousin. (Académie de Vaucluse, Mémoires, Avignon, 1913, série 2, tome 13, p. 193–207.) **DRR**

Account of work of two Frenchmen who made early studies in airplane stability. Also published as a separate, *VDY p.v.10, no.8.*

1629. **Dienstbach,** Carl. A. M. Herring's neue Flugversuche. illus. (Zeitschrift für Luftschiffahrt, Berlin, 1899, Jahrg. 18, p. 73–85.) **3–VDS**

Account of Herring's airplane experiments (1898).

1630. —— A. M. Herringsche Flugapparate. illus. (Illustrierte aëronautische Mitteilungen, Strassburg, 1904, Jahrg. 8, p. 256–258.) **VDS**

Herring's flying machine of 1898.

1631. —— Did Prof. S. P. Langley invent the first practical flying machine? (Scientific American, New York, 1914, v. 111, p. 59, 65–66.) **VA**

"To Langley belongs the merit of having anticipated, by the general trend of his mind and his intense sense of scientific responsibility, the most important achievement of today — knowledge of the effect of proportioning the different parts of an aeroplane to secure stability."

1632. —— The labors of Charles Matthews Manly. port. (American aeronaut, St. Louis, 1908, v. 1, June, p. 227–231.) **3–VDS**

Account of the work of S. P. Langley's first assistant.

1633. —— The perfect flying machine. First description of the marvelous invention which has given Wilbur and Orville Wright mastery of man's flight. — How they fly and how they learned the secret. illus. (American aeronaut, St. Louis, 1908, v. 1, Jan., p. 3–11.) **3–VDS**

1634. **Dime,** Eric Adolphus. What Europe is doing in aviation — Americans, please take notice! illus. (Scientific American, New York, 1920, v. 122, p. 489, 502.) **VA**

Account of Handley-Page power-saving airplane.

1635. **Doane,** Robert R. Aircraft demand, past, present and future. diagrs. (Aviation, New York, 1930, v. 28, p. 97–103.) **VDS**

Consumption since the war period, major factors of new demand and distribution, price, replacement demand, etc.

Heavier-Than-Air Craft — Airplanes, cont'd

1636. **Dollfus,** Charles. Clément Ader. illus. port. (L'Aéronautique, Paris, 1925, année 7, p. 208–209.) **VDS**

1637. —— Notes sur l'histoire des freins d'avions. illus. (L'Aéronautique, Paris, 1930, année 12, p. 310.) **VDS**

Brief history of brake development in France.

1638. The **Dreamer,** pseud. Armchair reflections. Cody. (Flight, London, 1913, v. 5, p. 907.) **VDS**

Appreciation of prominent British-American airplane designer.

1639. **Dreisch,** Th. Die Bremsung von Flugzeugen bei der Landung. illus. (Deutsche Luftfahrer-Zeitschrift, Berlin, 1917, Jahrg. 21, No. 11/12, p. 19–22.) **VDS**

Traces the development of brakes for landing gear of airplanes.

1640. **Drzewiecki,** Stepan. Le vol plané; essai d'une solution mécanique du problème. Paris: E. Bernard & Cie., 1891. 48 p. illus. 4°. **VDY p.v.8, no.1**

1641. **Dufaux,** Henri, and ARMAND DUFAUX. L'aéroplane Dufaux. illus. (L'Aérophile, Paris, 1908, année 16, p. 289–291.) **3–VDS**

1642. The **Dufaux** flying machine. illus. (Scientific American, New York, 1905, v. 93, p. 316.) **VA**

Invention of the Dufaux brothers of France.

1643. **Durand,** William Frederick. From the scientific standpoint. Wright's accomplishments constitute fine example of application of true scientific methods. (Airway age, New York, 1928, v. 9, Dec., p. 27–28.) **VDC**

1644. An **Early** air adventurer. The great flying hoax of 1876. illus. (Popular flying, London, 1933, v. 1, p. 716; illus. on p. 685.) **VDS**

Account of Ralph Stott who advertised to cross the English Channel in a machine equipped with secret power. Reproduction of his poster.

1645. An **Early** flight of fancy. illus. (Antiques, Boston, 1929, v. 15, p. 473–474.) **MAA**

Reproduction of a print on an English handkerchief, probably of Henson's aerial steam carriage projected in the 1840's.

1646. **Eberhardt,** J. C. The death of Wilbur Wright. illus. (Aero Club of America, Bulletin, New York, 1912, v. 1, July, p. 4–5.) **VDS (Flying)**

1647. The **Edward** Heath memorial completed. illus. (Popular aviation, Chicago, 1931, v. 9, Dec., p. 86.) **VDS**

Heath (1888–1931) was a well-known American builder of light airplanes, beginning his work in 1909.

1648. **Eiffel,** Alexandre Gustave. Projet d'aérodrome à la tour Eiffel. [Paris: L. Maretheux, 1904.] 12 p. illus. 4°. † **VBA p.v.7, no.23**

Proposal to install a cable from the Eiffel tower to a point near the ground for the purpose of sustaining airplanes in trial flights.

1649. **Esnault-Pelterie,** Robert Albert Charles. L'aéroplane et le moteur extra-léger Robert Esnault-Pelterie. illus. (Aérophile, Paris, 1907, année 15, p. 330–332.) **3–VDS**

1650. —— L'aviation et les aéroplanes. (Société industrielle de Mulhouse, Bulletin, Mulhouse, 1908, v. 78, p. 81–96.) **VA**

Brief history of heavier-than-air craft. Author describes machine of his invention.

1651. —— The trials of a pilot. (Flight, London, 1909, new series, v. 1, p. 60–61.) **VDS**

Abstract of lecture before the Aero Club, London. Includes a descriptive chronology of airplanes dating from 1879.

1652. **Espitallier,** Georges. L'état actuel de l'aviation. illus. (Le Génie civil, Paris, 1908, tome 52, p. 398–401, 409–412, 429–432, 455–457.) **VA**

History of the airplane, featuring French developments.

1653. —— Les progrès récents de l'aviation. Les expériences de MM. Wright, Farman et Blériot. illus. (Le Génie civil, Paris, 1908, v. 54, p. 33–37, 61–63.) **VA**

1654. **Étévé,** Albert. Les aéroplanes; leurs mouvements, leur stabilité. illus. (Revue du génie militaire, Paris, 1907, v. 33, p. 389–422, 505–542; v. 34, p. 21–52.) **VWA**

1655. **Evans,** Stanley H. Development of the slotted wing. illus. (Aero digest, New York, 1934, v. 24, June, p. 38–41.) **VDS**

1656. The **Evolution** of the Etrich "Taube." illus. (Flight, London, 1915, v. 7, p. 106–108.) **VA**

The form of this machine is stated to have been derived from the seed-leaf of the Zanonia palm. Also in *Scientific American supplement,* New York, 1915, v. 79, p. 284–285, *VA.*

1657. The **Evolution** of German aviation. (Flying, New York, 1912, v. 1, Oct., p. 7–9.) **VDS**

1658. **Experiments** of the brothers Wright. illus. (Aeronautical journal, London, 1904, v. 8, p. 37–42.) **VDS (Royal)**

Part 1: With gliding machines depending upon gravity for their motive power. Part 2: With the power flyer — communicated by Mr. Orville Wright.

1659. **Experiments** on flying machines. illus. (Engineering, London, 1890, v. 49, p. 687; v. 50, p. 769.) **VDA**

Illustrations and brief description of Hargrave's flying machine.

1660. **Experiments** with motor-driven aeroplanes. illus. (Scientific American, New York, 1903, v. 89, p. 204.) **VA**

Experiments of Gustave Whitehead, of Bridgeport, Conn.

1661. **Extrait** d'une lettre escrite à Monsieur Toynard sur une machine d'une nouvelle invention pour vôler en l'air. illus. (Journal des sçavans, Amsterdam, 1679, tome 6, p. 460–464.) **3–OA**

Quaint machine invented by Besnier Serrurier de Sablé. The illustration has been much copied.

Heavier-Than-Air Craft — Airplanes, cont'd

1662. The **Failure** of Langley's aerodrome. illus. (Scientific American, New York, 1903, v. 89, p. 272.) **VA**

Brief account of the mishap.

1663. A **Famous** Dutch aviation pioneer. The career of Mr. Anthony Fokker. illus. port. (Meccano magazine, Liverpool, 1932, Jan., v. 17, p. 28–29.) **VFA**

Fokker (born 1890) "at the age of 20 designed and constructed an aeroplane, and today, although he is only 41, his machines are built under license in 14 countries and 18 factories."

1664. Farman, T. F. Aviation in 1910. (Blackwood's magazine, London, 1911, v. 189, p. 161–176.) *** DA**

1665. —— French civil and military aviation in 1913. (Blackwood's magazine, London, 1914, v. 195, p. 1–21.) *** DA**

1666. The **Farman** aeroplane. Description of the flights made by this famous record-breaking machine. (Scientific American supplement, New York, 1908, v. 65, p. 67.) **VA**

1667. Farman's experiments with his triple-surface aeroplane. illus. (Scientific American, New York, 1908, v. 99, p. 468.) **VA**

1668. Faurote, Fay Leone. Airplane types. illus. (Society of Automotive Engineers, Transactions, New York, 1918, v. 13, part 2, p. 137–202.) **TOL**

Brief early history of the airplane and the effects of the World war on development. With a different title the paper is partly reprinted in *Aviation*, New York, 1918, v. 4, p. 760–764, *VDS; Automotive industries*, New York, 1918, v. 38, p. 1184–1186, *TOL.*

1669. Favaro, Antonio. Intorno alla vita ed ai lavori di Tito Livio Burattini, fisico agordino del secolo XVII. 140 p. (Reale istituto veneto di scienze, lettere ed arti, Memorie, Venezia, 1896, v. 25, no. 8.) *** ER**

See p. 55–57 for reference to flying machine model made by Burattini about 1648.

1670. Felgate, Russell A. Wing dope from historic, technical side. (U. S. air service, New York, 1920, v. 3, July, p. 20.) **VDS**

1671. Ferber, Louis Ferdinand. The aeroplane experiments of M. Louis Blériot. illus. (Scientific American supplement, New York, 1907, v. 64, p. 168–169.) **VA**

1672. —— Die Kunst zu fliegen, ihre Anfänge — ihre Entwickelung... Berechtigte Übersetzung mit Nachträgen von A. Schönig. Berlin: R. C. Schmidt & Co., 1910. 215 p. illus. 8°. (Bibliothek für Luftschiffahrt und Flugtechnik. Bd. 4.) **VDY**

Illustrated accounts of the pioneers in flying, including the author.

1673. —— Les progrès de l'aviation par le vol plane. illus. (Revue d'artillerie, Paris, 1905, tome 66, p. 317–375.) **VWA**

Illustrated account of the development of heavier-than-air machines during the early years of the twentieth century.

1674. —— Que valent des brevets Wright? illus. (L'Aérophile, Paris, 1908, année 16, p. 252–253.) **3–VDS**

Discussion of the value of the Wright brothers' patents.

For references to writings by and about Ferber see Paul Brockett, *Bibliography of aeronautics*, Washington, 1910, p. 305–309, *VDT*. Continued in United States. — National Advisory Committee for Aeronautics, *Bibliography of aeronautics*, 1909–1931, Washington, 1921–35, volume for 1909/16, p. 470–472, *VDT*.

1675. First airplane passenger. (Domestic air news, Washington, D. C., 1929, no. 51, April 30, p. 4.) **VDS (U. S. Commerce)**

Flight of Charles Furnas, an employee of the Wright brothers in 1908.

1676. The **First** flight of the Wright aeroplane at Fort Myer. illus. (Scientific American, New York, 1908, v. 99, p. 169.) **VA**

1677. The **First** Paris aeronautical salon. An interesting collection of novelties. illus. (Scientific American supplement, New York, 1909, v. 67, p. 52–54.) **VA**

Includes a second article entitled Aeroplane design and construction, in which some of the exhibits are described and commented on.

1678. First stainless steel, shot-welded airplane tested. illus. (U. S. air services, Washington, 1933, v. 16, Dec., p. 48.) **VDS**

Built by The Edward G. Budd Co., of Philadelphia.

1679. The **First** successful trial of a new American aeroplane. illus. (Scientific American, New York, 1908, v. 98, p. 200.) **VA**

Trial of airplane built and designed by members of the Aerial Experiment Association, Hammondsport, N. Y.

1680. First transcontinental airplane test. (Society of Automotive Engineers, Journal, 1920, v. 6, p. 422–424.) **TOL**

Reliability test of engines and planes, 1920.

1681. Fischer. Ueber Luftfahrzeuge mit Vorführung von Hofmann's Flieger im betriebsfähigen Modell. illus. (Verein zur Beförderung des Gewerbfleisses, Verhandlungen, Berlin, 1901, Jahrg. 80, p. 123–138.) **3–VA**

Brief history of airships and early airplanes, with a description of Hofmann's steam-driven airplane model.

1682. Flint, Charles Ranlett. Charles R. Flint's story of Great Britain's slowness in taking up aeroplanes. (Flying, New York, 1917, v. 6, p. 861, 869.) **VDS**

How the Wright machine was turned down.

1683. —— Memories of an active life: men, and ships, and sealing wax. New York: G. P. Putnam's Sons, 1923. xviii, 349 p. illus. 8°. **AN**

See p. 245–253 for account of Flint's relations with the Wright brothers. This selection is reprinted in *Aviation*, Highland, N. Y., 1923, v. 15, p. 797–798, *VDS*.

1684. Ein **Flugversuch** aus dem Jahre 1785. (Deutsche Luftfahrer-Zeitschrift, Berlin, 1916, Jahrg. 20, p. 8–9.) **VDS**

Flying experiments of Carl Friedrich Meerwein.

Heavier-Than-Air Craft — Airplanes, cont'd

1685. The **Flying** book, 1917 edition, edited by W. L. Wade. London and New York: Longmans, Green & Co., 1917. 291 p. illus. 8°. **3–VDS**

See p. 85–124 for details and pictures of airplanes no longer in use (in 1917), but of interest, "as in their day they assisted the development of military aviation."

1686. **Flying** machine. illus. (Scientific American, New York, 1871, new series, v. 24, p. 402.) **VA**

Flying machine with man-driven wings, the invention of W. F. Quimby, of Wilmington, Del.

1687. A **Flying** machine. illus. (Scientific American, New York, 1892, v. 66, p. 67.) **VA**

Invention of Edward P. Frost, of England.

1688. The **Flying** machine and its slow development. (Scientific American supplement, New York, 1903, v. 56, p. 23150–23151.) **VA**

Brief history.

1689. **Fokker,** Anthony Herman Gerard. Welded steel tube construction and the development of cantilever wings. illus. (Institution of Aeronautical Engineers, Proceedings, London [1923], no. 7, p. 7–26.) **VDS**

Author's experiences in these developments, with a brief autobiography. Abstract in *Flight,* London, 1923, v. 15, p. 739–742, *VDS.*

1690. **Fokker,** Anthony Herman Gerard, and B. Gould. Flying Dutchman; the life of Anthony Fokker... New York: H. Holt and Company [cop. 1931]. 3–282 p. illus. ports. 8°. **AN**

Reviewed in *Flight,* London, 1931, v. 23, p. 1139–1140, *VDS; Aviation,* New York, 1931, v. 30, p. 498, *VDS; Aeroplane,* London, 1931, v. 41, p. 826, 828, *VDS.*

1691. **Foolproof** airplane invented. (Science news-letter, Baltimore, 1928, v. 14, p. 163; illus. on p. 175.) **OA**

Invention of Albert A. Merrill that keeps the airplane always horizontal, even if the engine stalls. Also in *Literary digest,* New York, 1928, v. 99, Oct. 13, p. 76, * *DA.*

1692. **Ford,** W. B. Macy stabilizer patents. illus. (Aviation, New York, 1917, v. 1, p. 389.) **VDS**

1693. **Foreign** expressions concerning Wright brothers' plane. British air authority makes light of Wright gift. Discounts Wright's scientific contributions. illus. (Slipstream, Dayton, O., 1928, v. 9, April, p. 16–17, 19.) **VDS**

Arguments pro and con by C. C. Grey, J. H. Spottiswoode, and F. Handley Page.

1694. **Foreign** trade. charts. (Aviation, New York, 1932, v. 31, p. 129–130.) **VDS**

Statistics of international trade in aeronautical equipment, 1931.

1695. **Fournier,** Lucien. L'aéroplane Givaudan. illus. (La Nature, Paris, 1909, année 37, semestre 2, p. 117–119.) **OA**

1696. —— L'aéroplane V. Tatin. illus. (Cosmos, Paris, 1909, nouv. série, tome 60, semestre 1, p. 535–538.) **VA**

1697. —— Les deux nouveaux planeurs Blériot. illus. (Cosmos, Paris, 1909, nouv. série, v. 60, semestre 1, p. 172–174.) **VA**

1698. The **Fourth** aeroplane of the Aerial Experiment Association. illus. (Scientific American, New York, 1908, v. 99, p. 433–434.) **VA**

Illustrations and description of *The Silver dart.*

1699. **Franck,** François. Expériences sur le vol mécanique par M. V. Tatin. illus. (La Nature, Paris, 1877, année 5, semestre 1, p. 148–150.) **OA**

Experiments with model airplanes.

1700. **Freeburg,** Victor Oscar. Flying into the midnight sun. Sweden's progress in aviation is characteristic of that northern country's contribution to the world's industrial advancement. illus. (Aero digest, New York, 1924, v. 4, p. 298–301.) **VDS**

1701. **Fuller,** J. F. Bishop Wilkins on submarines and aeroplanes in 1648. (Kerry archaeological magazine, Cork, 1917, v. 4, p. 163–170.) **CT**

1702. **Fullerton,** J. D. Flight and flying machines — recent progress. (Aeronautical journal, London, 1897, v. 1, July, p. 1–9.) **VDS**

Deals with developments in scientific investigations, fuel, motors, materials, and flying machine design.

1703. **Further** flights with Langley's aeroplane. illus. (Scientific American, New York, 1914, v. 111, p. 307.) **VA**

1704. **Fyfe,** George. Flying prophets of the past. Hopes, fears and quaint devices of the days when aviation was very, very young. (Airways, London, 1928, v. 5, p. 135–138.) **VDS**

Qualifications of pilots. Stunting. Legislation. Opinions as to number of passengers that possibly might be carried. Safety devices. Land markings. Conjectures as to value of aviation in war.

1705. **Gabel,** S. L., and H. C. Knerr. The story of aircraft tubing. illus. (Aviation, New York, 1930, v. 28, p. 632–635, 763–766, 847–850.) **VDS**

1706. **Gall,** J. F. L'aviation au service de l'armée. Appareil aviateur Ader. (La Nature, Paris, 1898, année 26, semestre 2, p. 54–55.) **OA**

Interest of French army officials in Clément Ader's airplane.

1707. **Gammeter,** H. C. The Gammeter orthopter. illus. (Scientific American, New York, 1907, v. 97, p. 258.) **VA**

Illustration on p. 253.

1708. **Garber,** Paul Edward. The Henson airliner, 1840 historical model. illus. (U. S. air services, Washington, 1924, v. 9, Feb., p. 44.) **VDS**

Reconstruction based on patent specifications.

Heavier-Than-Air Craft — Airplanes, cont'd

1709. Gasnier, René. L'aéroplane Wright décrit par un de ses passagers. illus. (L'Aérophile, Paris, 1908, année 16, p. 470–476.) **3–VDS**

1710. —— Les progrès de l'aviation en 1909. Après Reims et l'Exposition. (L'Aérophile, Paris, 1909, année 17, p. 487–489.) **3–VDS**

1711. —— Les progrès de l'aviation en 1910. (L'Aérophile, Paris, 1910, année 18, p. 509–512.) **3–VDS**

Translation in *Scientific American supplement,* New York, 1911, v. 71, p. 6–7, *VA*.

1712. —— Les progrès de l'aviation en 1911. (L'Aérophile, Paris, 1911, année 19, p. 576–579.) **3–VDS**

1713. Gaudart, Louis. Victor Tatin. illus. port. (La Revue aërienne, Paris, 1912, année 5, p. 64–67.) **3–VDS**

1714. Gauthier, Aimé. The ambulance aeroplane. illus. (World's health, Paris, 1925, v. 6, p. 282–296.) **SPA**

Good history of developments during the World war and after.

1715. Genesis of the first successful aeroplane. (Scientific American, New York, 1906, v. 95, p. 442.) **VA**

Story of the Wright brothers.

1716. German court favors Wright patents. (Aero and hydro, Chicago, 1913, v. 5, p. 419.) **3–VDS**

1717. A **Giant** among airplanes. illus. (Scientific American, New York, 1924, v. 130, p. 107.) **VA**

Pictures and brief description of Barling bomber, 20-ton craft, built by the U. S. War Department.

1718. Gillette, Leslie S. Ford's progress in the air. What America's leading automobile manufacturer has accomplished in aerial development. illus. (Automotive industries, Philadelphia, 1926, v. 55, p. 1002–1006.) **VDS**

1719. Gillingham, James. A tribute to John Stringfellow. First man to make an engine-driven aeroplane which flew in 1848. ⌊By James Gillingham.⌋ Chard, 1914. 2 l. f°. **† VDT**

1720. Gilmore, W. L. The evolution of the modern racing airplane. (Society of Automotive Engineers, Journal, New York, 1925, v. 17, p. 476–478, 489.) **TOL**

1721. Giuseppe M. Bellanca has been designing successful planes since 1908. illus. (Aviation, Highland, N. Y., 1927, v. 22, p. 1434–1435, 1448.) **VDS**

1722. Glenn H. Curtiss. (Aviation yearbook, London, 1931, p. 65–68.) **VDS**

Curtiss (1878–1930), American pioneer in aeronautics, was "more than any other, directly responsible for the birth of public flying in America, and also for the inauguration of aircraft and aero-engine manufacture in the United States."

1723. Glenn L. Martin expands; a habit he has acquired during 23 years of extraordinary aeronautical activity. illus. (U. S. air services, Washington, 1930, v. 15, June, p. 33–35.) **VDS**

Career of the inventor of the famous Martin bomber.

1724. Gould, Bruce. Sky larking; the romantic adventure of flying. With seven illustrations by Cosmo Clark. New York: Horace Liveright, 1929. 259 p. illus. 8°. **VDY**

History of the airplane, beginning with the Wright brothers. Lindbergh's and other famous flights. Book has an unusual literary flavor. Selected bibliography.

1725. Goupil's aeroplane. illus. (Scientific American, New York, 1885, v. 52, p. 22.) **VA**

1726. Gradenwitz, Alfred. An interesting German flying machine. illus. (Scientific American, New York, 1907, v. 96, p. 252–253.) **VA**

Illustrated account of J. Hofmann's airplane.

1727. —— The new Némethy flying-machine and the principles of its construction. illus. (Scientific American, New York, 1904, v. 91, p. 31.) **VA**

Invention of E. Némethy, of Hungary.

1727a. —— A novel flying machine. illus. (American inventor, New York, 1904, v. 12, p. 197.) *** V**

Airplane invention of E. Némethy, of Hungary.

1728. Grahame-White, Claude. The story of the aeroplane. Boston: Small, Maynard and Co. ⌊cop. 1911.⌋ xii, 390 p. illus. port. 8°. **VDY**

Appreciation of pioneers. Author's experiences. Development of engine. Paulhan's London-Manchester flight. Notable flying records. Accidents.

1729. Grahame-White, Claude, and Harry Harper. Aeroplane triumphs; the amazing progress that has been made in flying in the last three years. illus. (Badminton magazine, London, 1911, v. 32, p. 489–504.) **MVA**

1730. Grahame-White Company, Limited, London. Milestones in aerial history; British aeronautics and the Grahame-White Company... ⌊London, 1919.⌋ 162 p. illus. ports. f°. **† VDY**

Brief résumé of aeronautical history, especially of the work of British pioneers. Numerous illustrations.

1731. The **Great** Wright lawsuits in America. illus. (Flight, London, 1910, v. 2, p. 284–285.) **VDS**

Abstracts from decisions, with comments, in the Wright-Curtiss and Wright-Paulhan suits in the United States courts.

1732. Green, F. M. The technical history of the aeroplane. (Aeronautical journal, London, 1918, v. 22, p. 3–16.) **VDS**

Tells how the various elements of the airplane have been gradually developed. Also in *Automotive industries,* New York, 1918, v. 38, p. 265–268, *TOL; American Society of Naval Engineers, Journal,* Washington, 1918, v. 30, p. 212–221, *VXA*.

Heavier-Than-Air Craft — Airplanes, cont'd

1733. Greenhill, Sir George. The Maxim flying machine. illus. (Nature, London, 1895, v. 52, p. 321–325.) **OA**

Interesting illustrations.

1734. Grey, Charles Grey. The death of a pioneer. [By Charles G. Grey.] (Aeroplane, London, 1917, v. 13, p. 567–568.) **VDS**

Brief biography of Harold Barnwell, British aviator, who built his first airplane in 1908.

1735. —— The death of S. F. Cody. port. (Aeroplane, London, 1913, v. 5, p. 186–187.) **VDS**

Appreciation of British-American pioneer aviator and experimenter.

1736. —— The evolution of the airplane. illus. (Aviation, New York, 1918, v. 4, p. 450–453, 523–524.) **VDS**

Pictures and descriptions of planes which have initiated fresh steps in development since the Wright machines.

1737. —— The late Horace Short. (Aeroplane, London, 1917, v. 12, p. 900, 902.) **VDS**

Appreciation of early English airplane constructor who was the inventor of the folding-wing seaplane.

1738. —— The late José Weiss. port. (Aeroplane, London, 1919, v. 17, p. 2048, 2050.) **VDS**

Weiss, an Alsatian who spent most of his life in England, was an early experimenter in the building of airplanes.

1739. —— On Australian aviation. illus. (Aeroplane, London, 1927, v. 32, p. 377–379, 382–383.) **VDS**

1740. —— On De Havilland history. ports. (Aeroplane, London, 1930, v. 38, p. 563–564.) **VDS**

Traces development of well-known firm of English airplane builders. This article is followed by a detailed history of the Moth machine, made famous by them.

1741. —— On the fifteenth birthday of "The Aeroplane." illus. (Aeroplane, London, 1926, v. 30, p. 633–634, 636, 638, 642–643, 646–647, 650–651.) **VDS**

Brief history of aviation (mainly in England) through 1911, and an account of the founding of the periodical *The Aeroplane.*

1742. —— On a great pioneer. port. (Aeroplane, London, 1929, v. 36, p. 37–38, 40.) **VDS**

"One may safely say that [George] Holt Thomas and the Aircraft Manufacturing Company did more than any one man or any one firm towards winning and holding that supremacy of the air which from the end of 1917 onwards did so much towards winning the War."

1743. —— On a matter of fraud. illus. (Aeroplane, London, 1921, v. 21, p. 357–360.) **VDS**

Discussion of Wright-Langley controversy.

1744. —— On a pioneer of aviation. illus. ports. (Aeroplane, London, 1930, v. 39, p. 273–274, 276, 278–280.) **VDS**

Excellent account of the work of Glenn Hammond Curtiss, American pioneer in aviation.

1745. —— On the pioneers of British aviation. (Aeroplane, London, 1916, v. 11, p. 130, 132, 134, 395–396, 398.) **VDS**

A list of some fifty names, with brief notes.

1746. —— The original Wright biplane. (Aeroplane, London, 1928, v. 34, p. 162, 228, 268.) **VDS**

Comments on the Langley-Wright controversy.

1747. —— Who made the first flying machine? (Aeroplane, London, 1925, v. 28, p. 536, 538.) **VDS**

Discussion of the Wright-Langley controversy.

1748. Grover Loening's statement [concerning Langley-Wright controversy]. (U. S. air services, Washington, 1925, v. 10, June, p. 34.) **VDS**

1749. Guggenheim, Harry Frank. Wright brothers and research. illus. (Airway age, New York, 1928, v. 9, Dec., p. 23–25.) **VDS**

1750. Guidoni, Alessandro. What has been done and is being done in Italian aeronautics. (U. S. air service, New York, 1920, v. 3, p. 34, 44.) **VDS**

1751. Gustave Whitehead's new machine. illus. (Aeronautical world, Glenville, O., 1903, v. 1, p. 225–226.) **VDS**

1752. Gyroscopic pilot declared success. (Aviation, New York, 1929, v. 27, p. 990.) **VDS**

Completion of tests on the invention of Elmer Sperry, jr.

1753. H. Die Flugmaschine des Ingenieurs W. Kress in Wien. illus. (Die Umschau, Frankfurt a. M., 1901, Jahrg. 5, p. 465–468.) **OA**

1754. —— Der verunglückte Versuch mit dem Drachenflieger des Ingenieurs Kress. illus. (Die Umschau, Frankfurt a. M., 1901. Jahrg. 5, p. 923–925.) **OA**

Account of mishap to Wilhelm Kress's airplane.

1755. Hänisch, Viktor. Konstruktion zur Ermöglichung der "intermittierenden Kraftausnützung" bei Fortbewegung von Massen in elastischen Mitteln unter spezieller Berücksichtigung des dynamischen Fluges. diagrs. (Oesterreichischer Ingenieur- und Architekten-Verein, Zeitschrift, Wien, 1904, Jahrg. 56, p. 529–533.) **3–† VDA**

Describes mechanism for reproducing wing motion.

1756. Hagen, Freiherr vom. Ueber dynamische Flugapparate. illus. (Zeitschrift für Luftschiffahrt, Berlin, 1882, Jahrg. 1, p. 66–73, 97–116.) **3–VDS**

Excellent history of heavier-than-air dreams and experiments from mediaeval times. Includes a section on aeronautical toys.

1757. Hall, Donald A. Technical preparation of the Ryan New York-Paris airplane. ports. illus. (Aero digest, New York, 1927, v. 11, p. 36, 38, 106.) **VDS**

How Lindbergh's *Spirit of St. Louis* was designed and built.

Heavier-Than-Air Craft — Airplanes, cont'd

1758. Handley-Page, Frederick. Developments in aircraft design by the use of slotted wings. illus. (Flight, London, 1921, v. 13, p. 844–846, 860–861.) **VDS**

1759. —— Wright brothers' contribution. illus. (Airway age, New York, 1928, v. 9, Dec., p. 29–30.) **VDS**

For lists of references concerning Handley-Page see United States. — National Advisory Committee for Aeronautics, *Bibliography of aeronautics,* 1909–1931, Washington, 1921–35, volumes for 1909/16, p. 607, 989; 1917/19, p. 186–188, 340; 1922, p. 63; 1923, p. 104; 1924, p. 43, 83; 1925, p. 74, 133; 1926, p. 59; 1927, p. 73–74; 1928, p. 85, 151–152; 1929, p. 97; 1930, p. 99; 1931, p. 119, *VDT.* Some of the entries are under the heading Page, Frederick Handley.

1760. Hargrave, Lawrence. Flying-machine memoranda. illus. (Royal Society of New South Wales, Journal, Sydney, 1889, v. 23, p. 70–74.) *** EC**

Experiments with power-driven models.

1761. —— Flying machines. (Engineering, London, 1890, v. 50, p. 758–759.) **VDA**

A reply to criticisms of his machine. See entry no. 1769.

1762. —— Notes on flying machines. illus. (Royal Society of New South Wales, Journal, Sydney, 1885, v. 19, p. 41–46.) *** EC**

Experiments with model airplanes.

1763. —— Nos. 13 and 14 compressed-air flying machines. illus. (Royal Society of New South Wales, Journal, Sydney, 1891, v. 25, p. 52–58.) *** EC**

Experiments with models.

1764. —— On a compressed-air flying machine. (Royal Society of New South Wales, Journal, Sydney, 1890, v. 24, p. 52–58, 253–256.) *** EC**

Author's experiments with model machines.

1765. —— On a form of flying machine. illus. (Royal Society of New South Wales, Journal, Sydney, 1885, v. 19, p. 47–50.) *** EC**

Experiment with model airplane propelled by rubber bands.

1766. —— Recent work on flying machines. illus. (Royal Society of New South Wales, Journal, Sydney, 1887, v. 21, p. 19–27.) *** EC**

Illustrated account of experiments.

1767. —— Rigid stable aeroplanes. illus. (Royal Society of New South Wales, Journal, 1909, Sydney, 1910, v. 43, p. 381–387.) *** EC**

Illustrated account of author's experiments with curved wings.
A second copy in *VDY p.v.6, no.11.*

1768. —— The trochoided plane. illus. (Royal Society of New South Wales, Proceedings, 1884, Sydney, 1885, v. 18, p. 61–72.) *** EC**

An early study of wing forms. Also in *Scientific American supplement,* New York, 1885, v. 19, p. 7826–7828, *VA.*

1769. Hargrave's compressed air flying machine. illus. (Engineering, London, 1890, v. 50, p. 769.) **VDA**

Also in *Scientific American,* New York, 1891, v. 64, p. 167. For Hargrave's reply see entry no. 1761.

1770. Hargrave's flying machine. illus. (Engineering, London, 1893, v. 55, p. 295.) **VDA**

Brief illustrated account of Lawrence Hargrave's steam-driven machine. See also *Scientific American supplement,* New York, 1893, v. 36, p. 14618–14619, *VA.*

1771. Harriman, John Emery, jr. Mechanical flight. illus. (Association of Engineering Societies, Journal, Philadelphia, 1904, v. 33, p. 43–53.) **3–VDA**

History of heavier-than-air craft beginning with J. Bell Pettigrew's book of 1874. Picture of the author's proposed machine.

1772. Harper, Harry. The personality of a great pioneer. Aviation owes much to men like the Hon. C. S. Rolls, who gave his life in its cause. port. (Airways, London, 1928, v. 4, p. 217–218.) **VDS**

1773. —— Twenty-five years of flying; impressions, recollections, and descriptions... London: Hutchinson & Co. [1929.] xviii, 19–292 p. ports. illus. 8°. **VDY**

"I have known personally most of those pioneers, from the Wright brothers onwards... [My purpose] is to give you the personal, and I hope vivid, impressions made upon an eye-witness by moving about here and there amid all the aerial doings of the past twenty-five years."
Reviewed in *Air,* London, 1929, v. 2, p. 241, *VDS.*

1774. Hayward, Charles Brian. The Gillespie aeroplane. illus. (Scientific American, New York, 1905, v. 92, p. 501.) **VA**

Invention of G. Curtis Gillespie of New York.

1775. Hegener, Henri. Belangrijke data uit het leven der gebroeders Wright. (Het Vliegveld, Amsterdam, 1928, jaarg. 12, p. 369–371.) **3–† VDS**

Chronological account of the Wright brothers, 1889–1912.

1776. Hegener, Henri, and others. De baanbrekers der dynamische luchtvaart. [By Henri Hegener, F. Reyneker, and Hydravion, pseud.] illus. (Het Vliegveld, Amsterdam, 1924, jaarg. 8, p. 3–9, 116–119, 145–149, 205–210, 269–271; 1925, jaarg. 9, p. 18–20, 50–53, 74–75; 1927, jaarg. 11, p. 126–131; 1928, jaarg. 12, p. 22–24, 79–80, 101–105, 124–127, 162–164, 339–340; 1929, jaarg. 13, p. 19–20, 63–66, 152–154.) **3–† VDS**

Careers of the Wright brothers, S. P. Langley, Wilhelm Kress, Louis Mouillard, Clément Ader, J. J. Montgomery, George Cayley, John Stringfellow, William Henson, Francis Wenham, Ferdinand Ferber, Percy S. Pilcher, Jean-Marie Le Bris, Gabriel and Charles Voisin, Octave Chanute, Jacob Ellehammer, Victor Tatin, Carl Friedrich Meerwein, Glenn H. Curtiss, and Sir Hiram Maxim.

1777. Heintz, Henry. Heintz's new flying machine. illus. (Aeronautical world, Glenville, O., 1903, v. 1, p. 171–172.) **VDS**

To start the machine from the ground "I will have 36 parachutes in operation." Striking illustration.

Heavier-Than-Air Craft — Airplanes, cont'd

1778. **Heinze,** Edwin P. A. Dr. Rumpler's trans-ocean airliner. port. illus. (Aero digest, New York, 1930, v. 17, Dec., p. 41, 152.) **VDS**

Portrait and brief account of career of Edmund Rumpler, Austrian pioneer aviation engineer, born 1872.

1779. —— Experiments in rocket power. port. (Popular aviation, Chicago, 1930, v. 6, May, p. 394, 410.) **VDS**

Deals mainly with the work of Fritz von Opel. This number of *Popular aviation* is published under the title of *Aeronautics*.

1780. —— Opel-Sander rocket plane. illus. (Aero digest, New York, 1929, v. 15, Nov., p. 158.) **VDS**

Describes and illustrates machine built by Fritz von Opel, with coöperation of Herr Sander.

1781. **Henry,** Louis Isidore. Étude du mouvement d'un aviateur-aéroplane. Nantes: M. Schwob & Cie, 1902. 4–124 p., 1 plan. 8°. **VDY**

1782. **Henry-Coüannier,** André. L'aviation est une découverte entièrement française. illus. (L'Illustration, 1912, tome 139, p. 12–14.) **†* DM**

"The writer goes so far as to claim for [Louis] Mouillard the invention not only of warping but that of the combination of wing working with the vertical rudder." Abstract in *Scientific American supplement*, New York, 1912, v. 73, p. 107–108, *VA*.

1783. **Henry** Ford buys Stout Metal Airplane Company. (U. S. air services, Washington, 1925, v. 10, Sept., p. 23–24.) **VDS**

1784. **Henson's** aerial carriage. illus. (Mechanics' magazine, London, 1843, v. 38, p. 257–263, 315–316.) **3–VA**

1785. **Henson's** aeroplane or "steam carriage" of 1843. illus. (Scientific American, New York, 1910, v. 103, p. 314, 316–317, 330–331.) **VA**

1786. **Henson's** machine and contemporary critics. illus. (Aeronautics, London, 1910, v. 3, p. 133–136.) **VDS**

Two criticisms, 1843.

1787. **Hermuth,** Paul. Geschichte der Flugmodelle. illus. (In: Hermann Hoernes, Buch des Fluges, Wien, 1911, Bd. 2, p. 393–401.) **VDW**

Illustrated history of heavier-than-air models.

1788. **Herring,** A. M. Dynamic flight. (Aeronautical annual, Boston, 1896, no. 2, p. 89–101.) **3–VDS**

For list of writings by Herring and references to his work see Paul Brockett, *Bibliography of aeronautics*, Washington, 1910, p. 414–415, *VDT*.

1789. **Hervé,** Henri. L'aéroplane de M. Maxim. illus. (Revue de l'aéronautique, Paris, 1892, année 5, p. 29–126.) **3–† VDS**

1790. **Hildebrandt,** Alfred Louis Heinrich. Die Brüder Wright: eine Studie über die Entwicklung der Flugmaschine von Lilienthal bis Wright. Berlin: O. Elsner Verlagsgesellschaft, 1909. 63 p. illus. 8°. **VDY**

1791. **Hill,** G. T. R. The tailless aeroplane. illus. diagrs. (Royal Aeronautical Society, Journal, London, 1926, v. 30, p. 519–544.) **VDS**

With discussion.
The inventor outlines the development of his machine.

1792. **Hill,** Thomas A. Status of the Wrights' suit. (Aeronautics, New York, 1909, v. 5, Oct., p. 122–123, 164.) **VDS**

Discussion of infringement suit against the Aeronautic Society, with references to anticipating claims.

1793. **Hirschauer,** Louis. L'aviation de transport; l'évolution de la construction de 1907 à 1919 et la réalisation des avions de transport; l'utilisation économique des appareils. Paris: Dunod, 1920. 233 p. illus. 4°. **VDY**

Technical discussion of the principal airplanes of various countries, 1907–1919. See p. 29–31 for table showing weights of planes and hydroplanes.

1794. **Historical** — the development of the man-carrying, motor-driven aeroplane from the Lilienthal glider of 1895 to the Wright biplane of 1903. illus. (Scientific American, New York, 1910, v. 103, p. 312–314.) **VA**

1795. The **History** of airplane dope. (Automotive industries, New York, 1918, v. 38, p. 691.) **3–† TOL**

1796. **History** of Fokker airplanes. illus. (Aero digest, New York, 1926, v. 9, p. 378–379, 390.) **VDS**

1797. **Hodges,** Lawrence K. Development of the aircraft spruce industry. (Automotive industries, New York, 1918, v. 39, p. 1037–1040, 1100–1101.) **TOL**

History of developments during the World war.

1798. **Hofmann,** Josef. Der Maschinenflug. Seine bisherige Entwicklung und seine Aussichten. Frankfurt a. M.: F. B. Auffarth, 1911. 232 p. illus. 8°. **VDY**

See p. 1–160 for well-illustrated history of aviation as based upon the study of birdflight. Descriptions of the author's various machines.

1799. —— Ueber das Flugproblem. illus. (Verein zur Beförderung des Gewerbfleisses, Verhandlungen, Berlin, 1896, Jahrg. 75, Sitzungsberichte, p. 216–236.) **3–VA**

Discussion of the problem is followed by a detailed description of the author's airplane. For brief English description of the machine see *Scientific American*, New York, 1901, v. 84, p. 281, *VA*.

1800. **Holland,** John P. Aerial navigation. illus. (Cosmopolitan, New York, 1892, v. 14, p. 89–96.) *** DA**

A study of power requirements for airplanes.

1801. —— Mechanical flight. illus. (Cassier's magazine, New York, 1893, v. 3, p. 243–254.) **VDA**

A study of the engineering problems involved in the flying machine.

Heavier-Than-Air Craft — Airplanes, cont'd

1802. Hollands, Sidney H. Motor-aviation of to-day and of recent years. illus. (Flying, London, 1901, Dec., no. 1, p. 8–18; 1902, March, no. 2, p. 53–61; June, no. 3, p. 116–119; Sept., no. 4, p. 156–162; 1903, Jan., no. 5, p. 203–208; April, no. 6, p. 251–253.) **VDS**

Features the work of Hargrave, Lilienthal, Chanute, and Frederick E. Brearey. Good portrait of Hargrave.

1803. —— The work of Lawrence Hargrave. illus. (Aero, London, 1912, v. 6, p. 274–277, 298–300.) **3–† VDS**

Illustrations and descriptions of his airplane models and kites.

1804. Horatio Phillips, doyen of British aviation. illus. port. (Aero, London, 1910, v. 2, p. 37, 39–40.) **3–† VDS**

Phillips (born 1845) was the inventor of the "Phillips entry," or dipping front edge.

1805. Horsfall, Jessie E. The first flight. (American Academy of Political and Social Science, Annals, Philadelphia, 1927, v. 131, May, p. 14–19.) **SA**

Brief story of the Wright brothers.

1806. How E. M. Laird built and flew a 12 horsepower plane in 1913. (Aerial age weekly, New York, 1921, v. 14, p. 233.) **VDS**

1807. Howard-Flanders, Richard Leonard. Reminiscences of the early days of aviation at Brooklands. (Institution of Aeronautical Engineers, Proceedings, London, 1923, no. 9, p. 31–42.) **VDS**

Recalls pioneer days at English aerodrome, commencing with the flying attempts of A. V. Roe.

1808. Howard Gill's death is great loss to aviation. port. (Aero and hydro, Chicago, 1912, v. 4, p. 542.) **3–VDS**

Portrait and brief biography of prominent American aviator and airplane designer.

1809. Hucks, B. C. A further three years' flying experience. diagrs. (Aeronautical journal, London, 1917, v. 21, p. 390–402, 426–429.) **VDS**

Discussion, p. 426–429.
Progress in flying knowledge and equipment as observed by the author, 1914–1917. Also in *Aviation*, New York, 1917, v. 2, p. 568–571, *VDS*.

1810. —— Three years' experience of flying. (Aeronautical journal, London, 1914, v. 18, p. 220–231.) **VDS**

With discussion.
Deals with early failures, improved skill of pilots, bad weather, winds, air-pockets, machine efficiency, and looping.

1811. Hugo Junkers 70 Jahre alt. port. (Illustrierte Flug-Woche, Berlin, 1929, Jahrg. 11, p. 1–3.) **VDS**

Portrait of Junkers and an account of his aeronautical work.

1812. Humphreys, F. E. The Wright flyer and its possible uses in war. illus. (United States Engineer Bureau, Professional memoirs. Washington, 1910, v. 2, p. 99–107.) **VDA**

1813. Hunter, H. Chadwick. Doom of the spherical balloon. (Fly, Philadelphia, 1909, v. 1, Feb., p. 10.) **3–† VDS**

Effect of the Wright brothers' invention on aeronautics.

1814. Hureau de Villeneuve, Abel. Les idées de M. Edison en aviation d'après le New York Herald. illus. (L'Aéronaute, Paris, 1880, année 13, p. 259–269.) **3–VDS**

1815. —— Les oiseaux mécaniques de M. Edison. illus. (L'Aéronaute, Paris, 1880, année 13, p. 123–147.) **3–VDS**

Comments on Edison's flying machine, with illustrations.

1816. Hurry, H. C. On the mechanical conditions of flight. (Engineering, London, 1867, v. 3, p. 419.) **VDA**

Abstract of paper read before the Aeronautical Society of Great Britain.

1817. Hyatt, Thaddeus. $1,000 reward — a flying machine wanted! (Scientific American, New York, 1860, new series, v. 3, p. 88.) **VA**

Well-known inventor offers prize for a practical machine. For comments see above magazine, 1860, new series, v. 3, p. 116.

1818. Igor I. Sikorsky. port. (U. S. air services, Washington, 1929, v. 14, Jan., p. 52–54.) **VDS**

Biography and excellent portrait.

1819. Igor Sikorsky's remarkable story. Designer of the famous Sikorsky amphibion is one of aviation's pioneers who has made history during war and peace. port. illus. (Airway age, New York, 1928, v. 9, Dec., p. 82.) **VDS**

Sikorsky (b. 1889) introduced the first successful multi-motored airplane.

1820. Ijserman, J. M. Het natuurlijk vliegen verklaard en het kunstmatig vliegen voorgesteld, in verband met de luchtvaart. Gevolgd door eene uitvoerige beschrijving der schaaktafel, die de despeelde partijen behoudt en aantoont... 's Gravenhage: Gebr. J. & H. van Langenhuysen, 1869. 70 p. illus. 8°. **VDY p.v.21, no.3**

1821. Important advance in aircraft design for military purposes. illus. (U. S. air services, Washington, 1930, v. 15, Oct., p. 42.) **VDS**

Description of the flying wing Fokker airplane *XO-27*.

1822. An **Improved** air ship. illus. (Scientific American, New York, 1889, v. 60, p. 152.) **VA**

Man-operated airplane invented by John P. Holmes of Kansas.

1823. "An **Industry** is born to-day." (Flight, London, 1909, v. 1, p. 154.) **VDS**

Discusses the significance of the Olympia exhibition, London, 1909.

1824. Ingenious ornithopter design. illus. port. (Fly, Philadelphia, 1910, v. 2, March, p. 9.) **3–† VDS**

Invention of Professor H. La V. Twining, of Los Angeles, California.

Heavier-Than-Air Craft — Airplanes, cont'd

1825. **Inglis,** William. The problem of flight solved. A flying-machine which automatically maintains its equilibrium has been invented by Augustus M. Herring... illus. (Harper's weekly, New York, 1908, v. 52, Oct. 24, p. 7–8, 27.) * DA

Describes the machine and has an account of the inventor.

1826. **Irish,** W. E. The aerial ship "Britannia." illus. (English mechanic, London, 1900, v. 72, p. 16–17.) VA

Author describes his jet-driven airplane.

1827. —— Feather aeroplanes. illus. (American inventor, New York, 1905, v. 14, p. 89–90.) * V

Significance of the author's experiments with feathered surfaces.

1828. **Irving,** H. B., and A. S. Batson. Some early model experiments on devices for improving lateral control near the stall. illus. (Great Britain. — Aeronautical Research Committee, Technical report of 1929/1930, London, 1931, v. 2, p. 819–823. Reports and memoranda. no. 1251.) VDW

Details of experiments begun in 1920.

1829. **Irwin,** R. Randall. Lockheed's gospel — speed. illus. (Western flying, Los Angeles, 1934, v. 14, Feb., p. 18–20.) VDS

Brief history of the Lockheed machines.

1830. **Jackson,** George Gibbard. ...The big book of aeroplanes. London: H. Milford [1928]. 160 p. illus. 8°. VDY

Popular illustrated history, including also balloons and airships.

1831. **Jacobs,** A. M. Twenty-five years of flight. illus. (St. Nicholas, New York, 1928, v. 56, p. 87–93, 147–148, 150–151.) **Children's Room**

1832. **James** H. Jones dead; airplane inventor. Built early biplane with his son in 1909, combining features of French and U. S. machines. (New York Times, New York, June 19, 1935, p. 19.) * A

Jones was also a pioneer motor builder.

1833. **Johnson,** G. R. Flying in the land of the Incas. illus. port., map. (Aero digest, New York, 1929, v. 15, Sept., p. 56–60, 352, 354.) VDS

History of aviation in Peru, with photographs by the author when associated with the Peruvian air force.

1834. **Johnson,** John Berlin. Relationship of metallurgy to the development of aircraft. illus. (American Society for Steel Treating, Transactions, Cleveland, O., 1926, v. 9, p. 517–538.) VIA

"Describes the development of the metal structure of the airplane in this country [United States] and the various types made and used on the Continent, especially in Germany and France, during the World War."

1835. **Johnson,** Julius F. The first steam airplane. (Popular aviation, Chicago, 1931, v. 8, June, p. 43, 55.) VDS

Unsuccessful attempts to fly airplane built for Hiram Maxim, 1889, the two engines for which were built by Harry A. House and his son.

1835a. **Johnson,** Robert. The use of dry ice in airplane construction. illus. (Ice and refrigeration, Chicago, 1931, v. 81, p. 145–146.) VOA

Experience of the Boeing Airplane Company in arresting hardening of rivets, etc.

1836. **Jones,** Bennett Melvill. Research on the control of aeroplanes. diagrs. (Nature, London, 1928, v. 121, p. 755–762.) OA

Has brief account of early work. For criticism by W. Lockwood Marsh see entry no. 1928.

1837. **Jotti,** da Badia, polesine. L'aeroplano razzo. illus. (Aeronautica, Milano, 1928, anno 2, p. 718–720, 1034–1035.) 3–† VDS

History of suggestions for rocket propulsion of aircraft.

1838. The **"June** bug" aeroplane — a competitor for the Scientific American trophy. illus. (Scientific American, New York, 1908, v. 99, p. 13.) VA

This machine, produced by the Aerial Experiment Association, was provided with movable wing tips.

1839. **Junkers,** Hugo. Eigene Arbeiten auf dem Gebiete des Metall-Flugzeugbaues. illus. (Zeitschrift für Flugtechnik und Motorluftschiffahrt, München, 1920, Jahrg. 11, p. 35–37.) VDS

English translation in *Aviation*, Highland, N. Y., 1920, v. 9, p. 219, *VDS*.

1840. **Kaemmerer,** W. Neuere Flugmaschinen. (Verein deutscher Ingenieure, Zeitschrift, Berlin, 1908, Bd. 52, p. 956–960.) 3–† VDA

Pictures and descriptions of contemporary airplanes.

1841. **Karos,** Willibald. Ein Universal-Drachenflieger mit rotierenden Tragflächen. illus. (Zeitschrift für Luftschiffahrt, Berlin, 1898, Jahrg. 17, p. 55–69.) 3–VDS

Author's combination of glider and helicopter principles.

1842. **Kellett,** Wallace. Farman developments. Record of achievement by three brothers who have given their whole lives to building aircraft. illus. (Aero digest, New York, 1924, v. 5, p. 339–340, 361.) VDS

1843. **Kérambrun,** Paul. Une invention nouvelle, récemment appliquée à l'aviation, va rendre les avions silencieux. (L'Aérophile, Paris, 1926, année 34, p. 186.) 3–VDS

The "Romino" for abating the noise of airplanes.

1844. **Kitty Hawk** 25 years after the event. Secretary of War and other officials honor the Wright brothers. illus. (U. S. air services, Washington, 1929, v. 14, Jan., p. 28–31.) VDS

Heavier-Than-Air Craft — Airplanes, cont'd

1845. Klemin, Alexander. All-metal cabin plane. illus. (Scientific American, New York, 1927, v. 137, p. 539.) **VA**

Small Hamilton monoplane.

1846. —— Byrd and the "America." illus. (Scientific American, New York, 1927, v. 137, p. 256, 258.) **VA**

Description and diagram of airplane used by Byrd in transatlantic trip.

1847. —— Present status of American airplane and seaplane construction. illus. (Aviation and aeronautical engineering, New York, 1919, v. 6, p. 136–140.) **VDS**

1848. Klemmer, Harvey. American aircraft builders. illus. port. (Sportsman pilot, Albany, N. Y., 1930, v. 4, July, p. 20–21.) **VDS**

Biography and portrait of Alfred V. Verville.

1849. Klinckowstroem, Carl Ludwig Friedrich Otto, Graf von. Emanuel Swedenborg und das Flugproblem. illus. port. (Geschichtsblätter für Technik, Industrie und Gewerbe, Berlin, 1916, Bd. 3, p. 207–217, 303–312.) **V**

German translation of Swedenborg's description of his flying machine, with pictures of a modern reconstruction.

1850. Kline, Gordon M. Fire-resistant doped fabric for aircraft. illus. (Industrial and engineering chemistry, Easton, Pa., 1935, v. 27, p. 556–561.) **VOA**

Has a bibliography of 86 titles.

1851. Koch, Gustav. Entwurf einer dynamischen Flugmaschine. illus. (Zeitschrift für Luftschiffahrt, Berlin, 1883, Jahrg. 2, p. 212–219, 238–243.) **3–VDS**

Design for steam-driven flying machine.

1852. —— Das Flugprincip und die Schaufelrad-Flugmaschine. illus. (Zeitschrift für Luftschiffahrt, Berlin, 1897, Jahrg. 16, p. 252–260, 275–285, 303–314.) **3–VDS**

Description and illustrations of the author's steam-driven paddle-wheel flying machine.

1853. —— Der freie menschliche Flug als Vorbedingung dynamischer Luftschiffahrt. Eine Denkschrift für Jedermann. München: G. Franz, 1891. 26 p. illus. 8°. **VDY p.v.2, no.1**

1854. —— Die Lösung des Flugproblems und das Luftschiff der Zukunft. München: J. Schön, 1893. 32 p. illus. 8°. **VDY p.v.2, no.4**

Drawing and description of author's flying machine.

1855. —— Die Lösung des Flugproblems in physikalischer und maschinentechnischer Hinsicht. Mit einem Anhang: Nutzen und Folgen praktischer Luftschiffahrt. München: H. Lukaschik, 1896. 72 p. 8°. **VDY p.v.2, no.5**

Description and drawings of author's flying machine.

1856. —— Ein Weg zur Lösung des Flugproblems. illus. (Zeitschrift für Luftschiffahrt, Berlin, 1887, Jahrg. 6, p. 55–59, 81–86.) **3–VDS**

Details and plans of the author's steam-driven flying machine.

1857. Konvalinka, Joseph George. Aerial navigation. Long Island City, N. Y., 1897. 11 p. illus. 8°. (Memoirs of an inventor and scientist. no. 5.) **3–OAZ**

Author's suggestion for a man-power airplane.

1858. Kosch, Rudolph. Latest inventions in aerial navigation. illus. (Scientific American supplement, New York, 1896, v. 41, p. 17026–17027.) **VA**

Author discusses Maxim's airplane, and a flapping wing machine of his own.

1859. Kress, Wilhelm. Aeroplanes and flapping flying machines. illus. (International Conference on Aerial Navigation, Chicago, 1893, Proceedings, 1894, p. 253–264.) **VDS**

Account of the author's models of glider, airplane, and ornithopter.

1860. —— Bericht über den Stand der Versuche mit einem Drachenflieger. illus. (Illustrierte aëronautische Mittheilungen, Strassburg, 1901, Jahrg. 5, p. 29–32.) **VDS**

History of the author's airplane experiments.

1861. —— Experimental-Vortrag über Luftschiffahrt in Wiener flugtechnischen Verein. (Zeitschrift für Luftschiffahrt, Berlin, 1892, Jahrg. 11, p. 186–191.) **3–VDS**

Description of the author's exhibition of power-driven models of airplanes.

1862. —— History of the evolution of my first flying model of a kite flyer in 1887. illus. (Aeronautics, New York, 1915, v. 16, p. 83.) **VDS**

1863. —— Der persönliche Kunstflug. illus. (Zeitschrift für Luftschiffahrt, Berlin, 1893, Jahrg. 12, p. 105–113.) **3–VDS**

Describes airplane model.

1864. —— Ueber den Bau des Kress'chen Drachenfliegers. (Oesterreichischer Ingenieur- und Architekten-Verein, Zeitschrift, Wien, 1900, Jahrg. 52, p. 390–391.) **VDA**

For list of references by and about Kress see Paul Brockett, *Bibliography of aeronautics*, Washington, 1910, p. 476–478, *VDT*; also United States. — National Advisory Committee for Aeronautics, *Bibliography of aeronautics*, 1909–1931, Washington, 1921–35, volume for 1909/16, p. 732–733, *VDT*.

1865. The **Kress** aeroplane. illus. (Scientific American, New York, 1901, v. 84, p. 137.) **VA**

Illustrations and brief description of Wilhelm Kress's invention.

1866–1867. L., J. H. Samuel Franklin Cody. (Aeronautical journal, London, 1913, v. 17, p. 207–208.) **VDS (Royal)**

Cody (1861–1913), a Texan, "summed up in his unique personality almost the entire history and development of aviation in this country [England]."

1868. Lahm, F. S. Wilbur Wright's first flights in France. illus. (U. S. air services, Washington, 1924, v. 9, Aug., p. 29–31.) **VDS**

1869. La Landelle, Guillaume Joseph Gabriel de. Curiosités bibliographiques. (L'Aéronaute, Paris, 1885, année 18, p. 26–31.) **3–VDS**

Reference to two Peruvian works on flight — that of Santiago de Cardenas (1762) and of Pedro Ruiz (1878).

Heavier-Than-Air Craft — Airplanes, cont'd

1870. **Langley,** Marcus. Metal aircraft construction. London: Sir Isaac Pitman and Sons, Ltd., 1934. x, 338 p. illus. 2. ed. 8°. **VDY**

Chapter 1 deals mainly with history. Library has a copy of the 1. ed., 1932.

1871. **Langley,** Samuel Pierpont. Experiments with the Langley aerodrome. illus. (Smithsonian Institution, Annual report for 1903/04, Washington, 1905, p. 113–125.) *** EA**

See also *Scientific American supplement,* New York, 1905, v. 60, p. 24765–24768, *VA.*

1872. —— The "flying machine." illus. port. (McClure's magazine, New York, 1897, v. 9, p. 647–660.) *** DA**

Description of the machine and a brief account of its flight as confirmed by Alexander Graham Bell, who witnessed the events.

1873. —— Mechanical flight. (Cosmopolitan, New York, 1892, v. 13, p. 54–58.) *** DA**

Believes that past mistakes will certainly lead to a successful flying machine.

1874. —— The possibility of mechanical flight. (Century magazine, New York, 1891, v. 42, p. 783–785.) *** DA**

1875. —— Story of experiments in mechanical flight. (Aeronautical annual, Boston, 1897, no. 3, p. 11–25.) **3–VDS**

Also in Smithsonian Institution, *Annual report for 1896–97,* Washington, 1898, p. 169–181, ** EA.*

1876. The **Langley** aerodrome. illus. (Smithsonian Institution, Annual report for 1899/1900, Washington, 1901, p. 197–216.) *** EA**

Largely from an article in *McClure's magazine,* for which see entry no. 1872. See also *Scientific American supplement,* New York, 1902, v. 54, p. 22494–22495, 22510–22512, ** DA.*

1877. **Langley** aerodrome flies well in new tests. (Aero and hydro, Chicago, 1914, v. 9, p. 25–26.) **3–VDS**

Account of test of Langley's historic airplane in June, 1914.

1878. The **Langley** aeroplane construction and control mechanism. illus. (Scientific American, New York, 1914, v. 110, p. 499–500.) **VA**

1879. **Langley's** aviation work reviewed. illus. (Aero and hydro, Chicago, 1914, v. 8, p. 119–120.) **3–VDS**

1880. The **Largest** airplane tire. illus. (Aviation, Highland, N. Y., 1923, v. 15, p. 249.) **VDS**

Brief reference to tire for the Barling bomber. Size, 54 by 12 in.

1881. The **Late** Fredk. W. Brearey. port. (front.) (Flying, London, 1903, April, no. 6, p. 254–257.) **VDS**

Frederick William Breary (1816–1896) was an early English experimenter with power-driven airplane models.

1882. The **Late** Mr. F. W. Brearey. (Aeronautical journal, London, 1897, v. 1, Jan., p. 9–11.) **VDS**

1883. The **Late** Samuel Pierpont Langley. port. (Scientific American supplement, New York, 1906, v. 61, p. 25229–25230.) **VA**

1884. The **Latest** French aeroplanes and their records. illus. (Scientific American, New York, 1907, v. 97, p. 358–359.) **VA**

Descriptions and illustrations of Esnault-Pelterie and Farman airplanes, and Bréguet helicopter.

1885. **Laurain,** Ernest. Un précurseur sabolien de l'aviation. illus. (Commission historique et archéologique de la Mayenne, Bulletin, Laval, 1908, série 2, v. 24, p. 464–468.) **DPF**

Account of Besnier's man-driven machine and its flights (seventeenth century).

1886. **Lautier,** Ferdinand. Étude comparative des aéroplanes de M. Samuel Langley et de MM. V. Tatin et Ch. Richet. (L'Aéronaute, Paris, 1897, année 30, p. 195–200.) **3–VDS**

1887. —— La machine volante de M. Arthur Stenzel. illus. (L'Aéronaute, Paris, 1897, année 30, p. 27–32.) **3–VDS**

Machine with beating wings.

1888. **Lawrance,** Charles Lanier. Reasons for the Wrights' success. illus. (Airway age, New York, 1928, v. 9, Dec., p. 19–20.) **VDS**

"With limited capital they proceeded step by step, logically and with extreme thoroughness."

1889. —— Where the industry stands. charts. (Aviation, New York, 1931, v. 30, p. 678–681.) **VDS**

A review of the ups and downs of the industry, chiefly since 1925.

1890. **Le Conte,** Joseph. New lights on the problem of flying. illus. (Popular science monthly, New York, 1894, v. 44, p. 744–757.) *** DA**

A study of bird flight. In the light of Langley's experiments the author modifies former statements concerning the impossibility of heavier-than-air flight.

1891. —— The problem of a flying machine. (Popular science monthly, New York, 1888, v. 34, p. 69–76.) *** DA**

Does not believe a successful heavier-than-air machine possible.

1892. **Lecornu,** Léon François Alfred. Sur le volant élastique. (Académie des sciences, Comptes rendus, Paris, 1900, tome 131, p. 253–255.) *** EO**

Calculations for Raffard's flying machine patented in 1890.

1893. **Ledeboer,** John Henry. "Warping": Professor Montgomery's claims. (Aeronautics, London, 1910, v. 3, p. 71.) **VDS**

Claims to have anticipated the Wright patent.

1894. **Leffmann,** Henry. A tribute: Samuel Pierpont Langley, pioneer in practical aviation. illus. (Franklin Institute, Journal, Philadelphia, 1919, v. 187, p. 49–63.) **VA**

"His aerodynamic experiments, some published and some as yet unpublished were complete enough to form a basis for practical pioneer aviation... He developed and built the first man-carrying aeroplane capable of sustained free flight."

Heavier-Than-Air Craft — Airplanes, cont'd

1895. The **Legend** of Ader's flight. illus. (Flying, New York, 1914, v. 3, p. 234–235.) **VDS**

Disproves claim that Ader was the first successfully to operate a motor-driven airplane.

1896. A **Light** and simple motor. illus. (Scientific American, New York, 1893, v. 68, p. 85.) **VA**

Man-operated motor invented by Theodore A. Stark, of Ottawa, Illinois.

1897. Lilienthal, Otto. Allgemeine Gesichtpunkt bei Herstellung und Anwendung von Flugapparaten. (Zeitschrift für Luftschiffahrt und Physik der Atmosphäre, Berlin, 1894, Jahrg. 13, p. 143–155.) **3–VDS**

Library has a reprint in *VDY p.v.3, no.3.*

1898. —— Die Flugmaschinen des Mr. Hargrave. illus. (Zeitschrift für Luftschiffahrt, Berlin, 1893, Jahrg. 12, p. 114–118.) **3–VDS**

Describes Hargrave's motor-driven airplane models.

1899. Limits of size in aeroplanes. (Engineering, London, 1910, v. 90, p. 403–404.) **VDA**

1900. Lindsay, Orrin. Orrin Lindsay's plan of aerial navigation, with a narrative of his explorations in the higher regions of the atmosphere, and his wonderful voyage round the moon! Edited by J. L. Riddell, M.D. New Orleans, 1847. 33 p. 8°. **OAZ p.v.1**

Story of highly magnetised alloy of steel and mercury, which afforded a screen against gravity!

1901. Loening, Grover Cleveland. Our wings grow faster... Garden City, New York: Doubleday, Doran & Co., 1935. vi, 203 p. illus. 4°. **VDY**

Author's personal episodes, enlivened by numerous photographs, give interesting side-lights on the history of American aeronautics, 1908 to date.

1902. Looking back at 1927. A resumé of airplane activities during the greatest year in the history of American aeronautics. illus. (Aviation, Highland, N. Y., 1928, v. 24, p. 22–25, 50, 52, 54.) **VDS**

1903. Louvriè, Ch. de. The advantage of beating wings. (International Conference on Aerial Navigation, Chicago, 1893, Proceedings, New York, 1894, p. 267–273.) **VDS**

1904. Low, Archibald R. Modern developments of aeroplane theory. illus. (Institution of Junior Engineers, Journal, London, 1913, v. 23, p. 461–496.) **VDA**

1905. Ludlow, Israel. Clement Ader's early experiments. French scientist studied birds to learn first principles. (Fly, Philadelphia, 1911, v. 3, March, p. 7.) **3–† VDS**

1906. —— The latest Ludlow aeroplane. illus. (Popular mechanics, Chicago, 1908, v. 10, p. 511.) **VFA**

1907. —— A new aeroplane. illus. (Scientific American, New York, 1905, v. 93, p. 45–46.) **VA**

Illustrations and description of author's invention.

1908. —— A successful flying machine. illus. (American inventor, New York, 1905, v. 14, p. 171–172.) *** V**

Author's experiments (1905) with towed motorless planes.

1909. —— The Wright Company is a menace to the development of aviation. (Aircraft, New York, 1910, v. 1, p. 94–95.) **VDS**

Statement of the counsel of Louis Paulhan.

1910. M., E. Un pioniere dell' aviazione. Achille Bertelli (1855–1925) e le sue macchine di volo. (Rivista aeronautica, Roma, 1929, anno 5, Oct., p. 189–195.) **VDS**

Heavier-than-air experiment (1902–1904).

1911. Macerone, Francis. Mechanical flying. (Mechanics' magazine, London, 1843, v. 38, p. 291–292.) **3–VA**

Author claims to have discovered airplane principle in 1824.

1912. MacKaye, Percy Wallace. The air voyage up the Hudson. (U. S. air service, Washington, 1922, v. 7, Dec., p. 11.) **VDS**

Poem inspired by Wilbur Wright's flight from Governor's Island to Grant's tomb, Oct. 4, 1909.

1913. McMahon, John Robert. An extra spectator at the first flight. illus. (Aero digest, New York, 1930, v. 17, July, p. 73, 202, 204.) **VDS**

Robert L. Westcott, a member of the Kitty Hawk life-saving crew, witnessed ascensions of the Wright brothers, Dec. 17, 1903, through a spyglass.

1914. —— The real fathers of flight. illus. (Popular science monthly, New York, 1929, v. 114, Jan., p. 17–19, 142–144; Feb., p. 42–44, 152–158; March, p. 42–44, 142–147; April, p. 42–44, 156–160; May, p. 48–50, 136–139; June, p. 52–53, 154.) *** DA**

Copiously illustrated account of the Wright brothers. The above articles form the basis of the author's work *The Wright brothers, fathers of flight,* Boston: Little, Brown & Co., 1930, *AN.*

1914a. Magoun, Frederick Alexander, and Eric Hodgins. The airplane establishes itself. illus. (In their: A history of aircraft, New York, 1931, p. 339–379.) **VDT**

Alberto Santos-Dumont. Louis Blériot. Hubert Latham. Cross-channel flights. Léon Delagrange. Pioneer flights of Farman. The Wright brothers in Europe. Early progress in theory and design. The Reims aviation meet. The work of Glenn H. Curtiss. The age of expositions. The Paulhan-Graham-White cross-country race. Other exploits and progress in 1910. Flights. Pilots in 1911. The impasse before the war.

1914b. —— Early experiments with heavier than air flight. illus. (In their: A history of aircraft, New York, 1931, p. 287–310.) **VDT**

Cayley as the originator of the airplane. The Aerial Transit Company of Henson and Stringfellow. The designs of Pénaud. Model of Victor Tatin. Work of Moy and Shill. Studies of Goupil. Theories and

Heavier-Than-Air Craft — Airplanes, cont'd

designs of Horatio Phillips. Ader and the "avions." Experiments of Sir Hiram Maxim. Hargrave's propeller models.

1914c. —— Langley and the Wright brothers. illus. (In their: A history of aircraft, New York, 1931, p. 311–338.) **VDT**

1914d. —— Ornithopters. illus. (In their: History of aircraft, New York, 1931, p. 189–220.) **VDT**

Theories of Borelli, Robert Hooke, Count d'Esterno, and F. H. Wenham. Machines of W. Miller, Karl Friedrich Meerwein, Jacob Degen, Thomas Walker, F. D. Artingstall, Struvé, Vincent de Groof, and others. References to original sources.

1915. Magrini, Effren. La navigazione aerea; gli areoplani, Teoria — applicazioni. Torino: G. Lavagnolo, 1909. 187 p. illus. 8°. **VDY**

History of the airplane, with descriptions of a large number of machines.

1916. Maigre, Étienne. La stabilisation des aéroplanes. Comment on pourrait l'obtenir à l'aide réflexes naturels. (Revue des idées, Paris, 1909, année 6, p. 155–162.) *** DM**

Discusses method of the Wright brothers and of Glenn Curtiss.

1917. Maitland, Lester J. Knights of the air ...with a foreword by General C. P. Summerall... Illustrations from photographs. Garden City, N. Y.: Doubleday, Doran & Company, Inc., 1929. xvi, 338 p. ports. illus. 8°. **VDY**

History of heavier-than-air apparatus in the United States.

1918. Manly, Charles Matthews. A history of aviation. (Aerial age weekly, New York, 1917, v. 5, p. 784–785.) **VDS**

Deals with the work of Cayley, Henson, and Langley. Author was assistant to the latter.

1919. —— Some recollections of the Langley airplane. (American machinist, New York, 1917, v. 47, p. 973–974.) **VFA**

1920. Man's first little wings. The sprouts of today that point to the possibilities of tomorrow. illus. (Collier's, New York, 1908, v. 42, Oct. 17, p. 18–19.) *** DA**

Double-page illustration showing twenty-one contemporary flying machines.

1921. Marmonnier, L. Un stabilisateur automatique pour aéroplanes: le pendule à gyroscope. illus. (L'Aérophile, Paris, 1909, année 17, p. 147–151.) **3–VDS**

1922. Marquis, Raoul. Les aéroplanes; historique, calcul et construction, par H. de Graffigny [pseud.]. Paris: B. Tignol [1908]. 138 p. illus. 8°. (Bibliothèque des actualités industrielles. no. 133.) **VDY**

1923. Marsh, William Lockwood. Clement Ader claimed to have forestalled the Wright brothers by thirteen years. (Airways, London, 1926, v. 3, p. 66.) **VDS**

1924. —— The first British airman. illus. ports. (Flight, London, 1928, v. 20, p. 417–418.) **VDS**

A. V. Roe is stated to have made the first flight in England (June 8, 1908); also that he was "the father" of the tractor airplane.

1925. —— The first flight of all. The original Wright biplane is now on view at the Science Museum, South Kensington. illus. (Airways, London, 1928, v. 4, p. 291–293.)

Brief description. **VDS**

1926. —— In 1843, [William Samuel] Henson designed a flying machine almost identical with the modern aeroplane. (Airways, London, 1927, v. 3, p. 232.) **VDS**

1927. —— John Stringfellow was the first man to design a power-driven, heavier-than-air machine. (Airways, London, 1926, v. 3, p. 132.) **VDS**

1928. —— Research on the control of aeroplanes. (Nature, London, 1928, v. 121, p. 939–940.) **OA**

A reply to Professor B. Melvill Jones. See entry no. 1836.

1929. —— The science and industry of aviation owes much to Ferdinand Ferber, a pioneer of 1899. (Airways, London, 1926, v. 3, p. 31.) **VDS**

Ferber (1862–1909) made early experiments in gliding and was one of the first to use an engine. He was widely known as Monsieur de Rue.

1930. Marshall, Chapman Frederick Dendy. The evolution of the "heavier than air" flying machine. (Engineer, London, 1920, v. 129, p. 83.) **VA**

Brief comprehensive history from early times.

1931. Marshall, J. Mechanical flight. (Aeronautical journal, London, 1898, v. 2, p. 57–60.) **VDS**

Author's experiments with airplanes equipped with flapping wings.

1932. Martin, Glenn L. The development of aircraft manufacture. illus. (Royal Aeronautical Society, Journal, London, 1931, v. 35, p. 894–927.) **VDS**

Traces development of the author's factories, the latest of which is situated near Baltimore, Md.

1933. Martin, James V. Clear statement of the bearing of the Martin aerodynamic stabilizer upon the Wright patent infringement controversy. (Flying, New York, 1917, v. 6, p. 19.) **VDS**

1934. Martin, Lajos. A repülő gépről. illus. (A Magyar mérnök- és építész-egylet közlönye, Budapest, 1894, kötet 28, p. 9–15.) **3–† VDA**

Experiments of the author with artificial flapping wings.

1935. The **Martinsyde** machines. illus. (Flight, London, 1920, v. 12, p. 637–641.) **VDS**

History of airplanes brought out by pioneer British firm of Martin and Handasyde.

Heavier-Than-Air Craft — Airplanes, cont'd

1936. Masfrand, Albert de. Expériences d'aviation de MM. Etrich et Wels. ports. illus. (L'Aérophile, Paris, 1908, année 16, p. 77–82.) **3–VDS**

1937. —— Les frères Voisin. ports. (L'Aérophile, Paris, 1908, année 16, p. 37.) **3–VDS**

Brief biographies of Gabriel and Charles Voisin, pioneer French airplane constructors. Portraits.

1938. —— Henri Farman. port. (L'Aérophile, Paris, 1907, année 15, p. 301.) **3–VDS**

Portrait and brief biography of pioneer French aviator and airplane designer.

1939. —— Le monoplan "Etrich 1910" et les records d'aviation autrichiens. illus. (L'Aérophile, Paris, 1910, année 18, p. 271–273.) **VDS**

1940. —— Paul Tissandier. port. (L'Aérophile, Paris, 1921, année 29, p. 135.) **3–VDS**

Portrait and brief biography of French aviator, who was the first pupil of Wilbur Wright. Like his father, Gaston Tissandier, he did much for the development of aeronautics in France.

1941. Mather, T. W. Flying machines. illus. (Popular science monthly, New York, 1885, v. 28, p. 1–18.) *** DA**

Ancient myths and early ideas. Bird flight suggests human possibilities. Machines of Henson (1842), La Landelle (1863), Stringfellow (1868), and Moy (1874).

1942. Matthias, Joachim, editor. Unsere Flieger erzählen... Berlin-Charlottenburg: C. J. E. Volckmann [1927?]. 175 p. illus. ports. 8°. **VDY**

Account of the development of German aviation from the time of Lilienthal, with reminiscences of the flyers Gottlob Espenlaub, Ernst Udet, Willy Polte, and Bruno Rodschinka.

1943. Maxim, Sir Hiram Stevens. Aerial navigation by bodies heavier than the air. (Aeronautical journal, London, 1902, v. 6, p. 2–7.) **VDS**

Author describes his experiments.

1944. —— Aerial navigation; the power required. illus. (Century magazine, New York, 1891, v. 42, p. 829–836.) *** DA**

Illustrated account of author's testing apparatus.

1945. —— The aeroplane. (Cosmopolitan, New York, 1892, v. 13, p. 202–208.) *** DA**

A statement of problems to be solved.

1946. —— Artificial and natural flight. London: Whittaker & Co., 1908. xv, 166 p. illus. 8°. **VDY**

1947. —— The development of aerial navigation. (North American review, New York, 1894, v. 159, p. 344–352.) *** DA**

Account of development of his engine and airplane at Baldwin Park.

1948. —— Experiments in aeronautics. (Society of Arts, Journal, London, 1894/95, v. 43, p. 21–38.) **VA**

With discussion.
Author's experiments with steam-driven airplane. Also in *Scientific American supplement,* New York, 1894, v. 38, p. 15838–15839, *VA.*

1949. —— The Maxim flying machine. illus. (Scientific American supplement, New York, 1891, v. 32, p. 13032–13033.) **VA**

As reported by him in the *New York Sun.* See also *Engineer,* London, 1891, v. 72, p. 46, *VA.*

1950. —— Mechanical flight. diagrs. (Engineer, London, 1893, v. 75, p. 226.) **VA**

Author's experiments, with some discussion of the work of Horatio Phillips.

1951. —— My early experiments. (Flight, London, 1910, v. 2, p. 161.) **VDS**

1952. —— Natural and artificial flight. illus. (Aeronautical annual, Boston, 1896, no. 2, p. 26–55.) **3–VDS**

1953. —— A new flying machine. Maxim's experiments in aerial navigation. illus. (Century magazine, New York, 1895, v. 49 [new series, v. 27], p. 444–456.) *** DA**

1954. —— The recent experiments conducted by the Wright brothers. (Aeronautical journal, London, 1906, v. 10, p. 37–39.) **VDS (Royal)**

For numerous references concerning Maxim's airplane experiments see Paul Brockett, *Bibliography of aeronautics,* Washington, 1910, p. 564–567, *VDT;* also United States. — National Advisory Committee for Aeronautics, *Bibliography of aeronautics,* 1909–1931, Washington, 1921–35, volumes for 1909/16, p. 853–854; 1917/19, p. 288, *VDT.*

1955. Maxim's flying machine. illus. (Engineer, London, 1894, v. 78, p. 130–132.) **VA**

Illustrated account shows details.

1956. Means, James. The problem of man-flight. illus. (Aeronautical annual, Boston, 1895, no. 1, p. 141–159.) **3–VDS**

Reprint of pamphlet published in 1894.
For references to Means' work and writings see United States. — Advisory Committee for Aeronautics, *Bibliography of aeronautics,* 1909–1930, Washington, 1921–1932, volume for 1909/16, p. 856–857, *VDT.*

1957. Mechanic, pseud. Aerial navigation. (Mechanics' magazine, London, 1843, v. 39, p. 46–47.) **3–VA**

Criticizes Henson's machine.

1958. Mechanical flight. illus. (Engineering, London, 1893, v. 55, p. 288, 653.) **VDA**

Description of Horatio Phillips' flying machine. Illustrations are on page 648.

1959. Mechanical flight [during 1908]. (Engineer, London, 1909, v. 107, p. 15–16.) **VA**

1960. Meerweins Gedanken über die Fliegkunst. illus. (Deutsche Zeitschrift für Luftschiffahrt, Berlin, 1909, Jahrg. 13, p. 1060–1064.) **VDS**

Quotations from Meerwein's paper of 1784.

1961. Melville, George Wallace. The engineer and the problem of aerial navigation. (North American review, New York, 1901, v. 173, p. 820–831.) *** DA**

Believes heavier-than-air flight practically impossible. For criticism see article by Carl E. Myers, entry no. 1988.

Heavier-Than-Air Craft — Airplanes, cont'd

1962. **Memento,** pseud. Clément Ader. port. illus. (L'Aérophile, Paris, 1927, année 35, p. 290–294.) 3–VDS

States that Ader made a flight in October, 1897, in an airplane equipped with two steam motors.

1963. **Merrill,** A. A. Some simple experiments with aerocurves. illus. (Aeronautical journal, London, 1899, v. 3, p. 65–67.) VDS

1964. **Merveilleuses** performances de Wilbur Wright. illus. (L'Aérophile, Paris, 1909, année 17, p. 12–13.) 3–VDS

How Wright won the Michelin cup in France in 1908. For another account see *L'Aérophile*, 1909, année 17, p. 38–39.

1965. **Mettler,** Hans. Automatisch stabiler Aeroplan konstruiert nach den Resultaten vergleichender Studien über Auftrieb, Fahrwiderstand, Geschwindigkeit und Energiebedarf der Luftfahrzeuge. Zürich: Gebr. Leeman & Co. [1909.] 61 p. 8°. VDY

1966. **Meyer,** E. Die Geschichte des freitragenden verspannungslosen Flugzeuges. illus. (Der Motorwagen, Berlin, 1923, Jahrg. 26, p. 247–250, 266–270, 279–284, 297–303.) TOL

History of unbraced cantilever wings, with a bibliography of 85 entries.

1967. **Meyer,** Willy. Von Wright bis Junkers, das erste Vierteljahrhundert Menschenflug, 1903–1928... Berlin: Deutsche Verlagsgesellschaft für Politik und Geschichte, 1928. xi, 51 p. illus. 8°. VDY

Deals principally with development of heavier-than-air craft in Germany from the time of Lilienthal.

1968. **Miles,** Russell H. How the New York to Paris plane was built. illus. ports. (Aviation, Highland, N. Y., 1927, v. 22, p. 1352–1353.) VDS

How Lindbergh's *Spirit of St. Louis* was ordered and designed.

1969. **Milestones.** illus. (Flight, London, 1915, v. 7, p. 22–24.) VDS

History of the celebrated Avro airplane, developed by Sir A. V. Roe.

1970. **Miller,** F. I regolatori di equilibrio per gli apparecchi volanti. (Rivista tecnica, Roma, 1906, anno 6, p. 59–75.) 3–VA

Author's suggestions for regulating stability of heavier-than-air craft.

1971. [**Miller's** flying machine.] (Universal magazine, London, 1784, v. 74, p. 23.) * DA

Involves flapping wings made of steel, stated to have been on exhibition in Hart Street, Covent Garden, London.

1972. **Mills,** Manley. Lightplaning in past years. The demand for light planes goes back over twenty years. Some of the tiny planes now long forgotten were excellent performing ships. illus. (Popular aviation, Chicago, 1931, v. 9, Oct., p. 23–24, 46.) VDS

Histories of some very small practical airplanes.

1973. **Mingos,** Howard. The man who sold the airplane to Ford. William B. Stout was a dreamer who stuck to his dreams. illus. ports. (World's work, Garden City, N. Y., 1925, v. 50, p. 423–428.) * DA

1974. —— Thus man learned to fly. illus. (Saturday evening post, Philadelphia, 1928, v. 201, July 7, p. 10–11, 105–106, 109–110; July 14, p. 18–19, 121–122, 124–126.) * DA

Story begins with the Wright brothers.

1975. **Mitchell,** John W. Langley's important work. (Aeronautics, New York, 1909, v. 5, p. 70.) VDS

1976. —— The real beginning of aviation. illus. (Aero, St. Louis, 1911, v. 3, p. 7–9, 39, 59, 63.) 3–VDS

Appreciation of the experiments of Samuel Pierpont Langley.

1977. **Mock,** Richard M. Progress in aeronautics [for 1929]: airplane design and construction. (Mechanical engineering, New York, 1930, v. 52, p. 4–5.) VFA

1978. The **Modern** Icarus; the newest of flying machines. illus. (English illustrated magazine, London, 1898, v. 20, p. 39–42.) * DA

Proposed airplane of George L. O. Davidson.

1979. The **Monospar.** illus. port. (Aeroplane, London, 1932, v. 42, p. 691–692, 694, 696, 698, 700, 704, 706, 708, 710.) VDS

System developed in 1928 by H. J. Stieger.

1980. **M.** Louis Blériot and his aeroplanes. Full description of the latest successful monoplanes of this celebrated French engineer. illus. (Scientific American supplement, New York, 1909, v. 68, p. 136–137.) VA

1981. **Montgomery,** John J. The origin of "warping." Professor Montgomery's experiments. (Aeronautics, London, 1910, v. 3, p. 63–64.) VDS

1982. The **Montgomery** aeroplane. illus. (Scientific American, New York, 1905, v. 92, p. 404, 406.) VA

1983. **Mort** tragique du capitaine Ferber. port. illus. (L'Aérophile, Paris, 1909, année 17, p. 440–441.) 3–VDS

Biography and portrait of pioneer French aviator and experimenter, Ferdinand Ferber.

1984. **Mortane,** Jacques. Adolphe Doutre — the inventor of the Doutre double automatic stabilizer. port. (Flying, New York, 1912, Oct., p. 6.) VDS

Portrait and brief biography of Doutre (born 1873).

1985. —— ...La vie des hommes illustres de l'aviation (depuis les origines jusqu'au 3 août 1914)... Paris: Roche d'Estrez, 1926. 208 p. 12°. VDY

Career of the pioneers. Claims that Clément Ader (b. 1840) was the "father of aviation." Flight records down to the time of the World War.

Heavier-Than-Air Craft — Airplanes, cont'd

1986. **Mouillard,** Louis Pierre. A programme for safe experimenting. (International Conference on Aerial Navigation, Chicago, 1893, Proceedings, New York, 1894, p. 310–315.) **VDS**

Advice on glider construction.

1987. **Munk,** Max M. Development of the "M" wing sections. diagrs. (Aviation, New York, 1930, v. 28, p. 12–16.) **VDS**

Author's model tests of mathematically derived sections.

1988. **Myers,** Carl E. Aerial navigation problems. (Scientific American supplement, New York, 1902, v. 53, p. 21859–21860.) **VA**

A reply to Admiral Melville. See entry no. 1961.

1989. **Myers,** Denys P. Collisions in the air. (Aircraft, New York, 1912, v. 3, p. 240, 244.) **VDS**

Accounts of accidents, 1910–1912.

1990. —— The patent situation. (Aircraft, New York, 1912, v. 3, p. 40, 48.) **VDS**

Status of various suits of Wright brothers for infringement.

1991. **Mytton,** R. G. V. Flying machine experiments. (Aeronautical world, Glenville, O., 1903, v. 1, p. 129–131.) **VDS**

Author's gasoline-driven airplane.

1992. **N.,** M. L'avion. illus. (L'Illustration, Paris, 1898, tome 112, p. 54–55.) *** DM**

Has large pictures of Ader's machine. Translation in *Scientific American,* New York, 1898, v. 79, p. 137–138, *VA.*

1993. **Nansouty,** Max Charles Emmanuel Champion de. Les machines volantes. illus. (Le Génie civil, Paris, 1894, tome 24, p. 279–280.) **VA**

Drawings and descriptions of flying machines of Wellner and of Quentin.

1994. **National** Aeronautic Association would re-label Langley machine. (U. S. air services, Washington, 1925, v. 10, June, p. 34.) **VDS**

1995. A **Natural** flying machine. illus. (Scientific American, New York, 1865, new series, v. 13, p. 195.) **VA**

Heavier-than-air craft kept aloft and propelled by eagles or swans.

1996. **Negro,** Piero. ..."Nidi d'aquila." Torino: C. Velentino & C., 1927. 179 p. illus. 4°. **VDY**

Illustrated history of Italian aviation.

1997. **Neilson,** Robert Morrison. Aeroplane patents. London: Constable & Company, Ltd., 1910. x, 91 p. illus. 8°. **VDY**

Lists British patents from 1882; American patents from 1896.

1998. **Némethy,** Emil. Die entgültige Lösung des Flugproblems. Leipzig: J. J. Weber, 1903–11. 2 v. illus. 4°. **VDY**

Description of author's airplane.

1999. **Nevill,** John T. Ford Motor Company and American aeronautic development. illus. (Aviation, New York, 1929, v. 26, p. 1871–1873, 2073–2076; v. 27, p. 41–44, 229–239, 306–308, 339–341, 568–570, 753–755.) **VDS**

2000. —— The story of Wichita. illus. (Aviation, New York, 1930, v. 29, p. 166–170, 291–295, 353–357.) **VDS**

A record of the city's development in airplane manufacture. It "has produced approximately one-fourth the commercial airplanes manufactured within the United States."

2001. **Neville,** Leslie E. 1929 design trends. A review of airplane and engine tendencies. charts. (Aviation, New York, 1930, v. 28, p. 327–330.) **VDS**

2002. —— The original Wright biplane. illus. (Aviation, New York, 1928, v. 25, p. 1724–1726.) **VDS**

2003. —— A review of 1928 design development. illus. (Aviation, New York, 1929, v. 26, p. 26–28, 58, 60, 62, 64.) **VDS**

2004. A **New** aerial machine. illus. (Scientific American, New York, 1885, v. 52, p. 291.) **VA**

Flying machine worked by human power and compressed air, the invention of W. O. Ayre of New Haven, Conn.

2005. **New** experiments on mechanical flight. illus. (Scientific American, New York, 1877, new series, v. 36, p. 179.) **VA**

Model experiments of Victor Tatin.

2006. A **New** flying machine. illus. (Scientific American, New York, 1885, v. 52, p. 374.) **VA**

Moveable wing apparatus of H. P. Booth, of Wisconsin. Also in *Knowledge,* London, 1885, v. 8, p. 50, *OA.*

2007. A **New** flying machine. illus. (Scientific American, New York, 1901, v. 84, p. 72.) **VA**

Power-driven airplane model of G. L. O. Davidson, of Scotland.

2008. **New** French aeroplanes. illus. (Scientific American, New York, 1907, v. 96, p. 269.) **VA**

Illustrated accounts of the Santos-Dumont and Delagrange airplanes.

2009. **Newcomb,** Simon. Is the airship coming? (McClure's magazine, New York, 1901, v. 17, p. 432–435.) *** DA**

2010. —— The outlook for the flying machine. (In his: Side-lights on astronomy, New York, 1906, p. 330–345.) **OMX**

In the above articles, Professor Newcomb is not optimistic.

2011. **Newell,** Joseph Shipley. Structural design progress in 1930. Curtailed production has made improved structural designs necessary. illus. (Airway age, New York, 1931, v. 12, p. 48–49.) **VDS**

Heavier-Than-Air Craft — Airplanes, cont'd

2012. **Newton,** Byron Rufus. One of the greatest achievements in the story of the human race. ports. (U. S. air services, Washington, 1931, v. 16, Dec., p. 19–24.) **VDS**

Author was aeronautical editor of the old *New York Herald* and was intimately acquainted with the Wright brothers.

2013. —— Recollections of the days when wings emerged. ports. (U. S. air services, Washington, 1932, v. 17, Jan., p. 26–30.) **VDS**

Recollections of the Wright brothers and of Glenn H. Curtiss.

2014. —— The scientists knew that human flight was impossible. illus. (U. S. air services, Washington, 1932, v. 17, Feb., p. 22–25.) **VDS**

Prophesies and pronouncements made concerning the future of the airplane.

2015. —— They said it was neither fact nor fiction and promptly turned it down. port. (U. S. air services, Washington, 1932, v. 17, July, p. 20–24.) **VDS**

Recollections of the Wright brothers at Kitty Hawk.

2016. —— Watching the Wright brothers fly. (Aeronautics, New York, 1908, v. 2, June, p. 6–10.) **VDS**

Author describes the flights of May, 1908, which he witnessed surreptitiously.

2017. **Nicolleau,** A. L'aéroplane Kapferer. illus. (L'Aérophile, Paris, 1908, année 16, p. 122–123.) **3–VDS**

2018. **Nimfuehr,** Raimund. New bird-like flying machine. illus. (Aeronautical world, Glenville, O., 1903, v. 1, p. 169–171.) **VDS**

Author's airplane is provided with flapping wings and a wing propeller.

2019. **1927** commercial production. Total estimated at 1653 with Advance Aircraft Co. heading list of 103 manufacturers with 454 planes. illus. chart. (Aviation, Highland, N. Y., 1928, v. 24, p. 35–38, 40.) **VDS**

2020. **1930** airplane design. charts. (Aviation, New York, 1931, v. 30, p. 22–25.) **VDS**

2021. **Noble,** V. d'Oyley. The aerial steam carriage 1842. illus. port. (Popular flying, London, 1934, v. 3, p. 115–117, 154.) **VDS**

Portrait and account of the author's grandfather, who was an associate of Henson and Stringfellow, and later a designer of an airship.

2022. **North,** J. D. The technical development of the aeroplane. (International Congress of Aerial Navigation, London, 1923, Report, London, 1923, p. 177–204.) **VDS**

With discussion.
Historical introduction. Partly reprinted in *Aviation*, Highland, N. Y., 1923, v. 15, p. 676–678, *VDS*.

2023. **Notes** on flying and flying machines. (Cornhill magazine, London, 1871, v. 24, p. 438–451.) *** DA**

Experiments of M. de Lucy, Besnier, and Charles Spencer. Discussion of problems to be solved. Also in *Scientific American*, New York, 1871, new series, v. 25, p. 384–385, 400–401, *VA*.

2024. **Ochoa,** V. L. Are the Wrights pirates? illus. (Aircraft, New York, 1910, v. 1, p. 55–57.) **VDS**

2025. **Ogilvie,** A. Some aspects of aeronautical research. (Aeronautical journal, London, 1922, v. 26, p. 381–389.) **VDS**

An appreciation of the experimental work of the Wright brothers.

2026. **O'Malley,** Frank Ward. When the Wrights joined the flying shows. illus. (U. S. air service, Washington. 1923, v. 8, Dec., p. 19–25.) **VDS**

Spicy accounts of pioneer Wright flyers.

2027. **On** aviation in 1925. illus. (Aeroplane, London, 1926, v. 30, p. 1–2, 4, 6, 53–54, 56, 105–106.) **VDS**

Developments in the British airplane industry during the year.

2028. **On** learning from the experience of others. (Aeroplane, London, 1926, v. 31, p. 517–518, 520, 522, 524, 526, 528, 530.) **VDS**

The development of airplane construction in England compared with that in America.

2029. The **Opel** rocket plane. A brief description based on authentic material received from Aviation's German correspondent. illus. (Aviation, New York, 1929, v. 27, p. 931.) **VDS**

2030. **Orcy,** Ladislas d'. How the war has modified the aeroplane. The passing of the military monoplane, and the development of the battle-plane. illus. (Scientific American, New York, 1915, v. 113, p. 196–197, 221–222.) **VA**

2031. —— Pioneers of aviation. illus. (Aircraft, New York, 1911, v. 2, p. 267; 1912, v. 3, p. 150, 161, 179, 317–318; 1913, v. 4, p. 199, 221–222; 1914, v. 5, p. 363–364.) **VDS**

Describes the work of Sir George Cayley, William Samuel Henson, John Stringfellow, Félix and Louis du Temple, Jean-Marie Le Bris, Count d'Esterno, Alphonse Pénaud, Thomas Moy, and Victor Tatin.

2032. The **Original** Avro triplane for South Kensington Museum. illus. (Flight, London, 1925, v. 17, p. 441.) **VDS**

Picture and history of plane constructed by British pioneer, Sir A. V. Roe.

2033. **Original** Langley machine flies. illus. (Aeronautics, New York, 1914, v. 14, p. 148–151.) **VDS**

Includes a biography of Langley.

2034. **Orville** Wright's new stabilizer almost perfected. (Aviation and aeronautical engineering, New York, 1916, v. 1, p. 331.) **VDS**

2035. **Orville** Wright's record flights at Fort Myer. illus. (Scientific American, New York, 1909, v. 101, p. 88, 99.) **VA**

2036. **Osborn,** Earl D. The industry's progress during 1928. illus. (Aviation, New York, 1929, v. 26, p. 24–25, 64.) **VDS**

Heavier-Than-Air Craft — Airplanes, cont'd

2037. An **Outline** of aviation. (Literary digest, New York, 1927, v. 93, June 25, p. 9–10.) *DA

Chronological from 1905 to Lindbergh, 1926.

2038. Pagé, Victor Wilfred. Airplane control; control methods of early airplanes. illus. (Everyday engineering magazine, 1918, v. 5, p. 172–174.) VA

2039. Parish, William Francis. Transportation. The flying machine. [By W. F. Parish.] illus. (Oil power, New York, 1927–28, v. 2, p. 99–110.) VFA

Brief account from time of Leonardo da Vinci, with numerous illustrations.

2040. Parrott, R. J. The history and evolution of the Avro training machine. illus. (Institution of Aeronautical Engineers, Proceedings, London, 1925, no. 14, p. 7–42.) VDS

History of early planes built by Sir A. V. Roe.

2041. Parsons, Bernard. The genesis of the aeroplane. illus. (Fry's magazine, London, 1910, v. 13, p. 83–90.) MVA

"Popular" article.

2042. Pawlowski, Felix Wladyslaw. Evolution of airplane wing-trussing. illus. (Society of Automotive Engineers, Transactions, New York, 1917, v. 12, part 1, p. 207–219.) TOL

With discussion.

"The fact that Henson at the early date of 1842 adopted the Fink truss for his wing construction must be considered and recognized as extraordinary."

Also in *Aerial age weekly,* New York, 1917, v. 4, p. 594–595, 619, *VDS.*

2043. Pay Wright brothers tribute in pilgrimage to Kitty Hawk. Ceremony follows International Conference. Harmon trophy to Lindbergh. illus. (Aviation, New York, 1928, v. 25, p. 2042–2043.) VDS

2044. Peyrey, François. L'idée aérienne — aviation. Avec une préface de Santos-Dumont. [v. 2.] Paris: H. Dunod & E. Pinat, 1909. 667 p. illus. 8°. VDY

Includes a chronology of the propeller.

Volume 2 is entitled *Les oiseaux artificiels.* It comprises an extensive, well-illustrated history of heavier-than-air machines.

2045. Pénaud, Alphonse, and PAUL GAUCHOT. Mémoire descriptif annexé au brevet d'invention de 15 ans pris le 18 février 1876. illus. (L'Aéronaute, Paris, 1877, année 10, p. 274–289.) VDS

2046. Percheron, Maurice. Les aéroplanes de la guerre: les aéroplanes de 1916 à 1920. Paris: E. Chiron, 1920. 183 p. illus. 4°. VDY

2047. Perkins, Frank C. A Swiss combination. illus. (Fly, Philadelphia, 1909, v. 1, March, p. 12.) 3–VDS

Illustrated account of the Dufaux airplane.

2048. Petit, Robert. Comment on construit un aéroplane. Paris: Librairie aéronautique [1909]. 136 p. illus. 8°. VDY

Includes a brief history of motors, and explains Drzewiecki's and Colliex's calculations for propellers. Library has also an English translation, London, 1910, *VDY.*

2049. Phélip, Gaston. Les aéroplanes de Chalais-Meudon. Le cerf-volant à moteur du capitaine Dorand. illus. (La Nature, Paris, 1909, année 37, semestre 2, p. 1–3.) OA

2050. Phillips, John Scott. Artificial flight. (Engineering, London, 1867, v. 4, p. 78.) VDA

Paper read before the Aeronautical Society of Great Britain. Discusses "by what mechanical means can we develop the principles developed in the flight of birds."

2051. —— Flying machines. illus. (Engineer, London, 1871, v. 32, p. 222–223.) VA

Author's machine equipped with flapping wings and a rotary engine.

2052. Phillips' flying machine. illus. (Scientific American supplement, New York, 1893, v. 35, p. 14528–14529.) VA

2053. Picard, Émile. Rapport sur le prix Osiris à décerner en 1909. (Revue générale des sciences, Paris, 1909, année 20, p. 570–574.) OA

The Osiris prize for 1909 was divided between Gabriel Voisin and Louis Blériot.

2054. Pictorial history of flight. 16 p. illus. (Flight, London, 1909, supplement of March 20.) VDS

Pictures begin with the flights of the Wright brothers.

2055. The **Pioneers.** illus. (Flight, London, 1913, v. 5, p. 730–734.) VDS

Pictures and brief accounts of pioneer airplanes and gliders.

2056. Pioneers of British aviation. ports. (Aeronautics, London, 1919–20, v. 17–v. 19.) VDS

v. 17, p. 112–113, José Weiss; p. 134–135, P. Y. Alexander; p. 160–161, A. V. Roe; p. 184–185, Horatio Barber; p. 208–210, J. W. Dunne; p. 232–233, T. O. M. Sopwith; p. 254–255, O. and H. Short; p. 274–275, R. Blackburn; p. 302–303, E. Holt Thomas; p. 326, Gertrude Bacon; p. 342–343, T. W. K. Clarke; p. 358–360, C. Grahame-White; p. 380–381, L. Howard Flanders; p. 402–403, E. P. Frost; p. 446–447, E. C. Gordon England; p. 466, J. T. C. Brabazon-Moore; p. 488, Griffith Brewer; p. 507–509, G. T. Richards; p. 524, G. B. Cockburn; p. 552, W. O. Manning; p. 575, R. F. Macfie.

v. 18, p. 22–23, Herbert Thomas; p. 58–59, F. Handley Page; p. 81–82, C. R. Fairey; p. 98–99, E. M. Maitland; p. 119, H. E. Watkins; p. 139, F. S. Barnwell; p. 158–159, J. D. Fullerton; p. 194–195, Robert Loraine; p. 213, S. D. A. Grey; p. 238, C. H. Pixton; p. 276, B. F. S. Baden-Powell; p. 295, S. V. Sippe; p. 308, F. H. Sykes; p. 325, F. W. Lanchester; p. 349, O. T. Gnosspelius; p. 375, G. H. Handasyde; p. 400–401, T. O'B. Hubbard; p. 430, E. V. Hammond; p. 464–465, A. Ogilvie; p. 479, H. M. Jullerot.

v. 19, p. 5, J. L. Travers; p. 24, J. G. Weir; p. 131–132, E. R. Calthrop; p. 144, O. C. Morison; p. 183–184, G. de Havilland; p. 239, F. Thurstan; p. 253–254, Leo Jezzi; p. 271–272, E. T. Willows.

2057. Plane that drops its engine to prevent crashes. illus. (Popular mechanics, Chicago, 1929, v. 52, p. 190–191.) VFA

2058. Platte, August. Aëronautische Betrachtungen. Wien: Lechner, 1879. 44 p. 8°. VDY

Discusses the possibility of mechanical flight with references to opinions of other writers.

Heavier-Than-Air Craft — Airplanes, cont'd

2059. **Pollner,** P. B. First European flight with craft heavier than air. (International Civil Aeronautics Conference, 1928, Proceedings, Washington, 1929, p. 105–106.) **VDS**

Experiments and flight of J. H. C. Ellehammer, in Denmark, Aug., 1906.

2060. **Pollog,** Carl Hanns. Hugo Junkers; ein Leben als Erfinder und Pionier... Dresden: C. Reissner, 1930. 205 p. port. illus. 8°. **VDY**

Junkers (b. 1859) stands in the forefront of German developers of the airplane.

2061. **Pompeïn-Piraud,** J. C. Aviation. Les secrets du coup d'ailes. Essai de construction d'une machine aérienne. Paris: F. L. Vivien, 1909. 366 p. illus. 2. ed. 4°. **VDY**

Author's experiments with flying-machine models based upon bird anatomy. His experiences as a balloonist. The disaster to the airship *L'Espérance.*

2062. —— Mémoire sur une machine aérienne à ailes artificielles articulées. Bâle: Henry George, 1890. 42 p. illus. 4°. (Exposition universelle, Paris, 1889. Congrès aéronautique.) **VDY p.v.3, no.4**

An endeavor to follow closely the laws of bird flight.

2063. **Ponton d'Amécourt,** Gustave, vicomte de, editor. Collection de mémoires sur la locomotion aérienne sans ballons, publiée par le Vte de Ponton d'Amécourt. [no.]1–6. Paris: Gauthier-Villars, 1864–67. 6 nos. 4°. † **VDY**

Memoirs by Ponton d'Amécourt, Emmanuel Liais, N. Landur, Henry Giffard, Babinet, L. Franchot, Ramon de Morènes, and A. Gerardin.

2064. **Post,** Augustus. Aeroplane records up to August 29, 1909. (American machinist, New York, 1909, v. 32, part 2, p. 564–565.) **VFA**

2065. —— A comparison of American aeroplanes. Both Curtiss and Wright machines exhibit individuality of their builders... (American machinist, New York, 1909, v. 32, p. 315–317.) **VFA**

2066. —— Details of the Antoinette monoplane. In one of these machines Latham made a record for long distance and duration flying at Rheims that was exceeded only by Farman. illus. (American machinist, New York, 1909, v. 32, no. 40, Oct. 7, p. 608–612.) **VFA**

2067. **Pouchet,** Paul. Édouard de Niéport, dit Nieuport. port. (L'Aérophile, Paris, 1911, année 19, p. 449.) **3–VDS**

Portrait and brief biography of French inventor and airplane constructor.

2068. **Practical** impossibility of human flight proved 22 years ago. Prof. Simon Newcomb, PH.D., LL.D., and Rear-Admiral Melville, U. S. Navy, wrote scientific reasons why — two years later Wilbur and Orville Wright gave aviation to world. (U. S. air service, Washington, D. C., 1923, Jan., p. 38–39.) **VDS**

2069. **Prepositi,** Clemente. La storia dell' aviazione. Firenze: Vallecchi [1931]. 3 v. 16°. **VDY**

History of airplane development in war and peace.

2070. **Pritchard,** John Laurence. The book of the aeroplane. London: Longmans, Green and Co., 1926. vii, 255 p. illus. 12°. **VDY**

Chapters 1 and 2 deal respectively with early history and with the work of the Wright brothers. There are accounts of famous flights, early engines, and airplanes in the World war.

2071. **Probert,** Richard J. How old is your airplane? (Western flying, Los Angeles, 1934, v. 14, March, p. 20–21.) **VDS**

A study of the life duration of airplanes.

2072. **Proctor,** Richard Anthony. Notes on flying and flying machines. (Knowledge, London, 1884, v. 5, p. 470–472; v. 6, p. 2–3, 25–28.) **OA**

Discusses the supporting and elevating power of the air as applicable to heavier-than-air machines.

2073. **Production** and licensing. charts. (Aviation, New York, 1931, v. 30, p. 136–141, 460–461, 514–515, 624.) **VDS**

Digest of statistics compiled for the years 1928–1931 by the United States Department of Commerce and by the Aeronautical Chamber of Commerce.

2074. **Production** and licensing. Who owns all the planes? And when will they buy more? (Aviation, New York, 1932, v. 31, p. 118–128.) **VDS**

Statistics for 1931.

2075. **Proves** Langley's first practical aeroplane. illus. (Aero and hydro, Chicago, 1914, v. 8, p. 117–118.) **3–VDS**

Brief account of test at Hammondsport, N. Y., June 28, 1914.

2076. A **Quaint** ornithopter. A Californian attempt at a wing-flapping machine. illus. (Aero, London, 1910, v. 2, p. 521.) **3–† VDS**

Bird-like apparatus of Professor H. L. Twining.

2077. **R.,** S. New system of aerostation. illus. (Mechanic's magazine, London, 1829, v. 11, p. 162–166, 177–183.) **3–VA**

Movable wing heavier-than-air machine.

2078. **Rameau,** Louis. Charles de Louvrié. (L'Aéronaute, Paris, 1894, année 27, p. 123–126.) **3–VDS**

Louvrié (1821–1894) was an early experimenter with the slightly inclined wing for airplanes.

2079. **Randolph,** Stella, and HARVEY PHILLIPS. Did Whitehead precede Wright in world's first power flight? illus. (Popular aviation, Chicago, 1935, Jan., p. 22–24, 55–57.) **VDY**

2080. **Rapid** development of aviation as shown by statistics. (Aircraft, New York, 1910, v. 1, p. 72–73.) **VDS**

Record of flights, 1903–1909.

Heavier-Than-Air Craft — Airplanes, cont'd

2081. **Rathbun,** John B. Some novelties of the early days. The radical designs of twenty years ago, furnish an interesting comparison with present day achievements in airplane construction. illus. (Popular aviation, Chicago, 1931, v. 9, Aug., p. 35–36, 57.) **VDS**

2082. —— The years go by but aviation stands still. illus. (Popular aviation, Chicago, 1931, v. 8, March, p. 20–22, 63.) **VDS**

"The airplane of today is no different from that of 10 years ago. The progress of aircraft design is largely horseradish."

2083. **Recent** foreign aeroplanes. illus. (Scientific American, New York, 1908, v. 99, p. 124.) **VA**

Airplanes of Captain Ferber, the Zens brothers, and Gastambide-Mangin; also a short account of Wilbur Wright's flights in France.

2084. **Recent** French aeroplanes and their performances. illus. (Scientific American, New York, 1909, v. 100, p. 481, 487.) **VA**

Antoinette, Blériot, and Voisin machines.

2085. **Recent** successful French aeroplanes. illus. (Scientific American, New York, 1909, v. 100, p. 352.) **VA**

Antoinette, Blériot, and Santos-Dumont machines.

2086. The **Red** letter days of aviation. Dates which ought to be remembered in the history of flying machines. (Aero, London, 1910, v. 3, p. 159–160.) 3–† **VDS**

Chronology beginning with Clément Ader, 1890.

2087. **Renard,** Charles. Aviation. (Scientific American supplement, New York, 1892, v. 34, p. 13819–13820.) **VA**

Reviews work of Pénaud, Langley, Piobert, and Morin, in addition to his own researches on heavier-than-air craft.

2088. —— La locomotion aérienne par les appareils plus lourdes que l'air. illus. (Revue de l'aéronautique, Paris, 1888, année 1, p. 118–130; 1889, année 2, p. 1–22.) **3–† VDS**

Author's studies of 1873.

2089. —— Note sur les expériences aérodynamiques de M. S. P. Langley. (Société française de physique, Séances, année 1892, Paris, 1892, p. 408–412.) **PAA**

Reviews airplane theories of Cayley, Wenham, Pénaud, Marey, Renard, and Langley.

2090. **Renard,** Paul. L'aéroplane Antoinette et les vols en hauteur. (Revue scientifique, Paris, 1909, année 47, semestre 2, p. 193–201.) **OA**

2091. **Report** of tests on reduction of airplane noise by use of mufflers. illus. (United States. — Commerce Department, Air commerce bulletin, Washington, 1932, v. 4, p. 277–290.) **VDS**

Account of experiments at the United States Bureau of Standards.

2092. **Reuterskiold.** The aerial ambulance in Sweden. illus. (World's health, Paris, 1925, v. 6, p. 406–411.) **SPA**

2093. A **Review** of the American aircraft industry. The steady growth of the industry a sign of stability. (Aviation, Highland, N. Y., 1926, v. 21, p. 180–181.) **VDS**

Statistics, 1914–1925.

2094. A **Revolution** in metallic design. illus. (Aeroplane, London, 1929, v. 37, p. 1248, 1252.) **VDS**

Describes Lepère-Weymann method of building up a standard wing section.

2095. **Richardson,** Holden Chester. Twenty-five years of airplane design. illus. (Aviation, New York, 1928, v. 15, p. 1716–1719.) **VDS**

A good review of technical developments.

2096. **Riddell,** M. R. The development and future of aviation in Canada. illus. (Engineering Institute of Canada, Journal, Montreal, 1919, v. 2, p. 200–209.) **VDA**

Developments during World war.

2097. **Riley,** C. V. Mr. Maxim's flying machine. illus. (Scientific American, New York, 1894, v. 71, p. 217.) **VA**

2098. **Roberts,** E. W. Reminiscences of early flight. illus. (Society of Automotive Engineers, Journal, New York, 1918, v. 2, p. 294–299.) **TOL**

"The first power flight of an airplane was not, as many suppose, that made by Ader in France in 1897. The first mechanical flight by any form of power was made by the large steam-powered machine designed and built by Sir Hiram Maxim...July 31, 1894." This flight is here described in detail.

2099. **Robertson,** Frederic Arthur de Vere. Aviation in Australia. illus. maps. (Flight, London, 1928, v. 20, p. 344–364.) **VDS**

This well-illustrated account has much of historical interest.

2100. —— The development and progress of flying in the [British] dominions. (Air annual of the British Empire, London, 1929, v. 1, p. 25–31.) **VDS**

2101. **Roché,** J. A. Engineering, 1930–31. A comprehensive review of development in the last year with some forecast as to its effect on achievement looked to in 1931. illus. (Airway age, New York, 1931, v. 12, p. 30–34, 88–89, 92, 98.) **VDS**

Developments in safety appliances, prevention of spinning, fog-flying aids, parachutes, low-power sport planes, gliders, structural design, engines, wheels, brakes, and airplane control.

2102. The **Rocket** airplane. (Mechanical engineering, New York, 1928, v. 50, p. 573–574.) **VFA**

Brief history of various attempts.

2103. **Roe,** Sir Alliot Verdon. Reminiscences of a pioneer. (Air annual of the British Empire, 1929, London, 1929, v. 1, p. 328–334.) **VDS**

Interesting account of author's early construction of airplanes.

2104. —— Twenty years of aviation. The reminiscences of a life-time devoted to British aviation. ports. illus. (Airways, London,

Heavier-Than-Air Craft — Airplanes, cont'd

1927–28, v. 4, p. 378–381, 419–422; 1928–29, v. 5, p. 14–16, 89–91, 133–134, 161–162, 237–238, 339–340.) **VDS**

For references to Roe's airplanes see United States. — National Advisory Committee for Aeronautics, *Bibliography of aeronautics*, 1909–1931, Washington, 1921–1935, volumes for 1909/16, p. 1145–1146; 1917/19, p. 383, *VDT*.

2105. Ross, Malcolm. The Henson "Aerial." illus. port. (Sportsman pilot, Albany, N. Y., 1930, v. 4, July, p. 12–15.) **VDS**

Article is based on original documents in the possession of Henson's daughter, Mrs. George W. Lipsham.

2106. Roux, Louis François. Nouvelles expériences d'aviation. illus. (L'Aéronaute, Paris, 1905, année 38, p. 41–48, 72–79, 95–103, 126–128, 140–152, 163–167.) **3–VDS**

Description of proposed flying machine of the author. Steam-driven beating-wing type. Includes a criticism by Paul Delaporte.

2107. The **Royal** Aircraft factory and the industry. illus. (Flight, London, 1914, v. 6, p. 112–113.) **VDS**

Account of three years' work.

2108. Rummel, Richard M. The birth and growth of Boeing. illus. (Western flying, Los Angeles, 1934, v. 14, March, p. 16–18.) **VDS**

History of the Boeing Aircraft Company of Seattle.

2109. —— Evolution of Boeing mail and passenger planes. illus. (Aero digest, New York, 1933, v. 23, Sept., p. 48–49.) **VDS**

2110. Rumpler, Edmund. Die Flugmaschine. Kritische Besprechung ausgeführter Flugmaschinen mit besonderer Berücksichtigung der geschichtlichen Entwicklung. Im Auftrage des Berliner Vereins für Luftschiffahrt ... Berlin: G. Braunbeck & Gutenberg-Druckerei A. G., 1909. 327 p. illus. 8°. **VDY**

Descriptions and discussions of a large number of contemporary airplanes.

2111. S., W. H. An ancestor of aviation. illus. (Aeroplane, London, 1919, v. 17, p. 1436, 1438.) **VDS**

Describes aerofoil patented by Horatio F. Phillips in England, 1884.

2112. S. F. Cody — the man and his work. port. (Flight, London, 1913, v. 5, p. 905–906.) **VDS**

There is also a good portrait of Cody in *Flight*, London, 1913, v. 5, p. 109.

2113. St. Martin, Henri Reda. A novel apparatus for aerial locomotion. (Engineering, London, 1867, v. 4, p. 85.) **VDA**

Paper read before the Aeronautical Society of Great Britain describes an airplane model of the author.

2114. Samson, Charles Rumney. When flying was young. illus. port. (Air, London, 1930, v. 3, p. 147–150.) **VDS**

Reminiscences of S. F. Cody and Horace Short.

2115. Samuel Franklin Cody. ports. (Aeronautics, London, 1913, v. 6, p. 331–332.) **VDS**

2116. Samuel P. Langley's work may affect the future history of aviation. (Aero and hydro, Chicago, 1914, v. 8, p. 120.) **3–VDS**

2117. Samuel Pierpont Langley. (Aeronautical annual, Boston, 1897, no. 3, p. 5–10.)

Brief biography. **3–VDS**

2118. Samuel Pierpont Langley. port. (Scientific American, New York, 1904, v. 90, p. 194.) **VA**

Biography and a good account of his works. Portrait, p. 189.

2119. Samuels, Frank E. Growth of the Douglas Company. illus. (Aero digest, New York, 1926, v. 8, p. 346, 392–393.) **VDS**

History of concern that has made many airplanes for the United States government.

2120. Samuelson, Arnold. Flight — velocity. Hamburg: Boysen & Maasch, 1906. 42 p. illus. sq. 8°. **VDY**

Experiments of the author with models of the beating-wing type.

2121. Sanderval, de. Recherches sur le vol plané. (Académie des sciences, Comptes rendus, Paris, 1886, tome 103, p. 648–650.) *** EO**

Author's experiments with gliders. Also in *L'Aéronaute*, Paris, 1886, année 19, p. 203–206, *3–VDS*.

2122. Santos Dumont's latest aeroplane. illus. (Scientific American, New York, 1908, v. 99, p. 433.) **VA**

Illustrations and brief description of machine, which weighs 297 pounds.

2123. Santos-Dumont's latest monoplane, with working drawings of the machine and all its parts. illus. (Scientific American supplement, New York, 1909, v. 68, p. 317–318.) **VA**

2124. Saunders, W. O. Then we quit laughing. illus. port. (Collier's, New York, 1927, v. 80, Sept. 17, p. 24, 56.) *** DA**

Author's experiences with the Wright brothers at Kill Devil Hill, N. C.

2125. Savage, E. W. Production and sales [1930]. Statistics tend to show improved state of industry. diagrs. (Airway age, New York, 1931, v. 12, p. 75–81.) **VDS**

Statistics for airplanes and for engines.

2126. Sayers, W. H. Fifteen years of the aircraft industry. (Aeroplane, London, 1926, v. 30, p. 674–676, 678, 680, 682, 684, 686, 688, 690, 692, 694, 696, 698, 700, 702.) **VDS**

Brief histories of some of the prominent British makers of airplanes, including accessories.

2127. —— Tail-less aeroplanes ancient and modern. illus. (Aeroplane, London, 1926, v. 30, p. 446, 448, 450, 452, 454.) **VDS**

Comparison of Hill's airplane with the early Weiss and Dunne machines.

2128. Says he warped wings in 1897. (Aeronautics, New York, 1910, v. 6, p. 52.) **VDS**

Statement of D. D. Wells, of Jacksonville, Florida.

2129. The **Scandal** of the first man-carrying aeroplane. illus. (Current opinion, New York, 1922, v. 72, p. 373–376.) *** DA**

Langley-Wright controversy.

Heavier-Than-Air Craft — Airplanes, cont'd

2130. **Schnepf,** John. Aerial mobile or flying machine. illus. (Aeronautical world, Glenville, O., 1902, v. 1, p. 79–80.) **VDS**

Electrically-driven flying machine provided with revolving sails.

2131. **Schnurmacher.** The fate of a pioneer. The colorful career of MacKensie-Kennedy embraces world famous achievements and disheartening disappointments. illus. port. (Popular aviation, Chicago, 1931, v. 8, Jan., p. 30, 58.) **VDS**

J. H. Chesborough MacKensie-Kennedy, prominent expert of the British Air Ministry during the World war, was "the daddy of the big plane."

2132. —— The inside history of the Christmas bullet. For the first time in thirteen years the designer of the much discussed "Christmas bullet" breaks his silence... illus. port. (Popular aviation, Chicago, 1931, v. 9, Aug., p. 37, 60.) **VDS**

Dr. W. W. Christmas "was the first man after the Wright brothers to fly." In 1918 he built his first cantilever airplane.

2133. **Schunck,** Helen Kay. The Fokker-Byrd controversy. The inside story of the beginnings of a famous dispute written by Mr. Fokker's former private secretary. ports. (Popular aviation, Chicago, 1931, v. 9, Oct., p. 13–14, 50.) **VDS**

2134. —— Working with a genius. illus. port. (Popular aviation, Chicago, 1929, v. 5, Aug., p. 14–15, 74, 76–80.) **VDS**

Author's experiences as stenographer to Anthony H. G. Fokker. This number of *Popular aviation* has the title of *Aeronautics*.

2135. The **Scientific** American award for plane improvement. illus. (Scientific American, New York, 1934, v. 150, p. 29, 34.) **VA**

First award to Luis de Florez.

2136–2137. **Scientific** research in aeroplane construction at the factory of the Curtiss Aeroplane and Motor Corporation. illus. (Aerial age weekly, New York, 1916, v. 3, p. 562–568.) **VDS**

2138. **Sée,** Alexandre. A l'appui des idées du capitaine Duchêne sur la stabilité latérale. illus. (Aérophile, Paris, 1913, année 21, p. 55–56.) **3–VDS**

2139. **Self** adjusting aeroplane constructed by Mr. Lawrence Hargrave, Sydney, N. S. W. illus. (Engineering, London, 1910, v. 89, p. 688.) **VDA**

2140. **Seux,** Edmond. L'aéroplane Edmond Seux. illus. (L'Aérophile, Paris, 1907, année 15, p. 165–166.) **3–VDS**

2141. —— Sur la stabilité des aéroplanes et la construction rationnelle des plans sustentateurs. (Académie des sciences, Comptes rendus, Paris, 1906, tome 142, p. 79–81.) *** EO**

For list of references to Seux's writings on airplanes and their stability see Paul Brockett, *Bibliography of aeronautics,* Washington, 1910, p. 765–766, *VDT.*

2142. **Shepherd,** William Gunn. They lifted the world with a toy. The intimate story of how the Wright boys, sons of a mid-western preacher, learned the mystery of flight. illus. (Collier's, New York, 1927, v. 80, Aug. 20, p. 8–9, 38–39.) *** DA**

2143. **Short,** Mac. Airplane design and construction. illus. (Mechanical engineering, New York, 1929, v. 51, p. 6–8.) **VFA**

Developments during 1928.

2144. A **Short** history of aviation in China. (Aviation, Highland, N. Y., 1921, v. 11, p. 422.) **VDS**

2145. **Short** survey of the history of Czechoslovak aviation prepared by Czechoslovak Aero Club. (International Civil Aeronautics Conference, Papers, Washington, 1928, Dec. 12–14, p. 151–162.) **VDY**

"The first man who in Czechoslovakia made successful attempts in flying was a Czechoslovak named Fučík. In the years 1760–1770 he constructed wings made from wood and tin. For these flights he also used bellows filled with fen-gas."

2146. **Silent,** invisible airplanes. (Literary digest, New York, 1928, v. 99, Nov. 24, p. 22.) *** DA**

Brief reference to researches in this direction.

2147. **Sir** Charles Wakefield. port. (Aeroplane, London, 1930, v. 38, p. 17.) **VDS**

Portrait and brief account of English "patron saint" of aviation.

2148. **Slosson,** Edwin Emery. What did Newcomb say? (Flying, New York, 1920, v. 9, p. 642.) **VDS**

Excuses Simon Newcomb for voicing the impossibility of human heavier-than-air flight.

2149. The **Smallest** successful American aeroplane. The Curtiss machine of the Aeronautic Society has made many flights... illus. (American machinist, New York, 1909, v. 32, p. 49–52.) **VFA**

2150–2151. **Smith,** William Harold. Airplane dopes and doping. Washington: Gov. Prtg. Off., 1920. 9 p. 4°. (National Advisory Committee for Aeronautics. Report no. 38.) **VDY (U. S.)**

Early history of compounds used. Also in *Aerial age weekly,* New York, 1919, v. 10, p. 208, 213, *VDS.*

2152. **Some** Fokker milestones. illus. (Flight, London, 1920, v. 12, p. 173–176.) **VDS**

Technical history of celebrated German airplanes.

2153. **Some** recent foreign flying machines. illus. (Scientific American, New York, 1906, v. 94, p. 252–253.) **VA**

Illustrated account of the airplanes of Vuia and J. H. C. Ellehammer; the revolving wing machine of George Clout; and the helicopter of Léger.

2154. The **Sopwith** machines. illus. (Flight, London, 1919, v. 11, p. 163–174.) **VDS**

Development of famous British airplane.

Heavier-Than-Air Craft — Airplanes, cont'd

2155. Soreau, Rodolphe. Développement de l'aviation. illus. charts. (Société des ingénieurs civils de France, Mémoires, Paris, 1913, année 1913, v. 2, p. 647–667.) **VA**

Development as shown by records and technical improvements.

2156. —— État actuel et avenir de l'aviation. illus. (Société des ingénieurs civils de France, Mémoires, Paris, 1908, année 1908, p. 13–234.) **VDA**

Extensive paper on the theory, design, construction, and future of airplanes, with much of historical interest.

2157. —— Note relative à la discussion sur l'aviation. (Société des ingénieurs civils de France, Mémoires, Paris, 1909, Aug., p. 113–131.) **VDA**

Comments on papers of Chaudy, Armengaud, Legrand, Boyer-Guillon, and Lumet, read before the society.

2158. —— Le problème générale de la navigation aérienne. illus. (Société des ingénieurs civils de France, Mémoires, Paris, 1897, série 5, année 50, partie 2, p. 119–191.) **VDA**

Largely a study of flight based on the principles of bird-flight. Details of Maxim's machines.

For references to the work of Soreau see United States. — National Advisory Committee for Aeronautics, *Bibliography of aeronautics*, 1909–1931, Washington, 1921–35, volumes for 1909/16, p. 1251–1252; 1917/19, p. 412; 1920/21, p. 381; 1922, p. 147; 1928, p. 184; 1930, p. 224, *VDT*.

2159. Spencer, G. K. 1926 — a turning point in American aviation. illus. (New York Times, Current history, New York, 1927, v. 25, p. 465–479.) **BAA**

"The progress of design was stimulated to such purpose that during 1926 more than 130 new or temporarily lost principles were introduced into newly manufactured aircraft."

2160–2161. Sperry, Elmer Ambrose. The automatic pilot. illus. (International Congress of Aerial Navigation, London, 1923, Report, London, 1923, p. 406–413.) **VDS**

Author recalls his early experiments.

2162. Sperry automatic pilot. illus. (Aero digest, New York, 1929, v. 15, Dec., p. 166.)

History of development. **VDS**

2163. The **"Spirit** of St. Louis." The Ryan monoplane described. illus. (Flight, London, 1927, v. 19, p. 376–379, 385.) **VDS**

Describes Lindbergh's machine.

2164. Squier, George Owen. Aeronautics in the United States, 1918. diagrs. (American Institute of Electrical Engineers, Proceedings, New York, 1919, v. 38, p. 53–114.) **VA**

Deals principally with airplane and engine development down to the Armistice, Nov. 11, 1918. Abridged in Society of Automotive Engineers, *Journal*, New York, 1919, v. 5, p. 402–414, *TOL*.

2165. —— The Wright brothers — a bit of history. (Flight, London, 1913, v. 5, p. 651–652.) **VDS**

History of the original army contract for airplanes (1907).

2166. Stabilizer qualifies Wright for trophy. (Aero and hydro, Chicago, 1914, v. 7, p. 179.) **3–VDS**

Orville Wright's device for maintaining automatic stability. Account of tests in 1913.

2167. Stearns, Myron Morris. The flying wing. illus. port. (Popular mechanics, Chicago, 1929, v. 52, p. 234–238.) **VFA**

Biography of W. W. Christmas, pioneer airplane designer who built the famous "Bullet" machine.

2168. Stevens, Truman. Amazing plane its own parachute. In emergency the wings drop off and the body unfolds. illus. (Popular science monthly, New York, 1925, v. 106, April, p. 30.) *** DA**

Invention of Captain Lepinte of France.

2169. Stockbridge, Frank Parker. Glenn Curtiss: air pilot no. 1. illus. (Popular science monthly, New York, 1927, v. 110, March, p. 20–22, 130–132; April, p. 22–24, 113–114; May, p. 24–26, 138–140.) *** DA**

2170. Story of the first flight. illus. ports. (Aero digest, New York, 1926, v. 9, p. 423–425, 476.) **VDS**

Story of how the Wright brothers designed their first machine and of the first flights.

2171. "Strasbourg," pseud. The influence of official control upon the design of aircraft, 1914–1918. (Aeroplane, London, 1919, v. 16, p. 53–54, 471–472, 573–574.) **VDS**

2172. Sturmey, Henry. A chapter in aviation history. (Motor, London, 1916–17, v. 30, p. 651–653.) **3–† TOL**

Account of Charles M. Manly's work with S. P. Langley.

2173. Suplee, Henry Harrison. The development of the aeroplane. illus. (Cassier's magazine, New York, 1909–10, v. 37, p. 174–181, 275–283.) **VDA**

2174. —— Plywood in aeroplane construction. illus. (Aerial age weekly, New York, 1918–19, v. 8, p. 945–947, 961.) **VDS**

Gives interesting early history.

2175. Swedenborg, Emanuel. Suggestions for a flying machine — a historical curiosity. illus. (Scientific American supplement, New York, 1911, v. 72, p. 286–287.) **VA**

Translation of manuscripts which, it is claimed, place Swedenborg as the discoverer of the airplane principle. Alfred Acton traces the history of the documents. See also *Aeronautical journal*, London, 1910, v. 14, p. 118–122, *VDS*.

2176. Swinehart, D. E. Aviation in north China. Little development up to 1925, but since then French assistance has made rapid progress possible. port. (Aviation, Highland, N. Y., 1928, v. 24, p. 636, 648.) **VDS**

2177. Swinton, Alan Archibald Campbell. An early aeroplane. (Engineering, London, 1910, v. 89, p. 120–121.) **VDA**

Model using steam power, constructed by Charles A. Parsons in 1893.

Heavier-Than-Air Craft — Airplanes, cont'd

2178. T. O. M. Sopwith. port. (Aeronautics, London, 1913, v. 6, p. 281.) **VDS**

Portrait and brief biography of a leading English pilot and airplane constructor.

2179. Tailless airplane is hailed as fool-proof. illus. (Popular mechanics, Chicago, 1926, v. 46, p. 366.) **VFA**

Invention of Capt. G. T. R. Hill, of England.

2180. Tanager wins grand prize in safe airplane contest. illus. (Aviation, New York, 1930, v. 28, p. 69–70.) **VDS**

Guggenheim prize, 1930, of $100,000, with a list of the severe requirements.

2181. Tarnowski, J. Der Flugwagen. illus. (Illustrierte aëronautische Mittheilungen, Strassburg, 1901, Jahrg. 5, p. 105–107.) **VDS**

Illustrated description of author's airplane.

2182. Tatin, Victor. Éléments d'aviation. Paris: H. Dunod & E. Pinat, 1908. vi, 65 p. illus. 4°. **VDW p.v.1, no.1**

Includes early history of the airplane.

2183. —— Étude et construction d'un aéroplane. illus. (L'Aérophile, Paris, 1906, année 14, p. 221–237.) **3–VDS**

2184. —— Expériences sur le vol mécanique. illus. (L'Aéronaute, Paris, 1877, année 10, p. 22–25, 46–49, 81–88, 141–145, 166–169.) **3–VDS**

Author's experiments in the Marey laboratory, 1874–1876.

2185. —— Navigation aérienne. Appareils plus lourds que l'air. illus. (La Nature, Paris, 1884, année 12, semestre 2, p. 328–331.) **OA**

Reviews briefly the experiments of Pénaud with helicopters, mechanical birds, and airplanes. Account of author's model airplane driven by compressed air. Translation in *Scientific American*, New York, 1884, v. 51, p. 342–343, *VA*.

2186. —— Nouvel aéroplane, mû par une machine à air comprimé; détermination expérimentale du travail nécessaire pour faire voler cet appareil. (Académie des sciences, Comptes rendus, Paris, 1879, tome 89, p. 1024–1027.) *** EO**

2187. —— Points d'histoire — l'aéroplane. (L'Aérophile, Paris, 1907, année 15, p. 281–284.) **3–VDS**

Discusses the work of Cayley, Wenham, Alphonse Pénaud, D. S. Brown, and Lawrence Hargrave.

For citations to Tatin's writings on experimental aeronautics see Paul Brockett, *Bibliography of aeronautics*, Washington, 1910, p. 812–814, *VDT*. Continued in United States. — National Advisory Committee for Aeronautics, *Bibliography of aeronautics*, 1909–1931, Washington, 1921–1935, volume for 1909/16, p. 1313–1314, *VDT*.

2188. Tatin, Victor, and Charles Richet. Expériences faites avec un aéroplane mû par la vapeur. (Académie des sciences, Comptes rendus, Paris, 1897, tome 125, p. 64–66.) *** EO**

Steam-driven model airplane flew 140 meters. Also in *L'Aéronaute*, Paris, 1897, année 30, p. 172–174, *3–VDS*.

2189. Teale, Edwin Way. Edison's early dream ships. illus. (Popular aviation, Chicago, 1933, v. 12, p. 360.) **VDS**

Edison's idea of 1880.

2190. Technical developments in aircraft construction. illus. (Aircraft yearbook, New York, 1920, p. 123–132; 1921, p. 139–157; 1922, p. 99–106; 1923, p. 194–200; 1924, p. 243–255; 1925, p. 197–214; 1926, p. 188–216; 1927, p. 205–216; 1928, p. 255–276; 1929, p. 233–246; 1930, p. 71–94; 1931, p. 117–144; 1932, p. 301–320.) **VDS**

Covers the years 1914–1931. Title varies considerably.

2191. The **Terredome.** illus. (English mechanic, London, 1902, v. 75, p. 262–263.) **VA**

Flying and leaping machine invented by William A. Eddy, of Bayonne, N. J.

2192. Test of the Bell tetrahedral-cell aeroplane in Nova Scotia. illus. (Scientific American, New York, 1909, v. 100, p. 205.) **VA**

2193. Tetens, Hans. Geschichte der deutschen Flugzeug-Industrie. (Jahrbuch für Luftfahrt, 1928, München, 1928, p. 1–11.) **3–VDS**

History of the German aircraft industry, beginning 1908.

2194. Thomas, H. J. The British aircraft industry in 1933–34. (Air annual of the British Empire, 1934/35, London, 1934, v. 6, p. 113–124.) **VDS**

Technical, military, and commercial developments.

2195. Thomson, G. P. The development of aeroplanes in the great war. illus. (Engineering, London, 1919, v. 107, p. 246–247.) **VDA**

2196. Todd, Frederick. The airship is here. How the aeroplane looks, how it moves, how it is built and managed. illus. (World's work, New York, 1908, v. 16, p. 10659–10680.) *** DA**

2197. The **"Totally** enclosed" aeroplane. illus. (Flight, London, 1917, v. 9, p. 363–365, 388, 416, 446, 473.) **VDS**

"One of the first, if not, indeed, the very first, full size machines in which the pilot and the passenger were totally enclosed in the *fuselage* covering, was the Piggot monoplane... 1911."

2198. Toulmin, H. A. Attacks on the Wright brothers wholly unjustified. (Aircraft, New York, 1910, v. 1, p. 93–94.) **VDS**

Statement of their counsel.

2199. Tournachon, Félix. Le droit au vol, par Nadar [pseud.]. Paris: J. Hetzel [1865]. iii, 115 p. 12°. **VDY**

Expresses hope for the heavier-than-air machine, with caustic comments on the opinions of members of the French Academy.

For references to Tournachon, writing under the pseudonym of Nadar, see United States. — National Advisory Committee for Aeronautics, *Bibliography of aeronautics*, 1909–31, Washington, 1921–35, volume for 1909/16, p. 936, *VDT*.

2200. Trend of German airplane design: a summary of features of captured enemy machines, issued by Technical Department (Aircraft Production), British Ministry of Munitions. illus. (Automotive industries, New York, 1919, v. 40, p. 262–265.) **TOL**

Heavier-Than-Air Craft — Airplanes, cont'd

2201. The **Trial** of Maxim's flying machine. illus. (Scientific American, New York, 1894, v. 71, p. 123–124.) **VA**

2202. **Trouvé's** aviator. illus. (Scientific American, New York, 1892, v. 66, p. 105.) **VA**

Model with flapping wings actuated by oxidation of hydrogen gas controlled by a Bourdon tube.

2203. **Turnbull,** W. R. Form and stability of aeroplanes. Researches supplementing those of Lilienthal and Langley. diagrs. (Scientific American supplement, New York, 1909, v. 67, p. 68–70.) **VA**

Account of the author's experiments.

2204. —— Researches on the forms and stability of aeroplanes. illus. (Physical review, Lancaster, Pa., 1907, v. 24, p. 285–302.) **PAA**

2205. **Turner,** George Kibbe. The men who learned to fly. The Wright brothers' story of their experiments, the sensations of flight, and their estimate of the future of the aeroplane. illus. ports. (McClure's magazine, New York, 1908, v. 30, p. 443–452.) *** DA**

2206. **Twenty-fifth** anniversary of the first airplane flight. Proceedings at the exercises at Kitty Hawk, N. C., on December 17, 1928, in commemoration of the twenty-fifth anniversary of the first flight of an airplane made by Wilbur and Orville Wright. Washington: Gov. Prtg. Off., 1929. xiii, 30 p. (United States. 70th cong., 2. sess. House document no. 520.) *** SBK**

2207. **Twenty-five** years of flight. (Nation, New York, 1928, v. 127, p. 674.) *** DA**

Brief account of the Wright-Langley controversy.

2208. **Types** of recent foreign flying machines. Some French and Danish aviation apparatus already built or building. illus. (Scientific American supplement, New York, 1908, v. 65, p. 172–174.) **VA**

Bertin helicopter, Coulomb ornithopter, Ellehammer airplane, Gilbert airplane.

2209. **United States.** — Air Service. Aircraft development since the armistice. Washington: Gov. Prtg. Off., 1923. 34 p. illus. 4°. (Air Service information circular no. 402.) **VDY**

Covers airplanes, armament, engines, and equipment.

2210. **United States.** — Military Affairs Committee (House, 70:1). First heavier-than-air flying machine. Hearing before subcommittee no. 8, Committee on Military Affairs, House of Representatives, 70th Congress, first session. April 27, 1928... Washington: Gov. Prtg. Off., 1928. ii, 56 p. diagrs., plates. 8°. *** SBF**

Hearings in regard to priority of Langley plane over that of the Wright brothers, and the question of placing the Wright plane in the Smithsonian Institution.

2211. **United States** has greatest year in world aviation. illus. (Aeronautical digest, New York, 1923, v. 2, p. 16–19.) **VDS**

Review of 1922.

2212. The **United States** Signal Corps aeroplane. The Wright machine that met the specifications by carrying two men for an hour's flight and making a speed of 42½ miles per hour. illus. (American machinist, New York, 1909, v. 32, part 2, p. 301–305.) **VFA**

2213. **Urlwin,** J. First motor-driven model. illus. (Flight, London, 1910, v. 2, p. 114.) **VDS**

Author claims that his model was the first to use a petrol engine.

2214. **Van Muffling,** Adrian. A victorious year of American aeronautics. (Aero digest, New York, 1927, v. 11, p. 642–643, 762–763.) **VDS**

Review of aviation for the year 1927.

2215. **Ventou-Duclaux,** L. L'aviation expliquée. Ouvrage de vulgarisation, renfermant un dictionnaire des termes couramment employés en aviation. Paris: F. L. Vivien, 1909. 183 p. 8°. **VDY**

Describes contemporary airplanes and motors. Good glossary.

2216. Die **Versuche** Wels-Etrich. illus. (Wiener Luftschiffer-Zeitung, Wien, 1908, Jahrg. 7, p. 281–282.) **VDS**

Illustrated description of the Wels-Etrich airplane.

2217. **Very,** Frank Washington. The rôle of viscosity in air support of a moving aeroplane. illus. (Technology quarterly, Boston, 1908, v. 21, p. 490–507.) **OA**

2218. **Veyrin,** E. De la pesanteur par rapport à l'aviation. illus. (L'Aéronaute, Paris, 1888, année 21, p. 143–156.) **3–VDS**

2219. The **Villard** flying machine. illus. (Scientific American, New York, 1902, v. 86, p. 173.) **VA**

Helicopter of Henri Villard, of Paris.

2220. **Violette,** Marcel. Les précurseurs de l'aviation. (In: L'Aéronautique, par Roland Garros [and others], Paris: Pierre Lafitte & Cie., 1914, p. 109–129.) **VDW**

Brief account of modern experimenters, with slight reference to their precursors.

2221. **Viscount** Carlingford's aerial machine. (Mechanic's magazine, London, 1857, v. 66, p. 538, 562–563.) **3–VA**

2222. **Voisin,** Gabriel. La naissance de l'aéroplane. [Paris? 1927?] 44 p. illus. 8°. **VDY p.v.43, no.6**

Autobiography of French pioneer, with numerous illustrations. Claims that his brother, Charles, was the first to accomplish flight in a plane equipped with an internal combustion engine.

Heavier-Than-Air Craft — Airplanes, cont'd

2223. **Vorreiter,** Ansbert. Drachen-, Schrauben- oder Schwingenflieger. illus. (Mitteleuropäischer Motorwagen Verein, Zeitschrift, Berlin, 1908, Jahrg. 7, p. 237–246.) **† TOL**

Descriptions of Santos-Dumont, Farman, Delagrange, Wright, Ellehammer, Jatho, Kress, Blériot, Kapferer, Ernault-Pelterie, de la Vaulx, and Gastambide-Mangin airplanes.

2224. —— Kritik der Drachenflieger... Berlin: R. S. Schmidt & Co., 1909. 120 p. illus. 8°. (Bibliothek für Luftschiffahrt und Flugtechnik. Bd. 1.) **VDY**

Describes and illustrates construction of important contemporary airplanes.

2225. —— Wilhelm Kress. port. (Zeitschrift für Flugtechnik, München, 1913, Jahrg. 4, p. 49–50.) **VDS**

Kress (1836–1913) was an early investigator of human flight.

2226. **Wade,** Walter Lewis. The development of German aeronautics. illus. (Aeroplane, London, 1916, v. 10, p. 50, 52, 54, 138, 140, 194, 196, 248, 250, 270, 384, 386, 558, 560, 908, 910.) **VDS**

2227. **Walcott,** Charles Doolittle. Orville Wright to give first plane to England. Mutilation of Langley's machine permitted by Smithsonian Institution, for purposes of private parties to a patent litigation, he charges. illus. (U. S. air services, Washington, 1925, v. 10, June, p. 31–33.) **VDS**

2228. —— Samuel Pierpont Langley and modern aviation. (American Philosophical Society, Proceedings, Philadelphia, 1926, v. 65, p. 79–82.) *** EA**

Brief account of experiments, with a copy of the label on the airplane exhibited at the U. S. National Museum.

2229. **Walker,** Thomas. A treatise upon the art of flying by mechanical means, with a full explanation of the natural principles by which birds are enabled to fly; likewise instructions and plans for making a flying car with wings, and, by working a small lever, cause himself to ascend and soar through the air with facility of a bird. Hull, 1810, 33 p. 8°. **3–VDS (Aero)**

Reprinted in *Aeronautical annual,* Boston, 1895, no. 1, p. 49–81. Library has also a reprint of the 2. ed. (1831), published by the Royal Aeronautical Society of Great Britain, London, 1910, xxi, 40 p. illus. 12°. (Aeronautical classics. no. 3.) *VDY (Royal).*

For an abridgment of the work see *Mechanic's magazine,* London, 1827, v. 7, p. 130–133, 154–157, 172–174, 205–207, *3–VA.*

2230. **Wallin,** B. H. Flugtechnik in Schweden. illus. (Illustrierte aëronautische Mitteilungen, Berlin, 1908, Jahrg. 12, p. 52–55, 150–153, 424–429, 452–456.) **VDS**

Aviation studies of the *Aktiebolaget Aviatorer,* 1906–1908.

2231. **Warner,** Edward Pearson. Is there a limit on speed? diagrs. (Aviation, New York, 1932, v. 31, p. 431–433.) **VDS**

Has an interesting chronological table, 1908–1930, showing airplane characteristics and speed of airplanes.

2232. **Watter,** Michael. Modern significance of the Wright patent. illus. (Aero digest, New York, 1928, v. 13, p. 1110–1112, 1284.) **VDS**

Includes a copy of the specifications for patent filed March 23, 1903.

2233. **Wead,** Frank Wilbur. Wings for men ... New York and London: The Century Company [1931]. xiii, 333 p. illus. 8°. **VDT**

Early history. War period. Famous flights.

2234. **Webb,** L. D. Improving the breed. (U. S. air services, Washington, 1931, v. 16, Dec., p. 13–17.) **VDS**

Developments for 1931 in engines, faster and stronger aircraft, elimination of vibrations, and automatic pilots.

2235. **Wellner,** Georg. Ueber dynamische Flugtechnik. (Oesterreichischer Ingenieur- und Architekten-Verein, Zeitschrift, Wien, 1893, Jahrg. 45, p. 49–53.) **3–† VDA**

Descriptions of flying machine models.

2236. —— Ueber Segelradflugmaschinen. illus. (Oesterreichischer Ingenieur- und Architekten-Verein, Zeitschrift, Wien, 1893, Jahrg. 45, p. 661–665.) **VDA**

Proposed flying machine of the author. For criticism see *L'Aéronaute,* Paris, 1894, année 27, p. 171–178, *3–VDS.*

2237. —— Ueber die Stabilität und Steuerung dynamischer Flugmaschinen. illus. (Zeitschrift für Luftschiffahrt, Berlin, 1893, Jahrg. 12, p. 64–69.) **3–VDS**

2238. —— Ueber den Weg zur Herstellung brauchbarer Flugmaschinen. (Zeitschrift für Luftschiffahrt, Berlin, 1896, Jahrg. 15, p. 207–218.) **3–VDS**

Author tells how he would construct a flying machine, using sail wheels.

For list of writings by and about Wellner see Paul Brockett, *Bibliography of aeronautics,* Washington, 1910, p. 883–886, *VDT.*

2239. **Wenham,** Francis Herbert. Aerial locomotion. London: Royal Aeronautical Society of Great Britain, 1910. x, 45 p. illus. port. 12°. (Aeronautical classics. no. 2.) **VDY (Royal)**

Reprint of paper read in 1866, with a biographical notice. Wenham (1824–1908) made close observations on bird flight and the above paper "contains...almost every principle, recognized to-day but unknown or unexpressed then, on which modern theory or practice in aviation is founded." Also in *Aeronautical annual,* Boston, 1895, no. 1, p. 82–113, *3–VDS;* Smithsonian Institution, *Annual report for 1888/89,* Washington, 1890, p. 303–323, ** EA.*

2240. —— First principles of aerial transport. (Popular science review, London, 1874, v. 13, p. 362–379.) **3–OA**

2241. —— On forms of surfaces impelled through air and their effects in sustaining weights. illus. (Aeronautical journal, London, 1900, v. 4, p. 134–137.) **VDS**

2242. —— Suggestions and experiments for the construction of aerial machines. illus. (International Conference on Aerial Navigation, Chicago, 1893, Proceedings, New York, 1894, p. 297–303.) **VDS**

For list of Wenham's writings on the airplane see Paul Brockett, *Bibliography of aeronautics,* Washington, 1910, p. 887, *VDT.*

Heavier-Than-Air Craft — Airplanes, cont'd

2243. Wer war der erste Flugpraktiker? port. illus. (Deutsche Luftfahrer-Zeitschrift, Berlin, 1913, Jahrg. 17, p. 211–213.) **VDS**

Brief account of Carl Friedrich Meerwein's projected flying machine (1782).

2244. Werthner, William. Personal recollections of the Wrights. (Aero Club of America, Bulletin, New York, 1912, v. 1, July, p. 13.) **VDS (Flying)**

The above bulletin is an early number of the periodical subsequently known as *Flying*.

2245. West, Rupert E. When the Wrights gave wings to the world. illus. (U. S. air services, Washington, 1927, v. 12, Dec., p. 19–23.) **VDS**

Reminiscences of Captain William J. Tate, in whose home the Wrights lived at Kitty Hawk, N. C. Picture of the house.

2246. Weyerbacher, Ralph D. Aviation's third decade. illus. (Saturday evening post, Philadelphia, 1929, v. 201, May 25, p. 18–19, 205–206.) *** DA**

Technical developments during twenty years.

2247. Weyher, Ch. Sur le vol direct de l'homme. illus. (L'Aéronaute, Paris, 1884, année 17, p. 123–131.) **3–VDS**

Possibilities based upon the principles of bird flight.

2248. What the factories are doing. Production records show growing dependence on military markets. illus. (Aviation, New York, 1933, v. 32, p. 77–81.) **VDS**

Production figures for five years shown by graphs.

2249. What Langley did for the science of aviation. His early experiments in aerodynamics at Allegheny Observatory. How he worked out fundamentals and developed basic principles before attempting to obtain free flight. port. illus. (Automotive industries, New York, 1918, v. 39, p. 660–662, 714–718, 738.) **TOL**

2250. When wings were young. illus. (Collier's, New York, 1927, v. 80, Sept. 24, p. 18–19.) *** DA**

Interesting pictures of the Wright brothers and of their machines.

2251. Whitehead, Gustave. The Whitehead flying machine. port. illus. (American inventor, Washington, 1902, v. 9, April 1, p. 1–2.) *** V**

Claims that his airplane flew successfully over Long Island Sound on January 17, 1902.

2252. Who was the first man to fly? (Flight, London, 1912, v. 4, p. 269.) **VDS**

Gives credit to the Wright brothers, with brief references to the work of their predecessors.

2253. Wilbur Wright, 1867–1912. port. (Flight, London, 1912, v. 4, p. 488–489.) **VDS**

Obituary notice and appreciation.

2254. Wilbur Wright — the passing of a great inventor. port. (Scientific American, New York, 1912, v. 106, p. 518.) **VA**

Appreciation of this great airplane pioneer (1867–1912). "It is not without significance that the flying machine should have been developed, not by an engineer or a trained physicist, but by two men who worked with their hands."

2255. Wild, Ralph Kirkby Bagnall. Safety in flight. (Aviation, Highland, N. Y., 1920, v. 9, p. 121–123, 150–154.) **VDS**

Effects of improvements in airplane construction.

2256. Wilhelm Kress. port. (Wiener Luftschiffer-Zeitung, Wien, 1904, Jahrg. 3, p. 143–146.) **VDS**

Biography of pioneer German airplane investigator.

2257. Wilhelm Kress. ports. illus. (Oesterreichische Flug-Zeitschrift, Wien, 1912, Jahrg. 6, p. 260–261; 1913, Jahrg. 7, p. 97–98, 118–119, 458–464.) **VDS (Flug)**

2258. Will the aeroplane ever assist us in everyday life? A symposium of the opinions of famous aviators and experts on aviation. (Fry's magazine, London, 1912, v. 4, p. 225–228.) **MVA**

2259. Williams, Bertram W. Freaks that flew. illus. (Popular aviation, Chicago, 1929, v. 4, May, p. 11–12, 84, 86.) **VDS**

Some of the earlier innovations.

2260. Williams, W. E. The stability of flying machines. (Royal Philosophical Society of Glasgow, Proceedings, Glasgow, 1903–04, v. 35, p. 181–188.) *** EC**

2261. Wilson, E. L'appareil de locomotion aérienne de M. Félix du Temple. illus. (L'Aéronaute, Paris, 1877, année 10, p. 223–234.) **3–VDS**

Patent specification of du Temple's flying machine of 1857.

2261a. —— Natural and artificial flight. illus. (Scientific American, New York, 1905, v. 93, p. 481.) **VA**

Author's experiments with airplanes at Wembly Park, England.

2262. Winchester, Clarence Arthur Charles. Flying men and their machines; a literary and photographic record of facts concerning flying with special reference to those pilots who have so bravely sacrificed their lives and to those who happily still practice the newest of arts... London: J. M. Dent & Sons, Ltd., 1916. xii, 242 p. illus. 8°. **VDY**

2263. Wing flight — some researches at Cambridge [England]. illus. (Automotor journal, London, 1905, v. 10, p. 648–651.) **3–† TOL**

Illustrated account of experiments with flapping wings by E. P. Frost, F. W. H. Hutchinson, and C. R. d'Esterre.

Heavier-Than-Air Craft — Airplanes, cont'd

2264. **Winters,** S. R. Life-saving airplanes. Aircraft are now employed for carrying life-lines to ships in distress. illus. (Scientific American, New York, 1927, v. 137, p. 497–499.) **VA**

Adopted by the United States Coast Guard. Invention of C. C. Van Paulsen.

2265. —— Sound proof airplanes. illus. (Popular aviation, Chicago, 1929, v. 4, May, p. 27–28.) **VDS**

Experiments by the United States Bureau of Standards. Reprinted in part in *Literary digest,* New York, 1929, v. 101, June 15, p. 19–20, * *DA.*

2266. **Wood,** George F. Campbell. Records and statistics. (Aircraft, New York, 1911, v. 1, p. 390–391.) **VDS**

Table of fatal airplane accidents, 1908–1910.

2267. —— What the Farmans have done. (Aircraft, New York, 1911, v. 2, p. 108–109.) **VDS**

2268. —— The Wright-Paulhan conflict. illus. (Aircraft, New York, 1910, v. 1, p. 50–55.) **VDS**

2269. **Woodhouse,** Henry. Evolution of aviation in 1911. (Aero Club of America, Bulletin, New York, 1912, Jan., p. 9–23.) **VDS (Flying)**

This bulletin is one of the early numbers of the periodical afterwards known as *Flying.*

2270. —— The trail of the aeroplane. illus. (Flying, New York, 1912, v. 1, Nov., p. 3–7; Dec., p. 6–9; 1913; Jan., p. 8–13; 1913, v. 2, Feb., p. 8–10.) **VDS**

"Account of how a few hardy pioneers introduced aviation in far-off countries...how the aeroplane was subsequently officially recognized and introduced in colonial service; also how it was used in Tripoli, and why it was not used in the Moroccan war."

2271. —— The world's tribute to Wilbur Wright. (Aero Club of America, Bulletin, New York, 1912, v. 1, July, p. 9–12.) **VDS (Flying)**

Eulogies in various publications are quoted. The above bulletin is an early number of the periodical subsequently known as *Flying.*

2272. The **Work** of Igo Etrich, and some comments on his machine. (Aeronautics, London, 1913, v. 6, p. 386–387.) **VDS**

Account of pioneer German experimenter with gliders and airplanes.

2273. The **World's** first aviators. (Domestic air news, Washington, D. C., 1929, no. 44, Jan. 15, p. 22.) **VDS (U. S. Commerce)**

List of thirty-six pioneer aviators in the order in which they flew. Wilbur Wright is first.

2274. **Wright,** Milton. Builders of the aviation industry. Back of the flyers are the men who are making of aeronautics a giant industry. ports. (Scientific American, New York, 1929, v. 140, p. 228–233.) **VA**

Brief biographies of A. H. G. Fokker, Sherman Mills Fairchild, Harry F. Guggenheim, Igor I. Sikorsky, Karl Arnstein, Charles H. Lawrance, Clement Melville Keys, William B. Stout, William E. Boeing, Richard F. Hoyt, Benjamin D. Foulois, Giuseppe M. Bellanca, Charles H. Colvin, Grover C. Loening, Sylvanus Albert Reed, Charles Sherman ("Casey") Jones.

2275. —— Rocking again the cradle of flight. Pioneers celebrate in Hammondsport the days that made aviation history. illus. (Scientific American, New York, 1928, v. 139, p. 231–233.) **VA**

Brief account of the early work of Glenn H. Curtiss.

2276. **Wright,** Orville. How we made the first flight. illus. (Flying, New York, 1913, Dec., v. 2, p. 10–12, 35–36.) **VDS**

Reprinted in *Aviation,* Highland, N. Y., 1923, v. 15, p. 737–741, *VDS.* Includes a biography of the author.

2277. —— Stability of aeroplanes. (Franklin Institute, Journal, Philadelphia, 1914, v. 178, p. 249–258.) **VA**

2278. **Wright,** Orville, and C. D. Walcott. The Wright-Langley controversy. Both sides presented by Orville Wright and Dr. Walcott. illus. (Aviation, Highland, N. Y., 1925, v. 18, p. 550–551.) **VDS**

2279. **Wright,** Orville, and Wilbur Wright. Our aeroplane tests at Kitty Hawk. illus. (Scientific American, New York, 1908, v. 98, p. 423.) **VA**

Tests at their original camp, 1908.

2280. —— The world's first 160 flights. illus. (U. S. air services, Washington, D. C., 1928, v. 13, Dec., p. 27–29.) **VDS**

Brief résumé of the authors' flights during the three years following their first flight at Kitty Hawk.

For numerous references to the Wright brothers see Paul Brockett, *Bibliography of aeronautics,* Washington, 1910, p. 900–905, *VDT.* Continued in United States. — National Advisory Committee for Aeronautics, *Bibliography of aeronautics,* 1909–1931, Washington, 1921–35, volumes for 1909/16, p. 1454–1470; 1917/19, p. 488–489; 1920/21, p. 443–444; 1922, p. 166; 1923, p. 142; 1924, p. 112–113; 1925, p. 186–187; 1926, p. 153; 1927, p. 182; 1928, p. 212; 1929, p. 239–240; 1930, p. 259, *VDT.*

2281. **Wright,** Wilbur. Experiments and observations in soaring flight. illus. (Western Society of Engineers, Journal, Chicago, 1903, v. 8, p. 400–417.) **VDA**

With discussion. Describes work of the Wright brothers after 1901.

Also in *Aero manual,* London, 1910, 2. ed., p. 22–33 (without the discussion).

2282. —— Some aeronautical experiments. illus. (Western Society of Engineers, Journal, Chicago, 1901, v. 6, p. 489–510.) **VDA**

Work of Lilienthal, Chanute, and Pilcher, with an account of the Wright brothers' experiments at Kitty Hawk, N. C. Reprinted in *Aeronautical journal,* London, 1916, v. 20, p. 86–97, *VDS.*

2283. —— The story of flight. (Aeronautics, New York, 1915, v. 16, p. 52–53, 62–63, 69, 79.) **VDS**

A record of the Wrights' work, with acknowledgment of what they owed to their predecessors. Also in *Aeronautics,* London, 1915, v. 8, p. 346–349, 414–415, *VDS.* Abstracted in *Aeroplane,* London, 1916, v. 10, p. 990, 992, 994, 1034, *VDS.*

Heavier-Than-Air Craft — Airplanes, cont'd

2284. —— What Clement Ader did. illus. (Aero Club of America, Bulletin, New York, 1912, v. 1, May, p. 17–19.) **VDS (Flying)**

Discussion of Ader's claim to successful flight. The above bulletin is an early number of the periodical subsequently known as *Flying.*

2285. —— What Mouillard did. (Aero Club of America, Bulletin, New York, 1912, April, p. 3–4.) **VDS (Flying)**

While doing justice to the studies of Mouillard, Wright criticizes French writers who "attempt to name their countryman as the 'father of aviation.'" The above bulletin is an early number of the periodical *Flying*. The article is reprinted in *Aeronautical journal*, London, 1916, v. 20, p. 107–110, *VDS.*

2286. The **Wright** aeroplane test in North Carolina. illus. (Scientific American, New York, 1908, v. 98, p. 393.) **VA**

2287. The **Wright** brothers. illus. (Popular aviation, Chicago, 1928, v. 3, Dec., p. 11–16, 116–117.) **VDS**

Experiments at Kitty Hawk. Interest of foreign governments. Apathy of the United States government. Business developments and the organization of the Wright Company.

2288. The **Wright** brothers' aeroplane in France and the United States. illus. (Scientific American, New York, 1908, v. 99, p. 140–141.) **VA**

Somewhat detailed description of the machine.

2289. The **Wright** brothers' flying machine and what it has accomplished. (Scientific American supplement, New York, 1906, v. 61, p. 25303.) **VA**

Includes a communication from the Wright brothers to the Aero Club of America.

2290. The **Wright** brothers' and Langley's aeroplanes. (Nature, London, 1928, v. 122, p. 930.) **OA**

Controversy as to disposition of the original machines.

2291. Wright Company. The Wright Company, complainant-appellee,-*vs.*-the Herring-Curtiss Company and Glenn H. Curtiss, defendants-appellants. In equity. Appeal from Judge Hazel's order. Brief for complainant-appellee, and abstract of evidence...H. A. Toulmin, counsel for complainant. Springfield, Ohio: The Young and Bennett Co. [1910.] 16, 189 p. diagrs. 4°. **VDY**

Appeal in the United States Circuit Court of Appeals to prove the pioneer character of the Wrights' invention. Much of historical interest.

2292. The **Wright Company** vs. Herring-Curtiss Company and Glenn H. Curtiss. Brief for complainant and abstract of evidence on motion for preliminary injunction. H. A. Toulmin, solicitor and of counsel for complainant. Springfield, Ohio: The Young and Bennett Co. [1909.] 163 p. diagrs. 4°. **VDY**

Much of historical interest in this law-suit in the United States Circuit Court.

2293. Wright declares patent question settled. (Aero and hydro, Chicago, 1914, v. 7, p. 285–286.) **3–VDS**

Brief account of final settlement of Wright suit against the Curtiss Company.

2294. Wright-Paulhan opinion. (Aeronautics, New York, 1910, v. 6, p. 144–146.) **VDS**

Judge Hand's opinion in the United States Circuit Court.

2295. The **Wrights'** first flights. illus. ports. (Aero digest, New York, 1928, v. 13, p. 1105–1109, 1290, 1292, 1294.) **VDS**

2296. Wrights get first Langley medal. Aviators' work honored by Smithsonian Institution. illus. (Fly, Philadelphia, 1910, v. 2, March, p. 7–8.) **3–† VDS**

Picture of the medal.

2297. A **Yorkshire** pioneer. port. (Aeronautics, London, 1930, v. 38, p. 332–334.) **VDS**

Account of Robert Blackburn (born 1885), early builder of airplanes.

2298. Young, Clarence Marshall. Airplanes of today. port. diagr. (Western flying, Los Angeles, 1929, v. 6, Sept., p. 64–65, 242.) **VDS**

Reprinted in United States. — Commerce Department, Aeronautics Branch, *Air commerce bulletin,* Washington, 1929, v. 1, Sept. 2, p. 1–3, under title of Trend of airplane design, *VDS.*

2299. Young, Edward H. Evolution of the curve. illus. (Aircraft, New York, 1910, v. 1, p. 286–287.) **VDS**

Early history of wing curves.

2300. —— Ft. Meyer and the Wrights. illus. (Fly, Philadelphia, 1909, v. 1, Sept., p. 8–9.) **3–† VDS**

2301. Young, Kennedy. Wings of the rising sun. illus. (Aero digest, New York, 1925, v. 7, p. 647–651; 1926, v. 8, p. 9–12, 46–48, 67–70, 106–107.) **VDS**

History of aviation in Japan.

2302. Zahm, Albert Francis. Some memories of Mr. Curtiss. port. (National aeronautic review, Washington, 1930, v. 8, Aug., p. 39, 62.) **VDS**

2303. —— Stability of aeroplanes and flying machines. illus. (International Conference on Aerial Navigation, Chicago, 1893, Proceedings, New York, 1894, p. 273–287.) **VDS**

2304. Zimmerman, Charles. Note on the Cycala flying machine. (Aeronautical journal, London, 1902, v. 6, p. 65–66.) **VDS**

Author's glider with vibrating wings. Author has a supplementary paper in *Aeronautical world,* Glenville, O., 1902, v. 1, p. 97–98.

2305. Zum zehnten Jahrestage der ersten Erfolge von Etrichs Taube. illus. (Luftfahrt, Berlin, 1919, Nov., p. 9–11.) **VDS**

Heavier-Than-Air Craft, continued

Seaplanes

2306. **A. S.** Janin dies; inventor was rival of Curtiss. Loser of 8-year court fight over hydro-airplane patent succumbs at New Brighton. (New York Herald Tribune, New York, Jan. 9, 1931, p. 21.) ***A**

2307. **Archdeacon,** Ernest. Ueber den Schwebeflug. illus. port. (Illustrierte aëronautische Mitteilungen, Strassburg, 1905, Jahrg. 9, p. 342–353.) **VDS**

Describes and illustrates author's pioneer work with seaplanes.

2308. **Armstrong,** E. R. The evolution of the flying boat. illus. (Aero and hydro, Chicago, 1913, v. 6, p. 41–42.) **3–VDS**

2309. **Baker,** George S. Ten years' testing of model seaplanes. illus. (Royal Aeronautical Society, London, 1923, v. 27, p. 224–243.) **VDS**

With discussion.

Account of work done at the William Froude National Experiment Tank, England. Abstracted in *Flight,* London, 1923, v. 15, p. 85, *VDS.*

2310. **Barton,** F. A. The first British dirigible and hydro-aeroplane. illus. (Aeronautics, London, 1911, v. 4, p. 248–249; 1912, v. 5, p. 8–9.) **VDS**

Description of pioneer craft built by the author and F. L. Rawson.

2311. **Benson,** S. R. Flying boats — old and new. illus. (Aeronautic review, Baltimore, 1929, v. 7, Sept., p. 30, 56, 58.) **VDS**

History beginning with crafts of Glenn Curtiss and Enea Bossi.

2312. [**Bibliography** of seaplanes.] (United States. — National Advisory Committee for Aeronautics, Bibliography of aeronautics, 1909–1931, Washington, 1921–35.) **VDT**

See volume for 1909/16, p. 660–663; 1917/19, p. 226, 401; 1920/21, p. 206–207; 371; 1922, p. 141–142; 1923, p. 123; 1924, p. 98; 1925, p. 158; 1926, p. 129–130; 1927, p. 156–157; 1928, p. 178–179; 1929, p. 201; 1930, p. 217–218; 1931, p. 254–255.

Some of the earlier headings read Hydroaeroplanes.

2313. **Blanke,** John H. D. Dornier flying boats. illus. (Popular aviation, Chicago, 1929, v. 4, Feb., p. 98–100.) **VDS**

History of the Dornier concern.

2314. **Blenk,** H., and F. Liebers. Das Wertungsverfahren im deutschen Seeflug-Wettbewerb 1926. diagrs. (Deutsche Versuchsanstalt für Luftfahrt, Jahrbuch 1927, München, 1927, p. 31–34.) **VDS**

Methods of rating in the German seaplane contest of 1926. For English translation see United States. — Advisory Committee for Aeronautics, Technical memorandum no. 454, Washington, 1928, p. 18–31, *VDS.*

2315. **Carter,** W. G. Flying boat development for speed research. illus. (Flight, London, 1930, v. 22, p. 1068a–1068c.) **VDS**

2316. **Chanute,** Octave. [William Kress.] (Aeronautics, New York, 1915, v. 16, p. 83, 85.) **VDS**

Kress, in 1901, experimented with a motor-driven seaplane.

2317. **Cléry,** A. Henri Fabre. port. (L'Aérophile, Paris, 1912, année 20, p. 529.) **3–VDS**

Fabre was prominent in the development of the seaplane.

2318. **Collected** reports on British high speed aircraft for the 1927 Schneider trophy contest, with an introduction by W. L. Cowley. (Aeronautical Research Committee, Reports and memoranda, London, 1931, no. 1300, 372 p. illus. 8°.) **VDW**

Detailed accounts of tests made before and after the contest.

2319. **Cross,** B. C. H. Evolution of seaplanes for war. illus. (Royal United Service Institution, Journal, London, 1926, v. 71, p. 765–776.) **VWA**

History, 1911–1926.

2320. **Curtiss** granted flying boat patent. illus. (Aerial age weekly, New York, 1918, v. 7, p. 208–209.) **VDS**

Copy of patent granted February 19, 1918, to Glenn H. Curtiss. Also in *Aviation,* New York, 1918, v. 4, p. 228–231, *VDS.*

2321. **Curtiss's** experiments in rising from the water. How the winner of the Scientific American trophy developed a satisfactory float for his biplane. illus. (Scientific American supplement, New York, 1911, v. 71, p. 132.) **VA**

2322. **Development** of the N. C. seaplanes. illus. (Aviation, New York, 1919, v. 6, p. 468–474.) **VDS**

2323. The **Development** of the seaplane. illus. (Aeroplane, London, 1915, v. 9, p. 842, 844, and supplementary p. iv–v.) **VDS**

Evolution is shown by 33 illustrations with brief descriptive texts.

2324. The **Development** of the seaplane. (Aerial age weekly, New York, 1916, v. 3, p. 627–628.) **VDS**

Twenty-six illustrations with brief descriptive texts.

2325. **Dornier,** Claude. Lessons of the Do-X. illus. (Royal Aeronautical Society, Journal, London, 1933, v. 37, p. 757–782.) **VDS**

With discussion.

2326. —— Vorträge und Abhandlungen aus dem Gebiete des Flugzeugbaues und Luftschiffbaues, 1914–1930. [Berlin-Lichterfelde: Verlag für deutsches Flugwesen, G.m.b.H., 1930.] 145 p. illus. port. 4°. **† VDY**

In addition to several technical papers, relates the history of the flying boat *Do X,* of which the author was designer.

Heavier-Than-Air Craft — Seaplanes, cont'd

2327. **First** flying boat service. (Aeronautics, London, 1919, v. 17, p. 163.) **VDS**

Established in 1919 by the Supermarine Aviation Works, of Southampton, England.

2328. The **First** light flying boat. illus. (Aviation, Highland, N. Y., 1924, v. 16, p. 593.) **VDS**

Description and drawing of light type with duralumin hull. Built by Short Brothers of England.

2329. **First** successful hydro-triplane tested. (Aero and hydro, Chicago, 1914, v. 7, p. 238.) **3–VDS**

Description of Euler plane made in Germany.

2330. **Garner,** H. M. Seaplane research. illus. (Royal Aeronautical Society, Journal, London, 1933, v. 37, p. 830–863.) **VDS**

With discussion.
Bibliography, p. 855.

2331. **Greer,** John Fulton. First flight of an American aeroplane from the water. illus. (Scientific American, New York, 1911, v. 104, p. 132.) **VA**

Exploit of Glenn H. Curtiss on Jan. 26, 1911.

2332. **Grey,** Charles Grey. Alessandro Guidoni. [By Charles G. Grey.] (Aeroplane, London, 1928, v. 34, p. 620.) **VDS**

Obituary of Italian engineer pioneer. He was one of the first to experiment with flying boats.

2333. —— The pioneers of seaplanes. illus. (Aeroplane, London, 1925, v. 28, p. 516, 518, 520.) **VDS**

History of Short Brothers, of England.

2334. **Grover** Loening granted patent on single wheel amphibian. illus. (U. S. air services, Washington, 1927, v. 12, May, p. 40.) **VDS**

2335. **Guidoni,** Alessandro. Development of the hydroplane. illus. (Engineering, London, 1913, v. 95, p. 170–172.) **VDA**

Technical developments of seaplanes with graphs and illustrations.

2336. **Gustosa,** Corrado. Particolari costruttivi dell' aerovascello "Do X." illus. (Rivista aeronautica, Roma, 1929, anno 5, no. 10, Oct., p. 84–113.) **VDS**

Translation in United States. — Advisory Committee for Aeronautics, Technical memorandum no. 546, Washington, 1930, 25 p., *VDS.*

2337. The **History** of the flying boat. illus. (Aeronautics, New York, 1913, v. 12, p. 5–15.) **VDS**

2338. **Hunsaker,** James Clarke. How the modern seaplane was born. (Aviation, Highland, N. Y., 1925, v. 18, p. 154.) **VDS**

History starts with Curtiss craft of 1911.

2339. —— Progress in naval aircraft. illus. (Society of Automotive Engineers, Journal, New York, 1919, v. 5, p. 31–44.) **TOL**

History of the N. C. flying boats of the United States Navy. "The development of really large flying craft before 1917 was held back because no suitable engine had been designed." Also in *Automotive industries,* New York, 1919, v. 41, p. 68–72, 120–123, 172–176, *TOL.*

2340. **Janin** vs. Curtiss law suit. (Aeronautics, London, 1916, v. 11, p. 425.) **VDS**

Albert S. Janin wins suit against the Curtiss Company for infringement of seaplane patent.

2341. **Kean.** Racing seaplanes. A designer's experience with their development in the United States. illus. (Aviation, New York, 1931, v. 30, p. 519–521.) **VDS**

2342. **Kress,** Wilhelm. Aëroveloce lenkbare Flugmaschine. Wien: Selbstverlag, 1880. 26 p. plate. 8°. **VDY p.v.27, no.11**

Description and drawing of proposed seaplane, with articulated wings.

2343. **Loening,** Grover Cleveland. An American designer's review of the Schneider race [of 1929]. illus. (Aviation, New York, 1929, v. 27, p. 650–652.) **VDS**

For writings concerning Loening and his amphibian planes see United States. — National Advisory Committee for Aeronautics, *Bibliography of aeronautics,* 1909–1931, Washington, 1921–35, volumes for 1909/16, p. 803–804; 1917/19, p. 268–269; 1920/21, p. 249; 1922, p. 87; 1923, p. 77; 1924, p. 61; 1925, p. 99; 1927, p. 103–104; 1928, p. 115; 1929, p. 130, *VDT.*

2344. **Loening** develops single-wheel amphibian. illus. (Aviation, Highland, N. Y., 1927, v. 22, p. 626.) **VDS**

2345. **Massy,** H. S. The seaplane and its development. (Royal United Service Institution, Journal, London, 1913, v. 57, p. 1452–1467.) **VWA**

2346. **Orcy,** Ladislas d'. Progress of the seagoing flying boat. illus. (Scientific American, New York, 1915, v. 113, p. 555, 565.) **VA**

History of the *America* built in the United States and sold to Great Britain in 1914. Account of building of a huge military plane at Buffalo, N. Y.

2347. **Penny,** R. E. Seaplane development. illus. (Royal Aeronautical Society, Journal, London, 1927, v. 31, p. 844–885.) **VDS**

Good résumé of early types.

2347a. **Poincaré,** L. L'évolution du matériel d'aviation maritime depuis la guerre. illus. (In: L'Aéronautique militaire, maritime, coloniale et marchande, Paris, 1931, p. 113–128.) **† VDY**

Technical advances in seaplanes since the World War.

2348. Le **Premier** aéroplane marin. illus. (L'Aérophile, Paris, 1910, année 18, p. 400–401.) **3–VDS**

Description and pictures of Henri Farman's machine.

2349. The **Ravaud** aero-hydroplane. illus. (Scientific American supplement, New York, 1909, v. 67, p. 280.) **VA**

2350. **Richardson,** Holden Chester. Developments in naval aeronautics. illus. (Society of Automotive Engineers, Journal, New York, 1924, v. 15, p. 412–417.) **TOL**

United States Navy developments. For discussion see above journal, 1925, v. 16, p. 361–366.

Heavier-Than-Air Craft — Seaplanes, cont'd

2351. —— Naval development of floats for aircraft. illus. (Society of Naval Architects and Marine Engineers, Transactions, New York, 1926, v. 34, p. 15–28.) **VXA**

Good early history.
With discussion. Illustrations at end of volume.

2352. —— The trend of flying boat development. illus. (American Society of Naval Engineers, Journal, Washington, 1926, v. 38, p. 233–253.) **VXA**

2353. Rivière, Pierre. Les hydroaéroplanes. Étude historique, technique et descriptive de l'aéroplane marin. illus. (Revue générale de l'aéronautique militaire, Paris, 1912, v. 2, p. 434–457; v. 3, p. 53–54, 138–155, 203, 230.) **3–VDS**

2354. Samson, Charles Rumney. The operation of flying boats in the Mediterranean. (Royal Aeronautical Society, Journal, London, 1925, v. 29, p. 114–131.) **VDS**

With discussion.
A record of 22 post-war flights and of the difficulties encountered. Abridged in *Flight*, London, 1925, v. 17, p. 101–102, *VDS*.

2355. Sayers, W. H. The work of the Marine Experimental Aircraft Depot, Isle of Grain. (Aeroplane, London, 1919, v. 17, p. 1589–1594, 1661–1666, 1743–1746, 1911–1912, 2013–2016.) **VDS**

2356. Seewald, F. Erfahrungen aus dem Deutschen Seeflug-Wettbewerb 1926. diagrs. (Deutsche Versuchsanstalt für Luftfahrt, Jahrbuch 1927, München, 1927, p. 26–30.) **VDS**

Lessons taught from the 1926 German seaplane contest. For English translation see United States. — Advisory Committee for Aeronautics, Technical memorandum no. 454, Washington, 1928, p. 1–17, *VDS*.

2357. Short, H. Oswald. Twenty-one years' progress of seaplane design. illus. (Flight, London, 1930, v. 22, p. 67–70.) **VDS**

2358. Sikorsky S-41 is the largest amphibian in the world; but a larger one is being built. illus. (U. S. air services, Washington, 1930, v. 15, Oct., p. 41.) **VDS**

2358a. Some Dornier "milestones." illus. (Flight, London, 1920, v. 12, p. 1269–1273, 1289–1292.) **VDS**

History of the seaplanes turned out by German firm, beginning with 1914.

2359. Stockbridge, Frank Parker. How sea flying was made safe. The first hydroplane. Ship-to-shore flights. Other triumphs of Glenn Curtiss, air pioneer. illus. (Popular science monthly, New York, 1927, v. 110, June, p. 32–33, 141–143.) *** DA**

2360. Tilgenkamp, Erich. Do X, das grösste Flugschiff der Welt. Zürich: Orell Füssli [cop. 1931]. 16 p. port. plates. 8°. **VDY**

Excellent photographs of the famous six-motored seaplane.

2361. Tuttle, Thomas Taft. Epoch-making achievements of Glenn H. Curtiss. Performances of hydro-aeroplane cause sensation in military circles. illus. (Fly, Philadelphia, 1911, v. 3, March, p. 4–5.) **3–† VDS**

2362. Woodhouse, Henry. The evolution of the hydroaeroplane. illus. (Aero Club of America, Bulletin, New York, 1912, v. 1, April, p. 5–9.) **VDS (Flying)**

2363. Zimmermann, Paul G. America builds and operates the first all metal hull flying boat. illus. (Aeronautical digest, New York, 1923, v. 3, p. 422–423.) **VDS**

Gliders and Gliding

2364. An **American** pioneer of soaring flight: John J. Montgomery. illus. (Aviation, New York, 1918, v. 4, p. 302–303.) **VDS**

Montgomery, whose experiments began in the early eighties, died in 1911.

2365. Ashwell-Cooke, J. R. Motorless flight. London: John Hamilton, Ltd. [1932.] vii, 149 p. illus. 12°. **VDY**

Brief history of developments, chiefly in England, during post-war period.

2366. Berchem, Beda von. Gustav [Lilienthal] carries on. An octogenarian seeks to prove that the world took the wrong cue from his brother's pioneer role. illus. (Sportsman pilot, Albany, N. Y., 1931, v. 6, Dec., p. 34–35, 42.) **VDS**

2367. [**Bibliography** of gliders and gliding.] (United States. — National Advisory Committee for Aeronautics, Bibliography of aeronautics, 1909–1931, Washington, 1921–35.) **VDT**

See volumes for 1909/16, p. 545–548; 1917/19, p. 187; 1920/21, p. 169; 1922, p. 57–58; 1923, p. 49–50; 1924, p. 38–39; 1925, p. 70; 1926, p. 54; 1927, p. 67; 1928, p. 77; 1929, p. 90; 1930, p. 88–89; 1931, p. 106–107.

2368. Blackden, L. S. Experiments relative to equilibrium and angle of fall in gliding flight. illus. (Aeronautical journal, London, 1903, v. 7, p. 53–61.) **VDS**

2369. —— Observations and experiments relative to equilibrium in air of a body heavier than air. illus. (Aeronautical journal, London, 1903, v. 7, p. 28–41.) **VDS**

Author's gliding experiments. Also in *Aeronautical world*, Glenville, O., 1903, v. 1, p. 241–247, 272–277, *VDS*.

2370. Blanke, John H. D. Pioneers in gliding. illus. (Popular aviation, Chicago, 1928, v. 2, May, p. 74–77; June, p. 38, 40, 86–87.) **VDS**

Lilienthal, Pilcher, Wright brothers, Chanute, development of the sport in Europe.

2371. Boyer, Jacques. The soaring flight of birds attained mechanically. illus. (Scientific American, New York, 1908, v. 99, p. 63.) **VA**

Experiments of Marcel Deprez.

Heavier-Than-Air Craft — Gliders, etc., cont'd

2371a. The **British** sailplane meet at Lewes. Peyret tandem monoplane piloted by A. Maneyrol makes new world's duration record of 3 hr. 21 min. 7 sec. illus. (Aviation, Highland, N. Y., 1922, v. 13, p. 684–686.) **VDS**

Daily Mail contest, October, 1922.

2372. **Bryan,** George Hartley, and W. E. Williams. The longitudinal stability of aerial gliders. (Royal Society of London, Proceedings, London, 1904, v. 73, p. 100–116.) *** EC**

"To show that the longitudinal stability of aeroplane systems *can* be made the subject of mathematical calculation." This paper is an aeronautical classic.

2373. —— The longitudinal stability of aëroplane gliders. illus. (Aeronautical journal, London, 1904, v. 8, p. 12–19.) **VDS**

This paper, read before the Aeronautical Society of Great Britain, is not the same as that read before the Royal Society (see preceding entry).

2374. **Burton,** Walter E. Sailboats of the air. illus. (Popular aviation, Chicago, 1930, v. 6, Jan., p. 26–27, 62, 64.) **VDS**

History of glider development since the world war. This number of *Popular aviation* is published under the title of *Aeronautics.*

2375. **Chanute,** Octave. American gliding experiments. illus. (Illustrierte aëronautische Mitteilungen, Strassburg, 1898, Jahrg. 2, p. 4–12.) **VDS**

Text in German and English.

2376. —— Experiments in flying. An account of the author's own inventions and adventures. (McClure's magazine, New York, 1900, v. 15, p. 127–133.) *** DA**

2377. —— Gliding experiments. illus. (Western Society of Engineers, Journal, 1897, v. 2, p. 593–628.) **VDA**

Also in *Scientific American supplement,* New York, 1898, v. 45, p. 18368–18370, 18380–18382, 18389–18391, *VA.*

2378. —— La navigation aérienne aux États-Unis. illus. (L'Aérophile, Paris, 1903, année 11, p. 171–183.) **3–VDS**

Gliding experiments of Chanute, Herring, and the Wright brothers.

2379. —— Recent experiments in gliding flight. (Aeronautical annual, Boston, 1897, no. 3, p. 30–53.) **3–VDS**

Author's experiences.

2380. **Chanute's** original models found. illus. (Popular aviation, Chicago, 1928, v. 2, Jan., p. 52–53.) **VDS**

2381. The **Death** of [Otto] Lilienthal. illus. (Aeronautical journal, London, 1897, v. 1, Jan., p. 11–13.) **VDS (Royal)**

Discussion as to the cause of fatal accident.

2382. **Death** of Professor John J. Montgomery. (Aeronautics, New York, 1911, v. 9, p. 151–154.) **VDS**

Claims to have used warped surfaces between 1883 and 1886.

2383. **Degoul,** M. Albert Bazin. port. (L'Aérophile, Paris, 1910, année 18, p. 145.) **3–VDS**

Brief biography of French pioneer in the scientific study and construction of gliders.

2384. **Deprez,** Marcel. Soaring flight. Models for its mechanical investigation. illus. (Scientific American supplement, 1908, v. 66, p. 165–166.) **VA**

Account of author's experiments.

2385. **Desmarest,** Léopold. La mort d'Otto Lilienthal. (L'Aéronaute, Paris, 1896, année 29, p. 239–245.) **3–VDS**

Account of Lilienthal's fatal accident.

2386. —— Les nouvelles expériences de M. Otto Lilienthal. illus. (L'Aéronaute, Paris, 1895, année 28, p. 27–39.) **3–VDS**

2387. **Dienstbach,** Carl. Ein Flug von 20 Minuten mit dem Gleitapparat von [J. J.] Montgomery. illus. (Illustrierte aëronautische Mitteilungen, Strassburg, 1905, Jahrg. 9, p. 254–259.) **VDS**

Flight of Montgomery's glider, with passenger, to a height of 1,000 meters, 1905.

2388. **Dowd,** R. E. Elmira, America's Wasserkuppe. The first national soaring competition in the United States. illus. (Aero digest, New York, 1930, v. 17, Nov., p. 48–49, 158, 160.) **VDS**

2389. —— Second national soaring meet. illus. (Aero digest, New York, 1931, v. 19, Sept., p. 38–39, 102.) **VDS**

Again held at Elmira, N. Y.

2390. —— Soaring and gliding at Elmira. The third national contest proves to be the greatest ever. illus. (Aero digest, New York, 1932, v. 21, Aug., p. 26–27, 78.) **VDS**

2391. —— Fifth annual soaring contest at Elmira. illus. (Aero digest, New York, 1934, v. 25, Aug., p. 18–20.) **VDS**

2392. The **Fatal** accident to Mr. Pilcher. port. illus. (Aeronautical journal, London, 1899, v. 3, p. 86–89.) **VDS (Royal)**

Percy S. Pilcher (1866–1899) began experimenting with soaring machines about 1894.

2393. **Ferber,** Louis Ferdinand. Expériences d'aviation par De Rue [pseud. of Louis Ferdinand Ferber.] illus. (L'Aérophile, Paris, 1903, année 11, p. 36–40.) **3–VDS**

Glider experiments of author.

2394. —— Les progrès de l'aviation depuis 1891 par le vol plané. Paris: Berger-Levrault & Cie., 1905. 53 p. illus. 8°. **VDY p.v.7, no.6**

Gliding experiments of Lilienthal, Pilcher, Chanute, the Wright brothers, Langley, and the author.

2395–2396. **Fitzgerald,** George Francis. Experimental soaring. (Aeronautical journal, London, 1898, v. 2, p. 75–76.) **VDS**

Experiments of distinguished British physicist.

Heavier-Than-Air Craft — Gliders, etc., cont'd

2397. **George** H. Madelung, aeronautical engineer, inventor of the world's record glider. port. (Aeronautical digest, New York, 1922, v. 1, p. 266.) **VDS**

2398. **Georgii,** Walter. Bericht über den 11. Rhön-Segelflug-Wettbewerb. illus. (Zeitschrift für Flugtechnik und Motorluftschiffahrt, München, 1931, Jahrg. 22, p. 129–140.) **VDS**

English translation in National Advisory Committee for Aeronautics, Technical memorandum no. 623, Washington, 1931, 16 p., *VDS.*

2399. —— Ten years' gliding and soaring in Germany. illus. (Royal Meteorological Society, Quarterly journal, London, 1930, v. 56, p. 141–149.) **PRA**

Reprinted in Smithsonian Institution, *Annual report for 1929/30,* Washington, 1931, p. 273–283, * *EA.*

2400. **Glenville,** I. E. Novel aerial sailing craft. illus. (American inventor, New York, 1904, v. 12, p. 449.) *** V**

Experiments of W. E. Irish, of Ohio, with glider constructed of light wood and turkey feathers.

2401. **Gliders** and motorless planes. illus. (Popular aviation, Chicago, 1927, v. 1, Aug., p. 20–22, 45.) **VDS**

Brief history from time of Leonardo da Vinci.

2402. **G[ustav]** Lilienthal, 84, German glider expert, is dead. With brother, who was killed in 1896, sought to develop bird-like plane. port. (New York Herald Tribune, New York, Feb. 2, 1933, p. 15.) *** A**

2403. **Hargrave,** Lawrence. The possibility of soaring in horizontal wind. illus. (Royal Society of New South Wales, Journal and proceedings, Sydney, 1897, v. 31, p. 207–213.) *** EC**

Also in *Aeronautical journal,* London, 1898, v. 2, p. 29–33, *VDS.*

2404. —— Soaring machines. illus. (Royal Society of New South Wales, Journal and proceedings, Sydney, 1898, v. 32, p. 209–222.) *** EC**

Further development of experiments explained in entry no. 2403.

2405. **Heitman,** Charles. Lilienthal. port. illus. (Aircraft, New York, 1910, v. 1, p. 10–11.) **VDS**

Portrait and brief account of Lilienthal's experiments.

2406. **Herring,** A. M. Recent advances toward a solution of the problem of the century. (Aeronautical annual, Boston, 1897, no. 3, p. 54–74.) **3–VDS**

Author's experiences with kites and gliders.

2407. **Hirth,** Wolfram K. E., and others. Elmira soaring contest, 1930, by W. K. E. Hirth, M. H. Schempp, and Jack Herrick. Washington, 1931. 15 p. illus. 4°. (United States. — National Advisory Committee for Aeronautics. Technical memorandum no. 613.) **VDS**

2408. **Howard-Flanders,** Richard Leonard, and C. F. Carr. Gliding and motorless flight. London: Sir Isaac Pitman & Sons, Ltd., 1932. xiii, 145 p. illus. 2. ed. 8°. **VDY**

Early history, and progress in recent times.

2409. **Hureau de Villeneuve,** Abel. L'homme volant: M. Otto Lilienthal. (L'Aéronaute, Paris, 1894, année 27, p. 5–10.) **3–VDS**

States that Lilienthal's machine is little more than a gliding parachute.

2410. **John** J. Montgomery falls to death in glider. Noted authority meets Lilienthal's fate. (Fly, Philadelphia, 1911, v. 4, Dec., p. 19.) **3–† VDS**

2411. **Josselyn,** Winsor. Red-head Dan Maloney. How fame flashed into the life of an unknown Irishman and then deserted him in history. illus. (Sportsman pilot, Albany, N. Y., 1931, v. 6, Nov., p. 14–15, 46.) **VDS**

Maloney meets death in Montgomery's glider, Santa Clara, California, 1905.

2412. **Kensche,** Heinz. Der 14. Rhönsegelflug-Wettbewerb 1933. illus. (Verein deutscher Ingenieure, Zeitschrift, Berlin, 1933, Bd. 77, p. 1187–1189.) **VDA**

2413. **Kreiss,** Eugen. Ueber die Mechanik im Dienste der Flugtechnik. (Zeitschrift für Luftschiffahrt, Berlin, 1893, Jahrg. 12, p. 33–43.) **3–VDS**

Discussion of Lilienthal's ideas.

2414. **Kress,** Wilhelm. Ueber Wind und Segelflug. (Zeitschrift für Luftschiffahrt, Berlin, 1888, Jahrg. 7, p. 257–264.) **3–VDS**

2415. **Kronfeld,** Robert. Kronfeld on gliding and soaring; the story of motorless human flight, translated by J. Manchot. Preface by Colonel the Master of Sempill. London: John Hamilton, Ltd. [1932.] xiv, 379 p. illus. 8°. **VDY**

Excellent, well-illustrated account of development from early days down to the present. Reviewed in *Air and airways,* London, 1932, v. 9, p. 287–288, *VDS.*

2416. **Krotki,** Rudolf. 14 hours motorless. illus. (Western flying, Los Angeles, 1930, v. 7, Jan., p. 67–68.) **VDS**

Story of record-breaking flight by Dinort in 1929, told in very quaint English.

2417. **L.,** W. J. S. Lilienthal's experiments on flying. illus. (Nature, London, 1896, v. 53, p. 300–301.) **OA**

2418. **Lachmann,** Gustav. The development of soaring flight in Germany. (Aeroplane, London, 1923, v. 25, p. 76.) **VDS**

2419. —— Neuere Versuchsergebnisse mit Spaltflügeln. diagrs. (Zeitschrift für Flugtechnik und Motorluftschiffahrt, München, 1924, Jahrg. 15, p. 109–116, 173–176, 181–185.) **VDS**

Heavier-Than-Air Craft — Gliders, etc., cont'd

2420. **Langsdorf,** Werner von. Das Segelflugzeug. München: J. F. Lehmann, 1931. 230 p. illus. 2. ed. 12°. VDY

Good historical introduction. Reviewed in *Aviation*, New York, 1931, v. 30, p. 660.

2421. The **Late** Percy S. Pilcher. port. (front.) (Flying, London, 1903, Jan., no. 5, p. 210.) VDS

2422. **Lauriol,** P. Les expériences de M. Lilienthal. illus. (Revue de l'aéronautique, Paris, 1895, année 8, p. 1–10.) 3–† VDS

2423. **Lilienthal,** Gustav. The early experiments of Otto and Gustav Lilienthal. illus. (Aeronautical journal, London, 1909, v. 13, p. 85–87.) VDS (Royal)

2424. —— ...Vom Gleitflug zum Segelflug: Flugstudien auf Grund zahlreicher Versuche und Messungen. Berlin-Charlottenburg: C. J. E. Volckmann Nachf., 1923. 159 p. illus. 16°. (Volckmann's Bibliothek für Flugwesen. Bd. 15.) VDY (Volckmann's)

Lilienthal's studies on bird flight and a record of his gliding experiments.

2425. **Lilienthal,** Otto. La découverte des meilleures formes d'ailes. illus. (L'Aéronaute, Paris, 1896, année 29, p. 5–18.) 3–VDS

For comments see *L'Aéronaute*, Paris, 1896, année 29, p. 27–29, *3–VDS*.

2426. —— Practical experiments for the development of human flight. illus. (Aeronautical annual, Boston, 1896, no. 2, p. 7–22.) 3–VDS

2427. —— Praktische Erfahrungen beim Segelflug. illus. (Prometheus, Berlin, 1893, Jahrg. 5, p. 161–162, 182–186.) 3–VA

Author's experiments. Translation of the second installment is in Smithsonian Institution, *Annual report for 1892/93*, Washington, 1894, p. 195–199, * *EA*.

2428. —— Die Profile der Segelflächen und ihre Wirkung. illus. (Zeitschrift für Luftschiffahrt, Berlin, 1895, Jahrg. 14, p. 42–57.) 3–VDS

2429. —— Die Tragfähigkeit gewölbter Flächen beim practischen Segelflug. illus. (Zeitschrift für Luftschiffahrt, Berlin, 1893, Jahrg. 12, p. 259–272.) 3–VDS

2430. —— Ueber die Ermittelung der besten Flügelformen. illus. (Zeitschrift für Luftschiffahrt, Berlin, 1895, Jahrg. 14, p. 237–245.) 3–VDS

2431. —— Ueber die Mechanik im Dienste der Flugtechnik. (Zeitschrift für Luftschiffahrt, Berlin, 1892, Jahrg. 11, p. 180–186.) 3–VDS

Discusses possibility of imitating bird flight.

2432. —— Ueber meine diesjährigen Flugversuche. (Zeitschrift für Luftschiffahrt, Berlin, 1891, Jahrg. 10, p. 286–291.) 3–VDS

2433. —— Ueber die Möglichkeit des freien Fluges. illus. (Verein zur Beförderung des Gewerbfleisses, Verhandlungen, Sitzungsberichte, Berlin, 1890, Bd. 69, p. 114–127.) 3–VA

Discusses wing form and describes apparatus used in measuring lifting power.

2434. —— Ueber Theorie und Praxis des freien Fluges. illus. (Zeitschrift für Luftschiffahrt, Berlin, 1891, Jahrg. 10, p. 153–164.) 3–VDS

2435. —— Zur Flugfrage. illus. (Prometheus, Berlin, 1893, Jahrg. 4, p. 753–756, 769–774.) 3–VA

The first part deals with Hargrave's flying machine; part 2, with the author's experiments. Translated abstract of part 2 is in Smithsonian Institution, *Annual report for 1892/93*, Washington, 1894, p. 189–194, * *EA*.

For numerous references concerning Otto and Gustave Lilienthal and their work see Paul Brockett, *Bibliography of aeronautics*, Washington, 1910, p. 520–523, *VDT*. Continued in United States. — National Advisory Committee for Aeronautics, *Bibliography of aeronautics*, 1909–1931, Washington, 1921–35, volumes for 1909/16, p. 798–799; 1917/19, p. 267–268; 1920/21, p. 247; 1922, p. 87; 1923, p. 76; 1924, p. 61; 1925, p. 97; 1926, p. 80; 1930, p. 138; 1931, p. 166, *VDT*.

2436. The **Lilienthal** and Pilcher gliders compared. illus. (Flight, London, 1910, v. 2, p. 6–8, 19–22.) VDS

2437. **Lilienthal's** experiments in flying. illus. (Nature, London, 1894, v. 51, p. 177–179.) OA

2438. **Lippish,** A. The development, design and construction of gliders and sailplanes. illus. (Royal Aeronautical Society, Journal, London, 1931, v. 35, p. 532–578.) VDS

With discussion.

Well-illustrated account of ten years' progress in practice and theory. Reprinted in United States. — National Advisory Committee for Aeronautics, Technical memorandum no. 637, Washington, 1931, 38 p., *VDS*.

2439. —— Technischer Bericht des Rhön-Segelflug-Wettbewerbs 1928. illus. (Zeitschrift für Flugtechnik und Motorluftschiffahrt, München, 1929, Jahrg. 20, p. 65–74.) VDS

Translated in United States. — National Advisory Committee for Aeronautics, Technical memorandum no. 514, Washington, May, 1929, 12 p., *VDS*.

2439a. **Magoun,** Frederick Alexander, and Eric Hodgins. Heavier-than-air man-carrying flight without power. illus. (In their: A history of aircraft, New York, 1931, p. 249–283.) VDT

Historic beginnings. The pioneering of Sir George Cayley. Adventures of Le Bris. Wenham's designs. Montgomery's gliders. Gliders of Otto Lilienthal. Hargrave's influence on gliding. The work of Pilcher. Bell's tetrahedral kites. Contributions of Chanute. The Wright gliders. Post-war glider progress.

2440. **Marsh,** William Lockwood. Octave Chanute, a pioneer of gliding, was also aviation's greatest historian. (Airways, London, 1926, v. 3, p. 98.) VDS

2441. **Maxim,** Hudson. Tribute to Octave Chanute. Father of aviation merited high honors. (Fly, Philadelphia, 1911, v. 3, Feb., p. 13.) 3–† VDS

Heavier-Than-Air Craft — Gliders, etc., cont'd

2442. Means, James. Octave Chanute's work in aviation. port. (Aeronautics, New York, 1911, v. 8, p. 3, 35.) **VDS**

2443. —— The problem of manflight. Boston: W. B. Clarke & Co., 1894. 20 p. illus. 8°. **VDY**

Description of author's experimental glider model.

2444. Montgomery, John J. Some early gliding experiments in America. illus. (Aeronautics, New York, 1909, v. 4, p. 47–50.) **VDS**

Account of flights by author in California, 1884–1905.

For references to the work of Montgomery see United States. — National Advisory Committee for Aeronautics, *Bibliography of aeronautics,* 1909–1931, Washington, 1921–35, volumes for 1909/16, p. 921; 1917/19, p. 315, *VDT.*

2445. Motorless flight; early experiments and the remarkable trials of 1922. (Aircraft yearbook, New York, 1923, p. 201–207.) **VDS**

2446. Müllenhoff, Karl. Éloge funèbre d'Otto Lilienthal. (L'Aéronaute, Paris, 1897, année 30, p. 51–59.) **3–VDS**

2447. —— Zur Erinnerung an Otto Lilienthal. port. (Zeitschrift für Luftschiffahrt und Physik der Atmosphäre, Berlin, 1896, Jahrg. 15, p. 289–295.) **3–VDS**

Account of Lilienthal's glider experiments. An English translation in *Aeronautical annual,* Boston, 1897, no. 3, p. 75–83, *3–VDS.*

2448. Needham, Cecil Hugh Latimer. The conquest of soaring flight. (Air annual of the British Empire, London, 1931–32, v. 3, p. 25–32.) **VDS**

History of motorless flight.

Abridged in *Aeronautic review,* Washington, 1932, v. 10, Jan., p. 13–16, *VDS.*

2449. Nimfuehr, Raimund. Das Segelflugproblem. Eine flugtechnische Studie in gemeinverständlicher Darstellung. (Oesterreichische Flug-Zeitschrift, Wien, 1912, Jahrg. 6, p. 211–217.) **VDS**

Has a bibliography.

2450. Oakes, Edward. Soaring flight. (Aeronautical journal, London, 1897, v. 1, April, p. 4–5.) **VDS**

Discussion of curved wings.

2451. Octave Chanute, 1832–1910. (Flight, London, 1910, v. 2, p. 984.) **VDS**

Brief outline of his career.

2452. Pagé, Victor Wilfred. Henley's ABC of gliding and sailflying. New York: The Norman W. Henley Publishing Co., 1930. x, 294 p. illus. 8°. **VDY**

Chapter 1: History of early glider development.

2453. Parseval, August von. Ueber das Segelproblem. illus. (Zeitschrift für Luftschiffahrt, Berlin, 1896, Jahrg. 15, p. 269–274; 1898, Jahrg. 17, p. 175–183.) **3–VDS**

2454. Pénaud, Alphonse. L'homme peut-il voler sans moteur auxiliaire? (L'Aéronaute, Paris, 1871, année 4, p. 81–88.) **3–VDS**

2455. Pilcher, Percy Sinclair. Experiments with soaring machines. illus. (Nature, London, 1896, v. 53, p. 365–366.) **OA**

2456. —— Gliding, to which is added the aeronautical work of John Stringfellow. London: Aeronautical Society of Great Britain, 1910. 39 p. illus. ports. 12°. (Aeronautical classics. no. 5.) **VDY (Royal)**

Account of author's experiments and those of Lilienthal. For abstract see *Aeronautical journal,* London, 1897, v. 1, April, p. 1–4, *VDS (Royal).* Includes a biography.

For references concerning Pilcher and his work see United States. — National Advisory Committee for Aeronautics, *Bibliography of aeronautics,* 1909–1931, Washington, 1921–35, volumes for 1909/16, p. 1042–1043; 1917/19, p. 354; 1920/21, p. 325; 1928, p. 157, *VDT.*

2457. Sanderval, de. Expériences d'aviation sur le vol plané. illus. (La Nature, Paris, 1886, année 14, semestre 2, p. 404–406.) **OA**

Author's experiences with gliders equipped with adjustable wings.

Translated in *Scientific American supplement,* New York, 1887, v. 23, p. 9175, *VA.*

2458. —— Sur un aéroplane pouvant contribuer aux progrès de la navigation aérienne. (Académie des sciences, Comptes rendus, Paris, 1883, tome 97, p. 1273–1274.) *** EO**

2459. Sayers, W. H. Gliders: their history and uses. (Royal Aeronautical Society, Journal, London, 1924, v. 28, p. 315–322.) **VDS**

2460. Slawson, H. H. The cradle of American aviation. port. illus. (Popular aviation, Chicago, 1928, v. 3, Nov., p. 80–81, 107–111.) **VDS**

Biography of Octave Chanute (1832–1910).

2461. Snow, V. G. Gliding — modern airman's sport — an early bird in flying history. illus. (Aeronautic review, Washington, 1930, v. 8, March, p. 10–15.) **VDS**

Brief history from early times.

2462. Soaring flight and soaring aircraft. A review of the progress achieved in soaring flight with considerations of its practical possibilities. illus. (Aviation, Highland, N. Y., 1922, v. 12, p. 195–198, 224–225.) **VDS**

History of modern developments.

2463. Speight, W. L. A South African gliding pioneer. illus. (Air, London, 1931, v. 4, p. 94–95.) **VDS**

It is stated that in 1875, a farmer's boy, Goodman Household, flew 200 yards in a home-made glider.

2464. Steinmetz, Charles Proteus. America's first glider club. illus. (Sportsman pilot, Albany, N. Y., 1930, v. 4, Dec., p. 26–27, 46–47.) **VDS**

Club was founded by the author at Schenectady, N. Y., in 1894.

2465. Tidswell, Ella. The work and experiments of Percy S. Pilcher. illus. (Aeronautical journal, London, 1909, v. 13, p. 87–89.) **VDS (Royal)**

Pilcher's sister presents a brief account of his experiments.

Heavier-Than-Air Craft — Gliders, etc., cont'd

2466. **Van Dusen,** L. Charlie Lindbergh, glider pilot. illus. (Western flying, Los Angeles, 1930, v. 7, May, p. 50–53, 143.) **VDS**
How Lindbergh got his glider license.

2467. **Vernon.** "The flying man." Otto Lilienthal's flying machine. illus. port. (McClure's magazine, New York, 1894, v. 3, p. 323–331.) *** DA**

2468. **Walker,** Donald F. A glider year [1930]. illus. (Airway age, New York, 1931, v. 12, p. 58–61.) **VDS**
Records and developments.

2469. —— Growth of the glider movement. illus. (Aero digest, New York, 1929, v. 14, June, p. 48–49, 224, 226.) **VDS**

2470. **Warner,** Edward Pearson. Gliding experiments in Europe — 1922. An eyewitness's review of the results achieved at the recent French and German gliding meets. illus. (Aviation, Highland, N. Y., 1922, v. 13, p. 376–380.) **VDS**

2471. **Weiss,** Joseph Bernard. Gliding and soaring flight; a survey of man's endeavour to fly by natural methods, with a preface by C. G. Grey, and an appendix by W. H. Sayers. London: S. Low, Marston & Co., Ltd. [1923.] xx, 164 p. illus. 12°. **VDY**
Describes experiments of Cayley, Le Bris, Mouillard, Lilienthal, Pilcher, Chanute, Langley, J. J. Montgomery, Wright brothers, Voisin brothers, José Weiss, and Igo Etrich.

2472. **Wilhelm** Kress. port. (L'Aérophile, Paris, 1913, année 21, p. 153.) **3–VDS**
Portrait and brief biography of Austrian pioneer in the scientific study of gliders.

2473. **Wright,** Wilbur. Otto Lilienthal. illus. (Aero Club of America, Bulletin, New York, 1912, v. 1, Sept., p. 20–21.) **VDS (Flying)**
Appreciation of Lilienthal's work.

2474. —— [The work of Octave Chanute.] port. (Aeronautics, New York, 1911, v. 8, p. 4.) **VDS**

2475. **Zahm,** Albert Francis. Octave Chanute, his work and influence in aeronautics. port. (Scientific American, New York, 1911, v. 104, p. 463–488.) **VA**
Excellent portrait of Chanute.

Parachutes

2476. **Angerstein,** Wilhelm. Die neueren Fallschirmversuche. illus. (Zeitschrift für Luftschiffahrt, Berlin, 1889, Jahrg. 8, p. 192–208.) **3–VDS**
Good history of parachutes.

2477. **Balloon** mad. (All the year round, London, 1862, v. 7, p. 500–504.) *** DA**
Account of Cocking's fatal parachute accident.

2478. [**Bibliography** of parachutes.] (In: Paul Brockett, Bibliography of aeronautics, Washington, 1910, p. 652; continued in United States. — National Advisory Committee for Aeronautics, 1909–1931, Washington, 1921–35, volumes for 1909/16, p. 995–998; 1917/19, p. 342–344; 1920/21, p. 315–317; 1922, p. 120–121; 1923, p. 104; 1924, p. 84; 1925, p. 133–134; 1926, p. 111; 1927, p. 136; 1928, p. 152; 1929, p. 173–174; 1930, p. 185; 1931, p. 216–217.) **VDT**

2479. **Carelli,** Jules. Petite expérience de parachute dirigeable. illus. (Illustrierte aëronautische Mittheilungen, Strassburg, 1899, Jahrg. 3, p. 55–57.) **VDS**

2480. **Collier** trophy awarded to Major Hoffman. (U. S. air services, Washington, 1927, v. 12, Feb., p. 45.) **VDS**
Edward L. Hoffman, of the United States Army, is noted for his parachute improvement. The article contains a list of previous awards of the trophy, beginning 1911.

2481. **Daedalus,** pseud. Ballooning. (Mechanic's magazine, London, 1826, v. 6, p. 4–5.) **3–VA**
Suggestion for steering a parachute.

2482. **Description** of Mr. Cocking's parachute, and narrative of the ascent and fatal descent. (Mechanic's magazine, London, 1837, v. 27, p. 258–264, 297–302.) **3–VA**

2483. **Description** of Mr. Cocking's parachute and a narrative of the circumstances attending his fatal descent. illus. (Franklin Institute, Journal, Philadelphia, 1837, new series, v. 20, p. 219–227.) **VA**

2484. **Dixon,** Charles. Parachutes for airmen. London and New York: Isaac Pitman & Sons, 1930. xii, 124 p. illus. 12°. **VDY**
Historical preface. Reviewed in *Aviation,* New York, 1930, v. 29, p. 371, *VDS.*

2485. **Dollfus,** Charles. Nouveaux parachutes français. illus. (L'Aéronautique, Paris, 1929, année 11, p. 23–27.) **VDS**

2486. **Fallschirmversuche** von Thomas S. Baldwin. illus. (Zeitschrift für Luftschiffahrt, Berlin, 1887, Jahrg. 6, p. 336–339.) **3–VDS**

2487. **Frankenberg und Ludwigsdorf,** Alex Victor von. Der Fallschirm, seine geschichtliche Entwicklung und sein technisches Problem, von Gustav von Falkenberg [pseud.]. Berlin: R. C. Schmidt & Co., 1912. 189 p. illus. 8°. **VDW**
History of the parachute with numerous illustrations and caricatures.

2488. **Glassman,** Don. The black sheep of aviation. Our aerial life preservers are deserving of a genealogical record — and here it is. illus. (Aeronautic review, Washington, D. C., 1929, v. 7, Nov., p. 23–24, 30.) **VDS**
Brief history of the parachute.

Heavier-Than-Air Craft — Parachutes, cont'd

2489. Grey, Charles Grey. E. R. Calthrop. [By C. G. Grey.] (Aeroplane, London, 1927, v. 32, p. 364.) **VDS**

Obituary of the inventor of the famous "Guardian Angel" parachute, used extensively during the World war.

2490. Hagen, Freiherr vom. Praxis und Theorie der Fallschirme. illus. (Zeitschrift für Luftschiffahrt, Berlin, 1883, Jahrg. 2, p. 143–154, 177–180.) **3–VDS**

History of famous exploits.

2491. Harmel, Falk. Bam! You're now a caterpillar! illus. ports. (Popular aviation, Chicago, 1931, v. 9, Oct., p. 17–18, 49, 51.) **VDS**

Experiences of some of the members of famous parachute club.

2492. Hengler, Lorenz. Verbesserung des Fallschirms. (Dingler's polytechnisches Journal, Stuttgart, 1832, Bd. 43, p. 102–104.) **3–VA**

2493. Hopson, William L. My five years as a parachute jumper. illus. (Popular aviation, Chicago, 1932, v. 11, p. 11–12, 56–57.) **VDS**

2494. Hubbard, Thomas O'Brien. Parachutes. illus. (Aeronautics, London, 1910, v. 3, p. 32–34.) **VDS**

Good history with a full-page illustration of Alessandro Volante's descent.

2495. La Landelle, Guillaume Joseph Gabriel de. De l'invention du parachute. Dans la période moderne, est-elle due à l'illustre Joseph de Montgolfier? (L'Aéronaute, Paris, 1883, année 16, p. 195–198.) **3–VDS**

2496. —— La marche rationnelle en aviation. (L'Aéronaute, Paris, 1884, année 17, p. 150–155.) **3–VDS**

History of the parachute.

2497. Lees, T. Orde. Parachutes. illus. (Aviation, Highland, N. Y., 1921, v. 11, p. 451–453, 485–488.) **VDS**

Has an historical retrospect, with records of mortalities.

2498. McDarment, Corley. Leap and live to leap again. Parachute history made in America. illus. (Saturday evening post, Philadelphia, 1925, v. 198, July 11, p. 10–11, 173–177.) *** DA**

Thrilling experiences of some American aviators.

2499. —— Stepping out into space. illus. (Saturday evening post, Philadelphia, 1924, v. 197, Nov. 29, p. 12–13, 118, 121–122.) *** DA**

History of parachutes in the United States since the World war. Experiences of John A. Macready and of H. R. Harris.

2500. Maceroni, Francis. Colonel Maceroni's parachutes. illus. (Mechanic's magazine, London, 1837, v. 27, p. 305–307.) **3–VA**

2501. Mackenzie, George. Mackenzie's improved parachute. illus. (Mechanic's magazine, London, 1837, v. 27, p. 307–308.) **3–VA**

2502. Macready makes first night parachute jump. port. (U. S. air services, Washington, 1924, v. 9, July, p. 43.) **VDS**

2502a. Magoun, Frederick Alexander, and Eric Hodgins. Parachutes. illus. (In their: A history of aircraft, New York, 1931, p. 87–98.) **VDT**

Beginnings. Early experiments. Arnold's attempted jump. Garnerin's development. The Michaud-Robertson parachute. Fatal attempt of Mosment. The Cocking disaster. Hampton and those who followed.

2503. Martell, P. Zur Geschichte des Fallschirms. (Der Luftweg, Berlin, 1924, July 27, p. 6–8.) **VDS**

2504. Mewes, Rudolf. Lorenz Hengler, der Verbesserer des Fallschirmes. (Zeitschrift für Luftschiffahrt, Berlin, 1886, Jahrg. 5, p. 252–255.) **3–VDS**

Includes a copy of Hengler's article in *Dingler's polytechnisches Journal.* See entry no. 2492.

2505. —— Ueber den Fallschirm. illus. (Zeitschrift für Luftschiffahrt, Berlin, 1887, Jahrg. 6, p. 65–73.) **3–VDS**

Discusses problems of the parachute, citing the machines of Cocking and Hengler.

2506. Milla, Karl. Der Drachenfallschirm. (Wiener Luftschiffer-Zeitung, Wien, 1902, Jahrg. 1, p. 4–5.) **VDS**

Discusses possibilities of the kite-parachute.

2507. Miller, H. B. Parachutes! illus. (Aero digest, New York, 1929, v. 15, Oct., p. 57–59, 264–268.) **VDS**

Good history.

2507a. Mr. Cocking's new parachute. illus. (The Mirror, London, 1838, v. 30, p. 65–69.) *** DA**

2507b. Mr. Hampton's descent in a parachute at Cremorne House, King's Road, Chelsea. illus. (The Mirror, London, 1839, v. 33, p. 376, 377.) *** DA**

Illustration on p. 369.

2508. Moedebeck, Hermann W. L. Die Entwicklung der Fallschirmtechnik. illus. (Prometheus, Berlin, 1889, Bd. 1, p. 137–140.) **3–VA**

2509. Moy, Thomas. The parachute and aërial engineering. (English mechanic, London, 1889, v. 48, p. 391–392.) **VA**

Discusses the backward state of parachute development.

2510. Murillo, Abraham, jr. Confessions of a Mexican parachute jumper. illus. (Popular aviation, Chicago, 1932, v. 10, p. 305, 329, 330.) **VDS**

2511. Murphy, Charles J. V. Parachute. New York: G. P. Putnam's Sons, 1930. viii, 275 p. illus. 8°. **VDS**

Traces development with exciting episodes. Reviewed in *Aviation,* New York, 1930, v. 29, p. 371, *VDS; Aeroplane,* London, 1930, v. 39, p. 344, *VDS.*

2512. Murphy, Marvin. Caterpillar Club. illus. (Popular aviation, Chicago, 1928, v. 2, Jan., p. 19–21, 68–69.) **VDS**

History of famous organization of parachute jumpers. Stories about Lindbergh and others.

Heavier-Than-Air Craft — Parachutes, cont'd

2513. A **New** parachute. illus. (Scientific American, New York, 1892, v. 67, p. 134.) **VA**
Illustration and brief account of Capazza's combined balloon and parachute.

2514. **Parachute** carries airplane to earth. illus. (Scientific American, New York, 1926, v. 135, p. 329.) **VA**
Successful landing at the San Diego Naval Air Station, 1926.

2515. **Particulars** of the parachute, with a circumstantial account of M. Garnerin's ascent in the balloon, and his descent with the parachute, on Tuesday, Sept. 21, 1802. [London, 1803.] 515–525 p. illus. 8°. **VDY p.v.41, no.5**

2516. **Paulus,** Käthe. Der Doppelfallschirm. port. (Illustrierte aëronautische Mittheilungen, Strassburg, 1900, Jahrg. 4, p. 75–77.) **VDS**
Invention of the German aeronaut, Lattemann.

2517. **Prieur,** C. A. Note historique sur l'invention et les premiers essais des parachutes. (Annales de chimie, Paris, an VII [1798], tome 31, p. 269–273.) **PAA**
Largely a quotation from a letter by Joseph Montgolfier, describing the latter's experiment.

2518. —— Réclamation relative à l'invention des parachutes. (Annales de chimie, Paris, an IX [1800], tome 36, p. 94–99.) **PAA**
Quotes from a memoir of Sébastien Le Normand, describing latter's experiences in 1784.

2519. **Prohaska,** Franz. Theorie der Fallschirmbewegung auf Grund des Newton'schen Luftwiderstandgesetzes. diagrs. (Zeitschrift für Luftschiffahrt, Berlin, 1892, Jahrg. 11, p. 5–14, 45–54.) **3–VDS**
Author's theory of parachute behavior.

2520. **Sala,** George Augustus Henry. On coming down in a parachute. A peculiarly personal experience. (Belgravia, London, 1874, v. 24, p. 166–176.) *** DA**
How the author descended from a sausage balloon in 1851.

2521. **Smith,** Floyd. The Floyd Smith aerial life pack. (Aviation, New York, 1920, v. 8, p. 67–70.) **VDS**
Introductory paragraphs trace the development of the modern parachute.

2522. **Stern,** Kenneth D. Aerial life savers. illus. (Aero digest, New York, 1927, v. 10, p. 175.) **VDS**
Successful results of parachute developed by Major E. S. Hoffman of the United States Army.

2523. **Tissandier,** Gaston. Histoire du parachute. illus. (La Nature, Paris, 1892, année 20, semestre 2, p. 337–339, 371–372.) **OA**
Also in *Scientific American supplement*, New York, 1892, v. 34, p. 14135–14136, *VA*.

2524. —— Le parachute de M. Capazza. illus. (La Nature, Paris, 1892, année 20, semestre 2, p. 195–196.) **OA**

2525. Der **Todessturz** Lattemanns. port. (Wiener Luftschiffer-Zeitung, Wien, 1904, Jahrg. 3, p. 197–199.) **VDS**
Portrait and biography of German aeronaut and inventor of the double parachute.

2526. **United States.** — War Department: Air Corps. Roster of the Caterpillar Club, prepared by the Information Division, Office of the Chief of the Air Corps. Washington [1934]. 9 p. 4°. **† VDW**
Lists 667 members, beginning July 1, 1919.

2527. **Veranzio** [vere Wranczy], Fausto. Machinae novae cum declaratione Latina, Italica, Hispanica, Gallica, et Germanica. Venetiis [1617?]. variously paged. 49 pl. f°.
See plate 38 for a much copied engraving entitled Homo volans, showing a parachute descent.

2528. **Wampler,** Rollin H. Famous parachute jumps. illus. (Popular aviation, Chicago, 1932, v. 11, p. 223–224, 269.) **VDS**
Account of American records since 1921.

2529. —— The inside story of the Caterpillar Club. illus. (Popular aviation, Chicago, 1932, v. 11, p. 153–154.) **VDS**
Includes a history of Lindbergh's jumps.

2530. **White,** Bert. How I broke the world's record jump. How it feels to walk out of doors with five miles of empty space yawning below you. ports. (Popular aviation, Chicago, 1932, v. 10, p. 85–86, 122.) **VDS**

2531. **Wilde,** Henry. On aerial locomotion. illus. (Manchester Literary and Philosophical Society, Memoirs, Manchester, 1900, v. 44, no. 11. 16 p.) *** EC**
Believes airplane impossible of success, but suggests what he calls a vibratory parachute. Accounts of his early experiments on the discharge of elastic fluids.

2532. **Winters,** S. R. A plane-size parachute. illus. (Popular aviation, Chicago, 1928, v. 3, Oct., p. 24–25.) **VDS**
Parachute 84 feet in diameter made for the U. S. Air Corps.

Helicopters and Autogiros

2533. **Albert I,** prince of Monaco. Expériences d'enlèvement d'un hélicoptère. (Académie des sciences, Comptes rendus, Paris, 1905, tome 140, p. 1311–1312.) *** EO**
Brief account of Leger's helicopter.

2534. The **Berliner** helicopter. illus. (Aeronautics, New York, 1908, v. 3, Oct., p. 19, 47.) **VDS**

2535. [**Bibliography** of helicopters.] (In: Paul Brockett, Bibliography of aeronautics, Washington, 1910, p. 404–405; continued in United States. — Advisory Committee for Aeronautics, Bibliography of aeronautics, 1909–1931, Washington, 1921–35, volumes for 1909/16, p. 627–629; 1917/19, p. 215; 1920/21, p. 193–195; 1922, p. 65–66; 1923, p. 57–58; 1924, p. 45; 1925, p. 77; 1926, p. 62; 1927,

Heavier-Than-Air Craft — Helicopters, etc., continued

p. 77; 1928, p. 88; 1929, p. 100; 1930, p. 103; 1931, p. 124.) **VDT**

2536. [**Bibliography** of gyroscopic action and of gyroplanes.] (In: Paul Brockett, Bibliography of aeronautics, Washington, 1910, p. 387.) **VDT**

2537. **Bobrovsky,** T. N. de. What's the matter with the helicopter? While this is one of the earliest flying machines proposed for conquest of the air, yet the helicopter has shown little progress during all these years. illus. (Popular aviation, Chicago, 1932, v. 11, p. 231–232, 268.) **VDS**

2538. **Bothezat,** Georges de. The helicopter. illus. port. (Aeronautical digest, New York, 1923, v. 3, p. 23.) **VDS**
Brief description of the author's machine, with his portrait.

2539. **Bourne,** John Frederick. [Experiments in South America.] (Engineering, London, 1867, v. 4, p. 84.) **VA**
Experiments with model helicopter in 1843.

2540. **Bréguet,** Louis Charles. Le giroplane Bréguet-Richet. illus. (L'Aérophile, Paris, 1907, année 15, p. 258–260.) **3–VDS**
For references concerning Bréguet and his gyroplane see Paul Brockett, *Bibliography of aeronautics*, Washington, 1910, p. 151–152, *VDT*.

2541. **Brie,** Reginald A. C. The autogiro of today. The new wingless, direct control autogiro now challenges comparison with any normal aircraft of similar power. illus. (Air and airways, London, 1933, v. 10, April, p. 17–18.) **VDS**

2542. **Caron,** Félix. Bienvenu et Launoy. (L'Aéronaute, Paris, 1877, année 10, p. 246–249.) **3–VDS**
References to the helicopter of Bienvenu and Launoy, 1784.

2543. **Castel,** P. Un hélicoptère à air comprimé. illus. (L'Aéronaute, Paris, 1878, année 11, p. 190–199.) **3–VDS**
Invention of the author.

2544. **Chalmers,** C. H. The helicopter. illus. (Aeronautics, New York, 1909, v. 4, Feb., p. 58–61.) **VDS**
Account of the author's machine.

2545. **Cierva,** Juan de la. The development of the autogyro. illus. (Royal Aeronautical Society, Journal, 1926, v. 30, p. 8–29.) **VDS**
With discussion.

2546. —— The evolution of the autogiro. A brief history of the development of this type of aerodyne from its inception. illus. (Aircraft engineering, London, 1934, v. 6, p. 132–133.) **VDS**

2547. —— Wings of tomorrow; the story of the autogiro. [New York:] Brewer, Warren & Putnam [cop. 1931]. 300 p. illus. ports. 8°. **VDY**

2548. **Cierva,** Juan de la, and D. F. Rose. From glider to autogiro. The inventor of the famous "flying windmill," latest sensation in aviation, tells for the first time the story of his aeronautical achievements. (North American review, New York, 1929, v. 228, p. 530–535.) *** DA**

2549. **Collier** trophy for 1930 awarded [to Harold F. Pitcairn] for autogiro development. illus. (National aeronautic magazine, Washington, 1931, v. 9, April, p. 33–34.) **VDS**

2550. **Cornu,** Paul. Hélicoptère Cornu et fils à propulseur spécial. illus. (L'Aérophile, Paris, 1906, année 14, p. 145–147.) **3–VDS**

2551. —— L'hélicoptère Paul Cornu. illus. (L'Aérophile, Paris, 1908, année 16, p. 138–141.) **3–VDS**
For list of writings by and about Cornu see Paul Brockett, *Bibliography of aeronautics*, Washington, 1910, p. 213–214, *VDT*.

2552. The **Cornu** helicopter. illus. (Aeronautics, New York, 1908, v. 2, May, p. 7–10.) **VDS**

2553. The **Cornu** helicopter. How this novel type of flying machine was developed. illus. (Scientific American supplement, New York, 1908, v. 65, p. 316–317.) **VA**

2554. The **D'Ascanio** helicopter. A successful Italian experiment. illus. (Flight, London, 1930, v. 22, p. 1249.) **VDS**

2554a. **Davidson,** George L. O. The problem of flight. (Engineering, London, 1908, v. 86, p. 207.) **VDA**
Author's experiments with a steam-driven helicopter.

2555. The **Davidson** "gyropter." illus. (Scientific American supplement, New York, 1909, v. 67, p. 353–354.) **VA**
Machine of George L. O. Davidson, of Scotland, measured tip-to-tip, 67 feet. Wrecked by explosion of the steam boiler.

2556. **Direct** control American wingless autogiros. illus. (Aero digest, New York, 1934, v. 25, Dec., p. 50–52.) **VDS**
Describes Pitcairn and Kellett machines.

2557. La **Distribution** des prix de l'Institut Lombard de Milan en 1879. (L'Aéronaute, Paris, 1880, année 13, p. 33–40.) **3–VDS**
Prize awarded to Enrico Forlanini for his helicopter.

2558. **European** developments in helicopters. illus. (Aviation, Highland, N. Y., 1921, v. 10, p. 492–495.) **VDS**
Petroczy-Karman and Lacoin-Damblanc inventions.

2559. The **Evolution** of the Cierva "autogiro." illus. (Flight, London, 1925, v. 17, p. 733.) **VDS**
Development shown by six photographs.

2560. **Fair,** Ernest W. The story of vertical flight. illus. (Popular aviation, Chicago, 1929, v. 5, Sept., p. 15–16, 67–68.) **VDS**
Cierva's autogyro and the helicopters of Pescara, Berliner, Karman, De Bothezat, and Damblanc. This number of *Popular aviation* has the title of *Aeronautics*.

Heavier-Than-Air Craft — Helicopters, etc., continued

2561. **First** fatal autogyro accident. (Flight, London, 1932, v. 24, p. 1233.) **VDS**

Death of Pierre Martin in France.

2562. **First** helicopter to fly a circular kilometer. illus. (Aviation, Highland, N. Y., 1924, v. 17, p. 888–889.) **VDS**

Account of Oehmichen machine, which met the French Air Service requirements.

2563. **For** the greatest achievement in American aviation. illus. (National aeronautic magazine, Washington, 1931, v. 9, May, p. 31–33.) **VDS**

Presentation of Collier trophy for 1930 to Harold F. Pitcairn.

2564. **Forlanini,** Enrico. Tentatives infructueuses avant le succès. (L'Aéronaute, Paris, 1879, année 12, p. 123–126.) **3–VDS**

Author's experiences with steam-driven helicopter.

2565. **Fournier,** Lucien. Hélicoplane Paul Cornu. illus. (Cosmos, Paris, 1909, nouv. série, tome 60, semestre 1, p. 200–202.) **VA**

2566. L'**Hélicoptère** Julian Felipe. illus. (L'Aérophile, Paris, 1907, année 15, p. 284–285.) **3–VDS**

2567. **Hovering** police of the air. illus. (Scientific American, New York, 1934, v. 150, p. 8.) **VA**

Experiments in the pursuit of criminals with the autogiro.

2568. **Hureau de Villeneuve,** Abel. Gabriel de la Landelle. port. (L'Aéronaute, Paris, 1886, année 19, p. 23–36.) **3–VDS**

Biography of eminent French scientist and writer who was interested in the development of the helicopter.

2569. —— Gustave de Ponton d'Amécourt. (L'Aéronaute, Paris, 1888, année 21, p. 47–51.) **3–VDS**

D'Amécourt, 1826–1888, French pioneer in helicopters.

2570. —— Le premier appareil d'aviation à vapeur qui ait quitté le sol. illus. (L'Aéronaute, Paris, 1879, année 12, p. 39–47.) **3–VDS**

Describes helicopter of Enrico Forlanini.

2571. **Hyatt,** Thaddeus. The dragon-fly; or, Reactive passive locomotion. A vacuum theory of aerial navigation, based on the principle of the fan-blower. To which is appended some remarks on water and its navigation. By a disciple of Bacon [T. H., i. e. T. Hyatt]. London: C. Whittingham & Co., 1882. 83 p. 4°. **VDY**

2572. **Isacco,** V. The heliogyre. illus. (Royal Aeronautical Society, Journal, London, 1929, v. 33, p. 573–614.) **VDS**

With discussion.
Author describes his airplane equipped with propellers on the wings. Brief discussion of earlier machines built on the helicopter principle.

2573. **Jotti,** da Badia, polesine. L'elicottero Forlanini nella sua genesi scientifica. illus. port. (Aeronautica, Milano, 1927, anno 1, p. 78–79.) **3–VDS**

Brief helicopter history; the Forlanini machine, and a portrait of its inventor.

2574. **Karos,** Willibald. Das Kreiselprincip und der Universal-Flugapparat. illus. (Zeitschrift für Luftschiffahrt, Berlin, 1900, Jahrg. 19, p. 39–46, 63–67.) **3–VDS**

Author's suggestion for lifting helicopter combined with a driving propeller.

2575. **Klemin,** Alexander. The autogiro. illus. (Scientific American, New York, 1931, v. 144, p. 192–194.) **VA**

Story of La Cierva's invention.

2576. —— The helicogyre. illus. (Scientific American, New York, 1929, v. 141, p. 70–71.) **VA**

Invention of V. Isacco.

2577. —— An introduction to the helicopter. A review of the aerodynamic and construction data thus far available, together with descriptions of a number of modern helicopters... (Mechanical engineering, New York, 1924, v. 46, part 2, p. 739–751.) **VFA**

2578. **Kreiss,** Eugen. Das Flugproblem und die Erfindung der Flugmaschine. (Patent angemeldet.) Hamburg: Hanseatische Druck- und Verlags-Anstalt [1908]. 57 p. diagr. 8°. **VDY**

Author's theory of flight, with a list of his writings, 1892–1899. Insertion describes and illustrates his patent for a helicopter.

2579. **Kress,** Wilhelm. Captiv-Schraube. illus. (Zeitschrift für Luftschiffahrt, Berlin, 1900, Jahrg. 19, p. 125–131.) **3–VDS**

Description of the author's captive helicopter for military purposes.

2580. **Lamé,** Maurice Luc Valère. Le vol vertical. Théorie générale des hélicoptères. Les appareils à voilures tournantes de leurs origines à 1934. Paris: Blondel la Rougery, 1934. 242 p. illus. 8°. **VDY**

History and theory of the autogiro and the helicopter.

2581. **Larsen,** Agnew E. Development of the autogiro. illus. (Aero digest, New York, 1930, v. 17, Oct., p. 54–56.) **VDS**

2582. **Léger,** M. Un nouvel essai de navigation aérienne par le plus lourd que l'air. illus. (Revue générale des sciences, Paris, 1905, année 16, p. 939–947.) **OA**

Author's combined helicopter and airplane.

2583. —— Nouvelles expériences d'enlèvement de l'hélicoptère "M. Léger," au Musée océanographique de Monaco. (Académie des sciences, Comptes rendus, Paris, 1905, tome 140, p. 1529–1531.) *** EO**

2584. **Liptrot,** R. N. Modern developments in the helicopter. (Royal Aeronautical Society, Journal, London, 1931, v. 35, p. 624–632.) **VDS**

Brief account of modern experiments, beginning with those of Oehmichen, 1923.

Heavier-Than-Air Craft — Helicopters, etc., continued

2585. **Lucas-Girardville,** P. Les appareils d'aviation expérimentes en 1905 en Europe. (Revue d'artillerie, Paris, 1906, v. 67, p. 369–381.) **VWA**
Experiments with the Dufaux and Léger helicopters.

2586. **Luyties,** Otto G. Experiments with a helicopter. (Scientific American, New York, 1908, v. 99, p. 26–27.) **VA**
Results obtained by the author.

2586a. **Magoun,** Frederick Alexander, and Eric Hodgins. The problem of vertical flight. illus. (In their: A history of aircraft, New York, 1931, p. 223–246.) **VDT**
First experimenters. Cayley's top. Remarks of David Mayer. The Phillips steam top. Cossus and Henry Bright. Gustave Ponton d'Amécourt. Researches of Wenham. Alphonse Pénaud. Pomés and de la Pauze. Helicopter of Renoir. Experiments of Dieuaide. Proposal of Mélikoff. Forlanini's successful method. Experiments of Castel. Trouve's propeller researches. Experiments of Thomas Alva Edison. Work of Bréguet. Helicopter of Paul Cornu. Experiments of Lorenzen. Helicopter development in the war. Damblanc's attempt. Helicopters in America. Work of de Bothezat. British helicopter competition. The Oehmichen helicopter. Present day experiments. The helicopter's present status.

2587. **Masfrand,** Albert de. L'hélicoptère des frères Dufaux. ports. illus. (L'Aérophile, Paris, 1905, année 13, p. 97–104.) **3–VDS**

2588. **Melikoff,** J. Projet d'un hélicoptère à vapeur d'éther. illus. (L'Aéronaute, Paris, 1879, année 12, p. 24–30, 103–110.) **3–VDS**
Details of helicopter using an ether turbine.

2589. **Moore,** Alvin Edward. Early setbacks in autogiro development. illus. (Popular aviation, Chicago, 1933, v. 12, p. 21–22, 59–61.) **VDS**
"The author, connected with the U. S. Patent Office, is thoroughly conversant with the many tribulations encountered by Cierva."

2590. **Mott,** S. D. Aerodromes. illus. (Scientific American supplement, New York, 1902, v. 54, p. 22416–22418.) **VA**
Author's flying machine built on the helicopter principle.

2591. The **Oehmichen** helicopter. illus. (Aviation, Highland, N. Y., 1921, v. 11, p. 319.) **VDS**
Describes machine of Étienne Oehmichen, which was built in France. Oehmichen is the author of a study of bird-flight entitled *Nos maîtres les oiseaux,* Paris: Dunod, 1920, of which the Library has a copy, *VDY.*

2592. **Osborn,** Earl D. The Pitcairn autogiro; a non-technical explanation of its operation and performance. illus. (Aviation, New York, 1929, v. 26, p. 2187–2189.) **VDS**

2593. **Pesce,** G. L. L'ingénieur Forlanini. port. illus. (L'Aérophile, Paris, 1902, année 10, p. 1–4.) **3–VDS**
Forlanini invented a steam-driven helicopter in 1877.

2594. **Pitcairn,** Harold F. The autogiro: its characteristics and accomplishments. illus. (Franklin Institute, Journal, Philadelphia, 1930, v. 209, p. 571–584.) **VA**
Account of Cierva's early work. Reprinted in Smithsonian Institution, *Annual report for 1929/30,* Washington, 1931, p. 265–271, ** EA.*

2595. **Platte,** August. Neue Erfahrungen in der Luftschiffahrt. (Neueste Erfindungen und Erfahrungen, Wien, 1893, Jahrg. 20, p. 535–538.) **3–VA**
Author believes that bird-flight can be successfully imitated by using helicopter driven horizontally by a propeller.

2596. Le **Premier** vol en hélicoptère. illus. (La Nature, Paris, 1921, année 49, semestre 1, p. 303–304.) **OA**
Credit given to Émile Berliner and his son, 1919.

2597. **Progress** in helicopters. (Aviation, Highland, N. Y., 1924, v. 16, p. 615.) **VDS**
Oehmichen's circuit flight. British Air Ministry's prize.

2598. **Ramakers,** L. The helicoptere: Santos-Dumont's latest flying machine. illus. (Scientific American, New York, 1906, p. 129–130.) **VA**

2599. **Renard,** Charles. Sur la possibilité de soutenir en l'air un appareil volant du genre hélicoptère en employant les moteurs à explosion dans leur état actuel de légereté. (Académie des sciences, Comptes rendus, Paris, 1903, tome 137, p. 843–846.) *** EO**

2600. **Renoir.** Expériences exécutées avec une hélice à collerettes. illus. (L'Aéronaute, Paris, 1873, année 6, p. 73–78.) **3–VDS**
Description of helicopter invented by Pillet, professor at the Cherbourg arsenal.

2601. —— Hélicoptère avec transmission de mouvement sans engrenage, et inclinaison variable de l'axe des hélices pendant la marche. diagrs. (L'Aéronaute, Paris, 1872, année 5, p. 17–24.) **3–VDS**

2602. **Sellers,** M. B. Some notes on the helicopter. illus. (Aviation, Highland, N. Y., 1922, v. 12, p. 228–230.) **VDS**
Brief review of modern developments.

2603. **Sohlman,** J. Om kraftbehofvet vid flygmaskiner, konstruerade enligt s. k. "heliokopter principen." (Teknisk Föreningen i Finland, Forhandlingar, Helsingfors, 1907, arg. 27, p. 185–190.) **VDA**
Power requirements for flying machines constructed on the helicopter principle.

2604. **Successful** helicopter trials at McCook Field. The de Bothezat helicopter lifts 3600 lb. with 170 Hp. illus. (Aviation, Highland, N. Y., 1923, v. 14, p. 97.) **VDS**

2605. **Taffoureau,** Edgar. Sur le coefficient d'utilisation des hélicoptères. (Académie des sciences, Comptes rendus, Paris, 1905, tome 141, p. 878–880.) *** EO**

2606. **Volumard,** Maurice. Le gyroplane Bréguet-Richet. illus. (Société des anciens élèves des écoles nationales d'arts et métiers, Bulletin technologique, Paris, 1908, p. 215–230.) **3–VA**

Kites

2607. **Abbe,** Cleveland. The development of the kite by European scientists. [By Cleveland Abbe.] (Monthly weather review, Washington, 1897, v. 25, p. 58–61.) † PRS

Describes the experiments of de Romas, P. Musschenbroek, Dr. Thomas Young, Heinrich Wilhelm Brandes, and Tiberius Cavallo.

2608. —— Kites in America and Europe. [By Cleveland Abbe.] (Monthly weather review, Washington, 1897, v. 25, p. 165–166.) † PRS

History of development for meteorological purposes.

2609. —— Sir Isaac Newton and his kites. (Monthly weather review, Washington, 1896, v. 24, p. 458–459.) † PRS

For a list of Abbe's writings on kites and meteorology see Paul Brockett, *Bibliography of aeronautics*, Washington, 1910, p. 1–2, *VDT*.

2610. **Baden-Powell,** Baden Fletcher Smyth. Kites: their theory and practice. illus. (Society of Arts, Journal, London, 1898, v. 46, p. 359–370.) VA

With discussion.
Author describes his experiments. Also in *Scientific American supplement*, New York, 1898, v. 45, p. 18718–18720, 18742–18743, *VA*.

2611. —— Die Verwendung von Drachen zum Aufheben von Menschen. illus. (Illustrierte aëronautische Mitteilungen, Strassburg, 1898, Jahrg. 2, p. 42–43.) VDS

Author began his experiments in 1892.

2612. **Bell,** Alexander Graham. The tetrahedral principle in kite structure. illus. (National geographic magazine, Washington, 1903, v. 14, p. 219–251.) KAA

Also in *Scientific American supplement*, New York, 1903, v. 55, p. 22947–22950, *VA*.

2613. **Bellet,** Daniel. Les cerfs-volants météorologiques de Blue Hill. illus. (La Nature, Paris, 1899, année 27, semestre 2, p. 408–410.) OA

2614. [**Bibliography** of kites.] (In: Paul Brockett, Bibliography of aeronautics, Washington, 1910, p. 463; continued in United States. — National Advisory Committee for Aeronautics, Bibliography of aeronautics, 1909–1931, Washington, 1921–35, volumes for 1909/16, p. 719–722; 1917/19, p. 243; 1920/21, p. 227; 1922, p. 78; 1924, p. 54; 1926, p. 73.) VDT

2615. **Blair,** William Richards. The methods and apparatus used in obtaining upper air observations at Mount Weather, Va. illus. (Mount Weather Observatory, Bulletin, Washington, 1908, v. 1, p. 12–19.) PRS

In this and subsequent bulletins are results of kite observations.

2616. **Brandes,** Heinrich Wilhelm. Drache, elektrischer. illus. (In: Gehler's physikalisches Wörterbuch, neu bearbeitet, Leipzig, 1826, Bd. 2, p. 583–591.) PAC (Gehler)

Developments during the eighteenth century of kites for investigation of atmospheric electricity. Illustration in Atlas, Bd. 2, figure 178. For translation of a part of the article see *Monthly weather review*, Washington, 1897, v. 25, p. 59–60, † *PRS*. See entry no. 2607.

2617. **Burgoyne,** Sir Alan Hugh. Notes on an aluminium kite. (Aeronautical journal, London, 1905, v. 9, p. 36–37.) VDS

2617a. **Cavallo,** Tiberius. A complete treatise on electricity in theory and practice... 4th ed. London, 1795. 3 v. 8°. PGE

Chapters 1 and 2 of volume 2 relate the author's experiments with kites. See also entry no. 2607.

2618. **Claudy,** Carl Harry. The tetrahedral kite. port. illus. (American inventor, New York, 1904, v. 12, p. 221–222.) * V

Experiments of Alexander Graham Bell.

2619. **Clayton,** Henry Helm. Professor Alexander Graham Bell on kite construction. illus. (Science, New York, 1903, v. 18, p. 204–208.) OA

For a list of Clayton's writings, principally on the meteorological use of kites, see Paul Brockett, *Bibliography of aeronautics*, Washington, 1910, p. 194–195, *VDT*.

2620. **Colladon,** Daniel. Expériences sur les cerfs-volants. Projet d'un flotteur pour les ballons dans les traversées maritimes. illus. (La Nature, Paris, 1887, année 15, semestre 2, p. 97–99.) OA

Experiences of the author in 1844 with a man-lifting kite provided with a parachute.

2621. **Davis,** John Woodbridge. Some experiments with kites. illus. (International Conference on Aerial Navigation, Chicago, 1893, Proceedings, New York, 1894, p. 320–328.) VDS

Applies his studies to suggestions for airplanes.

2622. The **Death** of Lawrence Hargrave. (Aeronautics, London, 1915, v. 9, p. 149.) VDS

Brief obituary of famous Australian kite expert.

2623. **Decker,** Horace M. On the mechanics of the kite. illus. (Monthly weather review, Washington, 1897, v. 25, p. 349–351.) † PRS

Illustrations are opposite p. 381.

2624. The **Development** of the kite and the forthcoming kite display. (Aeronautical journal, London, 1906, v. 10, p. 54–56.) VDS (Royal)

Includes a bibliography, 1897–1906.

2625. **Dines,** William Henry. The method of kite-flying from a steam-vessel, and the meteorological observations obtained thereby off the west coast of Scotland. illus. port. (Royal Meteorological Society, Quarterly journal, London, 1903, v. 29, p. 65–85.) PRA

Heavier-Than-Air Craft — Kites, continued

2626. —— Scientific kite flying. illus. (Nature, London, 1903, v. 68, p. 154–155.) **OA**
Brief account of recent development for meteorological purposes.
For a list of Dines' scientific articles, mainly on kites, see Paul Brockett, *Bibliography of aeronautics,* Washington, 1910, p. 247, *VDT.*

2627. Eine **Drachenfahrt.** (Die Gartenlaube, Leipzig, 1854, No. 40, p. 478–479.) *** DF**
Brief account of George Pocock's kite-driven road carriage.

2628. Eddy, William A. The Eddy Malay tailless kite. illus. (Scientific American, New York, 1894, v. 71, p. 169–170.) **VA**
Used for raising thermograph apparatus at the Blue Hill Observatory.

2629. —— Experiments with hexagon and tailless kites. (International Conference on Aerial Navigation, Chicago, 1893, Proceedings, New York, 1894, p. 315–320.) **VDS**

2630. —— A record of some kite experiments. (Monthly weather review, Washington, 1898, v. 26, p. 450–452.) **† PRS**
Author began his experiences in 1863.

2631. —— Some kite records in the United States. (Aeronautical journal, London, 1899, v. 3, p. 15–16.) **VDS**
Author recounts his experiences, 1891–1898.
For a list of Eddy's papers on kites see Paul Brockett, *Bibliography of aeronautics,* Washington, 1910, p. 271, *VDT.*

2632. Fergusson, Sterling Price. The early use of wire in kite flying. (Monthly weather review, Washington, 1897, v. 25, p. 135.) **† PRS**

2633. —— Exploration of the air by means of kites. illus. (Harvard College. Astronomical Observatory, Annals, Cambridge, 1897, v. 42, part 1, Appendix B, p. 41–67.) **† OPA**
Details of equipment used at the Blue Hill Observatory.

2634. —— The highest kite ascensions at Blue Hill. (Monthly weather review, Washington, 1897, v. 25, p. 392.) **† PRS**
Height of 11,847 feet above sea level reached Oct. 15, 1897.

2635. —— The international kite ascensions. illus. (Scientific American, New York, 1907, v. 97, p. 97–98.) **VA**
Account of the ascensions and equipment, July, 1907.

2636. —— Progress of experiments with kites in 1897–98 at Blue Hill Observatory. illus. (Scientific American supplement, New York, 1899, v. 47, p. 19375–19377.) **VA**

2637. —— Recent kite experiments at the Blue Hill Observatory. illus. (Scientific American, New York, 1899, v. 80, p. 123–124.) **VA**
For a list of Fergusson's writings on kites see Paul Brockett, *Bibliography of aeronautics,* Washington, 1910, p. 309, *VDT.*

2638. Frankenfield, Harry Crawford. The kite work of the United States Weather Bureau. illus. (Nature, London, 1900, v. 63, p. 109–111.) **OA**
"Seventeen stations were established in the spring of the year 1898."

2638a. Frantzen, L. P. Les progrès accomplis dans la construction et le rendement des planeurs captifs cerf-volants depuis 1914. illus. (La Science moderne, Paris, 1925, v. 2, p. 626–634.) **OA**

2639. Grosvenor, Gilbert Hovey. The tetrahedral kites of Dr. Alexander Graham Bell. illus. (Popular science monthly, New York, 1903, v. 64, p. 131–151.) *** DA**

2640. Hammon, W. H. Experiments with kites at San Francisco, Cal. (Monthly weather review, Washington, 1896, v. 24, p. 288–289.) **† PRS**
Work of the author and of Alexander McAdie at the Weather Bureau station.

2641. Hargrave, Lawrence. Aeronautics. illus. (Royal Society of New South Wales, Journal and proceedings, Sydney, 1898, v. 32, p. 55–65.) *** EC**
Celebrated author's experiments with soaring kites. Also in *Aeronautical journal,* London, 1898, v. 2, p. 80–83, *VDS.*

2642. —— Cellular kites. illus. (Engineering, London, 1893, v. 56, p. 523–524.) **VDA**
Includes photographs of six of the author's kites.

2643. —— Flying-machine engines and cellular kites. illus. (Royal Society of New South Wales, Journal, Sydney, 1893, v. 27, p. 75–81.) *** EC**
Illustrations at end of volume. Article is reprinted in *English mechanic,* London, 1893, v. 58, p. 238–239, *VA.*

2644. —— [Kites.] illus. (Aeronautical journal, London, 1899, v. 3, p. 49–58.) **VDS**
With discussion.
Author describes some of his work.

2645. —— On the cellular kite. illus. (Royal Society of New South Wales, Journal, Sydney, 1897, v. 30, p. 144–147.) *** EC**
Author describes and illustrates one of his kites. Also in *Aeronautical journal,* London, 1897, v. 1, April, p. 9–11, *VDS.*

2646. —— Paper on aeronautical work. illus. (Royal Society of New South Wales, Journal, Sydney, 1895, v. 29, p. 40–47.) *** EC**
Describes model motor using spirit-air vapor; also some experiments with man-carrying kites.
For lists of writings by and about Hargraves see Paul Brockett, *Bibliography of aeronautics,* Washington, 1910, p. 395–396, *VDT.* Continued in United States. — National Advisory Committee for Aeronautics, *Bibliography of aeronautics,* 1909–1931, Washington, 1921–35, volumes for 1909/16, p. 615; 1924, p. 44; 1930, p. 100, *VDT.*

2647. Julien, Stanislas. Notes sur l'emploi militaire des cerfs-volants... tirées des livres chinois. (Académie des sciences, Comptes rendus, Paris, 1847, tome 24, p. 1070–1071.) *** EO**
Ancient military use of kites in China. English translation in *Eureka,* New York, 1847, v. 2, p. 87–88, *VA.*

2648. The **Kite** in meteorology. Exploring the air to a height of 15,000 feet by means of kites. The instruments used in this interesting work. illus. (American inventor, Washington, 1902, v. 8, March 1, p. 1–3.) *** V**
Work at the Blue Hill Observatory, with a brief account of earlier investigations.

Heavier-Than-Air Craft — Kites, continued

2649. **Köppen**, Wladimir Peter. Flug eines ungefesselten Hargrave-Drachens. illus. (Prometheus, Berlin, 1901, Jahrg. 12, p. 257–262, 273–276.) 3–VA
Use of the Hargrave kite in meteorological work; also a discussion of its applicability to flying.

2650. **Lamson's** man-lifting kite. (Aeronautical journal, London, 1897, v. 1, Oct., p. 21.) VDS

2651. **Lawrence** Hargrave, a pioneer in aviation. (Popular aviation, Chicago, 1932, v. 10, Jan., p. 78, 126.) VDS

2652. **Lecornu**, J. Les cerfs-volants. Paris: Nony & Cie., 1902. iv, 240 p. illus. 8°. VDY
Historical introduction.

2653. The **Lecornu** cellular kite. illus. (Scientific American, New York, 1899, v. 81, p. 360–361.) VA

2654. **Lyon**, Albert B. The kite equipment of the Arlington Observatory. (Aeronautical journal, London, 1898, v. 2, p. 88–90.) VDS
Describes work at the U. S. Weather Bureau, Arlington, Va.

2655. **Man-lifting** war kites. illus. (Aeronautical journal, London, 1897, v. 1, April, p. 5–8.) VDS
History of various experiments.

2656. **Marvin**, Charles Frederick. Kite experiments at the Weather Bureau. (Monthly weather review, Washington, 1896, v. 24, p. 113–123, 156–167, 199–206, 239–255.) † PRS
Preceded by an account of earlier experiments. Illustrations at the end of each number.

2657. —— The mechanics and equilibrium of kites. illus. (Monthly weather review, Washington, 1897, v. 25, p. 136–161.) † PRS
This paper was awarded the "Chanute prize."
For citations to Marvin's writings on scientific kite flying see Paul Brockett, *Bibliography of aeronautics*, Washington, 1910, p. 557–558, *VDT*.

2658. **Means**, James. The kite considered as an instrument of value [to aeronautics]. (Aeronautical annual, Boston, 1896, no. 2, p. 111–118.) 3–VDS

2659. **Miller**, Charles M. The construction of kites. illus. (Manual training magazine, Peoria, Ill., 1909, v. 10, p. 200–221.) SSM
Has a bibliography of 54 entries.

2660. **Millet**, J. B. Scientific kite-flying, with special reference to the Blue Hill experiments. illus. (Century magazine, New York, 1897, v. 54, p. 66–77.) * DA

2661. —— Some experiences with Hargrave kites. illus. (Aeronautical annual, Boston, 1896, no. 2, p. 127–132.) 3–VDS

2662. **Moffett**, Cleveland. Scientific kite-flying. illus. (McClure's magazine, New York, 1896, v. 6, p. 379–392.) * DA
Tells of the experiments of W. A. Eddy, especially for photographic and meteorologic purposes.

2663. **Musschenbroek**, Petrus van. Introductio ad philosophiam naturalem. Lugduni Batavorum, 1762. 2 v. illus., plate 13 at back of volume. 4°. PAD
See v. 1, p. 177 for author's theory of the kite. There is a translation of this section in *Monthly weather review*, Washington, 1897, v. 25, p. 59, † *PRS*. See entry no. 2607.

2664. A **New** aerial apparatus. illus. (Scientific American, New York, 1890, v. 62, p. 117.) VA
Combination kite, balloon, and drag patented by Dr. David Thayer of Boston, Mass. A spirited picture of the apparatus in action.

2665. **New** kite record. (Aeronautical journal, London, 1898, v. 2, p. 90.) VDS
12,124 feet attained by Clayton and Fergusson at Blue Hill, Mass., Aug. 26, 1898.

2666. The **New** observation kites invented by S. F. Cody. illus. (Scientific American supplement, New York, 1903, v. 55, p. 22804.) VA

2667. **Nikel**, Hugo L. Versuche mit neuen Registrir-Drachen. illus. (Illustrierte aëronautische Mittheilungen, Strassburg, 1899, Jahrg. 3, p. 1–2.) VDS
Description of the author's kite.

2668. **Poujoula**, R. Le cerf-volant de sauvetage; historique, construction, fonctionnement. Paris: Librairie aéronautique [1910]. 43 p. illus. 8°. VDY
Notes some of the ideas concerning the use of kites in life-saving at sea.

2669. **Professor** Hargrave. port. (Flying, London, 1902, no. 2, p. [53.]) VDS
Good portrait of famous Australian kite and airplane pioneer.

2669a. **Romas**, Jacques de. Mémoire sur les moyens de se garantir de la foudre dans les maisons; suivi d'une lettre sur l'invention du serf-volant électrique... Bordeaux, 1776. 156 p. illus. 12°. PRK
Photostatic copy of original owned by the Franklin Institute. Author claims to have anticipated Franklin in use of the lightning kite. See also entry no. 2607.

2670. **Salier**, Cecil W. Lawrence Hargrave, a prelude to biography. port. map. (Royal Australian Historical Society, Journal, Sydney, 1929, v. 15, p. 142–188.) BHD
Hargrave (1850–1915), an Englishman who spent most of his life in Australia, made important researches in the science of flight.

2671. **Tissandier**, Gaston. Cerf-volants chinois. illus. (La Nature, Paris, 1888, année 16, semestre 1, p. 44–46.) OA
Interesting pictures of Chinese kites.

2672. **Wenz**, Émile. Bibliographie des applications du cerf-volant. (L'Aéronaute, Paris, 1904, année 37, p. 238–240.) 3–VDS
Twenty-four references, mostly contemporary.

2673. **Wise**, Hugh D. Experiments with kites. Including an account of the writer's ascent from Governor's Island, New York harbor. (Century magazine, New York, 1897, v. 54, p. 78–86.) * DA

2674. **Zimmerman**, Charles. A new form of kite or aeroplane. illus. (Aeronautical journal, London, 1898, v. 2, p. 64–65.) VDS
Man-carrying kite of the author's design.

ENGINES

2675. **Abell,** C. F. Airship machinery, past experience and future requirements. (Aeronautical journal, London, 1920, v. 24, p. 250–268.) **VDS**

With discussion.
Describes engines used in wartime on airships. Also in *Aeronautics,* London, 1920, v. 18, p. 244, 260–261, 273–275, *VDS.*

2676. —— Some recent progress in air-cooled aero-engine development. illus. (Engineering, London, 1930, v. 130, p. 308–311.) **VDA**

2677. The **Adams-Farwell** aero motor. illus. (Aeronautics, New York, 1908, v. 2, June, p. 43–45.) **VDS**

2678. **Aeronautical** motors in Paris. illus. (Engineer, London, 1909, v. 108, p. 342–343, 346, 364–365.) **VA**

Pictures and descriptions of motors shown at the International Exhibition of Aerial Locomotion, 1909.

2679. **Allen,** J. W. Starters for airplane engines. The progress of aviation starting equipment. illus. (Aviation engineering, Albany, N. Y., 1929, v. 2, Jan., p. 15–18.) **VDS**

2680. **Angle,** Glenn Dale. Developments in high-powered aircraft engines. illus. (Aero digest, New York, 1933, v. 23, Oct., p. 41–44, Nov., p. 46–47.) **VDS**

Has brief early history of large types.

2681–2682. **Army** Air Corps writes "finis" on Liberty engine's career. (Automotive industries, Philadelphia, 1934, v. 70, p. 163.) **TOL**

2683. **Baer,** Herbert Christian. Die Entwicklung des Flugmotors seit dem Kriege. illus. (Verein deutscher Ingenieure, Zeitschrift, Berlin, 1924, Bd. 68, p. 539–544.) **VDS**

Developments since the World war.

2684. **Besançon,** Georges. Louis Seguin. port. (L'Aérophile, Paris, 1912, année 20, p. 73.) **3-VDS**

Seguin established the Gnome motor industry in 1895.

2685. [**Bibliography** of engines.] (United States. — National Advisory Committee for Aeronautics, Bibliography of aeronautics, 1909–1931, Washington, 1921–35, volumes for 1909/16, p. 403–428; 1917/19, p. 132–148; 1920/21, p. 125–135; 1922, p. 43–47; 1923, p. 37–39; 1924, p. 28–31; 1925, p. 52–56; 1926, p. 40–44; 1927, p. 49–53; 1928, p. 59–62; 1929, p. 69–73; 1930, p. 67–72; 1931, p. 79–86.) **VDT**

2686. [**Bibliography** of the "Gnome" engine.] (United States. — National Advisory Committee for Aeronautics, Bibliography of aeronautics, 1909–1931, Washington, 1921–35, volumes for 1909/16, p. 549–551; 1917/19, p. 188; 1920/21, p. 170; 1924, p. 39.) **VDT**

2687. [**Bibliography** of the Green engine.] (United States. — National Advisory Committee for Aeronautics, Bibliography of aeronautics, 1909–1931, Washington, 1921–35, volumes for 1909/16, p. 583; 1917/19, p. 198; 1920/21, p. 179.) **VDT**

2688. [**Bibliography** of the "Gyro" engine.] (United States. — National Advisory Committee for Aeronautics, Bibliography of aeronautics, 1909–1931, Washington, 1921–35, volume for 1909/16, p. 595–596.) **VDT**

2689. [**Bibliography** of the Liberty engine.] (United States. — National Advisory Committee for Aeronautics, Bibliography of aeronautics, 1909–1931, Washington, 1921–35, volumes for 1917/19, p. 263–266; 1920/21, p. 245–246; 1922, p. 86; 1923, p. 75; 1924, p. 59; 1926, p. 79; 1927, p. 98; 1929, p. 127.) **VDT**

2690. [**Bibliography** of lubricants and lubrication.] (United States. — National Advisory Committee for Aeronautics, Bibliography of aeronautics, 1909–1931, Washington, 1921–35, volumes for 1909/16, p. 813; 1917/19, p. 271; 1920/21, p. 252; 1922, p. 88; 1923, p. 78; 1924, p. 62; 1925, p. 100; 1926, p. 81; 1927, p. 104; 1928, p. 116; 1929, p. 132; 1930, p. 141; 1931, p. 169.) **VDT**

2691. [**Bibliography** of the Mercedes engine.] (United States. — National Advisory Committee for Aeronautics, Bibliography of aeronautics, 1909–1931, Washington, 1921–35, volumes for 1909/16, p. 861; 1917/19, p. 291–292; 1920/21, p. 271; 1927, p. 111.) **VDT**

2692. [**Bibliography** of motors and power.] (In: Paul Brockett, Bibliography of aeronautics, Washington, 1910, p. 607.) **VDT**

Extensive list of reference citations down to July, 1909.

2693. [**Bibliography** of the Packard engine.] (United States. — National Advisory Committee for Aeronautics, Bibliography of aeronautics, 1909–1931, Washington, 1921–35, volumes for 1909/16, p. 988; 1917/19, p. 339; 1920/21, p. 312; 1922, p. 120; 1923, p. 103; 1925, p. 133; 1926, p. 110; 1927, p. 135; 1930, p. 183–184.) **VDT**

2694. [**Bibliography** of the Renault motor.] (United States. — National Advisory Committee for Aeronautics, Bibliography of aeronautics, 1909–1931, Washington, 1921–35, volumes for 1909/16, p. 1125–1126; 1917/19, p. 377; 1923, p. 114; 1924, p. 91; 1925, p. 146; 1926, p. 122; 1928, p. 167; 1929, p. 189; 1930, p. 201.) **VDT**

2695. [**Bibliography** of spark plugs.] (United States. — National Advisory Committee for Aeronautics, Bibliography of aeronautics, 1909–1931, Washington, 1921–35, volumes for 1909/16, p. 1255; 1917/19, p. 413–414; 1920/21, p. 383; 1923, p. 126; 1925, p. 163; 1929, p. 207; 1930, p. 225.) **VDT**

Engines, continued

2696. **Blanchet,** Georges. Alexandre Anzani. port. (L'Aérophile, Paris, 1911, année 19, p. 5.) **3–VDS**

Portrait and brief biography of inventor of aeronautical motor bearing his name.

2697. —— Pierre Clerget. port. (L'Aérophile, Paris, 1913, année 21, p. 241.) **3–VDS**

Brief account of pioneer designer of aeronautical motors.

2698. **Bradley,** W. F., and H. W. Perry. Power generation and transmission in aeroplanes. (Engineering magazine, New York, 1909, v. 36, p. 630–644.) **VDA**

Describes and illustrates Wright, Farman, Delagrange, Antoinette, Renault, Esnault-Pelterie, Curtiss, and Adams-Farwell engines.

2699. **Chanute,** Octave. Motors for aerial machines. (Scientific American supplement, New York, 1893, v. 35, p. 14281–14282.) **VA**

Discusses steam engines, gas engines, and electric motors.

2700. **Chenoweth,** Opie. Engines in 1930. Year extremely fruitful in research and experiment, though no great number of engines nor great profits were forthcoming. illus. (Airway age, New York, 1931, v. 12, p. 44–46, 86.) **VDS**

2701. **Chorlton,** Alan Ernest Leofric. Aero engines. illus. (Royal Society of Arts, Journal, London, 1921, v. 69, p. 689–705, 707–724, 725–740.) **VA**

Briefly describes internal combustion engines before the Wright brothers' machine of 1903, and gives detailed data concerning those developed since that date. Abstracted in *Engineering,* London, 1921, v. 111, p. 81–82, 110–111, 143–144, *VDA.*

2702. **Colvin,** Fred Herbert. How Ford built Liberty motors. illus. (American machinist, New York, 1919, v. 51, p. 1037–1041.) **VFA**

Illustrations and descriptions of technical operations.

2703. **Completion** of Liberty engine. (Air service journal, New York, 1917, v. 1, p. 338.) **VDS**

Brief history.

2704. **Crocé-Spinelli,** J. Considérations générales théoriques et pratiques sur les moteurs légers applicables à la navigation aérienne. illus. (L'Aéronaute, Paris, 1873, année 6, p. 135–140, 151–157, 166–176, 181–186, 214–217.) **3–VDS**

2705. —— Des condenseurs de vapeur d'eau dans les appareils de locomotion aérienne. illus. (L'Aéronaute, Paris, 1872, année 5, p. 89–98.) **3–VDS**

Author's suggestion for reclaiming waste steam in aeronautical motors.

2706. —— Le moteur Foillard. illus. (L'Aéronaute, Paris, 1871, année 4, p. 65–74.) **3–VDS**

2707. **Dacy,** G. H. Uncle Sam tests engines. illus. (Popular aviation, Chicago, 1928, v. 3, Aug., p. 27–28, 88.) **VDS**

Work at the U. S. Bureau of Standards.

2708. **Dantin,** Charles. Les moteurs à explosion légers pour dirigeables et aéroplanes. illus. (Génie civil, Paris, 1909, v. 55, p. 111–114, 125–129, 150–154.) **VA**

A review of motor development in France, Germany, Belgium, and Denmark.

2709. The **Death** of a pioneer engine designer. (Aeroplane, London, 1918, v. 15, p. 2470.) **VDS**

Biography of George Frederick Mort.

2710. **Denham,** Athel F. Cam is used instead of crank train in radial airplane engine. New four-cylinder, air-cooled powerplant constructed for commercial use by the Fairchild-Caminez Corp. has other unusual features. Large cylinders give output of 150 hp. illus. (Automotive industries, Philadelphia, 1926, v. 54, p. 891–893.) **VDS**

2711. —— Wright "J" type engine is result of extensive service tests. illus. (Automotive industries, Philadelphia, 1926, v. 55, p. 380–381.) **TOL**

2712. **Development** of aircraft engine design and lubrication. illus. (Lubrication, New York, 1928, v. 14, p. 73–84.) **VOA**

Contemporary practice with little of historical note.

2713. **Development** of an American pursuit engine. illus. port. (Aviation, Highland, N. Y., 1921, v. 11, p. 735–738.) **VDS**

Developments from the American-built Hispano-Suiza to the Wright "Superfighter."

2714. **Development** of the cam type aviation engine. illus. port. (Aero digest, New York, 1927, v. 11, p. 158, 160.) **VDS**

Invention of Harold Caminez.

2715. **Development** of the Stromberg carburetor. illus. (Aviation, Highland, N. Y., 1927, v. 23, p. 247.) **VDS**

2716. **Dinger,** Henry Charles. The development of the Wright air-cooled aviation engine. illus. (American Society of Naval Engineers, Journal, Washington, 1926, v. 38, p. 856–878.) **VXA**

Detailed account of the experiments of Charles L. Lawrance, beginning in 1916.

2717. **Dow,** J. H. The elastic-fluid turbine, a possible motor for aeronautical purposes. illus. (International Conference on Aerial Navigation, Chicago, 1893, Proceedings, New York, 1894, p. 120–125.) **VDS**

Steam-turbine invented by the author.

2718. **Eight** cylinder "Antoinette" motor for aeroplane. illus. (Engineering, London, 1906, v. 82, p. 703.) **VDA**

Two views. For text see p. 697.

2719. **Émile** Berliner, a man of achievements. port. (Flying, New York, 1914, v. 3, p. 331–332.) **VDS**

Brief biography of the inventor of the gyro motor.

2720. **Emmons,** H. H. Outline of history of aviation engine production. illus. (Aerial age weekly, New York, 1918, v. 8, p. 662–665.) **VDS**

American wartime developments, with statistics of production and horsepower.

Engines, continued

2721. **Enemy** aircraft engines; a detailed description of certain of the engines in use. illus. (Automobile engineer, London, 1916, v. 6, p. 350–357; 1917, v. 7, p. 32–34, 216–223, 246–247, 308–318, 351–364; 1918, v. 8, p. 162–169, 194–199, 254–260, 285–295, 316–319; 1919, v. 9, p. 178–184.) **TOL**

German engines used in the World war.

2722. The **Evolution** of aero engine weight. chart. (Aerial age weekly, New York, 1918, v. 7, p. 671.) **VDS**

Chart shows average values obtained from aircraft engines, other than the Liberty, during the period from 1903 to 1917.

2723. **Evolution** of aircraft engines. chart. (United States. — Advisory Committee for Aeronautics, Annual report, 1918, Washington, 1920, p. 28–29.) **VDY (U. S.)**

This brief account is reprinted in *Aviation,* New York, 1918, v. 4, p. 667.

2724. **Fedden,** A. H. R. The supercharging of aircraft and motor-vehicle engines. illus. (Royal Aeronautical Society, Journal, London, 1927, v. 31, p. 933–972.) **VDS**

Brief historical introduction.

2725. **Fell,** L. F. R. Twenty-one years of water-cooled engine development. illus. (Flight, London, 1930, v. 22, p. 81–83.) **VDS**

2726. **Ferber,** Louis Ferdinand. Léon Levavasseur. port. (L'Aérophile, Paris, 1908, année 16, p. 57.) **3–VDS**

Portrait and brief biography of prominent French engineer, a pioneer in the designing of light motors. For history of his "Antoinette" engine see article by Capt. Ferber, *L'Aérophile,* Paris, 1908, année 16, p. 58–61, *3–VDS.*

2727. **First** Liberty planes shipped. American-built battle-planes on way to France. (Automotive industries, New York, 1918, v. 38, p. 422.) **3–† TOL**

2728. **Foreign** aeronautic motors. illus. (Scientific American supplement, New York, 1908, v. 67, p. 108–110, 116–118, 132–133.) **VA**

2729. **Four-motored** airplanes. illus. (Scientific American, New York, 1926, v. 134, p. 207.) **VA**

Introduced in France by the Farman and Blériot companies.

2730. **Fournier,** Lucien. Le moteur d'aviation Clément-Bayard. illus. (Cosmos, Paris, 1909, nouv. série, tome 60, semestre 1, p. 284–287.) **VA**

2731. **Franklin Institute.** — Committee on Science and the Arts. Wright "whirlwind" engine. illus. (Franklin Institute, Journal, Philadelphia, 1928, v. 206, p. 681–687.) **VA**

Traces development begun by Charles L. Lawrance in 1916. Mr. Lawrance was awarded the Elliott Cresson medal for this work.

2732. **Gasterstädt,** Johannes. Die Entwicklung der Junkers-Diesel-Flugmotoren. illus. (Automobiltechnische Zeitschrift, Berlin, 1930, Bd. 33, p. 2–5, 41–46.) **TOL**

Translation in United States. — Advisory Committee for Aeronautics, *Technical memorandum no. 565,* Washington, 1930, 20 p., *VDS.*

2733. **Geisse,** J. H. Twenty-five years of engine design. illus. (Aviation, New York, 1928, v. 25, p. 1720–1723.) **VDS**

Good review of technical developments.

2734. **Goupil,** Albert Louis. La locomotion aérienne. Étude. Charleville: P. Pouillard, 1884. 109 p. illus. 8°. **VDY**

A study of light-weight motors for airships and airplanes. For a brief biography of Goupil, see *L'Aérophile,* Paris, 1909, année 17, p. 231, *3–VDS.*

2735. **Green,** F. M. Twenty-one years of aircraft engine development. illus. (Flight, London, 1930, v. 22, p. 71–74.) **VDS**

2736. **Grey,** Charles G. On a great pioneer. [By Charles G. Grey.] (Aeroplane, London, 1922, v. 22, p. 221–222.) **VDS**

Account of the work of Levavasseur, inventor of the Antoinette airplane and engine.

2737. **Haenig,** Alfred. Ballon- und Flugmotoren. Ihre technische Entwicklung und gegenwärtige Gestaltung. Rostock i. M.: C. J. E. Volckmann, 1910. 196 p. illus. 8°. **VDY**

Brief historical account of early experiments.

2738. **Hargrave,** Lawrence. Flying machine motors and cellular kites. illus. (International Conference on Aerial Navigation, Chicago, 1893, Proceedings, New York, 1894, p. 287–297.) **3–VDS**

Engines, designed by the author, for model airplanes.

2739. —— Flying-machine work and the ⅙ I. H. P. steam motor weighing 3¼ lbs. illus. (Royal Society of New South Wales, Journal, Sydney, 1892, v. 26, p. 170–175.) *** EC**

2740. **Harlaut.** L'évolution des moteurs d'aviation allemands pendant la guerre. illus. (L'Aéronautique, Paris, 1921, année 3, p. 310–315.) **VDS**

2741. **Hartson,** Joseph Tracy. The hero of all ocean flights. illus. port. (American review of reviews, New York, 1927, v. 76, p. 174–176.) *** DA**

Brief account of the development of the Wright "Whirlwind" engine. Portrait of its designer, Charles Lanier Lawrance.

2742. **Heinz,** F. Grundlagen zur Fluglehre. (Dingler's polytechnisches Journal, Stuttgart, 1899–1900, Bd. 313, p. 28–29, 132–134; Bd. 314, p. 76–78; Bd. 315, p. 207–209, 223–226.) **3–VA**

Investigation of the power requirements of dynamic flight.

2743. **Heldt,** Peter Martin. Engine performance at high altitudes studied by the Bureau of Standards. illus. (Automotive industries, Philadelphia, 1930, v. 63, p. 256–258, 263.) **TOL**

Engines, continued

2744. The **Hispano-Suiza** airplane engines. illus. (Aviation and aeronautical engineering, New York, 1918, v. 5, p. 549–553.) **VDS**
Originated by Marc Birkigt, a Swiss mining engineer.

2745. The **History** of the Scintilla magneto. illus. (Aviation engineering, East Stroudsburg, Pa., 1933, v. 8, Jan., p. 27.) **VDS**
Company was incorporated in 1921.

2746. **Hollands,** Sidney H. Aerial motor design. (Aëronautical Society of Great Britain, 20th annual report, London, 1885, p. 29–37.) **3–VDS**
Discusses requisites of steam engine and propeller.

2747. **Houston,** George H. Developing the Hispano-Suiza motor for the air service. Story of the invention of this engine by a Swiss engineer [Marc Birkigt] widely used by the French. illus. (Aerial age weekly, New York, 1918, v. 7, p. 1115.) **VDS**

2748. **Hureau de Villeneuve,** Abel. Rapport sur l'Exposition aéronautique de 1868 [London]. (L'Aéronaute, Paris, 1868, année 1, p. 51–53, 67–75, 83–89, 99–105, 115–120, 131–135; 1869, année 2, p. 3–6, 19–22, 35–39, 51–55, 84–88, 115–118, 131–136, 161–166; 1870, année 3, p. 65–69, 150–155.) **3–VDS**
Describes exhibits, especially of light motors.

2749. **Hynes,** G. B. A survey of recent work carried out in the engine research and experimental department of the Royal Aircraft Establishment, South Farnborough. (International Congress of Aerial Navigation, London, 1923, Report, London, 1923, p. 431–440.) **VDS**

2750. **Insley,** Robert. Air-cooled engines for aircraft, an historical survey of the design and development of these engines in America. illus. (Scientific American, New York, 1927, v. 137, p. 148–151, 182–183.) **VA**
"It was with a 24 horsepower Anzani air-cooled engine that Blériot first crossed the English Channel in 1909."

2751. **Jones,** E. L., and S. Y. Beach. Table of American aeronautic motors. Complete specifications of the motors of thirty-four manufacturers. (Scientific American, New York, 1911, v. 105, p. 6.) **VA**
Also in *Flight,* London, 1911, v. 3, p. 922, *VDS.*

2752. **Jotti,** da Badia, polesine. Il motore a vapore ed il suo impiego in aeronautica. illus. (L'Aerotecnica, Roma, 1931, v. 11, p. 1555–1564.) **VDS**
Historical review of the steam engine as applied to aeronautics.

2753. **Klemin,** Alexander. A new cooling system. (Scientific American, New York, 1929, v. 141, p. 70.) **VA**
Ethylene glycol (Prestone) proposed.

2754. —— A steam driven airplane engine. illus. (Scientific American, New York, 1933, v. 149, p. 124–125.) **VA**
Describes successful engine of William J. and George Besler.

2755. —— Superaviation. illus. (Scientific American, New York, 1933, v. 148, p. 232–233.) **VA**
Description of the Lorin jet-engine of 1913.

2756. **Krüger,** Richard. Ueber den geeignesten Motor für die Lenkbarkeit des Luftschiffes und für den Bau einer Flugmaschine. (Verein deutscher Ingenieure, Zeitschrift, Berlin, 1877, Bd. 21, col. 559–570.) **3–† VDA**
Discusses relative advantages of airships and airplanes, and believes the Otto gas engine will simplify the problem of flight.

2757. **Lawrance,** Charles Lanier. Air-cooled engine development. illus. (Society of Automotive Engineers, Transactions, New York, 1922, v. 17, part 1, p. 431–477.) **TOL**
With discussion.
Development was slow until the British began experiments during the World war.

2758. —— Air-cooled engine development. illus. port. (Aviation, Highland, N. Y., 1927, v. 23, p. 723–725.) **VDS**

2759. —— The development of the airplane engine in the United States. (International Civil Aeronautics Conference, 1928, Papers, Washington, 1928, p. 411–421.) **VDS**
Also in *Mechanical engineering,* New York, 1929, v. 51, p. 186–192, *VDS,* with added illustrations. Includes a good account of the pioneer work of Charles Matthews Manly (1874–1927).

2760. **Lescarboura,** Austin Celestin. Bringing the Gnome engine to America. illus. (Scientific American, New York, 1917, v. 116, p. 374–375.) **VA**

2761. **Light-weight** gasoline motors for aeronautical work. illus. (Scientific American supplement, New York, 1906, v. 62, p. 25833.) **VA**
Description of motors manufactured by Levavasseur.

2762. **Lindbergh's** Wright Whirlwind a result of seven years' development. Work began on Feb. 28, 1920, and since that time seven successive models of air-cooled radial engines have been produced. illus. (Aviation, Highland, N. Y., 1927, v. 22, p. 1358–1359, 1396.) **VDS**

2763. **MacCoull,** Neil. The mechanical development of aviation. illus. (Aerial age weekly, New York, 1917, v. 5, p. 256–257, 326–327, 354–356.) **VDS**
History begins with the steam-driven plane of Hiram Maxim, and the steam and gasoline driven planes of Langley. The first air-cooled revolving gasoline engine ever built was completed for Langley early in 1899. Abridged in American Society of Mechanical Engineers, *Journal,* New York, 1917, v. 39, p. 292–296, *VFA.*

2764. **McDarment,** Corley. The latest sky engine. illus. (Popular mechanics, Chicago, 1929, v. 52, p. 177–178.) **VFA**
Advent of the Diesel engine in aviation.

2765. **Machol,** Morris R. Spark gaps in series with spark plugs. (Aviation, New York, 1917, v. 2, p. 498.) **VDS**
The introduction traces spark gap development.

Engines, continued

2766. **McLaughlin,** George F. The story of the airplane engine. (American Academy of Political and Social Science, Annals, Philadelphia, 1927, v. 131, May, p. 34–37.) **SA**

Brief general article.

2767. **McMahon,** T. C. Development of aviation engines. illus. (Popular aviation, Chicago, 1928, v. 3, Dec., p. 64, 66, 70, 74, 76, 78, 107–108.) **VDS**

Wright brothers, Charles M. Manly, Glenn Curtiss, French engines, Daimler, Renault, Antoinette, and subsequent engines.

2768. **Macready,** John A. Functioning of supercharger in altitude flight. Experience of Lieut. John A. MacReady on record breaking altitude flight shows difficulties met in exceeding 40,000 feet. (Aviation, Highland, N. Y., 1922, v. 12, p. 51.) **VDS**

2769. The **Makers** of Napier engines. illus. ports. (Aviation, Highland, N. Y., 1925, v. 19, p. 586–587.) **VDS**

Firm, established in 1808, entered the aero engine field in 1898.

2770. **Manning,** William Oke. An echo of the 1868 exhibition. (Royal Aeronautical Society, Journal, London, 1933, v. 37, p. 288–290.) **VDS**

Refers to the exhibition at the Crystal Palace, London, and cites some of the observations of Doctor Abel Hureau de Villeneuve, especially concerning motors of the time.

2771. **Marquis,** Raoul. Les moteurs légers applicables à la navigation aérienne [par H. de Graffigny, pseud.]. illus. (L'Aérophile, Paris, 1894, année 2, p. 128–134, 175–182.) **3–VDS**

A review of contemporary light motors — steam, gas, electric, and compressed air.

2772. **Martinot-Lagarde,** Pierre Léonard Camille. L'évolution du moteur d'aviation en France depuis l'armistice jusqu'en 1930. illus. (L'Aéronautique militaire, maritime, coloniale, et marchande, Paris, 1931, p. 248–268.) **† VDY**

2773. **Mayo,** Robert H. The development of high-speed aircraft. illus. (Royal Aeronautical Society, Journal, London, 1924, v. 28, p. 158–188.) **VDS**

With discussion.
Developments date from the first Gordon Bennett race, 1909. Abridged in *Flight,* London, 1923, v. 15, p. 743–745, *VDS.*

2774. **Mead,** George J. The development of fixed radial air-cooled engines. illus. (Society of Automotive Engineers, Journal, New York, 1929, v. 25, p. 110–114.) **TOL**

2775. —— Historical development of air-cooled engines. illus. (Aero digest, New York, 1929, v. 14, Feb., p. 42–44, 224, 226, 228.) **VDS**

2776. —— Some aspects of aircraft-engine development. illus. (Society of Automotive Engineers, Journal, New York, 1925, v. 17, p. 496–507.) **TOL**

Contemporary trends, with a history of the air-cooled engine.

2777. **Mr.** Perkin's new steam engine. (Mechanic's magazine, London, 1823, v. 1, p. 36–37.) **3–VA**

Although never used on aircraft, this engine is interesting for the power developed in a cylinder only two inches in diameter.

2778. **Moedebeck,** Hermann W. L. Gottlieb Wilhelm Daimler. port. (Illustrierte aëronautische Mitteilungen, Strassburg, 1900, Jahrg. 4, p. 53.) **VDS**

Daimler (1834–1900), "father of the automobile," was also a pioneer in the manufacture of aeronautical engines.

2779. Le **Moteur** d'aviation. Ce qu'il était avant la guerre. Les progrès accomplis. Ce qu'il est aujourd'hui. Demain? illus. (L'Aérophile, Paris, 1919, année 27, p. 361–364.) **VDS**

2780. **Navy** changes in Liberty motor responsible for improvement in Navy plane operations. (Aerial age, New York, 1922, v. 15, p. 455, 478.) **VDS**

2781. The **1931** Collier trophy award. illus. (Aeronautic review, Washington, 1932, v. 10, April, p. 17–18.) **VDS**

Awarded to the Packard Motor Car Company for the development of the Diesel aircraft engine.

2782. **Nutt,** Arthur. Progress in aircraft-engine design. illus. (Society of Automotive Engineers, Journal, New York, 1926, v. 19, p. 239–247.) **TOL**

Ten years' review.

2783. **Oertel,** Walter. Die Entwicklung der Motor-Aëronautik. (Mitteleuropäischer Motorwagen-Verein, Zeitschrift, Berlin, 1907, Jahrg. 6, p. 93–98, 146–151.) **TOL**

Comprehensive history of aeronautical motors.

2784. **Pagé,** Victor Wilfred. Modern aviation engines — design, construction, operation, and repair. New York: Norman W. Henley Pub. Co., 1929. 2 v. illus. 8°. **VDY**

See historical introduction, p. xviii–xxvii; also chapter 24, Early and pre-war aviation engines; chapter 25, Typical war-time aviation engines. Brief items of historical interest in some of the other chapters. Reviewed in *Aeroplane,* London, 1929, v. 37, p. 1246, *VDS.*

2785. **Parsons,** Sir Charles Algernon. Flying engines. illus. (Nature, London, 1896, v. 54, p. 148–149.) **OA**

Account of steam model made by the author in 1893. Reprinted, with a postscript regarding his experiments with helicopters, in *Aeronautics,* London, 1918, v. 14, p. 38–39, *VDS.*

2786. **Pénaud,** Alphonse. Aéroplane automotor. illus. (L'Aéronaute, Paris, 1872, année 5, p. 2–9.) **3–VDS**

Use of india-rubber for propelling a model.

2787. **Peterson,** C. G. Three-engine planes for air transport. Reliability and punctuality insured by three-engine principle in commercial air transportation. illus. (Aviation, Highland, N. Y., 1926, v. 21, p. 354–357.) **VDS**

Engines, continued

2788. Powell, O. A. Development of the aircraft engine. (University of Nebraska Engineering Society, Nebraska blue print, Lincoln, 1919, v. 19, p. 15–20.) **VDA**

Brief account of engines at the beginning of the World war and subsequent developments.

2789. Pye, David Randall. The evolution of the modern aero engine. (Aeroplane, London, 1930, v. 38, p. 521–522.) **VDS**

Abstract of lecture at Cambridge University.

2790. —— The origin and development of heavy-oil aero engines. illus. (Royal Aeronautical Society, Journal, London, 1931, v. 35, p. 266–298.) **VDS**

With discussion.
Abridged in *Flight,* London, 1930, v. 22, p. 1491–1492, *VDS.*

2791. Renard, Paul. Ein historischer Luftschiffmotor. illus. (Deutsche Zeitschrift für Luftschiffahrt, Berlin, 1910, Jahrg. 14, Heft 2, p. 18–21.) **VDS**

History of the rotary motor of Charles Renard, about 1885.

2792. Ricardo, Harry Ralph. The development and progress of the aero engine. (Royal Aeronautical Society, Journal, London, 1930, v. 34, p. 1000–1015.) **VDS**

With discussion.
Abstracted in *Flight,* London, 1930, v. 22, p. 616–620, *VDS; Aeroplane,* London, 1930, v. 38, p. 1075–1080, *VDS.*

2793. —— Recent research work on the internal-combustion engine. illus. (Society of Automotive Engineers, Journal, New York, 1922, v. 10, p. 305–336, 347.) **TOL**

Account of work in the author's laboratory in England.

2794. Rumpler, Edmund. Motoren für Luftfahrzeuge. illus. (Verein deutscher Ingenieure, Zeitschrift, Berlin, 1909, Bd. 53, p. 441–448, 487–492, 532–538, 578–584.)

Describes a large number of types. **3–† VDA**

2795. Sayers, W. H. The first British inverted engine. illus. (Aeroplane, London, 1927, v. 32, p. 100, 102.) **VDS**

2796. Schwager, Otto. Development of German aircraft engines. diagrs. (Aviation, Highland, N. Y., 1920, v. 9, p. 161–164, 186–189.) **VDS**

Covers the war period.

2797. —— Die Entwicklung der Flugmotoren im In- und Auslande seit 1914. illus. (Illustrierte Flug-Welt, Leipzig, 1920, Jahrg. 2, p. 353–358, 381–385, 402–408, 464–469, 482–488; Illustrierte Flug-Woche, Leipzig, 1921, Jahrg. 3, p. 8–12.) **VDS**

2798. Sheldon, William. Aerial navigation and the patent laws. Boston: Thurston, Torry, & Co., 1850. 46 p. diagrs. 8°. **VFB p.v.21, no.12**

Describes a hot-air engine to be used on flying machines.

2799. Sherbondy, E. H. Aviation engine development. illus. (Society of Automotive Engineers, Journal, New York, 1917, v. 1, p. 194–203.) **TOL**

Good account of early engines. Also in *Aerial age weekly,* New York, 1917–18, v. 6, p. 472–474, 485, 564–567, 608–611, *VDS.*

2800. —— The Hispano-Suiza aviation motor. illus. (Aerial age weekly, New York, 1918, v. 8, p. 766–773.) **VDS**

Has a brief early history.

2801. Sherbondy, E. H., and G. D. Wardrop. Textbook of aero engines. New York: F. A. Stokes Company [cop. 1920]. 363 p. illus. 4°. **† VDY**

Describes American, British, French, and German types. See p. 48–49, 89–90 for historical résumé.

2802. Shoemaker, James M. Progress in aeronautics [for 1929]; power plants. (Mechanical engineering, New York, 1930, v. 52, p. 3–4.) **VFA**

2803. Sir Henry Royce. port. (Aeroplane, London, 1933, v. 44, p. 711.) **VDS**

Portrait and biography of prominent British manufacturer of airplane engines. He was one of the founders of Rolls-Royce, Ltd.

2804. Stentzel, A. Flüssige Kohlensäure, ein Mittel zur Lösung des Flugproblems. (Neueste Erfindungen und Erfahrungen, Wien, 1897, Jahrg. 24, p. 63.) **3–VA**

Author describes his motor using liquid carbonic acid gas.

2805. The **Story** of a remarkable endurance test. Wright E4 engine passes 300 hr. test with full power. illus. (Aviation, Highland, N. Y., 1923, v. 14, p. 475.) **VDS**

2806. Stout, R. Cheyne. The development of the Cirrus engine. illus. (U. S. air services, Washington, 1929, v. 14, April, p. 53–54.) **VDS**

2807. Swan, Andrew. Recent progress and developments in British engines. (Air annual of the British Empire, 1934/35, London, 1934, v. 6, p. 127–134.) **VDS**

2808. Symons, William L. The trade-mark "Liberty" registered by the United States. (Scientific American, New York, 1919, v. 121, p. 116.) **VA**

The government owns trade-mark as applied to engines. Slightly abridged in *Aerial age weekly,* New York, 1919, v. 9, p. 897, *VDS.*

2809. Taylor, Charles Fayette. History of the aeronautical engine. Basic features almost unchanged through development years. illus. (Aviation, Highland, N. Y., 1926, v. 21, p. 284–286.) **VDS**

2810. —— Power plants in 1933. illus. (Aviation, New York, 1934, v. 33, p. 19–20.) **VDS**

"The year 1933 was marked by further increases in the rated output of many models of American airplane engines...made possible largely by improvement in cooling through better fin design, cowling, and baffling of cylinders, and by improvement in the anti-knock value of available aircraft fuels."

Engines, continued

2811. **Taylor,** Charles E. Building the original Wright motor. World's first airplane mechanic relates how unique power plant was developed. illus. port. (Slipstream, Dayton, O., 1928, v. 9, May, p. 9–11.) **VDS**

2812. **Thomas,** W. E. The old and new aircraft engines. illus. (Western flying, Los Angeles, 1930, v. 7, Feb., p. 70–71, 150.) **VDS**

2813. **Tilley,** N. N. Progress in aeronautics. Power plants. (Mechanical engineering, New York, 1929, v. 51, p. 5–6.) **VFA**

Engine development during 1928.

2814. **Tissandier,** Gaston. Propulseur dynamo-électrique pour aérostat allongé. illus. (L'Aéronaute, Paris, 1883, année 16, p. 83–88.) **3–VDS**

Author's apparatus consisting of a Siemens dynamo and a bichromate of potassium battery.

2815. —— Sur l'application des moteurs électriques et des piles secondaires de M. G. Planté à la direction des aérostats. (Académie des sciences, Comptes rendus, Paris, 1881, tome 93, p. 254–256.) ***EO**

Also in *L'Aéronaute,* Paris, 1881, année 14, p. 178–180, *3–VDS.*

2816. A **Trade** change and some history. (Aeroplane, London, 1932, v. 42, p. 330, 332.) **VDS**

Chronological history of the firm of D. Napier and Son, Ltd., airplane engine builders.

2817. The **True** story of the Liberty motor; the lightest and most powerful airplane engine ever produced on a quantity basis. (Scientific American, New York, 1918, v. 118, p. 500, 515.) **VA**

2818. **Vincent,** J. G. The Liberty aircraft engine. illus. (Society of Automotive Engineers, Journal, New York, 1919, v. 4, p. 383–401.) **TOL**

With discussion.
"We had to start from absolutely nothing."

2819. **Walford,** Eric Weston. The German aeroplane engine trials. A detailed description of the equipment and proceedings. [By E. W. Walford.] illus. (Automobile engineer, London, 1916, v. 6, p. 318–321, 370–372; 1917, v. 7, p. 2–7, 38–42.) **TOL**

Tests were conducted in Germany during 1913.

2820. **Webb,** L. D. The Diesels take the air. illus. (U. S. air services, Washington, 1932, v. 17, March, p. 20–23.) **VDS**

2821. **Wheat,** George Seay. The air-cooled engine. Giant gains in efficiency mark the development of this type of power plane. illus. (Aero digest, New York, 1924, v. 5, p. 367–370.) **VDS**

2822. **Willgoos,** A. V. D. Pratt and Whitney fuel injection system. illus. (U. S. air services, Washington, 1932, v. 17, Feb., p. 35–36.) **VDS**

2823. **Wilson,** E. E. The trend of aircraft engine development. illus. (American Society of Naval Engineers, Journal, Washington, 1926, v. 38, p. 130–143.) **VXA**

2824. **Woolson,** L. M. The Packard Diesel aircraft engine. illus. (Society of Automotive Engineers, Journal, New York, 1930, v. 26, p. 431–442, 449; v. 27, p. 279–281, 319.) **VDS**

With discussion.

2825. The **Wright** J.-5 "Whirlwind." An American engine which has made history. illus. (Flight, London, 1927, v. 19, p. 390–392.) **VDS**

Description of engine used on the planes of Lindbergh and Chamberlin.

2826. **Wright** Whirlwinds have played important parts in making aeronautical history. illus. (Aviation, Highland, N. Y., 1927, v. 22, p. 1432–1433, 1449.) **VDS**

2827. **Zur** Geschichte des Luftschiff-Motors. port. illus. (Deutsche Luftfahrer-Zeitschrift, Berlin, 1916, Jahrg. 20, p. 10–11.) **VDS**

Account of Wilhelm Maybach, German pioneer designer.

PROPELLERS

2828. **Amans,** P. C. Perfectionnement des hélices aériennes. illus. (L'Aéronaute, Paris, 1892, année 25, p. 219–231.) **3–VDS**

Author's experiments to produce a propeller of maximum force.

2829. **Arson.** Étude sur le propulseur pour l'aviation. illus. (Société des ingénieurs civils de France, Mémoires, Paris, 1879, p. 342–365.) **VDA**

2830. **Balaban,** Karl. Ein Beitrag zur Entwicklungsgeschichte des Hubschraubers. illus. (Zeitschrift für Flugtechnik und Motorluftschiffahrt, München, 1922, Jahrg. 13, p. 299–303, 309–315.) **VDS**

History of the lifting airscrew.

2831. [**Bibliography** of helices.] (In: Paul Brockett, Bibliography of aeronautics, Washington, 1910, p. 404.) **VDT**

Numerous references to writings down to July, 1909.

2832. [**Bibliography** of propellers.] (United States. — National Advisory Committee for Aeronautics, Bibliography of aeronautics, 1909–1931, Washington, 1921–35, volumes for 1909/16, p. 1070–1084; 1917/19, p. 364–368; 1920/21, p. 18, 334–338; 1922, p. 127–128; 1923, p. 6–7, 109–110; 1924, p. 4, 88; 1925, p. 7–8, 141–142; 1926, p. 117–118; 1927, p. 142–143; 1928, p. 160–162; 1929, p. 182–183; 1930, p. 194–196; 1931, p. 227–230.) **VDT**

Some of the headings read *Airscrews.*

Propellers, continued

2833. Boyer-Guillon, Alfred. Essai des hélices aériennes au Laboratoire d'essais du Conservatoire national des arts et métiers. illus. (Revue de mécanique, Paris, 1908, tome 23, semestre 2, p. 242–249.) **3–VFA**

2834. —— Essais d'hélices au point fixe. illus. (Société des ingénieurs civils de France, Mémoires, Paris, 1909, Aug., p. 65–91.) **VDA**

Experimental study of propellers at the Laboratoire d'essais du Conservatoire national des arts et métiers.

2835. Brown, Harold S. A defense of the Wright system of propellers. (Scientific American, New York, 1908, v. 99, p. 471.) **VA**

2836. Caldwell, Frank Walker. Aircraft-propeller development and testing summarized. illus. (S. A. E. journal, transactions section, Philadelphia, 1934, v. 35, p. 297–310, 349–358.) **TOL**

With discussion.

"It is the purpose of this paper to outline in the briefest possible way some of the experiments... carried out in the United States during the last few years."

2837. Camus, Gaston. La technique des hélices aériennes; notions élémentaires sur le tracé, l'utilisation et la construction des hélices aériennes. Paris: F. L. Vivien, 1909. 71 p. 8°. **VDY**

2838. Cohn, Georg. Zur Geschichte der Luftschraube (Treibschraube). illus. (Deutsche Luftfahrer-Zeitschrift, Berlin, 1916, Jahrg. 20, p. 153–155.) **VDS**

2839. Crocé-Spinelli, J. Du meilleur propulseur applicable à la navigation aérienne. (L'Aéronaute, Paris, 1870, année 3, p. 49–61.) **3–VDS**

2840. —— Hélice à pas variable en marche. illus. (L'Aéronaute, Paris, 1871, année 4, p. 161–176.) **3–VDS**

Suggestion for variable pitch propeller.

2841. Dallwitz-Wegner, Richard von. Die Aeroplane und Luftschrauben der statischen und dynamischen Luftschiffahrt schwerer und leichter als Luft... Rostock i. M.: C. J. E. Volckmann Nachfolger, 1908. viii, 45 p. illus. 8°. **VDY p.v.5, no.9**

A study of the application of propellers to aircraft.

2842. Decazes, vicomte. Expériences d'allégement au moyen d'une hélice à axe vertical. illus. (L'Aérophile, Paris, 1903, année 11, p. 8–11.) **3–VDS**

2843. Deprez, Marcel. Calculs et théorèmes relatifs à l'hélice propulsive des aérostats. (Société des ingénieurs civils de France, Mémoires, Paris, 1902, année 1902, v. 1, p. 40–42.) **VDA**

2844. Drzewiecki, Stepan. Des hélices aériennes. Théorie générale des propulseurs hélicoïdaux et méthode de calcul de ces propulseurs pour l'air. Paris: F. L. Vivien, 1909. 62 p. 8°. **VDY**

2845. Fage, Arthur. Airscrews in theory and experiment. London: Constable & Co., Ltd., 1920. ix, 198 p. illus. diagrs. 4°. **VDY**

Bibliography, p. 191–195.

2846. First magnesium propeller meets test well. illus. (U. S. air services, 1925, v. 10, April, p. 47.) **VDS**

Magnesium-aluminum alloy used by Curtiss Aeroplane and Motor Company.

2847. Gerlach, Edm. Notiz über die Hubkraft von Schiffsschrauben mit senkrechter Axe. (Zeitschrift für Luftschiffahrt, Berlin, 1887, Jahrg. 6, p. 12–15.) **3–VDS**

Discusses the lifting power of vertical screws.

2848. Govi, Gilberto. Sur une très ancienne application de l'hélice comme organe de propulsion. (Académie des sciences, Comptes rendus, Paris, 1881, tome 93, p. 400–402.) **VA**

Refers to Leonardo da Vinci's invention of the air screw and of the parachute. Abstract in *Scientific American,* New York, 1881, new series, v. 45, p. 232, *VA.*

2849. Gustave Trouvé. (Wiener Luftschiffer-Zeitung, Wien, 1902, Jahrg. 1, p. 158.) **VDS**

Biography of prominent French investigator of gliders and of propellers.

2850. Haenlein, Paul. Ueber den Treibapparat bei Luftschiffen. illus. (Zeitschrift für Luftschiffahrt, Berlin, 1882, Jahrg. 1, p. 240–244.) **3–VDS**

Discussion of propellers and paddle-wheels.

2851. Hammond, E. V. The problem of flight. illus. (Engineering, London, 1908, v. 86, p. 206–207.) **VDA**

Account of author's experiments with multiple propellers.

2852. Hansen. Berechnung des Wirkungsgrades der Flügelschraube. illus. (Zeitschrift für Luftschiffahrt, Berlin, 1890, Jahrg. 9, p. 177–181.) **3–VDS**

2853. Hastings, C. W. On the problem of aerial navigation. illus. (International Conference on Aerial Navigation, Chicago, 1893, Proceedings, New York, 1894, p. 11–66.) **VDS**

Lengthy paper of one of Chanute's assistants. For discussion of the paper see entry no. 2890.

2854. Hauvel, Charles du. Conditions d'un élément héliçoidal pour l'effet utile maximum d'un propulseur. diagr. (L'Aéronaute, Paris, 1885, année 18, p. 7–14; 1887, année 20, p. 83–90.) **3–VDS**

2855. Havill, C. H. Development of aircraft propellers. (Aviation, New York, 1928, v. 25, p. 1190, 1206–1208.) **VDS**

2856. Hermann Hoernes. port. (Wiener Luftschiffer-Zeitung, Wien, 1907, Jahrg. 6, p. 270–271.) **VDS**

Portrait of pioneer Austrian aeronautical engineer and writer. He was especially interested in the development of the air propeller.

2857. Hervé, Henri, and Henri, comte de la Vaulx. Sur une nouvelle hélice. (Académie des sciences, Comptes rendus, Paris, 1904, tome 138, p. 1688–1690.) *** EO**

Propellers, continued

2858. Hoernes, Hermann. Wellner's Versuche mit grösseren Luftschrauben. (Zeitschrift für Luftschiffahrt, Berlin, 1897, Jahrg. 16, p. 5–8.) **3–VDS**

2859. Hofrat Georg Wellner. (Wiener Luftschiffer-Zeitung, Wien, 1909, Jahrg. 8, p. 323–324.) **VDS**
Wellner (1846–1909) is well known for his investigations of the air propeller.

2860. Hollands, Sidney H. Aerial propellers. Old and new forms. (Scientific American supplement, New York, 1908, v. 66, p. 387.) **VA**

2861. Jarolimek, Anton. Die Segelradversuche Professor Wellner's. (Oesterreichischer Ingenieur- und Architekten-Verein, Zeitschrift, Wien, 1895, Jahrg. 47, p. 89–91.) **3–VDA**

2862. —— Ueber Luftschrauben. (Oesterreichischer Ingenieur- und Architekten-Verein, Zeitschrift, Wien, 1894, Jahrg. 46, p. 547–549.) **3–†† VDA**
Discussion of Wellner's investigations.

2863. Kadarz, Theodor. Der Luftpropeller-Flügel und seine Eignung für Luftschiffahrts-Zwecke. illus. (Zeitschrift für Luftschiffahrt, Berlin, 1896, Jahrg. 15, p. 103–113, 145–153, 177–185.) **3–VDS**
Air-propeller calculations.

2864. Klassohn, Eduard. Ueber die willkürliche Bewegung von Luftschiffen. illus. (Der Civilingenieur, Leipzig, 1865, N. F. Bd. 11, p. 435–474.) **VDA**
A study of the application of the screw propeller to airships.

2865. Klemin, Alexander. The paddle wheel airfoil system. illus. (Aviation engineering, East Stroudsburg, Pa., 1933, v. 8, Feb., p. 5–8, 30.) **VDS**
Revival of an old idea shown in the patents of H. H. Platt and I. B. Laskowitz.

2866. Kress, Wilhelm. Note on the elastic air-screw. illus. (International Conference on Aerial Navigation, Chicago, 1893, Proceedings, New York, 1894, p. 264–266.) **VDS**
Screw for navigable balloon.

2867. Lefebvre, Ed. La première application de l'hélice aérienne. (L'Aérophile, Paris, 1909, année 17, p. 235.) **3–VDS**
Briefly refers to the patent of Frédéric Sauvage, 1836; also to the ideas of Meunier, 1792.

2868. The **Lifting** power of a screw propeller for aeronautical work. illus. (Scientific American, New York, 1905, v. 93, p. 461.) **VA**
Experiments of Louis Goddard, Félix Faure, and Hockengos.

2869. McCauley, E. G. Metal propeller development. illus. (Aviation, Highland, N. Y., 1927, v. 22, p. 1127–1130, 1144.) **VDS**
Describes characteristics of early metal propellers.

2870. Mallet, Maurice. Sur un essai de l'hélice à propulsion verticale. (Académie des sciences, Comptes rendus, Paris, 1893, tome 117, p. 300–302.) ***EO**
Vertical propeller for use on balloon. Also in *L'Aérophile,* Paris, 1893, année 1, p. 144–146, *3–VDS.*

2871. Marsh, William Lockwood. The first screw propeller. (Engineer, London, 1927, v. 144, p. 415–416.) **VA**
"We find that the use of the air propeller, both in conception and in actual practice, quite considerably antedates the marine propeller."

2872. Maxim, Sir Hiram Stevens. Mechanical flight. (Knowledge, London, 1895, v. 18, p. 14–15.) **OA**
A reply to Thomas Moy, concerning the aerial screw propeller. See entry no. 2877.

2873. —— Screw-propellers working in air. (Aeronautical annual, Boston, 1897, no. 3, p. 142–144.) **3–VDS**

2874. Mewes, Rudolf. Luftschraubenversuche von v. Bradsky. illus. (Dinglers polytechnisches Journal, Berlin, 1902, Bd. 317, p. 64–65.) **3–VA**

2875. Milestones in the development of the metal air-screw. illus. (Flight, London, 1924, v. 16, p. 445–446.) **VDS**
Seven years' development beginning in 1917.

2876. Moedebeck, Hermann W. L. Luftpropellerversuche auf dem Wasser und der erste Vorschlag der Luftschraube im Jahre 1784. (Illustrierte aëronautische Mitteilungen, Strassburg, 1900, Jahrg. 4, p. 74–75.) **VDS**
First use of propeller by the Robert brothers under the patronage of the Duke of Chartres.

2877. Moy, Thomas. Mechanical flight. illus. (Knowledge, London, 1894, v. 17 [new series, v. 9], p. 274–275.) **OA**
Does not believe that a balloon can be satisfactorily stiffened; also that the ordinary air propeller will not work. He took out a patent on a feathering screw. For reply by Hiram Maxim, see entry no. 2872.

2878. The **Multiple** air propeller. Its aeronautic possibilities for dirigible airships, aeroplanes, and helicopters. illus. (Scientific American, New York, 1908, v. 99, p. 244.) **VA**
Separate inventions of E. V. Hammond and Wilbur R. Kimball.

2879. Phillips, Donald B. War thrills in peace time. illus. (Popular aviation, Chicago, 1928, v. 3, Sept., p. 15–19.) **VDS**
Has account of the early use of duralumin in America, and of the duralumin propeller.

2880. Pillet, Louis. Expériences pratiques sur le rendement de différentes hélices aériennes et description de l'hélice à surfaces concavo-convexes. (L'Aéronaute, Paris, 1870, année 3, p. 33–43.) **3–VDS**
Experiments were begun by the author in 1848.

2881. Les **Premières** hélices aériennes. (Le Génie civil, Paris, 1928, tome 93, p. 98.) **VA**
Refers to Joseph Bramah, Leonardo da Vinci, Paucton, Meusnier, David Bourgeois, and Vallet.

2882. Professor Georg Wellner. port. (Wiener Luftschiffer-Zeitung, Wien, 1904, Jahrg. 3, p. 85–86.) **VDS**

Propellers, continued

2883. **Rateau,** Auguste Camille Edmond. Théorie des hélices propulsives. illus. (Académie des sciences, Comptes rendus, Paris, 1900, tome 130, p. 486–492, 702–705.) * EO

Also in Société d'encouragement pour l'industrie nationale, *Bulletin,* Paris, 1900, série 5, tome 5, p. 497–500, *3–VA.* Library also has author's *Théorie des hélices propulsives, marines et aériennes et des avions en vol rectiligne,* Paris, 1920, 159 p., *VDY p.v.23.*

2884. **Renard,** Charles. Sur un nouveau mode de construction des hélices aériennes. illus. (Académie des sciences, Comptes rendus, Paris, 1904, tome 139, p. 721–724.) * EO

2885. —— Sur la qualité des hélices sustentatrices. (Académie des sciences, Comptes rendus, Paris, 1903, tome 137, p. 970–972.) * EO

2886. **Riabouschinsky,** D. Recherches sur l'hélice aérienne se mouvant dans un courant aérien dirigé perpendiculairement à l'axe de l'hélice. illus. (Institut aérodynamique de Koutchino, Bulletin, St. Pétersbourg, 1906, fasc. 1, p. 13–17.) VDS

2887. —— Recherches sur les hélices aériennes. illus. (Institut aérodynamique de Koutchino, Bulletin, Moscou, 1909, fasc. 2, p. 3–102.) VDS

2887a. **S. A. Reed** dies; invented metal plane propeller. Mining engineer designed duralumin vanes that were used by Lindbergh. Succumbs here at 81. (New York Herald Tribune, New York, Oct. 2, 1935, p. 18.) * A

2888. **Sachse,** H. Der Kirsten-Boeing propeller. illus. (Zeitschrift für Flugtechnik und Motorluftschiffahrt, München, 1926, Jahrg. 17, p. 1–4.) VDS

Swivelling propellor for airships. Brief abstract in *Scientific American,* New York, 1926, v. 134, p. 354, *VA.*

2889. **Scoble,** Walter A. Some anchored tests of aerial propellers. diagrs. (Engineering, London, 1908, v. 86, p. 843–845.) VDA

Experiments at the Blythswood laboratory.

2890. **Soreau,** Rodolphe, and Edmond Henry. Sur l'hélice aérienne. Analyse critique de la théorie de M. C. W. Hastings. (L'Aéronaute, Paris, 1895, année 28, p. 51–63.) 3–VDS

For Hastings' paper see entry no. 2853.

2891. **Svehla,** George. The duralumin propeller. illus. (Popular aviation, Chicago, 1929, v. 4, March, p. 48, 50, 52, 54, 97.) VDS

Good history of development.

2892. **Taffoureau,** Edgar. Sur les hélices sustentatrices. (Académie des sciences, Comptes rendus, Paris, 1904, tome 139, p. 356–358.) * EO

2893. **Tastes,** de. Sur un nouveau propulseur. (Académie des sciences, Comptes rendus, Paris, 1871, tome 73, p. 1371.) * EO

Suggests rapidly vibrating discs.

2894. **Tatin,** Victor. Essais d'une hélice suspensive. (L'Aérophile, Paris, 1905, année 13, p. 138–141.) 3–VDS

2895. **Valer,** Paul. Note sur les propulseurs héliçoidaux. (L'Aéronaute, Paris, 1885, année 18, p. 188–192.) 3–VDS

Endeavors to adapt Antoine's marine propeller formulas of 1880 to aerial propellers.

2896. **Veyrin,** Émile. Les hélices à axe oblique s'avançant horizontalement. illus. (L'Aéronaute, Paris, 1892, année 25, p. 243–248.) 3–VDS

Description of model made by the author. For comments see above magazine, 1893, année 26, p. 61–62.

2897. **Vogt,** H. C. The air propeller. illus. (International Conference on Aerial Navigation, Chicago, 1893, Proceedings, New York, 1894, p. 113–119.) VDS

Conclusions drawn from experiments of several investigators.

2898. —— On the action of propellers and aeroplanes. (Engineering, London, 1909, v. 87, p. 11.) VDA

Author's experiments, using apparatus of Captain Rung, of the Danish artillery.

2899. **Walker,** William George. The lifting power of air propellers. illus. (Engineering, London, 1900, v. 69, p. 233–236.) VDA

Includes a list of other papers by the author.

2900. **Watts,** Henry C. Airscrew research in Great Britain. (International Congress of Aerial Navigation, London, 1923, Report, London, 1923, p. 422–430.) VDS

With discussion.

2901. **Weick,** Fred Ernest, and D. H. Wood. The twenty-foot propeller research tunnel of the National Advisory Committee for Aeronautics. illus. plans. (United States. — National Advisory Committee for Aeronautics. Technical report no. 300.) VDY

In the 14th annual report (1928) of the Committee, Washington, 1929, p. 429–442.

2902. **Wellner,** Georg. Die Segelradversuche. illus. (Oesterreichischer Ingenieur- und Architekten-Verein, Zeitschrift, Wien, 1894, Jahrg. 46, p. 561–566, 573–575.)

Investigation on sail propellers. 3–†† VDA

2903. —— Ueber Luftschrauben. Beitrag zur Kenntniss ihrer Betriebsverhältnisse und ihrer Propulsionswirkung. illus. (Oesterreichischer Ingenieur- und Architekten-Verein, Zeitschrift, Wien, 1894, Jahrg. 46, p. 405–410, 417–421, 537–540.) 3–†† VDA

For criticism see *L'Aéronaute,* Paris, 1895, année 28, p. 101–108, *3–VDS.*

2904. —— Versuche mit grösseren Luftschrauben. illus. (Oesterreichischer Ingenieur- und Architekten-Verein, Zeitschrift, Wien, 1896, Jahrg. 48, p. 504–508, 514–517.) 3–†VDA

Also in *Zeitschrift für Luftschiffahrt,* Berlin, 1897, Jahrg. 16, p. 98–110, 152–154, *3–VDS.*

2905. **Werner,** A. Die Fahnenschraube mit verstellbaren Flächen. illus. (Zeitschrift für Luftschiffahrt, Berlin, 1882, Jahrg. 1, p. 337–340.) 3–VDS

Describes adjustable vane-propellers.

2906. **Woodward,** Calvin Milton. A discussion of the problems connected with air-ship propellers. St. Louis, 1907. 10 p. 8°. VDY p.v.6, no.20

AIRPLANE FLIGHTS

2907. **Acosta,** Bertram B., and others. Fifty-one hours in the air! by Bert Acosta, Giuseppe Bellanca, and Clarence Chamberlin. ports. (Aero digest, New York, 1927, v. 10, p. 371.) VDS

Breaking of the world's endurance record, using Bellanca plane, 1927.

2908. **Across** Australia in two days. map. (Flight, London, 1926, v. 18, p. 38.) VDS

Colonel Brinsmead's flight, Dec., 1925.

2909. L'**Activité** aéronautique en 1924 [and 1925]. illus. ports. (L'Aérophile, Paris, 1925, année 33, p. 53–55, 119–120.) 3–VDS

Attempts for French and world's records for these years.

2910. The **Aerial** Derby, 1914. illus. (Flight, London, 1914, v. 6, p. 619–623.) VDS

Third Aerial Derby was won by W. L. Brock.

2911. The **Aero** loop-the-loop accomplished. (Aero and hydro, Chicago, 1913, v. 6, p. 431.) 3–VDS

Account of Pégoud's exploit.

2912. The **Aeroplanes** at Rheims. (Engineer, London, 1909, v. 108, p. 247–248, 266, 268.) VA

Descriptions of the competing machines.

2913. The **Air** not yet conquered. illus. ports. (Literary digest, New York, 1927, v. 94, July 16, p. 5–7.) * DA

Newspaper comments on Byrd's trans-Atlantic flight of 1927.

2913a. **Aircraft** year book; issued by the Manufacturers Aircraft Association, Inc., 1919–1934. New York: Manufacturers Aircraft Assoc., 1919–1934. 8°. VDS

Has yearly chronologies and records.

2914. **Alan** Cobham's Everest flight. (Flight, London, 1925, v. 17, p. 70.) VDS

2915. The **Alaska** flying expedition. illus. maps. (Flying, New York, 1920, v. 9, p. 626–630.) VDS

U. S. Army Air Service flight of 1920.

2916. **America** first in the air. The transcontinental non-stop flight. illus. ports. (Aeronautical digest, New York, 1923, v. 2, p. 403–409.) VDS

Flight of Oakley G. Kelly and John A. Macready, 1923.

2917. The **America** flies through fog to France. illus. ports. (Aviation, Highland, N. Y., 1927, v. 23, p. 68–70.) VDS

Flight of Commander Byrd and companions, 1927.

2918. **American** marathons of the sky. ports. (Literary digest, New York, 1929, v. 102, Aug. 10, p. 11.) * DA

Refueling flight of Dale Jackson and Forest O'Brine, 1929. Their record was 420 hours, 21 minutes, 30 seconds.

2919. **America's** first aviation meet at Los Angeles. Detailed account of the flights made by the American and French aviators. illus. (Scientific American, New York, 1910, v. 102, p. 105–106.) VA

2920. The **"America's"** flight to France. illus. (Aero digest, New York, 1927, v. 11, p. 7, 111.) VDS

Airplane flight commanded by Richard E. Byrd, 1927.

2921. **Amsterdam-Batavia** and back in 27 days. illus. (Flight, London, 1927, v. 19, p. 564–565.) VDS

Flight of Van Lear Black in 1927.

2922. **Another** four continent trip. illus. map. (Popular aviation, Chicago, 1928, v. 2, April, p. 15–16.) VDS

Brief account of flight of Costes and Le Brix in Europe, Africa, South America, and North America, 1927.

2923. **Anthony** (Tony) Jannus is dead. (Aviation, New York, 1916, v. 1, p. 263.) VDS

Jannus, an American aviator, was killed in the Russian service, Oct. 12, 1916. He featured in seaplane flights. For numerous references concerning him, see United States. — National Advisory Committee for Aeronautics, *Bibliography of aeronautics*, 1909–1931, Washington, 1921–35, volume for 1909/16, p. 690–692, *VDT*.

2924. **Aracena,** Diego. Across South America by the air route. port. (Aeronautical digest, New York, 1923, v. 2, p. 336.) VDS

Author's flight across the Andes.

2925. The **Army** flight to Hawaii. illus. ports. (Aero digest, New York, 1927, v. 11, p. 16–18.) VDS

United States Army flight by Lester J. Maitland and Albert F. Hegenberger, 1927.

2926. **Arnold,** Henry H. Contact and refueling now accomplished in the air. illus. (Scientific American, New York, 1925, v. 133, p. 90–91.) VDS

Experiences of Lowell Smith and J. P. Richter, of the United States Air Service, 1923.

2927. **Atlantic** flight chronology. (Aero digest, New York, 1931, v. 19, Dec., p. 33–37, 50.) VDS

Excellent list of 118 entries, 1873–1931.

2928. **Atwood** [Harry N.] and his St. Louis-New York flight. illus. (Scientific American, New York, 1911, v. 105, p. 225.) VA

2929. Une **Audacieuse** expérience de Pégoud. illus. (L'Aérophile, Paris, 1913, année 21, p. 420–421.) 3–VDS

Description of loop-the-loop first performed by him.

2930. **Aviation** in Australia. The first real flight in the Antipodes. illus. (Aero, London, 1910, v. 2, p. 478.) 3–† VDS

Claims that Harry Houdini was the first successful aviator in the country.

Airplane Flights, continued

2931. L'**Aviation** triomphante, par MM. d'Estournelles de Constant, C. Bouchard, E. Lavisse, P. Painlevé, L. Blériot, Paul Rousseau, capitaine Ferber, le comte de Lambert, Pierre Mille, et divers collaborateurs. Paris: Librairie aéronautique, 1910. 381 p. illus. ports. 12°. **VDT**

Blériot's Channel flight. Famous aeronautical meet at Reims, 1909. Flights of Captain Ferber, Comte de Lambert, and others.

2932. **Aviation's** victory over fog. (Literary digest, New York, 1929, v. 103, Oct. 5, p. 15.) *** DA**

Account of James H. Doolittle's blind flying over Mitchel Field, New York, 1929.

2933. L'**Avion** transatlantique du capitaine Fonck. illus. (L'Aérophile, Paris, 1926, année 34, p. 215–216.) **3–VDS**

Sikorsky plane equipped with three Gnome-Jupiter motors.

2934. Les **Avions** Farman et l'histoire de l'aviation. illus. (L'Aérophile, Paris, 1924, année 32, p. 31–32.) **3–VDS**

Brief history of the Société Henri et Maurice Farman.

2935. **Award** of the Langley medal. (Science, New York, 1927, v. 66, p. 9.) **OA**

Awarded to Charles A. Lindbergh, June 11, 1927.

2935a. **Balbo,** Italo. ...La centuria alata. Milano: A. Mondadori [1934]. 392 p. illus. 8° **VDYC**

North Atlantic flight, 1933. Library also has English translation, entitled *My air armada,* London, 1934.

2936. —— Da Roma a Odessa, sui cieli dell' Egeo e del Mar nero; note di viaggio. Milano: Fratelli Treves, 1929. 217 p. maps, ports. illus. 8°. **VDYC**

Well-illustrated account of author's airplane flight from Italy to the Aegean and Black Seas.

2936a. —— ...Stormi d'Italia sul mundo. Introduzione di V. Beonio-Brocchieri. Quatra edizione. Milano: A. Mondadori [1935]. 227 p. illus. 8°. **VDYC**

Mediterranean, South American, and North American flights commanded by the author.

2936b. —— ...Stormi in volo sull' oceano ... Verona: A. Mondadori [cop. 1931]. 256 p. plates. 8°. **VDYC**

Flight of Italian squadron, Italy to South America, 1931.

2937. **Balbo** shows the way. illus. (U. S. air services, Washington, D. C., 1933, v. 18, Aug., p. 11–14.) **VDS**

Italian trans-Atlantic flight of 1933.

2938. **Bans,** Georges. Léon Delagrange. port. (L'Aérophile, Paris, 1908, année 16, p. 137.) **3–VDS**

Portrait and brief biography of French sculptor-aviator.

2939. **Barker,** Fowler Wesley. Air pioneers of the West. illus. ports. (Aero digest, New York, 1931, v. 18, Feb., p. 40–43, 120.) **VDS**

Accounts of early fliers on the Pacific coast.

2940. **Beach** in Travel Air wins second reliability tour. illus. (Aviation, Highland, N. Y., 1926, v. 21, p. 362–366.) **VDS**

Walter Beach wins the Ford contest, 1926, with a score of 4043.3 points.

2941. **Beamish,** Richard Joseph. The boy's story of Lindbergh, the Lone Eagle... Philadelphia: The John C. Winston Company, 1928. 320 p. illus. ports. 8°. **AN**

2942. **Beery,** L. L. A pioneering flight through Central America. Navigation log of the Army Air Service's flight from Panama to Guatemala. (Aviation, Highland, N. Y., 1924, v. 16, p. 534–535.) **VDS**

Flight made by U. S. Army officers in February, 1924.

2943. **Bellanca** monoplane sets world endurance record. Pilots Chamberlin and Acosta hold Whirlwind engined plane in flight for over 51 hr. breaking previous record by wide margin. illus. (Aviation, Highland, N. Y., 1927, v. 22, p. 819–821.) **VDS**

Flight over Long Island, 1927.

2944. **Bellanca** plane arrives at German capital. illus. (Aviation, Highland, N. Y., 1927, v. 22, p. 1360–1361.) **VDS**

Final hop of the Chamberlin-Levine Transatlantic flight in 1927.

2945. **Benson,** Stuart. The Gordon Bennett cup. ports. (Aero Club of America, Bulletin, New York, 1912, Jan., p. 24–25.) **VDS (Flying)**

Brief history of contests for aviation trophy, 1909–1912. The above bulletin is one of the early numbers of the periodical afterwards known as Flying.

2946. **Bergner,** Hanns Heinz, compiler. Die Ozeanfluege bis Mai 1928, von Leutnant Read bis Hauptmann Köhl; zusammengestellt von Hanns Heinz Bergner. München: Südost-Verlag A. Dresler, 1928. 64 p. illus. ports. 8°. **VDY p.v.46, no.3**

Concise history of ocean flights, with a chronology, 1919–1928.

2947. **Bernard,** Marc. En hydravion au-dessus du continent noir. Paris: B. Grasset, 1927. xvi, 197 p. illus. ports., map. 12°. **VDY**

Account of trans-African flight (1926) by the author.

2948. **Besançon,** Georges. Louis Blériot. port. (L'Aérophile, Paris, 1908, année 16, p. 1.) **3–VDS**

2949. **Biard,** Henry Charles. Wings. London: Hurst & Blackett, Ltd. [1934.] x, 11–288 p. illus. ports. 8°. **VDYC**

Adventures and war career of British flying pioneer and winner of the Schneider trophy.

2950. [**Bibliography** of altitude flying.] (United States. — National Advisory Committee for Aeronautics, Bibliography of aeronautics, 1909–1931, Washington, 1921–35, volumes for 1909/16, p. 55–59; 1917/19, p. 25; 1920/21, p. 23–25; 1922, p. 9; 1923, p. 8; 1924, p. 6; 1925, p. 10–11; 1926, p. 7; 1927, p. 8; 1928, p. 12; 1929, p. 21; 1930, p. 14–15; 1931, p. 20.) **VDT**

Airplane Flights, continued

2951. [**Bibliography** of Claude Grahame-White.] (United States. — National Advisory Committee for Aeronautics, Bibliography of aeronautics, 1909–1931, Washington, 1921–35, volumes for 1909/16, p. 566–569; 1917/19, p. 192; 1920/21, p. 174.) **VDT**

2952. [**Bibliography** of flights in the Sahara desert.] (United States. — National Advisory Committee for Aeronautics, Bibliography of aeronautics, 1909–1931, Washington, 1921–35, volumes for 1909/16, p. 1177–1178; 1917/19, p. 394; 1920/21, p. 362; 1925, p. 154; 1928, p. 174; 1930, p. 211.) **VDT**

2953. [**Bibliography** of Gustave Hamel.] (United States. — National Advisory Committee for Aeronautics, Bibliography of aeronautics, 1909–1931, Washington, 1921–35, volumes for 1909/16, p. 604–605; 1917/19, p. 207.)
References to early English flier. **VDT**

2954. [**Bibliography** of Jacques Schneider and of the Schneider cup.] (United States. — National Advisory Committee for Aeronautics, Bibliography of aeronautics, 1909–1931, Washington, 1921–35, volumes for 1909/16, p. 1196; 1917/19, p. 398; 1920/21, p. 367–368; 1922, p. 140–141; 1923, p. 122; 1924, p. 98; 1925, p. 156–157; 1926, p. 128; 1927, p. 154–155; 1928, p. 176–177; 1929, p. 198; 1930, p. 214–215; 1931, p. 250–252.) **VDT**

2955. [**Bibliography** of Jules Vedrines.] (United States. — National Advisory Committee for Aeronautics, Bibliography of aeronautics, 1909–1931, Washington, 1921–35, volumes for 1909/16, p. 1374; 1917/19, p. 457.)
References to famous French aviator. **VDT**

2956. [**Bibliography** of Léon Delagrange.] (In: Paul Brockett, Bibliography of aeronautics, Washington, 1910, p. 233.) **VDT**
Cites references to Delagrange's airplane exploits.

2957. [**Bibliography** of looping the loop.] (United States. — National Advisory Committee for Aeronautics, 1909–1931, Washington, 1921–35, volume for 1909/16, p. 808–809.) **VDT**

2958. [**Bibliography** of the Michelin cup.] (United States. — National Advisory Committee for Aeronautics, Bibliography of aeronautics, 1909–1931, Washington, 1921–35, volumes for 1909/16, p. 871–873; 1920/21, p. 276; 1923, p. 86; 1927, p. 112.) **VDT**

2959. [**Bibliography** of night flying.] (United States. — National Advisory Committee for Aeronautics, Bibliography of aeronautics, 1909–1931, Washington, 1921–35, volumes for 1909/16, p. 955–956; 1917/19, p. 332; 1920/21, p. 301; 1922, p. 116–117; 1923, p. 100; 1924, p. 81; 1925, p. 129; 1926, p. 106–107; 1927, p. 132; 1928, p. 147; 1930, p. 178; 1931, p. 210.) **VDT**

2960. [**Bibliography** of records.] (United States. — National Advisory Committee for Aeronautics, Bibliography of aeronautics, 1909–1931, Washington, 1921–35, volumes for 1909/16, p. 1101–1117; 1917/19, p. 374–376; 1920/21, p. 346; 1922, p. 131–132; 1923, p. 112; 1924, p. 90; 1925, p. 144–145; 1926, p. 120–121; 1927, p. 145; 1928, p. 165–166; 1929, p. 187–188; 1931, p. 234.) **VDT**

2961. [**Bibliography** of transatlantic flights.] (United States. — National Advisory Committee for Aeronautics, Bibliography of aeronautics, 1909–1931, Washington, 1921–35, volumes for 1909/16, p. 88–89, 1340–1343; 1917/19, p. 443–447; 1920/21, p. 31; 1922, p. 155; 1926, p. 9, 141; 1927, p. 11, 170–171; 1928, p. 16–17; 1929, p. 24, 220; 1930, p. 19, 239.) **VDT**

2962. **Bierbaum,** Paul Willi. Im Aeroplan über die Alpen; Geo. Chavez' Simplonflug. Zürich: Art. Institut O. Füssli, 1910. 123 p. map, port. 12°. **VDY**
Account of fatal airplane flight over the Simplon Pass in 1910.

2963. **Bingham,** Hiram. History of the National Air Races. illus. (Aeronautic review, Baltimore, 1929, v. 7, Sept., p. 27, 46, 66.) **VDS**

2964. **Blackburn,** T. W. Lindbergh's record as a flying student. (Popular aviation, Chicago, 1930, v. 7, Aug., p. 42, 75.) **VDS**
Author was one of Lindbergh's instructors.

2965. **Blacker,** Latham Valentine Stewart. The Mount Everest flight. The special equipment and plan of campaign of the Houston-Mount Everest expedition described. illus. (Air and airways, London, 1933, v. 9, p. 423–424, 426.) **VDS**

2966. **Blake,** Wilfred Theodore. Flying round the world. London: H. Cranston, Ltd. [1923.] 238 p. illus. ports. 8°. **VDY**
Account of author's flight in 1922, London to India. Projected flight around the world was abandoned.

2967. —— Some difficulties in flying round the world. illus. (Geographical journal, London, 1923, v. 61, p. 273–279.) **KAA**
Observations on flight 1922, England to India, on world attempt.

2968. **Blériot,** Louis. Les expériences de Pégoud. illus. (Technique aéronautique, Paris, 1913–14, tome 8, p. 327–332, 363–369; tome 9, p. 16–22.) **3–VDS**

2969. **Blériot,** Louis, and E. Ramond. ...La gloire des ailes; l'aviation de Clément Ader à Costes. Paris: Les éditions de France, cop. 1927. vi, 335 p. 12°. **VDY**
Enthusiastic French account of notable events from Ader (1895) to Costes and Le Brix (1927). Chapters on Pégoud, the first aviator to "loop the loop," and Lindbergh. Altitude, speed, distance, and duration records.

Airplane Flights, continued

2970. **Blériot's** cross-Channel flight. illus. (Flight, London, 1909, new series, v. 1, p. 457–461.) **VDS**

2971. **Bonnefon,** Paul. Le premier "as" Pégoud. Préface du colonel aviateur Girod. Paris: Berger-Levrault, 1918. xiii, 140 p. illus. port. 12°. **BTZY**

2972. **Bonnet,** G. Adolphe Pégoud. port. (L'Aérophile, Paris, 1913, année 21, p. 409.) **3–VDS**

2973. —— Maurice Prévost. port. (L'Aérophile, Paris, 1913, année 21, p. 313.) **3–VDS**

Portrait and brief account of prominent French aviator.

2974. **Botved,** Anders Peter. København — Tokio — København gennem Luften. København: Gyldendalske Boghandel, 1926. 157 p. illus. port. 4°. **VDY**

Author's account of his flight (1926), Copenhagen to Tokio and return.

2975. **Bowen,** Robert Sidney. Ten years of air race programs. illus. (Aviation, New York, 1930, v. 29, p. 130–133.) **VDS**

2976. —— Twenty-five years of famous flights. illus. (Aviation, New York, 1928, v. 25, p. 1702–1715.) **VDS**

Account begins with the flights of the Wright brothers.

2977. **Bregeault,** Henry. Le premier atterrissage au Mont Blanc, 30 juillet 1921. Souvenirs d'un témoin. illus. (La Montagne, Paris, 1921, année 17, p. 177–196.) **PSL**

Author was a member of the first airplane expedition to land on Mt. Blanc.

2978. The **"Bremen's"** epic flight. illus. ports., map. (Literary digest, New York, 1928, v. 97, April 28, p. 5–7.) *** DA**

Newspaper comments on the flight of Captain Koehl, Major Fitzmaurice, and Baron von Huenefeld across the Atlantic ocean, 1928.

2979. The **"Bremen's"** flight to America. ports. illus. (Aero digest, New York, 1928, v. 12, p. 742–744.) **VDS**

Flight of 1928.

2980. **Bremer,** V. E. Across Africa on 80 h. p. The eventful story of a flight from Finland to the Cape and back on a light aeroplane with a British engine of only 80 h. p. illus. (Air and airways, London, 1932, v. 9, p. 160–161.) **VDS**

Author's flight, 1932.

2981. A **Brief** biography of Lindbergh. ports. (Aero digest, New York, 1927, v. 10, p. 534.) **VDS**

2982. **Brief** history of the round the world flight. World fliers covered 22,000 miles from Seattle to Boston in 153 days, using 300 flying hours. illus. map. (Aviation, Highland, N. Y., 1924, v. 17, p. 1015–1018.) **VDS**

United States Army Air Service flight of 1924.

2983. **Britain** pushes up the ceiling. (Aviation, New York, 1932, v. 31, p. 452.) **VDS**

Cyril Unwins reaches 43,976 feet in airplane, 1932.

2984–85. **Britain's** record-breaking air carnival. (Literary digest, New York, 1929, v. 102, Sept. 28, p. 46, 48–53.) *** DA**

High speeds made at Schneider Cup meeting of 1929, especially of A. H. Orlebar and H. R. D. Waghorn.

2986. **Brown,** Sir Arthur Whitten. Flying the Atlantic in sixteen hours, with a discussion of aircraft in commerce and transportation, by Sir Arthur Whitten Brown, K. B. E., assisted by Captain Alan Bott, R. F. C., with 21 illustrations from photographs. New York: Frederick A. Stokes Company [cop. 1920]. 178 p. illus. 12°. **VDYC**

2987. **Brown,** Robert J., jr. America girdles the globe in the air. illus. map. (New York Times current history magazine, New York, 1924–25, v. 21, p. 189–198.) **BAA**

1924 flight of United States Army fliers.

2988. —— American airmen first to cross the Pacific. illus. map. (New York Times current history magazine, New York, 1924, v. 20, p. 545–560.) **BAA**

United States Army flight around the world, 1924.

2989. —— Pacific first crossed by U. S. army fliers. World flight progress from April 19 to May 22. illus. (U. S. air services, Washington, 1924, v. 9, June, p. 51–52.) **VDS**

United States Army flight around the world, 1924.

2990. **Buchanan,** J. S. The Schneider cup race, 1925. (Royal Aeronautical Society, Journal, London, 1926, v. 30, p. 434–452.) **VDS**

With discussion.
Has a brief history of the earlier races.

2991. **Buist,** H. Massac. The flying races at Rheims. Being a gossip about an historical meeting that marks an epoch in the development of aerial locomotion. illus. (Flight, London, 1909, new series, v. 1, p. 537–541, 551–552.) **VDS**

Tabulated record of performances, p. 536.

2992. **Buranelli,** Albert. Pioneer air pilots. illus. (Popular aviation, Chicago, 1928, v. 2, June, p. 27–28, 83–85; v. 3, July, p. 40, 42, 44, Aug., p. 38, 40, 42, 44, Sept., p. 46, 48, 50, 91–92, Oct., p. 78, 80, 82, 117–118.) **VDS**

Biographies and feats of Adolphe Pégoud, John B. Moisant, Charles K. Hamilton, Thomas S. Baldwin, and A. Hoxsey.

2993. **Byrd,** Richard Evelyn. Our transatlantic flight. ports. illus. (Aero digest, 1927, v. 11, p. 8–10, 12.) **VDS**

Flight of the *America* under command of Richard E. Byrd.

2994. —— Skyward. Man's mastery of the air as shown by the brilliant flights of America's leading air explorer. His life, his

Airplane Flights, continued

thrilling adventures, his North pole and trans-Atlantic flights, together with his plans for conquering the Antarctic by air. New York: G. P. Putnam's Sons, 1928. xv, 359 p. illus. 8°. **VDY**

2995. Byrd's transatlantic trip. illus. (Popular aviation, Chicago, 1927, v. 1, Sept., p. 5–8.) **VDS**

2996. C., D. L. La traversée de la Manche en aéroplane par M. Blériot, avec son monoplan no. XI (25 juillet 1909). illus. (Le Génie civil, Paris, 1909, tome 55, p. 257–263.) **VA**

2997. Caldwell, C. S. Free lancing in the air, as told by C. S. Caldwell to E. L. Chicanot. port. illus. (Popular aviation, Chicago, 1928, v. 2, Jan., p. 10–15, 70–75.) **VDS**

Airplane experiences in the Canadian wilds.

2998. Caldwell, Cy. Evolution of the pilot. illus. (Aero digest, New York, 1931, v. 18, April, p. 42–43, 240, 242, 244.) **VDS**

Comparison of the old pilots with the new.

2999. Candelaria, Luis C. Memoria de la primera travesía de la cordillera de los Andes en aeroplano. [By Luis C. Candelaria.] Buenos Aires: Compañía argentina de tabacos, 1918. 127 p. illus. map, ports. 8°. **VDY p.v.47, no.8**

Airplane flight across the Andes in 1918.

3000. Captain [Frank M.] Hawks holds all transcontinental speed records. (U. S. air services, Washington, 1930, v. 15, Sept., p. 41–42.) **VDS**

3001. Captain W. N. Lancaster's fine flight. map. (Flight, London, 1928, v. 20, p. 39–40.) **VDS**

London to Dutch East Indies, 14,000 miles, in 1927.

3002. Carlstrom's achievement. illus. (Aerial age weekly, New York, 1916–17, v. 4, p. 225, 227.) **VDS**

Flying time Chicago-New York (1916) was 8 hours, 28 minutes, 30 seconds.

3002a. Cathcart-Jones, Owen. Aviation memoirs, including Australia and back and other record flights...with an introduction by Rt. Hon. Sir Philip Sassoon, bart... London: Hutchinson & Co. [1934.] 265 p. illus. 8°. **VDYC**

3003. Cave-Browne-Cave, H. M. The cruise of the Royal Air Force Far East flight. (Royal Aeronautical Society, Journal, London, 1929, v. 33, p. 542–572.) **VDS**

Also in *Aeroplane*, London, 1929, v. 36, p. 366–368, 369–370, 372, 374, *VDS*. Abridged in *Flight*, London, 1929, v. 21, p. 197–199, *VDS*.

3004. The **Challenge** de tourisme international. illus. map. (Aeroplane, London, 1930, v. 39, p. 177–178, 180–181.) **VDS**

An account of the preparations for this important meeting, 1930.

3005. Chamberlin, Clarence Duncan. Record flights. Philadelphia: Dorrance and Company [cop. 1928]. 286 p. illus. ports. 8°. **VDYC**

Biography of American aviator and an account of his trans-Atlantic flight with Charles A. Levine in 1927.

3006. Chamberlin flies Bellanca plane to Germany. Powered by a Wright "Whirlwind" engine the "Columbia" carries pilot and passenger to new world's non-stop distance record. illus. ports., map. (Aviation, Highland, N. Y., 1927, v. 22, p. 1276–1277.)

Flight of June 4–5, 1927. **VDS**

3007. Charles Keeney Hamilton. port. (Aircraft, New York, 1910, v. 1, p. 251.) **VDS**

Brief biography and portrait of daring American flier and aeronaut (born 1881).

3008. Charles Stewart Rolls. port. (Aircraft, New York, 1910, v. 1, p. 250.) **VDS**

Portrait and brief biography of prominent British aviator, who with Royce developed the Rolls-Royce automobile. He was the first British subject to cross the English Channel in an airplane.

3009. Charlie Lindbergh gets a job. port. (Literary digest, New York, 1928, v. 97, June 9, p. 34, 36, 41.) *** DA**

Becomes chairman of the technical committee of Transcontinental Air Transport, Inc.

3010. Chavez makes most daring flight of history, sailing over the Simplon Pass to Italy. (Aeronautics, New York, 1910, v. 7, p. 180.) **VDS**

3011. Chichester, Francis C. Solo to Sydney; with an introduction by the Baron von Zedlitz. London: J. Hamilton, Ltd. [1930.] 208 p. illus. map, port. 8°. **VDYC**

Author's account of his flight from London to Sydney, 1930.

3012. Chronology and tabulation of airplane and seaplane records. charts. (Aviation, New York, 1929, v. 27, p. 425–427.) **VDS**

Tables and charts show development, 1908–1929, of American and world records.

3013. Circuit of Britain. The race for the "Daily Mail" £10,000 prize. map. (Aeronautics, London, 1911, v. 4, p. 130–132, 155.)

Won by André Beaumont, 1911. **VDS**

3014. Claudy, Carl Harry. How Helwin died. (Fly, Philadelphia, 1908, v. 1, Nov., p. 14, 18.) **3–VDS**

Account of fatal airplane accident to Maxwell Helwin, Colorado, 1902.

3015. —— Thomas E. Selfridge — an appreciation. illus. (Aeronautics, New York, 1909, v. 4, p. 143–144, 172.) **VDS**

Career of early American aviator.

3016. Cléry, A. Augustin Seguin. port. (L'Aérophile, Paris, 1913, année 21, p. 481.) **3–VDS**

Portrait and brief account of prominent French aviator.

Airplane Flights, continued

3017. —— Charles Weyman. port. (L'Aérophile, Paris, 1911, année 19, p. 497.) **3–VDS**

Portrait and brief account of early American aviator, most of whose flying was done in France.

3018. —— Georges Legagneux. port. (L'Aérophile, Paris, 1911, année 19, p. 25.) **3–VDS**

Portrait and brief account of pioneer French aviator.

3019. —— Jules Védrines. port. (L'Aérophile, Paris, 1911, année 19, p. 401.) **3–VDS**

Portrait and brief account of prominent French aviator.

3020. **Cleveland,** Reginald M. Scheduled Atlantic flying. illus. (Scientific American, New York, 1934, v. 150, p. 82–83.) **VA**

Discussion of proposed seadromes.

3021. **Cobham,** Sir Alan John. Australia and back, with a foreword by the Rt. Hon. S. M. Bruce. London: A. & C. Black, Ltd., 1927. 124 p. illus. map. 8°. **VDY**

Account of author's flight in 1926.

3022. —— London-Rangoon-London. illus. map. (Flight, London, 1925, v. 17, p. 180–188.) **VDS**

Flight of 1924–1925.

3023. —— [My Australian flight.] (Royal Aeronautical Society, Journal, London, 1927, v. 31, p. 5–16.) **VDS**

Account of flight England-Australia and return.

3024. —— My flight to the Cape and back. London: A. & C. Black, Ltd., 1926. 70 p. illus. map, port. 8°. **VDY**

Airplane flight, 1925–1926, London-Cairo-Cape Town.

3025. —— Skyways; with a foreword by W. Sefton Brancker. London: Nisbet & Co., Ltd. [1925.] x, 314 p. ports. illus. 8°. **VDY**

Flights of noted British aviator in Europe and Africa.

3026. —— Twenty thousand miles in a flying-boat; my flight round Africa. London: G. G. Harrap & Co., Ltd. [1930.] 249 p. illus. ports. 8°. **VDY**

Reviewed in *Air*, London, 1930, v. 3, p. 133, *VDS*.

3027. **Coleman,** Charles H. Given up as lost, Navy Hawaiian flight plane found after nine days. (U. S. air services, Washington, 1925, v. 10, Oct., p. 15–18.) **VDS**

Rescue of Commander Rodgers and his crew of the *PN-9, no. 1,* 1925.

3028. —— The two thousand mile flight of navy planes from San Francisco to Honolulu. illus. ports. (U. S. air services, Washington, 1925, v. 10, Sept., p. 13–17.) **VDS**

Flight of 1925.

3029. **Collier** trophy. (Aeronautic review, Washington, 1926, v. 4, p. 178.) **VDS**

List of recipients, 1911–1925.

3030. **Collier** trophy committee to meet this month. (Aeronautic review, Baltimore, 1929, v. 7, Jan., p. 3, 19.) **VDS**

Cites awards, 1911–1926.

3031. **Collins,** John. Ely—the father of ship-shore flying. illus. (Aero digest, New York, 1927, v. 11, p. 330, 332.) **VDS**

Account of the adventurous career of Eugene Ely.

3032. **Collinson,** Clifford Whiteley, and Francis McDermott. Through Atlantic clouds. The history of Atlantic flight. Foreword by Lord Sempill. London: Hutchinson & Co. [1934.] 287 p. illus. maps. 8°. **VDYC**

Appendix has a chronological list of successful and unsuccessful flights, 1910 to 1934.

3033. **Colonel** Lindbergh's homecoming. illus. ports. (Aero digest, New York, 1927, v. 11, p. 31–32, 34.) **VDS**

3034. **Complete** official results, National Air Races, Cleveland, Ohio, August 24 – September 2 [1929], held under the auspices of the N. A. A. (Aeronautic review, Baltimore, 1929, v. 7, Oct., p. 46–51.) **VDS**

3035. The **Complete** story of the Rheims meeting. illus. (Aero and hydro, Chicago, 1913, v. 7, p. 41–42.) **3–VDS**

3036. **Coney,** William D. Flying across the continent in twenty-two hours. Army pilot's own story of historic trip. port. (U. S. air service, New York, 1921, v. 5, April, p. 12–15.) **VDS**

3037. **Connor,** Harry P. McLean. Crossing the north Atlantic. The story of how the navigation was done on the last successful flight. illus. (Airway age, New York, 1931, v. 12, p. 560–564, 602.) **VDS**

Flight of the author with Captain J. Erroll Boyd in the *Columbia,* 1930.

3038. —— Navigation on the non-stop flight from New York to Bermuda and return. ports., maps. illus. (Aero digest, New York, 1930, v. 17, Aug., p. 40–43, 182, 186.) **VDS**

Roger Q. Williams' flight of 1930.

3039. **Cooper,** Mabel C. An early bird goes to Paris. illus. (U. S. air services, Washington, 1931, v. 16, p. 21–24.) **VDS**

Career of Colonel Frank S. Lahm, balloonist and pioneer airplane flyer, who afterwards became associated with United States Army aviation.

3040. **Costes,** Dieudonné. Deux records du monde, préface de R. Peyronnet de Torres. Paris: La Nouvelle revue critique [1930]. 228 p. illus. map, ports. [24. ed.] 12°. (Collection "La vie d'aujourd'hui." no. 9.) **VDY**

Record of author's flight, Paris to China (1929) comprises the greater part of the book.

3041. —— The first non-stop flight from Paris to New York. Costes' own thoughts and impressions on flying across the Atlantic from France to the United States. illus. (Aero digest, New York, 1930, v. 17, Oct., p. 50–52.) **VDS**

Airplane Flights, continued

3042. Costes, Dieudonné, and J. M. Le Brix. Notre tour de la terre. [Paris:] Hachette [1928]. 261 p. illus. maps, ports. 12°. **VDY**

Account of the authors' famous flight (1927), Paris, South America, North America, across the Pacific by steamer, and again by airplane to Paris.

3043. Costes, Dieudonné, and M. Bellonte. Paris-New York; préface de R. Peyronnet de Torres. [Paris:] Nouvelle revue critique [1930]. 77 p. illus. map, ports. 12°. (Collection "La vie d'aujourd'hui." no. 13.) *** C p.v.2275**

3044. Costes and Le Brix. An account of their outstanding flight from France to the United States. illus. ports. (Aero digest, New York, 1928, v. 12, p. 332–334.) **VDS**

Four continent flight of 1927.

3045. The **Crew** of the "America." ports. (Aero digest, New York, 1927, v. 11, p. 14.) **VDS**

Portraits and brief biographies of Richard E. Byrd, Bertram B. Acosta, George O. Noville, and Bernt Balchen.

3046. Curtiss, Glenn Hammond. Albany-New York. (Aircraft, New York, 1910, v. 1, p. 177–178.) **VDS**

Brief account of flight, to which is appended a chronological record by Augustus Post, the official timekeeper.

3047. Curtiss Aeroplane and Motor Corporation. The flight across the Atlantic. New York City: Issued by the Department of Education, Curtiss Aeroplane and Motor Corporation, 1919. 88 p. illus. 4°. **† VDY**

Well-illustrated, detailed account of the flights of the *NC1, NC3,* and *NC4* in 1919.

3048. Curtiss makes $10,000 prize flight. Successful journey from Albany to New York. illus. (Fly, Philadelphia, 1910, v. 2, July, p. 9–10.) **3–VDS**

3049. The **Curtiss** marine trophy race. ports. (Popular aviation, Chicago, 1928, v. 3, July, p. 18–21.) **VDS**

Event of 1928.

3050. The **Daniel Guggenheim International Safe Aircraft Competition.** Final report, January 31, 1930. New York: The Daniel Guggenheim Fund for the Promotion of Aeronautics, Inc. [1930.] 147 p. illus. 4°. **VDY**

Records of airplanes in competition, 1930. The first prize was won by the Curtiss Company's *Tanager.*

3051. Davis, Elmer Holmes. On the wings of the army. A three thousand mile jaunt by the cloud route from New York to Texas [1924]. illus. (Collier's, New York, 1924, v. 74, Oct. 25, p. 8–9, 45–46.) *** DA**

3052. Death of [Charles K.] Hamilton. (Aeronautics, New York, 1914, v. 14, p. 23.) **VDS**

"The most daring exhibition flier in this country."

3053. Degoul, M. La coupe Michelin 1909 à Henri Farman. (L'Aérophile, Paris, 1910, année 18, p. 29.) **3–VDS**

3054. —— Louis Paulhan. port. (L'Aérophile, Paris, 1909, année 17, p. 337.) **3–VDS**

Portrait and brief biography of well-known French aviator.

3055. Delagrange's aeroplane flights in Italy. (Scientific American, New York, 1908, v. 98, p. 443.) **VA**

3056. Detroit race impressions. Great progress indicated in American design and construction. illus. (Aviation, Highland, N. Y., 1922, v. 13, p. 550–558.) **VDS**

Interesting account of the Curtiss and Pulitzer trophy races, Detroit, 1922.

3057. Dew, Arthur W. The first nonstop trans-Atlantic flight. (U. S. air services, Washington, 1929, v. 14, June, p. 42, 44.) **VDS**

Flight of Alcock and Brown, 1919.

3058. "Dick" Byrd as a Mercury of modern flying science. illus. (Literary digest, New York, 1927, v. 94, July 23, p. 34, 36, 38, 40, 42.) *** DA**

Biography.

3059. Dijk, Evert van. Over den oceaan, onze Oost-West vlucht Europa-Amerika, door Evert van Dijk, met talrijke foto's van den tocht, inleiding van A. H. G. Fokker, voorwoord van Kingsford Smith. Amsterdam: Scheltens & Giltay [1930]. 233 p. illus. ports. 8°. **VDYC**

Flight of the *Southern Cross,* Europe-America, in 1930.

3060. Dixon, Charles. The conquest of the Atlantic by air... London: S. Low, Marston & Co., Ltd. [1931.] ix, 246 p. incl. table. plates, port. 8°. **VDY**

History of attempts by airships and airplanes, with a chronological table of particulars.

For review see *Flight,* London, 1931, v. 23, p. 385, *VDS.*

3061. The **Dole** Pacific race. illus. port. (Aero digest, New York, 1927, v. 11, p. 250–252, 254, 366.) **VDS**

Race, 1927, won by Art Goebel and W. V. Davis.

3062. Dollfus, Charles. André Michelin. port. (L'Aéronautique, Paris, 1931, année 13, p. 203.) **VDS**

Michelin (1853–1931), famous tire manufacturer, was the originator of Les coupes Michelin (Tour de France).

3063. —— Fernand Laporte. port. (L'Aéronautique, Paris, 1925, année 7, p. 395–396.) **VDS**

Biography of well-known French aviator.

3064. Donovan, Thomas B. Bider's flight across the Alps reviewed. illus. (Aero and hydro, Chicago, 1913, v. 7, p. 13.) **3–VDS**

Brief account of flight in 1913.

3065. Doolittle does it. (Aviation, New York, 1932, v. 31, p. 419.) **VDS**

Brief account of flight averaging 294.39 miles per hour; also of James Haizlip's flight, Los Angeles-

Airplane Flights, continued

New York, in 10 hours, 19 minutes; also of Amelia Earhart's Los Angeles-New York flight in 19 hours, 5 minutes.

3066. **Du Bois,** H. M. The Times aerial derby around Manhattan. ports. (Flying, New York, 1913, v. 2, Nov., p. 24–25.) **VDS**

Race, 1913, won by W. S. Luckey.

3067. **Duckworth,** J. Herbert. Pioneer personalities. illus. (Popular aviation, Chicago, 1930, v. 6, April, p. 310–311, 326, 365.) **VDS**

Author was associated with the New York *Evening Sun* as early as 1909. He relates experiences of some notable early fliers. This number of *Popular aviation* is published under the title of *Aeronautics.*

3068. **Dunn,** H. H. Latin-America's first bird men. illus. (Popular aviation, Chicago, 1929, v. 5, July, p. 38, 42, 46.) **VDS**

Brief accounts of Roland Garros, René Barrier, Edmond Audemars, and René Simon. This number of *Popular aviation* has the title of *Aeronautics.*

3069. **Early** British flights. (Flight, London, 1929, v. 21, p. 132–133.) **VDS**

Committee of the Royal Aero Club of Great Britain gives the honor of the first flight to Lt. Col. J. T. C. Brabazon. Interesting evidence is given.

3070. **Earth's** highest peak conquered by British flyers. map. (U. S. air services, Washington, 1933, v. 18, May, p. 28–29.) **VDS**

Mt. Everest flight, 1933.

3071. The **East** to West flight. illus. ports. (Popular aviation, Chicago, 1928, v. 2, p. 13–16.) **VDS**

Transatlantic flight of the *Bremen* airplane, 1928.

3072. **Eastward** ho! Some comparative notes on the big flights to the East. chart. (Flight, London, 1928, v. 20, p. 9.) **VDS**

Interesting chart shows mileage and times of flights during four years.

3073. **Ehrhardt,** Gustav. Mittelmeerflug, mit 120 Fliegeraufnahmen und einer Einleitung von Walter Mittelholzer; Text von Gustav Ehrhardt. Zürich: Rascher & Cie., A. G., 1930. 164 p. illus. maps. 8°. **VDY**

Authors' flight, 1928, through Mediterranean countries with some remarkable aerial photographs.

3074. **Eighteen** years of world's records. illus. tables. (Flight, London, 1924, v. 16, p. 73–75.) **VDS**

Speed, altitude, distance, and duration records are tabulated and diagramed, 1906–1923.

3075. The **11,323** mile MacRobertson race from London to Melbourne within 71 hours. illus. (Aero digest, New York, 1934, v. 25, Nov., p. 20–22.) **VDS**

3076. **England-Australia-England.** illus. (Flight, London, 1926, v. 18, p. 650–660.) **VDS**

Sir Alan J. Cobham's flight of 28,000 miles in 1926.

3077. **England** to South America. illus. map. (Flight, London, 1933, v. 25, p. 149–150.) **VDS**

Flight of J. A. Mollison, 1933, England to Rio de Janeiro.

3078. An **Epic** of the air. Champion's great feat. illus. port. (Popular aviation, Chicago, 1927, v. 1, Sept., p. 11–12.) **VDS**

Carlton C. Champion ascended 47,000 feet in 1927.

3079. **Epochal** flights of 1927. ports. illus. map. (Aircraft yearbook for 1928, New York, 1928, p. 5–46.) **VDS**

Lindbergh's flight to Paris. Chamberlin's flight to Germany. Maitland and Hegenberger fly to Hawaii. Byrd's flight to France. More Hawaiian flights. Brock and Schlee, Detroit-Tokio. Haldeman and Elder, to the Azores. Pan-American good-will flight. Lindbergh's tour of U. S. and Latin America. Wilkins' Arctic flight.

3080. **Erickson,** H. A. The flight of the Question Mark. illus. (Popular aviation, Chicago, 1929, v. 4, March, p. 35–36, 38, 40, 104.) **VDS**

U. S. Army's tri-motored Fokker airplane remained aloft more than 150 hours with refueling.

3081. **Esteve,** Rafael M. Una aventura en el desierto (del vuelo Madrid-Manila). Madrid: Espasa-Calpe, s. a., 1926. 205 p. illus. maps. 12°. **VDY**

Story of the wrecking of the author's plane in the Syrian desert on the ill-fated Madrid-Manila flight of 1926. For a brief account of the expedition see *Aviation,* Highland, N. Y., 1926, v. 20, p. 750–751, *VDS.*

3082. **Eugene** Ely and Cromwell Dixon have fatal falls. (Fly, Philadelphia, 1911, v. 4, Nov., p. 9.) **3–† VDS**

3083. The **Events** of the New York air races. illus. ports. (Aviation, Highland, N. Y., 1925, v. 19, p. 533–543.) **VDS**

Among the ten contests were the Glenn Curtiss trophy, the Liberty engine builders trophy race, and the Pulitzer trophy race.

3084. **Everest** conquered. (Aeroplane, London, 1933, v. 44, p. 576, 578.) **VDS**

Successful British expedition of 1933.

3085. Les **Expériences** de Pégoud et la stabilisation des aéroplanes. (Revue du mois, Paris, 1913, tome 16, p. 641–646.) *** DM**

3086. **Famous** flights. Around the world. illus. (Popular aviation, Chicago, 1927, v. 1, Aug., p. 31–35.) **VDS**

Seaplane flight, U. S. Army, 1924.

3087. **Famous** flights. The NC–4 trip. ports. illus. (Popular aviation, Chicago, 1927, v. 1, Dec., p. 27–31, 62.) **VDS**

United States Navy plane's flight, of 1919, with accounts of Harry Hawker's and John Alcock's flights of the same year.

3088. **Famous** flights. The non-stop flight across America. map. illus. (Popular aviation, Chicago, 1927, v. 1, p. 19–21, 64.) **VDS**

Account of flight by Lt. John A. Macready and Lt. Oakley G. Kelly in 1923.

Airplane Flights, continued

3089. Famous flights of history. The first Channel crossing. illus. (Popular aviation, Chicago, 1927, v. 1, Sept., p. 13–15.) **VDS**

Flight of Blériot, 1909, and attempts of Hubert Latham.

3090. The **Far East** flight. illus. (Aeroplane, London, 1928, v. 35, p. 264, 266, 268, 270.) **VDS**

Royal Air Force flight, 1927–1928. Log of the Karachi-Singapore section.

3091. Farman's aeroplane flights in Belgium. illus. (Scientific American, New York, 1908, v. 98, p. 443.) **VA**

3092. Fatal accident to Calbraith P. Rodgers. illus. port. (Scientific American, New York, 1912, v. 106, p. 356.) **VA**

Brief account of transcontinental flight in 1912.

3093. The **Fatal** accident to M. Léon Delagrange. (Flight, London, 1910, v. 2, p. 30.) **VDS**

Brief account of famous French pioneer aviator.

3093a. Fawcett, Edward Douglas. From Heston to the high Alps; a chat about joy-flying... London: Macmillan & Co., Ltd., 1936. xiv, 73 p. illus. 8°. **VDYC**

Author's airplane flights, 1934. Excellent photographic reproductions.

3094. Ferrarin, Arturo. Voli per il mondo; con presentazione di Benito Mussolini. Milano: A. Mondadori, 1929. xv, 282 p. illus. maps, ports. 8°. **VDYC**

Author's flight, Rome-Tokio. Also account of his flight with Carlo del Prete, Rome-Brazil.

3095. Fifty-one hours in the air. illus. ports. (Literary digest, New York, 1927, v. 93, April 30, p. 13.) *** DA**

Newspaper comments on flight of Bert Acosta and Clarence D. Chamberlin, 1927.

3096. A **Fine** American altitude flight. illus. (Flight, London, 1926, v. 18, p. 68–69.) **VDS**

John A. Macready's ascent to 35,900 feet in 1926.

3097. First cross-country aeroplane flights. The long-distance flights of Farman and Blériot in France. illus. (Scientific American supplement, New York, 1908, v. 66, p. 381–382.) **VA**

3098. The **First** crossing of the Alps in an aeroplane. illus. (Scientific American supplement, New York, 1910, v. 70, p. 257–259.) **VA**

Chavez's fatal flight, 1910.

3099. First flight in Australia. Harry Houdini's success near Melbourne. illus. (Fly, Philadelphia, 1910, v. 2, June, p. 5.) **3–† VDS**

3100. First flight in Ireland. (Flight, London, 1910, v. 2, p. 23.) **VDS**

Short flight of H. G. Ferguson, Jan. 31, 1909.

3101. The **First** flight from shore to ship. How [Eugene] Ely flew from Selfridge Field, flew across country and landed on the U. S. S. Pennsylvania. illus. (Scientific American, New York, 1911, v. 104, p. 108.) **VA**

3102. First flights of the Aerial Experiment Association's second aeroplane. (Scientific American, New York, 1908, v. 98, p. 392.) **VA**

Flight of Glenn H. Curtiss in the *White Wing.*

3103. The **First** hydro-aeroplane meeting at Monaco [1912]. (Scientific American, New York, 1912, v. 106, p. 356, 368–369.) **VA**

3104. First non-stop flight from Chicago to New York. (Aircraft journal, New York, 1919, v. 4, April 26, p. 5.) **VDS**

Made by Capt. E. F. White, April 19, 1919, in 6 hours, 50 minutes.

3105. First non-stop flight to India. Fine performance of Fairey monoplane with Napier engine. Record attempt only frustrated by bad weather. illus. (Flight, London, 1929, v. 21, p. 351.) **VDS**

Flight in 1929 by A. G. Jones-Williams and N. H. Jenkins.

3106. First one-stop flight across the United States. illus. (Aviation, Highland, N. Y., 1921, v. 10, p. 332–333.) **VDS**

Flight of Lieutenant William D. Coney in 1921.

3107. First seaplane flight to Europe. illus. ports. (Aero digest, New York, 1928, v. 13, p. 39, 173.) **VDS**

Flight of Wilmer Stultz, Louis Gordon, and Amelia Earhart, 1928.

3108. The **First** successful cross-Channel flight. illus. (Scientific American, New York, 1909, v. 101, p. 88, 99–100.) **VA**

Flight of Louis Blériot, July 25, 1909.

3109. First successful flights of Blériot's no. 8 monoplane. illus. (Scientific American, New York, 1908, v. 99, p. 44.) **VA**

3110. The **First** trans-continental aeroplane flight. Account of the record-breaking aerial journey by Calbraith P. Rodgers on a Wright biplane. (Scientific American, New York, 1911, v. 105, p. 449.) **VDS**

For references to Rodgers' flight see United States. — National Advisory Committee for Aeronautics, *Bibliography of aeronautics*, 1909–1931, Washington, 1921–35, volume for 1909/16, p. 1143–1144, *VDT*.

3111. First transcontinental airplane test. (Society of Automotive Engineers, Journal, New York, 1920, v. 6, p. 422–424.) **TOL**

Fifty-nine planes competed (1920), 44 starting from New York, 15 from San Francisco.

3112. The **First** trans-Pacific flight. (Aeroplane, London, 1928, v. 34, p. 834.) **VDS**

Brief account of Kingsford Smith's flight of 1928.

3112a. Fischer von Poturzyn, Friedrich Andreas. Afrika von Oben; mit 3 Junkers nach Kapstadt. Berlin: Union deutsche Verlagsgesellschaft [1935]. 199 p. illus. maps. 8°. **VDYC**

Flight of three Junker airplanes to South Africa, 1934.

Airplane Flights, continued

3113. **Five** race around Manhattan in high wind. map. (Aero and hydro, Chicago, 1913, v. 7, p. 27.) 3–VDS

Won by William S. Luckey in 52 minutes, 54 seconds, Oct. 13, 1913.

3114. **Flies** to Bermuda and return non-stop. Roger Williams employs "Columbia" in 17 hr. trip. (Aviation, New York, 1930, v. 29, p. 29.) VDS

3115. The **Flight** of the Curtiss aeroplane for the Scientific American trophy. illus. (Scientific American, New York, 1909, v. 101, p. 62.) VA

Flight of Glenn H. Curtiss, Mineola, N. Y., 1909.

3116. The **Flight** from London to Manchester. The aviators' own accounts. illus. (Scientific American, New York, 1910, v. 102, p. 397–398.) VA

Flights of Grahame-White and Louis Paulhan, 1910.

3117. **Flights,** competitions, and records [during 1930]. illus. (Aviation, New York, 1931, v. 30, p. 50–53.) VDS

3118. The **Flights** over Everest. The success of the Houston expedition [1933] affords proof of efficiency of aeroplanes, engines and equipment. illus. map. (Aircraft engineering, London, 1933, v. 5, p. 99–100, 115.) VDS

3118a. **Flug-** und Erfahrungsberichte über den Amerika-Flug des Dornier-Wal D 2053 auf der Strecke List-Chicago über Island-Grönland-Labrador. [Berlin, 1934.] 88 f. incl. tables. 8°. VDYC

Flight of Dornier plane to Greenland and Chicago, under command of Wolfgang von Gronau, 1931.

3119. The **Flying** Scot in new fields. (Aviation, New York, 1932, v. 31, p. 419.) VDS

Brief account of James Allan Mollison's flight, Ireland-America, 1932; also of Professor Piccard's second ascension (52,500 feet) into the stratosphere.

3120. **Fontaine,** Charles. Comment Blériot a traversé la Manche; récit de la traversée par L. Blériot, conclusions de P. Painlevé. Paris: Librairie aéronautique [1909]. 154 p. illus. ports. 4°. VDY

3121. **Forty** persons attempted long ocean flights. (U. S. air services, Washington, 1927, v. 12, Oct., p. 58, 60.) VDS

Attempts during spring and summer of 1927. Twenty-one were lost.

3122. **Four** hundred miles an hour. map. (Aeroplane, London, 1931, v. 41, p. 860, 862.) VDS

Flight of G. H. Stainforth, Sept. 29, 1931, Southampton, England.

3123. **Fox,** Albert Whiting. Ely's flight from the "Birmingham." illus. (Air-scout, New York, 1910, v. 1, no. 2, p. 12–13, 45.) 3–VDS

How Eugene B. Ely flew from deck of United States cruiser in 1910.

3124. **Franco,** Ramón, and J. Ruiz de Alda. De Palos al Plata. Madrid: Espasa-Calpe, 1926. 294 p. illus. maps. 12°. VDY

Account of flight from Palos, Spain, to Buenos Aires (1926).

3125. **French,** Joseph Lewis, editor. Wings over the world. Introduction by Capt. Frank M. Hawks. Springfield, Mass. [cop. 1931.] 302 p. illus. 8°. VDY

Miscellany by various authors includes account of Lieut. Soucek's flight to 43,166 feet in airplane; John K. Montgomery's flight to South America (1929); biography of Ivan Sikorsky; pioneer women aviators; Kingsford-Smith's Atlantic flight.

3126. The **French** world tour. ports. map. (Flight, London, 1928, v. 20, p. 262–264.) VDS

Four continent flight of Dieudonné Costes and Joseph Le Brix, 1927–1928.

3127. **Gallarza,** Eduardo G., and J. Loriga Taboada. El vuelo Madrid-Manila. Madrid: Espasa-Calpe, s. a., 1927. 320 p. illus. maps, ports. 12°. VDY

Account of flight in 1926.

3128. **Garber,** Paul Edward. 6,000 miles from Spain to Argentina by air. illus. map. (U. S. air services, Washington, 1926, v. 11, March, p. 13–15.) VDS

Flight of Commander Franco and crew, 1926.

3129. —— Stinson "Detroiter" wins National Air Tour. (U. S. air services, Washington, 1927, v. 12, Aug., p. 33.) VDS

Won by Edward A. Stinson, 1927.

3130. **Gardner,** Lester Durand. A flying tour of Europe. illus. maps. (Aviation, Highland, N. Y., 1926, v. 21, p. 450–453, 494–498, 539–541, 582–585, 622–625.) VDS

Record of flights covering nearly 20,000 miles in 1926 over commercial lines. The last two installments are entitled An air tour of the Near East.

3131. —— The Schneider trophy race through American eyes. illus. (Aviation, Highland, N. Y., 1927, v. 23, p. 1050–1053.) VDS

Held at Venice, 1927.

3132. **Garros,** Roland. Guide de l'aviateur. Préface d'Alfred Leblanc. Biographie de Garros par Jacques Mortane. Paris: Pierre Lafitte & Cie. [cop. 1913.] 95 p. illus. 16°. VDY

Biography of Garros, famous French aviator, p. 13–32.

For lists of writings relating to Garros see United States. — National Advisory Committee for Aeronautics, *Bibliography of aeronautics*, 1909–1931, Washington, 1921–35, volumes for 1909/16, p. 517; 1917/19, p. 179; 1923, p. 47; 1925, p. 67; 1926, p. 51, *VDT*.

3133. **General** Mitchell's maximum speed record. How the great performance was officially observed. (Aviation, Highland, N. Y., 1922, v. 13, p. 717.) VDS

Brigadier General Mitchell, Assistant Chief of the United States Air Service, made a record of 224.38 miles per hour at the Detroit meeting in 1922.

3134. The **German** Atlantic flight. illus. (Flight, London, 1928, v. 20, p. 259–261.) VDS

Flight of the *Bremen* in 1928.

Airplane Flights, continued

3135. **German** plane crosses North Atlantic East to West. Junkers "Bremen" flies 2,125 miles non-stop from Ireland to Labrador. (Aviation, Highland, N. Y., 1928, v. 24, p. 1150–1151.) **VDS**

Flight of Hermann Koehl, James Fitzmaurice, and Gunther von Huenefeld, 1928.

3136. The **German** seaplane competition described by an eyewitness. (Aeroplane, London, 1926, v. 31, p. 221–222.) **VDS**

Meeting of 1926.

3137. **Gile-Nicaud,** Gilbert. Paris-Tokio en avion; le raid merveilleux de Pelletier Doisy, avec soixante-six photographies et cartes. Paris: Plon-Nourrit et Cie [cop. 1924]. ii, 261 p. illus. 12°. **VDY**

Account of flight made in 1924.

3138. **Glenn** Curtiss wins the Scientific American trophy. The first aeronautical trophy to be offered for competition in America. illus. (Scientific American, New York, 1911, v. 104, p. 29.) **VA**

Flight from Albany to New York in 1910.

3139. A **"Glidden** tour" of the air. illus. map. (Literary digest, New York, 1926, v. 90, Sept. 4, p. 12–13.) *** DA**

Account of Ford reliability tour of 1926.

3140. **Gray,** George A. Early barnstorming. port. (U. S. air services, Washington, 1929, v. 14, Feb., p. 49–50, 52, 54.) **VDS**

Author's experiences.

3141. The **Great** Australian flights. illus. (Aeroplane, London, 1930, v. 39, p. 918, 920.) **VDS**

Comparison of the records of Kingsford-Smith and Bert Hinkler.

3142. **Great** British flights, 1918 to 1932. An enduring record. (Air annual of the British Empire, London, 1932–33, v. 4, p. 29–41.) **VDS**

Also includes other than British flights, but using British engines.

3143. The **Great** Pacific flight. map. (Flight, London, 1928, v. 20, p. 437.) **VDS**

Flight of Kingsford-Smith, San Francisco to Sydney, 1928.

3144. **Great** world flights of 1928. map. (Air, London, 1929, v. 2, p. 120–121.) **VDS**

Double-page map.

3145. The **Greatest** seaplane contest. The Jacques Schneider cup won for second time by United States. illus. (Aviation, Highland, N. Y., 1925, v. 19, p. 620–626.) **VDS**

Won by Lieutenant James Doolittle of the Army Air Service, 1925.

3146. **Green,** Fitzhugh. Dick Byrd — air explorer: an intimate story of a great air explorer whose spectacular flights to the North and South poles, across the Atlantic Ocean and other adventures have thrilled red-blooded men and boys the world over. New York: G. P. Putnam's Sons, 1928. viii, 267 p. plates, ports. 8°. **AN**

3147. **Gregg,** Willis Ray. The first successful non-stop trans-Atlantic flight. (Monthly weather review, Washington, 1919, v. 47, p. 416.) **† PRS**

Weather conditions during flight of Alcock and Brown.

3148. —— The first trans-Atlantic flight. map. (Monthly weather review, Washington, 1919, v. 47, p. 279–282.) **† PRS**

Study of weather conditions during U. S. Navy flight of seaplanes in 1919.

3148a. **Grey,** Charles Grey. The Baltimore seaplane meet. (Aviation, Highland, N. Y., 1924, v. 17, p. 1221–1222.) **VDS**

Meeting of 1924.

3149. —— A Belgian pioneer. (Aeroplane, London, 1929, v. 36, p. 542.) **VDS**

Obituary notice of Major Georges Nélis, first Belgian military officer to fly, and the organizing genius of the Belgian Air Lines.

3150. —— Captain Lindbergh arrives. illus. (Aeroplane, London, 1927, v. 32, p. 644, 646, 648.) **VDS**

Account of strenuous welcome in London.

3151. —— The death of Captain Sir John Alcock, K. B. E., D. S. C. (Aeroplane, London, 1919, v. 17, p. 2144.) **VDS**

Account of accident to famous British aviator, 1919.

3152. —— The greatest air fighters of the war. (Air service journal, New York, 1918, v. 3, p. 190.) **VDS**

Brief accounts of aces of the Allies and of Germany.

3153. —— The late Harry Hawker. (Aeroplane, London, 1921, v. 21, p. 64.) **VDS**

Appreciation of prominent British flier and experimenter.

3154. —— The new world's record for speed. (Aeroplane, London, 1928, v. 34, p. 487.) **VDS**

318½ miles per hour, by Major de Bernardi, at the Lido, Venice, 1928.

3155. —— On a British pioneer. (Aeroplane, London, 1931, v. 40, p. 353–354.) **VDS**

Biography of George Bertram Cockburn, early English aviator.

3156. —— On the Far East flight. ports. (Aeroplane, London, 1929, v. 36, p. 69–70, 72.) **VDS**

Comments on the flight of the Royal Air Force, England-Singapore and return, 1928.

3157. —— On the first Deutsch cup contest. illus. (Aeroplane, London, 1921, v. 21, p. 293–296, 307.) **VDS**

Event of 1921.

3158. —— On flying boats. illus. (Aeroplane, London, 1928, v. 34, p. 313–314, 316, 318.) **VDS**

Comments on the Royal Air Force flight to Australia and the Far East, 1928.

Airplane Flights, continued

3159. —— On the German Rundflug. illus. (Aeroplane, London, 1925, v. 28, p. 557–558, 560, 562, 564, 566, 568, 570, 586, 588, 590, 592, 594.) **VDS**

Comments on the Round tour race of 1925.

3160. —— On the great flights. illus. (Aeroplane, London, 1927, v. 32, p. 601–604, 606, 608, 610, 612.) **VDS**

Flights of Lindbergh and futile attempt of the British Air Force expedition to India.

3161. —— On the history of the Schneider contest. illus. (Aeroplane, London, 1929, v. 37, p. 594, 596, 601–602, 606, 608, 613.) **VDS**

3162. —— On the King's cup race [of 1922]. illus. (Aeroplane, London, 1922, v. 23, p. 197–200, 202, 204, 206, 208.) **VDS**

Won by H. L. Barnard.

3163. —— On the King's cup race [of 1923]. illus. (Aeroplane, London, 1923, v. 25, p. 65–66, 68, 70, 72, 83.) **VDS**

3164. —— On the King's cup race [of 1926]. illus. (Aeroplane, London, 1926, v. 31, p. 45–46, 50, 52, 54, 56, 58, 60.) **VDS**

Won by Hubert Broad.

3165. —— On the King's cup race [of 1927]. illus. (Aeroplane, London, 1927, v. 33, p. 145–146, 150, 154, 156, 158, 160, 162, 164, 173–174, 176.) **VDS**

Won by Lawrence Hope.

3166. —— On the King's cup race [of 1928]. illus. (Aeroplane, London, 1928, v. 35, p. 181–182, 184–185, 188, 192, 194, 196, 198, 200.) **VDS**

Includes accounts by three other observers.

3167. —— On the King's cup race [of 1933]. illus. (Aeroplane, London, 1933, v. 45, p. 53–54, 56, 58.) **VDS**

Won by Captain Geoffrey de Havilland.

3168. —— On the Schneider picnic. illus. (Aeroplane, London, 1919, v. 17, p. 1045–1046, 1048–1050, 1052, 1054.) **VDS**

Seaplane competition of 1919 at Cowes, England.

3169. —— On the Schneider trophy competition. illus. (Aeroplane, London, 1923, v. 25, p. 329–330, 332, 334, 336, 338, 340.) **VDS**

Race of 1923 won by American team.

3170. —— On the Schneider trophy victory [of 1927]. illus. (Aeroplane, London, 1927, v. 33, p. 455–458, 460.) **VDS**

Won by the British team.

3171. —— On the trans-Atlantic flight. illus. (Aeroplane, London, 1928, v. 34, p. 545–546.) **VDS**

Flight of the *Bremen*, 1928.

3172. Grey, Charles Grey, and others. The King's cup competition [of 1924]. illus. (Aeroplane, London, 1924, v. 27, p. 157, 160, 164, 166, 168, 170, 172, 174, 178, 180, 182.) **VDS**

Won by H. W. G. Jones.

3173. Grey, Charles Grey, and WALTER KLEFFEL. On the start of the Tour of Europe. illus. (Aeroplane, London, 1930, v. 39, p. 217–218, 220, 222, 224, 226, 228, 230, 232, 234, 236, 238–239B, 241A–241B, 303–304, 306, 346, 350, 352, 354, 377–378, 380, 382, 384, 386.) **VDS**

Accounts of this important meeting (1930).

3174. Grierson, John. A flight in Russia. illus. (Flight, London, 1932, v. 24, p. 1043–1045, 1091–1092.) **VDS**

Author's flight, England to Samarkand and return.

3175. Gronau, Wolfgang von. Im Grönland-Wal; dreimal über den Atlantik und einmal um die Welt. Berlin: R. Hobbing [cop. 1933]. 176 p. illus. map. 8°. **VDYC**

3176. —— In a flying boat across the Greenland ice cap to U. S. A. illus. (Royal Aeronautical Society, Journal, London, 1932, v. 36, p. 918–944.) **VDS**

Describes flight of July – August, 1931.

3177. Hailing Bert Hinkler, "the Australian Lindbergh." port., map. (Literary digest, New York, 1928, v. 97, April 7, p. 54, 56, 58–61.) *** DA**

Newspaper comment on Hinkler's achievements.

3178. Haines, Lynn, and D. B. HAINES. The Lindberghs. New York: Vanguard Press [cop. 1931]. 307 p. port. 8°. **AN**

A biography of Charles A. Lindbergh, Senior, the father of the famous aviator.

3179. Hale, Albert. Crossing the Andes by aero and auto. (Flying, New York, 1914, v. 3, p. 125.) **VDS**

Tells of the early attempts.

3180. Haling, Elbert, and JAMES LOCKLEAR. Ormer Locklear — first aerial acrobat. illus. (Popular aviation, Chicago, 1931, v. 11, p. 355–356.) **VDS**

Account of early wing-walker and stunt man.

3181. Halliburton, Richard. The flying carpet. Indianapolis: The Bobbs-Merrill Company [cop. 1932]. 352 p. illus. 8°. **VDYC**

Author's partial flight around the world, with an account of his attempt on Mt. Everest. The chapter covering the latter is reprinted in *U. S. air services,* Washington, D. C., 1933, v. 18, June, p. 33–34, 36, *VDS.*

3182. Harper, Harry. The humour of the early flying days. (Aeronautics, London, 1919, v. 17, p. 19–21.) **VDS**

Author's experiences.

3183. —— Louis Paulhan — master of his craft. An appreciation of the winner of the first London-Manchester flight made sixteen years ago. port. (Airways, London, 1926, v. 3, p. 61–62.) **VDS**

3184. Hawker, Harry George, and K. MACKENZIE-GRIEVE. Our Atlantic attempt... With a preface by Major-General J. E. B. Seely... London: Methuen & Co., Ltd. [1919.] xi, 13–128 p. illus. 12°. **VDY**

Account of the authors' ill-fated flight and their rescue by the ship *Mary.*

Airplane Flights, continued

3185. **Hawker,** Muriel. H. G. Hawker, airman: his life and work, with a foreword by Lt.-Col. J. T. C. Moore-Brabazon. London: Hutchinson & Co. [1922.] xxii, 319 p. illus. ports. 8°. **VDYC**

Reviewed in *Aeroplane*, London, 1922, v. 23, p. 440, *VDS*.

3186. **Hegenberger** makes world's first blind solo flight. illus. port. (U. S. air services, Washington, 1932, v. 17, July, p. 13–14.) **VDS**

3187. **Henrickson,** H. B. Altitude flights [of 1928]. illus. (Aero digest, New York, 1929, v. 14, May, p. 39–42, 242, 244, 246.) **VDS**

3188. —— Altitude flights during 1929. illus. (Aero digest, New York, 1930, v. 17, July, p. 63–65, 190–192.) **VDS**

3189. —— Altitude flights [of 1930]. illus. (Aero digest, New York, 1931, v. 18, June, p. 42–43, 134–135; v. 19, July, p. 36–37, 128–130.) **VDS**

3190. —— Duration flights during 1929. illus. (Aero digest, New York, 1930, v. 16, p. 64–65, 272, 274, 276, 278.) **VDS**

3191. —— Endurance flights. port. illus. (Aero digest, New York, 1929, v. 14, April, p. 38–39, 266, 268, 270, 272, 274, 276, 278, 280.) **VDS**

Flights of 1928.

3192. —— Record altitude flights. illus. (Aero digest, New York, 1928, v. 12, p. 748–750, 882–885.) **VDS**

Flights of 1928.

3193. **Henry** Segrave. (Aeroplane, London, 1930, v. 38, p. 1155–1156.) **VDS**

Brief biography of distinguished British aviator, engineer, automobile and motor-boat racer.

3194. **Here** come the British — Bang! Bang! But the Dutch and Americans in order named, capture 2d and 3d prizes in the great air race from London to Melbourne, flying American transports. illus. (U. S. air services, Washington, 1934, v. 19, Nov., p. 16–20.) **VDS**

3195. The **Highest** aeroplane ascent. (Meteorological magazine, London, 1920, v. 55, p. 25.) **PRA**

"Major R. W. Schroeder, chief test pilot at Dayton, Ohio, U. S. A. ...in a Le Fère biplane...fainted, while his machine was nearly seven miles above the earth, and fell over five miles in a sheer nose dive."

3196. **Hinton,** Walter. The first trans-Atlantic flight. (American Academy of Political and Social Science, Annals, Philadelphia, 1927, v. 131, May, p. 94–100.) **SA**

Flight of the U. S. Navy's *NC-4* in 1919. The voyage was not a non-stop.

3197. **Hirth,** Hellmuth. 20,000 Kilometer im Luftmeer. Berlin: G. Braunbeck G.m.b.H., 1913. 240 p. illus. port. 12°. **VDY**

Experiences of German pioneer aviator.

3198. **History** making flights of 1929. (Aircraft yearbook for 1930, New York, 1930, p. 123–144.) **VDS**

The *Graf Zeppelin* girdles the globe. *Yellow bird* flies to Spain. Yancey and Williams span the Atlantic. Spaniards fly the South Atlantic. Moscow to New York flight. Orlebar makes new speed record. Schneider cup race. Hawks cuts transcontinental records. New York to Nome. *Southern Cross,* Australia to England. London-India non-stop. Flight of the *DO-X.* Byrd flies to South Pole.

3199. **History** making flights of 1930. illus. (Aircraft yearbook for 1931, New York, 1931, p. 95–116.) **VDS**

Transcontinental records broken. Latin-American flights. First Atlantic air mail. Kingsford-Smith completes world flight. Von Gronau's flying boat flight, Germany to New York. Coste-Bellonte, Paris to New York. *Graf Zeppelin's* four continent flight. *R-100,* London-Montreal and return. Refueling endurance flights. First flights to Bermuda.

3200. **History** making flights of 1931. illus. (Aircraft yearbook for 1932, New York, 1932, p. 105–136.) **VDS**

Post and Gatty around the world in 8 days. Boardman and Polando's non-refueling long distance record. First non-stop Pacific flight by Herndon and Pangborn. Italian South Atlantic flight. *DO-X* spans the Atlantic. Doolittle breaks transcontinental record. Hoiriis and Hillig fly North Atlantic. First New York-Hungary flight. England to Australia and Africa flights. Women's records broken. *"Graf"* spans Atlantic six times. Lindberghs fly to Tokyo. Hinkler spans South Atlantic.

3200a. **History** making flights of 1932. illus. (Aircraft year book for 1933, New York, 1933, p. 89–103.) **VDS**

Amelia Earhart's non-stop New York-Paris. Mollison's East to West Transatlantic; Mrs. Mollison's Cape flight. *"Do-X"*, New York-Germany. Mattern-Griffin, New York-Germany. Von Gronau's world flight. Altitude, speed, and transcontinental records.

3201. The **History** of the Pulitzer races, showing speed increases of almost 100 miles per hour. illus. (Aviation, Highland, N. Y., v. 19, p. 545.) **VDS**

3202. **History** of the Schneider trophy contests. illus. (Flight, London, 1931, v. 23, p. 895–903.) **VDS**

Records, 1913–1929.

3203. **History** of the Schneider trophy race. illus. (Flight, London, 1925, v. 17, p. 720–732.) **VDS**

3204. **History's** longest "hop" — across America. illus. map. (Literary digest, New York, 1923, v. 77, May 19, p. 38, 40, 42, 44.)

Flight of Macready and Kelly, 1923. *** DA**

3205. **Hoare,** Sir Samuel John Gurney. India by air. With an introduction by Lady Maude Hoare. London: Longmans, Green and Co., Ltd., 1927. xix, 155 p. illus. maps. 8°. **VDY**

Airplane flight, 1927, of author and his wife, London-India. Reviewed in *Aeroplane,* London, 1927, v. 33, p. 556, *VDS*.

3206. **Hoffman,** Frederick Ludwig. Scrap-books of material on aeronautics. **3–VDYB n.c.6**

Author's flying experiences as contributed chiefly to the *New York Times,* 1927–1928.

Airplane Flights, continued

3207. **Hoop,** Thomassen à T. van der. The Amsterdam-Batavia flight. (Flight, London, 1925, v. 17, p. 310–311.) **VDS**

Flight of author in 1924.

3208. **Hop,** skip, and jump across the Pacific. illus. ports., map. (Literary digest, New York, 1928, v. 97, June 16, p. 5–7.) *** DA**

Newspaper comments on flight of *Southern Cross*, under command of Charles E. Kingsford-Smith.

3209. **Horsfall,** Jessie E. Lindbergh's start for Paris. illus. port. (Aero digest, New York, 1927, v. 10, p. 503–504, 506, 508.) **VDS**

3210. The **Houston-Mount Everest** Expedition 1933. By Air-Commodore P. F. M. Fellowes and others...with a foreword by John Buchan, and an account of the filming of the flight by Geoffrey Barkas. London: John Lane [1933]. xix, 279 p. illus. maps. 8°. **VDY**

Copiously illustrated account of preparations, hazards, and accomplishments. Reviewed in *Aeroplane*, London, 1933, v. 45, p. 982, *VDS*.

3211. **Hubert** Latham meets death while on hunting trip. Pilot of the Antoinette killed by a buffalo. (Fly, Philadelphia, 1912, v. 4, Aug., p. 8.) **3–† VDS**

3212. **Huenefeld,** Ehrenfried Guenther, Freiherr von. Mein Ostasienflug: der erste Weltflug Berlin-Tokio. Vollendet und herausgegeben von Alexander Roechling auf Grund von Hünefelds Aufzeichnungen und Berichten von K. G. Lindner. Berlin: Union deutsche Verlagsgesellschaft [1929]. 159 p. illus. ports. 4. ed. 8°. **VDY**

Airplane flight of author in 1928.

3213. **Humble,** Richard. Cape Town to Clyde, with a foreword by Colonel the Master of Sempill. London: Longmans, Green & Co., 1932. xii, 91 p. maps. 8°. **VDY**

Airplane flight of author and wife in 1931. Reviewed in *Flight*, London, 1932, v. 24, p. 430, *VDS*.

3214. **Hurst,** Mont. Some pioneers of the air. illus. (Aero digest, New York, 1926, v. 9, p. 429.) **VDS**

A list of early American aviators.

3215. **Hydravion,** pseud. Historische luchtreizen over het Kanaal. illus. (Het Vliegveld, Amsterdam, 1927, jaarg. 11, p. 63–68.) **3–† VDS**

Balloon and airplane flights across the English Channel.

3216. **Ide,** John Jay. The Rheims aviation meeting and the Gordon Bennett cup. The winners and their machines. illus. (Scientific American supplement, New York, 1913, v. 76, p. 260–261.) **VA**

3217. **Into** the Grand Canyon, and out again, by airplane. (Literary digest, New York, 1922, v. 75, Oct. 7, p. 63–66.) *** DA**

Flight of Royal V. Thomas, 1922.

3218. **Italy.** — Delegazione al Congresso internazionale di aeronautica civile a Washington, 1928. Dall' Italia al Brasile; preparazione scientifica del volo per il record mondiale di distanza in linea retta. Memoria presentata al Congresso internazionale...a Washington dalla Delegazione governativa italiana... Roma: Senato, 1928. 43 p. illus. maps. 4°. **VBA p.v.186, no.9**

Preliminary scientific preparations for the flight of 1928.

3219. The **Japan Advertiser.** An account of the visit to Japan of Col. and Mrs. Charles A. Lindbergh, August to September, 1931. [Tokyo: The Japan Advertiser, 1931.] 32 p. illus. ports. f°. **† AN**

3220. **JL 6** makes new world's duration record. Piloted by E. Stinson and L. Bertaud, metal monoplane stays aloft 26 hr. 19 min. 35 sec. illus. (Aviation, Highland, N. Y., 1922, v. 12, p. 38–40.) **VDS**

Flight at Roosevelt Field, Dec., 1921.

3221. **Johnson,** George R. Peru from the air, with text and notes by Raye R. Platte... New York: American Geographical Society, 1930. xii, 159 p. illus. maps. obl. 4°. (American Geographical Society. Publication no. 12.) **HHY**

Interesting aerial photographs.

3222. **Jones,** Bradley. Shrinking the globe. ports. (U. S. air services, Washington, 1931, v. 16, Aug., p. 15–19.) **VDS**

Around-the-world flight of Wiley Post and Harold Gatty, 1931.

3223. **Jones,** Casey. All in a day's work. illus. (Saturday evening post, Philadelphia, 1929, v. 201, Feb. 2, p. 38, 41–42, 44.) *** DA**

Famous airman's experiences in newspaper work.

3224. **Jones,** Ernest. Chronology of world and American air records. (National Aeronautic Association review, Washington, 1925, v. 3, p. 13–14, 30–31, 46–47.) **VDS**

This chronology begins with the number for Dec., 1924, which the Library lacks.

3225. **Keyhoe,** Donald Edward. Flying with Lindbergh. New York: G. P. Putnam's Sons, 1928. xv, 299 p. illus. map. 8°. **VDY**

Author accompanied Lindbergh on 22,000 mile route across 48 states in 1927.

3226. **King's** cup air race. table. (Aeroplane, London, 1930, v. 39, p. 122–123.) **VDS**

Tabulation of results of 1930 race, won by Winifred Brown.

3227. **Kingsford-Smith,** Charles Edward, and C. T. P. Ulm. The flight of the Southern Cross. With a foreword by His Excellency, the Right Honorable Lord Stonehaven. New York: R. M. McBride & Co., 1929. xii, 295 p. illus. map, ports. 8°. **VDYC**

Airplane flight (1928), Oakland, Cal., to Australia.

Airplane Flights, continued

3228. **Kingsford-Smith** flies the Atlantic. Makes one-stop flight Ireland to New York [1930]. (Aviation, New York, 1930, v. 29, p. 28.) **VDS**

Flight, east to west, 1930.

3229. **Kingsford-Smith's** Atlantic flight; including some notes on the equipment of the "Southern Cross." illus. (Flight, London, 1930, v. 22, p. 757–761.) **VDS**

3230. **Klemin,** Alexander. Flying over Mount Everest. illus. (Scientific American, New York, 1933, v. 148, p. 286–287.) **VA**

Description of equipment for projected flight — electric heating, oxygen equipment, photography.

3231. —— Lessons of the Pacific flight. illus. (Scientific American, New York, 1925, v. 133, p. 340–341.) **VDS**

Seaplane flight, San Francisco to Honolulu, 1925.

3232. **Knauss,** Robert. Im Grossflugzeug nach Peking, der erste Weltflug der deutschen Luft Hansa. Berlin: Union deutsche Verlagsgesellschaft, Abt. Luftfahrtverlag, G.m.b.H. [cop. 1927.] 176 p. illus. port., map. 2. ed. 8°. **VDY**

Account of author's Berlin-Peking flight in 1926.

3233. **Köhl,** Hermann. The three musketeers of the air; their conquest of the Atlantic from East to West, by Captain Hermann Koehl, Major James C. Fitzmaurice, Baron Guenther von Huenefeld; German translation of Captain Koehl's and Baron von Huenefeld's versions by George F. Dunay. New York, London: G. P. Putnam's Sons, 1928. xi, 330 p. illus. ports. 8°. **VDYC**

Account of famous flight (1928) of the Junkers airplane *Bremen,* Germany to the United States.

3234. —— Unser Ozeanflug; Lebenserinnerungen von Hermann Köhl, James C. Fitzmaurice [und] E. G. Freiherr v. Hünefeld; der erste Ost-Westflug über den Atlantik in der "Bremen." Berlin: Union deutsche Verlagsgesellschaft [1928]. xi, 275 p. illus. diagr., ports. 9. ed., enl. 8°. **VDYC**

3235. **König von und zu Warthausen,** Friedrich Karl, Freiherr von. Wings around the world. New York: G. P. Putnam's Sons, 1930. xii, 185 p. illus. map, ports. 8°. **VDY**

Solo flight (1929) across Europe, Asia, and the United States. Reviewed in *Aviation,* New York, 1930, v. 29, p. 371–372, *VDS.*

3236. **Koppen,** G. A. Our flight from Amsterdam to Batavia and back. illus. map. (Flight, London, 1928, v. 20, p. 143–146.) **VDS**

Flight of 18,000 miles made in 1927 in record time of 28 days.

3237. **Lamb,** Dean Ivan. Across the Central American Andes. port. (Aviation, New York, 1921, v. 11, p. 107–108.) **VDS**

Flight, 1921, of the Chief of Aviation of the Republic of Honduras.

3238. **Lawson,** Alfred William. The joys of air travel. Remarkable story of the first flight of a huge airliner, American-designed and American-built. illus. (Aero digest, New York, 1924, v. 5, p. 362–366.) **VDS**

3239. **League,** A. W. Seventeen and a half days in the air. illus. ports. (Aero digest, New York, 1929, v. 15, Sept., p. 69, 314, 316.) **VDS**

Refueling flight of Dale Jackson and Forest O'Brine, 1929.

3240. **Ledeboer,** John Henry. Gustav Hamel: an appreciation. port. (Aeronautics, London, 1914, v. 7, p. 159–161.) **VDS**

Career of brilliant English pilot.

3241. **Le Grand,** G. Les records mondiaux de l'aéronautique. Paris: L'Aérophile, 1917. 63 p. 3. ed. 8°. **VDY**

List of records established in various countries through 1914. Also numerous prize winners.

3242. **Leighton,** B. G. Nine month's cruise of Atlantic Fleet Air Force without mishap. Six air boats make record trip to West Indies and return. illus. (U. S. air service, New York, 1920, v. 4, Sept., p. 8–10.) **VDS**

3243. **Léon** Delagrange. port. (Aeronautical journal, London, 1910, v. 14, p. 38–39.) **VDS (Royal)**

Brief biography of pioneer French aviator (1873–1910).

3244. **Léon** Delagrange. The latest victim to the progress of aviation. port. (Aero, London, 1910, v. 2, p. 21.) **3–† VDS**

Record of famous French airman (1873–1910).

3245. **Le Page,** Wynn Laurence. Are aviation records worth while? Experience shows that world aeronautical records foster technical development, increase international prestige. illus. charts. (Aviation, Highland, N. Y., 1927, v. 22, p. 767–771.) **VDS**

A study of records, with charts showing records in various countries.

3246. The **Lesson** of the flight to Australia. (Scientific American, New York, 1920, v. 122, p. 268.) **VA**

Ross Smith's flight of 1920.

3247. **Letellier,** R. Louis Blériot's work. (Aircraft, New York, 1912, v. 3, p. 270.) **VDS**

3248. **Lewis,** George M. First American piloted airplane to cross the Alps. (U. S. air service, New York, 1921, v. 5, March, p. 31.) **VDS**

Flight of Caproni biplane, July 21, 1918.

3249. **Lewis,** H. Latane, 2nd. The Curtiss marine trophy race. (U. S. air services, Washington, 1932, v. 17, May, p. 20–21.) **VDS**

History of various events, beginning in 1915.

Airplane Flights, continued

3250. Lieutenant Soucek's world altitude record. (U. S. air services, Washington, 1929, v. 14, June, p. 46, 48.) **VDS**

Flight of U. S. naval officer, 1929, to height of 39,140 feet.

3251. Lincoln Beachey. (Flying, New York, 1915, v. 4, p. 477, 488.) **VDS**

"With Beachey's death America loses her greatest flier, and the world one of its most daring aerial pioneers."

3252. Lindbergh, Anne Morrow. Flying around the north Atlantic, with a foreword by Charles A. Lindbergh. illus. maps. (National geographic magazine, Washington, 1934, v. 66, p. 259–337.) **KAA**

Copiously illustrated account of flight of the author and her husband, 1933.

3253. Lindbergh, Charles Augustus. My flight to Paris. (Aero digest, New York, 1927, v. 10, p. 514, 516, 518, 520, 522, 524.) **VDS**

3254. —— "We"; the famous flier's own story of his life and his transatlantic flight, together with his views on the future of aviation, with a foreword by Myron T. Herrick... New York and London: G. P. Putnam's Sons, 1927. 318 p. illus. 8°. **VDYC**

"No flaw marked any act or word, and he stood forth amidst clamor and crowds the very embodiment of fearless, kindly, cultivated American youth — unspoiled, unspoilable." — *Foreword.*

Reviewed in *Aeroplane,* London, 1927, v. 33, p. 556, *VDS.*

For numerous references to Lindbergh see United States. — National Advisory Committee for Aeronautics, *Bibliography of aeronautics,* 1909–1931, Washington, 1921–35, volumes for 1927, p. 99–102; 1928, p. 112–113; 1929, p. 128; 1930, p. 138; 1931, p. 166, *VDT.*

3255. Lindbergh ends Latin-American tour. illus. ports., map. (Aero digest, New York, 1928, v. 12, p. 340–341, 464–465.) **VDS**

3256. The **Lindbergh** survey. illus. map. (U. S. air services, 1934, v. 19, Jan., p. 10–13.) **VDS**

Around-the-Atlantic flight of Charles and Anne Lindbergh, 1933.

3257. Lindbergh welcomed in Europe. illus. (Aero digest, New York, 1927, v. 10, p. 530, 552.) **VDS**

3258. Lindbergh's Central American flight. illus. (Aero digest, New York, 1928, v. 12, p. 174, 301.) **VDS**

3259. Lindbergh's flight to Mexico and Central America. illus. map. (Aero digest, New York, 1928, v. 12, p. 14, 16, 18.) **VDS**

3260. Lindstrom, Siegfried F. White flyers who lead Chinese airmen. port. (Popular aviation, Chicago, 1932, v. 10, p. 165–166, 192.) **VDS**

Deals mainly with the aviator Bert Hall.

3261. Little, Ray. The passing of pioneer days. illus. port. (Aero digest, New York, 1929, v. 15, Aug., p. 77, 284.) **VDS**

Author's recollections of early days in the U. S. air postal service.

3262. Livingston, John H. How the 1929 air tour was won. illus. ports. (Aviation, New York, 1929, v. 27, p. 922–925.) **VDS**

Author, the winner, describes his preparations and how he ran up a score of 45,672.64.

3263. Loening, Grover Cleveland. An American designer's review of Schneider race [1929]. (Aviation, New York, 1929, v. 27, p. 650–652.) **VDS**

3264. Loewe, Karl F. Im Fluge nach dem Azoren; Flugerlebnisse und Reiseeindrücke. Berlin: Union deutsche Verlagsgesellschaft [1929?]. 44 p. illus. map. 4°. **VBA p.v.189, no.5**

Flight of hydro-airplane, Germany to the Azores, in October, 1927.

3265. London-Cairo-the Cape, by airplane. (Scientific American, New York, 1920, v. 122, p. 536.) **VA**

Lessons from the flight of Van Ryneveld and Brand in 1920.

3266. London-Cape Town-London. Alan Cobham's splendid achievement. illus. (Flight, London, 1926, v. 18, p. 155–161.) **VDS**

3267. The **London-Manchester** 10,000 flight prize. illus. (Flight, London, 1910, v. 2, p. 326–328.) **VDS**

Unsuccessful but brilliant flight of Claude Grahame-White. Won by Louis Paulhan.

3268. The **London** to Manchester flight [1910]. illus. (Aeronautics, London, 1910, v. 3, p. 60–62.) **VDS**

3269. London's welcome to Lindbergh. illus. (Flight, London, 1927, v. 19, p. 344–348.) **VDS**

3270. Lotti, Armeno, jr. Flying from Maine to Spain. illus. (Aero digest, New York, 1929, v. 15, July, p. 71, 246.) **VDS**

Flight of the *Yellow Bird,* 1929.

3271. Lund, Fred. Freddy Lund — exhibition thriller, as told to Legette Blythe. illus. port. (Popular aviation, Chicago, 1931, v. 9, Dec., p. 95–96, 126.) **VDS**

Experiences of famous American stunt flyer.

3272. McClure, Ivor H. Seven countries in eight days. map. (Flight, London, 1927, v. 19, p. 636–638.) **VDS**

Author's airplane flight through Europe, 1927.

3273. McDarment, Corley. Around the world by air. The details of the American army aviators' experience, and the lessons learned. illus. (Scientific American, New York, 1924, v. 131, p. 230–232, 292–293.) **VA**

Flight of 1924.

Airplane Flights, continued

3273a. **McElroy,** Clarence L. Seventeen days in the Mexican jungle. Greenfield, Ind.: The Mitchell Company, 1933. 172 p. illus. 12°. **VDYC**

Exciting adventures of the author, 1932.

3274. **McLaren,** A. S. C. S. The British world flight attempt. (Flight, London, 1924, v. 16, p. 641–643.) **VDS**

Account of the unsuccessful flight commanded by the author, in a Vickers-Napier seaplane, 1924.

3275. **MacLean,** Robert Alexander. Across the Arabian desert by aeroplane. (Canadian magazine, Toronto, 1923–24, v. 62, p. 435–442.) *** DA**

3276. **Macready,** John A. Exploring the earth's stratosphere. The holder of the American altitude record describes his experiences in reaching the "ceiling" of his plane at an elevation of nearly eight miles. (National geographic magazine, Washington, 1926, v. 50, p. 755–776.) **KAA**

3276a. **McReynolds,** Charles F. The refueling flight of the "Question mark." illus. (Aviation, New York, 1929, v. 26, p. 158–159, 162.) **VDS**

U. S. Army plane establishes world record, 1929, of 150 hours, 40 minutes, 14 seconds.

3277. **Magoun,** Frederick Alexander, and Eric Hodgins. Significant flights since the war. illus. (In their: A history of aircraft, New York, 1931, p. 397–425.) **VDT**

3278. **Mahgni-Eltten.** Night flying, ancient and modern. (Aeronautics, London, 1916, v. 10, p. 367–368.) **VDS**

"The first instance of night flying is Grahame-White's classic ascent at 2.30 a. m., at Roade, on April 28, 1910, during his second attempt in the London-Manchester race."

3279. **Maitland** and Hegenberger fly to Hawaii. illus. (Aviation, Highland, N. Y., 1927, v. 23, p. 72–73.) **VDS**

Flight of American aviators, 1927.

3280. **Major** Schroeder's true altitude 33,000 ft. illus. (Aviation, Highland, N. Y., 1920, v. 8, p. 361.) **VDS**

3281. **Man's** greatest altitude above the earth. Captain Gray, Army Air Corps, reaches height of more than 8 miles. port. (U. S. air services, Washington, 1927, v. 12, June, p. 49–50.) **VDS**

Balloon ascension of United States Army officer, 1927.

3282. **Marie,** Eugène Armand. Le Congo à six jours de Paris. Paris: Les Éticelles [1931]. 242 p. illus. ports., maps. 12°. **VDY**

3283. **Mason,** L. T. Air pioneers of the north Atlantic. illus. (Air, London, 1929, v. 2, p. 231–233, 283–285.) **VDS**

Early proposals, successes and failures, airships and airplanes.

3284. **Maxim,** Hudson, and W. J. Hammer. Chronology of aviation. [New York, 1912.] 5–23 p. 4°. **VDC p.v.23, no.10**

Tabular data on airplane records to date, with notes.

3285. **Mears,** John Henry. Racing the moon (and winning), being the story of the swiftest journey ever made, a circumnavigation of the globe by airplane and steamship in 23 days, 15 hours, 21 minutes and 3 seconds by two men and a dog. New York: Rae D. Henkle Co., Inc., 1928. vii, 11–320 p. illus. ports. 12°. **KBG**

Flight with C. B. D. Collyer (1928).

3286. La **Méditerranée** franchie d'un seul vol. Une grande date de l'aviation. map. (L'Aérophile, Paris, 1913, année 21, p. 446–447.) **3–VDS**

Describes flight of Roland Garros in 1913.

3287. **Minney,** Rubeigh James. Across India by air. Allahabad: A. H. Wheeler & Co., 1921. 91 p. illus. ports. 12°. **VDY p.v.25, no.3**

Author's account of his airplane flight in 1920.

3288. **Mr.** [C. W. A.] Scott's record flight to Australia. port. (Aeroplane, London, 1931, v. 40, p. 650.) **VDS**

Solo flight of 10,450 miles, 1931. For an appreciation of Scott, see *Aeroplane,* London, 1931, v. 40, p. 652, 654, *VDS.*

3289. **Mr.** Hinkler's triumph. port. (Aeroplane, London, 1928, v. 34, p. 274, 276, 295.) **VDS**

England-Australia flight, 1928.

3289a. **Mittelholzer,** Walter. Abessinienflug; mit dem dreimotorigen Fokker an dem Hof des Negus Negesti; mit einem Vorwort von Bundpräsident Pilet-Golaz. Zürich: A. G. Schweizer Aero-Revue [1934]. 95 p. illus. map. 8°. **VDYC**

Copiously illustrated account of author's Abyssinian flight, 1934.

3290. —— Kilimandjaro Flug. Zürich: O. Füssli [1930]. 114 p. illus. 8°. **VDY**

Flight of *Switzerland II* in 1927 from Zürich to central Africa and return. Some remarkable aerial photographs.

3291. —— Persienflug. Mit einem Nachwort von Prof. Dr. O. Flückiger. Zürich: O. Füssli [1926]. 212 p. illus. maps, port. 2. ed., enl. 8°. **VDY**

Zürich-Persia airplane flight, 1924–1925. Excellent illustrations.

3292. —— Tschadseeflug mit dem dreimotorigen Fokker der Swissair durch die Sahara zum Tschadsee, mit einem Vorwort von Prof. A. Piccard. Zürich: Schweizer Aero-Revue [1932]. 128 p. illus. map. 8°. **VDY**

Zürich-Lake Tchad airplane flight over the Sahara desert, 1930–1931. Numerous excellent aerial photographs.

3293. **Mittelholzer,** Walter, and H. Kempf. Alpenflug. Unter Mitarbeit von H. Kempf ...mit 191 Fliegerausnahmen und einer farbigen Tafel nach einem Gemälde von F.

Airplane Flights, continued

Hass. Zürich: O. Füssli [1928]. 155 p. illus. 8°. **VDY**

Experiences of the authors, beginning during the war period. Numerous aerial views, with descriptive text.

3294. **Mittelholzer,** Walter, and others. Afrikaflug im Wasserflugzeug "Switzerland" von Zürich über der dunkeln Erdteil nach dem Kap der guten Hoffnung [by Walter Mittelholzer, René Gouzy, and Arnold Heim]. Mit einem Vorwort von Professor Albert Heim. Zürich: O. Füssli, 1927. 235 p. illus. maps, port. 8°. **VDY**

Hydro-airplane flight, 1926, Zürich, Cairo, and Cape Town.

3295. **Mittelholzer,** Swiss pilot, makes great flight in "Dornier-Merkur" seaplane. illus. (U. S. air services, Washington, 1927, v. 12, May, p. 26.) **VDS**

Flight from Switzerland to South Africa in all-metal twin-float machine.

3296. **Mix,** Arthur. My 82,000 miles with Lincoln Beachey. port. (U. S. air services, Washington, 1929, v. 14, Jan., p. 62, 64.) **VDS**

Beachey's mechanician relates experience with noted pioneer flyer.

3297. **Mockler,** Don. The National Air Races in retrospect. (U. S. air services, Washington, 1931, v. 16, Oct., p. 36–40.) **VDS**

Official results of the 1931 events.

3298. —— National Air Tour effectively proves aircraft efficiency. (U. S. air services, Washington, 1931, v. 16, p. 21–22, 54.) **VDS**

Results of 1931 event.

3299. **Moffett,** William Adger. The Schneider seaplane trophy — symbol of speed supremacy in the air. illus. (National aeronautic magazine, Washington, 1931, v. 9, March, p. 45–48.) **VDS**

History of the various races.

3300. **Moisant** falls from machine. port. (Aeronautics, New York, 1911, v. 8, Feb., p. 61–62.) **VDS**

Biography of famous American aviator. He was the first to fly from London to Paris, also the first to carry a passenger over the same route. For references see United States. — National Advisory Committee for Aeronautics, *Bibliography of aeronautics*, 1909–1931, Washington, 1921–35, volume for 1909/16, p. 915, *VDT*.

3300a. **Mollison,** James Alan. The book of famous flyers. An interesting account of the history of aviation from the early pioneers to present-day aces of the air. London and Glasgow: Collins' Clear-type Press [1934]. 191 p. illus. ports. 8°. **VDYC**

3301. —— Death cometh soon or late. Foreword by Amy Johnson (Mrs. Mollison). London: Hutchinson & Co., Ltd. [1932.] 283 p. illus. map, ports. 8°. **VDYC**

Author's airplane flights: Australia to England; England to the Cape of Good Hope; and Ireland to America. Reviewed in *Air and airways*, London, 1932, v. 9, p. 325–326, *VDS*.

3302. **Mollison's** Atlantic flight. port. illus. (Flight, London, 1932, v. 24, p. 795–798.)

Ireland-New Brunswick in 30 hours. **VDS**

3303. **Monk,** F. V., and H. T. Winter. Great stories of the air. London: Blackie & Son, Ltd. [1934.] 63 p. illus. 16°. **VDYC**

Some heroes of the Caterpillar Club. Flight of G. H. Stainforth at 500 miles an hour. The Piccard stratosphere ascension. The 1933 Mt. Everest expedition.

3304. **Moore-Brabazon,** John Theodore Cuthbert. My first flight. port. illus. (Airways, London, 1930, v. 6, p. 243–244.) **VDS**

Author was officially adjudged the first Englishman to make a heavier-than-air flight above British soil.

3305. **Mortane,** Jacques. Marcel Brindejonc de Moulinais — free lance of the air. illus. (Flying, New York, 1913, v. 2, Aug., p. 12–13.) **VDS**

Career of youthful, but prominent, French aviator.

3306. —— Roland Garros, virtuose de l'aviation. Paris: L'Édition française illustrée, 1919. 126 p. 12°. **AN**

Biography of pioneer French civil and military aviator.

3307. —— The sixth Pommery contest. illus. (Flying, New York, 1914, v. 3, p. 22–23.) **VDS**

Of the ten events (1913), that of Brindejonc des Moulinais and of Eugène Gilbert were outstanding.

3308. **Murphy,** Charles J. V. The flight of Williams and Yancey. illus. ports. (Aero digest, New York, 1929, v. 15, Sept., p. 62–64, 312, 314.) **VDS**

Airplane flight, Old Orchard, Me., to Santander, Spain, 1929.

3309. The **"Mystery** ship" revealed. The fastest commercial plane yet flown, it has out-performed the best planes of the Army and Navy. illus. (Scientific American, New York, 1931, v. 144, p. 96–98.) **VA**

Used by Capt. Frank M. Hawks in transcontinental record flight, 1930.

3310. The **National** air races [of 1926]. (Aviation, Highland, N. Y., 1926, v. 21, p. 500–508.) **VDS**

Races contested in Philadelphia.

3311. The **National** air races [1928]. illus. ports. (Popular aviation, Chicago, 1928, v. 3, Nov., p. 11–15, 107.) **VDS**

Races held in Los Angeles, Cal.

3312. The **National** Air Tour. port. (Popular aviation, Chicago, 1928, v. 3, Sept., p. 38, 78, 80.) **VDS**

Brief account of 1928 event with table of scores.

3313. **Navy** sets 13 new records at Baltimore. Most successful naval seaplane race meet ever held. illus. ports. (Aviation, Highland, N. Y., 1924, v. 17, p. 1218–1220.) **VDS**

Baltimore Navy seaplane races of 1924.

3314. **Neely,** Frederick R. First civilian transcontinental air race. (U. S. air services, Washington, 1927, v. 12, Oct., p. 18–19.) **VDS**

National Air Derby and National Air Races, 1927.

3315. —— First pursuit group, using skis, makes great flight into Canada in midwinter... illus. (U. S. air services, Washington, 1927, v. 12, March, p. 16–19.) **VDS**

Flight of United States Army fliers, 1927.

Airplane Flights, continued

3316. **Nevill,** John T. The fifth annual National Air Tour [1929]. illus. map, ports. (Aviation, New York, 1929, v. 27, p. 670–673, 736–739, 844–848, 877–882.) **VDS**

3317. —— The national air races [1929] day by day and in summary. illus. (Aviation, New York, 1929, v. 27, p. 524–535.) **VDS**

Contests were held in Cleveland, Ohio.

3318. —— The National Air Tour [1928]. illus. (Aviation, Highland, N. Y., 1928, v. 25, p. 108–109, 180–181, 406–407, 422–424, 594, 611–614.) **VDS**

Accounts of the 4th reliability tour. Title varies.

3319. —— The tour in retrospect. illus. (Aviation, New York, 1930, v. 29, p. 278–280.) **VDS**

Comments on the 1930 National Air Tour.

3320. A **New** aeroplane passenger-carrying record. Louis Bréguet's feat of transporting 11 people 3 miles across country. Description of the machine. illus. (Scientific American, New York, 1911, v. 104, p. 407.) **VA**

3321. **New** altitude record: 40,800 ft. Lt. John A. MacReady, A. S., on a Lepere C. 11 airplane fitted with the Moss supercharger breaks all previous records. illus. port. (Aviation, Highland, N. Y., 1921, v. 11, p. 423–425.) **VDS**

3322. A **New** conquest. (Popular aviation, Chicago, 1928, v. 2, Feb., p. 7–9, 77.) **VDS**

Lindbergh's flight to Mexico and other Latin-American countries.

3323. The **New** endurance record. illus. (Aero digest, New York, 1930, v. 17, Aug., p. 116.) **VDS**

Flight of John and Kenneth Hunter, 1930, achieving 553 hours, 41 minutes, 30 seconds in the airplane *City of Chicago.*

3324. **New** records for the Cape flight. The story of Mrs. J. A. Mollison's record-making flights from England to South Africa and return. illus. (Air and airways, London, 1933, v. 9, p. 335–336.) **VDS**

3325. **New York-Berlin** (nearly). Clarence Chamberlin goes one better than Capt. Lindbergh. illus. (Flight, London, 1927, v. 19, p. 375.) **VDS**

3326. **New York-Paris** flight a reality. Capt. Charles A. Lindbergh makes first successful non-stop flight. illus. map, port. (Aviation, Highland, N. Y., 1927, v. 22, p. 1120–1122.) **VDS**

3327. The **New York Times.** [Scrapbook of newspaper clippings on Lindbergh's flights, May, 1927 – April, 1928. New York, 1929.] 86 p. illus. ports., maps. f°. **††† VDY**

3328. **Newcomb,** J. Arthur. Transatlantic flights, 1931. (U. S. air services, Washington, 1931, v. 16, Nov., p. 29–30.) **VDS**

3329. **Newton,** Byron Rufus. Charles K. Hamilton, one of the greatest early flyers. port. (U. S. air services, Washington, 1934, v. 19, Aug., p. 23–25.) **VDS**

3330. —— The meteoric Moisant and his all-star troupe. port. (U. S. air services, Washington, 1932, v. 17, March, p. 24–28.) **VDS**

3331. **Nine** days around the world. Mechanical aspects of the great flight. illus. ports. (Aviation, New York, 1931, v. 30, p. 468–469.) **VDS**

Flight of Wiley Post and Harold Gatty, 1931.

3332. The **1910** Rheims week. illus. (Flight, London, 1910, v. 2, p. 548–551.) **VDS**

Account of records made.

3333. The **1927** national air races. illus. (Aviation, Highland, N. Y., 1927, v. 23, p. 866–871.) **VDS**

Held at Spokane, Washington, Sept. 19–25, 1927.

3334. The **1934** national air races. (Western flying, Los Angeles, 1934, v. 14, Oct., p. 18–19.) **VDS**

Complete official results.

3335. **Notable** flights. (Aviation year book, London, 1931, p. 163–170.) **VDS**

Brief particulars of airplane and airship flights, 1903–1930.

3336. **Notable** flights, 1931–1932. 3 v. illus. maps. f°. **††† VDY**

Scrapbooks of clippings about the flights of Amelia Earhart, Griffin and Mattern, Post and Gatty, and others; mounted and bound by The New York Public Library, 1933.

3336a. **Notable** flights [of 1933]. (Aircraft year book for 1934, New York, 1934, p. 127–140.) **VDS**

Lindbergh's four-continent flight. Post's second world's flight. Balbo's transatlantic flight. British Mount Everest expedition. Codos and Rossi's New York-Syria flight. Speed and altitude records. Turner's trans-American flight. Hawks' trans-American flight. Australian flights.

3337. **Notes** on Major Schroeder's altitude flight. (Aviation, New York, 1920, v. 8, p. 293–294.) **VDS**

Technical discussion of the flight.

3338. The **Notes** of two tough hombres. ports. illus. (Popular aviation, Chicago, 1929, v. 5, Oct., p. 24–26, 78.) **VDS**

Refueling flight of the Buhl airplane *Angeleno,* by Loren Mendell and Roland Reinhart, 1929. Time aloft was 246 hours, 43 minutes and 32 seconds. This number of *Popular aviation* has the title of *Aeronautics.*

3339. **Nungesser** and Coli missing in Atlantic flight attempt. Notable French fliers unsuccessful in attempt to span Atlantic, Paris-New York, in Levasseur plane. illus. ports. (Aviation, Highland, N. Y., 1927, v. 22, p. 1041–1042.) **VDS**

Airplane Flights, continued

3340. [**Ocean** flight, May 20 – May 25, 1919. Clippings relating to the first non-stop flight across the Atlantic by John Alcock and Arthur Whitten Brown.] Mounted and bound, by The New York Public Library. 4°. **VDY**

3341. Ocean to ocean between lunch and breakfast. illus. map. (Literary digest, New York, 1928, v. 98, Sept. 8, p. 10–11.) *** DA**

Flight of Col. Arthur Goebel and Harry Tucker, Los Angeles-New York, in 18 hours, 50 minutes, 1928.

3342. Oelerich, Reinhold. How I flew five miles above the earth. (Aero and hydro, Chicago, 1914, v. 8, p. 221–222.) **3–VDS**

Brief account of flight in 1914, with a list of altitude records by others, beginning in 1908.

3343. Official results of the national air races. (Aviation, Highland, N. Y., 1927, v. 23, p. 814–816.) **VDS**

Held at Spokane, Washington, Sept. 19–25, 1927.

3344. On Bert Hinkler. port. (Aeroplane, London, 1933, v. 44, p. 165–167.) **VDS**

Appreciation of famous British aviator.

3345. On the best British aviator. port. illus. (Aeroplane, London, 1931, v. 41, p. 1257–1258, 1260, 1262, 1264.) **VDS**

Relates the career of H. J. L. (Bert) Hinkler.

3346. On the King's cup race. illus. (Aeroplane, London, 1929, v. 37, p. 41–76.) **VDS**

British race of 1929 as recorded by several observers.

3347. One year of flying and of "Flight." A 1909 retrospect. (Flight, London, 1910, v. 2, p. 2–4.) **VDS**

3348. Orcy, Ladislas d'. The editor's story of the Dayton air meet. illus. (Aviation, Highland, N. Y., 1924, v. 17, p. 1110–1121.)

Account of national air races of 1924. **VDS**

3349. —— The third Pulitzer trophy race. illus. port. (Aviation, Highland, N. Y., 1922, v. 13, p. 544–548.) **VDS**

Won by Lieut. R. L. Maughan at Detroit meeting, 1922.

3350. Orlebar, Augustus Henry. Schneider trophy, a personal account of high-speed flying & the winning of the Schneider trophy. London: Seeley, Service & Co. [1933.] 238 p. illus. 8°. **VDY**

Author, a wing-commander of the British Air Force, tells of his experiences in the events of 1929 and 1931.

Reviewed in *Aeroplane*, London, 1933, v. 45, p. 764, *VDS.*

3351. Our first air meet — 1910. (Aero digest, New York, 1928, v. 13, p. 468, 470.) **VDS**

Detailed account of the meet at Dominguez Field, Los Angeles, 1910.

3352. Our own story of the St. Louis air race meet [1923]. Three hundred thousand miles flown without serious mishap at world's biggest air meet. port. illus. (Aviation, Highland, N. Y., 1923, v. 15, p. 468–489.) **VDS**

International Air Meet. Among the events were the Pulitzer trophy races and the *Detroit News* air mail race.

3353. The **Outstanding** feat of the ages. illus. (Popular aviation, Chicago, 1927, Aug., p. 5–8.) **VDS**

Brief account of Lindbergh's flight, New York to Paris, 1927.

3354. Over the top of Mount Everest. The story of the conquest by air of the world's highest mountain told by the chief pilot and official observer of the Houston-Everest expedition. illus. map. (Air and airways, London, 1933, v. 10, May, p. 51–53.) **VDS**

3355. Ovington, Adelaide (Alexander). An aviator's wife. New York: Dodd, Mead & Co., 1920. 160 p. illus. ports. 12°. **VDY**

Experiences of Earle L. Ovington, early American aviator, as told by his wife.

3356. Owen, Russell. Lindbergh's historic Central American flight. (New York Times current history, New York, 1928, v. 28, p. 89–96.) **BAA**

Flight of 1927/28.

3357. Palmer, Paul J. Lincoln Beachey. port. (Aircraft, New York, 1915, v. 6, p. 448.) **VDS**

Biography of prominent American aviator, killed 1915.

3358. Parmelin, Agenor. Over Mount Blanc by aeroplane. illus. port. (Flying, New York, 1914, v. 3, p. 69.) **VDS**

Author was the first to make this flight (1914).

3359. The **Passing** of Harry Hawker. (Flight, London, 1921, v. 13, p. 494–495.) **VDS**

3360. Patrick, Mason Mathews. Engineering the round-the-world flight. illus. (U. S. air services, Washington, 1925, v. 10, June, p. 21–26.) **VDS**

How the planes and engines of the U. S. Army flight stood the strain.

3361. Pégoud, Adolphe Célestin. How I fly upside down. illus. (Royal magazine, London, 1913, v. 31, p. 132–137.) *** DA**

For references to Pégoud see United States. — National Advisory Committee for Aeronautics, *Bibliography of aeronautics*, 1909–1931, Washington, 1921–35, volumes for 1909/16, p. 1018–1021; 1917/19, p. 348; 1920/21, p. 321, *VDT.*

3362. Pégoud. Upside down flying at Brooklands. diagrs. (Flight, London, 1913, v. 5, p. 1066–1068.) **VDS**

Loop-the-loop stunt described with diagrams.

3363. Pégoud's remarkable performances. What coolness and nerve can accomplish. illus. (Scientific American, New York, 1913, v. 109, p. 301.) **VA**

3364. Pelletier Doisy, Georges Edme Charles Marie. Mon raid. 48 bois originaux de Jacques Boullaire. Paris: D. É. Champion, 1925. 127 p. illus. obl. 8°. **VDY**

Account of his Paris-Tokio flight, 1924.

Airplane Flights, continued

3365. **Philip** Parmelee, master aviator, falls to death. port. (Fly, Philadelphia, 1912, v. 4, July, p. 22–23.) 3–† VDS

Parmelee was a pupil of the Wright brothers.

3366. **Philos.** Chute mortelle de Léon Delagrange. illus. (L'Aérophile, Paris, 1910, année 18, p. 25–27.) 3–VDS

Famous French aviator was killed in airplane accident in France, Jan. 4, 1910.

3367. **Pinedo,** Francesco de. By seaplane to six continents. Cruising 60,000 miles, Italian argonauts of the air see world geography unroll, and break new sky trails over vast Brazilian jungles. illus. map. (National geographic magazine, Washington, D. C., 1928, v. 54, p. 247–302.) KAA

Copiously illustrated account of author's flight of 1927.

3368. —— A flight of 33,000 miles. (Royal Aeronautical Society, Journal, London, 1926, v. 30, p. 507–518.) VDS

Author's account of his airplane flight, Italy to Australia and the Orient, and return. Abstracted in *Flight*, London, 1926, v. 18, p. 249–250, *VDS; Aeroplane*, London, 1926, v. 30, p. 402, 404, *VDS*.

3369. —— Il mio volo attraverso l'Atlantico e le due Americhe, con un proemio di Gabriele d'Annunzio. Milano: U. Hoepli, 1928. 27 f., 279 p. illus. maps, ports. 8°. VDY

Account of author's hydro-airplane flight, 1927, Italy to South America, North America, and back to Italy, 43,820 kilometers. Library has a copy of the Spanish edition, Buenos Aires, 1928, *VDY*.

3370. —— Un volo di 55,000 chilometri. Milano: A. Mondadori [1926]. 237 p. illus. maps. 8°. VDY

Account of hydro-airplane flight by the author, 1925, Italy to Australia and the Orient, and return.

3371. **Pinedo** back in Rome. Successful conclusion to 25,000-mile flight. illus. port. (Flight, London, 1927, v. 19, p. 413–414.) VDS

Portrait of Pinedo.

3372. **Post,** Augustus. Aeroplane that flew to Philadelphia. illus. (American machinist, New York, 1910, v. 33, p. 1183.) VFA

Describes flight of Charles K. Hamilton, New York to Philadelphia, June 13, 1910.

3373. —— Columbus of the air. (North American review, New York, 1927, v. 224, p. 353–364.) * DA

Appreciation of Lindbergh's transatlantic flight.

3374. **Post,** Wiley, and Harold Gatty. Around the world in eight days; the flight of the Winnie Mae. Introduction by Will Rogers. New York, Chicago [etc.]: Rand, McNally & Company [cop. 1931]. 304 p. illus. ports. 8°. VDYC

Famous airplane flight, 1931. Reviewed in *Aviation*, New York, 1931, v. 30, p. 711, *VDS*.

3375. **Premiers** virages de l'aéroplane Henri Farman. Volte-face aérienne à pleines voiles. Le record de durée de vol mécanique. illus. (L'Aérophile, Paris, 1907, année 15, p. 302–303.) 3–VDS

3376. **Pritchard,** John Laurence. Pioneer flights. (International Civil Aeronautics Conference, Washington, 1928, Proceedings, Washington, 1929, p. 95–102.) VDS

Accounts of long flights from England, including first aeroplane transatlantic flight, 1919; the several Australian flights; Sir Alan Cobham's flights: London-Cape Town; around Africa.

3377. **Progress;** a pictorial review in "Flight" photographs. illus. (Flight, London, 1930, v. 22, p. 31–88.) VDS

A selection of *Flight* photographs of pilots, aircraft and events, covering a period of 21 years.

3378. **Pulford,** C. W. H. The Royal Air Force from Cairo to the Cape. (Royal United Service Institution, Journal, London, 1927, v. 72, p. 266–279.) VWA

Airplane flight, Cairo to Cape Town, 1926.

3379. The **Pulitzer** trophy race. illus. (Aviation, Highland, N. Y., 1920, v. 9, p. 378–381.) VDS

Won by Capt. Corliss C. Mosely with a Verville-Packard machine, 1920.

3380. The **R. A. F.** Cairo-Cape flight. Official report on the big 14,000 mile flight. (Flight, London, 1927, v. 19, p. 6–7, 21.) VDS

3381. **R. A. F.** Cape flight (official account, abridged). (Royal Aeronautical Society, Journal, London, 1927, v. 31, p. 142–158.) VDS

3382. **Raleigh,** Walter. Frank Hawks — the record breaker. illus. port. (Popular aviation, Chicago, 1931, v. 8, June, p. 23, 57.) VDS

Career of prominent American flier.

3383. **Raynham,** F. P. Some experiences during survey flying in the East. illus. (Air, London, 1928, v. 1, July, p. 13–16.) VDS

Seaplane work in India and Borneo.

3384. Le **Record** d'altitude en aéroplane. illus. (L'Aérophile, Paris, 1910, année 18, p. 30–31.) 3–VDS

Brief account of records established in 1910 by Hubert Latham and Louis Paulhan.

3385. The **Record** of the races. Who won what, and how. illus. (Aviation, New York, 1932, v. 31, p. 399–401.) VDS

Results of 1932 national air races at Cleveland.

3386. **Recouly,** Raymond. Across the Syrian deserts by airplane. illus. (Scribner's magazine, New York, 1922, v. 72, p. 387–400.) * DA

3387. **Redden,** Charles F. Over-water flying for 1923. illus. (U. S. air service, Washington, 1923, v. 8, Jan., p. 25–27.) VDS

Brief statistical account of operations in the United States.

3388. **Reed,** E. Stoy. Defying death in frozen Alaska. illus. (Popular aviation, Chicago, 1931, v. 8, May, p. 28–30, 60.) VDS

Thrilling adventures of several aviators.

Airplane Flights, continued

3389. **Refueling** endurance flights. illus. (Aircraft yearbook, New York, 1930, p. 145–159.) **VDS**

Some developments since 1923.

3390. **Reliability** demonstrated by Ford tour. illus. (Aviation, Highland, N. Y., 1925, v. 19, p. 481–485.) **VDS**

Detroit to Kansas City and return, 1925, with 17 entries.

3391. **Rescuing** "the three musketeers of the air." illus. (Literary digest, New York, 1928, v. 97, May 12, p. 32, 34, 36, 38, 40, 45, 47.) *** DA**

The plight of the *Bremen* at Greenly island, and the attempt of Lindbergh to carry serum to Floyd Bennett, 1928.

3392. **Results** of the Ford reliability tour. illus. (Aviation, Highland, N. Y., 1925, v. 19, p. 550–551.) **VDS**

Tabulation of results of 1925 tests.

3393. The **Rheims** aeronautical meeting. (Engineer, London, 1909, v. 108, p. 211, 237–238.) **VA**

3394. The **Rheims** aviation meeting. Details of the first great exhibition of dynamic flight. illus. (Scientific American supplement, New York, 1909, v. 68, p. 177, 180–182.) **VA**

For references to this meeting see United States. — National Advisory Committee for Aeronautics, *Bibliography of aeronautics*, 1909–1931, Washington, 1921–35, volume for 1909/16, p. 1130–1132, *VDT*.

3395. **Richardson,** Holden Chester. Some lessons of the transatlantic flight. (Aviation, Highland, N. Y., 1920, v. 8, p. 445–446.) **VDS**

Flight of the *NC1* and *NC2* of the U. S. Navy.

3396. **Robertson,** Frederic Arthur de Vere. Bert Hinkler — a memoir. A very great pilot. port. (Flight, London, 1933, v. 25, p. 151.) **VDS**

Appreciation of celebrated English long-distance flier.

3397. **Rodgers,** John. Rodgers tells his story of the flight. Wind and failure of fuel forced descent. ports. (U. S. air services, Washington, 1925, v. 10, no. 10, p. 19–21.) **VDS**

Commander Rodgers had charge of the *PN-9, no. 1,* in flight from San Francisco to Hawaii, 1925.

3398. **Roland** Rohlfs' altitude record. Climbs 34,610 feet in a Curtiss "Wasp" triplane with no supercharger. illus. charts. (Aircraft journal, New York, 1919, v. 5, Sept. 27, p. 3–4.) **VDS**

3399. The **Rome-Brazil** non-stop flight. The Atlantic again conquered. illus. map. (Flight, London, 1928, v. 20, p. 594–595.) **VDS**

Flight of Arturo Ferrarin and Major del Prete, 1928.

3400. **Rome** to Rio in formation. illus. (U. S. air services, Washington, 1931, v. 16, Feb., p. 35–38.) **VDS**

Flight of squadron under command of General Balbo, 1930/31.

3401. The **Roosevelt** field runway. illus. (Aviation, Highland, N. Y., 1928, v. 24, p. 442–443.) **VDS**

Deals with the starting point of several long distance flights.

3402. **Roster** of American aviation pilots. (Aero and hydro, Chicago, 1913, v. 6, p. 38–40.) **3–VDS**

Lists over 200 pilots in the order of their priority, with style of machine and motor used.

3403. **Round** Africa by light plane. illus. map. (Flight, London, 1931, v. 23, p. 137.) **VDS**

1930–1931 flight of Italian pilots covered 17,150 miles.

3404. **Round** Australia flights. illus. (Flight, London, 1924, v. 16, p. 289–290, 347, 644, 719.) **VDS**

Flights of Goble and MacIntyre, and H. C. Brinsmead, 1924.

3405. The **Round-the-world** flight. illus. (Aircraft yearbook, New York, 1925, p. 66–95.) **VDS**

United States Army Air Service's flight in 1924.

3406. **Round** the world in a flying-boat. illus. map. (Air and airways, London, 1933, v. 9, p. 339–340.) **VDS**

Wolfgang von Gronau's flight, 1932.

3407. The **Royal** Air Force Cape flight. illus. (Aeronautics, London, 1926, v. 31, p. 866, 868–870.) **VDS**

Flight of four planes, under command of C. W. H. Pulford, 1926.

3408. The **Royal** Air Force Far East flight. Log of Singapore-Australia-Singapore section. (Aeroplane, London, 1929, v. 36, p. 72–74, 76, 78, 80–82.) **VDS**

3409. The **Royal** Air Force flying-boat cruise. Log of the Far East flight from England [to Singapore]. illus. maps. (Flight, London, 1928, v. 20, p. 227–229, 695; 1929, v. 21, p. 52–54, 63–66.) **VDS**

Flight of four planes.

3410. **Rue,** Larry. I fly for news. New York: A. and C. Boni, Inc., 1932. viii, 13–307 p. illus. map, ports. 8°. **VDY**

Author's experiences, as correspondent for the Chicago *Tribune,* to Afghanistan, and through northern Africa.

3411. **Rulers** of three nations confer honors upon Captain Charles A. Lindbergh. illus. ports. (Aviation, Highland, N. Y., 1927, v. 22, p. 1216–1217.) **VDS**

3412. The **"St. Louis** robin" and its refueling flight. How the plane and engine were equipped and conditioned for their test. illus. (Aviation, New York, 1929, v. 27, p. 269–271, 310.) **VDS**

Established an endurance record of 420 hours, 1929.

Airplane Flights, continued

3413. **Samson,** Charles Rumney. A flight from Cairo to Cape Town and back. [London:] E. Benn, Ltd. [1931.] x, 196 p. illus. map, ports. 8°. **VDY**

Royal Air Force airplane flight, 1927, commanded by the author. Reviewed in *Aeroplane*, London, 1931, v. 40, p. 106B, *VDS*.

3414. **Sassoon,** Sir Philip Albert Gustave David, 3rd bart. The third route. London: W. Heinemann, Ltd., 1929. viii, 291 p. illus. maps. 8°. **VDY**

Author's flight to air stations in Egypt, the Soudan, Transjordania, Irak, India, and Malta, 1928. Reviewed in *Air*, London, 1930, v. 3, p. 268, *VDS*; *Aeroplane*, London, 1929, v. 36, p. 733–734, 735, 738, 777–778, 780, 784, *VDS*.

3415. **Sayers,** W. H. The flying boat and air operations at sea. (Aeroplane, London, 1924, v. 26, p. 136.) **VDS**

Experiences of the British Air Ministry and of the Chilian Naval Air Service.

3416. —— The lesson of the Schneider cup race [of 1923]. (Aeroplane, London, 1923, v. 25, p. 340, 342, 344, 346–348, 350.) **VDS**

Also in *Aviation*, Highland, N. Y., 1923, v. 15, p. 577–580, *VDS*.

3417. —— On the King's cup race [1925]. illus. map. (Aeroplane, London, 1925, v. 29, p. 41–42, 44, 46, 50, 52, 54, 56, 58, 60, 62.) **VDS**

Won by F. L. Barnard.

3418. **Scaroni,** Silvio. Colonel de Pinedo's long list of achievements. port. (U. S. air services, Washington, 1927, v. 12, April, p. 20–22.) **VDS**

3419. **Schaerowitz,** Ernst. Drei Jahre deutschen Flugsports! Chronik der wichtigsten flugsportlichen Ereignisse in Deutschland 1909–1912, mit einer kurzen Vorgeschichte. Berlin: Richard Carl Schmidt, 1913. 100 p. 8°. **VDY**

Record of flights and competitions in Germany, 1909–1912.

3420. The **Schneider** contest [1929] in retrospect. illus. (Aviation, New York, 1929, v. 27, p. 667–669.) **VDS**

3421. The **Schneider** trophy. (Aviation year book, London, 1931, p. 148–154.) **VDS**

Includes a table of the winners of this famous airplane contest, 1913–1929.

3422. The **Schneider** trophy race, Venice, 1927. illus. (Flight, London, 1927, v. 19, p. 684a–684r.) **VDS**

3423. **Schroeder,** Rudolph William. Capt. Schroeder's story of altitude record. port. (Aerial age weekly, New York, 1918, v. 8, p. 314–315.) **VDS**

28,900 feet by the author, an American.

3424. —— How I made the altitude record of 28,900 feet. ports. (Flying, New York, 1918, v. 7, p. 906–907.) **VDS**

3425. **Scott,** C. L. By floating boat to India. Sir Philip Sassoon's tour in the Blackburn "Iris." illus. map. (Flight, London, 1929, v. 21, p. 380–383.) **VDS**

Account of flight in 1928.

3425a. **Scott,** Charles William Anderson. Scott's book; the life and Mildenhall-Melbourne flight of C. W. A. Scott, told by himself. London: Hodder & Stoughton, Ltd., 1934. 263 p. illus. map. 8°. **VDYC**

3426. **Scott,** W. H. The refueling endurance record. illus. (Aero digest, New York, 1929, v. 15, July, p. 67, 266, 268.) **VDS**

Reginald L. Robbins and James Kelly made record of 172 hours, 32 minutes, at Fort Worth, Texas, May, 1929.

3427. **Scott,** Wisner Gillette. Christofferson's flight over Mt. Whitney. An epoch in California's aviation history. (Ace, Glendale, Cal., 1925, v. 7, Aug. 1, p. 5–6, 13.) **VDS**

Flight of Silas Christofferson, June 25, 1914.

3428. **Scott** succeeds. Butler's time for England-Australia flight reduced by 5 hours, 42 minutes. port. (Flight, London, 1932, v. 24, p. 389.) **VDS**

Flight made in 8 days, 20 hours, 47 minutes, by C. W. A. Scott.

3429. A **Second** flight over Everest. (Aeroplane, London, 1933, v. 44, p. 712.) **VDS**

Flight made on April 19, 1933, by the British Mt. Everest expedition.

3430. The **Second** great hop. illus. (Popular aviation, Chicago, 1927, v. 1, Aug., p. 10–12.) **VDS**

Brief account of Chamberlin-Levine flight, New York-Germany, 1927.

3431. The **Second** Hawaiian hop. illus. (Popular aviation, Chicago, 1927, v. 1, Sept., p. 9–10.) **VDS**

Smith-Bronte flight, 1927.

3432. **Shell,** first in aviation. (Popular flying, London, 1934, v. 3, p. 271.) **VDS**

List of great flights, 1909–1933, in which Shell fuel and oil were used.

3433. The **Ship** that crossed the continent from dawn to dusk. Curtiss PW8 pursuit plane — its characteristics and history. illus. (U. S. air services, Washington, 1924, v. 9, Aug., p. 19–22.) **VDS**

Piloted by Lieutenant Russell L. Maughan, 1924.

3434. **Ship-to-shore** flying. illus. (Aero digest, New York, 1927, v. 11, p. 330.) **VDS**

Clarence D. Chamberlin takes off from the steamship *Leviathan*, 1927.

3435. **Shoemaker,** James M. The Dornier Do. X. A flight experience and some remarks on the operation. illus. (Aviation, New York, 1929, v. 27, p. 1207–1209.) **VDS**

Account of author's flight over Lake Constance.

Airplane Flights, continued

3436. Sibley, Hi. California sets a pace in the air. illus. (Aero digest, New York, 1928, v. 13, p. 466, 614.) **VDS**

A history of activities in the state, beginning with the meet at Dominguez field, Los Angeles, in 1910.

3437. Silva, Louis de. The first flight to France. Seventeen years ago the Hon. C. S. Rolls made the first cross-Channel flight from Britain. (Airways, London, 1927, v. 3, p. 336–338.) **VDS**

3437a. The **Sky-crasher** peril of the Yellow bird. illus. (Literary digest, New York, 1929, v. 102, July 13, p. 48, 52, 54.) *** DA**

Flight of Assolant, Lotti, and Lefèvre, United States to Paris (1929), with account of the stowaway.

3438. Smedal, George. Alford J. Williams — speed king. illus. port. (Popular aviation, Chicago, 1931, v. 8, April, p. 31, 59.) **VDS**

"His death defying maneuvers in the air have done much to take the mystery out of flying and have been the means of discovering what a plane will do and why it does it."

3439. —— Jimmy Doolittle — air wizard. port. (Popular aviation, Chicago, 1931, v. 8, May, p. 23–24, 58.) **VDS**

Biography of prominent American post-war aviator.

3440. —— A tribute to Speed Holman. illus. ports. (Popular aviation, Chicago, 1931, v. 9, July, p. 21–22.) **VDS**

Biography of American aviator, Charles W. Holman, 1898–1931.

3441. Smith, Ernie. Ocean flights are the bunk. illus. (Popular aviation, Chicago, 1931, v. 8, Jan., p. 22–24, 58.) **VDS**

Tribulations of author's attempts to fly to Hawaii.

3442. Smith, Robert A. Temple hunting in Central America. illus. (Flight, London, 1932, v. 24, p. 3–6, 25–27.) **VDS**

3443. Smith, Sir Ross MacPherson. The first aeroplane voyage from England to Australia...with aeroviews by Capt. Frank Hurley. New South Wales edition... [Australia: Angus & Robertson, Ltd. [1921.] 32 p. illus. ports. obl. 12°. *** C p.v.1730, no.3**

3444. —— 14,000 miles through the air... illustrated with photographs. New York: The Macmillan Company, 1922. x, 136 p. illus. map. 8°. **VDY**

Account of the first flight from England to Australia.

3445. Smith-Bronte Hawaiian flight. illus. ports. (Aero digest, New York, 1927, v. 11, p. 130–131, 234.) **VDS**

Flight of Ernest L. Smith and Emory B. Bronte, Oakland, Cal., to Honolulu, in the *City of Oakland*, 1927.

3446. Soucek, Apollo. High boat. A tale of the realm above the earth in which no other American has trod. illus. port. (Aeronautic review, Baltimore, 1929, v. 7, July, p. 22–23.) **VDS**

Flight of author, 1929, establishing altitude record for seaplanes.

3447. —— My flight to 39,140 feet. port. (Aero digest, New York, 1929, v. 14, June, p. 52–53.) **VDS**

3447a. "Southern cross" makes a record flight. illus. (Flight, London, 1929, v. 21, p. 735–736.) **VDS**

Australia-London in less than 13 days, 1929 by Kingsford-Smith.

3448. Souvenir of the Schneider trophy contests. illus. (Flight, London, 1929, v. 21, p. 941–984.) **VDS**

Copiously illustrated accounts of the various contests, 1913–1927.

3449. Souvestre, Pierre, and AUGUST WIMILLE. La traversée de la Manche en aéroplane 1785–1909...avec une préface de L. Blériot. Paris: E. Flammarion [1909]. 314 p. illus. 12°. **VDY**

Account and chronological table (p. 36–37) of English Channel crossings, and a detailed account of Blériot's achievement.

3450. The **Spanish** transatlantic flight. illus. map. (Flight, London, 1926, v. 18, p. 59.) **VDS**

Franco's flight, Spain to Buenos Aires, 1926.

3451. Sprigg, Theodore Stanhope. Famous air trophies. Cups and prizes that have helped to make air history. illus. (Airways, London, 1927, v. 3, p. 189–191.) **VDS**

3452. Stack, T. Neville. By light aeroplane to India. (Royal Aeronautical Society, Journal, London, 1927, v. 31, p. 905–914.) **VDS**

Story of flight of Moth plane carrying the author and Mr. Leete in 1926. The enterprise was sponsored by the de Havilland Company.

3453. Standardization marks reliability tour. illus. (Aviation, Highland, N. Y., 1927, v. 23, p. 77–79.) **VDS**

Brief account of third tour — Detroit to Dallas, Texas, and return, 1927.

3454. The **Start** of the national air races [of 1926]. illus. (Aviation, Highland, N. Y., 1926, v. 21, p. 454–458.) **VDS**

Account of the first two days' racing.

3455. Stevens, Albert W. Exploring the valley of the Amazon in a hydroplane. Twelve thousand miles of flying over the world's greatest river and greatest forest to chart the unknown Parima river from the sky. illus. (National geographic magazine, Washington, 1926, v. 49, p. 353–420.) **KAA**

Hamilton Rice scientific expedition, 1924.

3456. Stokes, Allen. 46 years of aviation. illus. table. (Popular aviation, Chicago, 1933, v. 12, p. 297–298.) **VDS**

Airplane records, 1897–1911.

Airplane Flights, continued

3457. The **Story** of a great flight. port. (Aviation, Highland, N. Y., 1922, v. 13, p. 747.) **VDS**

Atlantic-Pacific one-stop flight by Lieut. James H. Doolittle in 1922.

3458. The **Story** of the Schneider cup seaplane race. How American aircraft piloted by Americans triumphed over the foreign competitors. illus. (Aviation, Highland, N. Y., 1923, v. 15, p. 574–576.) **VDS**

Race was held at Cowes, England, 1923.

3459. The **Story** of the trophy. The epic history of the Jacques Schneider trophy contest from 1913 to the present day. illus. (Air and airways, London, 1931, v. 8, p. 185–186.) **VDS**

3460. Streett, St. Clair. The first Alaskan air expedition. illus. map. (National geographic magazine, Washington, 1922, v. 41, p. 499–552.) **KAA**

United States Army Air Service flight of 1920. Copiously illustrated.

3461. Successful flights of Greene biplane. Machine does not infringe Wright patents. illus. (Fly, Philadelphia, 1910, v. 2, May, p. 3–4.) **3–† VDS**

Invention of Dr. William Greene, of Mineola, N. Y.

3462. Taylor, Charles E. The first transcontinental flight. Forty-nine day air journey recalls perseverance and courage of early flyers. illus. (Slipstream, Dayton, O., 1928, v. 9, June, p. 11–13.) **VDS**

Account of Calbraith Rodgers' flight in 1911, written by the Wright brothers' pioneer mechanic.

3463. Tennant, John Kingston. The first great transcontinental flight. (U. S. air services, Washington, D. C., 1928, v. 13, p. 28–29.) **VDS**

Made by Calbraith P. Rodgers in 1911, time 49 days.

3464. Terse biographies of leading aviators. ports. (Aero, Chicago, 1912, v. 4, p. 15–20, 47.) **3–VDS**

Accounts of Phil O. Parmelee, Glenn L. Martin, Horace B. Wild, Otto W. Brodie, Matilde Moisant, Blanche Stewart Scott, Hillery Beachy, Harold Kantner, William Somerville, Howard Gill, Andrew Drew, Paul Studensky, Tom W. Benoist, Walter E. Johnson, and Antony H. Jannus.

3465. Thaden, H. V. Impressions of the 1924 national balloon race. illus. (Aviation, Highland, N. Y., 1924, v. 16, p. 510–511.) **VDS**

Start was from San Antonio, Texas.

3466. Thomas, Lowell Jackson. European skyways; the story of a tour of Europe by airplane. Boston and New York: Houghton Mifflin Company, 1927. xv, 524 p. illus. ports. 8°. **VDYC**

3467. —— The first world flight; being the personal narratives of Lowell Smith, Erik Nelson, Leigh Wade, Leslie Arnold, Henry Ogden, John Harding. Boston and New York: Houghton Mifflin Company [cop. 1925]. xxii, 328 p. illus. ports. 8°. **VDY**

Reviewed in *Aeroplane*, London, 1926, v. 31, p. 324, *VDS*.

3468. Thomassen à Thuessink van der Hoop, A. N. J. Door de lucht naar Indie... Amsterdam: Scheltens & Giltay, 1926. 342 p. illus. 2. ed. 8°. **VDY**

Amsterdam-Batavia airplane flight in 1924, with numerous aerial photographs.

3469. Thornhill, Gertrude. Italy to Brazil. 6,450-mile formation flight. illus. ports. (Aero digest, New York, 1931, v. 18, Feb., p. 44–45.) **VDS**

General Italo Balbo's flight, 1930/31.

3470. Through the historic Caribbean. Remarkable record made by the PN–7 [U. S.] Navy plane powered with Wright T–3 engines. illus. (Aero digest, New York, 1924, v. 5, p. 291–295, 299.) **VDS**

Flight of 1924.

3471. The **Tragic** end of Moisant and Hoxsey. How two famous aviators met their death. ports. (Scientific American, New York, 1911, v. 104, p. 43.) **VA**

Accidents of John B. Moisant, at New Orleans, and Archie Hoxsey, at Los Angeles, both on Dec. 31, 1910.

3472. The **Transatlantic** flight. illus. (Flight, London, 1919, v. 11, p. 802–805.) **VDS**

Flight of J. Alcock and A. Whitten Brown, Newfoundland to Ireland, in 1919, the first non-stop.

3473. The **Transatlantic** flight. Hawker and Grieve retrieved. illus. ports., maps. (Flight, London, 1919, v. 11, p. 694–697.) **VDS**

Brief story of unsuccessful flight in 1919.

3474. The **Transatlantic** flight successfully completed. illus. (Aerial age weekly, New York, 1919, v. 9, p. 630–635.) **VDS**

Flight of the *NC4* in 1919.

3475. Transatlantic flights. illus. (Air, London, 1930, v. 3, p. 194–195.) **VDS**

Tabular data of 14 flights.

3476. Transocean fliers making history. illus. maps. (New York Times Current history, New York, 1927, v. 26, p. 741–752.) **BAA**

Byrd transatlantic flight. The flight of Hegenberger and Maitland to Honolulu.

3477. Transoceanic ventures. (United States. —Commerce Department, Aeronautics Branch. Air commerce bulletin, Washington, 1929, v. 1, July 15, p. 12–14.) **VDS**

Lists flights, both airships and airplanes, 1873 to 1929.

Airplane Flights, continued

3478. La **Traversée** des Alpes. illus. port. (L'Aérophile, Paris, 1910, année 18, p. 438–439.) 3–VDS

Brief account of flight and death of Georges Chavez, 1910.

3479. **Tucker,** Ray A. Honor Calbraith Perry Rodgers. illus. (Aviation, Highland, N. Y., 1926, v. 21, p. 915–916.) VDS

Rodgers was the first aviator to fly from the Atlantic ocean to the Pacific. Rodgers field, at Pittsburgh, is named for him.

3480. **Twenty** years after. Louis Blériot flies the Channel again. illus. (Flight, London, 1929, v. 21, p. 815–817.) VDS

3481. **Two** tough *hombres* in a second-hand "crate." ports. illus. (Literary digest, New York, 1929, v. 102, Aug. 3, p. 41, 44, 46, 48.) * DA

Refueling flight of Loren Mendell and R. B. Reinhart, 1929. Their record was 246 hours, 43 minutes, 2 seconds.

3482. **Uncle Sam's** second flying hop across the pond. illus. ports. (Literary digest, New York, 1927, v. 93, June 25, p. 37–38, 42, 44, 46, 48.) * DA

Flight of Clarence D. Chamberlin and Charles A. Levine, 1927.

3482a. **United States.** — Navigation Bureau (Navy Department). Morale; the navy's trans-Atlantic flight, United States Navy, 1919. [New York: Press U. S. Navy Recruiting Bureau, 1919.] 16 l. illus. obl. 24°. * C p.v.1652, no.10

Consists of brief text to accompany "sketches and paintings by Lieut. Comdr. Henry Reuterdahl...and Lieut. C. E. Ruttan."

3483. **United States.** — State Department. The flight of Captain Charles A. Lindbergh from New York to Paris, May 20–21, 1927, as compiled from the official records of the Department of State. Washington: Gov. Prtg. Off., 1927. 43 p. 8°. VDY p.v.42, no.2

3484. The **United States** wins the Schneider cup race. illus. ports. (Aviation, Highland, N. Y., 1923, v. 15, p. 436–437.) VDS

Race was held at Cowes, England, Sept. 28, 1923.

3485. **Van Every,** Dale, and M. D. Tracy. Charles Lindbergh, his life; introduction by Karl A. Bickel... New York: D. Appleton and Company, 1927. vi, 235 p. illus. ports. 12°. AN

3486. **Van Zandt,** John Parker. Flying from Spain to Brazil. ports. (U. S. air service, Washington, 1923, v. 8, May, p. 2–29.) VDS

Flight of Admiral Gago Coutinho and Captain Saccadura, both of the Portuguese navy, 1922.

3487. —— Greatest air meet in history of aviation. illus. (U. S. air service, Washington, 1923, v. 8, Nov., p. 25–28.) VDS

International air races, St. Louis, 1923.

3488. **Vaughan-Fowler,** R. London to China by light plane. "Cirrus-Avian's" 9,800 mile flight. illus. (Flight, London, 1929, v. 21, p. 545.) VDS

Flight, 1929, by Wen Lin Chen and Christian Johannsen.

3489. **Von Gronau's** world flight. map. (Flight, London, 1932, v. 24, p. 1150.) VDS

Flight in the Dornier flying boat *Grönlandwal,* 1932.

3490. **W.,** R. C. The Hanoi-Paris flight of Codos and Robida. illus. (Flight, London, 1932, v. 24, p. 107.) VDS

Indo-China to Paris in 3 days, 5 hours, and 40 minutes, 1932.

3491. **W.,** V. G. On our international challenge. illus. (Aeroplane, London, 1929, v. 37, p. 497–506.) VDS

Account of 1929 Challenge internationale de tourisme.

3492. **Wait,** William, jr. Development of high speed racing. illus. (Aviation, New York, 1929, v. 27, p. 428–431.) VDS

3493. —— The value of air races. illus. (Aero digest, New York, 1931, v. 18, p. 46–48, 132, 134, 136.) VDS

Effect on the development of speed.

3494. **Walmsley,** Leo. An airman's experiences in East Africa. (Blackwood's magazine, New York, 1919, v. 206, p. 633–652, 788–810; 1920, v. 207, p. 53–69, 189–209.) * DA

Author's experiences as a member of the Royal Flying Corps during the World war.

3495. **Walton,** Francis D. The flight of the "Southern cross" from California to Australia. map, ports. (Aero digest, New York, 1928, v. 13, p. 36–37.) VDS

Kingsford-Smith's flight of 1928.

3496. **Wead,** Frank Wilbur. Airplane racing. illus. (Popular aviation, Chicago, 1928, v. 2, May, p. 9–13, 86.) VDS

History of the Schneider seaplane trophy race and of the Pulitzer trophy race for landplanes.

3497. A **Week** in the sky. illus. (Literary digest, New York, 1929, v. 100, Jan. 19, p. 5–6.) * DA

Refueling flight of the *Question mark,* January, 1929.

3498. **Weems,** Philip Van Horn. The flight of the "Tingmissartoq." The authoritative account of the 1933 survey flights of the Lindberghs. illus. (Aviation, New York, 1934, v. 33, p. 33–36, 102–105.) VDS

Account of 30,000 mile trip embracing 21 countries.

3499. **Wells,** Linton. Around the world in twenty-eight days, with an introduction by

Airplane Flights, continued

Vilhjalmur Stefansson. Boston and New York: Houghton Mifflin Company, 1926. xxiv, 276 p. illus. ports. 8°. **KBG**

Voyage by the author and Edward S. Evans from New York in 1926, using aircraft over a large part of the route.

3499a. **Westervelt,** George Conrad, and others. The triumph of the N. C.'s, by Commander G. C. Westervelt, Commander H. C. Richardson and Lieut. Commander A. C. Read. Garden City, N. Y.: Doubleday, Page & Co., 1920. x, 308 p. illus. 8°. **VDY**

3500. **Whittaker,** W. E. de B. In memoriam — Hubert Latham, 1883–1912. (Aeroplane, London, 1912, v. 3, p. 78, 80.) **VDS**

Biography of pioneer British aviator.

3501. —— Sir John Alcock, K. B. E., 1892–1919. port. (Aeroplane, London, 1919, v. 17, p. 2050.) **VDS**

Brief biography of prominent British aviator who, with A. Whitten Brown, flew across the Atlantic ocean in 1919.

3502. **Wien,** Noel. Alaskan flying. illus. map. (Aero digest, New York, 1926, v. 9, p. 354–355, 409.) **VDS**

3503. —— Flying experiences in Alaska. (Aviation, Highland, N. Y., 1926, v. 20, p. 557–558.) **VDS**

Interesting account of pioneer work beginning in 1924.

3504. **Wiley** Post first to solo around the world. Also makes a new record for speed. illus. port. (U. S. air services, Washington, D. C., 1933, v. 18, Aug., p. 21–23.) **VDS**

3505. **Willard,** Charles F. Flutterings of an early birdling. illus. (Popular aviation, Chicago, 1932, v. 10, p. 95–96, 120, 129.) **VDS**

It is stated that the author was the first aviator to be taught by Glenn Curtiss.

3506. **Williams,** Alford Joseph. How I won the Pulitzer race, making an average speed of 4.05 miles per minute and 243.67 miles per hour. ports. (Aeronautical digest, New York, 1923, v. 3, p. 315–316.) **VDS**

Race of 1923.

3507. **Wines,** James P. Nine years of national air racing. illus. (Aviation, New York, 1929, v. 27, p. 404–416.) **VDS**

History of great series of American races, beginning at Selfridge field in 1921.

3508. The **Winning** flight of the "June bug" aeroplane for the Scientific American contest. illus. (Scientific American, New York, 1908, v. 99, p. 45.) **VA**

Flight by Glenn H. Curtiss, July 4, 1908 at Hammondsport, N. Y.

3509. **Wood,** George F. Campbell. History of the Michelin cup. illus. (Aircraft, New York, 1911, v. 1, p. 442–445.) **VDS**

3510. —— The International Aviation Meet. Belmont Park, New York, October 22d–30th [1910]. illus. (Aircraft, New York, 1910, v. 1, p. 319–323, 353–363.) **VDS**

Brief references to earlier meetings which led up to the above event. Tabular results.

3511. **Wood,** H. C. The 1934 Deutsch de la Meurthe cup race. illus. (National aeronautic magazine, Washington, 1934, v. 12, July, p. 8–9.) **VDS**

Won by Maurice Arnoux.

3512. **Woodhouse,** Henry. The evolution of great aeroplane circuits. illus. (Aero Club of America, Bulletin, New York, 1912, v. 1, June, p. 7–11.) **VDS (Flying)**

The above bulletin is an early number of the periodical subsequently known as *Flying*.

3513. —— The world's record men of 1910–1911. ports. (Air-scout, New York, 1911, v. 1, no. 5, p. 22–25.) **3–VDS**

Henry Farman, Maurice Tabuteau, Alfred Leblanc, and Arch Hoxsey.

3514. **"Woolaroc"** wins Dole race. illus. (Popular aviation, Chicago, 1927, v. 1, Oct., p. 5–8.) **VDS**

Flight of Art Goebel and William V. Davis, 1927, California to Hawaii.

3515. **World** records made at 1932 National Air Races. illus. (U. S. air services, Washington, 1932, v. 17, Oct., p. 9–14.) **VDS**

3516. The **World's** greatest aerial taxi-driver. port. (Literary digest, New York, 1926, v. 91, Dec. 4, p. 73–74.) *** DA**

Account of the flights of Sir Alan Cobham.

3517. **World's** height record. (Aeronautics, London, 1916, v. 10, p. 287.) **VDS**

23,000 feet by Harry G. Hawker, 1916.

3518. The **World's** records in F. A. I. categories and who holds them. (National aeronautic magazine, Washington, 1934, v. 12, July, p. 10–11.) **VDS**

Heavier-than-air records as established by the Fédération aéronautique internationale.

3519. **Wright** and Curtiss fly in Hudson-Fulton celebration. illus. (Aeronautics, New York, 1909, v. 5, Nov., p. 178–182.) **VDS**

3520. **Year's** [1913] progress reviewed shows step forward. chart. (Aero and hydro, Chicago, 1914, v. 7, p. 189–191.) **3–VDS**

Brief record of the year, with a chart showing distance, speed, duration, and altitude records, 1907–1914.

3521. **Zusammenstellung** der Rekordflüge von 1907 bis 1912. diagr. (Zeitschrift für Flugtechnik, München, 1913, Jahrg. 4, p. 47.) **VDS**

Chart shows airplane records for distance, duration, altitude, and speed, 1907–1912.

MILITARY AERONAUTICS

3522. **Adams,** Arthur A. The equipment of air forces. The Union of Soviet Socialist Republics. illus. (Aviation, New York, 1933, v. 32, p. 237–240.) **VDS**

Post-war developments in airplanes.

3523. **Ader,** Clément Agnès. L'aviation militaire. Paris: Berger-Levrault, 1909. xl, 161 p. illus. 12°. **VDW**

Describes his proposed military airplane. Library also has Paris editions of 1911 and 1916, *VDY.*

3524. **Admiral** Moffett reappointed. port. (Aviation, Highland, N. Y., 1925, v. 18, p. 325.) **VDS**

Brief biography and portrait of Chief of Aeronautics Bureau, U. S. Navy. He was killed in the *Akron* disaster of 1933.

3525. **Aeronautical** Board is reorganized. Secretaries of War and Navy agree on new functions of Board. (Aviation, Highland, N. Y., 1924, v. 17, p. 807.) **VDS**

Changes of the joint board designed to prevent duplication and to control development work.

3526. An **Aeroplane** machine gun that fires between the revolving propeller blades. illus. (Scientific American, New York, 1916, v. 115, p. 587.) **VA**

3527. L'**Aérostation** militaire en Allemagne. illus. (Revue militaire, Paris, 1908–09, v. 72, p. 497–520; v. 73, p. 28–50.) **VWA**

Development of Parseval, Gross, and Zeppelin airships.

3528. Les **Aérostats** militaires au Tonkin. map. (Revue de l'aéronautique, Paris, 1888, année 1, p. 64–71.) **3–† VDS**

Captive balloon operations in French-Chinese war.

3529. The **Air** demonstrates its command of the sea. The battleship bombing and the conference on the limitation of armament. illus. (Aircraft yearbook, New York, 1922, p. 45–59.) **VDS**

Accounts and pictures of the bombing of former German warships in 1921.

3530. **Aircraft** in warfare. (Aeronautic review, Washington, D. C., 1926, v. 4, p. 95.) **VDS**

Chronology, 1794–1914.

3531. The **Airplane-battleship** rumpus. illus. (Literary digest, New York, 1925, v. 84, March 7, p. 8–9.) *** DA**

Discussion of General William Mitchell's attacks on government's aeronautical policies.

3532. **American** military and naval aeronautics. (Aviation, New York, 1930, v. 28, p. 596–599.) **VDS**

Charts show expenditures, flying hours, research appropriations, accidents, numbers of airplanes, and numbers of qualified pilots, for the years 1921–1929.

3533. **American** military and naval development. charts. (Aviation, New York, 1929, v. 27, p. 712–713.) **VDS**

3534. **America's** drooping wings. illus. (Literary digest, New York, 1925, v. 84, Jan. 10, p. 14–15.) *** DA**

Backward condition of U. S. military aeronautics.

3535. **Andrews,** Solomon. Aerial navigation and a proposal to form an aerial navigation company. [New York: J. F. Trow,] 1865. 32 p. 8°. **VDY p.v.26, no.4**

The author presents evidence of the success of his airship. His unsuccessful efforts to have it accepted by the United States government during the Civil War.

3536. **Archen.** Les ballons dirigeables. Leur emploi comme arme offensive et les conditions du tir dirigé contre eux. Paris: R. Chapelot et Cie., 1908. 40 p. 8°. **VDW**

See also *Journal des sciences militaires,* Paris, 1908, série 12, v. 4, p. 404–414; v. 5, p. 45–57, 169–184, *VWA.*

3537. **Army** dirigible airships. (Scientific American supplement, New York, 1907, v. 64, p. 408.) **VA**

Specifications published by the Signal Corps of the United States Army, 1907.

3538. The **Army** and Navy carry on. illus. (Aviation, New York, 1933, v. 32, p. 82–83.) **VDS**

History of appropriations for United States forces shown by graphs.

3539. **Army's** first war plane. illus. (Aeronautics, New York, 1914, v. 15, Oct. 30, p. 117.) **VDS**

Brief description and drawings of the Burgess-Dunne no. 3, for the United States Army.

3539a. **Arnold,** Henry H. Wars have forced major developments in U. S. airplane design. illus. (Automotive industries, New York, 1925, v. 53, p. 297–300.) **TOL**

3540. **Ascent** of a squadron of war balloons in Russia. (Aeronautical journal, London, 1897, v. 1, April, p. 11–12.) **VDS**

3540a. **Aubé.** L'aviation de bombardement de 1919 à 1930. illus. (In: L'Aéronautique militaire, maritime, coloniale et marchande, Paris, 1931, p. 38–43.) **† VDY**

Development of ideas since the end of the World war.

3541. **Baden-Powell,** Baden Fletcher Smyth. Military ballooning. illus. (Royal United Service Institution, Journal, London, 1883, v. 27, p. 735–756.) **VWA**

A good historical review precedes a discussion of success, the relative advantages of captive and free balloons, the means of inflation, and the question of navigating balloons. Abridged (without the illustrations) in *Scientific American supplement,* New York, 1883, v. 16, p. 6339–6340, *VA.*

3542. —— The war balloon in South Africa. (Aeronautical journal, London, 1902, v. 6, p. 14–15.) **VDS**

3543. Les **Ballons** militaires et les dirigeables en Allemagne. (Nouvelle revue, Paris, 1909, série 3, v. 11, p. 145–163.) *** DM**

Military Aeronautics, continued

3544. The **Balloon** in modern warfare. (Scientific American supplement, New York, 1900, v. 50, p. 20800.) **VA**
Brief account of developments in various countries.

3545. **Bannerman,** Sir Alexander. The difficulty of aerial attack. (Royal United Service Institution, Journal, London, 1909, v. 53, p. 638–645.) **VWA**
Deals mainly with airship attack.

3546. **Bannerman-Phillips,** H. Achievements of military aircraft. Lessons taught by the European maneuvers and by the Tripolitan and Balkan campaigns. illus. (Scientific American, New York, 1913, v. 108, p. 306–307, 323.) **VA**

3547. **Barlow,** Lester P. Flying for Villa in Mexico. illus. port. (National Aeronautic Association review, Washington, 1926, v. 4, p. 51–52.) **VDS**
Brief account of adventures of aviator on Villa's staff in 1914.

3548. **Barnes,** G. M. Anti-aircraft ordnance — yesterday, today and tomorrow. illus. (Army ordnance, Washington, D. C., 1925, v. 6, p. 85–92.) **VWA**

3549. **Barrington-Kennett,** B. H. Military aeronautics in France. (Royal United Service Institution, Journal, London, 1912, v. 56, p. 171–178.) **VWA**
Describes organization, equipment, personnel, and service conditions of the French air service, with an account of the military airplane trials near Rheims.

3550. **Bayles,** J. C. Floating and flying navies; the military value of aerial navigation. (Cassier's magazine, New York, 1908, v. 35, p. 263–274.) **VDA**

3551. **Beaumont,** Frederick F. E. Balloon reconnaissance. (Royal United Service Institution, Journal, London, 1865, v. 8, p. 52–66.) **VWA**
Early history; also author's experiences during the American Civil war.

3552. —— On balloon reconnoissances as practised by the American army. (Royal Engineer Corps, Papers, Woolwich, 1863, new series, v. 12, p. 71–86.) **VDA (Gt. Britain, Army)**
Experiences of author during the American Civil war.

3553. **Berger.** Flugschiffabwehrartillerie. illus. (Schweizerische militärische Blätter, Frauenfeld, 1909, Jahrg. 45, p. 329–342.) **VWA**

3554. **Berlepsch,** Franz von. Bemerkungen über Ballonverwendung im Feldkriege — auf Grund von Manövererfahrungen. (Streffleurs militärische Zeitschrift, Wien, 1909, Jahrg. 50, Bd. 1, p. 417–426.) **VWA**

3555. **Bertrand,** Charles. État actuel de l'aéronautique militaire & navale en France et à l'étranger. Paris: Librairie aéronautique [1913]. 135 p. 8°. **VDY**
Status of the air equipment of the various European countries in 1913.

3556. **Besançon,** Georges. Commandant Bouttieaux. port. (L'Aérophile, Paris, 1907, année 15, p. 173.) **3–VDS**
Portrait and brief account of the career of French officer who had much to do with the development of military aviation. He was also a scientific author of distinction.

3557. **Béthuys,** G. Art militaire. Les ballons en Abyssinie. illus. (Le Génie civil, Paris, 1888, tome 12, p. 401–403.) **VA**
Describes and illustrates portable captive-balloon apparatus invented by Gabriel Yon for use in the Italian Abyssinian campaign. See also *Scientific American supplement,* New York, 1888, v. 25, p. 10135, *VA.*

3558. [**Bibliography** of military aeronautics.] (In: Paul Brockett, Bibliography of aeronautics, Washington, 1910, p. 582; continued in United States. — National Advisory Committee for Aeronautics, Bibliography of aeronautics, 1909–1931, Washington, 1921–35, volumes for 1909/16, p. 876–898; 1917/19, p. 301–311; 1920/21, p. 277–279; 1922, p. 97–98; 1923, p. 86–87; 1924, p. 69–70; 1925, p. 111–112; 1926, p. 90–91; 1927, p. 113; 1928, p. 126–128; 1929, p. 145–146; 1930, p. 153–156; 1931, p. 182–184.) **VDT**

3559. [**Bibliography** of naval aeronautics.] (United States. — National Advisory Committee for Aeronautics, Bibliography of aeronautics, 1909–1931, Washington, 1921–35, volumes for 1909/16, p. 942–944; 1917/19, p. 327; 1920/21, p. 295–296; 1922, p. 114–115; 1923, p. 98–99; 1924, p. 79–80; 1925, p. 127; 1926, p. 105; 1927, p. 130; 1928, p. 145.) **VDT**

3560. [**Bibliography** of the Royal Air Force.] (United States. — National Advisory Committee for Aeronautics, Bibliography of aeronautics, 1909–1931, Washington, 1921–35, volumes for 1917/19, p. 387–389; 1920/21, p. 357–358; 1922, p. 137–138; 1923, p. 117–118; 1924, p. 95–96; 1925, p. 151; 1926, p. 125; 1927, p. 150; 1928, p. 172; 1929, p. 194; 1930, p. 209; 1931, p. 243.) **VDT**

3561. [**Bibliography** of the United States Army Signal Corps.] (United States. — National Advisory Committee for Aeronautics, Bibliography of aeronautics, 1909–1931, Washington, 1921–35, volumes for 1909/16, p. 1225–1226; 1917/19, p. 406; 1920/21, p. 375; 1922, p. 143.) **VDT**

3562. **Blessing,** Arthur Reed. The development of air power policy. illus. (Aerial age, New York, 1923, v. 16, p. 165–168, 181.) **VDS**

3563. **Blind,** Karl. The siege of Paris and the airships. (North American review, New York, 1898, v. 166, p. 474–490.) *** DA**

3564. **Bombing** raid on Eastern cities. Sham attacks for practice on New York, Washington and other cities. (Aviation, Highland, N. Y., 1921, v. 11, p. 160–161.) **VDS**
Operations were in July, 1921.

Military Aeronautics, continued

3565. **Bombing** tests of the U. S. S. Alabama. illus. (Aviation, Highland, N. Y., 1921, v. 11, p. 396–397.) **VDS**
Account of tests, 1921.

3566. **Bombing** of warships proves air power. Martin bombers of Army Air Service sink destroyer G–102 and cruiser Frankfurt. illus. (Aviation, Highland, N. Y., 1921, v. 11, p. 96–98.) **VDS**
Account of operations during July, 1921.

3567. **Boone,** Andrew R. Reconnoitering over Nicaragua, as told in an interview with Capt. H. Denny Campbell, U. S. Marine Corps and winner of the Schiff trophy for safe flying. ports. (Aviation, Highland, N. Y., 1927, v. 23, p. 14–15, 28.) **VDS**
Thrilling experiences during the Nicaraguan rebellion.

3568. **Boynton,** Carroll A. Ballooning for the Federals. illus. (Sportsman pilot, Albany, N. Y., 1931, v. 5, Feb., p. 17, 44–45.) **VDS**
Experiences of T. S. C. Lowe during the American Civil war.

3569. **Brant,** Gerald. What really happened in Hawaiian maneuvers. (Aviation, Highland, N. Y., 1925, v. 19, p. 166–168.) **VDS**
Account of sham air operations against the Blue fleet in 1925.

3570. The **British** army aeroplane. illus. (Scientific American, New York, 1909, v. 100, p. 96.) **VA**
Remodeled airplane of S. F. Cody.

3571. **British** and foreign aircraft. Progress in naval aeronautics. (Naval annual, London, 1914, p. 134–149.) **VXA**
Developments in various countries during 1913.

3572. **Bruce,** Eric Henry Stuart. The aeroplane and its use in war. (United empire, London, 1916, new series, v. 7, p. 462–473.)
With discussion. **CX**
How the colonies helped Great Britain during the World war. Brief history of the airplane. German airplanes.

3573. —— Portable electric balloon signalling in war. (Aeronautical journal, London, 1900, v. 4, p. 148–150.) **VDS**

3574. —— The use of balloons in war. (Society of Arts, Journal, London, 1902, v. 50, p. 277–285.) **VA**
With discussion.
Deals mainly with use in the Boer war.

3575. —— War and ballooning. (Nature, London, 1887, v. 35, p. 259–262.) **OA**
Early history of the balloon. Capt. Renard's airship *La France.* Lhoste's floating brake. Author's proposal to use captive balloons for signal purposes.

3576. **Bryan,** John Randolph. Balloon used for scout duty. Terrible experiences of a Confederate officer who saw the enemy from dizzy heights. Rope cut as he ascended... (Southern Historical Society, Papers, Richmond, Va., 1905, v. 33, p. 32–42.) **IKLD**

3577. **Buchholtz,** Franz Hugo. Anforderung an Militär-Ballons und ihre Ausrüstung, sowie Bemerkungen über die Bedeutung der Luftschiffahrt für die Kriegsführung. illus. (Zeitschrift für Luftschiffahrt, Berlin, 1883, Jahrg. 2, p. 97–114.) **3–VDS**
Contains a tabular review of balloons used during the siege of Paris, 1870–1871, with a map showing where each landed.

3578. —— Die Geburt und erste Kindheit der preussischen Militär-Luftschiffer-Abtheilung. ports. (Illustrierte aëronautische Mitteilungen, Strassburg, 1900, Jahrg. 4, p. 106–110.) **VDS (Deutsche)**
Establishment of the Prussian Military Aeronautics Branch (1883). Good history of German military aeronautics.

3579. —— Der gegenwärtige Zustand der militärischen Aëronautik. (Zeitschrift für Luftschiffahrt, Berlin, 1883, Jahrg. 2, p. 65–78.) **3–VDS**
Results of contemporary experiments in France and England.
For list of Buchholtz's writings see Paul Brockett, *Bibliography of aeronautics,* Washington, 1910, p. 162, *VDT.*

3580. **Bülow,** Hilmer, Freiherr von. Geschichte der Luftwaffe; eine kurze Darstellung der Entwicklung der fünften Waffe mit einem Geleitwort des Reichsluftfahrt-Ministers Göring... Frankfurt am Main: Moritz Diesterweg, 1934. vii, 176 p. illus. ports. 8°. **VDYD**
History of military aeronautics, especially of Germany, during the twentieth century.

3581. **Bumpus,** F. A. The development of torpedo-carrying aircraft. diagrs. (Royal Aeronautical Society, Journal, London, 1928, v. 32, p. 206–226.) **VDS**
With discussion.
Good historical outline, p. 208–210. Abstract in *Flight,* London, 1927, v. 19, p. 872, *VDS.*

3582. **Capper,** Sir John Edward. Military ballooning. (Royal United Service Institution, Journal, London, 1906, v. 50, p. 890–909.) **VWA**
With discussion.
Discusses the use of captive balloons, kites, and free balloons; also the probable use of airships and airplanes.

3583. **Capt.** Henry C. Mustin dies. port. (Aviation, Highland, N. Y., 1923, v. 15, p. 218.) **VDS**
Brief account of work of United States naval officer, stated to be "father of naval aviation."

3584. **Chambers,** W. Irving. U. S. Navy's splendid aero record reviewed. (Aero and hydro, Chicago, 1914, v. 7, p. 228–229.)
Record for 1913. **3–VDS**

3585. **Chandler,** Charles de Forest. General [James] Allen: father of army aviation. (U. S. air services, Washington, D. C., 1933, v. 18, March, p. 11–12.) **VDS**
Served as Chief Signal Officer of the United States Army, 1906–1913.

3586. **Charlton,** Lionel Evelyn Oswald. War from the air; past, present, future. London, etc.: Thomas Nelson & Sons, Ltd. [1935.] vi, 183 p. 8°. **VDYD**
An attempt to describe "the effect on wars past, present, and to come of the advent of aircraft; from the time a hundred years ago, long before they were thought of, when their presence would have altered world events, to the time, soon to come, when they will control them."

Military Aeronautics, continued

3587. **Clarkson,** Coker Fifield. A pioneer in standardization. port. (Society of Automotive Engineers, Journal, New York, 1917, v. 1, p. 128–131.) **VDS**

Account of the work of Major Henry Souther (1865–1917), senior officer of the Aircraft Engineering Division of the U. S. Army Air Service.

3588. **Coontz,** Robert Edward. The wings of the fleet. illus. port. (Aeronautical digest, New York, 1923, v. 2, p. 75–77, 146–147.) **VDS**

United States naval developments from 1920.

3589. **Coupland,** R. C. Recent developments in aircraft armament. illus. (Aviation, New York, 1929, v. 26, p. 33–35, 47–48, 50, 52.) **VDS**

3590. **Crosara,** Leonardo. ...Gli aerostieri; notize storiche degli aerostieri militari dal 1793 al 1919. Roma: Stabilimento poligrafico editoriale romano [1924]. 155 p. illus. 4°. **VDY**

History of military aeronautics, especially during the World war.

3591. **Delaney,** Peter J. Telegraphing from a balloon in war. (Aeronautical journal, London, 1899, v. 3, p. 11–13.) **VDS**

Author's thrilling experience during the Spanish-American war.

3592. **Delmard,** Alaricus, and R. Blathwayt. European military ballooning. illus. (Pall Mall magazine, London, 1899, v. 17, p. 161–174.) *** DA**

A visit to the Krebs-Renard airship at Meudon. Experiences of the British Balloon Corps under Colonel Templer.

3593. The **Design** of weapons for attacking airships. (Engineer, London, 1909, v. 107, p. 505.) **VA**

3594. The **Development** of the military aeroplane. illus. (Aerial age weekly, New York, 1915–16, v. 2, p. 476–477.) **VDS**

Dates back to the Italian campaign in Tripoli. Twenty-nine small illustrations.

3595. **Development** of U. S. naval aviation. illus. (Aviation and aeronautical engineering, New York, 1919, v. 7, p. 424–428.) **VA**

3596. **Dickman,** Joseph Theodore. Balloons in war. A lecture delivered before the class of officers at the United States Infantry and Cavalry School, Fort Leavenworth, Kansas, March 27th, 1896... [Fort Leavenworth:] United States Infantry and Cavalry School, 1896. 14 p. 12°. **VDC p.v.10, no.8**

Brief history.

3597. **Dirigeable** balloon in warfare. (Scientific American, New York, 1907, v. 97, p. 286.) **VA**

Unencouraging editorial comment.

3598. **Engel,** Adolf. Ueber die militärische Verwendung der lenkbaren Ballons und der Aeroplane. (Mitteilungen über Gegenstände des Artillerie und Geniewesens, Wien, 1909, Jahrg. 1909, p. 369–374.) **VWI**

3599. The **Equipment** of air forces [Japan]. illus. (Aviation, New York, 1933, v. 32, p. 203–207.) **VDS**

Traces development of military aviation in Japan.

3600. **Espitallier,** Georges. Les ballons et leur emploi à la guerre. Paris: G. Masson [1889]. 92 p. illus. 12°. **VDY**

Account of development of balloons and airships for war purposes.

3601. **Estificief.** L'aérostation au Japon. (L'Aéronaute, Paris, 1904, année 37, p. 115.) **3–VDS**

Brief account of captive balloon for the Japanese army about 1890.

3602. The **Evolution** of the British naval airship. illus. (Engineering, London, 1919, v. 107, p. 757–759, 797–799.) **VDA**

Copiously illustrated.

3603. **Farcot,** Eugène. Voyage du ballon le Louis-Blanc. Paris: Le Chevalier, 1874. 80 p. illus. 12°. **VDC p.v.19, no.6**

Author's experiences during the Franco-Prussian war.

3604. **Farman,** T. F. Revolutionary rôle of aircraft in war. (Blackwood's magazine, New York, 1915, v. 198, p. 514–521.) *** DA**

3605. **Fechet,** James Edmond. The development of military aircraft from racing types. illus. (Aero digest, New York, 1923, v. 23, July, p. 15, 51.) **VDS**

3606. —— . Navy torpedo and bombing planes. illus. (Aero digest, New York, 1933, v. 22, May, p. 14–15.) **VDS**

Story of use in the United States Navy begins in 1917.

3607. —— Progress in military aviation. (Mechanical engineering, New York, 1929, v. 51, p. 11–12.) **VFA**

Developments during 1928.

3608. **Fingerhuth.** Die Luftschiffertruppe der I. französischen Republik. illus. (Illustrierte aëronautische Mitteilungen, Strassburg, 1898, Jahrg. 2, p. 33–37, 75–79.) **VDS**

3609. **First** Aero Squadron to Mexico. illus. (Aerial age weekly, New York, 1916, v. 3, p. 24–25.) **VDS**

Sent in pursuit of the rebel Villa and commanded by Benjamin D. Foulois.

3610. The **First** 3½ years of the 5 year expansion program. (U. S. air services, Washington, 1931, v. 16, p. 29–30.) **VDS**

Progress made by the United States Army Air Corps.

3611. **Fischer-Treuenfeld,** Richard Friedrich Eberhard von. Kriegs-Telegraphie. Geschichtliche Entwickelung, Wirkungskreis und Organisation derselben. Stuttgart: W. Kitzinger, 1879. ix, 374 p. illus. 8°. **TTC**

Modern development of military ballooning in North America, Brazil, France, and England, p. 172–193.

3612. **Fiske,** Bradley Allen. Defending America with torpedoplanes. illus. (Popular

Military Aeronautics, continued

science monthly, New York, 1917, v. 90, p. 690–692.) * DA
Author describes his invention for launching torpedoes from airplanes.

3613. **Fitzmaurice,** James C. The Irish Air Force. illus. (Aero digest, New York, 1926, v. 9, p. 190.) VDS
Organized in December, 1921.

3614. **Flores Alvarez,** Enrique. Historia de la aviación en Chile. [Santiago de Chile,] 1933–34. 2 v. 8°. VDY

3615. The **Flying** services of Denmark. illus. ports. (Aeroplane, London, 1932, v. 43, p. 909–910, 912, 914, 916–919.) VDS
Includes early history of the Danish military and naval flying forces.

3616. **Fonvielle,** Wilfrid de. De l'aérostation militaire. Paris: Auguste Ghio, 1876. 16 p. 8°. VDY p.v.7, no.5
Difficulties to be overcome, with suggestions.

3617. —— Les aérostats militaires en Angleterre. illus. (L'Aéronaute, Paris, 1879, année 12, p. 179–185.) 3–VDS
Describes organization and operations at the Woolwich arsenal.

3618. —— Balloon ascents for military purposes. (Nature, London, 1870, v. 3, p. 115–116, 134–135, 175–176.) OA
Experiences of the Franco-Prussian war, with criticism of Dupuy de Lôme's airship.

3619. —— Éloge de Dagron. (L'Aéronaute, Paris, 1900, année 33, p. 152–161.) 3–VDS
Dagron distinguished himself in the Franco-Prussian war by inventing a microphotographic means of sending long dispatches by pigeon post.

3620. —— How I came out of Paris in a balloon. (Temple bar, London, 1871, v. 31, p. 166–179.) * DA
Experience during the Franco-Prussian war.

3621. —— [Prudent René Patrice Dagron.] port. (L'Aérophile, Paris, 1900, année 8, p. 82–84.) 3–VDS

3622. The **French** captive war balloon. Its use at Casablanca. illus. (Scientific American supplement, New York, 1907, v. 64, p. 305, 312.) VA

3623. **G.,** C. von. Japanische Militärluftschiffahrt während der Belagerung von Port Arthur. illus. (Illustrierte aëronautische Mitteilungen, Strassburg, 1905, Jahrg. 9, p. 301–303.) VDS

3624. **Gamble,** Charles Frederick Snowden. The air weapon; being some account of the growth of British military aeronautics from the beginnings in the year 1783 until the end of the year 1929... v. 1. London: Oxford University Press, 1931. table. 8°. VDY
Bibliographical footnotes. Volume 1. covers the period Nov., 1783 – Aug., 1914.
Reviewed in *Aviation,* New York, 1931, v. 30, p. 712, *VDS; Aeroplane,* London, 1931, v. 40, p. 412, 414, 416, *VDS.*

3625. **General** Fechet. Col. James E. Fechet appointed Assistant Chief of the [United States] Air Service. port. (Aviation, Highland, N. Y., 1925, v. 18, p. 290.) VDS
Portrait and brief biography.

3626. **General** Patrick appointed C. of A. S. port. (Aviation, Highland, N. Y., 1921, v. 11, p. 395.) VDS
Mason M. Patrick becomes Chief of the U. S. Air Service, 1921. Portrait.

3627. **Gerlach,** Edmund. Ueber die Möglichkeit, den Drachen zu Rekognoscirungen zu benutzen. illus. (Zeitschrift für Luftschiffahrt, Berlin, 1883, Jahrg. 2, p. 257–273, 306–312.) 3–VDS
Calculations for man-carrying kites to be used for military observation purposes.

3628. **Geschichtlich-Aktuelles** vom Militärballon. (Wiener Luftschiffer-Zeitung, Wien, 1904, Jahrg. 3, p. 223–224.) VDS
Account of captive balloon work in the Austro-Franco campaign of 1794, according to Baron Selle de Beauchamp.

3629. **Gill,** Napier John. The aerial arm; its functions and development. With an introduction by W. Sefton Brancker. London: "Aeroplane" & General Publishing Co., 1919. 168 p. illus. sq. 8°. VDY
See p. 32–45 for specifications of early airships of various countries. See chapter 5 for evolution of airplane types; chapter 6 for use of airplanes during the World war. Reviewed in *Aviation,* New York, 1920, v. 8, p. 27, *VDS.*

3630. **Gillmore,** William E. Military aviation. A survey of the development of engineering materials for aviation uses, the air-cooled engine, and bombing equipment. illus. (Mechanical engineering, New York, 1928, v. 50, p. 839–845.) VFA

3631. **Glassford,** William Alexander. The balloon in the Civil war. (Military Service Institution of the United States, Journal, New York, 1896, v. 18, p. 255–266.) VWA

3632. —— Our army and aerial warfare. (American magazine of aeronautics, New York, 1908, v. 2, Jan., p. 18–21.) VDS
Brief early history of work in the United States, including that of the Spanish-American war.

3633. **Gradenwitz,** Alfred. Captive balloons of the German army and navy. illus. (Scientific American, New York, 1907, v. 97, p. 30–31.) VA
Description of the Parseval kite balloon.

3634. **Greely,** Adolphus Washington. Balloons in war. illus. (Harper's monthly magazine, New York, 1900, v. 101, p. 33–50.) * DA
History covers France, Great Britain, and the United States (especially during the Civil war).

3635. **Grey,** Charles Grey. Military and naval aviation. (English review, London, 1912, v. 11, p. 99–114.) * DA
Status of the European powers. Lessons of the 1911 maneuvres. British requirements and costs. Airplanes and airships as weapons.

Military Aeronautics, continued

3636. —— On the growth of British air power. illus. (Aeroplane, London, 1933, v. 44, p. 781–782, 784, 786–787, 791–792, 794, 796, 798.) **VDS**

3637. Gross, Hans. Das Kriegs-Luftschiffer-Material (System Yon und Lachambre). illus. (Zeitschrift für Luftschiffahrt, Berlin, 1889, Jahrg. 8, p. 33–41, 88–100.) **3–VDS**

Describes equipment of captive military balloons of Yon and Lachambre. Deals especially with gas generators.

3638. —— Ueber die Entwicklung der Militär-Luftschiffahrt in den Jahren 1894 und 95. (Zeitschrift für Luftschiffahrt, Berlin, 1896, Jahrg. 15, p. 81–89.) **3–VDS**

Developments in various nations.

3639. Grover, G. E. On the uses of balloons in military operations. (Royal Engineer Corps, Papers, Woolwich, 1863, new series, v. 12, p. 71–86.) **VDA (Gt. Britain, Army)**

Includes a good early history.

3640. Guidoni, Alessandro. Early experiments in launching torpedoes from an aeroplane. illus. (Flying, New York, 1915, v. 4, p. 539–540.) **VDS**

Work of the author, a captain in the Italian navy.

3641. H., E. Die Luftschiffertruppe im deutschen Heere. (Schweizerische militärische Blätter, Frauenfeld, 1904, Bd. 40, p. 141–152.) **VWA**

History of German Air Corps' organization.

3642. Hagen, Freiherr vom. Geschichte der militärischen Aëronautik. (Zeitschrift für Luftschiffahrt, Berlin, 1882, Jahrg. 1, p. 321–336, 353–366; 1883, Jahrg. 2, p. 12–21, 33–50.) **3–VDS**

History of military aeronautics with a brief bibliography.

3643. Hamlin, Percy G. Aerial observation, Army of the Potomac. port. (U. S. air services, Washington, 1934, v. 19, July, p. 16–19.) **VDS**

Account of the work of T. S. C. Lowe during the American Civil war.

3644. Hanst, Charles E. Marine aviation in Santo Domingo. illus. (Popular aviation, Chicago, 1928, v. 3, Oct., p. 26–28, 120–121.) **VDS**

3645. Harmel, Falk. A history of [United States] army aviation. illus. (Popular aviation, Chicago, 1928, v. 3, Dec., p. 17–27, 114–115.) **VDS**

"The declaration of war with Germany on April 6, 1917, found the United States with an air force, if such it could be called, of 65 officers (35 of whom were flyers), 1,087 enlisted men and 55 airplanes."

3646. Harmon, Benjamin F. The past and future of defence against aircraft. (Coast artillery journal, Fortress Monroe, Va., 1925, v. 63, p. 449–460, 555–566.) **VWA**

3647. Hawthorne, H. L. Balloons and dirigibles in war. (U. S. artillery journal, Fort Monroe, Va., 1909, v. 32, p. 107–115.) **VWA**

Discusses advantages of captive balloons and airships; also defense against aircraft.

3648. Hergesell, Hugo. The balloon in warfare. (Forum, New York, 1898, v. 26, p. 1–13.) *** DA**

Early history of military aeronautics. Modern German developments.

3649. Hinterstoisser. Die Anfänge der österreichisch-ungarischen Luftschiffertruppe. illus. (Illustrierte aëronautische Mitteilungen, Strassburg, 1900, Jahrg. 4, p. 1–5.) **VDS**

3650. A **History** of the U. S. Army Air Corps. illus. (Aviation, New York, Highland, N. Y., 1926, v. 21, p. 170–173.) **VDS**

3651. A **History** of U. S. naval aviation. Aviation in the navy from its earliest days. illus. (Aviation, Highland, N. Y., 1926, v. 21, p. 176–179.) **VDS**

3652. Hureau de Villeneuve, Abel. Sur un gaz pourrait substituer à celui qu'aujourd'hui on emploie d'ordinaire pour gonfler les ballons. (Académie des sciences, Comptes rendus, Paris, 1870, tome 71, p. 767–768.) *** EO**

Proposes to obtain balloon gas from green wood during the siege of Paris.

3653. Ide, John Jay. The equipment of air forces. France. illus. (Aviation, New York, 1933, v. 32, p. 101–105, 173–176.) **VDS**

Development of French heavier-than-air military and naval craft since the World war.

3654. Important events in military aeronautics. (Aircraft year book, New York, 1919, p. 315–316.) **VDS**

Chronology, 1794–1918.

3655. Jacques, P. L'Allemagne organise son aviation militaire. (L'Aérophile, Paris, 1926, année 34, p. 73–74.) **3–VDS**

Extensive post-war preparations.

3656. Un **Joli** geste du Conseil municipal. illus. (L'Aérophile, Paris, 1914, année 22, p. 64–65.), **3–VDS**

Proposal to erect tomb for Louis Godard, who built balloons for France during the Franco-Prussian war. Picture of interior of the Godard factory.

3657. Jones, Ernest Lester. The 15th anniversary of our Air Service. illus. (Aviation, Highland, N. Y., 1924, v. 17, p. 830–831, 863–865.) **VDS**

Recalls the work of the Wright brothers and the letting of the first army airplane contract, 1908.

3658. —— Our air force in the Civil war. (National Aeronautic Association review, Washington, D. C., 1925, v. 3, p. 100–103, 111.) **VDS**

"To the persistency of Thaddeus S. C. Lowe and the vision of Secretary Henry of Smithsonian we may give credit for the first American air force."

Military Aeronautics, continued

3659. **Jones,** H. B. Military ballooning. (Royal United Service Institution, Journal, London, 1892, v. 36, p. 261–279.) **VWA**

Results reached in England, with brief statements regarding foreign armies.

3660. —— Note on a performance of the "Bristol" war balloon during the South African campaign. (Aeronautical journal, London, 1902, v. 6, p. 65.) **VDS**

3661. **Jordanoff,** Assen. The first fighting squadron. illus. port. (Sportsman pilot, Albany, N. Y., 1932, v. 7, Feb., p. 14–15, 44–45.) **VDS**

"A veteran of the boiling Balkans of 1912 recalls the hangar-life and heroics of the initial air outfit in action against an enemy."

3662. **K.** Die erste Verwendung des Luftballons zu militärischen Zwecken — 1792. (Oesterreichische Flug-Zeitschrift, Wien, 1913, Jahrg. 7, p. 567–568.) **VDS (Flug)**

Experiment in Spain, 1792.

3663. **King,** Ernest J. Naval aviation [for the year 1933–34]. (Mechanical engineering, New York, 1934, v. 56, p. 613–615.) **VFA**

The author, a rear-admiral of the United States Navy, reports progress for the fiscal year.

3664. **King,** W. A. de C. Aerial reconnaissance, its possible effect on strategy and tactics. (Royal Engineers journal, London, 1913, v. 18, p. 147–154.) **VDA**

3665. **Kingston-McCloughry,** Edward James. Air control. (Air annual of the British Empire, 1934/35, London, 1934, v. 6, p. 16–33.) **VDS**

How Great Britain subdued risings in India, Iraq, the Egyptian Sudan, and Aden.

3666. **Lachambre,** Henri. Parc aérostatique pour ballons captifs. illus. (Le Génie civil, Paris, 1888, tome 13, p. 398.) **VA**

Description and illustrations of the author's portable captive-balloon equipment.

3667. **Lanchester,** Frederick William. Aircraft in warfare: the dawn of the fourth arm. (Engineering, London, 1914, v. 98, p. 312–313, 338, 367–368, 392–393, 422–423, 452–454, 481–482, 513–514, 539–540, 568–569, 596–598, 626–627, 680–681, 710–711, 762–765.) **VDA**

A study of possibilities in offense and defense.
Also issued under same title, New York: D. Appleton & Co., 1917, 222 p. 8°. *VDYD.*

3668. **Létonné,** F. Les aérostiers militaires pendant les guerres de la Révolution. (Revue du génie militaire, Paris, 1903, tome 25, semestre 1, p. 389–424.) **VWA**

3669. **Liggett,** Walter W. Feeding the favorites. The true aviation conditions in official Washington unmasked. illus. (Popular aviation, Chicago, 1932, v. 10, p. 282–285, 322, 324.) **VDS**

3670. —— Why America is a fifth rate air power. illus. (Popular aviation, Chicago, 1932, v. 10, p. 212–215, 257.) **VDS**

3671. **London's** "doom" in the air. illus. (Literary digest, New York, 1928, v. 98, Sept. 8, p. 9.) *** DA**

Four-day mimic invasion and defense tests.

3672. **Lowe,** Thaddeus Sobieski Coulincourt. [Balloon operations during the War of the Rebellion.] (United States. — War Department, The War of the Rebellion, a compilation of the official records of the Union and Confederate armies, series 3, v. 3, Washington, 1899, p. 252–319.) **IKC**

A report consisting of official documents. Lowe was Chief Aeronaut of the United States Army.
For good contemporary illustrations of these operations see *New York illustrated news,* New York, 1862, v. 6, p. 21, 24, opp. p. 105; 136, ** DA.*

3673. Die **Luftschiffahrt** zu militärischen Zwecken. (Zeitschrift für Luftschiffahrt, Berlin, 1888, Jahrg. 7, p. 312–315, 342–345.) **3–VDS**

3674. **Lutz,** Earle. Balloons in battle. Lighter-than-air warcraft of 1794 and after. illus. (Sportsman pilot, New York, 1929, v. 2, July, p. 39, 64–65.) **VDS**

3675. **Lynch,** Arthur. Du rôle des ballons militaires anglais dans la guerre de l'Afrique du Sud. (L'Aéronaute, Paris, 1902, année 35, p. 104–110.) **3–VDS**

3676. **M.,** E. Ai margini della storia. Il bombardamento aereo di Venezia nel 1849. illus. (Rivista aeronautica, Roma, 1927, anno 3, Nov., p. 173–176.) **VDS**

Use of bomb-dropping balloons.

3677. —— Il primo lancio di bombe dall' aeroplano. (Rivista aeronautica, Roma, 1929, anno 5, p. 606–607.) **VDS**

States that first bomb dropped from an airplane was during the Libyan campaign of 1911.

3678. **McAlery,** C. M. The first ten years. illus. (Aeroplane, London, 1928, v. 34, p. 472–472B, 474–476, 485–486.) **VDS**

History of the Royal Air Force, 1918–1928.

3679. —— The Royal Air Force in 1926 [–1932]. (Aeroplane, London, 1927, v. 32, p. 6, 8; 1928, v. 34, p. 6; 1929, v. 36, p. 8–10; 1930, v. 38, p. 24, 26; 1931, v. 40, p. 12–15; 1932, v. 42, p. 14–16; 1933, v. 44, p. 17–20.) **VDS**

3680. —— The years between. illus. (Aeroplane, London, 1933, v. 44, p. 823–824, 826, 828–832, 834, 836, 838–840, 843–844.) **VDS**

Post-war history of the Royal Air Force.

Military Aeronautics, continued

3681. **Mangin,** Théodore. Mes souvenirs du siège de Paris, 1870–1871. Historique du voyage en ballon Le Général Bourbaki. (L'Aéronaute, Paris, 1905, année 38, p. 235–250.) 3-VDS

3682. **Marie,** Félix. Les origines de l'aéronautique militaire (novembre 1909 – novembre 1910). Paris: Charles Lavauzelle et Cie., 1924. 121 p. port. 8°. VDY p.v.34, no.8

French developments for 1909–1910.

3683. **Maxfield,** Joseph E. Report on war-ballooning at Santiago de Cuba [during the Spanish-American war]. (United States. — War Department: Signal Office, Report for 1897–1898. Washington, 1898, p. 88–94.) VWZZ

3684. **Maxim,** Hudson. The warfare in the future. (Science, New York, 1908, new series, v. 28, p. 820–831.) OA

3685. **Meade,** F. N. Airships and their value in war. (Contemporary review, London, 1908, v. 94, p. 605–609.) * DA

3686. **Memorandum** concerning the use of the captive war balloon during the siege of Ladysmith. (Aeronautical journal, London, 1902, v. 6, p. 15.) VDS

3687. **Messimy,** A. Aéroplanes contre dirigeables. (Revue politique et littéraire, Paris, 1909, année 47, semestre 2, p. 521–523.) * DM

3688. **Metcalfe,** A. Wharton. Recent aeronautical progress in relation to sea power. (Empire review, London, 1913, v. 26, p. 303–312.) CA

3689. **Mettler,** C. G. Antiaircraft. illus. (Army ordnance, Washington, D. C., 1924, v. 6, p. 76–80.) VWA

Developments during the World war and after. Reprinted in *Coast artillery journal,* Fortress Monroe, Va., 1925, v. 63, p. 543–554, *VWA.*

3690. **Military** balloon experiments. illus. (Engineer, London, 1874, v. 38, p. 93.) VA

Trial of Bowdler's steering device at the Woolwich arsenal. Also in *Scientific American,* New York, 1874, v. 31, p. 167, *VA.*

3691. **Military** ballooning. (Engineering, London, 1882, v. 34, p. 429–430.) VDA

Brief history and discussion of the various difficulties in the use of balloons.

3692. **Military** ballooning evolutions in France. illus. (Scientific American supplement, New York, 1886, v. 22, p. 9031–9032.) VA

Pictures show how captive balloons are handled.

3693. **Military** ballooning at the Royal arsenal. (Engineer, London, 1879, v. 47, p. 431.) VA

Account of experiments by Captain J. Templer.

3694. **Military** and naval. charts. (Aviation, New York, 1931, v. 30, p. 171–172; 1932, v. 31, p. 131–134.) VDS

Brief review of developments in the United States and foreign countries during 1930.

3695. **Mills,** S. P. Progress in aeronautics [for 1929]. Military aircraft. (Mechanical engineering, New York, 1930, v. 52, p. 8–11.) VFA

3696. **Mitchell,** William. The collapse of the Coolidge air policy. port. illus. (Popular aviation, Chicago, 1929, v. 4, June, p. 11–15, 90.) VDS

Criticizes U. S. Government defense policy. This number of *Popular aviation* has the title of *Aeronautics.*

3697. —— Col. Mitchell's statements on Govt. aviation. These contain the assertions on which he expects to be court martialed. (Aviation, Highland, N. Y., 1925, v. 19, p. 318–320.) VDS

Daring indictment of the air policies of the United States War and Navy Departments.

3698. —— Crushing America by air. illus. (Popular aviation, Chicago, 1929, v. 5, July, p. 11–15.) VDS

Criticism of United States military air policies. This number of *Popular aviation* has the title of *Aeronautics.*

3699. —— General Mitchell attacks bomb test findings. (Aviation, Highland, N. Y., 1921, v. 11, p. 367.) VDS

Disagrees with government report on bombings off the Virginia capes in 1921.

3700. —— What I think about airplane bombers. illus. (Popular aviation, Chicago, 1932, v. 11, p. 348–350, 400.) VDS

Experiences in sinking old battleships, with comments.

3701. —— Winged defense. The development and possibilities of modern air power — economic and military. New York: G. P. Putnam's Sons, 1925. xxiv, 261 p. illus. port. 8°. VDYD

3702. **Mitchell** down but not out. illus. (Literary digest, New York, 1925, v. 84, March 21, p. 13–14.) * DA

Newspaper comments on Gen. Mitchell's criticisms of the U. S. air forces.

3703. The **Mitchell** trial. Colonel Mitchell faces court martial and produces evidence of his accusations. ports. (Aviation, Highland, N. Y., 1925, v. 19, p. 744–749, 770–772, 802–803.) VDS

Digest of evidence at famous trial of 1925.

3704. **Moedebeck,** Hermann W. L. Die Luftschiffahrt in Japan. illus. (Illustrierte aëronautische Mitteilungen, Strassburg, 1903, Jahrg. 7, p. 101–109.) VDS

Kites. Development of military aeronautics.

3705. —— Die russische Militärluftschiffahrt. illus. (Illustrierte aëronautische Mit-

Military Aeronautics, continued

teilungen, Strassburg, 1905, Jahrg. 9, p. 205–219.) **VDS**

Account of the author's visit to Russia.

3706. **Moffett,** William Adger. Aviation and the fleet. illus. port. (Aeronautical digest, New York, 1923, v. 2, p. 81–82, 148.) **VDS**

Post-war developments in the United States Navy.

3707. —— Five progressive years in naval aviation. illus. port. (Aero digest, New York, 1931, v. 18, March, p. 35–37, 124, 126.) **VDS**

Developments in the United States Navy, 1925–1930.

3708. —— Progress in aeronautics [for 1929]: progress of naval aviation. (Mechanical engineering, New York, 1930, v. 52, p. 7–8.) **VFA**

3709. —— The progress of naval aviation. Navy flying proves to be 99.99 per cent safe in 1930. Five year plan is approaching a successful and enlightening conclusion. Development in strategy made possible by improvement in planes and radio. illus. (Airway age, New York, 1931, v. 12, p. 52–55.) **VDS**

3710. —— Progress in naval aviation. (Mechanical engineering, New York, 1929, v. 51, p. 10–11.) **VFA**

Developments during 1928.

3711. —— Twenty years of naval aviation. illus. (Popular aviation, Chicago, 1928, Dec., v. 3, p. 28–34, 113–114.) **VDS**

History of United States naval airplanes and airships, 1908–1928.

3712. **Mortane,** Jacques. The evolution of military aviation in France. illus. (Flying, New York, 1912–13, v. 1, Dec., p. 26–29, Jan., p. 13–15, 30–32; 1913–14, v. 2, Feb., p. 11–14.) **VDS**

3713. **Neumann.** Die Verwendbarkeit von Ballon und Motorluftschiff in der Marine. map. (Marine-Rundschau, Berlin, 1908, Jahrg. 19, p. 831–867.) **VXA (Germany)**

Translation in Royal United Service Institution, *Journal,* London, 1908, v. 52, p. 1502–1517, 1653–1676, *VWA.*

3714. Le **Nouvel** établissement d'aéronautique militaire de Vienne. (L'Aéronaute, Paris, 1893, année 26, p. 272–274.) **3–VDS**

3715. **Oliver,** Robert Shaw. The development of aviation in the United States Army. (Aero Club of America, Bulletin, 1912, v. 1, July, p. 27–28.) **VDS (Flying)**

3716. **Orcy,** Ladislas d'. Early history of the French aeronautical corps, 1794–1802. (Air service journal, New York, 1918, v. 2, p. 133–134.) **VDS**

3717. —— New developments in military aeroplanes. Aeroplane destroyer versus battle aeroplane. illus. (Scientific American, New York, 1916, v. 114, p. 147, 164–166.) **VA**

3718. **Ostfriesland** sunk by 2000-lb. aircraft bombs. Martin bombers of Army Air Service sink the 23,000-ton battleship in fourteen minutes. illus. (Aviation, Highland, N. Y., 1921, v. 11, p. 128–132.) **VDS**

Operations in July, 1921.

3719. **Parkes,** Oscar. Japan's new navy. illus. (Scientific American, New York, 1929, v. 141, p. 396–399.) **VA**

Descriptions of new aircraft carriers.

3720. The **Part** the aeroplane played in bringing about the fall of Adrianople — how it saved life and money. illus. (Flying, New York, 1913, v. 2, May, p. 16–17.) **VDS**

3721. **Paullin,** Charles Oscar. Early use of balloons in war, 1782–1865. (United service magazine, London, 1909, new series, v. 39, p. 530–541.) *** DA**

3722. **Pierce,** G. Wyndham. Late developments in British aircraft. England has ruled the sea for centuries, and by intensive development, is rapidly approaching a similar rating in the air. illus. (Popular aviation, Chicago, 1932, v. 11, p. 83–84.) **VDS**

3723. **Pollock,** Granville A. The evolution of anti-aircraft artillery. illus. (Air power, Pensacola, Fla., 1918, v. 4, p. 129–136.) **VDS**

3724. **Postnikov,** F. A. Notes of a Russian military aeronaut on the application of ballooning to land and naval warfare. (American magazine of aeronautics, New York, 1907, v. 1, Nov., p. 25–27.) **VDS**

Discusses aeronautics in the Russo-Japanese war.

3725. **Pritchard,** Henry Baden. Military ballooning. (Nature, London, 1878, v. 18, p. 491–492.) **OA**

Developments in France. Brief reference to aerial camera invented by Walter Woodbury.

3726. The **Progress** of naval aeronautics. (Naval annual, London, 1913, p. 161–178.) **VXA**

Recent developments in the various countries.

3727. **Rameau,** Louis. Médaille commémorative de l'emploi des aérostats pendant le siége de Paris [1870–1871]. illus. (L'Aéronaute, Paris, 1876, année 9, p. 291–293.) **3–VDS**

3728. **Ramsey,** DeWitt C. The development of aviation in the fleet. front. (United States Naval Institute, Proceedings, Annapolis, Md., 1923, v. 49, p. 1395–1417.) **VXA**

Covers period immediately following the World war.

Military Aeronautics, continued

3729. **Ranney,** H. F. Colonel William Mitchell declared guilty. (U. S. air services, Washington, 1926, v. 11, Jan., p. 29–31.) **VDS**

Colonel Mitchell was courtmartialled by U. S. Army officers for severe criticisms of naval and military efficiency.

3730. **Reber,** Samuel. Recent progress in military aëronautics. (Franklin Institute, Journal, Philadelphia, 1915, v. 180, p. 437–448.) **VA**

Aeronautics in Russo-Japanese, Bulgarian, and Turko-Italian wars. Advantages of various airship types. Organization in France, Germany, and Great Britain.

3731. **Recent** military dirigeable balloons. (Scientific American, New York, 1908, v. 99, p. 135.) **VA**

Acceptance of Baldwin airship for the United States army. Brief reference to the German ships *Parseval* and *Gross II.*

3732. **Redempt.** Politique aéronautique des grandes puissances étrangères, 1920–1930. (In: L'aéronautique militaire, maritime, coloniale et marchande, Paris, 1931, p. 23–28.) **† VDY**

Discussion of the military air policies of France, England, Italy, United States, Russia, and Germany since the World war.

3733. **Reid,** Walter F. The use of explosives in aerial warfare, with some remarks on methods of defence. (Aeronautical journal, London, 1910, v. 14, p. 133–143.) **VDS (Royal)**

With discussion.
Brief history beginning 1794.

3734. **Renard,** Paul. L'aéronautique militaire en France. (Revue politique et parlementaire, Paris, 1911, v. 68, p. 18–37.) **SEA**

Reviews developments of 1910.

3735. **Rhees,** William Jones. Reminiscences of ballooning in the Civil war. ports. (Chautauquan, Meadville, Pa., 1898, v. 27 [new series, v. 18], p. 256–262.) *** DA**

3736. **Ritter,** Hans. Development of military aircraft design. charts. (Aviation, New York, 1931, v. 30, p. 626–629.) **VDS**

Based upon European developments since 1915. Author is a former officer of the German General Staff.

3737. **Robertson,** Frederic Arthur de Vere. A brief outline of the growth of British air power from the balloon company of the Royal Engineers to the present day Royal Air Force. illus. (Flight, London, 1931, v. 23, p. 581–598.) **VDS**

3738. —— Growth of the Air Service. (Flight, London, 1930, v. 22, p. 74–76.) **VDS**

History of the Royal Air Services of Great Britain, beginning in 1878.

3739. A **Roosevelt** administration in 1907 gave the United States army the first aeroplane ever used for military purposes. Another Roosevelt administration seems to be the only hope to give the U. S. army and navy their much needed air service. port. (Aerial age weekly, New York, 1916, v. 3, p. 261–263.) **VDS**

3740. **Rowell,** Ross E. Expeditionary movements of aircraft. illus. (U. S. air services, Washington, 1931, v. 16, March, p. 19–21.) **VDS**

Experiences of U. S. Marine Corps in Nicaragua.

3740a. **Ruggles,** C. L'H. Antiaircraft defense. illus. (Army ordnance, Washington, D. C., 1926, v. 6, p. 344–354.) **VWA**

3741. **Salvatores,** P. Il primo ampiego in guerra di mezzi aeronautica da parte delle forze italiane. maps. (Rivista aeronautica, Roma, 1929, anno 5, Oct., p. 195–201.) **VDS**

Use of observation balloons in Italian African campaign of 1887–1888.

3742. **Savorgnan di Brazzà,** Francesco, conte. La guerra nel cielo. Milano: Fratelli Treves, 1915. xi, 307 p. illus. 8°. **VDY**

First part of book deals with history of military aeronautics, especially of the Italian-Turkish war. Latter part treats of development during the World war.

3743. **Scaroni,** Silvio. The equipment of air forces. Italy. illus. (Aviation, New York, 1933, v. 32, p. 313–317, 341–343.) **VDS**

History of military aviation in Italy.

3744. **Schleiffarth,** W. L. Die Fortschritte der Militär-Luftschiffahrt in der Füllung von Ballons. illus. (Prometheus, Berlin, 1892, Jahrg. 3, p. 353–356.) **3–VA**

Progress in the filling of military balloons, with special reference to developments in England, and to the work of Latchinoff.

3745. **Scholle,** Howard A. Military balloons during the French revolution. (U. S. air services, Washington, 1927, v. 12, May, p. 41–42.) **VDS**

Advantages of balloons set forth in a report to the Committee of Public Safety, of which the above is a translation.

3746. **Serrell,** Edward Wellman. A flying machine in the army. (Science, New York, 1904, new series, v. 19, p. 952–955.) **OA**

Experiment with machine of the helicopter type by officers of the Northern army during the American Civil war.

3747. **Simonin,** L. L'aérostation pendant le siége de Paris. (Revue de deux mondes, Paris, 1870, période 2, année 40, tome 90, p. 612–630.) *** DM**

3748. **Some** interesting experiments in dirigible airships. illus. (Scientific American, New York, 1901, v. 84, p. 10.) **VA**

Description of Carl E. Myers' aerial torpedo.

Military Aeronautics, continued

3749. **Spaight,** James Molony. The beginnings of organized air power; a historical study... London: Longmans, Green and Co., Ltd., 1927. v, 317 p. 8°. **BTZY**

"It deals with the gradual development of the system of supply, design, and administration upon which our [British] enormous air strength of 1918 was based." Also discusses similar developments in France, United States, and Germany.

3750. **Spuller,** Eugène. Discours. (L'Aéronaute, Paris, 1893, année 26, p. 171–194.) **3–VDS**

Interesting experiences of the author during the Franco-Prussian war.

3751. **Squier,** George Owen. The present status of military aeronautics. illus. (American Society of Mechanical Engineers, Journal, New York, 1908, v. 30, p. 1571–1641.) **VFA**

See p. 1624–1639 for extensive bibliography of aeronautics in general.

3751a. **Squires,** J. Duane. Aeronautics in the Civil war. (American historical review, Richmond, Va., 1937, v. 42, p. 652–669.) *** R-IAA**

Good account with 81 references.

3752. **Statistique** des raids militaires en 1912. Martyrologe de la cinquième arme française. (In: L'Aéronautique, par Roland Garros [and others], Paris: Lafitte & Cie., 1914, p. 338.) **VDW**

List of fatalities to aviators in the French army, 1909–1913.

3753. **Stauber,** Josef. Ueber den militärischen Ballondienst. (Streffleurs militärische Zeitschrift, Wien, 1905, Jahrg. 46, p. 1092–1108, 1219–1232.) **VWA**

Discusses the value of the military balloon.

3754. **Stewart,** Oliver. The equipment of air forces. Great Britain. illus. (Aviation, New York, 1932, v. 31, p. 473–475; 1933, v. 32, p. 46–50.) **VDS**

Describes contemporary airplane equipment.

3755. **Stockhausen,** von. L'emploi militaire des dirigeables. (Internationale Revue über die gesamten Armeen und Flotten, Dresden, 1909, Jahrg. 27, supplément, p. 465–476.) **VWA**

Discusses relative advantages of dirigeables and captive balloons.

3756. **Stone,** Francis Gleadowe. Defence of harbours against naval airships. (Royal United Service Institution, Journal, London, 1909, v. 53, p. 559–577.) **VWA**

With discussion.

3757. **"Strafing"** New York from the clouds. illus. (Literary digest, New York, 1929, v. 101, June 8, p. 60, 62, 64–66.) *** DA**

Sham attack on New York City, 1929.

3758. **Sutner,** von. Le tir contre les ballons dirigeables. (Internationale Revue über die gesamten Armeen und Flotten, Dresden, 1908, Jahrg. 26, supplément, p. 257–272.) **VWA**

3759. **Sutton,** George W. The tenth anniversary of the world's military aviation. Looking back at 1909 when our war department secured the first war airplane and enthusiasm was worldwide. illus. (Scientific American, New York, 1920, v. 122, p. 107.) **VA**

3760. **Swoffer,** Frank Arthur. The war in the air in China. An eye-witness report from Shanghai. illus. (Air and airways, London, 1932, v. 9, p. 5–7.) **VDS**

3760a. **Talbot,** Frederick Arthur Ambrose. Aëroplanes and dirigibles of war. Philadelphia: J. B. Lippincott Company; London: W. Heinemann, 1915. xi, 283 p. 12°. **VDYD**

Describes equipment of the early part of the World war.

3761. **Templer,** D. [Balloon operations in South Africa.] illus. (Aeronautical journal, London, 1901, v. 5, p. 54–56.) **VDS**

3762. **Templer,** J. Military balloons. (Royal United Service Institution, Journal, London, 1879, v. 23, p. 173–184.) **VWA**

With discussion.

Balloons in the American Civil war and the siege of Paris, with some experiences of the author.

3763. **Thayer,** Russell. Dirigible balloons for war purposes. illus. (Military Service Institution of the United States, Journal, New York, 1886, v. 7, p. 177–194.) **VWA**

Author proposes jet propulsion, using air or carbonic acid gas.

3764. **Tissandier,** Gaston. Les aérostats captifs de l'armée française. illus. (La Nature, Paris, 1885, année 13, semestre 1, p. 196–198.) **OA**

Experiences during Franco-Prussian war and in China. Interesting copy of Chinese print. Somewhat abridged, without illustrations, in *L'Aéronaute*, Paris, 1885, année 18, p. 113–116, *3–VDS*.

3765. —— Les aérostats captifs de la marine française. illus. (La Nature, Paris, 1890, année 18, semestre 2, p. 311–314.) **OA**

Translation in *Scientific American supplement*, New York, 1890, v. 30, p. 12423, *VA*.

3766. —— Les aérostats militaires et les armées européennes. illus. (La Nature, Paris, 1887, année 15, semestre 1, p. 87–90.) **OA**

Brief record of contemporary developments. Translation of part of the article is in *Scientific American supplement*, New York, 1887, v. 23, p. 9287, *VA*.

3767. —— Les ballons captifs de l'armée chinoise. illus. (La Nature, Paris, 1887, année 15, semestre 1, p. 141–142.) **OA**

3768. —— Ballons captifs transportables pour le service des armées; système Gabriel

Military Aeronautics, continued

Yon. illus. (La Nature, Paris, 1885, année 13, semestre 2, p. 310–314.) **OA**

Translation in *Scientific American*, New York, 1885, v. 53, p. 328–329, *VA*.

3769. —— Le mousquet à ballons de M. Krupp. illus. (L'Aéronaute, Paris, 1876, année 9, p. 67–70.) **3-VDS**

Description of Krupp anti-aircraft gun.

3770. —— Nouveaux documents sur les aérostats militaires de la première république. illus. (La Nature, Paris, 1877, année 5, semestre 1, p. 145–146.) **OA**

Has a picture of Coutelle's apparatus for the manufacture of hydrogen (1794); also a description of de Selle de Beauchamp's process.

3771. —— Souvenirs et récits d'un aérostier militaire de l'armée de la Loire, 1870–1871 ... Paris: Maurice Dreyfous, 1891. x, 356 p. illus. 4°. **3-VDY**

Balloon experiences of the author during the Franco-Prussian war.

3772. **Torpedo** balloons. illus. (Scientific American supplement, New York, 1886, v. 22, p. 9031–9032.) **VA**

Inventions of George Rodeck, of Germany.

3773. **Tournachon,** Félix. A terre & en l'air. Mémoires du Géant, par Nadar [pseud.]. Avec une introduction par M. Babinet... Paris: E. Dentu, 1864. xx, 22–439 p. 12°. **VDW**

Tournachon (1820–1910), better known as Nadar, constructed the balloon *Le Géant* which did good service during the Franco-Prussian war. Library has also 2d edition, Paris, 1865.

3774. **Turbiaux,** E. Rapport du voyage de M. Turbiaux, aéronaute du ballon "La Poste de Paris" janvier 1871. (L'Aéronaute, Paris, 1902, année 35, p. 15–21.) **3-VDS**

Experiences during the Franco-Prussian war.

3775. **Two** pioneers of American naval aeronautics. Commander J. C. Hunsaker and Commander H. C. Richardson, U. S. N. ports. (Aviation, Highland, N. Y., 1923, v. 15, p. 710–711.) **VDS**

3776. **United States.** — Aeronautics Bureau (Navy Department). Annual report. (In: United States. — Navy Department, Annual report of the Secretary of the Navy, for the years 1922–1932, Washington, 1923–33.) **VYEB (US)**

Year 1922, p. 365–415; 1923, p. 527–569; 1924, p. 609–645; 1925, p. 591–623; 1926, p. 595–645; 1927, p. 615–666; 1928, p. 663–718; 1929, p. 587–635; 1930, p. 567–591; 1931, p. 333–356; 1932, p. 473–497. During these years William A. Moffett was chief of the bureau.

3777. **United States.** — Air Service (War Department). Aircraft development since the armistice. (Aviation, Highland, N. Y., 1923, v. 15, p. 6–9.) **VDS**

3778. —— Report of the director. 1918/19–1919/20, 1921/22. Washington, 1920–22. 3 v. 8°. **VDS**

3779. **United States.** — Army: Chief Signal Officer. Balloons. (In: United States. — War Department, Annual report, 1885/86, v. 4, part 1, p. 209; 1886/87, v. 4, part 1, p. 6; 1888/89, v. 4, part 1, p. 48; 1890/91, v. 4, p. 6; 1891/92, v. 1, p. 598–600, 618–620, 624; 1892/93, v. 1, p. 27, 648–649; 1893/94, v. 1, p. 493; 1894/95, v. 1, p. 581; 1895/96, v. 1, p. 601–603; 1897/98, v. 1, part 1, p. 888–891, 960–966; 1898/99, part 2, p. 747–748; 1902/03, v. 2, p. 343; 1903/04, v. 2, p. 397; 1904/05, v. 2, p. 240; 1905/06, v. 2, p. 196; 1906/07, v. 2, p. 179. Washington, 1886–1907.) **VWZZ**

Library lacks annual report for 1896/97; but has a separate copy of the Signal Officer's report (VWZZ) for that period, in which see p. 8–9. For continuation of the series, including airplanes and airships, see entry no. 3780.

3780. **United States.** — Army: Signal Office. Military aeronautics. (In: United States. — War Department, Annual report, 1907/08, v. 2, p. 210–214; 1908/09, v. 2, p. 252; 1909/10, v. 1, p. 650–653; 1910/11, v. 1, p. 739–742; 1911/12, v. 1, p. 964–971; 1912/13, v. 1, p. 781–802; 1913/14, v. 1, p. 514–524; 1914/15, v. 1, p. 740–748; 1915/16, v. 1, p. 882–891; 1916/17, v. 1, p. 839–841; 1917/18, v. 1, p. 1073–1084. Washington, 1908–18.) **VWZZ**

3781–3782. **United States.** — Army and Navy Board. Report...on results of aviation and ordnance tests held during June and July, 1921, and conclusions reached. Washington: Gov. Prtg. Off., 1926. 5 p. 8°. **VDY p.v.51, no.2**

Bombing tests on ex-German warships and submarines, also the United States warship *Iowa*, in 1921. Report is signed by John J. Pershing. Also in *Aviation*, Highland, N. Y., 1921, v. 11, p. 276–278, *VDS*.

3783. The **U. S.** aerial bombing experiments on warships. illus. (Flight, London, 1921, v. 13, p. 615–618.) **VDS**

Ships sunk were the *Iowa*, the German ships *Frankfurt* and *Ostfriesland*, and a German submarine.

3784. **U. S.** aerial bombing tests on battleships. illus. (Flight, London, 1923, v. 15, p. 657.) **VDS**

Bombing of obsolete ships *New Jersey* and *Virginia* in 1923.

3785. **United States** government and flying machines. (Aeronautical world, Glenville, O., 1902, v. 1, p. 55–56.) **VDS**

Government's interest in Langley's experiments.

3786. **Villiers du Terrage,** Marc, baron de. Les aérostiers militaires pendant la campagne d'Égypte. (L'Aéronaute, Paris, 1901, année 34, p. 85–91, 104–113.) **3-VDS**

Aeronautics in Napoleon's Egyptian campaign.

3787. **Vischer,** Peter. N. Y. World probes Government air situation. (Aviation, Highland, N. Y., 1924, v. 17, p. 1331–1333, 1366–1367.) **VDS**

Attempts to show that the United States has no definite aviation policies.

Military Aeronautics, continued

3788. **Vives y Vich,** Pedro. Aerostación militar. illus. (Memorial de ingeniores del ejército, Madrid, 1909, serie 4, v. 26, p. 205–223.) VWA

Developments in Spain beginning 1884.

3789. **Voisin,** André Paul. L'aviation de reconnaissance. illus. (In: L'aéronautique militaire, maritime, coloniale et marchande, Paris, 1931, p. 44–60.) † VDY

History of aerial reconnaissance during the World war and the post-war period.

3790. **War** balloons. (All the year round, London, 1869, new series, v. 1, p. 297–299.) * DA

History of ballooning during the French Revolution.

3791. **Warner,** Edward Pearson. The equipment of air forces in the United States of America. illus. (Aviation, New York, 1932, v. 31, p. 247–253, 290–294, 369–373, 411–415, 440–443.) VDS

Post-war developments.

3792. **Waters,** R. S. Ballooning in the French army during the Revolutionary wars. (Army quarterly, London, 1932, v. 23, p. 327–340.) VWA

3793. **Watkins,** John Elfreth. Air-scouts of other days. illus. (Air-scout, New York, 1911, v. 1, no. 3, p. 11–13.) 3-VDS

3794. **Watson,** Sir Charles Moore. Military ballooning in the British Army. (Royal Engineers' Institute, Occasional papers, Chatham, 1902, v. 28, p. 39–59.) 3-VDA

Chronological account, 1862–1901 inclusive.

3795. **Watson,** Geoffrey. The growth of guns on our aeroplanes. illus. (Air power, Pensacola, Fla., 1918, v. 4, p. 21–24.) VDS

3796. **Wead,** Frank Wilbur. The military value of airplane racing. illus. (Aviation, New York, 1929, v. 27, p. 418–422.) VDS

Author, a United States naval officer, traces developments in Europe and America.

3797. **Wells,** G. M. Development of anti-aircraft matériel. New types of anti-aircraft guns — improved methods for directing fire of guns and for locating and illuminating aircraft. illus. (Mechanical engineering, New York, 1931, v. 53, p. 427–432.) VFA

3798. **Werner,** A. Drachenballons für Kriegs- und Expeditionszwecke. illus. (Zeitschrift für Luftschiffahrt, Berlin, 1883, Jahrg. 2, p. 22–25.) 3-VDS

Discusses possibilities of the kite-balloon.

3799. **Where** the money goes. How $67,241,-327.95 was spent in 1922–23 for Government air services. (Aviation, Highland, N. Y., 1924, v. 16, p. 474–475, 509–510.) VDS

Army and navy expenditures.

3800. **Witzleben,** von. Ueber Geschütze zur Bekämpfung von Luftschiffen. (Schweizerische militärische Blätter, Frauenfeld, 1908, Jahrg. 44, p. 312–322.) VWA

3801. **Wolff,** Maurice. Bonaparte et les aérostiers militaires en Égypte. illus. (La Revue aërienne, Paris, 1912, année 5, p. 158–161.) 3-VDS

3802. **Woodhouse,** Henry. The evolution of the military aeroplane. illus. (Aero Club of America, Bulletin, New York, 1912, Feb./March, p. 3–7.) VDS (Flying)

History from the time of Clément Ader's airplane designed for the French government in 1900. The above bulletin is an early number of the periodical known as *Flying.*

3803. —— The evolution of military aviation in France. (Aircraft, New York, 1911, v. 2, p. 120–122.) VDS

3804. —— France's national movement to gain aerial supremacy. illus. (Aero Club of America, Bulletin, New York, 1912, v. 1, May, p. 13–16.) VDS (Flying)

The above bulletin is an early number of the periodical subsequently known as *Flying.*

3805. —— General Stephane Brun, father of French military aviation. (Aircraft, New York, 1911, v. 2, p. 149–150.) VDS

3806. —— Textbook of military aeronautics. New York: The Century Company, 1918. 298 p. illus. f°. † VDY

A well-illustrated account of how aeronautics developed technically during the World war and some years before. See especially chapter 17, History of United States Army aeronautics; and chapter 18, The evolution of military aviation.

3807. —— Textbook of naval aeronautics, with introduction by Bradley A. Fiske. New York: The Century Company, 1917. 288 p. illus. f°. † VDY

Copiously illustrated account of technical advances during the World war and some years preceding. See especially chapter 5, Aircraft mother ships; chapter 24, United States Navy aeronautics; chapter 28, Evolution of the seaplane and flying boat; chapter 32, Evolution of the aero motor.

3807a. **Woodman,** Dorothy. Les avions d'Hitler. [Paris:] Ernest Flammarion [1935]. xxxvi, 37–246 p. maps. 12°. VDYD

State of German military aeronautics, 1935.

3808. **Work** of the R. A. F. overseas in 1933. (Air annual of the British Empire, 1934/35, London, 1934, v. 6, p. 34–39.) VDS

Brief accounts of tribal raids, protectorate surveys, Palestine disturbances, and oversea flights.

WORLD WAR

3809. **Adams,** Briggs Kilburn. The American spirit. Letters of Briggs Kilburn Adams, Lieutenant of the Royal Flying Corps, with a preface by Arthur Stanwood Pier. Boston: Atlantic Monthly Press, Inc., 1918. 103 p. port. 8°. **BTZY**

Young aviator's letters mostly to his parents, dealing chiefly with his training period, are of high literary merit.

3810. **Aeronautics.** Retrospect of the year 1915. (Scientific American, New York, 1916, v. 114, p. 7.) **VA**

3811. **Aeronautics** and the war. A review of what aeroplanes have done and of their development during the war. (Scientific American supplement, New York, 1915, v. 79, p. 114–115.) **VA**

3812. L'**Aéronautique** dans la "grande flotte." En 1914–1915–1916. (L'Aéronautique, Paris, 1919, année 1, p. 213–215.) **VDS**

3813. The **African** cruise of the Zeppelin L 59. German naval airship made 4,250 mile voyage to succor beleagured German force in East Africa. (Aviation, Highland, N. Y., 1922, v. 13, p. 126.) **VDS**

Account of flight in 1917.

3814. **Air** raids on Paris and London. (The New York Times current history, New York, 1918, v. 8, p. 305–308.) **BAA (N. Y. Times)**

3815. **Aircraft** and the war. illus. (Naval annual, London, 1916, p. 88–107.) **VXA**

Lessons of the airship and the seaplane during the World war.

3816. **Aircraft** in the war. (Quarterly review, New York, 1916, v. 225, p. 334–348.)

Discussion of British developments. *** DA**

3817. **Aircraft** in warfare. (Aircraft year book, New York, 1920, p. 74–86.) **VDS**

Brief accounts of developments during the World war.

3818. **Alan** Ramsay Hawley. illus. port. (Flying, New York, 1918, v. 7, p. 1025–1027, 1087.) **VDS**

Biography and portrait of prominent balloonist and ardent advocate of aerial preparedness for war. He served several terms as president of the Aero Club of America.

3819. **Albret,** André. Au-dessus des batailles. Carnet de guerre d'un aviateur [by] C. H. A. André [pseud.]. Paris: Fayard & Cie [1917]. 316 p. 12°. **BTZY**

Experiences of a French aviator vividly described. Library has also an English translation by Mrs. Philip Duncan Wilson, with the title *Above the battles*, Garden City: Doubleday, Page & Co., 1919, 243 p. 12° *BTZY*.

3820. **America's** first achievement in providing bombing aeroplanes. (Aerial age weekly, New York, 1918, v. 7, p. 861–862.) **VDS**

Two machines, each equipped with Liberty motors, were made by the Standard Aircraft Corporation.

3821. **Armani,** Armando. Ex alto ad signum; aneddoti ed episodi di bombardamenti aerei (1915–1918). Roma: G. Bardi, 1925. 248 p. illus. 4°. **BTZY**

Exploits of Italian aviators during the World war.

3822–3823. **Ashmore,** Edward Bailey. Air defence. London: Longmans, Green and Co., 1929. vii, 179 p. illus. map. 8°. **VDY**

Air defence in England, with chapter on methods in Germany and France. Story of London raids. Statistical table of air raids on England with both airships and airplanes. Reviewed in *Aviation*, New York, 1930, v. 28, p. 184, *VDS*.

3824. **Aviaticus,** pseud. Boelcke der Held der Lüfte; ein deutsches Heldenleben. Chemnitz: Walther Berlinicke [1916?]. 36 p. illus. 12°. **AN p.v.249, no.10**

Small memorial pamphlet on the career of famous German military aviator.

3825. **Bairstow,** Leonard. Progress of aviation in the war period. Some items of scientific and technical interest. London: The Royal Aeronautical Society, 1919. 24 p. illus. 8°. (The Wilbur Wright lecture, 1919.) **VDY p.v.18, no.1**

Also in *Flight*, London, 1919, v. 11, p. 853–857, *VDS*; *Aeroplane*, London, 1919, v. 16, p. 2537–2538; v. 17, p. 60–62, *VDS*.

3826. **Baring,** Maurice. R. F. C. H. Q. 1914–1918. London: G. Bell & Sons, 1920. 315 p. 12°. **BTZE**

Observations and experiences of the author, who was an aide-de-camp attached to the Headquarters Staff of the Royal Flying Corps during the World war. Reviewed in *Aeroplane*, London, 1920, v. 18, p. 983–985, 1023–1026, 1067–1070, *VDS*.

3827. **Barnard,** William N. The U. S. Army School of Military Aeronautics at Cornell University. illus. (Sibley journal of engineering, Ithaca, 1919, v. 33, p. 63–65.) **VDA**

3828. **Barnes,** John K. The vindication of [George O.] Squier and [Edward A.] Deeds. What really happened to the billion dollar aircraft appropriation. (World's work, Garden City, N. Y., 1921, v. 42, p. 300–306.) *** DA**

History of aircraft production during the World war; also of the Liberty engine.

3829. **Barnett,** Gilbert. V. C.s of the air. The glorious record of men of the British Empire Air Force awarded the Victoria cross for valour, with an additional chapter on heroes of America. London: E. J. Burrow & Co., Ltd. [1918?] 36 p. illus. 8°. **BTZY**

3830. **Baumann,** A. Die Entwicklung des deutschen Riesenflugzeugbaues während des Krieges. illus. (Verein deutscher Ingenieure, Zeitschrift, Berlin, 1919, Bd. 63, p. 497–504, 1275–1281; 1920, Bd. 64, p. 78–83, 112–116, 129–134.) **VDA**

Development of giant airplanes in Germany during the World war.

World War, continued

3831. Baur, Fritz. Wir Flieger! 1914–1918. Der Krieg im Fliegerlichtbild. Wien: Selbstverlag, 1930. 116 p. plates. 8°. **BTZY**

Interesting series of aerial views taken on the Austrian front.

3832. Beecroft, David. America's ace of aces. Rickenbacker's record of twenty-six planes and how he made it. illus. (Motor age, Chicago, 1919, v. 35, Feb. 13, p. 7–10.) **TOL**

3833. Benn, Wedgwood. In the side shows. London: Hodder and Stoughton, 1919. xii, 310 p. illus. 8°. **BTZY**

Author's war experiences on the Italian front and in the Near-East.

3834. Bewsher, Paul. "Green balls." The adventures of a night-bomber. Edinburgh: William Blackwood & Sons, 1919. viii, 309 p. 12°. **BTZY**

"I have tried faithfully to describe the sensations, the strange inexplicable fears, the equally inexplicable fearlessness of a desk-bound London youth, pitch-forked...into a hitherto unknown, untried occupation — bombing at night from the air."

3835. Bibesco, Marthe Lucie (Lahovary), princesse. Lord Thomson of Cardigan: a memoir and some letters. London: J. Cape [1932]. 236 p. illus. ports. 8°. **AN**

Biography of British Air Minister.

3836. [**Bibliography** of George Marie Ludovic Guynemer.] (United States. — National Advisory Committee for Aeronautics, Bibliography of aeronautics, 1909–1931, Washington, 1921–35, volumes for 1909/16, p. 595; 1917/19, p. 203; 1922, p. 62; 1923, p. 55.) **VDT**

References to famous French war aviator.

3837. [**Bibliography** of the Royal Aircraft Factory.] (United States. — National Advisory Committee for Aeronautics, Bibliography of aeronautics, 1909–1931, Washington, 1921–35, volumes for 1909/16, p. 1157–1158; 1917/19, p. 389.) **VDT**

3838. [**Bibliography** of the Royal Flying Corps.] (United States. — National Advisory Committee for Aeronautics, Bibliography of aeronautics, 1909–1931, Washington, 1921–35, volumes for 1909/16, p. 1158–1159; 1917/19, p. 389–390.) **VDT**

3839. [**Bibliography** of the Royal Naval Air Force.] (United States. — National Advisory Committee for Aeronautics, Bibliography of aeronautics, 1909–1931, Washington, 1921–35, volumes for 1909/16, p. 1159; 1917/19, p. 390.) **VDT**

3840. Biddle, Charles John. The way of the eagle. New York: Charles Scribner's Sons, 1919. xvi, 297 p. illus. 8°. **BTZY**

Letters of a prominent member of the Lafayette Escadrille.

3841. Bishop, William A. Winged warfare. New York: Grosset & Dunlap [cop. 1918]. ix, 17–272 p. illus. port. 12°. **BTZY**

Major Bishop, of the British Flying Corps, destroyed 47 German machines during the Great war. Reviewed in *Aeroplane,* London, 1918, v. 15, p. 196, *VDS.*

3842. Blake, Wilfred Theodore. Over "Over there," by "Wing adjutant" [pseud. of Wilfred T. Blake]. London: Cassell & Co., Ltd. [1918.] viii, 209 p. 12°. **BTZY**

Stories of the Royal Air Force.

3843. —— Plane tales from the skies, by "Wing adjutant" [pseud. of Wilfred T. Blake]. London: Cassell & Co., Ltd. [1918.] x, 182 p. 12°. **BTZY**

Tales of the Royal Air Force.

3844. —— The Royal Flying Corps in the war, by "Wing adjutant" [pseud. of Wilfred T. Blake]. London: Cassell & Co., Ltd., 1918. 124 p. 12°. **BTZY**

3845. Blanchet, Georges. René Fonck. port. (L'Aérophile, Paris, 1921, année 29, p. 161–162.) **3–VDS**

Portrait and biography of France's famous "ace" in the World war.

3846. Bludworth, T. Francis. The battle of Eastleigh, England. [By T. F. Bludworth.] [New York,] 1918. 144 p. illus. obl. 8°. **BTZY**

Record and pictures of United States Navy Air Force quartered at Eastleigh, England, during the World war.

3847. Bölcke, Oswald. Hauptmann Bölckes Feldberichte, mit einer Einleitung von der Hand des Vaters und zwanzig Bildern. Vorwort von Joseph E. Ridder. New York: Deutschland Library Company, 1917. 220 p. illus. port. 12°. **BTZY**

Letters and reports of celebrated German ace of the Great war. Library has a copy of the English translation by R. R. Hirsch, entitled *An aviator's field book,* New York: National Military Pub. Co., 1917, *BTZY.*

3848. Bordeaux, Henry. Le chevalier de l'air. Vie héroique de Guynemer. Paris: Plon-Nourrit et Cie, 1918. 320 p. port. 12°. **BTZY**

Biography of celebrated French ace. Library has also a copy of the English translation entitled *Guynemer, knight of the air,* London: Chatto & Windus, 1918, *BTZY.* Reviewed in *Aeroplane,* London, 1919, v. 16, p. 1114, *VDS.*

3849. Bott, Alan John. Cavalry of the clouds, by "Contact" [pseud. of Alan Bott]. With an introduction by W. S. Brancker. Garden City, N. Y.: Doubleday, Page & Co., 1918. xvii, 266 p. port. 12°. **BTZY**

Author's experiences with the British Royal Air Force during the World war. Library has also a copy of the English edition, London, 1917, under the title of *An airman's outings, BTZY.* Extensive review in *Aeronautics,* London, 1917, v. 13, p. 1317–1320, 1322, *VDS.*

3850. Bradley, W. F. Airplane supply system of the American army in France. (Automotive industries, New York, 1919, v. 40, p. 987–989.) **TOL**

World War, continued

3851. —— Captain Georges Guynemer, France's greatest ace. The life story of the distinguished French aviator who had accounted for more than 100 enemy fliers. illus. port. (Automotive industries, New York, 1918, v. 38, p. 1030–1033, 1079–1081.) **3–† TOL**

Has an account of the development of the Hispano-Suiza engine.

3852. Bradshaw, Stanley Orton. Flying memories. London: John Hamilton, Ltd. [1934.] 52 l. 8°. **VDYD**

Forty-eight full-page illustrations with descriptive texts, covering deeds of the Royal Air Force during the World war and after.

3853. Bridgman, Leonard. The war, 1914–1918. (Aeroplane, London, 1933, v. 44, p. 808, 810–812, 814, 816, 818–818d, 822a, 822b.) **VDS**

Part played by Great Britain.

3854. Briscoe, Walter Alwyn, and H. S. Stannard. Captain Ball, V. C. The career of Flight Commander Ball, V. C., D. S. O. ... London: Herbert Jenkins, Ltd., 1918. xx, 320 p. ports. illus. 12°. **BTZY**

Battle exploits of celebrated English ace.

3855. Britain's air effort during the war. illus. (Engineer, London, 1919, v. 127, p. 461–463.) **VA**

3856. Bruce, Eric Henry Stuart. The German aeroplanes. (United service magazine, London, 1916, v. 174 [new series, v. 53], p. 266–275.) *** DA**

3857. Buddecke, Hans Joachim. El Schahin (der Jagdfalke). Aus meinem Fliegerleben. Berlin: A. Scherl [1918]. 124 p. illus. port. 12°. **BTZE p.v.740, no.1**

Author's experience as German aviator in the Turkish service during the World war.

3858. Building our aerial defenses. (Aerial age weekly, New York, 1916, v. 4, p. 165–166.) **VDS**

Preparations of United States army and navy.

3859. Bursey, Jack C. How we shot down Immelmann. This remarkable story is being told for the first time. It gives the inside story of how the British squadron set a trap for the great German ace, and sent him to his death. illus. (Popular aviation, Chicago, 1930, v. 7, Aug., p. 12–14, 76–77.) **VDS**

3860. Busset, Maurice. En avion, vols et combats, estampes et récits de la grande guerre, 1914–1918. Paris: Librairie Delagrave, 1919. 25 l. illus. f°. **††† BTZY**

Folio colored prints, with descriptive texts, showing combats and adventures during the World war.

3861. Bygott, John Ford, compiler. Two soldier brothers. London: Jarrolds [1918]. ix, 176 p. illus. 12°. **BTZY**

The story of Lieut. Walter B. Wood gives a good idea of airplane fighting.

3862. Campbell, G. L., compiler. Royal Flying Corps (military wing). Casualties and honours during the war of 1914-17. Compiled by Capt. G. L. Campbell, assisted by R. H. Blinkhorn. London: Picture Advertising Co., Ltd., 1917. 250 p. 12°. **BTZY**

3863. Carver, Leland M., and others. The Nineteenth Aero Squadron, American Expeditionary Forces. A history of its activities during the World war, from its formation to its return to the United States, by L. M. Carver, G. A. Lindstrom, and A. T. Foster. [Hinsdale, Ill.:] E. H. Greist [cop. 1920]. 91 p. illus. 8°. **BTZY**

3864. Castelnau, F. de. Les bombardements aériens de Paris. (Le Correspondant, Paris, 1923, tome 291 [nouv. série, tome 255], p. 3–29.) *** DM**

German attacks during the World war.

3865. Castle, Irene (Foote). My husband, by Mrs. Vernon Castle (Irene Castle). New York: Charles Scribner's Sons, 1919. xii, 264 p. illus. ports. 8°. **AN**

Letters of well-known professional dancer who became a captain in the Royal Flying Corps. He was killed in 1917.

3866. Cavara, Otello. Voli di guerra. Impressioni di un giornalista pilota. Milano: Fratelli Treves, 1918. 210 p. 12°. **BTZY**

Impressions and experiences of an Italian pilot during the World war.

3867. Clapp, Frederick Mortimer. A history of the 17th Aero Squadron... December, 1918. [Garden City, New York: Country Life Press, cop. 1920.] xi, 171, 10 pl. illus. 8°. **BTZY**

Record of United States unit in the World war, with detailed records of combats and bombing.

3868. Clark, Virginius Evans. Types of military airplanes. illus. (Scientific American supplement, New York, 1918, v. 85, p. 322.) **VA**

3869. Claudy, Carl Harry. British aircraft output during the war. A generous American tribute to Britain's past achievements and present position. (Aeronautics, London, 1919, v. 16, p. 453.) **VDS**

3870. Clifford, George Reid. My experiences as an aviator in the World war. Boston: Richard G. Badger [cop. 1928]. 276 p. illus. port. 8°. **BTZY**

An excellent account of service in the British Royal Air Force.

3871. Coburn, F. G. Problems of the [United States] Naval Aircraft Factory during the war. illus. (American machinist, New York, 1919, v. 51, p. 413–419.) **VFA**

Illustrated account of how the work was done.

3872. Col. Charles De F. Chandler retires. port. (Aviation, New York, 1920, v. 9, p. 353.) **VDS**

Portrait and brief biography. Chandler commanded the balloon section of the U. S. Signal Corps during the World war.

World War, continued

3873. Coppens, Willy. Jours envolés. Mémoires. Paris: Nouvelles Éditions Latines [1932]. 353 p. map. 12°. **BTZY**

Experiences of the author, a Belgian aviator, during the World war. Reviewed in *Aeroplane,* London, 1932, v. 43, p. 625–626, 628, *VDS.*

3874. Cowdin, Elliot C. How the famous Lafayette Escadrille was started. illus. (Flying, New York, 1918, v. 7, p. 239, 264.) **VDS**

3875. Crouvezier, Gustave. La guerre aérienne; le rôle de la cinquième arme. Paris: Berger-Levrault, 1916. 65 p. 16°. (Pages d'histoire. no. 98.) **BTZE (Pages)**

Equipment of nations during the World war, with a brief account of French operations.

3876. Cutlack, Frederic Morley. The Australian Flying Corps in the western and eastern theatres of war, 1914–1918. Sydney: Angus & Robertson, Ltd., 1923. xxvii, 485 p. maps. illus. 8°. (Official history of Australia in the war of 1914–1918. v. 8.) **BTZE (Official)**

3877. Delacommune, Charles. L'escadrille des éperviers. Impressions vécues de guerre aérienne. Préface de Maurice Barrès. Paris: Plon-Nourrit et Cie. [1918.] 310 p. 12°. **BTZY**

Experiences of the author, a French aviator, during the World war.

3878. Development of naval aeronautics. (Aviation, New York, 1918, v. 5, p. 434–435.) **VDS**

Excerpts from several United States government reports concerning developments during the World war.

3879. Dienstbach, Carl. Fighting the big guns from balloons. illus. (Popular science monthly, New York, 1917, v. 91, p. 50–51.) *** DA**

3880. —— Our enemies in the air. illus. (Scientific American, New York, 1918, v. 118, p. 133, 138–139.) **VA**

Account of German developments in the World war.

3881. Dollfus, Charles. Dirigeables et aérostats pendant la guerre. Les ballons dirigeables (1914–1918). illus. (L'Aérophile, Paris, 1920, année 28, p. 23–30, 121–123.) **3–VDS**

Brief accounts of airships of various countries, including Germany, during the World war.

3882. Donegan, James Francis. The uses of aeroplanes to the army medical services in the field. (Royal United Service Institution, Journal, London, 1914, v. 58, p. 321–333.) **VWA**

3883. Drake, Vivian. Above the battle, with an introduction by General C. G. Hoare, Royal Air Force. New York: D. Appleton & Co., 1918. vi, 323 p. 12°. **BTZY**

Gives excellent descriptions of the training and duties of aviators during the World war.

3884. Drew, George A. Canada's fighting airmen. Toronto: Maclean Publishing Co., Ltd., 1930. 305 p. 8°. **BTZY**

Individual records of Canada's aviators during the World war, with a chapter covering Canada's work in general. For review see *Flight,* London, 1931, v. 23, p. 385, *VDS; Aeroplane,* London, 1931, v. 40, p. 242, *VDS.*

3885. Driggs, Laurence La Tourette. Heroes of aviation. Boston: Little, Brown, and Co., 1918. xxvi, 301 p. ports. 12°. **BTZY**

Feats of Allies' aces (including American) during the World war, with appendix containing statistics of deeds of both Allied and German flyers.

3886. Eberhardt, Walter von, editor. Unsere Luftstreitkräfte 1914–18; ein Denkmal deutschen Heldentums... Berlin: C. A. Weller, 1930. ix, 467 p. illus. maps. f°. **† BTZY**

Copiously illustrated *de luxe* history of German aviation during the World war, by various contributors. Has a list of German air fatalities.

3887. Edmonds, Charles Humphrey Kingsman. Aerial cooperation with the navy. (Royal United Service Institution, Journal, London, 1921, v. 67, p. 237–257.) **VWA**

Account of British aerial operations at sea, during the World war, by a squadron leader of the Royal Air Force.

3888. L'Évolution de l'aviation de 1914 à 1920. illus. (L'Aérophile, Paris, 1919, v. 27, p. 354–358.) **VDS**

Aviation situation in France at the beginning of the war, and later progress.

3889. Ewart, Ernest Andrew. Air men o'war, by Boyd Cable [pseud. of Ernest Andrew Ewart]. New York: E. P. Dutton & Co. [1919.] x, 246 p. 12°. **BTZY**

Endeavors to show the value of cooperation of land and air forces in the World war.

3890. Famous planes of the World war. illus. (Popular aviation, Chicago, 1931, v. 8, April, p. 26–29, 58.) **VDS**

Numerous illustrations.

3891. Farman, T. F. Aeroplanes in war. (Blackwood's magazine, New York, 1916, v. 200, p. 788–795.) *** DA**

How the tide was turned in favor of the Allies during the World war. Some bombing exploits.

3892. Farré, Henry. Sky fighters of France. Aerial warfare, 1914–1918. Englished by Catharine Rush, with illustrations from paintings by the author. Boston: Houghton Mifflin Company, 1919. xviii, 143 p. illus. 4°. **BTZY**

Experiences of and pictures by French artist-aviator.

3893. Faulkner, G. L. Britain's bid for the control of the air. Training the British airmen for army and naval service. illus. (Scientific American, New York, 1917, v. 116, p. 422–423, 428–429, 544–545.) **VA**

World War, continued

3894. A **Few** ancestors of Germany's Graf Zeppelin. (Literary digest, New York, 1928, v. 99, Nov. 10, p. 62, 64, 66, 68, 70.) *** DA**

Deals mainly with the fates of German airships during the World war.

3895. The **First** aerial submarine hunt in American history. (Flying, New York, 1917, v. 6, p. 274.) **VDS**

False alarm in March, 1917.

3896. First American-made airplanes shipped. (Iron age, New York, 1918, v. 101, p. 564.) **3–† VIA**

Secretary of War Baker tells how many difficulties were overcome in producing military airplanes.

3897. The **First** coast patrol airship of the United States navy. illus. (Air service journal, New York, 1917, v. 1, p. 245.) **VDS**

3898. The **First** naval ace. New Assistant Secretary of the Navy for Aeronautics has brilliant war record. port. (Aeronautic review, Baltimore, 1929, v. 7, April, p. 13, 22.)

Account of David S. Ingalls. **VDS**

3899. Fischer, Johannes. Zwischen Wolken und Granaten. Berlin: E. S. Mittler & Sohn, 1932. 207 p. 8°. **BTZY**

Experiences of the author, a German aviator, during the World war.

3900. "Flight commander," pseud. Cavalry of the air. Illustrated by Geoffrey Watson. London: Ed. J. Burrow & Co., Ltd. [1918.] xii, 269 p. illus. 12°. **BTZY**

Author's experiences as aviator in the World war. Gives a brief account of development of aviation during that time.

3901. Fonck, René. Mes combats. Paris: Ernest Flammarion [1920]. 249 p. port. 12°. **BTZY**

Famous French ace recounts his experiences during the World war.

3902. Fracchia, Umberto. For the defence of Venice, published by the Special Department of the Ministry of Marine. Milan: Alfieri & Lacroix [1918]. 52 p. illus. 4°. **† BTZE p.v.403, no.15**

Operations during the World war.

3903. Frantzen, L. P. L'évolution du ballon captif pendant la guerre. illus. (La Nature, Paris, 1919, année 47, semestre 2, p. 212–218.) **OA**

Developments during the World war.

3904. Freeman, Lewis Ransome. To Kiel in the "Hercules." New York: Dodd, Mead & Co., 1919. 297 p. illus. 12°. **VYAF**

Record of trip into Germany at the close of the World war, by the Allied Armistice Commission. See especially chapter 5: Nordholz, the den of the Zeppelins.

3905. Freeston, Charles Lincoln. The aeroplane in warfare. illus. (Scribner's magazine, New York, 1915, v. 58, p. 1–18.) *** DA**

Developments and lessons of the first year of the World war.

3906. G., P. de. Évolution de l'aviation allemande pendant la guerre de 1914–1918. (L'Aérophile, Paris, 1919, année 27, p. 12–13.) **VDS**

Shows concisely the successive characteristics of military airplanes in chronological order.

3907. Gamble, Charles Frederick Snowden. Some notes on the activities of naval Zeppelin airships during the war, 1914–1918. illus. port. (Air, London, 1927, v. 1, Dec., p. 19–24.) **VDS**

Portrait of Peter Strasser, head of the German Naval Airship Service, 1913–1918.

3908. —— The story of a North Sea air station; being some account of the early days of the Royal Flying Corps (naval wing) and of the part played thereafter by the air station at Great Yarmouth and its opponents during the war, 1914–1918. London: Oxford University Press, 1928. xvi, 446 p. illus. map. 8°. **BTZY**

Includes a chapter on the development of the Zeppelin airship and the Imperial German airship service during the war. For an extensive review see *Aeroplane,* London, 1928, v. 35, p. 561–566, *VDS.*

3909. Geoffreys, Peter. Airships in the [World] war — and the future. illus. (Aeroplane, London, 1919, v. 17, p. 656B, 656D, 662, 664.) **VDS**

3910. Germany's naval airships and their war record. (Engineer, London, 1922, v. 133, p. 513–514.) **VA**

3911. Giglio, Vittorio. Le ali d'Italia nella nostra guerra. illus. (In: Luigi Segato, L'Italia nella guerra mondiale, Milano, 1927, v. 2, p. 1020–1126.) **BTZE**

Account of Italy's aerial operations during the World war, with the story of D'Annunzio's flight to Vienna.

3912. Goebel, Johannes. Afrika zu unsern Füssen. Lettow-Vorbeck entgegen und andere geheimnisvolle Luftschiffahrten... unter Mitarbeit des Schiffsmeteorologen Dr. Walter Förster. Leipzig: R. F. Koehler, 1925. xi, 124 p. illus. 8°. **BTZY**

Flights of German military airships into Africa and southeastern Europe during the World war.

3913. Gorrell, Edgar S. America's first World war appropriation. (U. S. air services, Washington, 1933, v. 18, May, p. 24–26.) **VDS**

3914. —— History of the [U. S.] Air Corps insignia. illus. (U. S. air services, Washington, D. C., 1933, v. 18, Feb., p. 31.) **VDS**

3915. Goualt, Émile. L'aéronautique maritime. illus. (La Nature, Paris, 1919, année 47, semestre 2, p. 86–91.) **OA**

Development of French seaplanes during the World war.

3916. Grace, Richard. Squadron of Death, by Dick Grace, Captain of the Squadron of

World War, continued

Death. Garden City, N. Y.: Doubleday, Doran & Co., Inc., 1929. viii, 304 p. illus. port. 12°. **AN**

Exciting adventures of moving picture actor who specialized in scenes depicting aerial warfare.

3917. Grahame-White, Claude, and HARRY HARPER. Heroes of the flying corps; a description of the work of the airmen during the early stages of the war...illustrated by Cyrus Cuneo. London: H. Milford [1919]. ix, 11–350 p. illus. 12°. **BTZY**

3918. —— Two years of aerial war. (Fortnightly review, London, 1916, new series, v. 100, p. 195–210.) *** DA**

Concise account of development of air strategy and equipment during the World war.

3919. —— Zeppelin airships: their record in the war [1914–1918]. (Fortnightly review, London, 1915, new series, v. 98, p. 542.) *** DA**

3920. Gran, Tryggve. Mellen Himmel og Jord. Oslo: Gyldendal Norsk Forlag, 1927. 195 p. illus. ports. 8°. **BTZY**

Experiences of the author, a Norwegian aviator, during the World war and after.

3921. Great Britain. — Air Ministry. ...Synopsis of British air effort during the war. Presented to Parliament by command of His Majesty. April, 1919. London: H. M. Stationery Off., 1919. 18 p. 4°. (Cmd. 100.) **† BTZE p.v.486, no.4**

Development of the service and accounts of its cooperation with the army and navy in the different theatres of war. Statistical tables.

3922. Great Britain's war work in the air. (Aeronautics, London, 1919, v. 16, p. 456–462.) **VDS**

3923. Greatest raid in aviation history. (Air power, New York, 1918, v. 4, p. 315.) **VDS**

American bombing squadron of more than 350 machines dropped 32 tons of explosives near Verdun, Oct. 9, 1918.

3924. Green, F. M. Development of the fighting aeroplane. illus. (Aeronautical journal, London, 1922, v. 26, p. 46–62.) **VDS**

Brief historical introduction shows developments beginning during the World war.

3925. Greene, Warwick. Letters of Warwick Greene, 1915–1928. Edited by Richard W. Hale. Boston: Houghton Mifflin Company, 1931. xxiv, 310 p. ports. illus. 8°. **BTZE**

Greene (1879–1929) was active in developing the American aviation organization in France during the World war.

3926. Grey, Charles Grey. Charles Rumney Samson. (Aeroplane, London, 1931, v. 40, p. 234–235.) **VDS**

Biography and appreciation of noted British World war aviator who was active in the organization of the Royal Flying Corps.

3927. —— The chief of the [British] Air Staff. port. (Aeroplane, London, 1933, v. 44, p. 924.) **VDS**

Biography and portrait of Sir Edward Ellington, appointed January, 1933.

3928. —— David Henderson. (Aeroplane, London, 1921, v. 21, p. 165.) **VDS**

Lieutenant-General Henderson was a leader in organizing the Royal Air Force of Great Britain.

3929. —— The defence of London. A remarkable article unveiling the mysteries of our anti-aircraft fights. illus. (London magazine, London, 1918, v. 39, p. 451–458.) *** DA**

Methods employed during the World war.

3930. —— The new chief of the Air Staff. port. (Aeroplane, London, 1929, v. 37, p. 518–519.) **VDS**

Portrait and brief biography of Sir John Salmond, who succeeded Sir Hugh Trenchard as Marshal of the Royal Air Force, Jan. 1, 1930.

3931. —— On the departing chief. ports. (Aeroplane, London, 1929, v. 37, p. 1401–1402, 1404, 1406, 1408, 1410, 1412, 1414, 1416.) **VDS**

Biography and account of the work of Sir Hugh Trenchard in building up the British Air Force during the World war and after.

3932. —— On the passing of the Commodore. (Aeroplane, London, 1917, v. 12, p. 429–430, 432, 434.) **VDS**

Account of Commodore Murray F. Sueter's work in founding the Royal Naval Air Service of Great Britain.

3933. —— On the tenth birthday of the R. A. F. illus. (Aeroplane, London, 1928, v. 34, p. 459–460, 462–464, 468–468B.) **VDS**

History of the Royal Air Force.

3934. —— On the work of the Royal Flying Corps in 1916. (Aeroplane, London, 1917, v. 12, p. 114, 116, 118, 122, 126, 128, 132, 134.) **VDS**

3935. —— Sefton Brancker. ports. (Aeroplane, London, 1930, v. 39, p. 798, 800.) **VDS**

Brancker, killed in the *R 101* disaster, did much to develop the Royal Flying Corps. In later years he became Director of Civil Aviation.

3936. —— Sir Geoffrey Salmond. port. (Aeroplane, London, 1933, v. 44, p. 745–746.) **VDS**

Salmond became chief of the British Air Staff in 1933. He has a distinguished war record and did much for the development of civil aviation.

3937. —— Sir John Salmond. (Aeroplane, London, 1933, v. 44, p. 565–568.) **VDS**

John Maitland Salmond, born 1881, became chief of the British Air Staff in 1930.

3938. —— Vincent Nicholl. port. (Aeroplane, London, 1927, v. 33, p. 576, 578.) **VDS**

Biography and portrait of distinguished British war-time flier.

3939. Hall, Bert. "En l'air!" (In the air.) Three years on and above three fronts. New York: The New Library [1918]. 153 p. illus. port. 12°. **BTZY**

Author's experiences with the Lafayette Escadrille and on the Russian front. Library has also a copy of the London edition entitled *In the air*, 1918, but not illustrated, *BTZY*.

World War, continued

3940. **Hall,** Bert, and J. J. Niles. One man's war; the story of the Lafayette Escadrille. New York: H. Holt and Company [cop. 1929]. vi, 353 p. illus. ports. 8°. **BTZY**

3941. **Hall,** James Norman. High adventure. London: Constable & Co., Ltd., 1918. xix, 232 p. 12°. **BTZY**

Exploits of the Lafayette Escadrille in the World war.

3942. **Hall,** James Norman, and C. B. Nordhoff, editors. The Lafayette Flying Corps ... Boston: Houghton Mifflin Company, 1920. 2 v. 4°. **BTZY**

Volume 1 comprises biographies; v. 2, letters from various members of the corps, describing their experiences. The work is copiously illustrated. Reviewed in *Aviation*, Highland, N. Y., 1920, v. 9, p. 263, *VDS*.

3943. **Hall,** Norman Shannon. The balloon buster, Frank Luke of Arizona. Garden City, N. Y.: Doubleday, Doran & Company, 1928. 191 p. illus. 8°. **BTZY**

Luke, American ace of the World war, won the Congressional medal of honor while under arrest.

3944. **Haslett,** Elmer. Luck on the wing. Thirteen stories of a sky spy. New York: E. P. Dutton & Co. [cop. 1920.] xxvii, 303 p. illus. 12°. **BTZY**

Author's experiences as an American aviator during the World war.

3945. **Hearne,** R. P. The romance of a new key industry. (Aeroplane, London, 1918, v. 15, p. 304.) **VDS**

Brief story of how Great Britain met the demand for magnetos during the World war.

3946. **Henkelburg,** Hans. Als Kampfflieger am Suez-Kanal. Berlin: August Scherl G.m.b.H. [cop. 1917.] 116 p. illus. 12°. **BTZY**

Author's experiences near the Suez Canal during the World war.

3947. **Heydemarck,** Georg. Doppeldecker "666" als Flieger im Westen. Berlin: A. Scherl, G.m.b.H. [cop. 1916.] 112 p. 12°. **BTZE p.v.849**

Experiences of author as German aviator during the World war. Library has also copy of English edition (illustrated), London, 1931, *BTZY;* of the Dutch edition, Amsterdam, 1917, *BTZY;* and of the Swedish edition, Stockholm, 1917, *BTZY.*

3948. —— Feldflieger. Berlin: Dom-Verlag [1930]. 220 p. illus. map. 12°. **BTZY**

Author's experiences as German aviator on the Macedonian front during the World war.

3949. **Hild,** Frederick C. War experiences of an air scout. illus. (Scientific American, New York, 1914, v. 111, p. 520, 530–531; 1915, v. 112, p. 20, 38–40, 51, 58–59.) **VA**

Diary of an American volunteer with the Aviation Corps of the French army.

3950. **Hildesheim,** Erik. German post-bellum reflections on war experiences. (Aeroplane, London, 1919, v. 17, p. 1852, 1854, 1863.) **VDS**

Development of German airplanes, engines, and airships during the World war.

3951. **Hoeppner,** Ernst von. Deutschlands Krieg in der Luft; ein Rückblick auf die Entwicklung und die Leistungen unserer Heeres-Luftstreitkräfte im Weltkriege... Leipzig: K. F. Koehler, 1921. vii, 184 p. tables. 8°. **BTZY**

Authoritative account of German organization and operations of aircraft during the World war. Library has also a copy of the French edition, entitled *L'Allemagne et la guerre de l'air,* Paris, 1923, *BTZY.*

3952. **How** dirigible airships are charged with gas on the battlefield. Portable plants for military use. illus. (Scientific American supplement, New York, 1914, v. 78, p. 155–156.) **VA**

3953. **How** powers are prepared for first aero war. illus. (Aero and hydro, Chicago, 1914, v. 8, p. 233–234.) **3–VDS**

Tabulation of Europe's aerial military equipment in 1914.

3954. **Huge** bombing machines America's most valuable contribution towards Allies' aerial supremacy. (Aerial age, New York, 1917, v. 6, p. 463.) **VDS**

3955. **Huisman,** Georges Maurice. Dans les coulisses de l'aviation 1914–1918... Paris: La Renaissance du livre [1921?]. 341 p. 12°. **BTZY**

Story of the organization and development of the French air service during the World war.

3956. **Hunt,** Frazier. The life story of Eddie Rickenbacker. How...one of America's greatest aces battled his way up from obscurity to become a national hero. port. (Popular aviation, Chicago, 1932, v. 11, p. 37–38.) **VDS**

3957. **Ide,** John Jay. The development of French naval aviation. illus. (Aviation, New York, 1919, v. 7, p. 340–345.) **VDS**

Seaplanes developed during the World war.

3958. **Identification** of German aeroplanes. illus. (Flight, London, 1917, v. 9, p. 368, 393, 415, 445, 475.) **VDS**

3959. L'**Illustrazione italiana.** Il volo della "Serenissima." Numero speciale. La gesta di Vienna, Km. 1100 in ore 6.40 9 agosto, 1918. 17 p. ports. illus. 4°. **† BTZX**

Gabriele d'Annunzio's flight to Vienna during the war. Proclamations were scattered by him over the city.

3960. **Immelmann,** Max Franz. Meine Kampfflüge, selbsterlebt und selbsterzählt. Berlin: August Scherl [1916]. 132 p. illus. ports. 12°. **BTZE p.v.717, no.4**

Letters of famous German ace during the World war, to his mother.

3961. **Italian** naval aircraft. illus. (Engineer, London, 1918, v. 125, p. 6–8.) **VA**

World War, continued

3962. **Jane,** Frederick Thomas. Submarines and aircraft. (Edinburgh review, New York, 1915, v. 221, p. 140–150.) *** DA**

3963. **Jeanjean,** Marcel. Sous les cocardes. Scènes de l'aviation militaire. Préface du capitaine Madon. Paris: Hachette [cop. 1919]. 38 p. plates. f°. **† BTZY**

Amusing colored caricatures of life in French aviation camps.

3964. **Johns,** William Earl. Fighting planes and aces. London: John Hamilton [1932]. 90 p. illus. ports. 8°. **VDYD**

Pictures and descriptions of planes of various countries during the World war. Portraits and brief accounts of exploits of aces.

3965. **Jones,** H. A. Over the Balkans and south Russia; being the history of No. 47 Squadron, Royal Air Force. London: Edward Arnold, 1923. xv, 176 p. illus. 8°. **BTZY**

3966. **Joynson-Hicks,** Sir William. The command of the air, or prophecies fulfilled. Being speeches delivered in the House of Commons. London: Nisbet & Co., Ltd. [1916.] 192 p. 12°. **BTZE**

Shows the unpreparedness of Great Britain for aerial defence and attack.

3967. **Jünke,** Arnold. Zeppelin im Weltkriege. Leipzig: Abel und Müller [1916]. 224 p. 12°. **BTZY**

Deeds of the German airships during the World war.

3968. **Kaempffert,** Waldemar Bernhard, and Carl Dienstbach. The armored flying tank. illus. (Popular science monthly, New York, 1918, v. 93, p. 37–40.) *** DA**

Account of the German Gothas.

3969. **Kaempffert,** Waldemar Bernhard. The eagle spreads his wings: how 23,625 American airplanes will be built at the rate of 3,500 a month to crowd the Germans out of the air. illus. (Popular science monthly, New York, 1917, v. 9, p. 894–898.) *** DA**

3970. **Kerr,** Mark Edward Frederic. Land, sea, and air. London and New York: Longmans, Green and Co., 1927. x, 406 p. illus. ports. 8°. **VYG**

Reminiscences of British admiral who was one of the originators of the Air Force. He attempted to fly the Atlantic in 1919. Reviewed in *Air*, London, 1928, v. 1, Feb., p. 47, *VDS*.

3971. **Kershaw,** John Baker Cannington. The European war from an engineer's standpoint. The conquest of the air and its results on warfare. illus. (Engineering magazine, New York, 1915, v. 48, p. 498–507.) **VDA**

3972. **Kingsford,** A. R. Night raiders of the air. Being the experiences of a night flying pilot, who raided Hunland on many dark nights during the war. London: John Hamilton, Ltd. [1930.] 222 p. illus. 8°. **BTZY**

3973. **Knappen,** Theodore Macfarlane. Wings of war; an account of the important contribution of the United States to aircraft invention, engineering, development and production during the World war, with an introduction by Rear-Admiral D. W. Taylor. New York and London: G. P. Putnam's Sons, 1920. xv, 289 p. illus. 8°. **VDYD**

3974. **Knowles,** James. The death of Quentin Roosevelt. illus. port. (Popular aviation, Chicago, 1931, v. 8, Feb., p. 10–12, 53.) **VDS**

3975. **Köhl,** Hermann. Bremsklötze weg! Das Lebensbuch eines deutschen Fliegers. Berlin: Sieben-Stäbe-Verlag, 1932. 228 p. illus. port., maps. 8°. **BTZY**

Autobiography of noted German aviator — his exploits in the World war and his trans-Atlantic flight on the *Bremen*, 1928.

3976. **Kranzler,** Wilhelm. Bezwinger der Luft im Weltkriege; siegreiche Fliegerkämpfe u. Luftschifffahrten unserer grossen Helden, Immelmann, Zeppelin, Boelcke, Parschau, Banfield u. a... Mit bunten Bildern von Kurt Schütz. Berlin-Charlottenburg: J. Hermann Michel [1916?]. 192 p. illus. 8°. **BTZY**

3977. **Lafont,** Bernard. Au ciel de Verdun. Notes d'un aviateur. Paris: Berger-Levrault, 1918. xii, 201 p. 12°. **BTZY**

Experiences of author, a French aviator during the World war.

3978. **La Frégeolière,** Renaud de. À tire d'ailes; carnet de vol d'un aviateur, et Souvenirs d'un prisonnier. Préface de M. René Bazin. Paris: Plon-Nourrit et Cie, 1916. ix, 311 p. [5. ed.] 12°. **BTZY**

Well-written diary of the French military pilot during the World war.

3979. **La Grange,** Amaury de. Early battles in the air. port. (Air service journal, New York, 1917, v. 1, p. 17.) **VDS**

3980. **Lagrange,** L. La fin d'un héros. Le capitaine Georges Guynemer tombe au champ d'honneur. port. (L'Aérophile, Paris, 1917, année 25, p. 337–338.) **3–VDS**

Obituary notice of French military aviator killed in 1917.

3981. **Lamb,** Dean Ivan. Evolution of the fighting airplane. illus. (Air service journal, New York, 1918, v. 3, p. 337–339.) **VDS**

3982. **Laredo de Mendoza,** Saverio. Gabriele d'Annunzio, aviatore di guerra; documenti e testimonianze... Milano: Impresa editoriale italiani [1930]. 518 p. illus. ports., facsims. 8°. **BTZY**

World war exploits of the Italian poet, as shown by correspondence and documents.

3983. The **Late** Captain Strasser. (Aeroplane, London, 1918, v. 15, p. 878, 880.) **VDS**

Brief account of Strasser's airship attacks on England during the World war.

World War, continued

3984. Lefranc, Jean Abel. L'évolution de l'aviation allemande. illus. (La Nature, Paris, 1918, année 46, semestre 1, p. 81–87, 161–170, 289–295; semestre 2, p. 17–24, 81–90.) **OA**

Development of German planes, instruments, accessories, respiratory apparatus, engines, and aerial photography during the period of the World war.

3985. —— German technique in aerial armament. illus. (Aviation, New York, 1918, v. 4, p. 95–98.) **VDS**

3986. Leigh, Howard. Planes of the Great war; with a foreword by Capt. W. E. Johns. London: J. Hamilton, Ltd. [1934.] 111 p. 8°. **VDYD**

Brief descriptions and mounted pen sketches of planes of England, Germany, and France.

3987. Lieutenant-Colonel George O. Squier. port. (Aviation, New York, 1916, v. 1, p. 114.) **VDS**

Portrait and brief biography of the Chief of Aviation in the United States Army during the World war.

3988. Lieut.-Col. George O. Squier, the man in charge of the aeronautical branch of our land defences. ports. (Flying, New York, 1916, v. 5, p. 331–333.) **VDS**

3989. Der **Luftkrieg,** 1914–1915, unter Verwendung von Feldpostbriefen und Berichten von Augenzeugen dargestellt von einem Flugtechniker. Leipzig: Hesse & Becker, 1915. 278 p. illus. 16°. **BTZE**

Equipment and operations of Germany and its enemies during the World war.

3990. McConnell, James Rogers. Flying for France. With the American Escadrille at Verdun. Garden City, N. Y.: Doubleday, Page & Co., 1917. xiv, 157 p. ports. illus. 12°. **BTZE**

3991. McCudden, James Thomas Byford. Five years in the Royal Flying Corps. London: "Aeroplane" & General Publishing Co., Ltd. [1918.] xvi, 348 p. illus. ports. 12°. **BTZY**

Author's experiences during the World war. Reviewed in *Aeroplane,* London, 1930, v. 39, p. 1216, 1218, *VDS.*

3992. Macmillan, Norman. Into the blue. London: Duckworth [cop. 1929]. 213 p. illus. port. 8°. **BTZY**

Well-written account of Captain Macmillan's experiences as aviator and instructor during the World war. Reviewed in *Air,* London, 1930, v. 3, p. 42–43, *VDS; Aeroplane,* London, 1929, v. 37, p. 1162, *VDS.*

3992a. Magoun, Frederick Alexander, and Eric Hodgins. The effect of the war on aviation. (In their: A history of aircraft, New York, 1931, p. 380–396.) **VDT**

The beginnings. Strategic uses. Reconnaissance. New aerial technique. Fokker and the new warfare. Engine evolution during the war. Speed and stability. Airplane weapons. Use of seaplanes. Aerial photography. The development of maneuvering. The record of aces. America's wartime aviation. Final results.

3993. Malkowsky, Emil Ferdinand. Vom Heldenkampf der deutschen Flieger. Ein Ruhmesbuch der deutschen Tapferkeit. Berlin: Askanischer Verlag, 1916. 198 p. illus. 8°. **BTZY**

Deeds of German aviators during the World war.

3994. Marben, Rolf. Ritter der Luft. Zeppelinabenteuer im Weltkrieg. Hamburg: Broschek & Co. [1931.] 191 p. port. illus. 12°. **BTZY**

Deeds of the Zeppelins during the World war, with a chronological record of their fates. Library has a copy of the English edition, translated by Claud W. Sykes, London [1932], under the title *Zeppelin adventures, BTZY.* The latter is reviewed in *Air and airways,* London, 1932, v. 9, p. 145, *VDS.*

3995. Marc. Notes d'un pilote disparu (1916–1917). Paris: Hachette, 1918. 223 p. 12°. **BTZY**

Diary of Pilot X, who was lost in combat, 1917.

3996. Markham, Christopher Alexander. Account of air raids in Northamptonshire by German Zeppelins during the great war of 1914–18. illus. (Associated Architectural Societies, Reports and papers, Lincoln, 1919, v. 35, part 1, p. 41–60.) **3–MQA**

3997. Marsh, William Lockwood. The aerial bombardment of the "Goeben" and "Breslau" in January, 1918. (Airways, London, 1927, v. 3, p. 160.) **VDS**

3998. —— The role of British airships in the war. illus. (Aeronautics, London, 1919, v. 16, p. 13–17.) **VDS**

"It was in February, 1915, that the real work of the airship service began."

3999. Martin, Robert E. Tony Fokker and the World war. Air battles given new fury when Dutch boy hooks machine gun to propeller. Rivals plan his ruin. illus. (Popular science monthly, New York, 1931, v. 118, June, p. 24–26, 148; v. 119, July, p. 38–39, 131–132.) *** DA**

4000. Massenet de Marancour, C. La chasse en avion. Paris: R. Chiberre [cop. 1921]. 112 p. 12°. **BTZE p.v.620, no.4**

A study of airplane tactics during the World war.

4001. Messrs. Henry A. Wise Wood and Alan R. Hawley force Secretary Daniels to publish the suppressed report of the General Board of the Navy. (Flying, New York, 1916, v. 4, p. 809–810.) **VDS**

Original report stated to have inadequately met the aerial needs of the nation.

4002. Meyer, Friedrich Albert. Immelmann und Boelke, deutsche Helden der Luft. Warendorf: C. Leopold [1917?]. 96 p. 16°. **BTZY**

Exploits of two famous German aviators during the World war.

4003. Middleton, Edgar Charles. Glorious exploits of the air. New York: D. Appleton & Co., 1918. 256 p. illus. 12°. **VDY**

Methods of aerial combat during the World war, with an account of the German air services.

World War, continued

4004. —— The great war in the air...with an introduction by Brigadier-General Lord Montagu of Beaulieu... London: The Waverley Book Co., Ltd., 1920. 4 v. illus. maps. 8°. **BTZY**

Critically reviewed in *Aeroplane*, London, 1920, v. 18, p. 130–132, *VDS*.

4005. —— The kingdom of the air. London: Ed. J. Burrow & Co., Ltd. [1919.] vii, 252 p. illus. 12°. **BTZY**

Airplanes and airships during the World war and after.

4006. —— Tails up. London: Simpkin, Marshall, Hamilton, Kent & Co., Ltd. [1918.] 314 p. 12°. **BTZY**

Dramatic incidents in the Royal Flying Corps during the World war.

4007. —— The way of the air. London: William Heinemann [1917]. viii, 184 p. 12°. **VDY**

How the aerial operations were carried on during the World war. Also diary of a British air pilot. Brief account of airship evolution. Reviewed in *Aeroplane*, London, 1917, v. 12, p. 1654, *VDS*.

4008. **Mr.** Godfrey L. Cabot's generous contribution to aeronautical development. port. (Aerial age weekly, New York, 1917–18, v. 6, p. 10.) **VDS**

Boston citizen gives $30,000 for the development of military airplane, 1917.

4009. **Mitchell,** William. Military aeronautics. illus. (Engineers' Club of Philadelphia, Proceedings, Philadelphia, 1917, v. 34, p. 324–330.) **VDA**

War preparations in the United States described by the Chief of Staff of the Aviation Corps. Mexican experiences.

4010. —— Our air force, the keystone of national defense. New York: E. P. Dutton & Co. [cop. 1921.] xxvi, 223 p. illus. map. 8°. **VDYD**

Contains excellent accounts of how the United States air forces met the problems of the World war.

4011. **Mixter,** George Webber, and H. H. Emmons. United States Army aircraft production facts, compiled at the request of the Assistant Secretary of War, January, 1919. Washington: Gov. Prtg. Off., 1919. 106 p. 8°. **VDY p.v.14, no.7**

Account of war-time activities, including engines, airplanes, machine guns, aerial bombs, the use of pyrotechnics, photography, oxygen for the aviator, the problem of aviators' clothing, radio, and the kite balloon.

4012. **Moffett,** William Adger. The [United States] Navy's record in aeronautics; statement of Admiral Moffett before the Naval Appropriation Committee, regarding post-war development of naval aviation. illus. (Aviation, Highland, N. Y., 1922, v. 12, p. 720–732.) **VDS**

4013. **Molter,** Bennett Arthur. Knights of the air. New York: D. Appleton & Co., 1918. 243 p. illus. 12°. **BTZY**

Describes methods of training and of fighting during the World war.

4014. **Morris,** Joseph. The German air raids on Great Britain, 1914–1918. London: Sampson Low, Marston & Co., Ltd. [1925.] xii, 306 p. illus. maps. 8°. **BTZE**

4015. **Morse,** Daniel Parmelee. The history of the 50th Aero Squadron; being the log and operations of the squadron while with the American Expeditionary Force in France, 1918–19, with other sundry items of interest to former members of the squadron and personal recollections of the author [D. P. Morse]. [New York: Privately printed, 1920.] 94 p. 8°. **BTZY**

4016. **Mortality** statistics of the air. (Automotive industries, New York, 1920, v. 42, p. 190–191.) **TOL**

Wartime fatalities of airmen of the Allied armies during the World war.

4017. **Mortane,** Jacques. Les ailes de la mort ... Paris: Éditions du Siècle [1932]. 252 p. 12°. **BTZY**

Efforts of Gen. Joffre to develop aviation during the early years of the World war. Aerial battles of July, 1918. Horrors of future aerial warfare.

4018. —— Les as, peints par eux-mêmes. Préface de M. Armand Rio. Étude sur les héros disparus suivie d'anecdotes... Paris: Alphonse Lemerre, 1917. 348 p. 12°. **BTZY**

Exploits of Allied aces. The preface gives a concise account of aerial operations in general. Reviewed in *Aeroplane*, London, 1918, v. 15, p. 194, *VDS*.

4019. —— Chasseurs de Boches. Paris: L'Édition française illustrée [cop. 1917]. 310 p. 12°. **BTZY**

Exploits of Allied aces during the World war. Reviewed in *Aeroplane*, London, 1918, v. 15, p. 194, *VDS*.

4020. —— La guerre des ailes (1914–1918). À travers les filets de l'ennemi. Paris: Éditions Baudinière [cop. 1929]. 270 p. 12°. **BTZY**

Exploits of Allies' aviators during the World war.

4021. —— La guerre des ailes (1914–1918). Navarre, sentinelle de Verdun. Paris: Éditions Baudinière [1930]. 319 p. 12°. **BTZY**

Biography of Jean Navarre, French ace.

4022. —— Guynemer, the ace of aces, translated by Clifton Harby Levy, together with transcripts from Guynemer's own note-book of flight, and photographic fac-similes of its pages. New York: Moffat, Yard & Co., 1918. xxxii, 267 p. ports. illus. 12°. **BTZY**

Reviewed in *Aviation*, New York, 1918, v. 5, p. 433, *VDS*.

4023. —— ...Les héros de l'air... Paris: Librairie Delagrave, 1930. 191 p. ports. illus. 8°. **VDY**

History of the heavier-than-air machine with accounts of famous flights.

4024. —— Special missions of the air. An exposition of some of the mysteries of aerial warfare. London: Aeroplane & General Publishing Co., Ltd., 1919. 131 p. 12°. **BTZY**

Stories of especially hazardous assignments during the World war. Library has also a copy of French edition, Paris [1929] entitled *Missions spéciales*. Reviewed in *Aeroplane*, London, 1919, v. 17, p. 1514, *VDS*.

World War, continued

4025. —— Les vols émouvants de la guerre. Paris: Éditions Pierre Lafitte [1917]. 287 p. 12°. **BTZE**

Tales of French heroism during the World war.

4026. Mortane, Jacques, and Jean Daçay. La guerre des nues racontée par ses morts. Préface du lieutenant Fonck. Paris: L'Édition française illustrée, 1918. viii, 306 p. 12°. **BTZY**

Letters of celebrated aces of the Allies during the World war. Reviewed in *Aeroplane,* London, 1918, v. 15, p. 1932, 1934, 1936, *VDS.*

4027. Munday, Albert. Sky high. The thrilling story of Canadian flying men who defeated the Hun. illus. (Canadian magazine, Toronto, 1927, v. 68, Nov., p. 8–11, 34; Dec., p. 10, 47–48.) *** DA**

4028. Munsey, Frank Andrew. The aeroplane a short cut to the winning of the war. (Munsey's magazine, New York, 1917, v. 61, p. 401–404.) *** DA**

4029. Nadaud, Marcel. Guynemer, l'as des as. Préface du lieutenant-colonel aviateur Girod. Paris: Albin Michel [1918]. 125 p. port. illus. 8°. **AN**

Life and deeds of French "ace of aces" during the World war.

4030. Naval aircraft. (Naval annual, London, 1919, p. 265–269.) **VXA**

Brief account of the work of British seaplanes during the World war.

4031. Naval Eight; a history of No. 8 Squadron R. N. A. S. — afterwards No. 208 Squadron R. A. F. — from its formation in 1916 until the Armistice in 1918. London: The Signal Press, Ltd., 1931. 207 p. illus. ports. 8°. **BTZY**

Reviewed in *Flight,* London, 1931, v. 23, p. 1222, *VDS.*

4032. Neon, pseud. The great delusion; a study of aircraft in peace and war; with a preface by A. H. Pollen. New York: The Dial Press, 1927. xxxix, 279 p. 8°. **VDY**

Deprecates the value of aeronautics in the World war. The defense of London. For criticism see entry no. 4090; also *U. S. air services,* Washington, 1927, v. 12, Sept., p. 47–48, *VDS.*

4033. Neumann, Georg Paul. Die deutschen Luftstreitkräfte im Weltkriege, unter Mitwirkung von 29 Offizieren und Beamten der Heeres- und Marine-Luftfahrt nach amtlichen Quellen. Berlin: E. S. Mittler und Sohn, 1920. x, 600 p. illus. 8°. **BTZY**

Comprehensive, authoritative account of development of aviation in Germany during the World war, especially of equipment. The Library has also a copy of the abridged English edition, translated by J. E. Gurdon, London: Hodder & Stoughton, Ltd., 1920, *VDYD.*

4034. —— In der Luft unbesiegt. Erlebnisse im Weltkrieg erzählt von Luftkämpfern. München: J. F. Lehmann [cop. 1923]. 324 p. ports. 8°. **BTZY**

Numerous German aviators contribute records of their experiences during the World war.

4035. New England aviators, 1914–1918, their portraits and their records. With an introduction by A. Lawrence Lowell. Boston: Houghton Mifflin Company, 1919–20. 2 v. ports. 8°. **AGZ (New England)**

4036. 1914–1918. L'aéronautique pendant la guerre mondiale. Paris: M. de Brunoff, 1919. viii, 735 p. illus. sq. f°. **† VDW (Nineteen)**

Copiously illustrated *de luxe* volume by various authors, covering French activities during the World war. Records of aces; colored emblems of the various escadrilles; military and naval operations; development of motors, instruments, and parachutes; statistics of manufacture, with accounts of the work of notable French firms.

4037. Noble, Walter. With a Bristol fighter squadron. With an introduction by the Right Hon. Winston Churchill. London: Andrew Melrose, Ltd., 1920. 186 p. port. illus. 16°. **BTZY**

Experiences of British aviator during the World war.

4038. Nordhoff, Charles Bernard. The fledgling. Boston: Houghton Mifflin Co., 1919. 201 p. 12°. **BTZY**

Diary of American aviator during the World war.

4039. Norman Prince, a volunteer who died for the cause he loved, with memoir by George F. Babbitt. Boston: Houghton Mifflin Company, 1917. 75 p. illus. ports. 8°. **BTZY**

Memorial volume concerning the career of American aviator of the World war, who was one of the organizers of the Lafayette Escadrille.

4040. Nunn, Wilfrid. Tigris gunboats; a narrative of the Royal Navy's co-operation with the military forces in Mesopotamia from the beginning of the war to the capture of Bagdad (1914–17), by Vice-Admiral Wilfrid Nunn. London: A. Melrose, Ltd. [1932.] 288 p. illus. maps, port. 8°. **VYAF**

Reviewed in *Air and airways,* London, 1932, v. 9, p. 326, *VDS.*

4041. Official history of aircraft production. Objects, problems, production of entire air program laid bare in conjunction with complete exhibition of planes and equipment. (Automotive industries, New York, 1918, v. 39, p. 968–969, 987–990.) **TOL**

World war developments in the United States.

4042. Ogilvie, Carl B. Battle tactics of German fighters. illus. (Popular aviation, Chicago, 1932, v. 11, p. 8–9.) **VDS**

4043. —— The career of Frank Luke — the balloon buster. illus. (Popular aviation, Chicago, 1932, v. 11, p. 280–282, 332.) **VDS**

War hero whose specialty was the bringing down of "sausage" balloons.

World War, continued

4044. —— A chat with Lothar von Richthofen. illus. port. (Popular aviation, Chicago, 1933, v. 12, p. 76–78, 127.) **VDS**

Brother of the famous ace, Manfred von Richthofen, tells of the plan to rescue the latter's body from the enemy.

4045. —— Combat methods used by leading aces. illus. (Popular aviation, Chicago, 1932, v. 10, p. 348–350, 402–403.) **VDS**

4046. —— Guynemer — the man of mystery. ports. (Popular aviation, Chicago, 1933, v. 12, p. 351–352, 398.) **VDS**

Author suggests that Guynemer may not have been killed by the Germans.

4047. On the aeronautical history of 1916. (Aeroplane, London, 1917, v. 12, p. 5–6, 10, 14, 16, 18, 20, 97–98, 102, 106, 108, 110.) **VDS**

Month by month record of naval and military operations, comparative losses, foreign and British events.

4048. Orthlieb. L'aéronautique, hier — demain. Préface du c[ommandant] Wateau. Paris: Masson & Cie., 1920. xii, 292 p. illus. maps. 12°. (Les leçons de la guerre.) **VDW**

History of aerial operations during the World war.

4049. —— Historique de l'emploi de l'aviation pendant la guerre 1914–1918. illus. (Aéronautique, Paris, 1919, année 1, p. 9–13, 49–52, 112–117, 177–179.) **VDS**

History of French operations.

4050. Osman, William H. Message carrying by pigeons from aircraft. illus. port. (Aeronautical digest, New York, 1923, v. 2, p. 347–349, 375–376, 381.) **VDS**

Interesting experiences of the Royal Air Force during the World war.

4051. Our eagle learns to fly. illus. (Scientific American, New York, 1917, v. 116, p. 170–171, 187.) **VA**

Account of training methods at the Mineola Army Aviation School.

4052. Our technical achievements in the great war. Providing the equipment and training for a great flying force. illus. diagrs. (Scientific American, New York, 1919, v. 121, p. 208–209.) **VA**

4053. Paine, Ralph Delahaye. The first Yale unit. A story of naval aviation, 1916–1919. Cambridge, Mass.: Riverside Press, 1925. 2 v. illus. ports. 8°. **STG (Yale)**

4054. Patrick, Mason Mathews. Cost of our wartime aircraft. (New York Times current history, New York, 1922–23, v. 17, p. 783–785.) **BAA**

United States' expenses during the World war.

4055. —— The United States in the air... introduction by F. Trubee Davison... Garden City, N. Y.: Doubleday, Doran and Company, Inc., 1928. xi, 191 p. illus. 8°. **VDYD**

"General Patrick discusses the many perplexing problems of wartime aviation with which he had to deal. He also paints one of the clearest pictures ever presented of American air operations over the front, behind the lines, and at home." — *Introduction.* Reviewed in *Air,* London, 1929, v. 2, p. 43, *VDS.*

4056. Pattinson, Lawrence Arthur. History of 99 Squadron, Independent Force, Royal Air Force, March, 1918 - November, 1918. Cambridge, England: W. Heffer & Sons, Ltd., 1920. 73 p. illus. map. 8°. **BTZY**

Some excellent aerial views of bombing operations. Reviewed in *Aeroplane,* London, 1921, v. 20, p. 128, 142, *VDS.*

4057. Pégoud est mort au champ d'honneur. port. (L'Aérophile, Paris, 1915, année 23, p. 195–196.) **3–VDS**

Famous loop-the-loop aviator killed in the French service, Aug. 31, 1915.

4058. Percheron, Maurice. Influence de la guerre sur la construction d'aviation en Europe et en Amérique. illus. (L'Aérophile, Paris, 1917, année 25, p. 122–125, 162–172.) **3–VDS**

Development of planes and engines during the World war.

4059. Perfetti, R. Italian airplane development. illus. port. (Air service journal, New York, 1917, v. 1, p. 89–91.) **VDS**

World war developments.

4060. Poirier, Jules. Les bombardements de Paris (1914–1918), avions, Gothas, Zeppelins, Berthas... Préface du général Niessel. Paris: Payot, 1930. 324 p. illus. 8°. **BTZY**

Includes history of development of military aviation in France and Germany. Bibliography.

4061. Pollard, Alfred Oliver. The Royal Air Force, a concise history. London: Hutchinson & Co., Ltd. [1934.] 288 p. illus. 8°. **VDYD**

4062. Porter, Harold Everett. Aerial observations. The airplane observer, the balloon observer, and the army corps pilot, by Harold E. Porter ("Holworthy Hall"). New York: Harper & Brothers [cop. 1921]. 356 p. port. illus. 12°. **VDY**

This book, for the lay reader, tells of development of the art during the World war.

4062a. Prepositi, Clemente. ...I cavalieri dell' aria; il primato italiano nella guerra aerea (1911–1912 — 1915–1918). Bologna: L. Cappelli [1933]. 253 p. 12°. **VDYD**

4063. —— La guerra nel cielo. Palermo: Remo Sandron, 1930. 192 p. illus. ports. 16°. **BTZY**

Brief history of use of aeronautics in war and the part played by Italy in the World war.

4064. The **Prospects** of aerial fighting in the present war. What may be expected of dirigibles and aeroplanes. illus. (Scientific American, New York, 1914, v. 111, p. 131–133.) **VA**

4065. Raleigh, Sir Walter Alexander, and H. A. Jones. The war in the air; being the story of the part played in the great war

World War, continued

by the Royal Air Force. Oxford: The Clarendon Press, 1922–31. 4 v. facsim., illus. (maps.) 8°. **BTZY**

Based on official documents. v. 2–3 are by H. A. Jones. Reviewed in *Flight*, London, 1929, v. 21, p. 15; 1931, v. 23, p. 611, *VDS; Aeroplane*, London, 1928, v. 34, p. 926, *VDS*. For a review of v. 1, see *Aeroplane*, London, 1922, v. 23, p. 121–123, *VDS*.

4066. Rawlinson, Alfred. The defence of London, 1915–1918, with introduction by Admiral Sir Percy Scott. London and New York: Andrew Melrose, 1923. xiii, 267 p. 2. ed. illus. 12°. **BTZY**

4067. Rear Admiral Peary showing us the way to attain air power just as Rear Admiral Mahan showed us the way to attain sea power. (Aerial age weekly, New York, 1917, v. 4, p. 453–454.) **VDS**

4068. Record of performance of American planes; this table serves to illustrate the numerous types of airplanes built by the U. S. government since June, 1917... (Automotive industries, New York, 1919, v. 40, p. 103.) **TOL**

4069. Reece, Robert Henry. Night bombing with the Bedouins, by one of the squadron, Robert H. Reece, lieut. D. F. C., R. A. F. Boston: Houghton Mifflin Company, 1919. 5 p.l., 99(1) p. illus. port. 12°. **BTZY**

How night bombing was effected during the World war.

4070. Reid, A. Cunningham. Planes and personalities, a pot-pourri, with an introduction by Lt.-Col. W. A. Bishop. London: Philip Allan & Co., 1920. 190 p. illus. 12°. **BTZY**

Descriptions of air-fighting methods during the World war. "The great war effected at least twenty years' progress in the history of flying."
Reviewed in *Aeroplane*, London, 1920, v. 18, p. 703–704, *VDS*.

4071. Renard, Paul. L'évolution de l'aéronautique pendant la guerre. illus. (Société d'encouragement pour l'industrie nationale, Bulletin, Paris, 1921, année 120, tome 133, p. 21–51.) **VA**

Developments during the World war.

4072. René Fonck, the French ace of aces. port. (Popular aviation, Chicago, 1932, v. 11, p. 148, 194.) **VDS**

4073. Richardson, James M. The rescue of the lost battalion. illus. (Popular aviation, Chicago, 1931, v. 8, May, p. 8–10, 62–63.) **VDS**

4074. —— The suicide patrol. The grim history of the day bombers. illus. (Popular aviation, Chicago, 1931, v. 9, Aug., p. 6–8, 62–63.) **VDS**

Story of "Major Brown's party" during the World war.

4075. —— The wartime diary of Clifford Allsopp — bomber pilot. A gripping account of the exciting experiences of an American pilot as revealed in his personal diary. illus. (Popular aviation, Chicago, 1931, v. 9, Oct., p. 8–10, 58, 60.) **VDS**

4076. Richthofen, Manfred Albrecht, Freiherr von. The red battle flyer. Translated by T. Ellis Barker, with a preface and notes by C. G. Grey. New York: Robert M. McBride & Co., 1918. vii, 222 p. 12°. **BTZY**

Germany's celebrated ace tells of his exploits.

4077. Rickenbacker, Edward Vernon. Fighting the flying circus, with maps and a foreword by Laurence La Tourette Driggs. New York: Frederick A. Stokes Co. [cop. 1919.] 371 p. port. 12°. **BTZY**

Exciting exploits of the United States "Hat-in-the-ring" squadron during the World war. Rickenbacker was commander of this organization and America's foremost ace.

4078. Roberts, E. M. A flying fighter. An American above the lines in France. New York: Harper & Brothers [cop. 1918]. 339 p. illus. port. 12°. **BTZY**

Deeds of an American member of the Royal Flying Corps during the World war.

4079. Roy, Jack Martin. What the war has done for aviation. illus. (Everyday engineering magazine, New York, 1918, v. 4, p. 213–214, 238–241, 242, 244, 246–247.) **VA**

4080. Samson, Charles Rumney. Fights and flights. London: E. Benn, Ltd. [1930.] xii, 372 p. illus. maps. 8°. **BTZE**

Experiences of the author, a British air commodore during the World war, on the Belgian coast and at the Dardanelles.
Reviewed in *Aeroplane*, London, 1930, v. 39, p. 1213, 1215–1216, *VDS*.

4081. Scaroni, Silvio. Impressioni e ricordi di guerra aerea. Torino: Stengrafica [pref. 1922]. 143 p. illus. ports., map. 8°. **BTZE p.v.707, no.11**

Letters from the author, who served as aviator on the Austro-Italian front during the World war.

4082. Schäffer, Ernst. Pour le mérite. Flieger im Feuer. Berlin: Union deutsche Verlagsgesellschaft [cop. 1931]. 178 p. 8°. **BTZY**

Deeds and portraits of German aces who achieved this medal.

4083. Schiller, Hans von. Acht Jahre Marineluftschiffahrt. illus. (Luftfahrt, Berlin, 1922, Jahrg. 26, p. 98–101.) **VDS**

History of German airships during the war period, 1914–1918, with tabular details. Includes statistics of disasters.

4084. Schofield, Harry Methuen. The high speed and other flights. London: John Hamilton, Ltd. [1932.] 270 p. illus. port. 8°. **VDY**

Author, a British aviator, tells of his training, experiences on the Eastern front during the World war, and of post-war days, especially the winning of the Schneider cup at Venice. Also briefly relates a flight with the autogyro.

World War, continued

4085. Scott, A. J. L. Sixty Squadron, R. A. F. A history of the squadron from its formation...with a preface by the Rt. Hon. Lord Hugh Cecil. London: William Heinemann, 1920. xx, 145 p. illus. 8°. **BTZY**

This was the squadron of Capt. A. Ball and Lieut.-Col. W. A. Bishop. Reviewed in *Aeronautics,* London, 1921, v. 20, p. 127–128, *VDS.*

4085a. Sherman, William Carrington. Air warfare. New York: The Ronald Press Company [cop. 1926]. ix, 307 p. illus. 8°. **VDYD**

Has much of interest concerning lessons learned during the World war.

4086. Smith, Earl Hamilton. Colonel John B. Bennet, officer in charge of Aeronautic Division, Signal Corps, United States Army. port. (Flying, New York, 1917, v. 6, p. 201–202.) **VDS**

4087. The **Spider** web. The romance of a flying-boat war flight. By P. I. X. [pseud.] Edinburgh: Wm. Blackwood & Sons, 1919. x, 278 p. illus. 8°. **BTZE (Spider)**

Early history of development of flying boats for warfare on submarines and excellent account of how attacks were carried out.

4088. Strahlmann, Fritz, editor. Zwei deutsche Luftschiffhäfen des Weltkrieges, Ahlhorn und Wildeshausen. Erinnerungen und Aufsätze... Oldenburg i. O., 1926. 306 p. illus. ports., plans. 8°. **BTZE**

Reminiscences of two German airship stations during the World war, by several officers.

4089. Strain, L. H. Aircraft versus submarines. (Flight, London, 1921, v. 13, p. 46–48.) **VDS**

Record of air defense during the World war.

4090. Sueter, Murray Fraser. Airmen or Noahs: fair play for our airmen. The great "Neon" air myth exposed. With illustrations in colour by W. Russel Flint, and illustrations in black and white by E. G. O. Beuttler. London: Sir Isaac Pitman & Sons, Ltd., 1928. xxvii, 448 p. illus. 8°. **VDYD**

A defense of British air operations during the World war. Post-war developments. The value of long-distance flights. The Schneider cup race of 1927. For the provocative book by "Neon" see entry no. 4032. Reviewed in *Air,* London, 1929, v. 2, p. 191, *VDS; Aeroplane,* London, 1928, v. 34, p. 878, *VDS.*

4091. Sullivan, Alan, compiler. Aviation in Canada, 1917–1918. Being a brief account of the work of the Royal Air Force, Canada, the Aviation Department of the Imperial Munitions Board and the Canadian Aeroplanes, Limited. Toronto: Rous and Mann, Ltd. [cop. 1919.] 318 p. illus. 8°. **VDY**

Deals mainly with history of organization and of training methods. Copiously illustrated.

4092. Sutton, B. E. Some aspects of the work of the Royal Air Force with the B. E. F. in 1918. (Royal United Service Institution, Journal, London, 1922, v. 67, p. 336–348.) **VWA**

Comments on artillery observation, photography, reconnaissance, communication, air-fighting, bombing, low flying, transportation, and cooperation with tanks and with cavalry.

4093. Sweetser, Arthur. The American air service; a record of its problems, its difficulties, its failures, and its final achievements ...with an introduction by Newton D. Baker ... New York: D. Appleton and Company, 1919. xxx, 384 p. illus. 8°. **BTZY**

Birth of American aviation and developments during the World war, including supplies of raw materials, aviation engines, balloons, and a list of aces. Reviewed in *Aircraft journal,* New York, 1919, v. 4, May 3, p. 9, *VDS.*

4094. Sykes, Sir Frederick Hugh. Aviation in peace and war. London: Edward Arnold & Co., 1922. 139 p. 8°. **VDY**

Deals mainly with the period of the World war. For a critical review see *Aeroplane,* London, 1922, v. 24, p. 417–419, *VDS.*

4095. A **Synopsis** of British air effort during the war. (Aeroplane, London, 1928, v. 35, p. 766, 768, 770, 772, 774, 776, 778, 780, 782.) **VDS**

4096. Tennant, John Edward. In the clouds above Baghdad; being the records of an air commander. London: Cecil Palmer, 1920. x, 289 p. illus. maps. 8°. **BTZY**

Experiences on the Turkish front during the World war.

4097. Thébault, Louis. L'Escadrille 210. Paris: Jouve & Cie., 1925. 250 p. illus. ports., map. 8°. **BTZY**

History of French air squadron in World war.

4098. Theilhaber, Felix A. Jüdische Flieger im Kriege; ein Blatt der Erinnerung. Berlin: Louis Lamm, 1919. 52 p. ports. 8°. **BTZE p.v.470, no.8**

A record of Jews who served as aviators in the German army during the World war.

4099. Thenault, Georges. The story of the Lafayette Escadrille, told by its commander Captain Georges Thenault, translated by Walter Duranty, with an introduction by André Tardieu. Boston: Small, Maynard & Co. [cop. 1921.] xv, 172 p. 8°. **BTZY**

4100. Thiéry, Maurice. Paris bombardé par Zeppelins, Gothas & Berthas. Paris: E. de Boccard, 1921. 296 p. 12°. **BTZY**

Bombardments of Paris during the World war.

4101. Three and a half years of aerial warfare. illus. (Scientific American, New York, 1918, v. 118, p. 517.) **VA**

Graphic representation shows increase in number of certified pilots, growth of horse-power, and flying speed.

4102. Tomlinson, Daniel Webb. The sky's the limit. Philadelphia: Macrae Smith Company [cop. 1930]. 289 p. port. illus. 8°. **VDY**

Author's experiences as a U. S. naval aviator during the World war; also as a commercial manager during the post-war period. Reviewed in *Aviation,* New York, 1931, v. 30, p. 254, *VDS.*

World War, continued

4103. The **Torpedoplane** — an American invention successfully used by Germans against British shipping. (Aerial age weekly, New York, 1917, v. 5, p. 415–416.) **VDS**

4104. **Toulmin,** Harry Aubrey. Air Service, American Expeditionary Force, 1918, with an introduction by Jacques Louis Dumesnil. New York: D. Van Nostrand Company, 1927. xxxii, 388 p. illus. port. 8°. **BTZS**

History of organization and operations during 1918.

4105. **Tousley,** Bennett E. Aeronautics again plays an important part in arousing interest in Liberty loan. illus. (Flying, New York, 1917, v. 6, p. 856–858.) **VDS**

4106. **Towers,** J. H. Operations of naval aircraft. illus. (Society of Automotive Engineers, Journal, New York, 1919, v. 4, p. 368–371.) **TOL**

"At the declaration of war the [U. S.] Navy had only one aircraft station, twenty-two training seaplanes, and a total of thirty-eight aviators. It had no seaplanes fit for service use. Its so-called lighter-than-air equipment was negligible."

4107. **Trenchard,** H. The work of the I[ndependent] A[ir] F[orce]. map. (Flight, London, 1919, v. 11, p. 52–55.) **VDS**

Account of British organization's work in bombing the industrial centers of Germany.

4108. **Treusch von Buttlar Brandenfels,** Horst Julius Ludwig Otto, Freiherr. Zeppeline gegen England. Zürich: Amalthea-Verlag [1931]. 215 p. illus. 12°. **BTZY**

Zeppelin operations against England during the World war, with appreciation of the record of Capt. Peter Strasser. Library also has copies of the American edition, New York: Harcourt, Brace & Co. [cop. 1932], *BTZY;* and of the English edition, London: G. C. Harrap & Co. [1931], *BTZY.* For review see ***Flight,*** London, 1931, v. 23, p. 1115–1116, *VDS.*

4109. **Turner,** Charles Cyril. The struggle in the air, 1914–1918. London: Edward Arnold, 1919. viii, 288 p. illus. 8°. **BTZY**

Excellent comprehensive work, including exploits of prominent aviators.

4110. **$225,000,000** in contracts cancelled. Secretary Baker's drastic treatment of aircraft production... War contracts thus far abrogated aggregate $1,336,000,000 and more will be ended. (Air service journal, New York, 1918, v. 3, Nov. 28, p. 2.) **VDS**

4111. **United States.** — Air Services, Select Committee of Inquiry into Operations of the. Inquiry into operations of the United States Air Services. Hearings... Washington: Gov. Prtg. Off., 1925. 6 parts. charts, tables. 8°. *** SBF**

4112–4113. —— Report... Washington: Gov. Prtg. Off., 1925. ii, 54 p. 8°. **VDY p.v.35, no.5**

4114. **United States.** — Aircraft Production Bureau. Report [for the fiscal year ending June 30, 1918]. (United States. — War Department, Annual reports, 1918, Washington, 1919, v. 1, p. 1407–1417.) **VWZZ**

Account of organization and war activities.

4115. **United States.** — Army General Staff, 1917–19. Summary of air information, Second Section, General Staff, General Headquarters... March, April, May, 1918. France: American Expeditionary Forces, 1918. 278 p. illus. f°. **† BTZY (U. S.)**

Progress of operations and enemy information as furnished by prisoners and captured documents.

4116. **United States.** — Army War College: Historical Section. The Signal Corps and Air Service, a study of their expansion in the United States, 1917–1918... July, 1922. Washington: Gov. Prtg. Off., 1922. viii, 128 p. diagrs., tables. 8°. (Monograph no. 16.) **VWE p.v.121, no.7**

4117. **United States.** — Director of Military Aeronautics. Report [for the fiscal year ending June 30, 1918]. (United States. — War Department, Annual reports, 1918, Washington, 1919, v. 1, p. 1381–1390.) **VWZZ**

Account of organization and war activities. Also in *Aerial age weekly,* New York, 1918, v. 8, p. 720–722, *VDS.*

4118. **United States.** — Expenditures in the War Department, Select Committee on. War expenditures. Hearings before Subcommittee No. 1 (Aviation) of the Select Committee... House of Representatives, 66th Congress... July 31 – December 19, 1919. Washington: Gov. Prtg. Off., 1919–20. 3 v. charts, maps. 8°. *** SBF**

Voluminous report on evidence and findings on expenditures during the World war. Abstracted in *Aerial age weekly,* New York, 1919–20, v. 10, p. 720–721, 778–781; 1920, v. 11, p. 14, 58, 89–90, 158, 192, 226, 260, 294, 332, 372, 406, 448, 514, 550, 584, 618, 656, 714, 752, 778; 1920–21, v. 12, p. 52, *VDS.*

4119. **United States.** — Justice Department. Report of the aircraft inquiry. October 25, 1918. Washington, 1918. 182 p. 8°. **BTZY**

This report, signed by Charles E. Hughes, is concerned with waste and mismanagement in aircraft production in the United States during the World war. Also in *Aerial age weekly,* New York. 1918–19, v. 8, p. 462–471, 511–516, 568–569, 630, 670–672, 820–821, 864–865, 872, 908–909, 952–953, 962, *VDS.* For comments see *Scientific American,* New York, 1918, v. 119, p. 386, *VA.*

4120. **United States.** — Military Affairs Committee (Senate, 65:2). Aircraft production. Hearings... Washington: Gov. Prtg. Off., 1918. 2 v. 8°. **VDY**

Voluminous record in two volumes has testimony concerning the management of production during the World war. The findings are in Senate reports, 65th Congress, 2d session, report no. 555 (serial no. 7306), 1918, 19 p., ** SBE;* also reprinted in *Aviation,* New York, 1918, v. 5, p. 161–169, *VDS.*

4121. **United States.** — Military Affairs Committee (House, 70:1). Pioneer aviators. Hearings before the Committee on Military Affairs, House of Representatives, 70th

World War, continued

Congress, first session, on. H. R. 11273... [Part 1.] Washington: Gov. Prtg. Off., 1928. 8° *SBF

Hearings on an amendment to National Defence Act providing for rewards to army aviators before the World war. Includes a history of the Air Service, 1862–1920.

4122. **United States.** — Naval Operations Bureau. Development of U. S. naval aviation. illus. (Aviation, New York, 1919, v. 7, p. 424–428.) VDS

Developments from 1911 through 1919.

4123. **United States.** — President's Aircraft Board. Aircraft. Hearings before the President's Aircraft Board. Washington: Gov: Prtg. Off., 1925. 4 v. tables. 8°. VDW

Voluminous testimony on the condition of government aeronautics during World war and after.

4124. —— Report of President's Aircraft Board. November 30, 1925. Washington: Gov. Prtg. Off., 1925. 30, 5 p. 8°. VDW

Bound with the hearings of the Board. See entry no. 4123a. For a summary of the report, see *Aviation*, Highland, N. Y., 1925, v. 19, p. 834–837, *VDS*.

4125. **United States Spruce Production Corporation.** History of Spruce Production Division, United States Army, and United States Spruce Production Corporation. [Portland, Ore.: Press of Kilham Stationery & Printing Co., 1920?] ix, 126 p. illus. ports., maps. 4°. BTZS

4126. **Vernillat.** L'évolution des avions militaires français depuis la guerre. illus. (In: L'Aéronautique militaire, maritime, coloniale et marchande, Paris, 1931, p. 61–75.)

Copiously illustrated. †VDY

4127. **Victors** over enemy aircraft. Air Service fliers who brought down one or more enemy aircraft. (Aviation, Highland, N. Y., 1926, v. 20, p. 185–188.) VDS

Records of United States aviators during World war.

4128. **"Vigilant,"** pseud. German war birds. London: John Hamilton, Ltd. [1932.] 264 p. ports. 8°. BTZY

Deeds of Richthofen, Plüschow, Immelmann, Boelcke, von Eschwege, and other German aviators during the World war.

4129. **Voisin,** André Paul. La doctrine de l'aviation française de combat au cours de la guerre (1915–1918). Paris: Éditions Berger-Levrault, 1932. xi, 172 p. illus. map. 8°. BTZY

A study of French organization, tactics, and equipment.

4130. —— Le rendement de l'aviation française de bombardement de jour au cours de la guerre de 1914–1918. (Revue militaire française, Paris, 1924, année 94, Jan./March, p. 109–128.) VWA

History of French aerial bombing operations.

4131. **Wade,** Walter Lewis. The aeroplane in the great war. A record of its achievements, with illustrations by Geoffrey Watson. London: Virtue & Co. [1919?] ix, 229 p. illus. 8°. BTZY

Has some spirited pictures; also a detailed summary year by year.

4132. **Walcott,** Stuart. Above the French lines; letters of Stuart Walcott, American aviator: July 4, 1917, to December 8, 1917. Princeton, N. J.: Princeton University Press, 1918. v, 93 p. port. 8°. VDS

Training and fighting in the French army described.

4133. The **War** in the air. illus. (Flying, New York, 1914–15, v. 3, p. 325–328, 346–351, 365–367, 378–382; 1915–16, v. 4, p. 407, 416–418, 440–441, 518, 525–528, 559, 562–565, 596–601, 634–637, 680–684, 717–720, 752–757, 796–798, 835–838; 1916–17, v. 5, p. 39–42, 78–82, 130, 133–137, 174–177, 209–211, 251, 257–258, 303, 307, 352, 356–359, 392–395, 429–431, 469–473, 515–517; 1917–18, v. 6, p. 160, 228–231, 309–311, 315–317, 413–415, 418–420, 511–513, 604–607, 700, 702, 712, 787–788, 792, 981, 986, 988, 1076–1078; 1918–19, v. 7, p. 76–77, 266–269, 356, 361–364, 370, 460–463, 554–559, 650–655, 746–751, 838–843, 934–939, 1064–1069.) VDS

A day-by-day record of aerial events in the World war.

4134. **War** airplanes of today. illus. (Automotive industries, New York, 1918, v. 39, p. 14–18.) TOL

4135. **War** Department anti-aircraft record. [United States] Coast Artillery gives official records made during war. (Aviation, Highland, N. Y., 1925, v. 19, p. 247.) VDS

A study of percentage of planes brought down compared to number of shots fired, during the World war.

4136. **Wasner,** Adolf. Rittmeister Manfred Freiherr von Richthofen, Deutschlands grösster Fliegerheld. Ein Lebensbild nach Zeitungsberichten... Diesdorf bei Gäbersdorf: Schreiberhau-Diesdorfer Rettungsanstalten [1918]. 88 p. 8°. AN p.v.255, no.7

Memorial volume to famous German aviator of the World war.

4137. **Weir,** William Weir, 1st baron. Some developments in aircraft design and application during the war. illus. (North-east Coast Institution of Engineers and Shipbuilders, Transactions, Newcastle-upon-Tyne, 1918, v. 35, p. 268–347.) VDA

With discussion.

Covers aerodynamical aspects, construction and application, engine design, navigation, and meteorology. Also in *Engineering*, London, 1919, v. 108, p. 59–61, 93–99, 108–112, *VDA; Flight*, London, 1919, v. 11, p. 955–959, 987–991, 1013–1016, 1048–1053, *VDS*. The section on engine design is also in *Automobile engineer*, London, 1919, v. 9, p. 263–266, *TOL*.

4138. **Wellman,** William Augustus. Go, get 'em! The true adventures of an American aviator of the Lafayette Flying Corps who

World War, continued

was the only Yankee flyer fighting over General Pershing's boys of the Rainbow Division in Lorraine, when they first went "over the top," with an introduction and notes by Eliot Harlow Robinson. Boston: The Page Company, 1918. xi, 284 p. illus. ports. 12°. **BTZY**

4139–4140. Wemp, Bert, and A. H. MUNDAY. Canadian pilots in the great war. illus. (Canadian magazine, Toronto, 1927, v. 68, Oct., p. 6–7.) *** DA**

4141. Williams, Bertram W. The classification of military aeroplanes. illus. (Scientific American, New York, 1917, v. 117, p. 248–249.) **VA**

4142. —— Recent enemy aircraft. How the Germans make the most of information obtained through captured Allied planes. illus. (Scientific American, New York, 1918, v. 119, p. 274.) **VA**

4143. Wilson, N. W. How the war in the air developed. illus. (Flying, New York, 1915, v. 4, p. 574–578, 611–615, 707.) **VDS**

Illustrated history of early operations at the front during the World war.

4144. "Wings" (Captain R. A. F.). Over the German lines, and other sketches illustrating the life and work of an artillery squadron of the R. A. F. in France. London: Hodder and Stoughton, 1918. xxiv, 184 p. 12°. **BTZY**

4145. Wings for our eagle. illus. (Scientific American, New York, 1916, v. 115, p. 474–486.) **VA**

American preparations for World war.

4146. Winslow, Carroll Dana. With the French flying corps. London: Constable [1917]. 190 p. illus. 12°. **BTZY**

Good account of training and fighting during the World war.

4147. Wood, Eric. Thrilling deeds of British airmen. London: George G. Harrap & Co., 1917. 318 p. illus. 12°. **BTZY**

Exploits during the World war.

4148. Woodhouse, Henry. The aircraft's part in beating the U-boat. illus. (United States Naval Institute, Proceedings, Annapolis, 1918, v. 44, p. 2727–2739.) **VXA**

4149. —— Every military aviator ought to know what his own and the enemy's machine can do and how they look. illus. (Aerial age weekly, New York, 1918, v. 7, p. 42–47, 112–114, 155, 257.) **VDS**

Illustrations and brief descriptions.

4150. —— Henri Farré, artist and aerial artilleryman — founder of a new school. illus. (Flying, New York, 1918, v. 7, p. 123–135.) **VDS**

Reproductions of some of the artist's war scenes.

4151. —— Revolutionary developments in naval aeronautics. illus. (Flying, New York, 1915, v. 4, p. 461–463, 487.) **VDS**

4152. —— Russian defeats due to lack of aeroplanes. illus. (Flying, New York, 1915, v. 4, p. 671.) **VDS**

4153. —— Submarine hunting by aircraft. illus. (Flying, New York, 1917, v. 6, p. 267–273.) **VDS**

4154. —— Twenty-eight different uses for aircraft in the present war. illus. (Flying, London, 1918, v. 7, p. 38–39.) **VDS**

4155. —— United States has the machinery needed to make America first in aeronautics and only needs cooperation of efforts. (Aerial age weekly, New York, 1916, v. 4, p. 325–327.) **VDS**

4156. The **Work** and training of the Royal Flying Corps. Published, by authority of the Royal Flying Corps, at the offices of "The Illustrated London news." London [1917?]. 46 p. illus. f°. **† VDY (Work)**

A series of excellent photographs showing training and operations during the World war.

4157. The **Work** and training of the Royal Naval Air Force. Published by authority of the Admiralty at the offices of the "The Illustrated London news" [1917]. 46 p. illus. f°. **† VDY (Work)**

Excellent photographs taken during the World war.

4158. The **Work** of the British Air Services — 1914–1915. From official despatches. (Aeronautics, London, 1916, v. 10, p. 15–19.) **VDS**

Deeds of the Royal Naval Air Force and of the Royal Flying Corps.

4159. The **World's** aces. (Aircraft yearbook, New York, 1919, p. 295–298.) **VDS**

War records of American, British, French, Italian, Russian, German, and Turkish aces, 1914–1918.

4160. Wortley, Rothesay Stuart. Letters from a flying officer. London: Oxford University Press, 1928. ix, 207 p. port. 12°. **BTZY**

Author, British aviator, presents well-written accounts of fighting methods during the World war. Reviewed in *Air*, London, 1928, v. 1, April, p. 49, *VDS; Aeroplane*, London, 1928, v. 34, p. 252, *VDS*.

4161. Zeigler, S. J. The naval aircraft factory. illus. (Aero digest, New York, 1925, v. 7, p. 65–69.) **VDS**

Brief history of United States establishment during the World war.

CIVIL AERONAUTICS

4162. **Aerial** newspaper delivery service in Florida. Curtiss flying boat maintains 100 per cent schedule. illus. (Aviation, Highland, N. Y., v. 14, p. 423.) VDS

4163. **Aeronautics** Branch reorganized. (United States. — Commerce Department. Air commerce bulletin, Washington, 1933, v. 5, p. 107.) VDS

Eugene L. Vidal heads the Branch with title of Director of Aeronautics.

4164. **Air** liner travels 682 miles first 10 days. illus. (Aero and hydro, Chicago, 1914, v. 7, p. 213–214.) 3–VDS

Account of operation of the St. Petersburg-Tampa line, the first commercial aero undertaking in America.

4165. **Air** traffic in Italy during 1926. illus. map. (Aviation, Highland, N. Y., 1927, v. 23, p. 934–935.) VDS

4166. **Air** transport aids mining in New Guinea. illus. (Engineering and mining journal, New York, 1930, v. 129, p. 594–596.) VHA

4167. **Air** transport. Development of Yugoslavia. map. (Flight, London, 1931, v. 23, p. 407.) VDS

4168. **Air** transport. How does it progress? And where most rapidly? charts, maps. (Aviation, New York, 1932, v. 31, p. 102–105.) VDS

Statistical account of development in the United States during 1931.

4169. **Air** transport progress. illus. (Aviation, New York, 1931, v. 30, p. 18–21.) VDS

Developments in the United States during 1930.

4170. **Aircraft** in commerce. (Aircraft yearbook, New York, 1920, p. 13–59.) VDS

After-the-war development.

4171. **Airport** construction in 1930. illus. (Aviation, New York, 1931, v. 30, p. 31–33.)

Progress in the United States. VDS

4172. **Airports** and airways. diagrs. (Aviation, New York, 1934, v. 33, p. 90.) VDS

Growth of airports in the United States and expenditures for same shown by diagrams.

4173. **American** commercial flying. (Aviation, Highland, N. Y., 1926, v. 20, p. 522, 524, 560.) VDS

Mileage statistics of commercial flyers operating throughout the United States.

4174. **Analysis** of civil flying in the United States. Total mileage for 1925, 9,203,165 miles. Figures discussed in detail. (Aviation, Highland, N. Y., 1926, v. 20, p. 709–710.) VDS

4175. **Anderson,** John W. Airport development in Oregon. illus. (Aviation, New York, 1928, v. 25, p. 697, 764, 766, 768, 770.) VDS

4176. **Automatic** airport lighting. illus. (Scientific American, New York, 1928, v. 138, p. 356, 378.) VA

General Electric Company's device for automatically lighting airport when an airplane approaches.

4177. **Aviation** in Alaska. (United States. — Commerce Department, Air commerce bulletin, Washington, 1934, v. 5, p. 225–229.) VDS

Development and present condition of commercial aeronautics.

4178. **Banks,** Charles A. Gold mining by air. By using air instead of road transport a New Guinea goldfield was able to save a year in development time and win a profit of nearly £6,000,000. illus. (Air and airways, London, 1932, v. 9, p. 205–208.) VDS

4179. **Bauer,** P. P. von. Commercial aviation in the American tropics; an economic analysis. illus. (Aviation, Highland, N. Y., 1926, v. 20, p. 407–410.) VDS

Development of the Scadta Company of Colombia, South America.

4180. **Bellah,** James Warner. Linking the Americas. illus. (Saturday evening post, Philadelphia, 1929, v. 201, April 20, p. 6–7, 82, 84.) * DA

Status of commercial aviation development between the two continents.

4181. [**Bibliography** of airports.] (In: United States. — National Advisory Committee for Aeronautics, Bibliography of aeronautics, 1909–1931, Washington, 1921–35, volumes for 1920/21, p. 18; 1922, p. 8; 1925, p. 7; 1927, p. 6; 1928, p. 8–9; 1929, p. 11–14; 1930, p. 8–10; 1931, p. 12–15.) VDT

4182. [**Bibliography** of airways.] (In: United States. — National Advisory Committee for Aeronautics, Bibliography of aeronautics, 1909–1931, Washington, 1921–35, volumes for 1917/19, p. 23; 1920/21, p. 21; 1922, p. 9; 1923, p. 7; 1924, p. 5; 1925, p. 9; 1926, p. 5; 1927, p. 7; 1928, p. 11; 1929, p. 17–18; 1930, p. 12–13; 1931, p. 17.) VDT

4183. [**Bibliography** of civil and commercial aeronautics.] (In: United States. — Advisory Committee for Aeronautics, Bibliography of aeronautics, 1909–1931, Washington, 1921–35, volumes for 1917/19, p. 87–88, 95–98; 1920/21, p. 78–81, 85–93; 1922, p. 28, 30–31; 1923, p. 25, 26–28; 1924, p. 18, 19–20; 1925, p. 35–36, 37–40; 1926, p. 26–27, 29–30; 1927, p. 32, 34–36; 1928, p. 41, 43–44; 1929, p. 50, 53–55; 1930, p. 44, 46–48; 1931, p. 53, 56–58.) VDT

Civil Aeronautics, continued

4184. **Bigelow,** A. A. Airport growth in 1930–1931. The over-development in the rest of the industry has not been experienced in the airport field. Some design, construction and administrative progress has been made. illus. (Airway age, New York, 1931, v. 12, p. 72–74.) **VDS**

4185. **Black,** Archibald. Air transport and other civil flying. (Mechanical engineering, New York, 1929, v. 51, p. 9–10.) **VFA**
Developments during 1928.

4186. —— Leviathans of the air. illus. (Aero digest, New York, 1926, v. 8, p. 19–21, 48–49.) **VDS**
Discusses the work of the German commercial airship lines before and shortly after the war.

4187. —— Progress in aeronautics [for 1929]: air transport and other civil flying. (Mechanical engineering, New York, 1930, v. 52, p. 6–7.) **VFA**

4188. **Blee,** Harry H. Aeronautic development. (Scientific monthly, New York, 1930, v. 31, p. 236–240.) ***DA**
Account of the activities of the Aeronautic Development Service of the United States Department of Commerce.

4189. **Bonforte,** John. Progress in aeronautics [for 1929]: airports. (Mechanical engineering, New York, 1930, v. 52, p. 12–13.) **VFA**

4190. **Bouché,** Henri. L'aviation marchande en Perse. illus. map. (L'Aéronautique, Paris, 1930, année 12, p. 275–278.) **VDS**
Development of civil aviation in Persia.

4191. **Brancker,** Sir Sefton. The lessons of six years' experience in air transport. (Royal Aeronautical Society, Journal, London, 1925, v. 29, p. 552–585.) **VDS**
With discussion.
Suggestions for increasing safety. Abstracted in *Flight,* London, 1925, v. 17, p. 658–661, *VDS.*

4192. **Brilioth,** Borge. Sweden's air industry. Swedish-built airliners now fly regularly to the principal European cities. illus. (Aero digest, New York, 1925, v. 7, p. 656–657.) **VDS**

4193. **Brittain,** Sir Harry Ernest. By air. With an introduction by...the marquess of Londonderry. London: Hutchinson & Co., Ltd. [1933.] xiv, 194 p. illus. 8°. **VDY**
Traces the rise of civil aviation, with accounts of long flights by the author, as well as of the pioneering work of the Royal Air Force.

4194. **Campaigning** for an airport. How the airminded of San Diego, Calif., succeeded in obtaining a $650,000 airport bond issue. illus. ports. (Aviation, Highland, N. Y., 1928, v. 24, p. 310–313.) **VDS**

4195. **Canada.** — Department of National Defence. Report on civil aviation. 1923–1931. Ottawa, 1924–32. 9 v. 8°. **VDY**
These volumes trace the development of civil and postal aviation in Canada; also of prospecting by airplanes.

4196. **Canadian** activities. charts. (Aviation, New York, 1929, v. 27, p. 714.) **VDS**
Development of commercial aviation shown by charts.

4197. **Carll,** George S. Congress provides for commercial aeronautics. (U. S. air services, Washington, 1926, v. 11, July, p. 22–24.) **VDS**

4198. **Chicanot,** E. L. Civil aviation in Canada. illus. (Popular aviation, Chicago, 1929, v. 4, Jan., p. 36–37, 110.) **VDS**

4199. —— Revolutionizing Canadian mining. illus. (Popular aviation, Chicago, 1928, v. 3, Aug., p. 29–32.) **VDS**

4200. **Civil** aeronautics in statistics. (United States. — Commerce Department: Aeronautics Branch, Air commerce bulletin, Washington, 1929, v. 1, Aug. 1, p. 8–9.) **VDS**
Detailed figures for the United States, 1926 to 1928 inclusive.

4201. **Civil** aviation abroad in 1926. chart. (Flight, London, 1927, v. 19, p. 415–416.) **VDS**
Developments in Great Britain, France, Germany, and the United States.

4202. A **Commercial** & historical atlas of the world's airways. Foreword by Sir W. Sefton Brancker... Birmingham, Eng.: F. J. Field, Ltd. [1925.] 85 p. maps, tables. 8°. **VDY p.v.36, no.3**
Brief references to early histories of air routes in the principal countries of the world, with maps.

4203. The **Coolidge-Ford** aviation alliance. illus. (Literary digest, New York, 1926, v. 90, Aug. 21, p. 17–18.) ***DA**
Discusses cooperation of United States government with Henry Ford.

4204. **Daniel** Guggenheim gives $2,500,000 more to foster aeronautics. (U. S. air services, Washington, 1926, v. 11, Feb., p. 13–15.) **VDS**

4205. **Day** and night transcontinental service. The Boeing system begins its big plane passenger line. illus. (Airway age, New York, 1930, v. 11, p. 658–659.) **VDS**

4206. **De Ritis,** B. The Italian aeronautical industry. (Aero digest, New York, 1929, v. 15, Dec., p. 74, 108.) **VDS**
Brief histories of airplane and engine companies.

4207. **Development** of air transport activities in the United States. charts. (Aviation, New York, 1929, v. 27, p. 704–705.) **VDS**
Charts show development of mail and commercial lines.

4208. **Development** of civil aeronautics in the United States [during 1928]. (United States. — Commerce Department: Aeronautics Branch, Air commerce bulletin, Washington, 1929, v. 1, Nov. 1, p. 1–6.) **VDS**
Developments in scheduled air transport, air-rail service, air marking, scenic tours, equipment design, aircraft production, and exports.

Civil Aeronautics, continued

4209. The **Development** of the Croydon aerodrome. (Aeroplane, London, 1923, v. 24, p. 182.) VDS

4210. **Dillon,** Charles. What airplanes are doing to railroads abroad. Days are saved, and tickets are not more costly than by train in some classes. illus. (Railway review, Chicago, 1925, v. 77, p. 231–233.) † TPB

Quoted in *Literary digest,* New York, 1925, v. 86, Sept. 26, p. 66–69, * *DA.*

4211. **Dresel,** A. Air transport of mine equipment, supplies, and personnel. illus. (Engineering and mining journal, New York, 1933, v. 134, p. 201–202.) VHA

Experiences of the Bulolo Gold Dredging, Ltd. of New Guinea.

4212. **Duval,** Albert B. Les vols de nuit et le développement de l'aéronautique commerciale. (La Technique aéronautique, Paris, 1924, année 15, p. 298–305.) 3–VDS

Development of night-flying in various countries.

4213. **Empire** commercial aviation. (Air annual of the British Empire, 1934/35, London, 1934, v. 6, p. 74–101.) VDS

Recent developments in Australia, Canada, India, and South Africa.

4214. **Fansler,** P. E. The first commercial airline. illus. (Aero digest, New York, 1929, v. 15, Dec., p. 58–59, 260, 262, 264, 266, 268.) VDS

St. Petersburg-Tampa line was inaugurated Jan. 1, 1914, under the direction of the author.

4215. **Faragher,** R. G. London-Paris. World's foremost airway. (Aeronautical digest, New York, 1923, v. 3, p. 149–151, 255–256.) VDS

History of cross-Channel commercial undertakings.

4216. **Faunce,** Cy Q. The airliner and its inventor, Alfred W. Lawson, with a summary of the entire aeronautical movement...containing appendices: A 2,000 mile trip in the first airliner, and Natural prophesies; aircraft — airology, by Alfred W. Lawson. Columbus, O.: Rockcastel Publishing Co. [cop. 1921.] 206 p. illus. 8°. VDY

Lawson, former editor of *Aviation,* was a pioneer in building and operating commercial airplanes.

4217. **Faurote,** Fay Leone. A survey of Canadian commercial aviation. illus. port. (Aviation, New York, 1929, v. 27, p. 1250–1254.) VDS

Interesting account of operations in northern Canada.

4218. The **First** aerodrome in the world. A race course for army flying machines. illus. (Scientific American supplement, New York, 1909, v. 67, p. 104.) VA

Picture of "Port Aviation" near Paris.

4219. The **First** aeroplane express. illus. (Scientific American, New York, 1910, v. 103, p. 464.) VA

Express matter carried from Dayton, O., to Columbus, O., Nov. 7, 1910.

4220. The **First** imperial air route. illus. ports. (Flight, London, 1926, v. 18, p. 862–863.) VDS

Sir Samuel Hoare inaugurates the Cairo-Karachi route, 1926.

4221. **First** private aerodrome in America. illus. (American magazine of aeronautics, New York, 1907, v. 1, July, p. 13.) VDS

Picture of building owned by Glenn H. Curtiss, Hammondsport, N. Y.

4222. **Fokker,** Anthony Herman Gerard. Trend of aeronautics in Europe. port. (Aero digest, New York, 1930, v. 16, April, p. 67, 226.) VDS

Effects of discontinuance of subsidies on commercial aviation.

4223. **Ford** Air Transport reports favorable year. (Aviation, Highland, N. Y., 1926, v. 20, p. 662.) VDS

"Over 1000 trips, carrying large loads of freight and express."

4224. **Foreign** activities. charts. (Aviation, New York, 1931, v. 30, p. 173–177.) VDS

Statistics for 1930.

4225. **Foreign** activities. How do Europe and America compare. charts. (Aviation, New York, 1932, v. 31, p. 136–139.) VDS

World statistics for 1931 and previous years.

4226. **Forman,** Harrison. Following the West. China becomes one of the world's great aeronautical frontiers and one of the greatest of markets. map. (Aviation, New York, 1934, v. 33, p. 148–149.) VDS

History of manufacture and transport.

4227. **France** leads the world in development of commercial air transportation. (Aviation, Highland, N. Y., 1926, v. 20, p. 57.) VDS

4228. **French** civil aviation. A review of French air transportation with statistics for past three years of operation. illus. (Aviation, Highland, N. Y., 1926, v. 21, p. 127–129.) VDS

4229. **Fullerton,** Frederick H. Flying in British Columbia. illus. (Popular aviation, Chicago, 1929, v. 4, Feb., p. 78–81, 104–105.) VDS

Development of commercial aviation.

4230. **Gale,** Charles H. The French air-boat line to South America. illus. map. (Aviation, New York, 1930, v. 28, p. 192–197.) VDS

The Latecoere system, beginning in 1918, has developed a combined boat and air line from France to Buenos Aires.

4231. **Gardner,** Lester Durand. The Deruluft airline to Moscow. In successful operation since 1922. Russian progress in commercial aviation. illus. (Aviation, Highland, N. Y., 1927, v. 22, p. 318–322, 371–376, 404–406.) VDS

History of Berlin-Moscow line.

Civil Aeronautics, continued

4232. —— German air transport. illus. (Aviation, Highland, N. Y., 1927, v. 23, p. 146–150, 253–254, 256, 258, 260, 365–369, 469–472, 495.) **VDS**
Brief history, with author's experiences on the routes of the Deutsche Luft Hansa.

4233. Gold transport by air. Loads on Africa and India routes increasing. (Engineering and mining journal, New York, 1934, v. 135, p. 297.) **VHA**

4234. Goldsborough, Paul. Are European air lines better than those of the United States? illus. (Aero digest, New York, 1931, v. 19, Nov., p. 36–37, 88, 90, 92–93.) **VDS**
With certain exceptions, the author believes not.

4235. Grayson, King Hamilton. The Russian air fleet. maps. illus. (Popular aviation, Chicago, 1930, v. 6, Feb., p. 15–16, 87.) **VDS**
This number of *Popular aviation* is published under the title of *Aeronautics.*

4236. Great Britain. — Civil Aviation Directorate. Annual report on the progress of civil aviation. 1919–1934. London, 1919–35. 8°, f°. **† VDY and VDY**

4237. Grey, Charles Grey. On the world's longest airway. illus. maps. (Aeroplane, London, 1931, v. 40, p. 265–266, 268, 318, 320–321, 356, 358, 360, 362.) **VDS**
History of the Compagnie générale aéropostale, operating the line from Toulouse to the southern part of South America.

4238. The **Growth** of an Australian airline. illus. (Aviation, Highland, N. Y., 1926, v. 20, p. 663–664.) **VDS**
History of operations of The Queensland and Northern Territory Aerial Services, Ltd.

4239. Gubbins, M. N. T. The development of civil aviation and night flying. illus. (Flight, London, 1930, v. 22, p. 1163–1164.) **VDS**

4240. Handley-Page, Frederick, and W. P. SAVAGE. The development of commercial aviation. (International Congress of Aerial Navigation, London, 1923, Report, London, 1923, p. 689–700.) **VDS**
Technical developments.

4241. Hanson, Earl. Armstrong seadrome project progresses. Plans complete for installing first drome within a year. illus. (Airway age, New York, 1930, v. 11, p. 353–357.) **VDS**

4242. —— Canadian aviation — 1930. illus. (Airway age, New York, 1931, v. 12, p. 68–71.) **VDS**

4243. —— Canadian Pacific Express experiments with air-rail service. map. (Airway age, New York, 1929, v. 10, p. 835–836.) **VDS**

4244. —— Floating airports. port. (Popular aviation, Chicago, 1927, v. 1, Oct., p. 50.) **VDS**
Author is the inventor of the radio piloting cable system, which he proposes to apply to floating airports.

4245. Harper, Harry. The romance of a modern airway...with an introduction by Air Vice-Marshal Sir Sefton Brancker... illustrated by 100 special photographs, selected and arranged by the author. London: S. Low, Marston & Co., Ltd. [1930.] xiii, 241 p. illus. 8°. (Romance series.) **VDY**
Section 1 deals with British airmail tests, the first London-Paris express, early traffic problems, and the establishment of the Imperial Airways.

4246. Henderson, Paul. Air transportation, its origin, growth and outlook for the future. illus. port. (Aeronautic review, Washington, 1930, v. 8, Jan., p. 16–17, 56–58, 65.) **VDS**
Developments in the United States.

4247. Henry Ford has started his first regular flying machine route. illus. (U. S. air services, Washington, 1925, v. 10, May, p. 18–20.) **VDS**
Chicago-Detroit line.

4248. The **History** of Canadian civil flying. (Aeroplane, London, 1927, v. 32, p. 524–525.) **VDS**

4249. Hoover, Herbert Clark. Civil aviation's rapid progress. port. (Aero digest, New York, 1928, v. 12, p. 509, 696–698.) **VDS**

4250. Italy builds civil air service. illus. (Airway age, New York, 1929, v. 10, p. 296–297.) **VDS**
Brief history of the eight main lines.

4251. Johnson, George R. Flying in the land of the Incas. illus. (Aero digest, New York, 1929, v. 15, Sept., p. 57–60, 352, 354.) **VDS**
Development of civil and military aviation in Peru.

4252. Johnson, Robert. Boeing starts Chicago-to-Coast passenger service. illus. (Aero digest, New York, 1930, v. 16, May, p. 102, 104.) **VDS**

4253. —— 50,000,000 miles of flying. illus. (Scientific American, New York, 1933, v. 149, p. 114–116.) **VA**
Story of the United Air Lines.

4254. Key West-Havana route inaugurated. illus. (Aviation, Highland, N. Y., 1920, v. 9, p. 226–227, 230.) **VDS**

4255. Klemin, Alexander. American passenger air transport. The passenger-carrying airlines of the United States have shown remarkable growth in the last two years. illus. maps. (Scientific American, New York, 1929, v. 141, p. 324–329, 404–407.) **VA**

4256. Knight, W. Commercial aviation developments in Europe. (Aerial age weekly, New York, 1922, v. 15, p. 30–32, 58–60, 107–109, 133–134.) **VDS**

4257. —— Commercial aviation developments in the United States. (Aerial age weekly, New York, 1922, v. 15, p. 294–296, 311.) **VDS**
A discussion of ways and means.

Civil Aeronautics, continued

4258. **Lafargue.** Aviation saharienne. (Société de géographie d'Alger et de l'Afrique du Nord, Bulletin, Alger, 1910, année 15, p. 437–448.) **KAA**

Detailed suggestions for the establishment of an airplane service across the Sahara desert.

4259. **Lamarche,** Paul E., jr. Belgium in the air. illus. (Popular aviation, Chicago, 1928, v. 3, Nov., p. 28–31, 120–121.) **VDS**

Commercial developments in Belgium and in the Congo.

4260. —— Le Bourget. illus. (Popular aviation, Chicago, 1929, v. 4, April, p. 23–27.) **VDS**

History of French airport.

4261. —— Britain in the air. illus. (Popular aviation, Chicago, 1928, v. 3, Aug., p. 18–22, 89–91.) **VDS**

Development of commercial aviation.

4262. —— Denmark in the air. illus. (Popular aviation, Chicago, 1929, v. 4, Jan., p. 22–25.) **VDS**

Development of commercial lines.

4263. —— France in the air. illus. (Popular aviation, Chicago, 1928, v. 2, March, p. 12–17.) **VDS**

Table shows development of commercial aviation, 1919–1928.

4264. —— Germany in the air. illus. (Popular aviation, Chicago, 1928, v. 3, July, p. 22–26, 80–81.) **VDS**

Development of commercial aviation.

4265. —— Mussolini pilots Italy. illus. (Popular aviation, Chicago, 1929, v. 4, May, p. 13–15, 92–93.) **VDS**

Italy's advance in commercial aviation under Mussolini's direction.

4266. —— Switzerland in the air. illus. (Popular aviation, Chicago, 1928, v. 3, Sept., p. 28–31, 92–93.) **VDS**

Development of commercial aviation.

4267. **Langley,** Marcus. The Royal Dutch Indian Airways. illus. map. (Flight, London, 1931, v. 23, p. 53–55.) **VDS**

The Koninklijke Indische Luchtvaart Maatschappij was established in 1928.

4268. The **Lehigh** airports competition. illus. (Aeronautic review, Washington, 1930, v. 8, Feb., p. 18–23, 56.) **VDS**

Reproduction of some of the prize-winning designs. Competition was sponsored by the Lehigh Portland Cement Company.

4269. **Lehigh Portland Cement Company.** American airport designs, containing 44 prize winning and other drawings from the Lehigh airports competition, the first national contest for designs of modern airports held in the United States...with analysis of designs by Archibald Black... New York and Chicago: Published for the Lehigh Portland Cement Company by Taylor, Rogers & Bliss, Inc. [cop. 1930.] 96 p. illus. f°. **† VDY**

4270. **Lima,** Peru, to Para, Brazil. Inauguration of air line between Lima and Iquitos, Peru, is first step in South American transcontinental route. illus. (Aviation, Highland, N. Y., 1928, v. 24, p. 577, 593.) **VDS**

4271. **MacGregor,** John D. Development of inter-American air transport. (U. S. air services, Washington, 1931, v. 16, May, p. 54.) **VDS**

History of the South American West Coast service established by the Pan American-Grace Airways.

4272. **Macmillan,** Norman. The history and growth of commercial air transport. illus. (Flight, London, 1930, v. 22, p. 77–80.) **VDS**

4273. **McMullin,** C. R. Among the pagodas. After 4,000 years of crude transportation facilities China takes up air navigation. illus. (Aero digest, New York, 1924, v. 5, p. 80–83.) **VDS**

4274. **McReynolds,** Charles F. Air transport development in 1929. illus. (Aviation, New York, 1930, v. 28, p. 318–322.) **VDS**

Brief account of accomplishments in the United States.

4275. **Manning,** L. B. Aviation shows rapid advance in last four years. (New York Herald Tribune, New York, 1934, Jan. 2, p. 30.) *** A**

Chairman of the Board, American Airways, Inc., reviews United States transportation developments, 1930–1933 inclusive.

4276. **Mingos,** Howard. America takes the lead in aviation. maps, diagrs. (World's work, Garden City, N. Y., 1925–26, v. 51, p. 633–644.) *** DA**

Growth of commercial aviation shown by maps and charts.

4277. **Mr.** G. Holt Thomas. (Flight, London, 1929, v. 21, p. 25.) **VDS**

Brief biography of prominent pioneer of civil aviation in England.

4278. **Montagnes,** James. Canada's aerial expansion during 1929. illus. (Aero digest, New York, 1930, v. 16, March, p. 66–67, 216.) **VDS**

4279. —— Canadian aviation progress. illus. (Aero digest, New York, 1929, v. 14, March, p. 54–55.) **VDS**

Tells what the government has done in the promotion of civil and postal aviation.

4280. —— The modern prospector flies. He now covers in hours the trails it once took him days to travel. illus. (Airway age, New York, 1929, v. 10, p. 1410–1412.) **VDS**

Interesting accounts of work done.

4281. **Mourot,** N. O. Flying in Alaska. ports. illus. (Popular mechanics, Chicago, 1928, v. 2, April, p. 9–14, 78, 80, 82.) **VDS**

Pioneer flights of Ben Eielson and George Wilkins, and the establishment of the Fairbanks Airplane Corporation.

4282. **N. A. T.** planes flown over 700,000 miles during first year of operation. illus. (Aviation, Highland, N. Y., 1927, v. 23, p. 80–82.) **VDS**

National Air Transport, Inc. reports on its operation of the Chicago-Dallas air mail service.

Civil Aeronautics, continued

4283. **Navy** releases $4,000,000 worth of aircraft for commercial flying. (U. S. air service, New York, 1921, v. 4, Jan., p. 20.) **VDS**
Retirement of United States surplus aircraft and engines.

4284. The **New** German air transport combine. An amalgamation, in the interests of economy, of two of the largest air transport systems in Europe. (Aviation, Highland, N. Y., 1926, v. 20, p. 221–222.) **VDS**
Aero Lloyd and Junkers company unite to form the Deutsche Lufthansa A. G.

4285. **Noel,** Edmond Percy. Wings for Hawaii. illus. (Popular aviation, Chicago, 1930, v. 6, March, p. 22–23, 74.) **VDS**
Development of commercial aviation and of the John Rodgers airport. This number of *Popular aviation* was published under the title of *Aeronautics.*

4286. The **Northwest** Airways. Notable record of efficiency and safety in operation. Established first air-rail passenger service. illus. (Airway age, New York, 1929, v. 10, p. 19–21.) **VDS**
Established in 1926 for service between Chicago and Minneapolis-St. Paul.

4287. **Offermann,** Erich. Anfänge des europäischen Luftverkehrs. (Jahrbuch für Luftverkehr, 1924, München, 1924, p. 1–13.) **3–VDS**

4288. **Orcy,** Ladislas d'. The dawn of American commercial aviation. illus. (Scientific American, New York, 1920, v. 123, p. 546, 555, 557.) **VA**
Reasons why the United States has trailed behind Europe.

4289. —— French air expansion in South America. map. (Aviation, Highland, N. Y., 1925, v. 18, p. 241–242.) **VDS**

4290. —— International air transportation. map. (Aviation, Highland, N. Y., 1925, v. 18, p. 514–516.) **VDS**
History of the C. I. D. N. A. (Compagnie internationale de navigation aérienne), organized in 1920.

4291. The **Past** decade [of commercial aviation]. (Aviation year book, London, 1930, p. 5–25.) **VDS**
Developments in Belgium, the Belgian Congo, France, Germany, Great Britain, Holland, and the United States.

4292. **Perrin,** H. E. Air transport in the strike. illus. (Flight, London, 1926, v. 18, p. 295–297.) **VDS**
Incidents of the British general strike of 1926.

4293. **Pioneering** in East Africa. (Aeroplane, London, 1931, v. 40, p. 290, 292.) **VDS**
History of the Wilson Airway, Ltd., established in 1929.

4294. **Pollog,** Carl Hanns. Der Weltluftverkehr; seine Entwicklung, Geographie und wirtschaftliche Bedeutung... Leipzig: B. G. Teubner, 1912. 94 p. illus. (maps.) 8°. **VDY p.v.46, no.4**
Accounts of the development of commercial aeronautics in various countries since the World war.

4295. **Post,** Augustus. The evolution of Pan-American aeronautics. illus. ports. (Flying, New York, 1917, v. 6, p. 46–53.) **VDS**
A history of the organization of the Pan-American Aeronautic Federation, in which Santos-Dumont took an active part.

4296. **Progress** of air transport in Europe. Total traffic of six years aggregates over 25,000 tons. (Aviation, Highland, N. Y., 1925, v. 18, p. 270–271.) **VDS**
Development, 1919–1923.

4297. **Progress** of civil aeronautics in the United States. (United States. — Commerce Department: Aeronautics Branch, Air commerce bulletin, Washington, 1933, v. 4, p. 530–533.) **VDS**
Detailed statistics, including those of the air mail, 1926–1932 inclusive.

4298. The **Progress** of Imperial Airways, 1929–1930. illus. (Air annual of the British Empire, London, 1930, v. 2, p. 151–156.) **VDS**

4299. The **Railroads** take to the air. map. (Literary digest, New York, 1929, v. 102, July 20, p. 8.) *** DA**
Combined airplane and railway services across the United States.

4300. **Renvoise,** Ange. Le port aérien du Bourget-Dugny de 1919 à 1930. illus. (In: L'Aéronautique militaire, maritime, coloniale et marchande, Paris, 1931, p. 179–182.) **† VDY**
History of famous French aviation field.

4301. **Resumé** of commercial information. (Royal Aeronautical Society, Journal, London, 1933, v. 37, p. 708–722.) **VDS**
Statistics show operations of French companies and of the Imperial Airways since their inception.

4302. **Rosette,** Breese. Flying heavy freight into the Peruvian Andes. illus. (Compressed air magazine, Phillipsburg, N. J., 1933, v. 38, p. 4231–4232.) **VDS**
Remarkable transportation of mining machinery on a 60-mile flight from Cuzco.

4303. **Russia.** (Aeronautical digest, New York, 1923, v. 3, p. 87.) **VDS**
Post-war developments in commercial aviation.

4304. **Sayre,** Daniel. World airways, 1934. A survey of international airlines, their commercial and political significance, their aspirations and their governmental relationships. map. (Aviation, New York, 1934, v. 33, p. 235–239.) **VDS**
Histories of the principal services.

4305. **Scheduled** air transport operation statistics for the calendar year 1932. (United States. — Commerce Department, Air commerce bulletin, Washington, 1933, v. 4, p. 473–477.) **VDS**
Statistics of American companies engaged in domestic and foreign operations.

4306. **Schmeckebier,** Laurence Frederick. ...The Aeronautics Branch, Department of Commerce; its history, activities and organi-

Civil Aeronautics, continued

zation. Washington, D. C.: The Brookings Institution, 1930. x, 147 p. 8°. (Institute for Government Research. Service monographs of the United States government. no. 61.) **VDY**

The United States government entered the field of heavier-than-air aeronautics in 1907. First air mail was carried in 1911. The book reviews history, cites legislation affecting the service and gives a list of air-transport routes with dates of commencement. Bibliography.

4307. **Seadrome** corporation asks NRA for millions to bridge Atlantic. illus. map. (Iron age, New York, 1933, v. 132, Oct. 19, p. 41.) **VIA**

Seadrome Ocean Dock Corporation asks for $30,000,000 to establish five landing platforms.

4308. **Shannon,** Homer H. Breaks records with express traffic. Kohler Aviation Corp. handling 13,000 pounds monthly across Lake Michigan — overshadows passenger business. illus. (Airway age, New York, 1930, v. 11, p. 1195–1198.) **VDS**

4309. **Shaw,** B. Russell. Air transport values evolved from American duration flights. (U. S. air service, Washington, 1923, v. 8, June, p. 14–15.) **VDS**

4310. **Shepherd,** Edwin Colston. Commercial flying progress within the Empire. (Air annual of the British Empire, 1934/35, London, 1934, v. 6, p. 64–73.) **VDS**

4311. **Simmonds,** O. E. The development of civil marine aircraft. illus. (Institution of Aeronautical Engineers, Proceedings, London, 1926, no. 20, p. 7–37.) **VDS**

Includes a brief historical introduction concerning developments in Great Britain of marine air lines, beginning 1923.

4312. **Sinclair,** D. The development of civil aviation wireless. (International Congress of Aerial Navigation, III, Brussels, 1925, Reports, Paris, tome 1, p. 206–212.) **VDS**

4313. **Smedal,** George. Sweden turns to aviation. illus. (Popular aviation, Chicago, 1928, v. 2, June, p. 17–18.) **VDS**

Brief account of development of commercial aviation in the country.

4314. **Somerville,** Harry A. Canadian aviation in 1929. illus. (Airway age, New York, 1930, v. 11, p. 58–60.) **VDS**

4315. **Stoll,** Bert. Air transportation with safety. A brief history of a million miles of flying on Stout Air Lines. illus. (Aero digest, New York, 1930, v. 16, May, p. 55–56.) **VDS**

4316–4317. **Stout,** Wesley Winans. Aviation comes out of a tail spin. (Saturday evening post, Philadelphia, 1926, v. 198, May 8, p. 3–5, 137, 141–142, 145, 149.) *** DA**

Development of commercial aviation in the United States. History of the pony express and of the air mail.

4318. **Svehla,** George. A survey of civil aviation in the Southwest [United States]. illus. (Aero digest, New York, 1930, v. 17, August, p. 35–37, 58, 170, 172, 174, 176, 178; Sept., p. 52–54, 162, 164, 166.) **VDS**

4319. **Ten** years of air transport. port. (Aeroplane, London, 1929, v. 37, p. 1001–1004.) **VDS**

History of the Royal Air-Traffic Company of the Netherlands and the Colonies, with a portrait of its chief director, A. Plesman.

4320. **Ten** years of air transport. Progress, unique in the history of transport, has been made by air travel during the past decade. illus. (Airways, London, 1929, v. 6, p. 97–99.) **VDS**

4321. **Transport** goes steadily forward. charts. (Aviation, New York, 1934, v. 33, p. 71–78.) **VDS**

"The story of [United States] air transport in 1933 can be comprehensively summarized in two words — everything up. More miles flown, more passengers carried." Interesting colored charts.

4322. **Ullendorff,** Hans. Russia plans growth in the air. Soviet has five-year plan to increase service. illus. (Airway age, New York, 1930, v. 11, p. 206–208.) **VDS**

4323. **United States.** — Commerce Department: Aeronautics Branch. Chronology of U. S. air transport. Washington, 1928. 2 p. 4°. (Information bulletin. no. 17.) **TAKB**

Table shows mileage and dates of beginning of 27 services, 1919–1927.

4324. —— Civil aeronautics in America. Washington, 1927. 49 p. 4°. (Information bulletin. no. 1.) **TAKB**

Brief history.

4325. —— Civil air accidents and casualties. Washington, 1928. 4 l. 4°. (Information bulletin. no. 30.) **TAKB**

Analysis of accidents for the year 1927.

4326. —— Commercial aeronautics in Germany. Washington, 1927. 8 p. 4°. (Information bulletin. no. 22.) **TAKB**

Brief history.

4327. **United States** air mail and passenger transport. charts, maps. (Aviation, New York, 1930, v. 28, p. 574–581.) **VDS**

Statistics of development from 1918 shown by charts, statistics, and maps.

4328. **United States** has reciprocal agreements on civil aeronautics with nineteen nations. (United States. — Commerce Department, Air commerce bulletin, Washington, 1933, v. 5, p. 151–154.) **VDS**

4329. **U. S.** air mail and passenger transport. charts. (Aviation, New York, 1931, v. 30, p. 151–156, 461, 515, 624–625; 1932, v. 31, p. 35–36.) **VDS**

Statistics, charts, and maps show growth from the beginning.

4330. **Van Dusen,** W. I. The first million miles. A story of true aerial pioneering between North and South America which is

Civil Aeronautics, continued

advancing the progress of the Western Hemisphere. illus. map. (Aeronautic review, Baltimore, 1929, v. 7, Sept., p. 19–22, 47.) **VDS (National Aeronautics)**

History of Pan-American airways to Cuba, Mexico, Central America, and South America.

4331. **Van Zandt,** John Parker. Frontier facts in air transportation. map, diagrs. (Society of Automotive Engineers, Journal, New York, 1925, v. 17, p. 354–358.) **TOL**

Charts show development of air transportation (including air mail) in Europe and America, 1919–1924. Article is reprinted, with an analysis of the charts, in *Aviation*, Highland, N. Y., 1925, v. 19, p. 677–681, *VDS*.

4332. —— Subsidizing civil aviation in Europe and America. (U. S. air services, Washington, 1931, v. 16, Feb., p. 15–18.) **VDS**

Air mail contracts in America serve the same purpose as subsidies in Europe.

4333. **Warner,** Edward Pearson. Development of transportation by air. (Scientific monthly, New York, 1924, v. 18, p. 383–393.) *** DA**

Commercial aviation, air mail, effects of subsidies, rates, night-flying, and express service.

4334. —— Transport airplanes then and now. illus. (Aviation, New York, 1932, v. 31, p. 323–327.) **VDS**

Post-war history of mail and passenger airplanes.

4335. **Wenneman,** Joseph H. Municipal airports; history, development and legal aspect of municipal airports, text of federal acts and regulations, digests of state aviation laws and of state enabling acts... Cleveland, Ohio: The Flying Review Publishing Company [cop. 1931]. xxvii, 879 p. illus. maps. 8°. **VDYL**

Has histories of aeronautics, international conferences on jurisprudence, and the United States air mail.

4336. **Weyerbacher,** Ralph D. The common carrier hops off. illus. (Saturday evening post, Philadelphia, 1927, v. 200, July 30, p. 6–7, 93–94, 99.) *** DA**

Development of civil aviation in the United States.

4337. **White,** Thomas D. The first American commercial airplane in China. illus. (U. S. air services, Washington, 1929, v. 14, Feb., p. 32–34.) **VDS**

4338. **Willcox,** H. Case. Air transport development in Latin America. illus. map. (Aviation, New York, 1930, v. 29, p. 286–290.)

Brief accounts of the various companies. **VDS**

4339. —— Air transportation in Latin America. illus. map. (Geographical review, New York, 1930, v. 20, p. 587–604.) **KAA**

Good history of development.

4340. **Williams,** W. T. S. European air progress — 1928 in retrospect. (Air, London, 1929, v. 2, p. 25–26, 73–74.) **VDS**

4341. —— World aviation. illus. (Air, London, 1927–28, v. 1, Dec., p. 25–29, Jan., p. 35–40, March, p. 31–34, April, p. 27–32, May, p. 35–37, June, p. 31–34, July, p. 31–35, Aug., p. 27–32, Sept., p. 29–32, Oct., p. 33–38, Nov., p. 29–32, Dec., p. 31–34; 1929, v. 2, p. 221–224.) **VDS**

Development of commercial aviation, after the War, in Italy, United States, Australia, Russia, Afghanistan, Spain, Africa, South America, Central America, Canada, Holland, France, and India.

4342. **Wolfin,** Hugh M. Use of airplanes in mining and petroleum operations (abstract). Washington, Feb., 1934. 28 p. 8°. (United States. — Mines Bureau. Information circular. no. 6767.) **VHCA**

Relates use in Alaska, Canada, New Guinea, and the United States.

4343. **Woolley,** James G., and E. W. Hill. Airplane transportation...with special papers by Hon. Wm. P. MacCracken, jr... and Dr. Carl-Gustaf Rossby... [Hollywood, Calif.: Hartwell Publishing Corporation, cop. 1929.] xii, 353 p. illus. maps. 8°. **VDYE**

Brief history of the airplane and its wartime development. Postwar conditions leading to commercial aviation and development of the air mail service. Development of private lines.

4344. **World** air transportation. America's steadily lengthening lead in commercial aviation. illus. (Aviation, New York, 1934, v. 33, p. 86–87.) **VDS**

Growth and present condition shown by pictures.

4345. **Wright** Field is dedicated. Many leading army, navy and civilian air officials attend the ceremonies at Dayton, O. illus. (Aviation, Highland, N. Y., 1927, v. 23, p. 995–997.) **VDS**

4346. **Wronsky,** Martin. Air traffic in Germany. illus. diagrs. (Royal Aeronautical Society, Journal, London, 1927, v. 31, p. 690–714.) **VDS**

Growth of service is traced.

4347. —— Drei Jahre Luftverkehr. maps. illus. (Zeitschrift für Flugtechnik und Motorluftschiffahrt, München, 1922, Jahrg. 13, p. 17–22.) **VDS**

Three years' history of the Deutsche Luft-Reederei in the handling of passenger, commercial and postal traffic.

4348. **Young,** Clarence Marshall. Civil aviation and the trend of development. (United States. — Commerce Department: Aeronautics Branch, Air commerce bulletin, Washington, 1930, v. 2, p. 215–223.) **VDS**

4349. —— Present problems of the aircraft industry. (United States. — Commerce Department: Aeronautics Branch, Air commerce bulletin, Washington, 1932, v. 3, p. 493–497.) **VDS**

Economic influence, air transportation, manufacturing, air mail, and legislation.

4350. —— Progress of commercial aeronautics in the past 18 months. (United States.— Commerce Department: Aeronautics Branch, Air commerce bulletin, Washington, 1930, v. 2, p. 135–138.) **VDS**

4351. **Zarzar,** V. New airlines in Russia. illus. (Aero digest, New York, 1929, v. 15, July, p. 72–73.) **VDS**

AIR MAIL

4352. **Air** mail losses by fire negligible. (Aviation, New York, 1929, v. 27, p. 1043.) **VDS**
Statistics of United States Post Office Department, 1918–1929.

4353. **Air** mail. The record in the case. (Aviation, New York, 1934, v. 33, p. 64–67.) **VDS**
United States government cancels contracts with private concerns.

4354. **Air** mail. The Watres act in its workings. charts. (Aviation, New York, 1932, v. 31, p. 106–108.) **VDS**
Account of operations for 1931 and previous years.

4355. **Armstrong,** Douglas B. Air postage stamps. (Flight, London, 1923, v. 15, p. 524, 554, 568, 588, 614, 632, 664, 678, 692, 706, 718, 732, 746, 758, 772, 786; 1924, v. 16, p. 14, 40, 70, 82, 98, 112, 126, 140, 168, 200, 228, 258, 282, 376, 432, 504, 546, 582, 690, 726, 738, 796, 808; 1925, v. 17, p. 76, 90, 194, 206, 232, 266, 314, 364, 396, 514, 558, 574, 622, 740, 860; 1926, v. 18, p. 100, 172, 280, 302, 312, 588, 710, 798; 1927, v. 19, p. 196, 676, 786, 844; 1928, v. 20, p. 276, 680, 756, 786; 1929, v. 21, p. 18, 128, 308, 436, 1084, 1206, 1318, 1356; 1930, v. 22, p. 200, 314, 524, 1180, 1398, 1450; 1931, v. 23, p. 242, 304, 396, 514, 654, 798, 892, 1004, 1118, 1204; 1932, v. 24, p. 44, 124, 204, 314, 432, 522, 640, 836, 950, 1040, 1160; 1933, v. 25, p. 48, 114, 214, 332, 434, 528, 682, 792, 926, 1014, 1112, 1250; 1934, v. 26, p. 22, 116, 208, 344, 450, 574, 724, 826, 964, 1044, 1188, 1314.) **VDS**
A valuable running account of various issues.

4356. **Austin,** Warren Robinson. Air mail facts laid before American public. (National aeronautic magazine, Washington, 1934, v. 12, April, p. 6–7, 18–19.) **VDS**
Comments on the cancellation of United States air mail contracts.

4357. **Baldwin,** N. C., compiler. The air mails of British Africa, 1925–1932. Sutton Coldfield, England: Francis J. Field [1932]. 68 p. illus. 8°. ("Aero field" handbook. no. 2.) **VDS (Aero field)**

4358. The **Balloon** and pigeon posts. (Chambers' journal, London, 1871, v. 48 [series 4, v. 8], p. 129–134, 154–156.) *** DA**
As used in the Franco-Prussian war.

4359. **Barry,** R. A. Air mail stamps trace history of mankind's aerial conquests. illus. (New York Herald Tribune, New York, July 29, 1933, p. 7.) *** A**

4360. **Bernardi,** Mario de. The first permanent aero mail line. illus. (Flying, New York, 1917, v. 6, p. 684–685.) **VDS**
Author flew the first airplane on the Turin-Rome route, May 22, 1917.

4361. [**Bibliography** of air mail.] (In: United States. — National Advisory Committee for Aeronautics, Bibliography of aeronautics, 1909–1931, Washington, 1921–35, volumes for 1909/16, p. 828–830; 1917/19, p. 276–281; 1920/21, p. 256–262; 1922, p. 90–91; 1923, p. 80–82; 1924, p. 64–65; 1925, p. 104–105; 1926, p. 84–85; 1927, p. 106–107; 1928, p. 118–119; 1929, p. 136–137; 1930, p. 144–145; 1931, p. 172–173.) **VDT**

4362. **Brearley,** Norman. Air mail and transport in Australia. illus. (Aviation, New York, 1928, v. 25, p. 1480–1481, 1516, 1518.) **VDS**
First air mail in 1921.

4363. **Brunel,** Georges. La voie postale aérienne. illus. (L'Aérophile, Paris, 1920, année 28, p. 240–245.) **3–VDS**
Brief history of the air-mail stamps of the various countries.

4364. **Carrying** mail by aeroplane. The first experiments in the country [United States] at the International Aviation meet. illus. (Scientific American, New York, 1911, v. 105, p. 313.) **VA**

4365. **Chicago-Dallas** air mail opened. illus. (Aviation, Highland, N. Y., 1926, v. 20, p. 783–785.) **VDS**
Opened May 12, 1926.

4366. **Chronology** [of the United States air mail]. (National aeronautic magazine, Washington, 1934, v. 12, [March], special air mail issue, p. 6–8.) **VDS**

4367. **Cleveland,** Reginald M. Airmail. illus. (Scientific American, New York, 1934, v. 150, p. 290–291.) **VA**
Story of U. S. government cancellation of contracts, 1934.

4368. **Collins,** Gertrude L. My year-old air mail collection. illus. (Air, London, 1929, v. 2, p. 81–83.) **VDS**
Pictures of stamps and post-cards.

4369. A **Convincing** demonstration. The first round trip night air mail letter, New York, Chicago, New York. illus. (Aviation, Highland, N. Y., 1925, v. 19, p. 69.) **VDS**

4370. **Coyler,** D. B. The air mail service. (Western Society of Engineers, Journal, Chicago, 1927, v. 32, p. 161–164.) **VDA**
Brief history, beginning May 15, 1919.

4371. **Davis,** Alec. A postal history of aircraft. The evolution of the aeroplane is depicted on the stamp issues of the world. illus. (Air and airways, London, 1932, v. 9, p. 73–74.) **VDS**

4372. **Death** of Colonel Shaughnessy. Second Assistant Postmaster General, severely injured in collapse of Washington theatre, dies in Walter Reed Hospital. port. (Aviation, Highland, N. Y., 1922, v. 12, p. 199.) **VDS**
Shaughnessy (1882–1922) was in charge of the U. S. Air Mail Service.

Air Mail, continued

4373. **Delivery** of air mail to ocean liner successful. illus. (Aircraft journal, New York, 1919, v. 5, Aug. 23, p. 4.) **VDS**
Mail pouch dropped to steamship *Adriatic,* August 14, 1919.

4374. **Early** history of aerial mail carrying. illus. (Air service journal, New York, 1918, v. 2, p. 696–699.) **VDS**
Operations in the United States, beginning in 1910.

4375. **Eielson,** C. B. Blazing the air mail path in Alaska. Report to the Second Assistant Postmaster General on the first trip of the Alaskan Air Mail Service. illus. map. (Aviation, Highland, N. Y. 1924, v. 16, p. 394–396.) **VDS**

4376. —— First air mail service in Alaska. illus. map. (U. S. air services, Washington, 1924, v. 9, May, p. 12–14.) **VDS**

4377. **England-India-England.** First out and home air mail service completed in 15½ days. (Flight, London, 1929, v. 21, p. 312–313.) **VDS**

4378. The **Epoch** making night mail flights. ports., map. (Aviation, Highland, N. Y., 1923, v. 15, p. 266–269.) **VDS**
United States mail planes in 1923 maintained average of 28 hours for six flights between New York and San Francisco.

4379. **Field,** Francis J. Air mails and aerophilately. illus. (Air, London, 1928, v. 1, March, p. 37–40.) **VDS**
Reproductions of special stamps.

4380. **Florman,** Carl. Night air mails. maps. (Royal Aeronautical Society, Journal, London, 1931, v. 35, p. 460–488.) **VDS**
Development in Europe and the United States.

4381. **Ford** contract air mail opens. New service connects Detroit with transcontinental route. illus. (Aviation, Highland, N. Y., 1926, v. 20, p. 292.) **VDS**

4382. **Fortschritte** im Militärbrieftaubenwesen. (Zeitschrift für Luftschiffahrt, Berlin, 1890, Jahrg. 9, p. 234–239.) **3–VDS**
Progress in pigeon mail service.

4383. **Gale,** Charles H. Ship to shore aerial mail service. illus. (Aviation, New York, 1930, v. 28, p. 1084–1087.) **VDS**
The French line made its experiments in 1928, while the North German Lloyd established the service in 1929.

4384. **Glover,** Warren Irving. The eagle spreads its wings. illus. (Popular aviation, Chicago, 1928, v. 3, Aug., p. 11–15, 92.) **VDS**
History of the United States air mail.

4385. —— The [U. S.] air mail. map. (American Academy of Political and Social Science, Annals, Concord, N. H., 1927, v. 131, May, p. 43–48.) *** DA**
Brief history.

4386. **Gregor.** Die Entwicklung der Luftpost von 1922 bis 1926. (Germany. — Postamt, Archiv für Post und Telegraphie, Berlin, 1927, Bd. 55, p. 181–189, 216–224.) **TVA**
Developments in various countries, notably in Germany.

4387. —— Die Luftpost von 1920 bis 1922. (Germany. — Postamt, Archiv für Post und Telegraphie, Berlin, 1923, Bd. 51, p. 253–278.) **TVA**
Development of air service in various countries, notably in Germany.

4388. **Gross,** Hans. Die Ballon-Brieftaubenpost während der Belagerung von Paris im Jahre 1870–71. (Zeitschrift für Luftschiffahrt, Berlin, 1887, Jahrg. 6, p. 161–176.) **3–VDS**

4389. **Harper,** Harry, and Robert Brenard. The romance of the flying mail, a pageant of aerial progress. London: G. Routledge & Sons, Ltd., 1933. xvi, 240 p. illus. ports. 8°. **VDY**
History of British air mail service.
Reviewed in Royal Aeronautical Society, *Journal,* London, 1933, v. 37, p. 1049, *VDS.*

4390. **Henderson,** Paul. Review of the air mail night flying trials. illus. (Aviation, Highland, N. Y., 1923, v. 15, p. 521–523.) **VDS**

4391. **Hill,** Roderic. The Baghdad air mail, by Wing Commander Roderic Hill. London: E. Arnold & Co., 1929. x, 328 p. illus. maps. 8°. **VDY**
History of the service, which was opened in 1921. Experiences of the author as aviator. Reviewed in *Flight,* London, 1929, v. 21, p. 234, *VDS; Aeroplane,* London, 1929, v. 36, p. 364, *VDS.*

4392. **Howard,** W. H. L. Steps in the evolution of mail carrying. (Flying, New York, 1918, v. 7, p. 437, 439.) **VDS**
Development from Colonial times.

4393. **Inauguration** of the first aerial post of the United Kingdom. illus. (Flight, London, 1911, v. 3, p. 798–799.) **VDS**

4394. **Key West** to Havana. Some facts and figures about the organization and operating activities of the Pan American Airways, Inc. illus. (Aviation, Highland, N. Y., 1928, v. 24, p. 440–441.) **VDS**

4395. **Kropp,** H. Die Entwicklung der Luftfahrt und ihre Bedeutung für den Postverkehr. (Germany. — Postamt, Archiv für Post und Telegraphie, Berlin, 1920, Jahrg. 48, p. 212–237.) **TVA**
A history of aeronautics precedes an account of the early days of the German air postal service.

4396. **Lawrance,** Charles Lanier. Development and cost of the air mail service. (In his: Our national aviation problem, New York, 1932, p. 93–112.) **VDY**
This chapter appeared originally in *Western flying,* Los Angeles, Cal., 1931, v. 10, Oct., p. 21–24, *VDS.*

4397. **Liggett,** Walter W. The inside story of the airmail monopoly. illus. (Popular aviation, Chicago, 1932, v. 10, p. 351–352, 404–405.) **VDS**
Story of how four interlocked companies have controlled matters in the United States.

Air Mail, continued

4398. **Los Angeles-Salt Lake City** air mail opened. illus. (Aviation, Highland, N. Y., 1926, v. 20, p. 704–705.) VDS
Opened April 17, 1926.

4399. **Mead,** George J. Journeying with the Boeing air mail. illus. (Aviation, Highland, N. Y., 1927, v. 23, p. 990–994.) VDS

4400. **Mr.** Praeger honored. port. (Aviation, Highland, N. Y., 1921, v. 11, p. 280.) VDS
As Second Assistant Postmaster-General, Otto Praeger organized the United States Air Mail.

4401. **Moore,** Samuel Taylor. Over the perilous path of the air mail. (Literary digest, New York, 1925, v. 84, March 14, p. 73–78.) * DA
Author, associated with the New York *World,* relates his experiences.

4402. **Morris,** Woodbridge E. The pony express takes to the air. Once it took twenty-five days to carry a letter from New York to San Francisco, but now Uncle Sam's fearless pilots do the job in a day and a night. illus. (Light, Cleveland, O., 1926, v. 4, July, p. 5–7.) VGA
How the route between New York and Cleveland is lighted, with experiences of some of the pilots. Quoted in *Literary digest,* New York, 1926, v. 90, July 31, p. 18–19, * *DA.*

4403. **New York-Chicago** night air mail. illus. (Aviation, Highland, N. Y., 1925, v. 19, p. 38–40.) VDS
Service began July 1, 1925.

4404. **New York-Philadelphia-Washington** aerial mail great success. Extension to Pittsburgh, Cleveland and Boston proposed. (Aerial age, New York, 1918, v. 7, p. 531–532.) VDS
Inaugurated May 15, 1918.

4405. The **Night** flight of the air mail. illus. (Aeronautical digest, New York, 1923, v. 3, p. 223–231, 280–282.) VDS
Account of illumination of the Chicago-Cheyenne night airway, 1923.

4406. **Oppermann,** C. A. Les ballons-poste du siége de Paris. illus. (Portefeuille économique des machines, Paris, 1871, année 16, col. 65–67.) 3–† VFA
List of 54 balloons in the service. Excellent detailed drawings of the Godard balloon shown in plates 33–34 at end of volume.

4407. **Ovington,** Adelaide (Alexander). The first air mail flight. An inside story of the first airmail trip, flown by Earle Ovington [1911]... illus. (Popular aviation, Chicago, 1933, v. 12, p. 79–80.) VDS

4408. The **Paris** balloon post. illus. (Engineering, London, 1871, v. 12, p. 236.) VDA
Brief account of ascents (1870–1871), with description of the Godard balloon.

4409. **Pernicious** air mail bill becomes Federal law. (National aeronautic magazine, Washington, 1934, v. 12, June, p. 6–9.) VDS
Includes addresses before the United States Senate by Senators Warren B. Austin and Simeon D. Fess on the Black-McKellar bill.

4410. **Phillips,** Stanley. Postage stamps of the air mails. illus. (Aeronautical digest, New York, 1923, v. 3, p. 154–156.) VDS

4411. The **Pony** express takes to the air. illus. (Literary digest, New York, 1927, v. 94, Sept. 17, p. 10–11.) * DA
Newspaper comments on success of the United States air mail.

4412. **Postmaster General** Burleson on the Air Mail. port. (Aviation, New York, 1920, v. 9, p. 417–419.) VDS
Excerpts from his annual report on progress of the United States Air Mail during 1920.

4413. **Potter,** Leslie S. Cairo-Baghdad air mail route. How the British Royal Air Force inaugurated a desert airline. illus. map. (Aero digest, New York, 1930, v. 17, July, p. 56–57.) VDS

4414. **Premier** essai d'aviation postale. illus. (L'Aérophile, Paris, 1913, année 21, p. 484–485.) 3–VDS
This French test of the airplane in the postal service is stated to be first in any country.

4415. **Pritchard,** Robert J. Facts of the air mail. (Western flying, Los Angeles, 1934, v. 14, March, p. 10–13, 22.) VDS
A history of mail-carrying airlines in the United States and of legislation.

4416. **Rampont.** La poste aérienne pendant le siége de Paris. (L'Aéronaute, Paris, 1875, année 8, p. 28–33.) 3–VDS

4417. A **Review** of air mail law. (National aeronautic magazine, Washington, 1934, v. 12, [March], special mail issue, p. 14–15.) VDS
Text of the four laws, including the Kelly and Watres acts.

4418. **Route** history [of the United States air mail]. (National aeronautic magazine, Washington, 1934, v. 12, [March], special air mail issue, p. 11–13.) VDS

4419. **Sarigny,** René de. South Africa's air mail. illus. (Aero digest, New York, 1925, v. 6, p. 313–314.) VDS

4420. **Shaughnessy,** E. H. The United States Air Mail service. illus. (Aviation, Highland, N. Y., 1922, v. 12, p. 96–99.) VDS
"The air-mail service operated by the Post Office Department was started in May, 1918, the first step being the establishment of a mail route extending from the City of Washington to New York City." Also, somewhat condensed, in Society of Automotive Engineers, *Journal,* New York, 1922, v. 11, p. 435–437, *TOL.*

4421. **Statistical** picture of the air mail. An analysis of recent operations in the United States. (Aviation, New York, 1929, v. 27, p. 1027–1031.) VDS
Detailed account of operations, 1927–1929.

4422. **Texas-Chicago** air mail operates at 97 per cent perfect. (Aviation, Highland, N. Y., 1926, v. 21, p. 320.) VDS
What the National Air Transport accomplished during the first three months.

Air Mail, continued

4423. **Three** years of air mail service [in the United States]. illus. (Aviation, Highland, N. Y., 1921, v. 10, p. 559–561, 564.) **VDS**

4424. **Truby,** Harry A. The aerial post epoch. illus. (Aero digest, New York, 1925, v. 7, p. 416–418, 448, 450.) **VDS**
Has pictures of early air mail stamps.

4425. The **United States** air mail experiments. illus. (Flight, London, 1923, v. 15, p. 629–630.) **VDS**
Conducted in 1923 for determining practicability of 28-hour service between Atlantic and Pacific coasts.

4426. **United States** air mail operation and traffic. charts. (Aviation, New York, 1929, v. 27, p. 706–707.) **VDS**
Statistics and charts of development.

4427. **U. S.** aviation and the air mail. The journalistic nigger in the word pile remains: "Just what happened?" So Fortune considers the ten-year record, looks into the future. illus. ports., map. (Fortune, New York, 1934, v. 9, May, p. 85–91, 140, 142, 144, 153–154, 156, 158.) **† TMA**
Spicy account of the cancellation of the air mail contracts, 1934, with portraits of some of the actors.

4428. **Van Zandt,** John Parker. On the trail of the air mail. A narrative of the experiences of the flying couriers who relay the mail across America at a speed of more than 2,000 miles a day. illus. (National geographic magazine, Washington, 1926, v. 49, p. 1–61.) **KAA**

4429. **Woodhouse,** Henry. The first permanent aerial mail line. illus. (Flying, New York, 1918, v. 7, p. 703–704.) **VDS**
New York-Washington line.

4430. —— The inauguration of the New York-Philadelphia-Washington aerial mail line. illus. (Flying, New York, 1918, v. 7, p. 413–421.) **VDS**

POLAR EXPEDITIONS

4431. The **Airship** Norge flies across the Arctic. Amundsen-Ellsworth Arctic expedition succeeds in linking Spitzbergen and Alaska by air. illus. ports., map. (Aviation, Highland, N. Y., 1926, v. 20, p. 824–826.) **VDS**

4432. **Amundsen,** Roald Engelbregt Gravning. Gjennem luften til 88° nord, av Roald Amundsen, Hj Riiser-Larsen, Leif Dietrichson, Fredrik Ramm, Jakob Bjerknes. Oslo: Gyldendal norsk forlag, 1925. 278 p. illus. ports. 8°. **KBR**
Accounts of the Amundsen-Ellsworth Arctic expedition 1925.

4433. —— The hop-off; radiograms of the Amundsen-Ellsworth polar flight. [New York: W. Hyams & Co.,] 1925. 35 p. 4°. **KAT p.v.175, no.9**

4434. **Amundsen,** Roald Engelbregt Gravning, and L. W. Ellsworth. First crossing of the Polar sea; with additional chapters by other members of the expedition. New York: George H. Doran Company, 1927. x, 13–324 p. illus. ports., map. 8°. *** KF (1927)**

4435. —— The first flight across the Polar sea; with additional chapters by Joh. Höver, Hj. Rüser-Larsen, Gustav Amundsen, Finn Malmgren [and] B. L. Gottwaldt. London: Hutchinson & Co., Ltd. [1926.] 274 p. illus. ports., map. 8°. **KBR**

4436. The **Andrée** polar expedition. illus. (Scientific American supplement, New York, 1896, v. 42, p. 17260–17261.) **VA**
Account of equipment.

4437. **Arnesen,** Odd. The Polar adventure; the "Italia" tragedy seen at close quarters. [Translated from the Norwegian by Asta and Rowland Kenney.] London: V. Gollancz, Ltd., 1929. 158 p. illus. ports. 8°. **KBR**

4438. **Bahier,** Eugène. L'expédition aéronautique de M. Andrée au pôle nord. illus. (Le Génie civil, Paris, 1897, tome 31, semestre 2, p. 257–259.) **VA**

4439. **Balchen,** Bernt. Aviation lessons of the Antarctic. The Antarctic region, its extent, elevation, and climatic conditions — flight problems of the Byrd expedition as told by its chief pilot... illus. (Mechanical engineering, New York, 1931, v. 53, p. 489–496.) **VFA**

4440. **Baschin,** Otto. Die Arktisfahrt des Luftschiffes "Graf Zeppelin." illus. (Naturwissenschaften, Berlin, 1932, Jahrg. 20, p. 6–13.) **OA**
Early use of aeronautics in the Arctic regions. Scientific results of the *Graf Zeppelin* flight.

4441. **Běhounek,** František. Sieben Wochen auf der Eisschole; der Untergang der Nobile-Expedition. Leipzig: F. A. Brockhaus, 1929. 263 p. illus. maps, ports. 5. ed. 12°. **KBR**

4442. **Bennett,** Cora Lillian (Orkins). ... Floyd Bennett; with a foreword by Rear Admiral Richard E. Byrd. New York: W. F. Payson, 1932. xii, 168 p. illus. ports. 12°. **AN**
Bennett was airplane pilot on Byrd's flight to the North Pole.

4443. **Bennett,** Floyd. Our flight over the North Pole. illus. ports. (Aero digest, New York, 1926, v. 9, p. 175–177, 261–263.) **VDS**
Flight of the author and Admiral Byrd in the airplane *Josephine Ford.*

Polar Expeditions, continued

4444. [**Bibliographies** of aerial polar expeditions.] (In: Paul Brockett, Bibliography of aeronautics, Washington, 1910, p. 635; continued in United States. — National Advisory Committee for Aeronautics, Bibliography of aeronautics, 1909–1931, Washington, 1921–35, volumes for 1909/16, p. 962; 1917/19, p. 334; 1920/21, p. 304; 1922, p. 117; 1923, p. 101; 1924, p. 82; 1925, p. 130; 1926, p. 108–109; 1927, p. 133; 1928, p. 148–149; 1929, p. 169; 1930, p. 180; 1931, p. 225.) **VDT**

4445. [**Bibliography** of S. A. Andrée.] (In: Paul Brockett, Bibliography of aeronautics, Washington, 1910, p. 45–46.) **VDT**
List of writings by and about Andrée.

4446. Binney, George. With seaplane and sledge in the Arctic, by George Binney, leader Oxford University Arctic Expedition, with a preface by Professor W. J. Sollas. New York: G. H. Doran Co., 1926. 287 p. illus. map. 8°. **KBX**
Library has also London edition, 1925, *KBX*.

4447. The **British** Arctic air route expedition. (Geographical journal, London, 1930, v. 76, p. 67–68, 426–427.) **KAA**

4448. British Arctic Air Route Expedition to Greenland, 1930–1931. Northern lights; the official account of the...expedition, 1930–1931, by F. Spencer Chapman. With a foreword by Admiral Sir William Goodenough, and introduction by the late H. G. Watkins, leader of the expedition, and additional chapters by J. M. Scott, Capt. P. M. H. Lemon & Augustus Courtauld. London: Chatto and Windus, 1932. xv, 304 p. illus. ports., maps. 4°. **KBR**
Reviewed in *Air and airways*, London, 1932, v. 9, p. 325, *VDS*.

4449. Bruce, Eric Henry Stuart. The use of electric balloon signalling in Arctic and Antarctic expeditions. (British Association for the Advancement of Science, Report of the 68th meeting, Bristol, 1899, London, 1899, p. 948–949.) *** EC**
Author's suggestions. Also in *Electrical review*, London, 1898, v. 43, p. 505, *3–†† VGA*.

4450. Byrd, Richard Evelyn. Little America, aerial exploration in the Antarctic, the flight to the South Pole. New York: G. P. Putnam's Sons, 1930. xvi, 422 p. illus. ports., maps. 8°. **KBY**
Reviewed in *Aeroplane*, London, 1931, v. 40, p. 242, *VDS*.

4451. The **Byrd** flight to the North Pole. Commander Byrd gives details of his flight under the Polar regions. ports., map. (Aviation, Highland, N. Y., 1926, v. 20, p. 780–782.) **VDS**
An abstract of Byrd's accounts in the *New York Times*.

4452. Crocé-Spinelli, J., and A. Saco. Rapport de la commission chargée d'apprécier le projet d'exploration du pôle nord, à l'aide d'un aérostat présenté par M. Sivel. illus. (L'Aéronaute, Paris, 1872, année 5, p. 137–157.) **3–VDS**

4453. Crowding the North Pole. illus. (Literary digest, New York, 1926, v. 89, May 22, p. 8–10.) *** DA**
Preparation for the various expeditions of the year and an account of Byrd's airplane flight.

4454. The **Discovery** of Andrée's body in the Arctic ice: the pioneer of polar aeronautics found after 33 years. illus. (Illustrated London news, London, 1930, v. 177, p. 360.) *** DA**
Brief account of discovery, with reprinting of old illustrations showing preparations for the original expedition.

4455. Ekholm, Nils. S. A. Andrée's Polarfahrt im Luftballon. map. (Illustrierte aëronautische Mitteilungen, Strassburg, 1898, Jahrg. 2, p. 43–47, 65–71.) **VDS**
For list of Ekholm's papers on the Andrée expedition see Paul Brockett, *Bibliography of aeronautics*, Washington, 1910, p. 273–274, *VDT*.

4456. Ellsworth, Lincoln W. Air pioneering in the Arctic; the two polar flights of Roald Amundsen and Lincoln Ellsworth. Part I: The 1925 flight from Spitzbergen to 88° north; part II: The first crossing of the Polar sea, 1926. New York: National Americana Society, 1929. 126 p. illus. 4°. **† KBR**

4457. Espitallier, Georges. Au pôle nord en ballon dirigeable. L'expédition Wellman. illus. (Le Génie civil, Paris, 1906, tome 49, p. 246–250.) **VA**
Illustrated account of the airship.

4458. Essén, Axel. Polarhjälten; en bok om S. A. Andrée, sammanställd av Anders Eje [pseud.]. Stockholm: H. Schildts förlag [1930]. 175 p. illus. ports., maps. 8°. **KBR**

4459. The **First** hop over the earth's roof. ports., map. (Literary digest, New York, 1928, v. 97, May 5, p. 10–11.) *** DA**
Newspaper comments on flight of George H. Wilkins and Carl B. Eielson, Point Barrow to Spitzbergen, 1928.

4460. The **Flight** of the airship Norge. Italy-England-Norway-Russia flight carried out successfully as preliminary to flight into Arctic. illus. (Aviation, Highland, N. Y., 1926, v. 20, p. 624–626.) **VDS**
Amundsen-Ellsworth expedition of 1926.

4461. Fonvielle, Wilfrid de. L'exposition du "Pôle nord" au Champ-de-Mars. illus. (L'Aérophile, Paris, 1896, année 4, p. 104–111.) **3–VDS**
Description of Andrée's balloon.

4462. Freeburg, Victor Oscar. What of Andrée? Will Amundson find relics of Andrée? Possibility that Norwegian explorer may discover wreck of Swedish balloon lost twenty-seven years ago. Dramatic story of first attempt to fly to the Pole. illus. port. (Aero digest, New York, 1924, v. 5, p. 34–38.) **VDS**

Polar Expeditions, continued

4463. The **French** polar balloon expedition. (Aeronautical journal, London, 1897, v. 1, April, p. 8–9.) **VDS**

An interview with Louis Godard on the plans of the enterprise.

4464. **Giudici,** Davide. The tragedy of the Italia; with the rescuers to the red tent. London: E. Benn, Ltd. [1928.] viii, 215 p. illus. map, ports. 8°. **KBR**

4465. **Gouzy,** René. Dans le ciel des pôles; explorations d'autrefois — d'aujourd'hui. Genève: A. Jullien [1931]. 206 p. illus. maps, ports. 8°. **KBM**

Illustrated history of Arctic and Antarctic aerial expeditions.

4466. **Hanson,** Earl. The place of aircraft in Polar research. illus. (Airway age, New York, 1929, v. 10, p. 275–279.) **VDS**

Discusses the suitability of aircraft for scientific and practical purposes.

4467. **Hoenerssen,** G. A., pseud. Neue Erkentnisse; Amundsens missgluecter Flug 1925 und das Geheimnis des Nordpols. Obermenzing-München: Raumkraftverlag, 1925. 63 p. diagrs. 12°. **VDY p.v.36, no.1**

4468. **Italy.** — Commissione d'Indagini per la Spedizione Polare dell' Aeronave "Italia." ...Relazione... Roma, 1930. 138 p. maps. (Rivista marittima, anno 63, gennaio, 1930, supplemento.) **VXA**

Report of Italian commission appointed by Mussolini to investigate cause of disaster.

4469. **Jacobowsky,** C. Vilhelm. Andrée en man och en bragd... Lund: Borelius [1930]. 194 p. illus. map, port. 4°. **KBR**

Life and works of Salomon August Andrée, born 1854.

4470. **Joerg,** Wolfgang Louis Gottfried. Brief history of polar exploration since the introduction of flying, to accompany a physical map of the Arctic and a bathymetric map of the Antarctic. New York: American Geographical Society, 1930. 50 p. maps. 8°. (American Geographical Society. Special publication. no. 11.) **KBN**

Brief summaries of various expeditions, with a list of the polar publications of the Society. Reviewed in *Aviation,* New York, 1931, v. 30, p. 499, *VDS.*

4471. —— The work of the Byrd Antarctic expedition, 1928–1930... New York: American Geographical Society, 1930. 71 p. illus. maps. 8°. **KBY**

"A running account...derived...from radio messages covering the achievements and adventures of the expedition which were published in the *New York Times* in the course of the two years." — *Introduction.*

4472. **Kohl-Larsen,** Ludwig. Die Arktisfahrt des "Graf Zeppelin," im Auftrage der Internationalen Gesellschaft zur Erforschung der Arktis mit Luftfahrzeugen (Aeroarctic) herausgegeben. Vorwort von Dr. Hugo Eckener. Berlin: Union deutsche Verlagsgesellschaft [1931]. 202 p. illus. map, port. 8°. **KBR**

Account of Arctic flight, 1931.

4473. **Lachambre,** Henri, and Alexis Machuron. Andrée's balloon expedition in search of the North Pole. New York: Frederick A. Stokes Company, 1898. 306 p. 12°. **VDW**

The authors were Andrée's assistants in the preparation, equipment, and departure of the balloon.

4474. **Lavergne,** Gérard. Le ballon Andrée pour l'exploration des régions polaires arctiques. illus. (Le Génie civil, Paris, 1896, tome 29, p. 49–54.) **VDA**

Detailed account of equipment.

4475. —— L'expédition Andrée pour l'exploration aérostatique des régions polaires. illus. (Le Génie civil, Paris, 1897, tome 31, semestre 2, p. 184–187.) **VA**

4476. **Lawrence,** John, pseud. Bernt Balchen, viking of the air. New York: Brewer, Warren and Putnam, 1931. 165 p. ports. illus. 12°. **AN**

Story of Admiral Byrd's famous pilot in Polar and Transatlantic flights.

4477. **Lundborg,** Einar Paul Albert Muni. The Arctic rescue; how Nobile was saved, translated by Alma Luise Olson. New York: The Viking Press, 1929. 221 p. illus. ports. 8°. **KBR**

4478. —— Captain Einar Lundborg tells the truth about Nobile's rescue. illus. ports. (Popular aviation, Chicago, 1929, v. 4, June, p. 17–18, 82–83.) **VDS**

This number of *Popular aviation* has the title of *Aeronautics.*

4479. **Macdonald,** A. C. Over land and sea; or, Andrée's aerial voyage to the North Pole. illus. ports. (Royal Geographical Society of Australasia, Victoria, Transactions, Melbourne, 1897, v. 15, p. 53–67.) **KAA**

4480. **Marcillac,** P. L'aéronautique scientifique et les grandes expéditions polaires. map, plate. (Société scientifique industrielle de Marseille, Bulletin, Marseille, 1904, année 32, p. 21–91.) **3–VA**

Lengthy discussion of the possibilities of the balloon for Arctic exploration.

4481. **Martinet,** Ludovic. L'exploration du pôle en aérostat. (L'Aéronaute, Paris, 1875, année 8, p. 287–301.) **3–VDS**

Discusses problems involved.

4482. **Meyer,** Willy. Der Kampf um Nobile; Versuch einer objektiven Darstellung und Wertung der Leistungen des italienischen Luftschiffers, mit einem gemeinsamen Vorwort von Franz Behounek, Arthur Berson [und] Leonid Breitfuss. Berlin: Gebr. Radetzki, 1931. 360 p. maps. 8°. **KBR**

Bibliography, p. 349–354.

Polar Expeditions, continued

4483. Miethe, Adolf, and Hugo Hergesell. Mit Zeppelin nach Spitzbergen; Bilder von der Studienreise der deutschen arktischen Zeppelin-Expedition... Mitarbeiter: v. Bartenwerffer, v. Drygalski...Graf v. Zeppelin... Berlin: Deutsches Verlagshaus Bong & Co. [1911.] 291 p. illus. ports. 4°. **KBR**

Numerous illustrations.

4484. Miller, Francis Trevelyan. The fight to conquer the ends of the earth: Byrd's great adventure, with the complete story of all polar expeditions for one thousand years ...with forewords by General A. W. Greely ...and Dr. Henry Fairfield Osborn... Philadelphia: The John C. Winston Company [cop. 1930]. 383 p. illus. ports. 8°. **KBY**

4485. Mohr, Adrian. 33 Jahre verschollen im Packeis; die arktische Freiballon-Expedition des Schweden Salomon August Andrée, von Adrian Mohr, Carsten Borchgrevink, Tryggve Gran, G. V. E. Svedenborg und Otto Sverdrup... Leipzig: Grethlein & Co. [1930.] 160 p. front. (port.), map, plates. 12°. **KBR**

4486. Neely, Frederick R. To Siberia and return. Alaskan airmen blaze a commercial air trail across the Arctic wastes. illus. map. (Aeronautic review, Baltimore, 1929, v. 7, May, p. 14–15, 31.) **VDS**

Flight of Noel Wien and Calvin Cripe, Fairbanks, Alaska, to Cape North, 1929.

4487. New York Public Library. The Byrd Antarctic expedition [1928–1930]. A scrapbook of clippings, taken mainly from the New York Times... Prepared by the staff of the Science Room. New York, 1931. 2 v. illus. f°. **Desk (117)**

Volume 1, in two parts, covers text. Volume 2 contains the illustrations.

4488. Nobile, Umberto. L'"Italia" al Polo nord. [Verona:] A. Mondadori, 1930. xv, 475 p. illus. maps, ports. 2. ed. 4°. **KBR**

Library has also English translation by Fran Fleetwood entitled *With the 'Italia' to the North Pole,* London: G. Allen & Unwin, Ltd., 1930, *KBR.*
For references to Nobile see United States. — National Advisory Committee for Aeronautics, Bibliography of aeronautics, 1909–1931, Washington, 1921–35, volumes for 1917/19, p. 332; 1920/21, p. 302–303; 1922, p. 117; 1923, p. 101; 1924, p. 81–82; 1925, p. 129–130; 1926, p. 107; 1928, p. 148; 1929, p. 169; 1930, p. 179; 1931, p. 211, *VDT.*

4489. Il **"Norge"** al Polo nord; da Roma allo Stretto di Behring in dirigible, 13,000 km. in 172 ore di volo; note del comandante Y. H. Bologna: L. Cappelli, 1926. 218 p. illus. maps, ports. 8°. **KBR**

Norge expedition to North Pole in 1926.

4490. Pesce, G. L. Historique des divers projets d'exploration aérostatic du pôle nord. illus. (L'Aérophile, Paris, 1899, année 7, p. 51–55, 63–65, 75–79.) **3–VDS**

History begins with the project of Jules François Dupuis-Delcourt, 1845.

4491. The **Problem** of flight: ballooning in Arctic exploration. (Journal of science, London, 1879, series 3, v. 1, p. 163–168.) **OA**

Views of F. W. Brearey on mechanical flight and on the use of the balloon in Arctic regions.

4492. Progress of the Detroit Arctic expedition. Captain Wilkins explores North and flies nearer Pole than anyone has flown before. illus. (Aviation, Highland, N. Y., 1926, v. 20, p. 587–588.) **VDS**

4493. Proposed balloon voyage to the North Pole. illus. (Scientific American, New York, 1877, new series, v. 37, p. 375.) **VA**

Three balloons connected by cables, proposed by Henry Coxwell.

4494. Putnam, George Palmer. Andrée; the record of a tragic adventure. New York: Brewer & Warren, Inc., 1930. 6 p.l., 11–239 p. illus. ports., maps. 12°. **VDW**

"No 'flying machines,' no radio, in those days [1897], remember. The Arctic explorer, once beyond the rim of civilization, dropped from sight and sound, until he returned again — if he did return."

4495. Quattrini, Antonio G. Col "Norge" sulla via del Polo, con una nota illustrativa di U. Nobile. Firenze: A. Quattrini [1926]. 237 p. illus. maps, ports. 4°. **KBR**

Norge polar flight of 1926.

4496. Robertson, Hugh W. The eagle screams at the North Pole. illus. (U. S. air services, Washington, 1926, v. 11, June, p. 13–23.) **VDS**

Airplane flight of Byrd, 1926.

4497. Samoilovich, Rudol'f Lazarevich. S-O-S in der Arktis; die Rettungsexpedition des Krassin. Berlin: Union deutsche Verlagsgesellschaft [1929]. 410 p. illus. maps, port. 8°. **KBR**

Describes the rescue of the Nobile expedition by the Russian ice-boat *Krassin.* Translated from the Russian by H. Leuschner. Library has also Italian translation, Firenze: Bemporad & figlio, 1930, 2. ed., *KBR.*

4498. Shackleton, Sir Ernest Henry. The use of the balloon in the British Antarctic expedition. (Aeronautical journal, London, 1904, v. 8, p. 11–12.) **VDS**

4499. Smith, A. M. The Wilkins Arctic expeditions. illus. (Aero digest, New York, 1928, v. 12, p. 741, 874–875.) **VDS**

Accounts of expeditions of 1926, 1927, and 1928.

4500. Stefánsson, Vilhjálmur. The airplane and the Arctic. New pioneers dispel an old myth. (Harper's monthly magazine, New York, 1927, v. 155, p. 595–604.) *** DA**

Polar flight of Commander Byrd; also of Wilkins and Eielson.

4501. —— New polar trails. (Forum, New York, 1927, v. 77, p. 54–64.) *** DA**

"Focuses...the new knowledge of the North gained during the past twelve months."

Polar Expeditions, continued

4502. **Strindberg,** Nils. La fabrication de l'hydrogène pour le ballon de l'expédition polaire de M. Andrée. (L'Aérophile, Paris, 1896, année 4, p. 68–70.) **3–VDS**

4503. **Stuart,** John. Wellman — precursor of heroes. illus. port. (Sportsman pilot, Albany, N. Y., 1931, April, p. 14–15, 55–56.) **VDS**

Brief account of American Arctic explorer and of his airship, the *America.*

4504. **Suplee,** Henry Harrison. Lessons from the Wellman expedition. (Cassier's magazine, New York, 1910–11, v. 39, p. 79–80.) **VA**

4505. **Surcouf,** Édouard. Expédition française au pôle nord en ballon. (Société des ingénieurs civils de France, Mémoires, Paris, année 1897, v. 1, p. 174–192.) **VDA**

A study of the problems confronting the author and his associate, Louis Godard, for their proposed voyage.

4506. **Svenska sällskapet för antropologi och geografi,** Stockholm. The Andrée diaries; being the diaries and records of S. A. Andrée, Nils Strindberg and Knut Fraenkel written during their balloon expedition to the North Pole in 1897 and discovered on White Island in 1930, together with a complete record of the expedition and discovery. Authorized translation from the official Swedish edition by Edward Adams-Ray... London: J. Lane, Ltd. [1931.] xx, 471 p. illus. maps, ports. 8°.

For Swedish original see entry no. 4508. **KBR**

4507. —— Andrée's story; the complete record of his polar flight, 1897, from the diaries and journals of S. A. Andrée, Nils Strindberg, and K. Fraenkel, found on White Island in the summer of 1930 and edited by the Swedish Society for Anthropology and Geography; translated from the Swedish by Edward Adams-Ray. New York: The Viking Press, 1930. xvi, 389 p. illus. ports. 8°. **KBR**

4508. —— Med Örnen mot Polen, av. S. A. Andrée, Nils Strindberg och Knit Fraenkel. Andrées Polar-expedition år 1897... Stockholm: A. Bonniers förlag [1931]. 487 p. illus. facsims., table, ports., plates. 8°. **KBR**

Diaries of the Andrée polar expedition of 1897. The Library has a copy of the authorized English translation by Edward Adams-Ray, see entry no. 4506.

4509. **Tomaselli,** Cesco. L'inferno bianco; racconto della spedizione Nobile. Milano: Unitas, 1929. xvi, 349 p. 4. ed. enl. illus. maps, ports. 8°. **KBR**

Italia polar expedition under Nobile, 1928.

4510. **Vaniman,** Melvin. Story of the Wellman ocean trip. Vivid account of the trip. illus. (Scientific American, New York, 1910, v. 103, p. 340–341, 351.) **VA**

4511. **Waack,** Karl. Von Andree bis Zeppelin. Das Luftschiff im Dienste der Polarforschung. Rostock: E. Wette, 1910. 63 p. ports. 8°. **VDT**

4512. **Watkins,** Henry George. The British Arctic Air Route Expedition. illus. (Geographical journal, London, 1932, v. 79, p. 353–367, 466–501.) **KAA**

Survey (1930–1931) of air route, Winnipeg, Canada, to London. For account of aerial work see p. 473–480.

4513. **Wellman,** Walter. The aerial age; a thousand miles by airship over the Atlantic Ocean. Airship voyages over the Polar sea. The past, the present and the future of aerial navigation. New York: A. R. Keller & Company, 1911. 448 p. illus. ports. 8°. **VDY**

Wellman (1858–1934) endeavored in 1906–1907 to reach the North Pole by airship. Later he and his companions made a record trip by dirigible over the Atlantic Ocean.

4514. —— The polar airship. (National geographic magazine, Washington, 1906, v. 17, p. 208–219.) **KAA**

For references to Wellman see Paul Brockett, *Bibliography of aeronautics,* Washington, 1910, p. 882–883; continued in United States. — National Advisory Committee for Aeronautics, *Bibliography of aeronautics,* 1909–1931, Washington, 1921–35, volume for 1909/16, p. 1413–1414, *VDT.*

4515. The **Wellman** Polar airship expedition. illus. (Scientific American, New York, 1907, v. 96, p. 509–510.) **VA**

Description of the airship.

4516. **Wellman's** airship for his North Polar expedition. illus. (Scientific American, New York, 1906, v. 95, p. 7–8.) **VA**

Illustrated description of the airship.

4517. **Wilkins,** Sir George Hubert. Arctic flights. (Royal Aeronautical Society, Journal, London, 1928, v. 32, p. 885–900.) **VDS**

Lecture on the adaptation of airplanes to Arctic explorations.

4518. —— Flying the Arctic. New York: G. P. Putnam's Sons, 1928. xv, 336 p. illus. ports. 8°. **VDY**

Author's Polar airplane flight of 1928.

4519. —— The problems of Polar flying. Some interesting high lights on Sir Hubert Wilkins' aerial ventures into the Arctic and Antarctic regions are told here. illus. (Aviation, New York, 1930, v. 28, p. 1076–1079.) **VDS**

4520. **Wilkins'** arctic flight. illus. ports. (Popular aviation, Chicago, 1928, v. 2, June, p. 9–12.) **VDS**

Flight of George H. Wilkins and Carl B. Eielson, Point Barrow to Green Harbor, 1928.

4521. **Wisting,** Oscar. 16 år med Roald Amundsen, fra pol til pol. Oslo: Gyldendal norsk forlag, 1930. 206 p. illus. ports. 8°.

Author's experiences with Amundsen. **KBR**

4522. **Woodhouse,** Henry. Greatest race of all time. Expeditions of five nations vie for the honor of being the first to reach the North and South Poles by air. illus. (Illustrated world, Chicago, 1923, v. 38, p. 679–683, 786.) **VDA**

ASSOCIATIONS AND CONFERENCES

4523. Aeronautic Society. Epitome of the work of the Aeronautic Society from July, 1908, to December, 1909. 39 p. illus. 8°. (Aeronautical Society of America. Bulletin no. 1.) **VDS**

Aeronautic Society subsequently became the Aeronautical Society of America.

4524. The **Aeronautical** Society. (Aeronautical journal, London, 1897, v. 1, Jan., p. 3–4.) **VDS**

History of the society.

4525. Air Conference, London. Proceedings of Air Conference. Conferences 1–3, 1920–1923. London, 1920–23. 3 v. maps. 8°. **VDY**

no. 1 is shelved in *VBA p.v.70, no.9;* no. 2, in *VBA p.v.78, no.1.* In these conferences the development of civil air service and of air mail service is discussed.

4526. Air conquest from the early days of giant kites and birdlike gliders. The National Geographic Society has aided and encouraged the growth of aviation. illus. (National geographic magazine, Washington, D. C., 1927, v. 52, p. 233–242.) **KAA**

Record of the society.

4527. "The **Air** hath bubbles." (Mechanic's magazine, London, 1837, v. 26, p. 400.) **3–VA**

Unflattering paragraph on the "Aeronautical Association for Applying Aero-Station to the Purposes of Geographical Survey."

4528. Association française aérienne. Le vol à voile et l'Association française aérienne; rapport officiel du congrès de Combegrasse & programme du congrès de Vauville, 1922–1923; préface de M. Laurent-Eynac. Paris [1922?]. 178 p. illus. 8°. **VDS**

4529. [**Bibliographies** of aeronautical societies.] (In: Paul Brockett, Bibliography of aeronautics, Washington, 1910, p. 785; continued in United States. — National Advisory Committee for Aeronautics, Bibliography of aeronautics, 1909–1931, Washington, 1921–35, volumes for 1909/16, p. 1243–1247; 1917/19, p. 410–411; 1920–21, p. 380–381; 1922, p. 146–147.) **VDT**

4530. [**Bibliographies** of congresses and conferences.] (In: Paul Brockett, Bibliography of aeronautics, Washington, 1910, p. 207–211; continued in United States. — National Advisory Committee for Aeronautics, Bibliography of aeronautics, 1909–1931, Washington, 1921–35, volumes for 1909/16, p. 292–293; 1917/19, p. 99; 1920/21, p. 94; 1922, p. 31; 1923, p. 28; 1925, p. 41; 1926, p. 31; 1927, p. 37; 1928, p. 45–46; 1929, p. 56; 1930, p. 49; 1931, p. 58.) **VDT**

4531. Bordé. Congrès de météorologie aéronautique de St. Pétersbourg [1904]. (L'Aéronaute, Paris, 1904, année 37, p. 201–217.) **3–VDS**

4532. Bowen, Robert Sidney. The second National Aeronautics Conference. ports. (Aviation, Highland, N. Y., 1927, v. 23, p. 1458–1463.) **VDS**

Held at Washington, D. C., Dec. 5th to 9th, 1927.

4533. Bruce, Eric Henry Stuart. The Aeronautical Society of Great Britain. (Aeronautical journal, London, 1900, v. 4, p. 132–134.) **VDS**

Refers to work of several prominent members of the society; also to aeronautical operations in the Boer war.

4534. Congrès international d'aéronautique de 1889. Séances. (L'Aéronaute, Paris, 1889, année 22, p. 169–192, 197–214, 221–254, 261–276, 285–311; 1890, année 23, p. 5–25, 39–47, 67–73, 91–93.) **3–VDS**

4535. Congrès international d'aéronautique, 2d, Paris, 1900. ...Congrès international d'aéronautique tenu à Paris du 15 au 21 septembre 1900. Procès-verbaux sommaires. [Paris:] Imprimerie Wellhoff et Roche, 1906. 27 p. 4°. **VDS**

4536. Congrès international d'aéronautique, 3d, Milan, 1906. Rapports et mémoires, publiés par les soins de la Commission permanente internationale d'aéronautique. Paris: H. Dunod et E. Pinat, 1907. 239 p. illus. 4°. **VDS**

4537. Congrès international d'aéronautique, 4th, Nancy, 1909. IVe Congrès international d'aéronautique, Nancy, 18–23 septembre 1909. Procès-verbaux, rapports & mémoires. Paris: H. Dunod & E. Pinat, 1909. 473 p. 4°. **VDS**

4538. Congrès international d'aéronautique, 5th, Turin, 1911. V. Congrès international d'aéronautique... Turin, 25–31 octobre 1911. Procès-verbaux, rapports & mémoires. Paris: H. Dunod & E. Pinat [1911]. 437 p., 14 pl. illus. 4°. **VDS**

4539. Congrès international d'aéronautique, 6th, Ghent, 1913. [Report of proceedings.] (La Conquête de l'air, Bruxelles, 1913, année 10, p. 250–254, 265–268, 281–282, 336–340.) **3–† VDS**

Abstract.

4540. Congrès international de législation aérienne. [Compte rendu.] [nos.] 1–9 (1911–1913, 1921–1925, 1928, 1930). Paris, 1911–1931. 8°. **VDS**

None issued, 1914–1921.

4541. Congrès international de la sécurité aérienne, 1st, Paris, 1930. Rapports. Paris: Comité français de propagande aéronautique, 1930. 2 v. in 1. illus. 4°. **VDS**

Associations and Conferences, continued

4542–4543. Congresso giuridico internazionale per il regolamento della locomozione aerea, 1st, Verona, 1910. Atti e relazioni. Verona: Società tipografica cooperativa, 1910. 262 p. 8°. **VDW**

4544. Cooper, Bertram G. Meeting of the International Commission for Aeronautical Maps, Brussels, 26 and 27 May, 1911. [By B. G. Cooper.] (Aeronautical journal, London, 1911, v. 15, p. 123–124.) **VDS**

4545. Erk, Fritz. Die erste Konferenz der Internationalen aëronautischen Kommission. (Meteorologische Zeitschrift, Wien, 1898, Jahrg. 15, p. 241–248.) **PRA**

4546. Fonvielle, Wilfrid de. Aeronautics in France. (Aeronautical journal, London, 1901, v. 5, p. 38–43.) **VDS**

Some account of the achievements of the Aéro-club de France and of the Société française de navigation aérienne.

4547. —— La médaille aéronautique de l'Académie des sciences. illus. (L'Aérophile, Paris, 1910, année 18, p. 112–113.) **3–VDS**

4548. Gale, Charles H. The first National Airport Conference. (Aviation, New York, 1929, v. 26, p. 1879–1880.) **VDS**

Held at Cleveland, O., May 15–17, 1929.

4549. Hardecker, John F. Standardization in the air. (Automotive industries, Philadelphia, 1927, v. 56, p. 575–577.) **TOL**

History of Army-Navy Standards Conferences, begun in 1924.

4550. Hodgson, J. E. The Royal Aeronautical Society. A survey, 1866–1929. (Air annual of the British Empire, 1929, London, 1929, v. 1, p. 42–50.) **VDS**

4551. Ide, John Jay. International aeronautic organisations. (Flight, London, 1930, v. 22, p. 975–976, 1004–1005, 1027.) **VDS**

Brief histories of various conferences and congresses.

4552. International Aeronautic Federation. Bulletin de la Fédération aéronautique internationale. année 1–date (Jan., 1920–date). **3–VDS**

With année 6–7 are bound Pilotes brevetés — records mondiaux, Jan., 1913, Jan., 1914; and Commission internationale de la carte aéronautique, Bulletin trimestrielle no. 1 (Feb., 1914).

4553. —— Conférence de Paris, 10–15 juin 1930. Rapport de M. l'ingénieur en chef Hirschauer...sur l'exercice 1929–1930. [Paris, 1930.] 28, xlix p. 4°. **VDS**

4554. International Aeronautical Congress, New York, 1907. Summary of meeting. (American magazine of aeronautics, New York, 1907, Oct., p. 15–23.) **VDS**

4555. The **International Aeronautical Federation.** London congress. (Aeronautics, New York, 1908, v. 3, July, p. 19–20.) **VDS**

4556. International Civil Aeronautics Conference, Washington, D. C., 1928. Papers submitted by the delegates for consideration by the conference... Washington: Gov. Prtg. Off., 1928. viii, 669 p. 8°. **VDS**

4557. —— Proceedings of the conference. Washington: Gov. Prtg. Off., 1929. vii, 286 p. illus. 8°. **VDS**

4558. International Commission for Air Navigation. Bulletin officiel... Official bulletin. nos. 1–20. Paris, 1922–32. f°. **† VDS**

French and English texts on opposite pages.

4559. International Commission for the Investigation of the Upper Air. Report of the proceedings of the meeting. [nos.] 1, 7–[10] (1898, 1912, 1921, 1925, 1927). Strasbourg [etc.], 1898–1928. 8°, 4°. **VDS**

Another copy of the 1927 proceedings is in *PRD p.v.87, no.3.*

4560. The **International Commission for Scientific Aeronautics.** (Nature, London, 1909, v. 80, p. 354–356.) **OA**

Account of meeting at Monaco, 1909.

4561. International Conference on Aerial Navigation, Chicago, 1893. Proceedings. New York: American Engineer and Railroad Journal, 1894. iv, 429 p. illus. 8°. **VDS**

Numerous papers by experts.

4562. International Congress of Aerial Navigation. ...Rapports. 1921, 1923, 1925, 1927, 1930. Paris, etc., 1921–31. 5 v. 4°. **VDS**

For 1921, 1925, 1930 name appears as Congrès international de la navigation aérienne; 1923, International Air Congress. 1923, title in English, text in English and French. 1927, title in Italian, English, and French.

4563. Die **Internationale Aéronautische Conferenz** zu Strassburg i. E. vom 31. März bis 4. April 1898. (Zeitschrift für Luftschiffahrt, Berlin, 1898, Jahrg. 17, p. 133–159.) **3–VDS**

Account of the proceedings.

4564. Ledeboer, John Henry. The fourth International Aeronautical Congress [Nancy, 1909]. (Aeronautical journal, London, 1909, v. 13, p. 113–116.) **VDS (Royal)**

Résumé of the proceedings.

4565. Light airplane clubs. An historical review of English clubs and the part they have played in the progress of English aeronautics. (Aviation, New York, 1928, v. 24, p. 377–379.) **VDS**

4566. The **Manufacturers Aircraft Association.** ports. (Aircraft yearbook, New York, 1919, p. 31–41.) **VDS**

History of the association.

4567. Manufacturers Aircraft Association, Inc., organized. (Aviation, New York, 1917, v. 3, p. 43.) **VDS**

4568. Meyer, Alex. Der erste internationale Kongress für Luftrecht. (Deutsche Zeitschrift für Luftschiffahrt, Berlin, 1911, Jahrg. 15, No. 13, p. 25–26.) **VDS**

Held in Paris, 1911.

Associations and Conferences, continued

4569. Moedebeck, Hermann W. L. Der Deutsche Verein zur Förderung der Luftschiffahrt, seine Vergangenheit und Zukunft. (Zeitschrift für Luftschiffahrt, Berlin, 1887, Jahrg. 6, p. 354–364.) **3–VDS**

4570. —— 25 Jahre Geschichte des Berliner Vereins für Luftschiffahrt. illus. ports. (Illustrierte aëronautische Mitteilungen, Strassburg, 1906, Jahrg. 10, p. 329–380.) **VDS**

A history of German aeronautics, 1880–1925, with portraits and illustrations.

4571. National Aeronautic Association. (Aeronautical digest, New York, 1922, v. 1, p. 192.) **VDS**

Brief account of second Aeronautical Congress, Detroit, 1922.

4572. National Air Traffic Conference. Proceedings... [no.] 1 (1929). [New York, 1929.] 8°. **VDS**

4573. Nevill, John T. The first National Air Traffic Conference. ports. (Aviation, New York, 1929, v. 27, p. 653–658.) **VDS**

Held at Kansas City, Sept. 16, 17, 18, 1929.

4574. Noel, Edmond Percy. First American convention of aero clubs. (Aeronautics, New York, 1910, v. 6, p. 87–88.) **VDS**

Held in St. Louis, Mo., 1910.

4575. The **Paris** International Aeronautical Congress. (Aeronautical journal, London, 1901, v. 5, p. 2–5.) **VDS**

Brief account of proceedings.

4576. Pole, William. The Chicago conference on aerial navigation. (Institution of Civil Engineers, Proceedings, London, 1896, v. 124, p. 378–389.) **VDA**

Abstracts of papers read at the International Conference, 1893.

4577. Rotch, Abbott Lawrence. The International Conference at Strassburg [1898]. (Science, New York, 1898, new series, v. 7, p. 846–848.) **OA**

Also in *Aeronautical journal*, London, 1898, v. 2, p. 61–63, *VDS*.

4578. Royal Aeronautical Society, London. Annual reports. 1871, 1885–1886. London, 1871–[1886?]. 3 v. 16°. **3–VDS**

4579. Sayers, W. H. The International Air Congress [4th] at Rome. (Aeroplane, London, 1927, v. 33, p. 666.) **VDS**

4580. The **Second** National Aero Congress. Notable aeronautic gathering establishes National Aeronautic Association of U. S. A. illus. (Aviation, Highland, N. Y., 1922, v. 13, p. 586–590.) **VDS**

4581. Some celebrities of the Paris Congress. ports. (Aeronautical journal, London, 1901, v. 5, p. 9–14, 24.) **VDS**

Accounts of Pierre J. C. Janssen, Major F. C. Trollope, Wilfred de Fonvielle, Louis Triboulet, Jacques Faure, and Abbott Lawrence Rotch.

4582. Tolmachoff, I. Dissolution of the American section of the [International Society for the Exploration of the Arctic Regions by Means of Aircraft]. (Science, New York, 1933, v. 78, p. 277.) **OA**

Membership terminated by act of Congress.

4583. Turner, Charles Cyril. The Royal Aero Club. (Air annual of the British Empire, London, 1930, v. 2, p. 167–173.) **VDS**

Founded in 1901.

4584. Veit, Sidney B. The International Air Conference at Gothenburg [1923]. illus. (Aeronautical digest, New York, 1923, v. 3, p. 258–259, 287.) **VDS**

4585. Walcott, Charles Doolittle. Smithsonian — a pioneer in aeronautics. (National Aeronautic Association review, Washington, D. C., 1925, v. 3, p. 99, 111.) **VDS**

Brief record of the Smithsonian Institution's services.

JURISPRUDENCE

4586. Aero-Club von Deutschland. Gesetze und Entwürfe von Gesetzen betr. die Luftfahrt. Zusammengestellt von der Juristischen Kommission des Aero-Clubs von Deutschland. Berlin: G. Braunbeck, 1919–21. 3 v. 8°. **VDW**

Laws of the various countries in chronological order.

4587. Air legislation one hundred years ago. Congress petitioned to give monopoly of air to inventor. (Aviation, Highland, N. Y., 1922, v. 13, p. 251.) **VDS**

J. Bennet, of Philadelphia, asked for exclusive use of the air, 1822.

4588. American Bar Association. — Special Committee on the Law of Aviation. First preliminary report of the executive committee (with annexed list of publications). New York, 1921. 40 p. 8°. **VDY p.v.21, no.1**

Contains lists of references in the Library of Congress, The New York Public Library, and other libraries.

4589. [**Bibliography** of aerial law.] (In: United States. — National Advisory Committee for Aeronautics, 1909–1931, Washington, 1921–35, volumes for 1909/16, p. 767–772; 1917/19, p. 255–258; 1920/21, p. 237–240; 1922, p. 82–84; 1923, p. 72–73; 1924, p. 57–58; 1925, p. 94; 1926, p. 77–78; 1927, p. 95–96; 1928, p. 108–110; 1929, p. 123–125; 1930, p. 133–135; 1931, p. 158–163.) **VDT**

4590. Bielenberg, Johannes. Die Freiheit des Luftraums. Berlin: Franz Vahlen, 1911. viii, 75 p. 8°. **BTZY**

Discussion of the freedom of the air, with an historical background. Good bibliography.

4591. Blachère, André. L'air, voie de communication et le droit. Paris: Société du recueil Sirey, 1911. xii, 226 p. 8°. **VDWB**

A thesis on air law, with a brief historical introduction and a bibliography.

Jurisprudence, continued

4592. **Blanchard's** ascension in Holland, Hague, July 13, 1785. (New London magazine, London, 1785, v. 1, p. 47–48, 105.) *** DA**
Has an amusing account of probably the first aeronautical suit for damages.

4593. **Bogert,** George Gleason. Problems in aviation law. ⌊Ithaca, N. Y., 1921.⌋ 39 p. 8°. **VDY p.v.23, no.2**
History of development, with 167 references. Reprinted from the *Cornell law quarterly.*

4594. **Bouvé,** Clement Lincoln. Private ownership of airspace. (Air law review, New York, 1930, v. 1, p. 232–257, 376–400.) **3–VDS**
A history of conceptions beginning with Roman times.

4595. **Cogliolo,** Pietro. Codice aeronautico, con appendici di aggiornamento a tutto il 1930. 2. ed. Milano: Ulrico Hoepli, 1931. vii, 733 p. 16°. **VDWB**
Aerial laws of various countries, especially those of Italy. Text of the Paris convention of 1919. Bibliography, 1901–1926.

4596. **Colegrove,** Kenneth Wallace. International control of aviation. Boston: World Peace Foundation, 1930. 235, xx p. 12°. **VDY**
Progress before and after the World war. Reviewed in *Aeroplane,* London, 1931, v. 40, *VDS.*

4596a ...**Convention** for the regulation of aerial navigation. (13 October 1919.) London: H. M. Stationery Office, 1920. 45 p. f°. (Great Britain. Parliamentary papers. 1920. v. 30. Cmd. 670.) **†* SDD**
Text in French and English on opposite pages.

4597. **Cuthell,** Chester Welde. Development of aviation laws in the United States. (Air law review, New York, 1930, v. 1, p. 86–107.) **3–VDS**

4598. **Daus,** Edgar. Die Luftschiffahrt in staats- und völkerrechtlicher Hinsicht. Erlangen: E. Th. Jacob, 1908. 44 p. 8°. **VDW p.v.2, no.6**
Historical introduction and a bibliography of 39 titles.

4599. **Davis,** Warren Jefferson. Aeronautical law. Los Angeles: Parker, Stone & Baird Co. ⌊cop. 1930.⌋ xxxviii, 541 p. 8°. **VDY**
Bibliography, p. 473–482, supplements list of American Bar Association (1921). See entry no. 4588.

4600. —— The International Air Code. port. (Aeronautical digest, New York, 1923, v. 2, p. 103–107.) **VDS**
Meeting of the fifth congress of the Comité juridique internationale de l'aviation, Prague, 1922. Has an English translation of the code.

4601. **Du** régime juridique de la navigation aérienne dans les rapports internationaux. (Journal du droit international, Paris, 1909, v. 36, p. 1010–1015.) **XAA**

4602. **Eagleton,** Clyde. International law and aerial discovery at the South Pole. (Air law review, New York, 1930, v. 1, p. 125–127.) **3–VDS**

4603. **Ebel,** Martin. Die Haftpflicht des Luftschiffers. Greifswald: J. Abel, 1913. 60 p. 8°. **VDY p.v.11, no.6**
Bibliography, p. 5–7.

4604. **Elliott,** Shelden D. Unobstructed airport approaches. (Journal of air law, Chicago, 1932, v. 3, p. 207–225.) **VDS**
A survey of foreign and state legislation.

4605. **Ellis,** W. E. Aerial-land and aerial-maritime warfare. (American journal of international law, New York, 1914, v. 8, p. 256–273.) **XBA**
Comments on the work of the Hague Conferences.

4606. **Fackeldey,** Ferdinand. Luftschiffahrt und Schadenshaftung. Ein Beitrag zur Lehre vom Recht der Luftschiffahrt. Leipzig: R. Noske, 1915. 73 p. 8°.
Bibliography, p. vii–ix. **VDY p.v.41, no.9**

4607. **Fixel,** Rowland W. The law of aviation. Albany, N. Y.: Matthew Bender & Co., 1927. xv, 403 p. 8°. **VDY**
See chapter 3 for historical background.

4608. **Freeman,** Harry Joseph. Survey of state aeronautical legislation, 1928–1929. (Air law review, New York, 1930, v. 1, p. 61–84.) **3–VDS**

4609. **Freeman,** William Marshall. Air and aviation law (civil aviation), being an exposition of the statute and case law affecting air navigation. London: Sir Isaac Pitman & Sons, Ltd., 1931. xi, 163 p. 8°. **VDWB**
Has a brief historical introduction.

4610. **Garner,** James Wilford. International regulation of warfare. (Air law review, New York, 1932, v. 3, p. 104–125, 309–323.) **3–VDS**

4611. **German** aviation and allied restrictions. (Flight, London, 1925, v. 17, p. 457.) **VDS**
Germans protest the harshness of the revised rules.

4612. **Gordini,** L. Il I Congresso giuridico internationale per il regolamento della navigazione aerea (Verona 31 maggio–2 giugno 1910). (Rivista di diritto internazionale, Roma, 1910, anno 5, p. 269–280.) **XBA**

4613. **Greer,** Rowan A. Aeronautics from a legal point of view. (U. S. air services, Washington, 1929, v. 14, Jan., p. 32–38.) **VDS**
Law of the air had its inception with the Hague Peace Conference of 1899. Captain Greer traces developments since that time.

4614. **Hotchkiss,** Henry Greene. A treatise on aviation law, giving authorities with complete citations of Federal and State decisions, and including texts in full of the International Convention of 1919, the Havana Convention ⌊etc.⌋. New York: Baker, Voorhis & Co., 1928. xviii, 492 p. 8°. **VDY**
See chapters 1–2 for historical introduction. Bibliography.

4615. **Hudson,** Manley Ottmer. Aviation and international law. (Air law review, New York, 1930, v. 1, p. 184–210.) **3–VDS**
Excellent history of the various conventions, with a list of international agreements prior to 1930.

Jurisprudence, continued

4616. Ide, John Jay. The history and accomplishments of the International Technical Committee of Aerial Legal Experts (C. I. T. E. J. A.). (Journal of air law, Chicago, 1932, v. 3, p. 27–49.) **VDS**

Has accounts of the first six sessions.

4617. Kuhn, Arthur Kline. The beginnings of an aerial law. (American journal of international law, New York, 1910, v. 4, p. 109–132.) **XAA**

4618. Lacroix, Pierre. Le domaine aérien et la guerre. Essai sur les hostilités et la neutralité aériennes dans le conflit de 1914–1919. Toulouse: Bonnet, 1919. 442 p. 8°. **BTZY**

A study of the development of international law as related to aeronautics, especially during the World war. Bibliography.

4619. Ladrone, Francesco. Airspace rights in Roman law. (Air law review, New York, 1931, v. 2, p. 455–467.) **3–VDS**

4620. Lebon, Émile. De la guerre aérienne dans ses rapports avec le droit international. Gérardmer: G. Remy [1923]. vi, 234 p. 4°. **VDW**

International aerial law as affected by the World war.

4621. Leo, Kurt. Die zivilrechtliche ausserkontraktliche Haftung des Luftschiffers nach bestehenden Recht. Greifswald: J. Abel, 1912. 70 p. 8°. **VDY p.v.10, no.2**

Bibliography, p. 7–10.

4622. Lin, Wo-chiang. Aeronautical law in time of war. (Journal of air law, Chicago, 1932, v. 3, p. 79–92.) **VDS**

Deals with the laws regulating aerial warfare before 1914.

4623. —— The development of aviation and aeronautical law in China. (Journal of air law, Chicago, 1931, v. 2, p. 159–175.) **VDS**

4624. Logan, George Bryan. Recent developments in aeronautical law. (Journal of air law, Chicago, 1934, v. 5, p. 548–563.) **VDS**

4625. Lupton, George W., jr. Progress of aviation law. (Aviation, New York, 1932, v. 31, p. 463–467; 1933, v. 32, p. 43–45.) **VDS**

Developments as shown by significant law cases.

4626. MacCracken, William Patterson. The growth of aeronautical law in America. (Journal of air law, Chicago, 1930, v. 1, p. 415–421.) **VDS**

4627. McNair, Arnold Duncan. The beginnings and the growth of aeronautical law. (Journal of air law, Chicago, 1930, v. 1, p. 383–392.) **VDS**

4628. —— The law of the air (The Tagore law lectures of 1931). London: Butterworth & Co., 1932. xv, 249 p. 8°. **VDWB**

Introductory historical chapter and a brief bibliography. Acts and conventions in the appendices.

4629. Meili, Friedrich. Das Luftschiff im internen Recht und Völkerrecht. Zürich: Orell Füssli, 1908. 61 p. 8°. **VDW p.v.2, no.7**

Brief historical introduction and numerous footnote references.

4630. Meyer, Alex. Rechtliche Glossen zum Falle Nobile. (Archiv für Luftrecht, Königsberg, 1932, Bd. 2, p. 59–62.) **3–VDS**

Legal points involved in the disaster of the Nobile North Pole expedition.

4631. Ostrop, Wilhelm. Die zivilrechtlichen Ansprüche des Grundeigentümers, auf dessen Grundstück sich eine Ballonlandung vollzieht. Greifswald: J. Abel, 1913. 45 p. 8°.

Bibliography, p. 3–4. **VDY p.v.11, no.4**

4632. The **Prosecution** of Brindejonc des Moulinais. (Aeronautics, London, 1913, v. 6, p. 219.) **VDS**

Defendant cited for air trespass over the coastline of Great Britain, the first case of its kind.

4633. Roer, Max. Nationales und internationales Luftverkehrsrecht (Transportrecht). Osnabrück: Druckerei-Gesellschaft, 1928. xvii, 149 p. 8°. **VDY**

Bibliography, p. vii–xvii.

4634. Rohlfing, Charles Carroll. National regulation of aeronautics. Philadelphia: University of Pennsylvania Press, 1931. ix, 298 p. 8°. **VDWB**

History of Federal assistance and legislation, also of the development of state legislation. Bibliography.

4635. Rolland, Louis. Les pratiques de la guerre dans le conflit de 1914 et le droit des gens. (Revue générale de droit international public, Paris, 1916, tome 23, p. 497–604.) **XBA**

A study of aerial international law as developed during the World war.

4636. Roper, Albert. Recent developments in international aeronautical law. (Journal of air law, Chicago, 1930, v. 1, p. 395–414.) **VDS**

4637. Runkel, Peter. Die ausservertragliche Haftung des Luftschiffers. Geltendes Recht und Erwägungen de lege ferenda. Berlin: Franz Vahlen, 1913. 107 p. 8°. **VDY**

Bibliography, p. 7–11.

4637a. Spaight, James Molony. Air power and the cities. London: Longmans, Green & Co., 1930. ix, 244 p. 8°. **VDYD**

Treats of air bombing during the World war. A continuation of entry no. 4638.

4638. —— Air power and war rights. London: Longmans, Green & Co., 1924. ix, 493 p. 8°. **VDY**

A study of international military law in the light of events of the World war.

4639. —— Aircraft in peace and the law. London: Macmillan & Co., 1919. viii, 227 p. 8°. **VDY**

Excellent history, with an extensive bibliography.

4640. Thibout, André. Le domain aérien des états en temps de paix et essais de réglementation de la circulation aérienne. Bar-le-Duc, 1911. 171 p. 8°. **VDY p.v.10, no.12**

Bibliography.

Jurisprudence, continued

4641. **Trend** in state aeronautic legislation is toward uniformity. (United States. — Commerce Department, Air commerce bulletin, Washington, 1933, v. 4, p. 508–509.) **VDS**

4642. **United States.** — Interstate and Foreign Commerce Committee (House). Civil aeronautics. Legislative history of the Air Commerce Act of 1926... together with miscellaneous legal matters relating to civil air navigation. Revision of the 1923 edition of Law memoranda upon civil aeronautics. Corrected to August 1, 1928. Washington: Gov. Prtg. Off., 1928. v, 178 p. 8°. **VDY p.v.44, no.1**

4643. **United States** aviation reports... 1928–1931. Baltimore: United States Aviation Reports, Inc., 1928–31. 5 v. illus. diagrs. 8°. **VDWB**

Contains decisions relating to the law of aviation covering the period 1822–1932. Also the various state laws.

4644. **Verzeichnis** der luftjuristischen Literatur. (Jahrbuch für Luftverkehr, 1924, München, 1924, p. 369–382.) **3–VDS**

Extensive bibliography issued under the auspices of the Aero-Club of Germany.

4645. **Wäsch,** Curt. Der Schutz gegen Gefährdungen durch Luftfahrzeuge in zivilrechtlicher und öffentlichrechtlicher Hinsicht. Berlin: Ludwig Wröbel, 1915. 52 p. 8°.

Bibliography, p. 7–10. **VDY p.v.35, no.4**

4646. **Warner,** Edward Pearson. The International Convention for Air Navigation: and the Pan American Convention for Air Navigation: a comparative and critical analysis. (Air law review, Spencer, Mass., 1932, v. 3, p. 221–308.) **3–VDS**

4647. **Weck,** Hermann. Deutsches Luftrecht. Berlin: Carl Heymanns Verlag, 1913. xxiii, 334 p. 8°. **VDY**

Extensive bibliography, p. vii–xvii.

4648. **Wilhelm,** A. De la situation juridique des aéronautes en droit international. (L'Aeronaute, Paris, 1892, année 25, p. 171–183.) **3–VDS**

A discussion of the rights of belligerent aeronauts, citing experiences of the Franco-Prussian war.

4649. **Woodhouse,** Henry. Textbook of aerial laws and regulations for aerial navigation, international, national, and municipal, civil and military. New York: Frederick A. Stokes Company [cop. 1920]. 171 p. f°. **† VDW**

History and texts of international conventions. Group photograph of the first Interallied Aeronautical Commission.

4650. **Zollmann,** Carl Frederick Gustav. State control of aeronautics in 1931. (Journal of air law, Chicago, 1932, v. 3, p. 68–74.) **VDS**

AERONAUTICS AND SCIENCE

4651. The **Aerodynamic** experimental station at Göttingen. illus. (Engineering, London, 1921, v. 112, p. 218–220.) **VDA**

Erected 1907 under the direction of Professor L. Prandtl.

4652. **Aerodynamic** experiments of Duc de Guiche. illus. (Scientific American, New York, 1912, v. 107, p. 368, 375.) **VA**

Field experiments to determine wind pressure upon plane surfaces.

4653. **Aerodynamische Versuchsanstalt,** Göttingen. Ergebnisse. Lief. 1–4. München: R. Oldenbourg, 1921–32. 4 v. illus. diagrs. 4°. **PCP**

Has bibliographies. Library has also the 2. ed. of Lief. 1. For a review by Alexander Klemin see *Aviation,* Highland, N. Y., 1924, v. 16, p. 619, *VDS.*

4654. **Aeronautical** work at M. I. T. Aeronautics course at the Massachusetts Institute of Technology covers all fields. Extensive research and test work also carried out. illus. (Aviation, Highland, N. Y., 1926, v. 20, p. 630–631.) **VDS**

4655. **Aeronautisches Observatorium,** Lindenberg. Ergebnisse der Arbeiten des Königlich Preussischen Aeronautischen Observatoriums bei Lindenberg. Bd. 1–2 (1905–06). Braunschweig, 1906–10. f°. **† PRR**

Volumes contain accounts of balloon and kite work.

4656. **Agafonov,** V. K. Vozdukhoplavaniye v yevo proshlom i v nastoyashchem. Sostavleno po Lekornyu, Linke, Pomortzevu, Tisand'ye i dr. pod redaktzieĭ V. K. Agafonova. St. Petersburg: I. N. Skorokhodov, 1903. 236 p. illus. ports. 8°. *** QH**

Historical compilation from the works of Lecornu, Linke, Pomortzeff, Tissandier, and others.

4657. **Aimé,** Emmanuel. Gustave Hermite. port. (L'Aérophile, Paris, 1896, année 4, p. 1–5.) **3–VDS**

Portrait and brief biography of French aeronaut and instrument inventor. He and Georges Besançon were the first to suggest the use of "ballons sondes" for meteorological purposes.

4658. —— M. Wilfrid de Fonvielle. port. (L'Aérophile, Paris, 1894, année 2, p. 97–100.) **3–VDS**

Brief account of the career of distinguished French scientist, author, and aeronaut.

4659. **Albert I,** prince of Monaco. Sur les lancements de ballons sondes et de ballons pilotes au-dessus des océans. (Académie des sciences, Comptes rendus, Paris, 1905, tome 141, p. 492–493.) *** EO**

Brief account of experiments over the Mediterranean sea. Greatest altitude reached was 14,000 meters.

Aeronautics and Science, continued

4660. **Alexander,** A. Power in relation to weight in aerial navigation. (Engineering, London, 1868, v. 5, p. 318–320, 383.) **VDA**

Paper read before the Aeronautical Society of Great Britain. Has comments by F. H. Wenham.

4661. **Alexander,** Patrick Y. Sounding the air by flying machines controlled by Hertzian waves. (Aeronautical journal, London, 1901, v. 5, p. 59.) **VDS**

4662. **Ames,** Joseph Sweetman. Recent progress in the science of aeronautics. (Franklin Institute, Journal, Philadelphia, 1925, v. 199, p. 83–90.) **VA**

4663. —— A résumé of the advances in theoretical aeronautics made by Max M. Munk. (United States. — Advisory Committee for Aeronautics, Technical report no. 213, Washington, 1926. In: 11th annual report, 1925, p. 93–134.) **VDY**

4664. **Anderson,** S. Herbert. Fog penetration. (U. S. air services, Washington, 1931, v. 16, July, p. 20–25.) **VDS**

Early experiments; also those of the United States Government.

4665. **Arago,** Dominique François Jean. Du voyage aéronautique de MM. Bixio et Barral. (Académie des sciences, Comptes rendus, Paris, 1850, tome 31, p. 5–7.) *** EO**

4666. —— Sur les voyages aéronautiques exécutés dans l'intérêt de l'avancement des sciences. (In his: Oeuvres complètes, Paris, 1857, tome 9, p. 489–534.) **OAL**

"The first aeronautic voyage to which science was indebted for some useful indications was that performed at Hamburg, July 18, 1803, by the physicist Robertson, accompanied by his countryman, Lhoest." — *From the English translation* in Smithsonian Institution, Annual report for 1863, Washington, 1872, p. 331–349, * *EA*.

4667. **Ascension** scientifique exécutée par MM. J. Crocé-Spinelli, Jobert, A. Pénaud, le docteur Petard, et Sivel. (L'Aéronaute, Paris, 1873, année 6, p. 96–100.) **3–VDS**

4668. **Ascension** scientifique de longue durée; par MM. Sivel, Crocé-Spinelli, A. et G. Tissandier et Jobert. illus. (Académie des sciences, Comptes rendus, Paris, 1875, tome 80, p. 866–871.) *** EO**

Describes ascent of the balloon *Zenith,* March, 1875.

4669. **Assmann,** Richard. Die Erfordernisse einer Ballonfahrt zu wissenschaftlichen Zwecken. illus. (Zeitschrift für Luftschiffahrt, Berlin, 1892, Jahrg. 11, p. 33–44, 69–85.) **3–VDS**

Account of instruments developed by the author, with a comparison with those used by the British investigator Glaisher.

4670. —— Uebersicht über die vom dem "Deutschen Vereine zur Förderung der Luftschiffahrt in Berlin" ausgeführten wissenschaftlichen Ballonfahrten. (Zeitschrift für Luftschiffahrt, Berlin, 1895, Jahrg. 14, p. 83–94.) **3–VDS**

Includes a record of balloon ascensions sponsored by the above society, 1888–1895.

For list of Assmann's writings on aeronautical and meteorological subjects see Paul Brockett, *Bibliography of aeronautics,* Washington, 1910, p. 63–68, *VDT.*

4671. **Assmann,** Richard, and Arthur Berson. Wissenschaftliche Luftfahrten; ausgeführt vom Deutschen Verein zur Förderung der Luftschiffahrt in Berlin... Braunschweig: F. Vieweg und Sohn, 1899–1900. 3 v. diagr., pl., tables. 4°. **† VDW**

Bd. 1 deals with history of ascensions for scientific purposes, giving details of observations.

4672. **Austin,** A. O. Lightning investigation as applied to the airplane. illus. (U. S. air services, Washington, 1931, v. 16, p. 28–33.) **VDS**

Studies made at the high-voltage laboratory of the Ohio Insulator Company for Ward T. Van Orman.

4673. **Averly,** A. Le problème général du "vol" et la force centrifuge. fasc. 1. Paris: Vve C. Dunod, 1904. illus. 4°. **VDY**

Author's theory of flight.

4674. **Babinet,** Jacques. Navigation aérienne, avec ou sans ballon. (Les Mondes, Paris, 1863, tome 2, p. 117–124, 152–160, 181–185.) **3–OA**

History of early aeronautics. Discussion of the laws of heavier-than-air flight, with criticism of the formulas of Navier.

4675. **Bacon,** Francis. Sylva sylvarum, or a natural history in ten centuries... The ninth and last edition, with an alphabetical table of the principal things contained... London: Printed by J. R. for William Lee, 1670. 215,[19] p. f°. **NCF**

See Century 8, no. 791, p. 167 for "experiment solitary touching the flying of unequal bodies in the air"; Century 9, no. 886, p. 191, for "experiment solitary touching flying in the air." Also in the *Works* of Francis Bacon, edited by J. Spedding, R. L. Ellis, and D. D. Heath, London, 1872, v. 4, p. 151–483, *NCF.* Library has a copy of the Latin edition, Leyden, 1648, 24°, *Reserve Room.*

4676. **Bacon,** Gertrude. The acoustical experiments carried out in balloons by the late Rev. J. M. Bacon. (Aeronautical journal, London, 1906, v. 10, p. 5–8.) **VDS (Royal)**

4677. **Bacon,** John Mackenzie. The balloon as an instrument of scientific research. (Society of Arts, Journal, London, 1899, v. 41, p. 277–283.) **VA**

Early investigations. Glaisher. The author's own work.

4678. **Bacon,** Roger. Fr. Rogeri Bacon. Opera quaedam hactenus inedita. Edited by J. S. Brewer. London: Longman, Green, Longman, and Roberts, 1859. 2 v. 8°. (Great Britain. Public Record Office. Chronicles and memorials. v. 15.) **C (Gt. Br. Public)**

See v. 1, p. 533 for brief reference to the possibility of flight.

Aeronautics and Science, continued

4679. **Bacon,** Roger. Roger Bacon's letter concerning the marvelous power of art and of nature and concerning the nullity of magic. Translated from the Latin by Tenney L. Davis...together with notes and an account of Bacon's life and work. Easton, Pa.: The Chemical Publishing Co., 1923. 76 p. 12°. **OAD**

See p. 27 for brief reference to flight.

4680. **Bairstow,** Leonard. The scientific progress of aviation during the war. illus. (Engineering, London, 1919, v. 108, p. 493–496.) **VDA**

For abstract see British Association for the Advancement of Science, *Report of the 87th meeting, Bournemouth, 1919,* London, 1920, p. 266–267, * *EC.* For a discussion see entry no. 5026.

4681. —— Twenty-one years' progress in aerodynamic science. (Flight, London, 1930, v. 22, p. 28–30.) **VDS**

4682. The **Balloon** in meteorology. (Science, Cambridge, Mass., 1885, v. 5, p. 91–93.) **OA**

The first ascent of its kind in America, Jan. 19, 1885, under the direction of the chief signal officer of the United States Army, General Hazen.

4683. **Barbet,** A. Rapport fait...sur les expériences de M. Canovetti relatives à la résistance opposée par l'air aux corps en mouvement. illus. (Société d'encouragement pour l'industrie nationale, Bulletin, Paris, 1903, tome 104, semestre 1, p. 166–195.) **3–VA**

4684. —— Rapport au nom du Comité des arts mécaniques sur le prix proposé pour une étude des coefficients nécessaires au calcul mécanique d'une machine aérienne. Recherches de M. Canovetti et de M. l'abbé Le Dantec. (Société d'encouragement pour l'industrie nationale, Bulletin, Paris, 1899, série 5, tome 4, p. 946–950.) **3–VA**

4685. **Baroni Cavalcabo,** Clemente. L'impotenza del demonio di trasportare a talento per l'aria da un luogo all' altro i corpi umani ... Dove anche si dimostra l'impossibilità di volare con artifizio umano. Rovereto: Marchesani, 1753. xxiv, 141 p. 12°. **YLE**

Discusses the impossibility of human flight.

4686. **Barral,** Jean Augustin, and A. J. Bixio. Journal du voyage aéronautique fait le 27 juillet 1850. (Académie des sciences, Comptes rendus, Paris, 1850, tome 31, p. 126–135.) ***EO**

Translation, lacking the meteorological tables, in *Mechanic's magazine,* London, 1851, v. 54, p. 174–177, *3–VA.*

4687. **Bauer,** Louis Hopewell. Aviation medicine. Published by authority of the Surgeon-General. Baltimore: Williams and Wilkins Co., 1926. xv, 241 p. illus. 8°. **VDY**

Extensive bibliography, p. 179–201.

4688. —— The School of Aviation Medicine. illus. port. (Aeronautical digest, New York, 1924, v. 4, p. 37–39.) **VDS**

History of this service of the United States War Department.

4689. **Beach,** A. E. Dr. S. P. Langley's experiments in aerodynamics. [By A. E. Beach.] (Scientific American, New York, 1892, v. 66, p. 101.) **VA**

Based upon several of Langley's papers.

4690. **Béjeuhr,** Paul. Richard Assmann. port. (Deutsche Luftfahrer Zeitschrift, Berlin, 1913, Jahrg. 18, p. 173.) **VDS**

Portrait and account of pioneer aerological investigator and writer (1845–1918).

4691. **Berget,** Alphonse. Les applications de l'aéronautique à la géographie. illus. (Revue de géographie, Paris, 1909, v. 3, p. 551–571.) **KAA**

For list of Berget's writings on the science of aeronautics see Paul Brockett, *Bibliography of aeronautics,* Washington, 1910, p. 111–112, *VDT.*

4692. **Berson,** Arthur. Michael Pomortzeff: wissenschaftliche Ergebnisse von 40 Ballonfahrten in Russland. (Zeitschrift für Luftschiffahrt, Berlin, 1892, Jahrg. 11, p. 99–109.) **3–VDS**

Ascensions by a professor of meteorology in the airship station of St. Petersburg.

For list of Berson's writings on scientific aeronautics see Paul Brockett, *Bibliography of aeronautics,* Washington, 1910, p. 115–118, *VDT.*

4693. **Berson,** Arthur, and R. Süring. Ein Ballonaufstieg bis 10,500 m. (Illustrierte aëronautische Mitteilungen, Strassburg, 1900, Jahrg. 4, p. 117–119.) **VDS**

Account of authors' ascension in 1900 — stated to have been the record. See p. 119–120 for biographies and portraits of the aeronauts.

4694. **Bert,** Paul. [Obituary on Crocé-Spinelli and Sivel.] (L'Aéronaute, Paris, 1876, année 9, p. 6–16.) **3–VDS**

Includes brief biographies.

4695. **Besançon,** Georges. Commandant Paul Renard. port. (L'Aérophile, Paris, 1908, année 16, p. 97.) **3–VDS**

Portrait and biography of French student of aeronautics and a prolific writer on the subject.

4696. —— Rodolphe Soreau. port. (L'Aérophile, Paris, 1909, année 17, p. 169.) **3–VDS**

Portrait and brief notice of the works of well-known French aeronautical and scientific writer.

4697. [**Bibliographies** of air resistance.] (In: Paul Brockett, Bibliography of aeronautics, Washington, 1910, p. 34; continued in United States. — National Advisory Committee for Aeronautics, Bibliography of aeronautics, 1909–1931, Washington, 1921–35, volumes for 1909/16, p. 1128; 1917/19, p. 378; 1920/21, p. 349; 1922, p. 133; 1925, p. 146–147; 1926, p. 122; 1927, p. 146; 1928, p. 167; 1929, p. 189; 1930, p. 202; 1931, p. 235.) **VDT**

4698. [**Bibliographies** of meteorology in relation to aeronautics.] (In: Paul Brockett, Bibliography of aeronautics, Washington, 1910, p. 575; continued in United States. — National Advisory Committee for Aeronautics, Bibliography of aeronautics, 1909–1931, Washington, 1921–35, volumes for 1909/16,

Aeronautics and Science, continued

p. 864–868, 1408–1409; 1917/19, p. 293–299, 467; 1920/21, p. 273–275, 425; 1922, p. 95–96; 1923, p. 85; 1924, p. 69; 1925, p. 110–111; 1926, p. 89; 1927, p. 111–112; 1928, p. 124–125; 1929, p. 143–145; 1930, p. 151–152; 1931, p. 180–181.) **VDT**

Some of the early references have the heading *Weather*.

4699. [**Bibliography** of the Aerial Experiment Station.] (In: Paul Brockett, Bibliography of aeronautics, Washington, 1910, p. 6.) **VDT**

List of writings concerning work of the association.

4700. [**Bibliography** of aerodynamics.] (In: United States. — National Advisory Committee for Aeronautics, Bibliography of aeronautics, 1909–1931, Washington, 1921–35, volumes for 1909/16, p. 23–28; 1917/19, p. 7; 1920/21, p. 7–8; 1922, p. 2–3; 1923, p. 2–3; 1924, p. 2; 1925, p. 2–4; 1926, p. 2–3; 1927, p. 2–3; 1928, p. 4–5; 1929, p. 5–6; 1930, p. 4–6; 1931, p. 5–8.) **VDT**

4701. [**Bibliography** of aeronautical research.] (In: United States. — National Advisory Committee for Aeronautics, Bibliography of aeronautics, 1909–1931, Washington, 1921–35, volumes for 1909/16, p. 1128; 1917/19, p. 377–378; 1920/21, p. 349; 1922, p. 132–133; 1923, p. 114; 1924, p. 91; 1925, p. 146; 1926, p. 122; 1927, p. 146; 1928, p. 167; 1929, p. 189; 1930, p. 202; 1931, p. 235.) **VDT**

4702. [**Bibliography** of aeronautical work at the National Physical Laboratory of England.] (In: United States. — National Advisory Committee for Aeronautics, Bibliography of aeronautics, 1909–1931, Washington, 1921–35, volumes for 1909/16, p. 940–941; 1917/19, p. 326; 1920/21, p. 295; 1925, p. 127; 1926, p. 205; 1931, p. 207.) **VDT**

4703. [**Bibliography** of aeronautics at the Massachusetts Institute of Technology.] (In: United States. — National Advisory Committee for Aeronautics, Bibliography of aeronautics, 1909–1931, Washington, 1921–35, volumes for 1909/16, p. 847; 1917/19, p. 287; 1922, p. 93–94; 1925, p. 107; 1926, p. 86; 1927, p. 109.) **VDT**

4704. [**Bibliography** of Albert R. von Miller-Hauenfels.] (In: Paul Brockett, Bibliography of aeronautics, Washington, 1910, p. 584–585.) **VDT**

Citations to Miller-Hauenfel's writings on the theory of flight.

4705. [**Bibliography** of the atmosphere.] (In: United States. — National Advisory Committee for Aeronautics, Bibliography of aeronautics, 1909–1931, Washington, 1921–35, volumes for 1909/16, p. 89–91; 1917/19, p. 35; 1920/21, p. 32; 1922, p. 11; 1923, p. 10; 1925, p. 14–15; 1926, p. 9–10; 1928, p. 17; 1929, p. 24–25; 1930, p. 19; 1931, p. 26.) **VDT**

4706. [**Bibliography** of the ballon-sonde.] (In: Paul Brockett, Bibliography of aeronautics, Washington, 1910, p. 90.) **VDT**

4707. **Bibliography** of the Blue Hill Meteorological Observatory.] (In: Paul Brockett, Bibliography of aeronautics, Washington, 1910, p. 137–138.) **VDT**

List of writings concerning the institution.

4708. [**Bibliography** of Edward P. Warner.] (In: United States. — National Advisory Committee for Aeronautics, Bibliography of aeronautics, 1909–1931, Washington, 1921–35, volumes for 1909/16, p. 1405; 1917/19, p. 466; 1920/21, p. 421–423; 1922, p. 160; 1923, p. 137; 1924, p. 109; 1925, p. 179; 1926, p. 147; 1927, p. 176; 1928, p. 202; 1929, p. 228; 1930, p. 247–248; 1931, p. 297.) **VDT**

4709. [**Bibliography** of Elmer Ambrose Sperry.] (In: United States. — National Advisory Committee for Aeronautics, Bibliography of aeronautics, 1909–1931, Washington, 1921–35, volumes for 1909/16, p. 1259–1260; 1917/19, p. 415; 1920/21, p. 385–386; 1922, p. 147; 1923, p. 127; 1926, p. 135; 1927, p. 161; 1929, p. 208.) **VDT**

4710. [**Bibliography** of experiments at the Eiffel tower.] (In: Paul Brockett, Bibliography of aeronautics, Washington, 1910, p. 273.) **VDT**

4711. [**Bibliography** of the gyroscope in aviation.] (In: United States. — National Advisory Committee for Aeronautics, Bibliography of aeronautics, 1909–1931, Washington, 1921–35, volumes for 1909/16, p. 596–599; 1917/19, p. 203; 1923, p. 55; 1925, p. 74; 1930, p. 97; 1931, p. 117.) **VDT**

4712. [**Bibliography** of L. Teisserenc de Bort.] (In: Paul Brockett, Bibliography of aeronautics, Washington, 1910, p. 815–817.) **VDT**

Citations to Teisserenc de Bort's writings, principally on the use of sounding balloons in meteorology.

4713. [**Bibliography** of laboratories.] (In: United States. — National Advisory Committee for Aeronautics, Bibliography of aeronautics, 1909–1931, Washington, 1921–35, volumes for 1909/16, p. 740–742; 1920/21, p. 231; 1922, p. 80; 1924, p. 56; 1925, p. 91; 1926, p. 75; 1927, p. 93; 1929, p. 121.) **VDT**

4714. [**Bibliography** of Louis Paul Cailletet.] (In: Paul Brockett, Bibliography of aeronautics, Washington, 1910, p. 171.) **VDT**

List of Cailletet's writings concerning his various aeronautical instruments.

4715. [**Bibliography** of physiology in relation to aeronautics.] (In: United States. — National Advisory Committee for Aeronautics, Bibliography of aeronautics, 1909–1931, Washington, 1921–35, volumes for 1909/16, p. 1039; 1917/19, p. 354; 1920/21, p. 325; 1925, p. 137; 1926, p. 114; 1928, p. 156; 1930, p. 189.) **VDT**

Aeronautics and Science, continued

4716. [**Bibliography** of Pierre Jules César Janssen.] (In: Paul Brockett, Bibliography of aeronautics, Washington, 1910, p. 448–449.) **VDT**

List of Janssen's writings on balloon photography and other scientific aeronautical subjects.

4717. [**Bibliography** of practical navigation.] (In: United States. — National Advisory Committee for Aeronautics, Bibliography of aeronautics, 1909–1931, Washington, 1921–35, volumes for 1909/16, p. 944; 1917/19, p. 327; 1920/21, p. 297; 1923, p. 99; 1924, p. 80; 1925, p. 127; 1926, p. 106; 1927, p. 130; 1928, p. 145; 1929, p. 164–165; 1930, p. 174–175; 1931, p. 207–208.) **VDT**

4718. [**Bibliography** of Rudolf Mewes.] (In: Paul Brockett, Bibliography of aeronautics, Washington, 1910, p. 577–579.) **VDT**

Citations to Mewes' writings on the science of aeronautics.

4719. [**Bibliography** of scientific ballooning.] (In: Paul Brockett, Bibliography of aeronautics, Washington, 1910, p. 92.) **VDT**

List of writings on the subject.

4720. [**Bibliography** of Smithsonian Institution's publications on aeronautics.] (In: Paul Brockett, Bibliography of aeronautics, Washington, 1910, p. 779–783.) **VDT**

4721. [**Bibliography** of the stability of aircraft.] (In: Paul Brockett, Bibliography of aeronautics, Washington, 1910, p. 795; continued in United States. — Advisory Committee for Aeronautics, Bibliography of aeronautics, 1909–1931, Washington, 1921–35, volumes for 1909/16, p. 1263–1280; 1917/19, p. 417–418; 1920/21, p. 387–388; 1922, p. 148; 1923, p. 128; 1924, p. 102; 1925, p. 164–165; 1926, p. 162–163; 1927, p. 135; 1928, p. 186; 1929, p. 209; 1930, p. 227–228; 1931, p. 265–266.) **VDT**

4722. [**Bibliography** of Wilhelm Angerstein.] (In: Paul Brockett, Bibliography of aeronautics, Washington, 1910, p. 47–48.) **VDT**

List of some of Angerstein's writings on meteorology and aeronautics.

4723. [**Bibliography** of wind.] (In: United States. — National Advisory Committee for Aeronautics, Bibliography of aeronautics, 1909–1931, Washington, 1921–35, volumes for 1909/16, p. 1428–1430; 1917/19, p. 475; 1920/21, p. 431; 1922, p. 163; 1923, p. 140; 1924, p. 110; 1925, p. 182; 1926, p. 149–150; 1927, p. 179; 1929, p. 234; 1930, p. 253; 1931, p. 303.) **VDT**

4724. [**Bibliography** of wind tunnels.] (In: United States. — National Advisory Committee for Aeronautics, Bibliography of aeronautics, 1909–1931, Washington, 1921–35, volumes for 1909/16, p. 1431; 1917/19, p. 476; 1920/21, p. 432–433; 1922, p. 163; 1923, p. 140; 1924, p. 111; 1925, p. 182–183; 1926, p. 150; 1927, p. 179–180; 1928, p. 207–208; 1929, p. 234; 1930, p. 254; 1931, p. 303–304.) **VDT**

4725. **Biddlecomb,** C. H. Some scientific aspects of Commander Byrd's transatlantic flight. illus. (Outlook, New York, 1927, v. 146, p. 406–407.) *** DA**

4726. **Bilancioni,** Guglielmo. Svolgimento storico del concetto di "aria." illus. (International Congress of Aerial Navigation, Fourth, Rome 1927, Rapports, memorials, Rome, 1928, v. 4, p. 439–487.) **VDS**

History of the physics of the air and of its effects on the human body.

4727. **Biot,** Jean Baptiste. Extrait de la relation d'un voyage aérostatique, fait par MM. Gay-Lussac et Biot... (Journal de physique, Paris, 1804, tome 59, p. 314–320.) **3–OA**

German translation in *Annalen der Physik,* Halle, 1805, Bd. 20, p. 1–18, *PAA.*

4728. **Blanchet,** Georges. Lieutenant-colonel Georges Espitallier. port. (L'Aérophile, Paris, 1908, année 16, p. 285.) **3–VDS**

Portrait and brief biography of distinguished French scientist and prolific writer on aeronautical subjects.

4729. **Börnstein,** Richard. Wissenschaftliche Luftfahrten. (Deutsche Revue, Stuttgart, 1900, Jahrg. 25, Bd. 3, p. 143–154.) *** DF**

History of scientific results obtained in balloon ascensions.

For list of Börnstein's writings on scientific ballooning see Paul Brockett, *Bibliography of aeronautics,* Washington, 1910, p. 138–139, *VDT.*

4730. **Boothby,** F. L. M. The recent eclipse as seen from a balloon. illus. (Aviation, Highland, N. Y., 1927, v. 23, p. 248–250.) **VDS**

Solar eclipse of 1927 as seen in England.

4731. **Briggs,** Lyman J. Aeronautic research at the national Bureau of Standards. illus. port. (Aero digest, New York, 1929, v. 14, March, p. 47–49; April, p. 52–53, 290.) **VDS**

4732. **Brodetsky,** Selig. Prof. G. H. Bryan, F. R. S. (Nature, London, 1928, v. 122, p. 849–850.) **OA**

Bryan (1864–1928) is noted for his mathematical theory of aeroplane flight.

4733. **Brown,** Fay C. Aviation work of the Bureau of Standards. illus. port. (Aerial age, New York, 1923, v. 16, p. 115–119.) **VDS**

4734. **Bruno,** H. A. Flying explorers. Lieut. Walter Hinton and Capt. Stevens discover the source of the Parima river in unknown Brazil. illus. (Aero digest, New York, 1925, v. 7, p. 362–363, 396.) **VDS**

4735. **Bryan,** F. I. The Fund is ended but its work still carries on. ports. (Western flying, Los Angeles, 1930, v. 7, Jan., p. 64–66, 106.) **VDS**

Brief history of the Daniel Guggenheim Fund for the Promotion of Aeronautics.

Aeronautics and Science, continued

4736. Bryan, George Hartley. Artificial flight. (Science progress, London, 1897, v. 6 [new series, v. 1], p. 531–553.) 3–OA

Brief history of aeronautics. Considers "one by one the various qualifications which are required to make a successful flying machine and how they have been attained."

4737. —— The longitudinal stability of aërial gliders. (British Association for the Advancement of Science, Report of the 72d meeting, Belfast, 1902, London, 1903, p. 524–525.) * EC

4738. —— The resistance of the air. illus. (Nature, London, 1899, v. 61, p. 107–109, 248.) OA

Describes the experiments of Canovetti and of the Abbé Le Dantec. Also in *Scientific American supplement,* New York, 1900, v. 49, p. 20116–20117, *VA.*

For a list of Bryan's writings, mainly on the theory of flight, see Paul Brockett, *Bibliography of aeronautics,* Washington, 1910, p. 160–161, *VDT.*

4739. Budau, Arthur Adalbert Joseph. Die mechanischen Grundgesetze der Flugtechnik unter der vereinfachenden Annahme konstanten specifischen Volumens der atmosphärischen Luft. Anhang: Erwiderung auf die gegen die obigen Grundgesetze gemachten Einwendungen. Wien: Der Verfasser, 1904. 52 p. 8°. VDY

Author's theory of heavier-than-air flight.

4740. Bunnell, Charles Beecher. Leonardo da Vinci. (Aeronautics, New York, 1914, v. 15, p. 35.) VDS

Brief accounts of da Vinci's engineering works, including aeronautics.

4741. Busk, Mary (Acworth). E. T. Busk, a pioneer in flight, with a short memoir of H. A. Busk. London: John Murray [1925]. xi, 167 p. ports. illus. 8°. AN

Edward Teshmaker Busk (1886–1914) was prominently associated with the British Aircraft Factory and with the National Physical Laboratory as an investigator in scientific aeronautics. For a review see *Flight,* London, 1925, v. 17, p. 829, *VDS.*

4742. Cailletet, Louis Paul, and E. Colardeau. Expériences sur la résistance de l'air et de divers gaz au mouvement des corps. diagrs. (Académie des sciences, Comptes rendus, Paris, 1893, tome 117, p. 145–150.) * EO

Also in *L'Aérophile,* Paris, 1893, année 1, p. 141–144, *3–VDS.*

4743. —— Recherches expérimentales sur la chute des corps et sur la résistance de l'air à leur mouvement. Expériences exécutées à la tour Eiffel. illus. (L'Aéronaute, Paris, 1892, année 25, p. 159–165.) 3–VDS

4744. Canovetti, Cosimo. Expériences sur le coefficient de résistance de l'air. illus. (Société d'encouragement pour l'industrie nationale, Bulletin, Paris, 1899, série 5, tome 4, p. 1038–1048.) 3–VA

4745. —— Sur la résistance de l'air au mouvement des corps. (Académie des sciences, Comptes rendus, Paris, 1907, tome 144, p. 1030.) * EO

For a list of Canovetti's writings, mainly on air resistance, see Paul Brockett, *Bibliography of aeronautics,* Washington, 1910, p. 172–173, *VDT.*

4746. Canovetti's air-resistance experiments. illus. (Scientific American, New York, 1907, v. 96, p. 170–171.) VA

4747. Carpenter, Ford Ashman. Weather and air transport. illus. (Aero digest, New York, 1929, v. 15, Dec., p. 51–52, 276, 278.) VDS

History of upper air investigations since 1913.

4748. Cayley, Sir George. Aeronautical and miscellaneous note-book (ca. 1799–1826) of Sir George Cayley, with an appendix comprising a list of the Cayley papers. Cambridge, England: Published for the Newcomen Society by W. Heffer & Sons, Ltd., 1933. xx, 93 p. illus. 8°. (Newcomen Society for the Study of the History of Engineering & Technology. Extra publication. no. 3.) VA

4749. —— Extracts from the aeronautical and miscellaneous note-book (ca. 1799–1826). (Royal Aeronautical Society, Journal, London, 1933, v. 37, p. 547–568.) VDS

4750. —— On aerial navigation. (Aeronautical annual, Boston, 1895, no. 1, p. 16–48.) 3–VDS

Reprint of articles in *Nicholson's journal,* 1809–1810. Also published, with slight abridgment, in Royal Aeronautical Society of Great Britain, *Aeronautical classics,* no. 1, London, 1910, 40 p., *VDY.* French translation in *Revue générale de l'aéronautique,* Paris, 1912, v. 2, p. 270–304, *3–VDS.*

4751. —— On aërial navigation. (Philosophical magazine and journal, London, 1816, v. 47, p. 81–86, 321–329; 1817, v. 50, p. 27–35.) OA

Reprinted in *Journal* of the Royal Aeronautical Society, London, 1925, v. 29, p. 76–92, *VDS.*

4752. —— On the principles of aerial navigation. illus. (Mechanics' magazine, London, 1842, v. 38, p. 273–278.) 3–VA

Description of machine proposed by him, with comments on the work of Henson.

Also in Franklin Institute, *Journal,* Philadelphia, 1843, series 3, v. 6, p. 202–209, *VA.*

4753. —— Retrospect of the progress of aerial navigation, and demonstration of the principles by which it must be governed. illus. (Mechanics' magazine, London, 1843, v. 38, p. 263–265.) 3–VA

Also in Franklin Institute, *Journal,* Philadelphia, 1843, series 3, v. 6, p. 132–134, *VA.*

For a list of writings concerning Cayley and his investigations see United States. — National Advisory Committee for Aeronautics, *Bibliography of aeronautics,* 1909–1931, Washington, 1921–35, volumes for 1909/16, p. 247; 1917/19, p. 83; 1923, p. 24; 1925, p. 33, *VDT.*

4754. Chamberlin, Clarence Duncan. The weather conditions on my flight to Germany. port., map. (Aero digest, New York, 1927, v. 11, p. 513, 616–618.) VDS

Contradicts statements of Ramsey and Kincaid. See entry no. 4993.

Aeronautics and Science, continued

4755. Chatley, Herbert. Mechanical flight. illus. (Society of Engineers, Transactions, London, 1908, p. 241–280.) **VDA**

With discussion.

See p. 243–244 for brief history of investigations on the theory of air resistance.

4756. Chaudy, F. Sur le frottement des fluides contre les surfaces solides. illus. (Société des ingénieurs civils de France, Mémoires, Paris, 1896, partie 2, p. 24–30.) **VDA**

Develops formula for resistance of solids to gas flow.

4757. —— Théorie de la résistance à l'avancement des plaques dans l'air. illus. (Société des ingénieurs civils de France, Mémoires, Paris, 1909, Aug., p. 92–109.) **VDA**

Based on data of Eiffel.

4758. Clarence LeRoy Meisinger, 1895–1924. port. (U. S. air services, Washington, 1924, v. 9, July, p. 35–36.) **VDS**

As a member of the staff of the United States Weather Bureau, Dr. Meisinger made important contributions to the physics of the upper air.

4759. Cléry, A. Les fondateurs de l'aérodynamique expérimentale. Gustave Eiffel. port. (L'Aérophile, Paris, 1911, année 19, p. 265–266.) **3–VDS**

4760. Coe, C. C. Observations in balloons. (International Conference on Aerial Navigation, Chicago, 1893, Proceedings, New York, 1894, p. 354–359.) **VDS**

Author, who states that he is the oldest aeronaut in the United States, records his experiences in detecting submarine objects from a balloon.

4761. Collecting insects by airplane. Bureau of Entomology develops method of using airplanes for collecting insects at altitudes. illus. (Aviation, Highland, N. Y., 1927, v. 22, p. 1173.) **VDS**

4762. Condroyer, L. Le compas de navigation aérienne. illus. (L'Aéronautique, Paris, 1919–20, année 1, p. 387–395.) **VDS**

Has a brief historical introduction and a bibliography.

4763. Cordero di Montezemolo, Vittorio. Studio sulla navigazione aerea. Roma: E. Voghera, 1903. vii, 200 p. illus. 8°. **VDY**

Study of flight theory for heavier-than-air and lighter-than-air craft, with references to earlier investigations.

4764. Crocco, Gaetano Arturo. Sull' equilibrio delle macchine volanti. (Rivista di artiglieria e genio, Roma, 1905, annata 22, v. 2, p. 23–46.) **VWW**

For references to Crocco's writings on aeronautical theory see Paul Brockett, *Bibliography of aeronautics,* Washington, 1910, p. 219–220, *VDT.* See also Giuseppe Boffito, *Biblioteca aeronautica italiana,* Firenze, 1929, p. 139–143, † *VDT.*

4765. Crocé-Spinelli, J. Stabilité des appareils destinés a se mouvoir dans l'air. illus. (L'Aéronaute, Paris, 1869, année 2, p. 22–26, 73–77.) **3–VDS**

For references to Crocé-Spinelli's ascensions, also to his papers on engines, see Paul Brockett, *Bibliography of aeronautics,* Washington, 1910, p. 220–221, *VDT.*

4766. Crocé-Spinelli, J., and TH. SIVEL. Ascension scientifique à grande hauteur exécutée le 22 mars 1874. illus. (L'Aéronaute, Paris, 1874, année 7, p. 145–154.) **3–VDS**

Ascension of the authors.

4767. Dacy, George H. In the air on the ground. illus. (Popular aviation, Chicago, 1928, v. 3, Oct., p. 18–20, 121.) **VDS**

Account of research work at the United States Bureau of Standards.

4768. —— Navy pilot charts. illus. (Popular aviation, Chicago, 1928, v. 3, Nov., p. 16–19, 121.) **VDS**

Work of the Hydrographic Office of the U. S. Navy.

4769. Dallwitz-Wegner, Richard von. Hilfsbuch für den Luftschiff- und Flugmaschinenbau. Eine übersichtliche Darstellung der verschiedenen Konstruktionen, sowie Anleitung zur Berechnung der Leistungen und des Wirkungsgrades von Luftschiffen, Flugmaschinen aller Art und von Treibschrauben, nebst einem Anhang: Die Mechanik des Gleitbootes. Rostock: Volckmann Nachfolger, 1909. vii, 142 p. illus. 4°. **VDY**

Calculations for the construction of various aircraft.

4770. The **Daniel Guggenheim Fund for the Promotion of Aeronautics, Inc.** Equipment used in experiments to solve the problem of fog flying. New York: The Daniel Guggenheim Fund for the Promotion of Aeronautics, Inc., 1930. 57 p. illus. 8°. **VDY p.v.51, no.12**

4771. —— Solving the problem of fog flying. A record of the activities of the Fund's Full Flight Laboratory to date. New York: The Daniel Guggenheim Fund for the Promotion of Aeronautics, Inc. [pref. 1929.] 52 p. 8°. **VBA p.v.196, no.2**

4772. Daniel Guggenheim gives $2,500,000 to foster aviation. port. (Aviation, Highland, N. Y., 1926, v. 20, p. 106–107.) **VDS**

4773. The **Death** of a great pioneer. Gustave Eiffel passes away. (Flight, London, 1924, v. 16, p. 4.) **VDS**

4774. Decazes, vicomte. Théorie des flottements dynamiques et son application à la navigation aérienne. illus. diagrs. (L'Aérophile, Paris, 1894, année 2, p. 101–110, 134–142, 158–167, 210–214; 1895, année 3, p. 82–87, 137–142; 1896, année 4, p. 269–275; 1897, année 5, p. 27–33, 92–96.) **3–VDS**

4775. Degouy, Marcel. Étude sur la dynamique des ballons libres. (Revue du génie militaire, Paris, 1902–03, v. 24, p. 5–28, 325–348, 394–426, 481–506; v. 25, p. 125–150, 323–348.) **VWA**

Aeronautics and Science, continued

4776. Desaguliers, John Theophilus. An account of some experiments made on the 27th day of April, 1719, to find how much the resistance of the air retards falling bodies. (Royal Society of London, Philosophical transactions, 1717/18, London, 1720, p. 1071–1078.) * EC

Experiments with hollow glass vessels and air-filled bladders.

4777. Donalies, Hans. Leonardo da Vinci's Flugtheorie. illus. (Deutsche Luftfahrer-Zeitschrift, Berlin, 1912, Jahrg. 16, p. 8–10, 32–34, 54–55.) VDS

4778. Drzewiecki, Stepan. De la nécessité urgente de créer un laboratoire d'essais aérodynamiques destiné à fournir aux aviateurs les éléments nécessaires à la construction des aéroplanes et de la manière d'organiser ce laboratoire. Paris: F. L. Vivien, 1909. 15 p. 8°. VDY p.v.8, no.3

For a list of Drzewiecki's writings see Paul Brockett, *Bibliography of aeronautics,* Washington, 1910, p. 258–259, *VDT.*

4779. Dupont, Paul. Les travaux du Laboratoire aérodynamique. illus. (In: L'Aéronautique militaire, maritime, coloniale et marchande, Paris, 1931, p. 216–223.) † VDY

History of French institution at Saint-Cyr.

4780. Durand, William Frederick. The development of our knowledge of the laws of fluid mechanics. (Science, New York, 1933, v. 78, p. 343–351.) OA

4781. —— Historical sketch of the development of aerodynamic theory. (International Civil Aeronautics Conference, Washington, 1928, Papers, Washington, D. C., 1928, p. 447–462.) VDS

4782. Duroy de Bruignac, Albert, baron. Recherches sur la navigation aérienne. diagrs. (L'Aéronaute, Paris, 1876, année 9, p. 131–140, 198–202, 229–230, 271–280.) 3–VDS

Two chapters from de Bruignac's book of above title, dealing with action of plane surfaces, and with bird flight.

4783. —— Remarques sur le problème de la navigation aérienne. (Société des ingénieurs civils de France, Mémoires, Paris, 1898, année 1898, v. 1, p. 313–331.) VDA

Discusses the best form for an airship; also gliding of birds. For discussion see entry no. 5040.

4784. Duté-Poitevin, Adrien. Observations critiques sur les ascensions à grande hauteur. illus. (L'Aéronaute, Paris, 1876, année 9, p. 103–111.) 3–VDS

A record of high ascensions, with a chart showing comparative heights attained.

4785. Ebert, Hermann. Magnetische Messungen im Ballon. (Illustrierte aëronautische Mitteilungen, Strassburg, 1900, Jahrg. 4, p. 137–146.) VDS

History of aeronautical magnetic observations. The first were made in 1804 by Guy-Lussac and Biot.

4786. Edmond Seux. port. (L'Aérophile, Paris, 1907, année 15, p. 53–54.) 3–VDS

Brief account and portrait of pioneer student of aviation.

4787. Eiffel, Alexandre Gustave. New aerodynamical researches. Results of many important experiments. illus. (Scientific American supplement, New York, 1915, v. 80, p. 260–261.) VA

4788. —— Résumé des principaux travaux exécutés pendant la guerre au Laboratoire aérodynamique Eiffel, 1915–1918... Paris: E. Chiron [1921]. xx, 212 p. incl. tables. illus. diagrs. f°. † VDW

For a résumé of this important book see *Le Génie civil,* Paris, 1920, tome 76, p. 324–329, *VA.*
For a list of references concerning Eiffel and his investigations see United States. — National Advisory Committee for Aeronautics, *Bibliography of aeronautics,* 1909–1931, Washington, 1921–35, volumes for 1909/16, p. 394–397; 1917/19, p. 130; 1920/21, p. 123; 1922, p. 27–28, *VDT.*

4789. The **Eiffel** aero-dynamometric installation. illus. (Engineer, London, 1910, v. 110, p. 134–136.) VA

Also in *Scientific American supplement,* New York, 1910, v. 70, p. 265–266, *VA.*

4790. Espitallier, Georges. Le ballon anémométrique Julliot. illus. (La Nature, Paris, 1904, année 32, semestre 2, p. 353–354.) OA

Balloon has a device for measuring direction and strength of the wind at different altitudes. Also in *Scientific American supplement,* New York, 1905, v. 59, p. 24481, *VA.*

4791. —— Ueber das Lebenswerk von Oberst Renard. port. (Illustrierte aëronautische Mitteilungen, Strassburg, 1905, Jahrg. 9, p. 170–172.) VDS

Account of work of great French pioneer in the study of the upper air. Portrait on p. 169.

4792. Experiments with currents of air. illus. (Engineering, London, 1885, v. 40, p. 160–161.) VDS

Pioneer experiments of Horatio Phillips.

4793. Faccioli, Aristide. Teoria del volo e della navigazione aerea; ricerche sperimentali sulla resistenza dell' aria. Teorie dell' elice e del timone. Milano: U. Hoepli, 1895. viii, 309 p. illus. 8°. VDY

Author's experiments on air resistance and its relation to mechanical flight.

4794. Faraud, Louis. Le plus lourd que l'air. diagrs. (Société des ingénieurs civils de France, Mémoires, Paris, 1888, année 1888, v. 2, p. 618–643.) VDA

Calculations for airplanes, balloon-airplanes, and air resistance.

4795. Ferber, Louis Ferdinand. Les progrès de l'aviation depuis 1891 par le vol plané. (Revue d'artillerie, Paris, 1904–05, v. 63, p. 397–443; v. 66, p. 317–375; v. 67, p. 44–74, 81–105; v. 69, p. 133–148.) VWA

Mathematical discussion of flight.

Aeronautics and Science, continued

4796. Fergusson, Sterling Price. The exploration of the upper air by means of ballons sondes. illus. (Scientific American, New York, 1909, v. 100, p. 169–170.) **VA**

History and current practice.

4797. Fitzgerald, Maurice F. On flapping flight of aeroplanes. illus. (Royal Society, Proceedings, 1898/99, London, 1899, v. 64, p. 420–430.) *** EC**

See also *Aeronautical journal,* London, 1899, v. 3, p. 59–65, *VDS.*

4798. Flammarion, Camille. Wilfrid de Fonvielle. port. (Société astronomique de France, Bulletin, Paris, 1914, année 28, p. 339–343.) **OMA**

Account of prominent French balloonist, writer, and scientist, 1826–1914.

4799. Fonvielle, Wilfrid de. Les ballons-sondes de MM. Hermite et Besançon et les ascensions internationales; précédé d'une introduction par M. Bouquet de la Grye. Paris: Gauthier-Villars et fils, 1898. ix, 112 p. illus. 12°. **VDW**

Details of Hermite's and Besançon's experiments with sounding balloons, and of the International Commission. For review see *Science,* New York, 1898, new series, v. 7, p. 33–34, *OA.*

4800. —— Exploration of the upper atmosphere. illus. (International Conference on Aerial Navigation, Chicago, 1893, Proceedings, New York, 1894, p. 364–396.) **VDS**

An account of the results obtained by Hermite with his small exploring balloons. Includes a discussion by H. A. Hazen.

4801. —— Le général major Michel Alexandrowitch Rykatchew. port. (L'Aérophile, Paris, 1898, année 6, p. 149–152.) **3–VDS**

Portrait and account of the scientific balloon ascensions of the director of the Russian Meteorological Observatory.

4802. —— M. Cailletet. port. (L'Aérophile, Paris, 1898, année 6, p. 117–121.) **3–VDS**

Portrait and biography of famous French physicist.

4803. —— Thermometer soundings in the high atmosphere. (Nature, London, 1893, v. 48, p. 160–161.) **OA**

Account of Hermite's investigations.

4804. Fournier, Lucien. L'Institut aérodynamique de Koutchino. illus. (La Nature, Paris, 1909, année 37, semestre 2, p. 133–141.) **VA**

Description of laboratory, near Moscow, built by Riabouchinsky. Translation in *Scientific American supplement,* New York, 1909, v. 68, p. 244–246, *VA.*

4805. Friedrich Ritter von Loessl. port. (Wiener Luftschiffer-Zeitung, Wien, 1907, Jahrg. 6, p. 106–107.) **VDS**

Biography and portrait of prominent German experimenter in aerodynamics.

4806. Frion, O. Jules Jamin. port. (L'Aéronaute, Paris, 1886, année 19, p. 43–50.) **3–VDS**

Jamin (1818–1886), eminent French physicist, was president of the Société de navigation aérienne.

4807. Gammeter, H. C. Aerodynamics — development of mechanical flight. diagrs. (Cleveland Engineering Society, Journal, Cleveland, O., 1914–15, v. 7, p. 442–460.) **VDA**

Refers to some of the early studies on air pressure.

4808. Gay-Lussac, Joseph Louis. Bericht von seiner aerostatischen Reise, am 16ten Sept. 1804. (Annalen der Physik, Halle, 1805, Bd. 20, p. 19–37.) **PAA**

French abridgment in *Journal de physique,* Paris, 1804, tome 59, p. 454–462, *3–OA.*

4809. Genet, Edmond Charles. Memorial on the upward forces of fluids and their applicability to several arts, sciences, and public improvements, for which a patent has been granted by the government of the United States to the author. Albany: Packard and Van Benthuysen, 1825. 112 p. 8°. **PCW**

Describes his horse-driven airship. Has an appendix containing his plan for applying steam power to aeronautics.

4810. —— Reply to Dr. Jones, in the Franklin journal, on the subject of the memorial on the upward forces of fluids. (American journal of science, New Haven, 1827, v. 12, p. 310–325.) **OA**

For remarks by Benjamin Silliman see above magazine, 1827, v. 12, p. 325–330, *OA.*

4811. —— Vindication of the memorial on the upward force of fluids. (American journal of science, New Haven, 1827, v. 12, p. 94–103.) **OA**

Replies to criticisms of his airship.

4812. Gerlach, Edmund. Ableitung gewisser Bewegungsformen geworfener Scheiben aus dem Luftwiderstandsgesetze. illus. (Zeitschrift für Luftschiffahrt, Berlin, 1886, Jahrg. 5, p. 65–89.) **3–VDS**

Deductions drawn from the propulsion of certain forms of flat disks through the air.

4813. Gerstner, Ferdinand. Ueber die Versuche zur Lösung des Problemes der Luftschiffahrt. diagrs. (Oesterreichischer Ingenieur- und Architekten-Verein, Zeitschrift, Wien, 1903, Jahrg. 55, p. 229–235.) **VDA**

Discusses problems of aerial navigation and the different classes of balloons, airships, and heavier-than-air machines.

4814. Giacomelli, R. The aerodynamics of Leonardo da Vinci. illus. (Royal Aeronautical Society, Journal, London, 1930, v. 34, p. 1016–1038.) **VDS (Royal)**

"The purpose...is to collect all Da Vinci's writings on the matter in chronological order and to explain them suitably in modern language." Abstract in *Mechanical engineering,* New York, 1930, v. 52, p. 888, *VFA.*

4815. —— Leonardo da Vinci e il volo meccanico. illus. (L'Aerotecnica, Pisa, 1927, v. 7, p. 486–524.) **VDS**

4816. —— Les machines volantes de Léonard de Vinci et le vol à voile. illus. (International Congress of Aerial Navigation, 4th, Rome, 1927, Reports, Roma, 1928, v. 3, p. 230–236.) **VDS**

Aeronautics and Science, continued

4817. Giacomelli, R., and E. PISTOLESI. Historical sketch [of the development of aerodynamics]. (In: W. F. Durand, Aerodynamic theory, a general review of progress, Berlin, 1934, v. 1, p. 305–394.) **PCN (Durand)**

An excellent review from antiquity to the present, with numerous references.

4818. Glaisher, James. An account of meteorological and physical observations in eight balloon ascents, made under the auspices of the committee of the British Association for the Advancement of Science. (British Association for the Advancement of Science, Report of 32d meeting, Cambridge, 1862, London, 1863, p. 376–502.) *** EC**

Ascents were made in 1862.

4819. —— An account of meteorological and physical observations in five balloon ascents in the year 1863 (in continuation of eight made in the preceding year...) (British Association for the Advancement of Science, Report of 33d meeting, Newcastle-upon-Tyne, 1863, London, 1864, p. 426–516.) *** EC**

4820. —— The balloon and its application. illus. (Leisure hour, London, 1864, v. 13, p. 293–296, 316–320, 327–331, 343–346.) *** DA**

History of ballooning. Meteorological observations and the experiences of the author and his companion Coxwell.

4821. —— Balloon ascents and their scientific importance. illus. (Popular science review, London, 1865, v. 4, p. 403–418.) **3–OA**

Illustration shows arrangement of instruments in the car.

4822. —— On aerial navigation as applied to scientific research. (Good words, London, 1863, v. 4, p. 219–225.) *** DA**

Describes an ascension with Henry Coxwell to six miles altitude.

4823. —— On scientific experiments in balloons. (Royal Institution of Great Britain, Proceedings, London, 1862–66, v. 4, p. 65–72.) *** EC**

Results of author's ascensions and reference to work of previous experimenters. Also in *Engineer*, London, 1863, v. 15, p. 318, *VA*.

For citations to writings by and about Glaisher see Paul Brockett, *Bibliography of aeronautics*, Washington, 1910, p. 362–363, *VDT*.

4824. Glaisher, James, and others. Voyages aériens, par J. Glaisher, Camille Flammarion, W. de Fonvielle et Gaston Tissandier... Paris: L. Hachette et Cie, 1870. 612 p. illus. ports. 4°. **3–VDW**

Well-illustrated record of pioneer voyages by the above in the interest of science. Library has also English edition entitled *Travels in the air*, London, 1871, 398 p., 4°, *VDW*.

4825. Glazebrook, Sir Richard Tetley. The development of the aeroplane. diagrs. (Aeronautical journal, London, 1914, v. 18, p. 272–301.) **VDS**

Traces experiments and scientific research. Also in *Scientific American supplement*, New York, 1914, v. 77, p. 361–363, 382–384, *VA*; also in *Aeronautics*, London, 1914, v. 7, p. 169–176, *VDS*.

4826. Gold, Ernest. The international kite and balloon ascents. A dissertation which obtained the first prize awarded in 1912 by the German Meteorological Society... London: H. M. Stationery Office, 1913. 62–104 p. f°. (Meteorological Office. Geophysical memoirs. no. 5.) **PSA**

Discusses 12,000 observations made during fifteen years.

4827. Gold, Ernest, and W. A. HARWOOD. The present state of our knowledge of the upper atmosphere as obtained by the use of kites, balloons, and pilot balloons. (British Association for the Advancement of Science, Report of the 77th meeting, Winnipeg, 1909, London, 1910, p. 71–124.) *** EC**

Excellent historical introduction.

4828. Gore, Howard Mason. Aircraft and the Department of Agriculture. (National Aeronautic Association review, Washington, 1925, v. 3, p. 67–72.) **VDS**

Airplane used by the Bureau of Entomology, Bureau of Plant Industry, Weather Bureau, Forest Products Laboratory, and in the estimation of crops.

4829. Goupil, Albert Louis. Aviation. illus. (Bulletin technologique, Paris, 1904, Sept., p. 745–844.) **3–VA**

Presentation of flight theory, with references to some of the author's experiments.

4830. —— Aviation. Aérodynamique, les secrets du planement et du vol ramé; appareils aériens... Paris: J. Loubat & Cie. [1910.] 318 p. 8°. **VDY**

Résumé of author's experiments, 1884–1908.

4831. —— Lancement d'un aéroplane. (Société des anciens élèves des écoles nationales d'arts et métiers, Bulletin technologique, Paris, 1908, p. 956–968.) **3–VA**

Derives formulas for airplane take-off.

4832. —— La locomotion aérienne. Étude. Charleville: A. Pouillard, 1884. 112 p. illus. 8°. **VDY (Room 120)**

Sets forth the science of airplanes and airships and suggests the use of a light steam-engine for propulsion.

4833. —— Virage d'un aéroplane; nécessité de régulateurs de roulis. (Société des anciens élèves des écoles nationales d'arts et métiers, Bulletin technologique, Paris, 1907, p. 727–737.) **3–VA**

A study of the banking of airplanes.

For a list of Goupil's writings on the theory of flight see Paul Brockett, *Bibliography of aeronautics*, Washington, 1910, p. 370–372, *VDT*.

4834. Gradenwitz, Alfred. An aeronautical observatory. The elaborate equipment of a model institution. illus. (Scientific American supplement, New York, 1907, v. 64, p. 24–25.) **VA**

Illustrates and describes the Lindenberg laboratory, of which Dr. R. Assmann is director.

4835. Graham, W. H. Johnny Starr — first aerial weatherman. illus. ports. (Popular aviation, Chicago, 1932, v. 10, p. 355–356, 401–402.) **VDS**

United States Weather Bureau flyer's experiences.

Aeronautics and Science, continued

4836. Gregg, Willis Ray. The highest aerial sounding. (Monthly weather review, Washington, 1920, v. 48, p. 633.) † PRS

Discussion of the sounding balloon record made at Pavia, Italy, 1911, and remarks on some subsequent records.

4837. —— History of the application of meteorology to aeronautics with special reference to the United States. (Monthly weather review, Washington, 1933, v. 61, p. 165–169.) † PRS

4838. —— Recent advances in weather service in relation to aerial navigation. maps. (U. S. air services, Washington, 1932, v. 17, Aug., p. 28–33.) VDS

4839. —— Trans-Atlantic flight and meteorology. maps. (Aviation, New York, 1919, v. 6, p. 370–372, 422–425.) VDS

Has a bibliography.

4840. Gronau, Johann Friedrich Wilhelm. Historische Entwicklung der Lehre vom Luftwiderstande. (Naturforschende Gesellschaft in Danzig, Schriften, Danzig, 1868, N. F. Bd. 2, No. 5, 1–28 p.) 3–* EE

History of air resistance investigations from the sixteenth century.

4841. Günther, S. C. Kramp und die Aëronautik. (Illustrierte aëronautische Mittheilungen, Strassburg, 1899, Jahrg. 3, p. 69–72.) VDS

Christian Kramp (1760–1826) was a pioneer German investigator in theoretical aeronautics.

4842. Guggenheim, Harry Frank. The first year of the Guggenheim Fund. (Aviation, Highland, N. Y., 1927, v. 22, p. 18.) VDS

4843. Guggenheim Fund endows universities. Leland Stanford University and California Institute of Technology to become centers of aeronautical research and learning. (Aviation, Highland, N. Y., 1926, v. 21, p. 402.) VDS

4844. Guillaume, Charles Édouard. L'exploration des hautes régions. illus. (La Nature, Paris, 1892, année 21, semestre 1, p. 1–3.) OA

Discusses the use of small free balloons. Translation in *Scientific American supplement,* New York, 1893, v. 35, p. 14290–14291, *VA.*

4845. Haedicke, H. Grundzüge zu einer Theorie des Fluges. (Der Civilingenieur, Leipzig, 1879, N. F. Bd. 25, col. 561–586.) VDA

Presents theory of heavier-than-air flight.

4846. Hagen, Freiherr vom. Der Luftballon im Dienste der Wissenschaft. (Zeitschrift für Luftschiffahrt, Berlin, 1884, Jahrg. 3, p. 65–76, 97–112, 129–146.) 3–VDS

Contributions of aeronautics to science, 1783–1884.

4847. Hammon, W. H. Meteorological observations taken in four balloon voyages. (American meteorological journal, Ann Arbor, 1891, v. 7, p. 498–528.) PRA

Details of ascensions undertaken by the United States Signal Corps in 1885.

4848. Harding, Charles. Scientific balloon ascents. (Aeronautical journal, London, 1904, v. 8, p. 69–74.) VDS (Royal)

Principally a summarization of the results obtained by James Glaisher.

4849. Harrington, Mark Walrod. Systematic explorations of the upper air, with estimates of cost. (International Conference on Aerial Navigation, Chicago, 1893, Proceedings, New York, 1894, p. 349–354.) VDS

The Chief of the United States Weather Bureau discusses the use of kites, pilot balloons, and large balloons.

4850. Hart, Ivor Blashka. Leonardo da Vinci as a pioneer of aviation. illus. (Royal Aeronautical Society, Journal, London, 1923, v. 27, p. 244–269.) VDS

4851. —— The mechanical investigations of Leonardo da Vinci. London: Chapman & Hall, Ltd., 1925. vii, 240 p. illus. port. 8°. V

See chapters 7 and 8 for aviation and bird flight respectively.

4852. Hazen, H. A. Scientific results gained by balloons. (International Conference on Aerial Navigation, Chicago, 1893, Proceedings, New York, 1894, p. 361–364.) VDS

Brief record of investigations.

4853. —— Two balloon voyages. (Science, New York, 1887, v. 9, p. 591–592.) OA

By the author, chief signal officer of the United States Army, for meteorological purposes, at Providence, R. I., 1887.

4854. Hellmann, Gustav. Die erste Ballonfahrt zu wissenschaftlichen Zwecken. port. (Zeitschrift für Luftschiffahrt, Berlin, 1892, Jahrg. 11, p. 14–17.) 3–VDS

Ascensions of Jeffries, 1784 and 1785. Portrait of Jeffries.

4855. Helmholtz, Hermann Ludwig Ferdinand von. Theoretische Betrachtungen über lenkbare Luftballons. (Verein zur Beförderung des Gewerbfleisses in Preussen, Verhandlungen, Berlin, 1872, Jahrg. 51, p. 289–292; 1873, Jahrg. 52, p. 326–335.) 3–VA (Gewerb.)

The first installment is translated into French in *L'Aéronaute,* Paris, 1874, année 7, p. 203–207, *3–VDS.*

4856. —— Ueber ein Theorem, geometrisch ähnliche Bewegungen flüssiger Körper betreffend, nebst Anwendung auf das Problem, Luftballons zu lenken. (Preussische Akademie der Wissenschaften, Monatsberichte, Berlin, 1873, p. 501–514.) * EE

For citations to Helmholtz's writings on balloons see Paul Brockett, *Bibliography of aeronautics,* Washington, 1910, p. 405–406, *VDT.*

Aeronautics and Science, continued

4857. **Hergesell,** Hugo. Zur Geschichte der internationalen Ballonfahrten. (Illustrierte aëronautische Mittheilungen, Strassburg, 1898, Jahrg. 2, p. 22–24.) **VDS**

Inception and development of the International Commission for Scientific Aeronautics.

For list of Hergesell's writings on upper-air investigations see Paul Brockett, *Bibliography of aeronautics,* Washington, 1910, p. 409–411, *VDT.*

4858. **Hermite,** Gustave. L'exploration de la haute atmosphère. Ascension de ballon "Aérophile." illus. (L'Aérophile, Paris, 1893, année 1, p. 45–55, 165–170.) **3–VDS**

4859. —— L'exploration de la haute atmosphère. Troisième ascension du ballon L'Aérophile. illus. (L'Aérophile, Paris, 1896, année 4, p. 5–21.) **3–VDS**

Scientific results of ascension.

4860. —— Exploration des hautes régions de l'atmosphère à l'aide de ballons non montés, pourvus d'enregistreurs automatiques. (Académie des sciences, Comptes rendus, Paris, 1892, tome 115, p. 862–864.) *** EO**

Extract of report made to the Académie. Tabular data.

For a list of Hermite's writings on the exploration of the upper air see Paul Brockett, *Bibliography of aeronautics,* Washington, 1910, p. 411–414, *VDT.*

4861. **Hermite,** Gustave, and Georges Besançon. Sur les principaux résultats de la dernière ascension à grande hauteur du ballon explorateur *L'Aérophile* (22 mars 1895). (Académie des sciences, Comptes rendus, Paris, 1896, tome 122, p. 849–850.) *** EO**

4862. **Hervé,** Henri. Sur les ascensions maritimes appliquées à la météorologie. (L'Aéronaute, Paris, 1886, année 19, p. 144–146.) **3–VDS**

Suggests improvements in balloon equipment.

4863. **Hervé** Mangon. illus. (Revue de l'aéronautique, Paris, 1888, année 1, p. 78–84.) **3–† VDS**

Mangon (1821–1888), prominent meteorologist, was active during the siege of Paris, 1871–72, and subsequently contributed much to the development of aeronautics in France.

4864. **Hill,** Frederick Thomas. Developments in aeronautical science. (Nature, London, 1935, v. 135, p. 750–752.) **OA**

Survey covers 25 years.

4865. **Hoernes,** Hermann. Kritische Betrachtungen über die Wellner'schen Versuche über den Luftwiderstand gewölbter Flächen im Winde und auf Eisenbahnen. (Zeitschrift für Luftschiffahrt, Berlin, 1896, Jahrg. 15, p. 162–171.) **3–VDS**

4866. **Holland,** Maurice. Aviation's apostles. Jerome C. Hunsaker. illus. (Sportsman pilot, Albany, N. Y., 1930, v. 4, Sept., p. 14–15, 42, 44.) **VDS**

Career of noted American designer of airships, and scientific investigator.

4867. **Hooke,** Robert. An account of flying. Of a flying trip. (Royal Society of London, Memoirs, London, 1739, v. 2, p. 138–140.) *** EC**

These two papers (in abstract) cover discussions of Besnier's apparatus and Lana's airship.

4868. —— The posthumous works of Robert Hooke, containing his Cutlerian lectures, read at the meetings of the illustrious Royal Society... London: Published by Richard Waller, 1705. xxviii, 572, [11] p. illus. f°. **† OAD**

Preceded by a biographical sketch in which reference is made to author's experiments with models (p. iv) and to his observations on animal flight (p. xix). Possibility of flight is discussed on p. 56–57.

4869. **Hopkins,** Albert Allis. Leonardo da Vinci as aviation engineer. The man who first suggested the helicopter and the motor to drive flying machines. illus. (Scientific American monthly, New York, 1921, v. 3, p. 355–359.) **VA**

4870. **Hunter,** H. Chadwick. Samuel Pierpont Langley. port. (Fly, Philadelphia, 1909, v. 1, Sept., p. 3–4.) **3–† VDS**

4871. **Hureau de Villeneuve,** Abel. Babinet. port. (L'Aéronaute, Paris, 1872, année 5, p. 181–189.) **3–VDS**

Brief account of Jacques Babinet (1794–1872), French physicist and writer on aeronautics.

4872. —— Léonard de Vinci, aviateur. illus. (L'Aéronaute, Paris, 1874, année 7, p. 259–274.) **3–VDS**

4873. The **Influence** of aviation upon mathematical physics. (Scientific American supplement, New York, 1919, v. 87, p. 341.) **VA**

4874. **Jacob,** Emil. Der Luftwiderstand. (Zeitschrift für Luftschiffahrt, Berlin, 1895, Jahrg. 14, p. 143–156.) **3–VDS**

For a list of Jacob's writings on the theory of flight see Paul Brockett, *Bibliography of aeronautics,* Washington, 1910, p. 445–446, *VDT.*

4875. **Jarolimek,** Anton. Ueber den Einfluss fluchtuirender Windströmungen und regelmässiger Schwingungen auf die Grösse des Luftwiderstandes mit Beziehung auf den Bau des Vogelfluges. illus. (Oesterreichischer Ingenieur- und Architekten-Verein, Zeitschrift, Wien, 1894, Jahrg. 46, p. 296–301.) **†† VDA**

4876. —— Ueber die Möglichkeit des dynamischen Fluges mit Beziehung auf die Versuche Lilienthal's. (Zeitschrift für Luftschiffahrt, Berlin, 1892, Jahrg. 11, p. 145–159.) **3–VDS**

4877. —— Ueber das Problem dynamischer Flugmaschinen. illus. (Oesterreichischer Ingenieur- und Architekten-Verein, Zeitschrift, Wien, 1893, Jahrg. 45, p. 417–422, 429–436.) **3–†† VDA**

A discussion of the Wellner, Loessl, and Lilienthal formulas for heavier-than-air flight.

For a list of Jarolimek's writings on flight theory see Paul Brockett, *Bibliography of aeronautics,* Washington, 1910, p. 449–450, *VDT.*

Aeronautics and Science, continued

4878. Jeffries, John. A narrative of two aerial voyages of Doctor Jeffries with Mons. Blanchard; with meteorological observations and remarks... London: the author, 1786. 7–60 p., 1 pl. 4°. **VDT**

4879. Jobert, Claude. De l'adhérence de l'air en mouvement aux surfaces polies et de non adhérence aux surfaces rugueuses. Conséquences à en tirer pour la navigation aérienne. (L'Aéronaute, Paris, 1890, année 23, p. 194–199.) **3–VDS**

4880. Johnson, Harry Miles. Résumé of research in the psychology of aviation during the year 1919. (Science, New York, 1920, v. 51, p. 449–452.) **OA**

Results of tests in the United States Air Service Medical Research Laboratory on the effects of oxygen, and on general aptitude.

4881. Jones, Thomas P. "Upward force of fluids." Remarks on Mr. Genet's "Memorial on the upward forces of fluids," and also of Dr. F. Pascalis's communication on the above subject, published in Silliman's journal [American journal of science]. (Franklin journal, Philadelphia, 1827, v. 3, p. 33–37.) **VA**

4882. Jouguet, Émile. Indications historiques sur l'application des lois de la similitude mécanique au mouvement des fluides. (La Technique aéronautique, Paris, 1924, année 15, p. 70–78.) **3–VDS**

Numerous references to authorities.

4883. Joukovsky, N. De la chute dans l'air de corps légers de forme allongée, animés d'un mouvement rotatoire. illus. (Institut aérodynamique de Koutchino, Bulletin, St. Pétersbourg, 1906, fasc. 1, p. 51–65.) **VDS**

4884. Journal of the aeronautics ascension of MM. Barral and Bixio, made at Paris 27th July, 1850. (Franklin Institute, Journal, Philadelphia, 1851, v. 51, p. 34–39.) **VA**

4885. Kargl, Ludwig. Zur Mechanik des Fluges. illus. (Der Civilingenieur, Leipzig, 1870, N. F. Bd. 6, col. 452–468.) **VDA**

Author's calculations for heavier-than-air craft lead him to believe that the steam engine, so far as developed, is not suitable.

4886. Kassner, Carl Julius Arminius. Die Anfänge der Aerologie. (Illustrierte aëronautische Mitteilungen, Berlin, 1909, Jahrg. 13, p. 632–634.) **VDS**

Chronology of the study of the upper air, 1749–1885.

4887. Kimball, James Henry. Trans-Atlantic flights of scientific value. maps. (Aero digest, New York, 1930, v. 17, Nov., p. 40–42.) **VDS**

4888. —— Trans-Atlantic weather. port., map. (Aero digest, New York, 1930, v. 17, Oct., p. 35–36, 144.) **VDS**

Reference to the care taken by noted transatlantic fliers to avoid bad weather.

4889. Kircher, Athanasius. Ars magna lucis et umbrae, editio altera, priori multo auctior. Amstelodami, 1671. 810 p. f°. **†† PEB**

For picture and description of his flying machine see p. 724. For comments by Gaspar Schott see entry no. 5028.

4890. Klein, Felix. Aeronautics at Göttingen University. (Scientific American supplement, New York, 1909, v. 67, p. 287.) **VA**

Outlines program.

4891. Klemin, Alexander. Aerodynamics. (Mechanical engineering, New York, 1929, v. 51, p. 8–9.) **VFA**

Developments during 1928.

For a list of Klemin's writings see United States. — National Advisory Committee for Aeronautics, *Bibliography of aeronautics,* 1909–1931, Washington, 1921–35, volumes for 1909/16, p. 722; 1917/19, p. 244-245; 1920/21, p. 227–228; 1923, p. 68; 1924, p. 54; 1925, p. 89; 1926, p. 73; 1927, p. 91; 1928, p. 103; 1929, p. 117–118; 1930, p. 126; 1931, p. 150, *VDT.*

4892. Köppen, Wladimir Peter. Beiträge zur Mechanik des Fluges und schwebenden Falles. diagrs. (Illustrierte aëronautische Mittheilungen, Strassburg, 1901, Jahrg. 5, p. 149–159.) **VDS**

4893. Koppe, Heinrich. Verfahren zur Messung der Geschwindigkeitsleistung von Luftfahrzeugen. illus. (Zeitschrift für Flugtechnik und Motorluftschiffahrt, München, 1923, Jahrg. 14, p. 17–25.) **VDS**

Has 42 footnote citations.

4894. Koppen, Otto C. Progress of research and invention. (Aviation, New York, 1931, v. 30, p. 39.) **VDS**

Brief account of work in the United States during 1930.

4895. Kousnetzow, V. Premiers lancers de ballons-sondes à l'Institut aérodynamique de Koutchino. illus. (Institut aérodynamique de Koutchino, Bulletin, St. Pétersbourg, 1906, fasc. 1, p. 67–86.) **VDS**

4896. Kremser, Victor Felix Karl. The study of the upper atmosphere by means of balloons. (United States. — Weather Bureau, Bulletin no. 11, Washington, 1896, part 3, p. 725–733.) **PRS (U. S.)**

Much of historical interest.

4897. Kress, Wilhelm. Wind und Wellenflug. illus. (Zeitschrift für Luftschiffahrt, Berlin, 1887, Jahrg. 6, p. 232–239.) **3–VDS**

States that the solution of dynamic flight lies in the speed of the plane.

4898. Krouchkoff, Michel. Un mémoire provenant du recueil de M. Dubois-Reymond. illus. (L'Aéronaute, Paris, 1879, année 12, p. 67–72.) **3–VDS**

Illustration of Hans Strasser's apparatus for studying the action of human muscles as power for flight.

4899. Labadié, Jean. Dans la stratosphere, à 16,000 mètres. illus. (Science et la vie, Paris, 1931, tome 40, p. 89–96.) **VA**

Account of Auguste Piccard's equipment for studying cosmic rays during his flight into the stratosphere in 1931.

Aeronautics and Science, continued

4900. Lancaster, I. Observations in aeronautics. [Fairhope, Ala., 1907?] 1 l. nar. 4°. **VDW p.v.1, no.5**

Discusses the relation of gravity to air pressure.

4901. —— Transformation of gravity. [Fairhope, Ala., 1907.] 1 l. 8°. **VDW p.v.2, no.4**

4902. —— Travel in air. diagrs. (Engineer, London, 1901, v. 92, p. 623–624.) **VA**

Discussion of forces involved in bird flight and heavier-than-air machines.

4903. —— Travel in air. [Fairhope, Ala., 1907.] 2 l. nar. 4°. **VDW p.v.1, no.4**

Author's explanation of soaring flight. Not same paper as the preceding.

4904. Lanchester, Frederick William. Aerodynamics, constituting the first volume of a complete work on aerial flight. [Aeronautics, constituting the second volume]... London: A. Constable & Co., Ltd., 1907–08. 2 v. illus. 8°. **VDW**

For a criticism by Herbert Chatley see *Engineering,* London, 1909, v. 88, p. 148–149, *VDA.*

For a list of Lanchester's writings see United States. — National Advisory Committee for Aeronautics, *Bibliography of aeronautics,* 1909–1931, Washington, 1921–35, volumes for 1909/16, p. 751–753; 1917/19, p. 251–252; 1920/21, p. 233–234; 1924, p. 56; 1925, p. 92; 1926, p. 76, *VDT.*

4905. Langley, Samuel Pierpont. Experiments in aerodynamics. Washington: Smithsonian Institution, 1902. ii, 115 p. illus. 2. ed. f°. (Smithsonian contributions to knowledge. no. 801.) **† PCP**

4906. —— The internal work of the wind. Washington: Smithsonian Institution, 1893. iii, 23 p. illus. f°. (Smithsonian contributions to knowledge. no. 884.) *** EA**

Also in *American journal of science,* New Haven, 1894, series 3, v. 47, p. 41–63, *OA.*

4907. —— Langley memoir on mechanical flight. Part 1. 1887 to 1896, by S. P. Langley ... Part 2. 1897 to 1903, by C. M. Manly. Washington: Smithsonian Institution, 1911. x p., 1 l., 320 p., 104 pl. f°. (Smithsonian contributions to knowledge. v. 27, no. 3.) **†† VDY**

Reviewed in *Scientific American supplement,* New York, 1912, v. 74, p. 95–96, *VA.*

4908. —— Locomotion aérienne. Recherches expérimentales aérodynamiques et données d'expériences. (Académie des sciences, Comptes rendus, Paris, 1891, tome 113, p. 59–63.) *** EO**

Translation in *Nature,* London, 1891, v. 44, p. 277–279, *OA; Scientific American supplement,* New York, 1891, v. 32, p. 13050–13051, *VA.* For comments see *L'Aéronaute,* Paris, 1891, année 24, p. 243–247, *3–VDS.*

4909. —— Researches and experiments in aerial navigation. Washington: Government Prtg. Off., 1908. v. p. illus. 8°. **VDW p.v.2, no.10**

Reprints of Langley's papers in the *Smithsonian reports* for 1897, 1900, 1901, and 1904, with a bibliography of Langley's writings on aeronautics.

For other lists of writings by and about Langley see Paul Brockett, *Bibliography of aeronautics,* Washington, 1910, p. 492–496, *VDT;* continued in United States. — National Advisory Committee for Aeronautics, *Bibliography of aeronautics,* 1909–1931, Washington, 1921–35, volumes for 1909/16, p. 758–760; 1917/19, p. 254; 1920/21, p. 235; 1922, p. 82; 1925, p. 93; 1926, p. 76; 1927, p. 95; 1928, p. 107; 1930, p. 132, *VDT.*

4910. The **Late** Mr. James Glaisher. (Aeronautical journal, London, 1903, v. 7, p. 22–24.) **VDS (Royal)**

Glaisher (1809–1903) is famous for his pioneer work in aeronautical meteorology. His experiences are set forth in his well-known book *Travels in the air.*

4911. Launay, L. de. Un précurseur en aviation: Léonard de Vinci. illus. (La Nature, Paris, 1909, année 37, semestre 1, p. 402–404.) **OA**

History of Da Vinci's manuscripts on aviation, with reproductions of some of the drawings.

4912. La Vaulx, Henri, comte de. L'aérostation scientifique. (In: L'Aéronautique, par Roland Garros [and others], Paris: Pierre Lafitte & Cie. [1914], p. 47–54.) **VDW (Aéronautique)**

Brief account of service of the balloon to science.

4913. Lavoisier, Antoine Laurent. Extraits des mémoires de Lavoisier concernant la météorologie et l'aéronautique, publiés par les soins de l'Office national météorologique de France. Paris: E. Chiron [1917]. 229 p. incl. tables, front. (port.) 8°. **PRD**

4914. Lecornu, Joseph. Considérations sur l'aérostation au point de vue des études astronomiques. (Revue scientifique, Paris, 1883, tome 31 [série 3, tome 5], p. 532–534.) **OA**

4915. Le Dantec, Louis Marie. Expériences faites dans la chapelle des Arts et Métiers, en septembre 1898. illus. (Société d'encouragement pour l'industrie nationale, Bulletin, Paris, 1899, série 5, tome 4, p. 1024–1037.) **3–VA**

Experiments on air resistance.

4916. Ledeboer, John Henry. The greatest pioneer. (Aeronautics, London, 1911, v. 4, p. 54.) **VDS**

Appreciation of the pioneer work of Alphonse Pénaud (1851–1880), "the father of aviation in the true sense of the word, for he was the first to combine theory and practice, thus becoming the first aeronautical engineer."

4917. Lepère, G. Recherches expérimentales d'aéronautique aux États-Unis. illus. (La Technique aéronautique, Paris, 1924, année 15, p. 210–225, 246–260.) **3–VDS**

Illustrated review of researches in the United States during 1923.

4918. Lippert, P. W. Ueber Arbeitsökonomie in der Flugbewegung. illus. (Zeitschrift für Luftschiffahrt, Berlin, 1888, Jahrg. 7, p. 11–17, 40–49, 79–85, 114–119, 148–155, 186–187.) **3–VDS**

Develops his theory of flight.

Aeronautics and Science, continued

4919. **Loessl,** F., Ritter von. Aërodynamische Betrachtung über das Verhalten einer in wagrechter Stellung durch die Luft fallenden dünnen Platte. illus. (Zeitschrift für Luftschiffahrt, Berlin, 1900, Jahrg. 19, p. 222–227, 238–242, 265–279.) 3–VDS

Study of thin plates in horizontal position falling through the air.

4920. —— Der aërodynamische Schwebezustand eine dünnen Platte und deren Sinkgeschwindigkeit... illus. (Oesterreichischer Ingenieur- und Architekten-Verein, Zeitschrift, Wien, 1898, Jahrg. 50, p. 454–456, 471–475, 481–485.) 3–† VDA

Also in *Zeitschrift für Luftschiffahrt,* Berlin, 1899, Jahrg. 18, p. 10–20, 25–31, 58–68, 91–95, *3–VDS.*

4921. —— Biographische Notizen über meine aërodynamischen Studien und Erlebnisse. port. (Wiener Luftschiffer-Zeitung, Wien, 1904, Jahrg. 3, p. 3–8.) VDS

4922. —— Einige Beobachtungen über das (mechanische) Verhalten der Luft gegen bewegte Körper. illus. (Zeitschrift für Luftschiffahrt, Berlin, 1888, Jahrg. 7, p. 129–136, 169–178.) 3–VDS

With discussion.
A study of air pressure against moving bodies.

4923. —— Erwiderung auf die kritischen Bemerkungen des Herrn Ingenieurs J. Popper über die Loessl'sche Formel der Sinkgeschwindigkeit einer in der Luft schwebenden dünnen Platte. (Oesterreichischer Ingenieur- und Architekten-Verein, Zeitschrift, Wien, 1899, Jahrg. 51, p. 264–267, 284–288, 506–509.) 3–† VDA

4924. —— Der Luftwiderstand in Allgemeinen und in seiner besondern Beziehung auf Luftschiffahrt. illus. (Zeitschrift für Luftschiffahrt, Berlin, 1886, Jahrg. 5, p. 161–173, 193–202.) 3–VDS

4925. —— Ein neues experimentelles Verfahren sur Messung von Luftwiderstand. illus. (Oesterreichischer Ingenieur- und Architekten-Verein. Zeitschrift, Wien, 1894, 3–† VDA

4926. —— Nouveau procédé expérimental pour la mesure de la résistance de l'air. (L'Aéronaute, Paris, 1894, année 27, p. 77–90.) 3–VDS

4927. —— Wiederholte Erläuterung des Schwebefluges. (Illustrierte aëronautische Mittheilungen, Strassburg, 1903, Jahrg. 7, p. 250–260.) VDS

Presents the laws of hovering flight.

4928. —— Winddruckmesser. illus. (Illustrierte aëronautische Mittheilungen, Strassburg, 1903, Jahrg. 7, p. 141–162.) VDS

Illustrated description of author's invention.
For citations to writings of von Loessl see Paul Brockett, *Bibliography of aeronautics,* Washington, 1910, p. 526–529, *VDT.*

4929. **Louis** Paul Cailletet. port. (Wiener Luftschiffer-Zeitung, Wien, 1906, Jahrg. 5, p. 96.) VDS

Portrait and brief account of the work of French physicist who was a pioneer in the use of sounding balloons for the study of the upper air.

4930. **Lyons,** Sir Henry George. Meteorology during and after the war. (Royal Society of Arts, Journal, London, 1919, v. 67, p. 167–180.) VA

With discussion.
How meteorology and aeronautics mutually helped each other. Mostly reprinted in *Nature,* London, 1919, v. 103, p. 12–16, *OA.* Abridged in *Monthly weather review,* Washington, 1920, v. 47, p. 81–83, † *PRS.*

4931. **McAdie,** Alexander George. The attainment of high levels in the atmosphere. (Science, New York, 1920, v. 51, p. 287–289.) OA

Account of Major Schroeder's ascent, with brief reference to some of the earlier exploits, beginning with Dr. John Jeffries in 1785.

4932. —— The discovery of the stratosphere. port. (American Meteorological Society, Bulletin, Worcester, Mass., 1934, v. 15, p. 174–177.) PRA

History of the work of Léon Teisserenc de Bort, with a portrait.

4933. **McCurdy,** Edward. Leonardo da Vinci and the science of flight. (Nineteenth century and after, London, 1910, v. 68, p. 126–142.) * DA

4934. **Mach,** Ludwig. Ueber die Sichtbarmachung von Luftstromlinien. illus. (Zeitschrift für Luftschiffahrt, Berlin, 1896, Jahrg. 15, p. 129–139.) 3–VDS

Author's method of photographing stream-lines. Illustrations are between p. 60 and 61.

4935. **Makower,** Walter. Note relating to experiments carried out at Glossop Observatory. (Great Britain. — Advisory Committee for Aeronautics. Report for 1909–1910, London, 1910, p. 127–128.) VDW

Deals with the electrification of balloons.

4936. **Maleire,** E. Le problème de l'aviation et la résistance de l'air. illus. diagrs. (Le Génie civil, Paris, 1907, tome 51, p. 245–248, 264–267.) VA

Describes the experiments of Canovetti.

4937. **Mangon,** Hervé, and Léon Durand-Claye. Expériences sur la résistance au mouvement des ballons dans l'air. illus. (Société d'encouragement pour l'industrie nationale, Bulletin, Paris, 1872, année 71 [série 2, tome 19], p. 17–22.) 3–VA

Experiments during the siege of Paris. See also *L'Aéronaute,* Paris, 1878, année 11, p. 184–190, *3–VDS.*

4938. **Mannesmann,** O. Luftwiderstandmessungen mit einem neuen Rotationsapparat. illus. (Annalen der Physik und Chemie, Leipzig, 1899, N. F. Bd. 67, p. 105–131.) PAA

Air resistance studied with a rotation apparatus. References to previous investigators. See also *Dingler's polytechnisches Journal,* Stuttgart, 1899, Bd. 311, p. 147–148, *3–VA.*

Aeronautics and Science, continued

4939. **Marguet,** Frédéric Philippe. L'inventeur de la direction aérienne. (Revue de Paris, Paris, 1912, v. 6, p. 309–334.) * DM
Story of the experiments of Pierre Louis Mouillard.

4940. **Marriott,** William. The balloon ascents made by James Glaisher, F.R.S., for scientific purposes, 1862–69. port. (Aeronautical journal, London, 1904, v. 8, p. 19–23.)
Frontispiece portrait. VDS (Royal)

4941. —— James Glaisher, F.R.S., 1809–1903. port. (Royal Meteorological Society, Quarterly journal, London, 1903, v. 29, p. 115–120.) PRA

4942. **Marsh,** William Lockwood. Aeronautical experiments. Wind tunnel tests have solved many of the most baffling problems known to aeronautical science. illus. (Air and airways, London, 1932, v. 9, p. 173–175.)
History of wind tunnels. VDS

4943. **Martienssen,** Oscar. Theoretische Grundlagen für die Construction eines Schraubenfliegers. diagrs. (Illustrierte aëronautische Mittheilungen, Strassburg, 1902, Jahrg. 6, p. 125–133.) VDS

4944. The **M**[**assachusetts**] I[nstitute of] [Technology] aeronautical course. The pioneer American course in aeronautical engineering. ports. (Aviation, Highland, N. Y., 1925, v. 19, p. 42–43.) VDS
Established in 1913 under the directorship of Jerome C. Hunsaker.

4945. **Mendeleyeyv,** D. O soprotivlyeniyi szidkostyey i o vozdukhoplavaniyi. St. Petersburg: V. Demakova, 1880. 160, 80 p. illus. 8°. VDY p.v.3, no.5
A study of liquid and air resistance.

4946. **Miller,** William H. Research and the airplane. illus. (Aero digest, New York, 1928, v. 13, p. 1128, 1130, 1132; 1929, v. 14, Jan., p. 50, 52, 174–176.) VDS
Account of investigations from the time of Sir George Cayley.

4947. **Millikan,** Robert Andrews. Some scientific aspects of the meteorological work of the United States army. diagrs. (American Philosophical Society, Proceedings, Philadelphia, 1919, v. 58, p. 133–149.) * EA

4948. **Mr.** [J. M.] Bacon's scientific ascents. (Aeronautical journal, London, 1898, v. 2, p. 78–80.) VDS
Ascents of 1898.

4949. **Moedebeck,** Hermann W. L. Anleitung zum wissenschaftlichen Ballonfahren auf Grund der Renard'schen Prinzipien. (Zeitschrift für Luftschiffahrt, Berlin, 1886, Jahrg. 5, p. 313–322, 363–366.) 3–VDS
Discusses the adaptability of balloons to scientific investigations.

4950. —— Die Benutzung verschiedener Luftströmungen. (Zeitschrift für Luftschiffahrt, Berlin, 1884, Jahrg. 3, p. 353–366.) 3–VDS
A history of improvements to render balloons adaptable for meteorological purposes.

4951. —— Unser Hochfahrer. ports. (Illustrierte aëronautische Mittheilungen, Strassburg, 1901, Jahrg. 5, p. 119–120.) VDS
Portraits and brief accounts of A. Berson and R. Süring, German meteorological aeronauts, who made record for altitude.

4952. **M.** Claude Jobert. (L'Aéronautique, Paris, 1904, année 37, p. 19–20.) 3–VDS
Chronological account of Jobert's aeronautical inventions and investigations.

4953. **Moore,** R. F. Report on experiments to ascertain the power and means necessary for flight with wings. illus. (Aeronautical journal, London, 1897, v. 1, Oct., p. 2–17.)
Includes an airplane design of the author. VDS

4954. **Mora,** Alfred. Aérolocomotion & aéro-automobiles; étude. Paris: F. Bouchy & Cie., 1901. 47 p. 8°. VDY p.v.7, no.7
A study of balloon, airship, and airplane principles, with a description of the Roze combined airship and airplane.

4955. **Morin,** Arthur Jules. Note sur un moyen de déterminer la loi du mouvement d'ascension et de transport horizontal des ballons. (Académie des sciences, Comptes rendus, Paris, 1868, tome 67, p. 635–639.) * EO

4956. **Mort** de L. P. Cailletet. port. (L'Aérophile, Paris, 1913, année 21, p. 40–44.) 3–VDS
Brief biography and appreciation of French pioneer in the use of balloons for meteorological purposes, for which he invented several recording instruments.

4957. **Mort** de M. Léon Teisserenc de Bort. (L'Aérophile, Paris, 1913, année 21, p. 44–45.)
Biography of French meteorologist. 3–VDS

4958. **Mott,** S. D. Electro-aerostatics. Proposed meteorological machines. Raising self-registering instruments into the higher atmosphere by orthogonal flight. illus. (Electrical world, New York, 1896, v. 27, p. 737–739.) VGA
Light-weight electrically-operated planes suggested by the author.

4959. **Myers,** Carl E. Balloon meteorology. (International Conference on Aerial Navigation, Chicago, 1893, Proceedings, 1894, p. 359–361.) VDS
Author's experiences in exploding balloons to produce rainfall.

4960. **Navier,** Claude Louis Marie Henri. Sur l'évaluation approximative de la quantité d'action nécessaire pour le vol des oiseaux, et pour la direction des aérostats. (Académie royal des sciences, Mémoires, Paris, 1832, tome 11, p. lxxvi–cxviii.) * EO
Famous early calculations.

4961. **Nayler,** J. L. High flights above the earth. Some scientific aspects of the recent height record and a survey of the problems confronting the designer of high altitude aircraft. illus. (Air and airways, London, 1932, v. 9, p. 269–270.) VDS
Flight of C. F. Unwins, 1932, to 43,976 feet.

Aeronautics and Science, continued

4962. **Neue Augsburger Zeitung.** Professor Piccards Forschungsflug in die Stratosphäre; Verlauf des Stratosphärenfluges und dessen wissenschaftliches Ergebnis. Mit Beiträgen von Professor Dr. A. Piccard, Ingenieur P. Kipfer und anderen Sachverständigen... Augsburg: Literar. Institut von Haas & Grabherr, 1931. 128 p. illus. 8°. **VDW**

Detailed account of famous free-balloon ascent of 16,000 meters, May, 1931. Biographies of Piccard and Kipfer.

4963. **New** system of schedules for broadcast of airways weather information now in effect. map. (United States. — Commerce Department: Aeronautics Branch, Air commerce bulletin, Washington, 1932, v. 3, p. 438–440.) **VDS**

4964. **Nimfuehr,** Raimund. Flugtechnische Betrachtungen. (Zeitschrift für Luftschiffahrt, Berlin, 1899, Jahrg. 18, p. 266–275, 293–299; 1900, Jahrg. 19, p. 14–20.) **3–VDS**

Discussion of the theories of Buttenstedt, Kreiss, von Loessl, Jacob, and Platte.

4965. **Pannell,** John Robert. The wind channel: its design and use. illus. (Aeronautical journal, London, 1918, v. 22, p. 211–248.) **VDS**

A history of installations in various countries, with a discussion. Also, without discussion and the illustrations, in *Aviation,* New York, 1918, v. 4, p. 91–95, 160–163, *VDS.*

4966. **Parkin,** J. H. The new aerodynamic laboratory of the University of Toronto. illus. (Engineering journal, Montreal, 1927, v. 10, p. 390–399.) **VDA**

A description of the laboratory with the results of air flow studies and power consumption tests.

4967. **Pascalis,** Felix. Genet's memorial. Fifth experiment, or application of the upwards aerostatic forces to aeronautics. illus. (American journal of science, New Haven, 1826, v. 11, p. 346–348.) **OA**

Illustration and description of E. C. Genet's horse-driven airship.

4968. **Pawlowski,** Felix Wladyslaw. Aeronautical research. (International Civil Aeronautics Conference, 1928, Papers, Washington, 1928, p. 435–443.) **VDS**

Brief references to noted workers in aerodynamical research.

4969. **Pénaud,** Alphonse. Aviation. Appareils de vol mécanique. illus. (Journal de physique, Paris, 1875, tome 4, p. 289–296.) **PAA**

The author's experiments with model helicopters and model airplanes. Translated in *Popular science monthly,* New York, 1876, v. 8, p. 453–460, * *DA;* also in Franklin Institute, *Journal,* Philadelphia, 1876, v. 101, p. 193–201, *VA.*

4970. **Peppler,** Albert Gustav. Richard Assmann. port. (Deutsche Luftfahrer-Zeitschrift, Berlin, 1918, Jahrg. 22, No. 11/12, p. 14–16.) **VDS**

Biography of distinguished German aerologist and writer.

4971. **Pérez de Guzmán,** Juan. El primer ensayo científico de aerostación en España. (España moderna, Madrid, 1910, v. 259, July, p. 58–75.) *** DR**

Ascension of Domingo Badía y Leblich, 1795.

4972. **Petliakov,** V. M. Aviation progress under the Soviet. A description of the work of the Russian Central Aero-Hydro-Dynamic Institute, with some of its practical results. illus. map. (Aviation, New York, 1930, v. 28, p. 108–112.) **VDS**

4973. **Phillips** Flugversuche. illus. (Illustrierte aëronautische Mitteilungen, Berlin, 1908, Jahrg. 12, p. 429–431.) **VDS**

Experiments with models and machines, 1885–1907.

4974. **Philos.** A. Goupil. port. (L'Aérophile, Paris, 1909, année 17, p. 231.) **3–VDS**

Brief biography of French writer and investigator whose aeronautical career began with the Franco-Prussian war.

4975. —— Louis Capazza. port. (L'Aérophile, Paris, 1908, année 16, p. 181.) **3–VDS**

Portrait and brief biography of Italian aeronaut who did much for the advancement of the study of aerial navigation and of the upper atmosphere. With Adolphe Clément he built the famous French airship *Bayard-Clément.*

4976. **Piccard,** August. A 16,000 metri; i miei viaggi nella stratosfera. Prefazione di Italo Balbo. Milano: A. Mondadori [cop. 1933]. xx, 311 p. illus. 8°. **VDW**

Author's ascent to the stratosphere, 1932.

4977. **Pichault,** S. Aéronautique. Réflexions et calculs à propos de la navigation aérienne sans ballons. diagrs. (Annales industrielles, Paris, 1873, année 5, tome 2, col. 615–625, 658–666, 679–686, 720–727, 747–755; 782–791; 1874, année 6, tome 1, col. 48–55, 109–113, 171—175, 240–243, 298–302, 370–374, 437–442, 538–542, 600–606, 631–635, 666–670.) **3–† VA**

Investigations show that heavier-than-air navigation is impossible until lighter and more powerful motors are available. The first installment contains a bibliographical account of work by the author's predecessors.

4978. **Pierre** Jules César Janssen. port. (Wiener Luftschiffer-Zeitung, Wien, 1908, Jahrg. 7, p. 4–5.) **VDS**

Portrait and biography of French astro-physicist, an early user of balloons in scientific investigations.

4979. **Platte,** August. Die Flugtechniker und die Mechanik. (Zeitschrift für Luftschiffahrt, Berlin, 1893, Jahrg. 12, p. 10–20.)

Author criticizes ideas of Lilienthal. **3–VDS**

4980. —— Professor Miller v. Hauenfels. (Zeitschrift für Luftschiffahrt, Berlin, 1898, Jahrg. 17, p. 1–2.) **3–VDS**

Brief biography of German scientist (1818–1898) who developed a theory of gliding flight.

4981. —— Ueber die Berechnung der Leistungsfähigkeit von Luftschiffen. (Zeitschrift für Luftschiffahrt, Berlin, 1885, Jahrg. 4, p. 177–182, 203–210.) **3–VDS**

Exposition of von Loessl's formulae for determining lifting power of an airship.

For citations to Platte's writings on aeronautical theory see Paul Brockett, *Bibliography of aeronautics,* Washington, 1910, p. 675–678, *VDT.*

Aeronautics and Science, continued

4982. Pomortzev, Mikhail Mikhailovich. Vozdukhoplavaniye i izsledovaniye atmosfery. Pod redaktziei predsedatelya otdela M. M. Pomortzev. vyp 1–4, 7–9. St. Petersburg: Tip. Imp. Ak. Nauk, 1897–1905. 3 vol. (7 vozdukhoplavatel'nyi otdel Imper. Rus. Tekh. Obshchest.) **VDY p.v.4, no.1–7**

Treats of aeronautics and its relation to the study of the upper atmosphere.

4983. Popper, Josef. Kritische Bemerkungen zu der Abhandlung des Herrn Ober-Ingenieurs F. R. v. Loessl: "Der aërodynamische Schwebezustand einer dünnen Platte und deren Sinkgeschwindigkeit... illus. (Oesterreichischer Ingenieur- und Architekten-Verein, Zeitschrift, Wien, 1899, Jahrg. 51, p. 51–57, 73–76.) **3–† VDA**

Also in *Zeitschrift für Luftschiffahrt*, Berlin, 1899, Jahrg. 18, p. 86–91, 108–115, 133–145, *3–VDS*.

4984. —— Ueber die Quelle und den Betrag der durch Luftballons geleisten Arbeit. diagrs. (Akademie der Wissenschaften, Sitzungsberichte, Mathematisch-naturwissenschaftliche Classe, Wien, 1875, Bd. 71, Abt. 2, p. 809–845.) *** EF**

Develops equations for the physics of balloon ascensions.

4985. Porta, Giovanni Battista della. Joh. Baptistæ Portæ Neapolitani Magiæ naturalis libri viginti... Lugd. Batavorum: Ap. H. de Vogel, 1644. 670 p. illus. 24°. **PKD**

On the last page the author describes a dragon kite and prophesies that man may learn to fly by attaching wings to his body. Library has a later edition, Leyden, 1651, *YMD;* also an English translation entitled *Natural magick,* London, 1658, *PKD.*

4986. Poulter, R. M. Ballooning. (Meteorological magazine, London, 1934, v. 69, p. 35–37.) **PRA**

Brief account of Glaisher's ascents.

4987. Prandtl, Ludwig. Aerodynamic and aerostatic studies. Experiments with models of airships and aeroplanes. illus. (Scientific American supplement, New York, 1909, v. 68, p. 293–294.) **VA**

Brief account of experiments at the Goettingen laboratory.

4988. —— Die Bedeutung von Modellversuchen für die Luftschiffahrt und Flugtechnik und die Einrichtungen für solche Versuche in Göttingen. illus. (Verein deutscher Ingenieure, Zeitschrift, Berlin, 1909, Bd. 53, p. 1711–1719.) **3–VDA**

For a list of references to Prandtl's writings see United States. — National Advisory Committee for Aeronautics, *Bibliography of aeronautics,* 1909–1931, Washington, 1921–35, volumes for 1909/16, p. 1062; 1917/19, p. 360; 1920/21, p. 331–332; 1922, p. 126; 1923, p. 109; 1924, p. 88; 1925, p. 139–140; 1926, p. 116; 1927, p. 141–142; 1928, p. 159; 1929, p. 181; 1930, p. 194; 1931, p. 226, *VDT.*

4989. Primbault, H. La navigation aérienne. (La Revue générale, Bruxelles, 1902, année 38, v. 76, p. 102–126, 164–182.) *** DM**

Problems of the airplane and airship. Discussion of kites.

4990. Le **Professeur** Piccard atteint en ballon l'altitude de 16,000 mètres. illus. ports. (La Nature, Paris, 1931, année 59, p. 570–571.) **OA**

Remarkable ascension of Piccard and Kipfer in 1931.

4991. Professor Zahm's experiments. (Aeronautical annual, Boston, 1897, no. 3, p. 148–149.) **3–VDS**

Experiments to determine resistance of air for speeds upwards of one hundred feet a second.

4992. Radio weather maps. port. (Aero digest, New York, 1927, v. 11, p. 220.) **VDS**

Account of C. Francis Jenkins' device for sending daily weather maps by radio.

4993. Ramsey, Logan Carlisle, and E. H. Kincaid. An analysis of recent trans-Atlantic flights. maps. (United States Naval Institute, Proceedings, Menasha, Wis., 1927, v. 53, p. 986–999.) **VXA**

A discussion of the weather charts of the flights of Nungesser and Coli, Lindbergh, Chamberlin, and Byrd. Also in *Aero digest,* New York, 1927, v. 11, p. 402–405, *VDS.* For discussion by C. D. Chamberlin see entry no. 4754.

4994. Randers-Pehrson, Nils Hendrik. Pioneer wind tunnels. Washington: Smithsonian Institution, 1935. 20 p. illus. 8°. (Smithsonian miscellaneous collections. v. 93. Publication no. 3294.) *** EA**

Work of John Smeaton, F. H. Wenham, Horatio Phillips, Ludwig Mach, Johann Irminger and H. C. Vogt, Charles Renard, Sir Hiram Maxim, Paul Lacour, Étienne Marey, A. F. Zahm, the Wright brothers, T. E. Stanton, G. A. Crocco, D. Riabouchinski, Ludwig Prandtl, A. Rateau, Gustave Eiffel, and the National Physical Laboratory.

4995. Rayleigh, John William Strutt, 3d baron. The mechanical principles of flight. 26 p. diagr. (Manchester Literary and Philosophical Society, Memoirs and proceedings, Manchester, 1899–1900, v. 44, paper no. 5.)

A second copy in *VDY p.v.27, no.6.* *** EC**

4996. Reed, Thomas R. Meteorological aspects of the San Francisco-Hawaii airplane flight. maps. (Monthly weather review, Washington, 1925, v. 53, p. 384–387.) **† PRS**

U. S. Navy flight, 1925.

4997. Reissner, H. Die Seitensteuerung der Flugmaschinen. illus. (Zeitschrift für Flugtechnik und Motorluftschiffahrt, Berlin, 1910, Jahrg. 1, p. 101–106, 117–123.) **VDS**

Calculations for airplane stability.

4998. Renard, Charles. Nouvelles expériences sur la résistance de l'air. (L'Aéronaute, Paris, 1889, année 22, p. 73–81.) **3–VDS**

4999. —— Recherches relatives à la résistance de l'air au moyen d'un nouvel appareil appelé balance dynamométrique. illus. (Académie des sciences, Comptes rendus, Paris, 1904, tome 138, p. 1201–1204.) *** EO**

5000. —— Résistance de l'air. Comparaison des résistances directes de divers carènes aériennes. Résultats numériques. (Aca-

Aeronautics and Science, continued

démie des sciences, Comptes rendus, Paris, 1904, tome 138, p. 1264–1266.) * EO

5001. —— Sur l'emploi des ballons perdus pour l'exécution des observations météorologiques à grandes hauteurs. illus. (Revue de l'aéronautique, Paris, 1893, année 6, p. 1–19.) 3–† VDS

Abstracted in Académie des sciences, *Comptes rendus,* Paris, 1892, tome 115, p. 1049–1053, * EO; also in *L'Aéronaute,* Paris, 1893, année 26, p. 5–9, *3–VDS.*

For citations to Renard's writings on the various phases of scientific aeronautics see Paul Brockett, *Bibliography of aeronautics,* Washington, 1910, p. 706–708, *VDT;* and United States. — National Advisory Committee for Aeronautics, *Bibliography of aeronautics,* 1909–1931, Washington, 1921–35, volume for 1909/16, p. 1122–1123, *VDT.*

5002. Renard, Paul. Conférences sur l'aviation. illus. (Société d'encouragement pour l'industrie nationale, Bulletin, Paris, 1909, année 108, p. 72–91, 300–350, 755–784, 1112–1182.) 3–VA

A series of papers on aerodynamical principles, covering airships, airplanes, helicopters, and propellers. Also published in book form: Paris: H. Dunod et E. Pinat, 1909. vii, 183 p. 4°. *VDY.*

5003. —— L'œuvre du colonel Charles Renard. (Société d'encouragement pour l'industrie nationale, Bulletin, Paris, 1905, tome 107, p. 1202–1229.) 3–VA

5004. —— L'œuvre de M. Léon Teisserenc de Bort. (La Technique aéronautique, Paris, 1913, tome 7, p. 97–110.) 3–VDS

An account of the work of famous French meteorologist.

5005. A **Review** of the work of the Guggenheim Fund. port. illus. (Aeronautic review, Washington, 1930, v. 8, Feb., p. 42–43.) VDS

5006. Rey, Robert. Un grand ancêtre, Léonard de Vinci. illus. port. (La Revue aërienne, Paris, 1912, année 5, p. 363–366.) 3–VDS

5007. Rosenhain, Walter, and Guy Barr. Report from the National Physical Laboratory on the determination of the rate of leakage or diffusion of hydrogen through balloon fabrics. illus. (Great Britain. — Advisory Committee for Aeronautics, Report for 1909–1910, London, 1910, p. 86–89.) VDW

5008. Rotch, Abbott Lawrence. The chief scientific uses of kites. (Aeronautical journal, London, 1901, v. 5, p. 56–59.) VDS

Has early history.

5009. —— The exploration of the free air by means of kites, at Blue Hill Observatory, Mass., U. S. A. illus. (Royal Meteorological Society, Quarterly journal, London, 1898, v. 24, p. 250–261.) PRA

Includes references to some of the early work with kites, beginning with that of Dr. Alexander Wilson, 1749.

5010. —— The first observations with "ballons-sonde" in America. (Aeronautical journal, London, 1905, v. 9, p. 51.) VDS

Experiments were conducted by S. P. Fergusson at St. Louis, Mo., 1905.

5011. —— On the first observations with registration balloons in America. (American Academy of Arts and Sciences, Proceedings, Boston, 1905/06, v. 41, p. 347–350.) * EA

5012. —— On obtaining meteorological records in the upper air by means of kites and balloons. (American Academy of Arts and Sciences, Proceedings, Boston, 1897, v. 32, p. 245–251.) * EA

History of various experiments in America and Europe.

5013. —— Progress in the exploration of the air with kites at the Blue Hill Observatory, Massachusetts. (Monthly weather review, Washington, 1898, v. 26, p. 355–356.) † PRS

Also in *Aeronautical journal,* London, 1899, v. 3, p. 17–19, *VDS.*

5014. —— The relation of the wind to aeronautics. (Aeronautical annual, Boston, 1896, no. 2, p. 105–110.) 3–VDS

Experiences at the Blue Hill Observatory.

5015. —— Sounding the ocean of air, being six lectures delivered before the Lowell Institute of Boston in December, 1898. London: Society for Promoting Christian Knowledge, 1900. viii, 184 p. illus. 16°. VDW

Ancient and modern knowledge of the atmosphere; notable balloon ascensions; and the history of kites. Reviewed in *Aeronautical journal,* London, 1901, v. 5, p. 29–31, *VDS.*

For lists of Rotch's writings, principally on the use of kites and balloons in meteorology, see Paul Brockett, *Bibliography of aeronautics,* Washington, 1910, p. 725–729, *VDT.* Continued in United States. — National Advisory Committee for Aeronautics, *Bibliography of aeronautics,* 1909–1931, Washington, 1921–35, volumes for 1909/16, p. 1152–1153; 1917/19, p. 385; 1920/21, p. 136, *VDT.*

5016. Roy, Maurice. A propos de recherches théoriques sur l'aérodynamique. (L'Aérophile, Paris, 1925, année 33, p. 368–370; 1926, année 34, p. 76–77.) 3–VDS

Reviews the researches of Dmitri Ryabushinski and C. Witoszynski.

5017. Ryabushinski, Dmitri. Études du courant aérien artificiel dans un tube destiné à des expériences sur la résistance de l'air. illus. (Institut aérodynamique de Koutchino, Bulletin, St. Pétersbourg, 1906, fasc. 1, p. 6–12.) VDS

Describes wind tunnel experiments of the author. See also *Nature,* London, 1907, v. 75, p. 609–610, *OA.*

5018. —— Recherches sur la pression que supportent des plaques rectangulaires en rotation dans un courant aérien. illus. (Institut aérodynamique de Koutchino, Bulletin, Moscou, 1909, fasc. 3, p. 30–45.) VDS

5019. —— Recherches sur la rotation de plaques symétriques dans un courant aérien et sur la détermination de la pression qu'elles supportent. illus. (Institut aérodynamique de Koutchino, Bulletin, St. Pétersbourg, 1906, fasc. 1, p. 18–50.) 3–VDS (Moskovski)

Aeronautics and Science, continued

5020. —— Spectres aérodynamiques. illus. (Institut aérodynamique de Koutchino, Bulletin, Moscou, 1909, fasc. 3, p. 59–66.) **VDS**

Registers air-currents around a solid body by means of imprint left by lycopodium powder.

For references to Ryabushinski's researches see United States. — National Advisory Committee for Aeronautics, *Bibliography of aeronautics*, 1909–1931, Washington, 1921–35, volumes for 1909/16, p. 1132–1134; 1920/21, p. 350; 1922, p. 115; 1926, p. 122; 1930, p. 203, *VDT*.

5021. Saint-Loup. Expériences sur la résistance opposée par l'air au mouvement d'une surface. illus. (L'Aéronaute, Paris, 1879, année 12, p. 255–267.) **3–VDS**

Author's experiments.

5022. Saladini, Girolamo. Riflessioni circa la memoria intorno la salita delle macchine aereostatiche nell' aria di Leonardo Eulero. (Società italiana delle scienze, Memorie, Modena, 1803, tomo 10, parte 1, p. 264–284.) *** ER**

Exposition and discussion of Euler's investigations of balloons.

5023. Samuelson, Arnold. Einige Gesetze des Widerstandes der Flüssigkeiten. illus. (Zeitschrift für Luftschiffahrt, Berlin, 1895, Jahrg. 14, p. 261–274, 291–303; 1896, Jahrg. 15, p. 3–9, 90–103.) **3–VDS**

5024. —— Zur physikalischen Grundlage des Fluges. illus. (Zeitschrift für Luftschiffahrt, Berlin, 1895, Jahrg. 14, p. 210–215.) **3–VDS**

For a list of Samuelson's writings on the theory of flight see Paul Brockett, *Bibliography of aeronautics*, Washington, 1910, p. 743–744, *VDT*.

5025. Sanson, A. J. Preuves sur preuves d'une nautique aérienne. Inductions logiques tirées de faits patents d'aérostation, de marine, d'ornithologie, d'ichthyologie, de physique, de statique, de dynamique... Paris: Ledoyen, 1857. 32, 16 p. illus. 8°. **VDY p.v.6, no.15**

5026. Sayers, W. H. The scientific progress of aviation. (Aeroplane, London, 1919, v. 17, p. 1309–1310.) **VDS**

Discussion of a paper by Leonard Bairstow. See entry no. 4680.

5027. Schellbach, K. H. Ueber einen Apparat zur Ermittelung der Gesetze des Luftwiderstandes. illus. (Annalen der Physik und Chemie, Leipzig, 1871, Bd. 143, p. 1–14.) **PAA**

Schellbach was the first to use a rotating device for the determination of air resistance.

5028. Schott, Gaspar. Magia universalis naturæ et artis, sive recondita naturalium et artificialium rerum scientia. Opus quadripartitum... Herbipoli, 1657–59. 4 v. 4°. **OAD**

See v. 3, p. 571–573 for comments on Kircher's ideas (see entry no. 4889).

5029. Scientific äerostation. (Mechanics' magazine, London, 1865, new series, v. 14, p. 32–34.) **3–† VA**

Brief references to work of Robertson and Lhoest, Biot and Gay-Lussac, Humboldt, Barral and Bixio, with a more detailed account of James Glaisher. Reprinted in Franklin Institute, *Journal*, Philadelphia, 1866, v. 81, p. 388–393, *VA*.

5030. Scientific ballooning and the higher strata of our atmosphere. (Engineering, London, 1897, v. 64, p. 460–461, 486–487, 554–555, 609, 612.) **VDA**

An excellent history of various investigations.

5031. Second scientific balloon ascent of MM. Barral and Bixio. (Dublin University magazine, Dublin, 1850, v. 36, p. 304–311.) *** DA**

5032. Selby, Francis James. Report as to existing knowledge on the subject of the accumulation of electrostatic charges on balloons, and the precautions to be adopted to avoid the dangers arising therefrom. (Great Britain. — Advisory Committee for Aeronautics, Report for 1909–1910, London, 1910, p. 119–121.) **VDW**

History of early investigations.

5033. Shaw, Sir William Napier. Contributions of balloon investigations to meteorology. (Aeronautical journal, London, 1903, v. 7, p. 8–13.) **VDS**

5034. —— Illusions of the upper air. diagrs. (Royal Institution of Great Britain, Proceedings, London, 1918, v. 21, part 3, p. 603–624.) *** EC**

A review of progress of meteorological theory in England since 1866. Reprinted for the most part in *Nature*, London, 1916, v. 97, p. 191–194, 210–213, *OA*; *Scientific American supplement*, New York, 1917, v. 83, p. 124–126, *VA*.

5035. —— Scientific ballooning. illus. (Nature, London, 1902, v. 65, p. 224–226.) **OA**

Deals mainly with developments in Germany.

5036. Sir George Cayley's work in aeronautics. Early pioneer originated airplane flight theory and anticipated metal-hull airship. (Aviation, New York, 1923, v. 15, p. 744–747.) **VDS**

5037. Soreau, Rodolphe. Aéronautique. (Société des ingénieurs civils de France, Mémoires, Paris, 1898, série 5, année 5, Cinquantenaire 1848–1898, partie 1, p. 421–436.) **VDA**

A brief exposition of airship and airplane principles.

5038. —— La navigation aérienne. illus. (Le Génie civil, Paris, 1900, tome 36, semestre 1, p. 264–266, 278–280, 298–301.) **VA**

Effects of wind. Notable efforts to steer balloons. History of development of heavier-than-air craft.

5039. —— Navigation aérienne. illus. (Société des ingénieurs civils de France, Mémoires, Paris, 1902, année 1902, v. 2, p. 507–583.) **VDA**

A study of the wind, airship stability, the airships *La France* and the *Santos-Dumont 6*, the airplane, and aerodynamics. The section on the airplane is abstracted in *Scientific American supplement*, New York, 1904, v. 58, p. 23949–23950, *VA*.

5040. —— Réponse aux "Remarques sur le problème de la navigation aérienne." (Société des ingénieurs civils de France, Mémoires, Paris, 1898, série 5, année 51, p. 332–349.) **VDA**

Discusses paper of Duroy de Bruignac; see entry no. 4873.

Aeronautics and Science, continued

5041. Sprat, Thomas. The history of the Royal Society of London, for the Improving of Natural Knowledge. London: F. Martyn, 1667. 438 p. illus. 8°. *** EC (Royal)**

On p. 218 reference is made to the flotation of glass balls in air; but no mention is made of the experimenter or of the results.

5042. Squier, George Owen. Scientific research for national defense as illustrated by the problems of aeronautics. (Franklin Institute, Journal, Philadelphia, 1917, v. 183, p. 35–40.) **VA**

A list of technical problems awaiting solution, including those of aerodynamics, engines, physics of the air, and others.

5043. Stade, H. Professor Dr. Victor Kremser. port. (Illustrierte aëronautische Mitteilungen, Berlin, 1909, Jahrg. 13, p. 1055–1058.) **VDS**

Kremser (1858–1909) was a prominent German investigator of the upper atmosphere.

5044. Stewart, Charles John. Modern developments in aircraft instruments. illus. (Royal Aeronautical Society, Journal, London, 1928, v. 32, p. 425–481.) **VDS**

With discussion.

5045. The **Stratosphere** expedition. (U. S. air services, Washington, 1934, v. 19, Sept., p. 22.) **VDS**

Ascension of Capt. Albert W. Stevens, United States Army Air Corps, 1934.

5046. Süring, Reinhard Joachim. Aerology in Germany. (American magazine of aeronautics, New York, 1907, v. 1, Nov., p. 27–29.) **VDS**

Historical development of the science, of which Dr. Assmann was the pioneer.

5047. —— Die Beziehungen zwischen Meteorologie und Luftschiffahrt. (Illustrierte aëronautische Mitteilungen, Strassburg, 1900, Jahrg. 4, p. 49–52.) **VDS**

A brief history of the relations of meteorology and aeronautics, with a bibliography.

5047a. —— Wissenschaftliche Ballonhochfahrten. illus. (In: Karl Bröckelmann, Wir Luftschiffer, Berlin, 1909, p. 48–65.) **VDW**

History of scientific balloon ascensions.

5048. —— Zum Gedächtniss an James Glaisher. port. (Illustrierte aëronautische Mitteilungen, Strassburg, 1903, Jahrg. 7, p. 132–136.) **VDS**

5049. Surcouf, Édouard. Le colonel Charles Renard. port. (L'Aérophile, Paris, 1905, année 13, p. 83–85.) **3–VDS**

Portrait and appreciation of famous French aeronautical engineer.

5050. Sverdrup, Harald Ulřik, and Einar Lönnberg. Scientific results of the Andrée expedition. (Geografiska annaler, Stockholm, 1931, arg. 13, p. 121–153.) **KAA**

Separate papers by each of the above authors. Bibliography, p. 140.

5051. Sykes, William H. On the desirableness of renewing balloon ascents in England for meteorological objects. (British Association for the Advancement of Science, Report of the 18th meeting, Leeds, 1858, London, 1859, Notices and abstracts, p. 39.) *** EC**

Also in *Mechanic's magazine,* London, 1858, v. 69, p. 372–373, *3–VA.*

5052. Teisserenc de Bort, Léon Philippe, and A. L. Rotch. General results of the meteorological cruises of the "Otaria" on the Atlantic in 1905, 1906, and 1907. map. illus. (Nature, London, 1909, v. 80, p. 219–221.) **OA**

5053. Thomas, Pierre. Considérations élémentaires sur les conditions théoriques de la navigation aérienne. (Société des ingénieurs civils de France, Mémoires, Paris, 1870, p. 543–553.) **VDA**

A study of air resistance. Believes heavier-than-air flying impossible with known mechanical means.

5054. Tissandier, Gaston. L'étude des courants aériens par les ballons libres. map. (La Nature, Paris, 1891, année 19, semestre 2, p. 259–260.) **OA**

Map shows the course of sixty sounding balloons liberated at Amiens, France, 1888–1890.

5055. Tissandier, Paul. Marc Seguin (1786–1875). Ses travaux sur l'aéronautique. port. (L'Aérophile, Paris, 1923, année 31, p. 205–206.) **3–VDS**

Biography of eminent French scientist and engineer, with extracts from his *Mémoire sur l'aviation,* 1866.

5056. Touche, P. E. Projet d'étude de la résistance de l'air. illus. (L'Aéronaute, Paris, 1893, année 26, p. 126–128.) **3–VDS**

5057. Tziolkovski, K. Voprosy vozdukhoplavaniya (Po povodu Trudov po vozdukhoplav. 7 Otd. Imp. Russkago Tekhnich. Obshch. za... 1895–1900 g.) St. Petersburg: Tip. P. P. Soikina, 1901. 18 p. 8°. **VDY p.v.4, no.10**

From the transactions of the Imperial Russian Technical Society, 1895–1900. Deals with the various problems of aeronautics.

5058. United States. — National Advisory Committee for Aeronautics. Annual report, including technical reports. nos. 1–20. Washington, 1915–34. 4°. **VDY**

Record of United States government activities in aeronautical research for the period covered.

5059. United States. — Forest Products Laboratory: Decennial Celebration Committee. The Forest Products Laboratory, a decennial record, 1910–1920. [Madison, Wis.:] Published by The Decennial Committee, 1921. 196 p. 8°. **VQY**

Account of work done in the kiln drying of aircraft woods, on the design of aircraft parts, development of water-resistant glues, and propellers.

5060. United States. — Standards Bureau. War work of the Bureau of Standards. Washington: Gov. Prtg. Off., 1921. 299 p. illus. 8°. (United States. — Standards Bureau. Miscellaneous publications. no. 46.) **VBDA**

Researches on instruments, engines, airplane construction, dopes, and balloon gases.

Aeronautics and Science, continued

5061. Van Everdingen, E. Aerial navigation and meteorology. (Nature, London, 1920, v. 105, p. 637–638.) **OA**

Comments on Annexe G of the Convention for the Regulation of Aerial Navigation, 1919.

5062. Van der Maas, H. J. The development of aviation and aeronautical research in Holland in recent years. illus. diagrs. (Royal Aeronautical Society, Journal, London, 1929, v. 33, p. 965–979.) **VDS**

5063. Vinci, Leonardo da. I manoscritti di Leonardo da Vinci. Codice sul volo degli uccelli e varie altre materia. Publicato da T. Sabachnikoff. Trascrizioni et note di G. Piumati. Traduzione in lingua francese di C. Ravaisson-Mollien. Parigi: E. Rouveyre, 1893. 156 p., 15 facsims. f°. **† VDY**

Transcriptions and sketches of da Vinci's studies in the mechanics of flight. For history of this translation see *L'Aéronaute,* Paris, 1894, année 27, p. 101–105, *3–VDS.*

5064. Voyer. Die aëronautischen Arbeiten des Generals Meusnier. (Illustrierte aëronautische Mitteilungen, Strassburg, 1905, Jahrg. 9, p. 137–144.) **VDS**

A résumé of the aerodynamical laws of General Meusnier, published in 1783–84.

5065. —— Histoire de l'aérostation; les lois de Meusnier. illus. (Revue du génie militaire, Paris, 1902, année 16, v. 23, p. 421–430, 521–532; v. 24, p. 135–156.) **VWA**

In 1784 Meusnier presented to the French Academy his paper on the laws of equilibrium as applied to balloons. He also made detailed suggestions for an airship and for governing the same. See entry no. 1180.

5066. Walker, G. W. Reports on experiments made at the Eskdalemuir Observatory with regard to the electrification of bodies in the air. (Great Britain. — Advisory Committee for Aeronautics. Report for 1909–1910, London, 1910, p. 122–127.) **VDW**

5067. Watter, Michael. Engineering aspects of Lindbergh's transatlantic flight. diagrs. (Aero digest, New York, 1927, v. 11, p. 396–397, 483–485.) **VDS**

5068. Wellner, Georg. Die Ausführbarkeit dynamischer Flugmaschinen. illus. (Zeitschrift für Luftschiffahrt, Berlin, 1891, Jahrg. 10, p. 70–77, 164–177, 257–272; 1892, Jahrg. 11, p. 61–69.) **3–VDS**

Discusses laws of flight and their relation to several proposed types of heavier-than-air machines.

5069. —— Friedrich Ritter von Lössl. port. (Illustrierte aëronautische Mitteilungen, Strassburg, 1907, Jahrg. 11, p. 235–236.) **VDS**

Account of aerodynamical researches of Austrian pioneer (1817–1907).

5070. —— Versuche über den Luftwiderstand gewölbter Flächen im Winde und auf Eisenbahnen mit Rücksicht auf das Problem dynamischer Flugmaschinen. illus. (Oesterreichischer Ingenieur- und Architekten-Verein, Zeitschrift, Wien, 1893, Jahrg. 45, p. 353–357, 366–369, 377–384, 389–396.) **3–VDA**

5071. What science owes to Wright brothers. A complete resumé of the early Wright experiments. illus. (Slipstream, Dayton, O., 1928, v. 9, April, p. 9–15, 18.) **VDS**

Includes also a pictorial review of airplane development from the time of Leonardo da Vinci to the World war.

5072. Whipple, F. J. W. Progress in knowledge of the upper air. (Nature, London, 1935, v. 135, p. 698–702.) **OA**

History begins with Teisserenc de Bort in 1910.

5073. Wilfrid de Fonvielle. (Aeronautical journal, London, 1914, v. 18, p. 168.) **VDS (Royal)**

Brief biography of French pioneer (1826–1914) who figured prominently in the siege of Paris during the war of 1870–1871.

5074. Wilfrid de Fonvielle. port. (L'Aérophile, Paris, 1914, année 22, p. 231–232.) **3–VDS**

Obituary of prominent French scientific writer who specialized in aeronautical subjects.

5075. Wilkins, John, bishop of Chester. ...The discovery of a new world; or, A discourse tending to prove, that ('tis probable) there may be another habitable world in the moon. With a discourse concerning the possibility of a passage thither. The fifth impression; corrected and enlarged. London: Printed for John Nicholson, 1707. 135 p. 12°. (In his: Mathematical and philosophical works, London, 1708.) **OAD**

Library has also the collected works, London, 1802, 2 v. *OAD.* For reference to flight possibilities see proposition no. 14.

5076. —— Mathematical magick, or, The wonders that may be performed by mechanicall geometry. In two books. Concerning mechanicall powers, motions. Being one of the most easie, pleasant, usefull, (and yet most neglected) part of mathematics. Not before treated of in this language. London, 1648. 295 p. illus. 16°. **3–PBC**

Chapter 6: Of the volant automata, Archytas his dove; and Regiomontanus his eagle. The possibility and great usefulnesse of such inventions. Chapter 7: Concerning the art of flying. The severall ways whereby this hath been or may be attempted. Chapter 8: A resolution of the two chief difficulties that seem to oppose the possibility of a flying chariot.

Library has also the edition of 1691; also the author's collected works, both the 1708 and 1802 editions.

5077. Willson, Robert W. Professor Abbott Lawrence Rotch, the pioneer explorer of the upper air. (Aero Club of America, Bulletin, New York, 1912, v. 1, May, p. 29.) **VDS (Flying)**

5078. Wilson, Edwin Bidwell. The mathematics of aërodynamics. (American mathematical monthly, Lancaster, Pa., 1918, v. 25, p. 292–297.) **OEA**

Brief history of the subject, beginning with the work of G. H. Bryan. Also in *Aviation,* New York, 1918, v. 4, p. 157–158, *VDS.*

Aeronautics and Science, continued

5079. **Wimperis,** Harry Egerton. The relationship of physics to aeronautical research. (Royal Aeronautical Society, Journal, London, 1926, v. 30, p. 668–675.) **VDS**

Work of the Aeronautical Research Committee, the National Physical Laboratory, Ludwig Prandtl, Juan de la Cierva, Osborne Reynolds, and Bertram Hopkinson.

5080. **Wise,** John. On the balloon as an instrument in meteorological research. (Franklin Institute, Journal, Philadelphia, 1871, v. 91, p. 345–349.) **VA**

5081. **Wyatt,** B. H. The eclipse of the sun. Observations entirely impossible on account of climatic conditions, except from the airplanes at high elevations. (Aeronautical digest, New York, 1923, v. 3, p. 344–345.) **VDS**

United States Navy observations, southern California, 1923.

5082. **Zahm,** Albert Francis. Atmospheric friction with special reference to aeronautics. Communicated to the International Aeronautic Congress of 1904. [Washington:] The Catholic University of America, 1904. 30 p. 8°. **VDW p.v.2, no.9**

5083. —— Atmospheric gusts and their relation to flight. illus. (International Conference on Aerial Navigation, Chicago, 1893, Proceedings, New York, 1894, p. 163–172.) **VDS**

For a list of Zahm's writings on experimental aeronautics see Paul Brockett, *Bibliography of aeronautics,* Washington, 1910, p. 908–910, *VDT.*

PHOTOGRAPHY AND SURVEYING

5083a. **Abrams,** Talbot. Mapping Michigan for geologists. illus. (Engineering and mining journal, New York, 1935, v. 136, p. 564–565.) **† VHA**

5084. **Aerial** mapping of Burmese forests. (Engineering, London, 1926, v. 122, p. 360.) **VDA**

Survey made by R. C. Kemp of 1000 square miles.

5085. **Aerial** survey of central Australia. illus. (Flight, London, 1930, v. 22, p. 1257–1258.) **VDS**

Expedition under command of Donald Mackay mapped 67,000 square miles of territory.

5086. **Aerial** surveys for the revision of Coast and Geodetic Survey charts. (United States. — Commerce Department: Aeronautics Branch, Aircraft commerce bulletin, Washington, 1929, v. 1, Aug. 15, p. 12–16.) **VDS**

5087. **Aerial** photography. Danger of government competition. illus. (Aircraft year book, New York, 1926, p. 87–95.) **VDS**

A brief review of United States government and private developments.

5087a. **Alcock,** F. J. Airplanes an aid in geologic studies. illus. (Engineering and mining journal, New York, 1935, v. 136, p. 546–547.) **† VHA**

A comparison of field work conditions in 1914 and 1935.

5088. **Beazeley,** George Adam. Reconnaissance survey from aircraft. Dehra Dun: Geodetic Branch Off., Survey of India, 1927. 34 p. front. 4°. (India. Survey Dept. Professional papers. no. 20.) **† KCV (India)**

Record of work in the Punjab.

5089. [**Bibliographies** of aerial photography.] (In: Paul Brockett, Bibliography of aeronautics, Washington, 1910, p. 670; continued in United States. — National Advisory Committee for Aeronautics, Bibliography of aeronautics, 1909–1931, Washington, 1921–35, volumes for 1909/16, p. 1036–1038; 1917/19, p. 351–353; 1920/21, p. 323–325; 1922, p. 124–125; 1923, p. 107; 1924, p. 85–86; 1925, p. 136; 1926, p. 113; 1927, p. 139; 1928, p. 156; 1929, p. 177–178; 1930, p. 188–189; 1931, p. 221–222.) **VDT**

5090. [**Bibliography** of aerial mapping.] (In: United States. — National Advisory Committee for Aeronautics, Bibliography of aeronautics, 1909–1931, Washington, 1921–35, volumes for 1909/16, p. 255, 836–838; 1917/19, p. 85, 284; 1920/21, p. 264–265; 1922, p. 92; 1923, p. 82; 1924, p. 65–66; 1925, p. 105–106; 1926, p. 85; 1927, p. 108; 1928, p. 119; 1929, p. 138; 1930, p. 145–146; 1931, p. 175–176.) **VDT**

Some of the earlier entries have the heading *Charts.*

5091. [**Bibliography** of aerial surveying.] (In: United States. — National Advisory Committee for Aeronautics, Bibliography of aeronautics, 1909–1931, Washington, 1921–35, volumes for 1909/16, p. 1304; 1917/19, p. 429; 1920/21, p. 396; 1922, p. 151; 1923, p. 130; 1924, p. 103–104; 1925, p. 168–169; 1926, p. 137–138; 1927, p. 166; 1928, p. 190; 1929, p. 214; 1930, p. 232–233; 1931, p. 273.) **VDT**

5092. **Blon,** John L. von. Alaska is mapped from the air. With three airplanes, a government expedition did more mapping in Alaska in fifteen days than ordinary surveyors could have accomplished in fifteen years. illus. (Scientific American, New York, 1927, v. 136, p. 303–305.) **VA**

Work of the United States Aerial Survey Expedition.

5093. —— Mapping Alaska from the air. illus. (Flight, London, 1927, v. 19, p. 172–175.) **VDS**

40,000 square miles were mapped by aviators of the United States Navy.

5094. **Bourne,** Ray. Aerial survey in relation to the economic development of new countries, with special reference to an investi-

Photography and Surveying, continued

gation carried out in Northern Rhodesia. illus. maps. (Oxford forestry memoirs. Oxford, 1928, no. 9, p. 1–35.) † VQP

5095. Bouttieaux, Victor Paul. La télephotographie en ballon. illus. (Revue de l'aéronautique, Paris, 1894, année 7, p. 99–130.)
Historical introduction, p. 100–104. 3–† VDS

5096. Brock, Norman H. Brock development of aerial mapping. illus. (Military engineer, Washington, 1925, v. 17, p. 504–509.) † VWA
Tells of the development of the Brock camera and process.

5097. Bruno, H. A. A job for one in a million. Why only a few men can photograph cities, chart forests, explore jungles from aircraft. illus. (Popular science monthly, New York, 1927, v. 110, April, p. 11–13, 104.) * DA
Notable experiences. Also in *Literary digest,* New York, 1927, v. 92, March 26, p. 62–67, * *DA.*

5098. Burkett, E. F. Mapping 13,000 square miles of Alaska. illus. (Aviation, New York, 1929, v. 27, p. 1210–1213.) VDS
Account by executive officer of the expedition, 1929.

5099. Canada. — Air Board: Library. Selective bibliography on air photography, generally. [Ottawa, 1922.] 5, 1, 3, 2 f. f°. (Its: Bibliography. no. 3.) † VQP
Mimeographed.

5100. Canada. — Interior Department: Northwest Territories and Yukon Branch. Preliminary report on the aerial mineral exploration of northern Canada, by G. H. Blanchet. Ottawa: F. A. Acland, 1930. 32 p. illus. maps. 8°. VHN
Deals briefly with operations during 1928 and 1929.

5101. Carriel, Cruse. Hawaii from the air. An account of the first aerial survey of our possessions in the Pacific, involving the complete circumnavigation by naval flying-boats of all the islands in the archipelago. illus. (Flying, New York, 1921, v. 10, p. 127–130.) VDS

5102. Carrier pigeons as photographers. illus. (Scientific American, New York, 1909, v. 100, p. 4.) VA
Apparatus of Dr. Neubronner, of Germany.

5103. Cassinis, Gino. Lo stato attuale dei metodi di rilevamento topografico dagli aerei. illus. (Ingegneria, Milano, 1924, anno 3, p. 306–317.) VDA

5104. Close, Sir Charles Frederick. Surveys from the air; the present position. diagrs. (Nature, London, 1924, v. 114, p. 195–197.) OA

5105. Cochran-Patrick, Charles Kennedy. Air surveys in Burma. illus. maps. (Royal Aeronautical Society, Journal, London, 1925, v. 29, p. 603–633.) VDS
Abstracted in *Aeroplane,* London, 1925, v. 29, p. 482, 484, *VDS.*

5106. Conquest of the Arabian desert by aircraft. (Aviation, Highland, N. Y., 1921, v. 11, p. 535.) VDS
Brief account of survey by the Royal Air Force.

5106a. Corbin, Paul, and Nicolas Oulianoff. La photographie aérienne au service de la géologie. (Académie des sciences, Comptes rendus, Paris, 1934, tome 199, p. 431–432.)
Authors' experiences in the Alps. * EO

5107. The **Death** of Nadar. port. (Aeronautical journal, London, 1910, v. 14, p. 76.) VDS (Royal)
Nadar was the pseudonym of Félix Tournachon (1820–1910), who took out the first patents on aerial photography.

5108. Death Valley mapped under severe conditions. (Aviation, New York, 1931, v. 30, p. 710–711.) VDS
Brief account of work done under direction of the United States Geological Survey, 1931.

5109. The **Development** of aerial surveying. illus. (Engineering, London, 1926, v. 122, p. 595–596.) VDA
Account of developments by the Air Survey Co., Ltd., of London, England.

5110. Dollfus, Charles. Historique de la photographie aérienne. illus. (L'Aéronautique, Paris, 1922, année 4, p. 177–178.) VDS
Has a photograph taken by "Nadar" in 1868.

5111. Doughty, John G. Balloon experience of a timid photographer. illus. (Century magazine, New York, 1886, v. 32 [new series, v. 10], p. 679–694.) * DA
Experiences and views, 1885, by the author, who was companion to the aeronaut, A. E. Moore.

5112. Eddy, William A. Photographing from kites. illus. (Century magazine, New York, 1897, new series, v. 32 [whole no., v. 54], p. 86–91.) * DA
Author's experiences.

5113. Exploration by air. illus. map. (Aircraft yearbook, New York, 1930, p. 161–180.) VDS
Expeditions of 1929. Byrd Antarctic expedition. Wilkins-Hearst Antarctic expedition. Lindbergh Pueblo and Maya survey. Alaskan aerial survey. Canadian prospecting survey.

5114. Fairchild, Sherman M. Photographing New York from the airplane. (Aeronautical digest, New York, 1923, v. 3, p. 434–435.) VDS
Account of surveys made by the Fairchild Camera Corporation.

5115. Finsterwalder, Sebastian. Eine Grundaufgabe der Photogrammetrie und ihre Anwendung auf Ballonaufnahmen. map. illus. (Königlich Bayerische Akademie der Wissenschaften, Denkschriften: Abhandlungen der Mathematisch-physikalischen Klasse, München, 1906, Bd. 22, p. 223–260.) * EE

5116. —— Ueber die Konstruktion von Höhenkarten aus Ballonaufnahmen. illus. (Königlich Bayerische Akademie der Wissenschaften, Sitzungsberichte, Mathematisch-physikalische Klasse, München, 1900. Bd. 30, p. 149–164.) * EE

Photography and Surveying, continued

5117. **Gill,** Donald. Air survey in relation to economic geology. illus. (Institution of Mining and Metallurgy, Transactions for 1932–1933, London, 1933, v. 42, p. 81–175.) **VHA**

With discussion.
Extensive bibliography, p. 128–136. The paper is slightly abridged in Royal Aeronautical Society, *Journal*, London, 1933, v. 37, p. 227–287, *VDS*.

5118. **Grey,** Charles Grey. On the progress of air survey. illus. (Aeroplane, London, 1927, v. 33, p. 601–602.) **VDS**

Recent technical developments.

5119. **Haferkorn,** Henry Ernest. Aerial photography. (United States. — Engineer Department, Professional memoirs, Washington, 1918, v. 10, p. 421–436, 569–590, 855–884.) **VDA**

Extensive annotated bibliography on aerial photography and surveying, containing numerous early titles.

5119a. **Heiland,** C. A. Geophysical mapping from the air; its possibilities and advantages. (Engineering and mining journal, New York, 1935, v. 136, p. 609–610.) **† VHA**

5120. **Heinsen,** Walter A. I photograph the enemy from the air. illus. (Popular aviation, Chicago, 1931, v. 9, July, p. 24–25, 54.) **VDS**

Experiences of a German aerial observer during the World war.

5121. **Hemming,** H. Air surveying in Rhodesia. Two aeroplanes have surveyed nearly 40,000 square miles of virgin country in twelve months. illus. (Airways, London, 1927, v. 4, p. 135–137, 150.) **VDS**

5122. —— Wonders of aerial survey: aviation, photography, astronomy, geology and wireless in collaboration. illus. (Illustrated London news, London, 1926, v. 169, p. 1232, 1254; illus. on p. 1212–1213.) *** DA**

Interesting illustrations show how work is done.

5123. **Hotine,** Martin. Surveying from air photographs. London: Constable and Co., Ltd., 1931. xi, 250 p. illus. 4°. **VDE**

Bibliography, p. 243.

5124. **Jennings,** W. Nicholson. Balloon photography. illus. (Fly, Philadelphia, 1909, v. 1, June, p. 4–5.) **3–VDS**

Brief history.

5125. **Jones,** Bennett Melvill, and J. C. Griffiths. Recent experiments in aerial surveying by vertical photographs. illus. (Nature, London, 1923, v. 111, p. 707–709, 745–748.) **OA**

Experiments at University of Cambridge beginning 1920.

5126. **Judd,** Neil Merton. The airplane aids the archaeologist in reconstructing an ancient civilization. illus. (National aeronautic magazine, Washington, 1931, v. 9, Jan., p. 21–25, 27, 29.) **VDS**

Account of author's work in collaboration with the United States Army Air Corps in Arizona.

5127. **Kahle.** Die Bedeutung der Luftschiffahrt mit lenkbaren Fahrzeugen für Städtebau, Kartographie und Erdkunde. (Zeitschrift für Vermessungswesen, Stuttgart, 1909, Bd. 38, p. 57–62.) **OXA**

5128. **Klemin,** Alexander. Developing motion pictures in the air. illus. (Scientific American, New York, 1928, v. 139, p. 448.) **VA**

5129. —— Photographing a state from the air. illus. (Scientific American, New York, 1934, v. 150, p. 90–91.) **VA**

How an air picture of Massachusetts was made.

5130. **Lallemand,** Charles. International air map and aeronautical marks. (Geographical journal, London, 1911, v. 38, p. 469–483.) **KAA**

Proposals of the Permanent Committee for Aerial Navigation of the Public Works Department of the French Government. Abstract in British Association for the Advancement of Science, *Report of meeting, Portsmouth, 1911,* London, 1912, v. 81, p. 447–448, ** EC*.

5131. The **Latest** African air survey. map. (Aeroplane, London, 1932, v. 42, p. 382.) **VDS**

Survey of Uganda and the Sudan, 1930.

5132. **Laussedat,** Aime. La photographie en ballon. illus. (In his: Recherches sur les instruments, les méthodes, et le dessin topographiques, Paris, 1898–1901, tome 2, p. 157–208.) **VDE**

Accounts of various apparatus for balloons and kites. For a portrait and brief biography of Laussedat see *L'Aérophile,* Paris, 1894, année 2, p. 153–157, *3–VDS*.

5133. **Lewis,** C. G. I: Air-survey in the Irrawaddy delta, 1923, by Major C. G. Lewis ... II: Reconnaissance survey in Bhutan and South Tibet, 1922, by Captain H. R. C. Meade ... Dehra Dun: Geodetic Branch Off., Survey of India, 1925. 49 p. diagrs., map. 4°. (India. Survey Dept. Records. v. 21.) **† VDE**

5134. **Literature** on aerial photography. (Photographic journal, London, 1921, new series, v. 45, p. 396–399.) **MFA**

5135. **Loeffler,** H. C. Aerial photography. illus. (World's health, Paris, 1926, v. 7, p. 32–39.) **SPA**

Brief history of use in surveying and archaeology.

5136. **Lomet,** Antoine François, baron. Sur l'emploi des machines aérostatiques aux reconnaissances militaires et à la construction des cartes géographiques. illus. (L'École polytechnique, Journal, Paris, 1801, tome 4, cahier 11, p. 252–258.) **OEA (France)**

English translation in *Nicholson's journal,* London, 1803, v. 6, p. 194–200, *OA*. For a biography and portrait of Lomet see *Revue de l'Agenais,* Agen, 1918, tome 45, p. 73–94, 141–159, 265–279, ** EN*.

Photography and Surveying, continued

5137. Maatschappij voor Landometing. Photographic mapping. Munich: Dietz & Lüchtrath, 1923. 41 p. illus. maps. 4°. **Map Room**
Bibliography, p. 14–15.

5138. McKinley, Ashley Chadbourne. Applied aerial photography, with a chapter on oblique aerial surveys [Canada] by A. M. Narraway. New York: J. Wiley & Sons, Inc., 1929. xiii, 341 p. illus. maps. 8°. **MFP**
Bibliography, p. 331–332.

5139. MacLeod, Malcolm Neynoe. Aeroplane photographic survey. (Chartered Surveyors' Institution, Transactions, London, session 1923–1924, v. 56, p. 25–59.) **3–VDA**
With discussion.
Discussion has some reminiscences of Sir Charles Close concerning balloon surveying in Nova Scotia and in India.

5140. The **Making** of Greater New York's air map. Planes flew 3,000 miles and photographed 625 miles on largest city mapping project ever attempted. illus. (Aviation, Highland, N. Y., 1924, v. 16, p. 16–17.) **VDS**

5141. Mason, Ethel I. Alaska is mapped by the "polar bears." illus. ports. (U. S. air services, Washington, 1926, v. 11, Aug., p. 9–12.) **VDS**
U. S. Navy expedition, 1926, using Loening amphibian planes.

5142. Montagnes, James. Charting Canada's wilderness from the air. Rich mineral areas amid lakes and forests of northern Canada mapped accurately for the first time. illus. (Scientific American, New York, 1929, v. 141, p. 214–215.) **VA**

5143. Narraway, Athos Maxwell. Aerial surveying as applied to engineering problems. illus. maps. (Engineering Institute of Canada, Journal, Montreal, 1930, v. 13, p. 171–178.) **VDA**
Three years' experiences on large Canadian surveys.

5144. Norman, William James. Notes on air survey in India. Dehra Dun: Geodetic Branch Off., Survey of India, 1929. v, 45 p. diagrs. 8°. (India. Survey Dept. Professional paper. no. 24.) **† KCV**

5145. Page, William H. Pillars of light. Army Air Service revives surveying method found successful in India a century ago. illus. (Aero digest, New York, 1925, v. 7, p. 424–425.) **VDS**
Night surveying with searchlights.

5146. Painter, K. M. Mapping Hispaniola from the air. illus. (Popular aviation, Chicago, 1928, v. 3, July, p. 13–17.) **VDS**

5147. Photographing Mount Everest; some details of the photographic aspects of the recent Houston-Mt. Everest expedition. illus. (Air and airways, London, 1933, v. 10, p. 136.) **VDS**

5148. Pitts, Nelson F., jr. What aerial mapping has done for Syracuse. An indispensable adjunct to many departments of the municipality. illus. (American city, New York, 1927, v. 37, Sept., p. 354–356.) **SERA**

5149. Poidebard, A. Méthode aérienne de recherches en géographie historique. illus. (Terre, air, mer. Paris, 1932, tome 57, p. 1–16.) **KAA (Société)**
Bibliography, p. 16.

5150. Radford, A. W. Three months at eleven thousand feet above Alaska. A human-interest story of the second Naval Alaskan Aerial Survey expedition, and also about things leading up to and after the remarkable accomplishment. illus. ports. (Aeronautic review, Baltimore, 1929, v. 7, Dec., p. 19–30.) **VDS**

5151. Reed, Henry Albert. Photography applied to surveying. New York: John Wiley & Sons, 1888. 68 p. illus. 4°. **VDE**
See p. 53–61 for account of development of balloon photographic surveying.

5152. Riddle, Karl. Aerial surveys lend confidence on hastily organized job. Plat of 18,000 acres of undeveloped land surveyed in minimum time. Conclusions drawn from aerial photographs. illus. (Engineering news-record, New York, 1926, v. 97, p. 1072–1073.) **VDA**
Survey at Boca Raton, Florida, made by the Fairchild Aerial Surveys, Inc.

5153. Roussilhe, H. Résultats obtenus en 1921 et 1922 par l'application de la photographie aérienne aux levès de précision à grande échelle. (Académie des sciences, Comptes rendus, Paris, 1922, tome 175, p. 944–946.) *** EO**

5154. Sagendorph, Kent. Aerial mapping in Mexico. illus. (Popular aviation, Chicago, 1929, v. 4, April, p. 20–22, 94.) **VDS**
Work of Fairchild Aerial Surveys Company, 1928.

5155. —— The hazards of aerial mapping. illus. port. (Popular aviation, Chicago, 1933, v. 12, p. 155–156, 190–191.) **VDS**
Account of the adventures of Sydney O. Bonnick.

5156. Sarnetzky, Heinrich. Grundzeuge der Luft- und Erdbildmessung. Berlin: Gebrüder Borntraeger, 1928. 236 p. illus. 8°. (Sammlung Borntraeger. Bd. 14.) **VDE**
Bibliography, p. 218–230.

5157. Smith, George Otis. Geological Survey in the air. (National Aeronautic Association review, Washington, 1925, v. 3, p. 74–75.) **VDS**
Use of the airplane in photographic surveying.

5157a. Smith, Philip S. Airplanes expedite Alaskan survey. illus. (Engineering and mining journal, New York, 1935, v. 136, p. 565–566.) **† VHA**

Photography and Surveying, continued

5158. Somerville, Harry A. Geodetic surveying by air. Successful use is made of planes for this work in Canada. illus. (Airway age, New York, 1930, v. 11, p. 1061–1062.) **VDS**

5159. Stewart, N. C. Surveying from the air. illus. (Royal Astronomical Society of Canada, Journal, Toronto, 1931, v. 25, p. 153–165.) **OMA**

Developments in Canada since the World war.

5159a. The **Survey** of Rio de Janeiro. Aircraft Operating Company, Ltd., completes important contract. (Flight, London, 1932, v. 24, p. 452.) **VDS**

Account of survey of the Federal District, taking three and a half years.

5160. Taylor, H. W. Mapping Porto Rico from the sky. illus. (National aeronautic magazine, Washington, 1931, v. 9, Nov., p. 12–21.) **VDS**

5161. Thomas, Hugh Hamshaw. Aircraft photography in war and peace. (Royal Society of Arts, Journal, London, 1919–20, v. 68, p. 749–753, 763–766.) **VA**

"Aerial photography may be said to date back to the year 1858, when Paris was photographed from a balloon."

5162. —— Some features in the present position of aerial photographic survey. illus. (Royal Aeronautical Society, Journal, London, 1924, v. 28, p. 475–500.) **VDS**

With discussion.
Deals with the advances made in the art by Major J. C. Griffiths. Somewhat abridged in *Flight,* London, 1924, v. 16, p. 121–122, 134–136, *VDS.*

5163. Tissandier, Gaston. La photographie en ballon. Paris: Gauthier-Villars, 1886. vii, 52 p. illus. map. 8°. (Bibliothèque photographique.) **MFF p.v.36**

Has accounts of work of early experimenters.

5164. —— La photographie en cerf-volant. illus. (La Nature, Paris, 1889, année 17, semestre 1, p. 257–258.) **3–VDS**

Apparatus of Arthur Batut.

5165. United States. — Coast and Geodetic Survey. Topography. Aerial survey of the Mississippi river delta, by G. C. Mattison... Washington: Gov. Prtg. Off., 1924. 16 p. illus. 8°. (Special publication. no. 105.) **OXA**

5166. United States. — Library of Congress: Division of Bibliography. List of references on aerial photography and the applicability of telephotography to the same... March 27, 1918. 5 f. 4°. (Select list of references. no. 238.) *** SAB**

Typewritten.

5167. Use of aircraft for surveying. illus. (Hydrographic review, Cannes, 1927, v. 4, Nov., p. 53–103.) **VXA**

Accomplishments of the United States Hydrographic Office.

5168. Vest, J. R. W. Charting Cuba's coastline from the sky. illus. (National aeronautic magazine, Washington, 1931, v. 9, Feb., p. 9–20.) **VDS**

Work of the Hydrographic Office of the United States Navy.

5169. Walker, George L. Surveying the Witwatersrand from the air. illus. (Engineering and mining journal, New York, 1934, v. 135, p. 456–457.) **VHA**

5170. Webber, John E. Notable achievements in aerial photography. illus. (Photo-era magazine, Wolfeboro, N. H., 1924, v. 52, p. 310–315.) **MFA**

5171. Wenz, Émile. Résumé historique de l'invention de la photographie aérienne par cerf-volant. (L'Aéronaute, Paris, 1901, année 34, p. [264–265.]) **3–VDS**

Tabulation of results obtained by several kite operators. See also Société française de photographie, *Bulletin,* Paris, 1902, série 2, tome 18, p. 210–211, *MFA.*

5172. Whitehead, R. F. Problems of aerial photography in Alaska. illus. (Aviation, New York, 1930, v. 28, p. 1024–1027.) **VDS**

Work of the Alaska Mapping Expedition of 1929.

5173. Wilkins, H. F. Competing with sunshine. The origin and development of night photography in the [U. S.] Army Air Corps. illus. (Aero digest, New York, 1929, v. 14, April, p. 46–47, 280, 282.) **VDS**

5174. Willcox, H. Case. Value of aerial photographic surveying and mapping to petroleum companies and their geologists. Papers by H. Case Willcox and Gerard H. Matthes read at the New York meeting of the Petroleum Division of the American Institute of Mining and Metallurgical Engineers, Feb., 1925. New York [1925]. 11 p. illus. 4°. (Fairchild Aerial Surveys, Inc. Bulletin. no. 3.) **† VBA p.v.165**

5175. Williams, C. L. Looking for holes in the sky. Photographing the Great Smoky and Blue Ridge mountains. illus. (Aero digest, New York, 1926, v. 9, p. 184–185.) **VDS**

5176. Wilson, Elwood. Forest mapping and estimating from the air. Brief account of the aerial timber survey carried out in the Canadian woods by a crew of four men. illus. (Aviation, Highland, N. Y., 1922, v. 12, p. 538–539.) **VDS**

5177. Wyatt, B. H. Mapping our roughest region. port. illus. (Aero digest, New York, 1925, v. 7, p. 604–605, 636.) **VDS**

Navy pilot describes flying experiences on Colorado oil shale survey.

ANIMAL FLIGHT

5178. **Airy,** Hubert. The soaring of birds. (Nature, London, 1883, v. 27, p. 590–592.) **OA**

5179. **Amans,** P. C. Physiologie du vol d'après Léonard de Vinci. (L'Aéronaute, Paris, 1892, année 25, p. 195–210.) **3–VDS**

5180. **Argyll,** George Douglas Campbell, 8th duke of. The reign of law. London: Alexander Strahan, 1867. vi, 435 p. 2. ed. illus. 12°. **YAR**
See chapter 3 for a discussion of bird flight.

5181. **Basté,** J. F. Mémoire relatif au problème de la locomotion dans l'air. illus. (L'Aéronaute, Paris, 1887, année 20, p. 165–178, 183–196, 203–218.) **3–VDS**
For comments see *L'Aéronaute,* Paris, 1888, année 21, p. 7–10, *3–VDS.*

5182. **Bazin,** Albert. Mouillard. (L'Aérophile, Paris, 1910, année 18, p. 255–257.) **3–VDS**
Characterizes the investigations of prominent pioneer of bird-flight study.

5183. —— Mouillard et les mathématiciens. (L'Aérophile, Paris, 1910, année 18, p. 350–351.) **3–VDS**
Mouillard, in spite of his comparative ignorance of mathematics, developed his scientific theories of bird flight.

5184. **Berthenson,** Georg. Elasticität und Bewegung. (Zeitschrift für Luftschiffahrt, Berlin, 1895, Jahrg. 14, p. 215–224.) **3–VDS**

5185. [**Bibliographies** of bird flight.] (In: Paul Brockett, Bibliography of aeronautics, Washington, 1910, p. 129; continued in United States. — National Advisory Committee for Aeronautics, Bibliography of aeronautics, 1909–1931, Washington, 1921–35, volumes for 1909/16, p. 159–160; 1917/19, p. 57; 1920/21, p. 49–50; 1922, p. 17–18; 1923, p. 16; 1924, p. 11–12; 1925, p. 22; 1926, p. 16; 1927, p. 18; 1928, p. 25; 1929, p. 31; 1930, p. 27–28; 1931, p. 35.) **VDT**

5186. [**Bibliographies** of soaring flight.] (In: Paul Brockett, Bibliography of aeronautics, Washington, 1910, p. 783; continued in United States. — National Advisory Committee for Aeronautics, Bibliography of aeronautics, 1909–1931, Washington, 1921–35, volumes for 1909/16, p. 1241–1243; 1920/21, p. 379–380; 1922, p. 144–146; 1923, p. 125; 1924, p. 101; 1925, p. 162; 1926, p. 134; 1928, p. 183; 1929, p. 206; 1930, p. 222–224; 1931, p. 261–262.) **VDT**

5187. [**Bibliography** of Fr. Ahlborn.] (In: Paul Brockett, Bibliography of aeronautics, Washington, 1910, p. 30–31.) **VDT**
Lists Ahlborn's writings on bird flight.

5188. [**Bibliography** of insect flight.] (In: United States. — National Advisory Committee for Aeronautics, Bibliography of aeronautics, 1909–1931, Washington, 1921–35, volumes for 1909/16, p. 671; 1917/19, p. 228; 1920/21, p. 210; 1925, p. 82; 1931, p. 135.) **VDT**

5189. [**Bibliography** of Marcel Deprez.] (In: Paul Brockett, Bibliography of aeronautics, Washington, 1910, p. 235–236.) **VDT**
Cites Deprez's writings, principally on bird flight.

5190. The **Bird** as a model for the aeroplane. illus. (Scientific American, New York, 1905, v. 93, p. 7–8.) **VA**
Briefly reviews the experiments of E. P. Frost, F. W. H. Hutchinson, and C. R. D'Esterre.

5191. **Borelli,** Giovanni Alfonso. The flight of birds. [London:] Printed and published for the Aëronautical Society of Great Britain, 1911. 3 p.l., [v]–viii p., 1 l., 40 p. illus. 12°. (Aeronautical classics. no. 6.) **VDY (Royal)**
A translation of the section "De Volatu," omitting a few propositions which have no general interest in the argument. The first edition of this work was published in Rome, 1680–81.

5192. **Bretonnière,** J. Étude sur le vol plané. illus. (L'Aéronaute, Paris, 1889, année 22, p. 124–133, 148–155; 1890, année 23, p. 93–96, 115–121.) **3–VDS**
Study of the airplane and bird flight.

5193. —— Sailing flight. From observations made at Constantine, Algeria. illus. (International Conference on Aerial Navigation, Chicago, 1893, Proceedings, New York, 1894, p. 172–205.) **VDS**
With discussion.

5194. —— Le vol plané. Paris: H. Dunod et E. Pinat, 1909. 29 p. illus. 8°. **VDY**

5195. —— Le vol plané. illus. (Revue scientifique, Paris, 1898, série 4, tome 9, p. 35–40.) **OA**
Author's reply to criticisms of his theory.

5196. **Bryan,** George Hartley. On the sailing flight of birds. (British Association for the Advancement of Science, Report of the 66th meeting, Liverpool, 1896, London, 1896, p. 726–728.) ***EC**

5197. **Bull,** Lucien. Mécanisme de mouvement de l'aile des insectes. illus. (Académie des sciences, Comptes rendus, Paris, 1904, tome 138, p. 590–592.) ***EO**
Brief account of the author's continuation of the work of Marey.

5198. **Buttenstedt,** Karl. Meine Spannungstheorie. illus. (Zeitschrift für Luftschiffahrt, Berlin, 1893, Jahrg. 12, p. 272–277.) **3–VDS**

5199. —— Naturstudien zu einer Segeltheorie. Berlin: der Verfasser, 1894. 21 p. illus. 8°. **VDY p.v.1, no.2**

5200. —— Perpetuum mobile und Schwerkraft-Spannung. (Zeitschrift für Luftschif-

Animal Flight, continued

fahrt, Berlin, 1899. Jahrg. 18, p. 243–249.) **3–VDS**

Reply to some criticisms of Dr. Emil Jacob. See entry no. 5277.

For list of Buttenstedt's writings see Paul Brockett, *Bibliography of aeronautics*, Washington, 1910, p. 166–168, *VDT*.

5201. Chanute, Octave. Sailing flight. (Aeronautical annual, Boston, 1896, no. 2, p. 60–76.) **3–VDS**

Has a bibliography of 29 titles.

5202. Collins, Thomas Byard. The action of a bird's wing and its bearing on the problem of mechanical flight. illus. (Scientific American, New York, 1903, v. 88, p. 171–172.) **VA**

5203. Coues, Elliott. The mechanism of the flight of birds. (Science, New York, 1887, v. 10, p. 321–322.) **OA**

Discusses the mechanism of flexion and extension of wings, with history of earlier investigations.

5204. Coughtrie. Aërial locomotion. Pettigrew versus Marey. (Journal of science, London, 1875, new series, v. 5, p. 234–251.) **OA**

"Professor Marey has added comparatively little to the science of aërostation."

5205. Cousin, Joseph. Le vol à voile. Étude historique, critique, anatomo-physiologique et théorique du vol des oiseaux voiliers et son application à l'homme. Paris: F. Louis Vivien, 1910. xvi, 224 p. illus. 4°. (Bibliothèque Avia. v. 1.) **VDW**

Historical study of bird flight with numerous illustrations.

5206. Davidson, George L. O. Flight, and how birds get a rise out of the wind when soaring. illus. (Aeronautical journal, London, 1898, v. 2, p. 15–20.) **VDS**

5207. Drzewiecki, Stepan. ⌊Une explication rationnelle du vol des oiseaux.⌋ illus. (L'Aéronaute, Paris, 1889, année 22, p. 221–254.) **3–VDS**

5208. —— Les oiseaux considérés comme des aéroplanes animés. Clermont (Oise): Daix frères, 1889. 36 p. illus. 8°. **VDY p.v.8, no.2**

5209. Dutczynski, Alfred R. von. Der Insectenflug. illus. (Zeitschrift für Luftschiffahrt, Berlin, 1893, Jahrg. 12, p. 166–183.) **3–VDS**

5210. Dutczynski, L. von, and A. R. von Dutczynski. Zum Insektenflug. illus. (Zeitschrift für Luftschiffahrt, Berlin, 1895, Jahrg. 14, p. 57–67, 94–103.) **3–VDS**

5211. Esclangon, Ernest. La navigation aérienne et le planement des oiseaux. illus. (Revue des idées, Paris, 1909, année 6, p. 1–22.) *** DM**

5212. Esterno, Henri Philippe Ferdinand, comte d'. Du vol des oiseaux. Indication des sept lois du vol ramé et des huit lois du vol à voile. Paris: Librairie nouvelle, 1864. 61 p. illus. 8°. **VDY**

Author's suggestion for flying machine based upon the laws of bird flight. For Esterno's remarkable predictions as to the future of aviation see *L'Aérophile*, Paris, 1918, année 26, p. 371, *3–VDS*.

5213. Étévé, Albert. Le vol de l'oiseau et les ornithoplanes. illus. (Le Génie civil, Paris, 1909, v. 55, p. 282–285, 294–298.) **VA**

5214. Funcke, H. Ueber Vogelflug und Flug-Apparate. (Zeitschrift für Luftschiffahrt, Berlin, 1894, Jahrg. 13, p. 206–213, 235–240.) **3–VDS**

5215. Gatke, Heinrich. Fifty years observation on bird flight. (Aero, Chicago, 1912, v. 4, p. 259, 265, 278, 284, 297, 302, 319, 324, 339, 342, 359, 379, 382.) **3–VDS**

Chapters from the author's work entitled *Heligoland, an ornithological observatory*, 1890.

5216. Gerlach, Edmund. Beitrag zur Erklärung des Segelfluges der Vögel. (Zeitschrift für Luftschiffahrt, Berlin, 1886, Jahrg. 5, p. 281–292.) **3–VDS**

5217. Guarini, Emile. The flight of birds mechanically studied. illus. (Scientific American, New York, 1903, v. 89, p. 256–257.) **VA**

Apparatus of Achille Bertelli.

5218. Hargrave, Lawrence. Sailing birds are dependent on wave-power. (Royal Society of New South Wales, Journal and proceedings, Sydney, 1899, v. 33, p. 125–128.) *** EC**

Reprinted for the most part in *Aeronautical journal*, London, 1900, v. 4, p. 122–123, *VDS*.

5219. Hart, Ivor Blashka. Leonardo da Vinci's manuscript on the flight of birds. illus. (Royal Aeronautical Society, Journal, London, 1923, v. 27, p. 289–317.) **VDS**

Complete English translation.

5220. Hauvel, Ch. du. Théorie mathématique du vol des oiseaux. illus. (L'Aéronaute, Paris, 1878, année 11, p. 225–235, 260–268, 288–299, 316–324, 344–355, 369–382; 1879, année 12, p. 12–24, 47–59, 72–84, 126–137, 159–166, 185–193, 308–313; 1880, année 13, p. 288–296.) **3–VDS**

5221. Huber, François. Observations sur le vol des oiseaux de proie...accompagnées de figures, dessinées par l'auteur. Genève: P. Barde, 1784. 51 p. illus. sq. f°. **QMD**

5222. Huffaker, E. C. On soaring flight, with introduction by S. P. Langley. (Smithsonian Institution, Annual report for 1896/1897, Washington, 1898, p. 183–206.) *** EA**

5223. —— Soaring flight. illus. (International Conference on Aerial Navigation, Chicago, 1893, Proceedings, New York, 1894, p. 208–219.) **VDS**

5224. —— The way of an eagle in the air. illus. (Aeronautical annual, Boston, 1897, no. 3, p. 128–141.) **3–VDS**

Describes artificial birds made by the author.

Animal Flight, continued

5225. **Hutton,** Frederick Wollaston. On the mechanical principles involved in the sailing flight of the albatross. illus. (Philosophical magazine, London, 1869, series 4, v. 38, p. 130–136.) **OA**

5226. **Institut égyptien.** À la mémoire de Louis Pierre Mouillard, hommage de l'Institut égyptien. Le Caire, 1912. 5–46 p. illus. port. 4°. **† VBA p.v.6, no.14**

Appreciation of the work of Mouillard, French scientist, who studied bird-flight in Egypt, and who in 1881 published *L'Empire de l'air,* of which the Library has a copy, *VDY.*

5227. **Jacob,** Emil. Ueber Buttenstedt's Flugtheorie. (Zeitschrift für Luftschiffahrt, Berlin, 1899, Jahrg. 18, p. 169–175.) **3–VDS**

For Buttenstedt's reply see entry no. 5200.

5228. **Janet,** Charles. Sur le mécanisme du vol chez les insectes. illus. (Académie des sciences, Comptes rendus, Paris, 1899, tome 128, p. 249–253.) *** EO**

5229. **Jarolimek,** Anton. Der Insektenflug. (Zeitschrift für Luftschiffahrt, Berlin, 1894, Jahrg. 13, p. 9–16.) **3–VDS**

5230. —— Ueber den Flug der Vögel mit Beziehung auf die Luftschiffahrtsfrage. (Zeitschrift für Luftschiffahrt, Berlin, 1882, Jahrg. 1, p. 225–240.) **3–VDS**

5231. **Korf,** Georg. So werden wir fliegen! Die natürliche Lösung des Flugproblems. Der Mensch als Segler der Lüfte in naher Zukunft. Mit einem II. Teil: Wenn wir fliegen. Oranienburg: Orania-Verlag, 1909. 87 p. 8°. **VDY**

Discusses bird-flight and the possibility of flying.

5232–5233. **Kotzauer,** Wenzel. Die Luftschiffahrt und ihre Zukunft. Wien: J. B. Wallishausser, 1895. 40 p. illus. 8°. **VDY p.v.2, no.3**

Deals with principles of bird flight as applied to aviation.

5234. **Kress,** Wilhelm. A theory of sailing flight. illus. (International Conference on Aerial Navigation, Chicago, 1893, Proceedings, New York, 1894, p. 206–208.) **VDS**

5234a. **L.,** L. Mechanical flying — examination of the theory of M. Chabrier. diagr. (Mechanics' magazine, London, 1843, v. 38, p. 430–433.) **3–VA**

5235. **Lancaster,** I. The flight of birds. (Engineer, London, 1883, v. 55, p. 172–173.) **VA**

For editorial comments see above magazine, 1883, v. 55, p. 167–168.

For list of Lancaster's writings on bird flight see Paul Brockett, *Bibliography of aeronautics,* Washington, 1910, p. 489, *VDT.*

5236. **Lanchester,** Frederick William. The laws of flight. illus. (Engineer, London, 1908, v. 106, p. 303–305, 329–331.) **VA**

A study of bird flight.

For a list of Lanchester's writings on animal and mechanical flight see Paul Brockett, *Bibliography of aeronautics,* Washington, 1910, p. 490, *VDT.*

5237. **Langley,** Samuel Pierpont. The greatest flying creature; introducing a paper on the pterodactyl, by F. A. Lucas. illus. (Smithsonian Institution, Annual report for 1901, Washington, 1902, p. 649–659.) *** EA**

5238. **Lendenfeld,** Robert J., Edler von Lendlmayr, Reichsritter von. Relation of wing surface to weight. (Smithsonian Institution, Annual report for 1903/1904, Washington, 1905, p. 127–130.) *** EA**

Data for a large number of birds.

5239. **Lilienthal,** Otto. Der Vogelflug als Grundlage der Fliegekunst. Ein Beitrag zur Systematik der Flugtechnik auf Grund zahlreicher von O. und G. Lilienthal ausgeführter Versuche. Berlin: R. Gaertner, 1889. viii, 187 p. illus. 8°. **VDY**

Library has a copy of the English translation, London: Longmans, Green & Co., 1911, *VDY.* For criticism by Émile Simon see *L'Aéronaute,* Paris, 1892, année 25, p. 6–12, *3–VDS.*

5240. —— Our teachers in sailing flight. (Aeronautical annual, Boston, 1897, no. 3, p. 84–91.) **3–VDS**

5241. **Lochet,** G. L'inauguration du monument Mouillard. illus. (La Revue aërienne, Paris, 1912, année 5, p. 202–206.) **3–VDS**

5242. **Louvrié,** Charles de. Erreur de Navier. (L'Aéronaute, Paris, 1868, année 1, p. 89–92, 106–111, 120–127.) **3–VDS**

Takes exceptions to Navier's calculations on bird flight.

5243. —— Les mystère du vol à voiles. (L'Aéronaute, Paris, 1884, année 17, p. 83–89.) **3–VDS**

5244. —— Stabilité absolue de l' oiseau dans l'air. Conditions de cette stabilité. illus. (L'Aéronaute, Paris, 1878, année 11, p. 199–210; 1880, année 13, p. 56–64.) **3–VDS**

5245. —— Suspension et propulsion des oiseaux vivants ou mécaniques. (L'Aéronaute, Paris, 1880, année 13, p. 166–177.)

With comments by Charles du Hauvel. **3–VDS**

5246. —— Theory of soaring flight. (International Conference on Aerial Navigation, Chicago, 1893, Proceedings, Chicago, 1894, p. 220–226.) **VDS**

For citations to the writings of de Louvrié see Paul Brockett, *Bibliography of aeronautics,* Washington, 1910, p. 531–532, *VDT.*

5247. **Lucanus,** Friedrich von. Die Höhe des Vogelzuges, und seine Richtung zum Winde auf Grund aeronautischer Beobachtung. Zwei Vorträge. Neudamm: J. Neumann, 1904. 24 p. 8°. **QMD p.v.12, no.7**

5248. **Lucas,** Frederic Augustus. The greatest flying creature, the great pterodactyl ornithostoma. illus. (Smithsonian Institution, Annual report for 1901, Washington, 1902, p. 654–659.) *** EA**

Animal Flight, continued

5249. Lucy, de. On the flight of birds, of bats, and of insects, in reference to the subject of aerial locomotion. (Engineering, London, 1868, v. 5, p. 486–488, 508–509.) **VDA**

5250. Marey, Étienne Jules. Analyse des mouvements du vol des oiseaux par la photographie. illus. (L'Aéronaute, Paris, 1883, année 16, p. 144–151.) **3–VDS**

5251. —— Animal mechanism: a treatise on terrestrial and aërial locomotion. New York: D. Appleton and Co., 1874. xvi, 283 p. illus. 12°. (International science series. no. 11.) **QCL**

5252. —— Détermination des inclinaisons du plan de l'aile aux différents instants de sa révolution. illus. (L'Aéronaute, Paris, 1872, année 5, p. 76–79.) **3–VDS**

5253. —— Emploi de la photographie instantée pour l'analyse des mouvements chez les animaux. illus. (Académie des sciences, Comptes rendus, Paris, 1882, tome 94, p. 1013–1020.) *** EO**

Also in *L'Aéronaute,* Paris, 1882, année 15, p. 123–131, *3–VDS.*

5254. —— Expériences sur la résistance de l'air pour servir à la physiologie du vol des oiseaux. Paris: G. Masson, 1875. 39 p. illus. 8°. (École des hautes études, Bibliothèque: Section des sciences naturelles, tome 13, article no. 1.) *** EN (École)**

5255. —— Lectures on the phenomena of flight in the animal kingdom. illus. (Smithsonian Institution, Annual report for 1869, Washington, 1872, p. 226–285.) *** EA**

5256. —— Le mécanisme du vol chez les animaux. Cours professé au Collège de France. (L'Aéronaute, Paris, 1870, année 3, p. 134–148, 187–189; 1871, année 4, p. 23–27, 34–42, 54–58, 75–80, 88–94, 102–104.) **3–VDS**

5257. —— Le mécanisme du vol des oiseaux éclairé par la photochronographie. illus. (La Nature, Paris, 1888, année 16, semestre 1, p. 8–14.) **OA**

Instantaneous photographs of bird flight taken by the author. Translation in *Nature,* London, 1888, v. 37, p. 369–374, *OA.*

5258. —— Le mécanisme du vol des oiseaux étudié par la chronophotographie. illus. (Académie des sciences, Comptes rendus, Paris, 1887, tome 104, p. 210–215.) *** EO**

Also in *L'Aéronaute,* Paris, 1887, année 20, p. 103–109, *3–VDS.*

5259. —— Les mouvements de l'aile de l'oiseau représentés suivant les trois dimensions de l'espace. illus. (L'Aéronaute, Paris, 1887, année 20, p. 123–127, 143–148.) **3–VDS**

5260. —— Physiologie du mouvement. Le vol des oiseaux. Paris: G. Masson, 1890. xvi, 394 p. illus. 8°. **VDY**

Extensive, well-illustrated study of the laws of bird flight and of their application to aviation.

5261. —— Physiologie du vol des oiseaux; du point d'appui de l'aile sur l'air. (Académie des sciences, Comptes rendus, Paris, 1874, tome 78, p. 117–121.) *** EO**

Also in *L'Aéronaute,* Paris, 1874, année 7, p. 39–45, with comments of Alphonse Pénaud.

5262. —— Le problème mécanique du vol. illus. (Revue scientifique, Paris, 1888, année 25, semestre 2, p. 289–300.) **OA**

5263. —— Valeurs relatives des deux composantes de la force déployée dans le coup d'aile de l'oiseau, déduites de la direction et de l'insertion des fibres du muscle grand pectoral. (Académie des sciences, Comptes rendus, Paris, 1888, tome 107, p. 549–551.) *** EO**

For citations to Marey's writings see Paul Brockett, *Bibliography of aeronautics,* Washington, 1910, p. 550–553, *VDT.*

5264. Masfrand, Albert de. Victor Tatin. port. (L'Aérophile, Paris, 1906, année 14, p. 219–220.) **3–VDS**

Portrait and brief account of Tatin's experiments in the application of the principles of bird flight.

5265. Means, James. Leonardo da Vinci. [By James Means.] port. illus. (Aeronautical annual, Boston, 1895, no. 1, p. 7–15.) **3–VDS**

Has illustrations and excerpts from Leonardo's treatise on bird flight.

5266. Mewes, Rudolf. Der mechanische Flug einst und jetzt (Leonardo da Vinci und Karl Buttenstedt). illus. (Dinglers polytechnisches Journal, Stuttgart, 1901, Jahrg. 82, p. 29–33, 46–51.) **3–VA**

Discussion of the bird-flight theories of da Vinci and Buttenstedt.

5267. Milla, Karl. Die Flugbewegung der Vögel. Leipzig: F. Deuticke, 1895. 93 p. illus. 8°. **VDY p.v.2, no.2**

5268. Miller-Hauenfels, A. R. von. Das Mysterium des Vogelfluges. illus. (Zeitschrift für Luftschiffahrt, Berlin, 1896, Jahrg. 15, p. 259–268, 308–314.) **3–VDS**

5269. Mott, S. D. An analysis of the functions of a bird's wing during flight and its mechanical imitation. illus. (Scientific American supplement, New York, 1894, v. 38, p. 15518–15519.) **VA**

Suggestion for a flying machine.

5270. Mouillard, Louis Pierre. L'empire de l'air; essai d'ornithologie appliquée à l'aviation. Paris: G. Masson, 1881. 284 p. illus. 8°. **VDY (120)**

Important work by famous French investigator. Translated abstract in Smithsonian Institution, *Annual report for 1892–1893,* Washington, 1893, p. 397–463, ** EA.*

For extensive review see *L'Aéronaute,* Paris, 1881, année 14, p. 217–230, *3–VDS.*

5271. —— Gliding flight. illus. (Cosmopolitan, New York, 1894, v. 16, p. 459–466.) *** DA**

"What the bird does in a wind, man can do."

5272. —— Le vol sans battement; ouvrage posthume inédit de L. P. Mouillard; reconstitué et précédé d'une étude sur l'œuvre

Animal Flight, continued

ignorée de L. P. Mouillard par A. Henry-Coüannier. Paris: Librairie aéronautique, 1912. 484 p. illus. port. 8°. **VDY**

Papers on bird-flight and air flight apparatus by almost forgotten French pioneer (1834–1897), preceded by an extensive biography. For review by Alexandre Sée, see *L'Aérophile,* Paris, 1912, année 20, p. 108–109, *3-VDS.*

For references concerning Mouillard see Paul Brockett, *Bibliography of aeronautics,* Washington, 1910, p. 608, *VDT.* Continued in United States. — National Advisory Committee for Aeronautics, *Bibliography of aeronautics,* 1909–1931, Washington, 1921–35, volumes for 1909/16, p. 930–931; 1923, p. 89, *VDT.*

5273. **Müllenhoff,** Karl. Die Grösse der Flugarbeit. (Pflüger's Archiv für die gesammte Physiologie des Menschen und der Thiere, Bonn, 1885, Bd. 36, p. 548–572.) **QCA**

A study of the comparative flying abilities of large and small flying animals. Reprinted in *Zeitschrift für Luftschiffahrt,* Berlin, 1885, Jahrg. 4, p. 243–251, 259–269, *3-VDS.*

5274. —— Die Grösse der Flugflächen. diagrs. (Pflüger's Archiv für die gesammte Physiologie des Menschen und der Thiere, Bonn, 1885, Bd. 35, p. 407–453.) **QCA**

Comprehensive study of wing size, with history of earlier investigations; also tabular data for 396 animals. Reprinted (without diagrams and tables) in *Zeitschrift für Luftschiffahrt,* Berlin, 1885, Jahrg. 4, p. 161–169, 210–216, 228–233, *3-VDS.*

5275. —— Ueber den Einfluss des Windes auf den fliegenden Vogel. (Zeitschrift für Luftschiffahrt, Berlin, 1891, Jahrg. 10, p. 116–124.) **3–VDS**

5276. —— Ueber das Schweben und Kreisen der Vögel. (Zeitschrift für Luftschiffahrt, Berlin, 1894, Jahrg. 13, p. 256–269.) **3–VDS**

For citations to Müllenhoff's writings see Paul Brockett, *Bibliography of aeronautics,* Washington, 1910, p. 609–611, *VDT.*

5277. **Murie,** James. Recent researches on [animal flight]. (Nature, London, 1871, v. 4, p. 516.) **OA**

Appreciation of the work of Dr. J. Bell Pettigrew.

5278. **Natural** and artificial flight. illus. (Quarterly journal of science, London, 1872, new series, v. 2, p. 27–37, 182–208.) **OA**

Describes studies of Marey and of Pettigrew.

5279. **Nees,** Chr. Om Luftseilads, baseret paa Fugleflugt. Helsingör: A. Schnipp, 1869. 82 p. 8°. **VDC p.v.10, no.4**

Discusses bird flight and aviation.

5280. **Parseval,** August von. Die Mechanik des Vogelflugs. Wiesbaden, 1889. vii, 138 p. plates. 8°. **VDY p.v.3, no.2**

5281. **Peal,** S. E. The soaring of birds. illus. (Nature, London, 1891, v. 44, p. 56.) **3–VDS**

5282. **Pénaud,** Alphonse. Du vol à voiles, ou vol continu sans battements. illus. (L'Aéronaute, Paris, 1875, année 8, p. 75–98, 111–116.) **3–VDS**

5283. —— Lois du glissement dans l'air. (L'Aéronaute, Paris, 1873, année 6, p. 4–18.) **3–VDS**

5284. —— Sur la force des ètres volants. (L'Aéronaute, Paris, 1876, année 9, p. 322–329.) **3–VDS**

For citations to Pénaud's writings on the theory of flight, see Paul Brockett, *Bibliography of aeronautics,* Washington, 1910, p. 658, *VDT.* See also list in Royal Society of London, *Catalogue of scientific papers,* London, 1879, v. 8, p. 585; 1894, v. 10, p. 1020; 1921, v. 17, p. 775, *OAB.*

5285. **Pettigrew,** James Bell. Animal locomotion; or, Walking, swimming, and flying, with a dissertation on aëronautics. New York: D. Appleton & Co., 1874. xiii, 264 p. illus. 12°. (International scientific series.) **VDY**

Library also has a copy of the London edition, H. S. King & Co., 1873, *QGF;* also of the Paris edition, Germer Baillière, 1874, *QCL.*

5286. —— On the various modes of flight in relation to aeronautics. (Royal Institution of Great Britain, Proceedings for 1866–1869, London, 1869, v. 5, p. 94–107.) *** EA**

Also in Smithsonian Institution, *Report for 1867,* Washington, 1868, p. 325–334, ** EA.*

For list of Pettigrew's writings see Paul Brockett, *Bibliography of aeronautics,* Washington, 1910, p. 665–666, *VDT.*

5287. **Prechtl,** Johann Joseph, Ritter von. Untersuchungen über den Flug der Vögel. Wien: Carl Gerold, 1846. vi, 259 p. illus. 8°. **QMD**

5288. **Rayleigh,** John William Strutt, 3d baron. Lord Rayleigh on flight. (Aeronautical journal, London, 1900, v. 4, p. 113–117.) **VDS**

Discusses bird flight and the possibilities of artificial flight.

5289. —— The soaring of birds. (Nature, London, 1883, v. 27, p. 534–535.) **OA**

5290. **Reichenbach,** Oscar. Das Problem des Flugs. (Zeitschrift für Luftschiffahrt, Berlin, 1892, Jahrg. 1, p. 304–312, 340–345.) **3–VDS**

5291. **Saco,** A. Le cours professé au Collège de France. Le mécanisme du vol chez les animaux. illus. (L'Aéronaute, Paris, 1869, année 2, p. 6–9, 26–30, 39–43, 60–63, 67–72, 88–94, 118–121, 137–143, 153–157, 183–188; 1870, année 3, p. 3–14, 29–32, 44–47, 62–64.) **3–VDS**

Describes experimental methods of Étienne Jules Marey.

5292. **Samuelson,** Arnold. Fischschwanz und Flügelschütteln. illus. (Zeitschrift für Luftschiffahrt, Berlin, 1897, Jahrg. 16, p. 191–198.) **3–VDS**

Study of fish and bird movements.

5293. —— Zum Vogelflug. illus. (Zeitschrift für Luftschiffahrt, Berlin, 1896, Jahrg. 15, p. 218–229.) **3–VDS**

5294. **Sée,** Alexandre. Le vol à voile et la théorie du vent louvoyant. (L'Aérophile, Paris, 1909, année 17, p. 242–254.) **3–VDS**

Study of wing flight, with a review of early investigators.

Animal Flight, continued

5295. Silberschlag, Esaias. Von dem Fluge der Vögel. illus. (Gesellschaft naturforschender Freunde, Schriften, Berlin, 1781, Bd. 2, p. 214–270.) 3–* EE

For illustrations see plates 8 and 9 at end of volume.

5296. Smyth, W. Experiments practically demonstrating the laws by which birds fly, and their application to an aerial machine. (Engineering, London, 1867, v. 4, p. 56.) VDA

Paper read before the Aeronautical Society of Great Britain. Discussion on p. 78.

5297. Snodgrass, Robert Evans. How insects fly. illus. (Smithsonian Institution, Annual report for 1928/29, Washington, 1930, p. 383–421.) * EA

See p. 420–421 for bibliography of the more important papers. The article is reprinted, with abridgment, in Royal Aeronautical Society, *Journal*, London, 1933, v. 37, p. 113–114, *VDS*.

5298–5299. Soreau, Rodolphe. Aéronautique. Le vol à voile et l'aviation. illus. (Revue scientifique, Paris, 1895, série 4, tome 3, p. 395–400, 426–431.) OA

Discusses the investigations of Langley, Mouillard, Marey, and Bretonnière.

5300. Sykes, H. M. Thoughts upon the present position of aëronautical science. (Aëronautical Society of Great Britain, Sixth annual report, London, 1871, p. 29–45.) 3–VDS

Discusses relation of bird flight to aeronautics, with comments on the ideas of Wenham and of de Lucy.

5301. Tatin, Victor. Expériences sur la reproduction mécanique du vol de l'oiseau. (Académie des sciences, Comptes rendus, Paris, 1876, tome 83, p. 457–459.) * EO

5302. —— Recherches et expériences sur le mécanisme du vol des oiseaux. illus. (L'Aéronaute, Paris, 1880, année 13, p. 211–229.) 3–VDS

Author's experiments with model airplanes.

5302a. Weisse, H. Das Flug-Gesetz als Grundlage zur Lösung des Flug-Problems im Sinne des Buttenstedt'schen Princips. Kiel: Selbstverlag, 1897. [45] p. illus. 8°.

Aviation and the law of bird-flight. VDY

5303. —— Ueber den heutigen Stand der Luftschiffahrt und die Einführung von Flug-Apparaten nach dem Buttenstedt'schen Princip in unsere Verkehrs-Praxis. (Zeitschrift für Luftschiffahrt, Berlin, 1898, Jahrg. 17, p. 184–197.) 3–VDS

Appreciation of what Buttenstedt did for aviation.

5304. Werner. Der eigentliche Flugmotor der Vögel. illus. (Zeitschrift für Luftschiffahrt, Berlin, 1888, Jahrg. 7, p. 17–21, 49–54, 85–89, 120–126.) 3–VDS

5305. Weyher, Charles. Observations sur le vol plané par orbes. illus. (L'Aéronaute, Paris, 1890, année 23, p. 235–240.) 3–VDS

5306. Winter, Wilhelm. Der Vogelflug. Erklärung der wichtigsten Flugarten der Vögel, mit Einschluss des Segelns und Kreisens. München: Theodor Ackermann, 1895. viii, 172 p. illus. 8°. VDY p.v.1, no.1

5307. Zachariae, August Wilhelm. Die Elemente der Luftschwimmkunst. Wittenberg: Zimmermann, 1807. vi, 282 p. diagr. 8°. VDY

Discusses bird and fish locomotion as related to the problem of steering balloons.

For comments by J. S. C. Schweigger see *Journal für Chemie und Physik*, Nürnberg, 1823, Bd. 38, p. 353–361, *PAA*. Schweigger also refers to the steam engines of Reichenbach and Perkins, and discusses aeronautical problems to be solved.

WOMEN IN AERONAUTICS

5308. Abbott, Mary. Woman's share in the conquest of the air. Some pioneers. ports. (Air, London, 1928, v. 1, June, p. 17–18.) VDS

5309. Aerial excursion of Madamoiselle Garnerin. illus. (Mechanic's magazine, London, 1823, v. 1, p. 33–36.) 3–VA

5310. The **American** girl. Ruth Elder's transatlantic flight. ports. illus. (Aero digest, New York, 1927, v. 11, p. 515, 612.) VDS

Flight with George Haldeman in 1927.

5311. The **American** super-girl and her critics. port. illus. (Literary digest, New York, 1927, v. 95, Oct. 29, p. 52, 54, 56–57.) * DA

Newspaper comments on Ruth Elder's trans-Atlantic flight, 1927.

5312. Amy Johnson's triumph, England-Australia. illus. port. (Flight, London, 1930, v. 22, p. 577–579.) VDS

Log of flight in 1930.

5312a. Ascent of Biggin and Mrs. Sage, June 29, 1785. illus. (In: J. E. Hodgson, History of aeronautics, London, 1924, p. 125–127.) VDT

5313. Bailey, Ethel H. Katharine Wright helped brothers. Close companion and real inspiration to them in their research and experimental work. port. illus. (Airway age, New York, 1928, v. 9, Dec., p. 35–36.) VDS

5314. Bans, Georges. Mademoiselle Marie Marvingt. port. (L'Aérophile, Paris, 1909, année 17, p. 505.) 3–VDS

Brief account of ascensions of famous French woman aeronaut. Portrait.

5315. Beinhorn, Elly. Flying girl. London: G. Bles [1935]. 256 p. front., plates, ports. 8°. VDYC

Flying adventures of the author, a young German woman-flier, in Africa, the Far East and South America, 1932.

Women in Aeronautics, continued

5316. **Bergstrom,** Florence O. Women's City Club of Washington goes aeronautic. First ground school class graduates at ceremonies in honor of girl flyers in Dixie Air Derby. (U. S. air services, Washington, 1930, v. 15, Nov., p. 60.) **VDS**

5317. **Brown,** Margery. Why women pilots have organized. The formation of the "99" Club is a tribute to the growing strength of women's activities in aviation. ports. (Popular aviation, Chicago, 1930, v. 6, March, p. 35–36.) **VDS**

This number of *Popular aviation* is published under the title of *Aeronautics.*

5318. **Bruce,** Mildred Mary (Petre). The Bluebird's flight, with a foreword by Colonel the Master of Sempill. London: Chapman & Hall, Ltd., 1931. xii, 292 p. illus. maps, ports. 8°. **VDYC**

Author's solo flight across Europe, Asia, and America in 1930. Reviewed in *Flight,* London, 1931, v. 23, p. 1140, *VDS.*

5319. **Burtnett,** Gerald B. Bessica Raiche, America's first aviatrix. illus. (Popular aviation, Chicago, 1931, v. 9, July, p. 44, 55.) **VDS**

First flight made Sept. 16, 1910.

5320. **Camac,** Harriet Julia Metcalfe. From India to England by air, with a foreword by Henry Prather Fletcher. New York: Privately printed, 1929. 20 p. illus. ports. 12°. **VDY**

Brief account of author's airplane flight, 1929.

5321. **Chubb,** I. William. Women in airplane construction. illus. (American machinist, New York, 1917, v. 47, p. 705–708; 1918, v. 48, p. 221–225.) **VFA**

Account of women's work in British factories during the World war.

5322. **Cléry,** A. La première femme-aviateur. port. (L'Aérophile, Paris, 1908, année 16, p. 309.) **3–VDS**

Portrait and biography of Mme. Thérèse Peltier.

5323. The **Crossing** of the English Channel by an American aviatress. illus. port. (Scientific American, New York, 1912, v. 106, p. 492.) **VA**

Brief account of Harriet Quimby's flight in 1912.

5324. **Diard,** Ennemonde. Coups d'ailes de l'oiseau à l'avion. Préface de M. Laurent Eynac, Ministre de l'Air. Saint-Étienne: Imprimerie industrielle, 1930. xiv, 532 p. 8°. **VDW**

Various reminiscences of a woman parachute operator. Her impressions of World war aviation. Bird-flight studies of Pierre Lucien Buisson. History of the parachute.

5325. **Dixon,** Charles. Amy Johnson — lone girl flyer. London: S. Low, Marston & Co., Ltd. [1930.] vii, 151 p. illus. map, ports. 12°. **VDYC**

Account of Miss Johnson's airplane flight, England to Australia, in 1930; also of Hinkler's flight over practically the same route in 1928. Reviewed in *Air,* London, 1930, v. 3, p. 308, *VDS; Aeroplane,* London, 1930, v. 38, p. 1138, *VDS.*

5326. **Duchess** of Bedford's record flight. England-India-England in eight days. illus. (Flight, London, 1929, v. 21, p. 878–880.) **VDS**

5327. The **Duchess** of Bedford's record flight. England to Cape Town and back in twenty days. illus. (Flight, London, 1930, v. 22, p. 499–500.) **VDS**

5328. **Earhart,** Amelia. The fun of it. Random records of my own flying and of women in aviation. New York: Brewer, Warren & Putnam, 1932. 218 p. illus. 8°. **VDYC**

In May, 1932, the author made the first transatlantic airplane flight by a woman.

5329. —— 20 hrs. 40 min.; our flight in the Friendship. The American girl, first across the Atlantic by air, tells her story. New York: G. P. Putnam's Sons, 1928. 314 p. illus. ports. 8°. **VDYC**

5330. Die **Ersten** Luftschifferinnen. (Wiener Luftschiffer-Zeitung, Wien, 1904, Jahrg. 3, p. 143–146.) **VDS**

List of women aeronauts, 1784–1848.

5331. **Espitallier,** Georges. Les femmes aéronautes. ports. (Illustrierte aëronautische Mitteilungen, Strassburg, 1904, Jahrg. 8, p. 309–322.) **VDS**

Portraits and brief accounts of women balloonists from 1784 to 1904.

5332. **Etzdorf,** Marga von. Kiek in die Welt. Als Deutsche Fliegerin über drei Erdteilen. Geleitwort von Hugo Junkers. Berlin: Union deutsche Verlagsgesellschaft [cop. 1931]. 172 p. ports. illus. 8°. **VDY**

Flights of German woman aviator in Europe, Africa, and Asia.

5333. The **First** aviatress in Great Britain. port. (Aero, London, 1910, v. 3, p. 3.)

Brief reference to Mrs. Gavin. **3–†VDS**

5334. **Fonvielle,** Wilfrid de. Mlle. Dorothée Klumpke. port. (L'Aérophile, Paris, 1900, année 8, p. 1–3.) **3–VDS**

Portrait and biography of French astronomer and aeronaut.

5335. **Fradiss,** O. Une femme ingénieur d' aéronautique. port. (L'Aérophile, Paris, 1925, année 33, p. 266.) **3–VDS**

The author, a woman engineer of note, discusses the adaptability of women for aeronautical engineering.

5336. **Gibson,** Ada. Women who fly. ports. (Aircraft, New York, 1910, v. 1, p. 134–136.) **VDS**

Accounts of some of the well-known aviators of 1910.

5337. **Goldstrom,** John. Modern Valkyries. ports. (Popular aviation, Chicago, 1929, v. 4, June, p. 28–29, 100–104.) **VDS**

Brief accounts of some contemporary women fliers. This number of *Popular aviation* has the title *Aeronautics.*

5338. **Gould,** Bruce. Girls have wings, too! port. (St. Nicholas, New York, 1929, v. 56, p. 631, 674.) **Children's Room**

How Elinor Smith, 17, set an endurance record.

Women in Aeronautics, continued

5339. **Graham,** Lloyd S. When ladies step out. Miss Kauffman was thrown from her plane while looping. Mrs. Fisher was lost in the fog and was forced to bail out. Others have jumped for various reasons. ports. (Popular aviation, Chicago, 1931, v. 8, Feb., p. 23–24, 60.) **VDS**

5340. **Grey,** Charles Grey. On Lady Bailey's achievement. [By George G. Grey.] illus. (Aeroplane, London, 1929, p. 105–106, 108.) **VDS**
Flight around Africa, 1928.

5341. —— Winifred Spooner. port. (Aeroplane, London, 1933, v. 44, p. 94, 96, 98.) **VDS**
Appreciation and portrait of prominent English woman aviator.

5342. **Grosse,** Margarete. Die Frau im Ballonsport. ports. (In: Adolf Mehl, Der Freiballon im Theorie und Praxis, Stuttgart: Franckh'sche Verlagshandlung, 1911, Bd. 1, p. 236–246.) **VDW**
In 1784 Mme. Tible made the first woman's ascent.

5343. **Hattoom,** Fred L. Teaching women to fly. ports. (Aero digest, New York, 1930, v. 16, March, p. 63–64.) **VDS**
Experiences at the Standard Flying Schools, Los Angeles.

5344. **Heath,** Sophie Mary (Peirce-Evans) Elliott-Lynn, lady. Airwomen. ports. (Air, London, 1927, v. 1, Dec., p. 14–18.) **VDS**
Accounts of contemporary English women aviators.

5345. —— Woman and flying, by Lady Heath and Stella Wolfe Murray. London: J. Long, Ltd. [1929.] 223 p. illus. ports. 8°. **VDY**
Introductory historical chapter is followed by accounts of flights of the authors. Reviewed in *Air,* London, 1929, v. 2, p. 451, *VDS.*

5346. —— Women who fly. port. (Popular aviation, Chicago, 1929, v. 4, May, p. 32, 34, 86.) **VDS**
Deals chiefly with some contemporary fliers.

5347. **Hewlett,** Hilda Beatrice (Herbert). London to Batavia by air. port. (Aeroplane, London, 1933, v. 44, p. 142, 144, 146.) **VDS**
Author's flight (1932?).

5347a. **Hodgson,** J. E. George Graham and Mrs. Graham. illus. (In his: History of aeronautics, London, 1924, p. 224–229.) **VDT**
Mrs. Graham and her husband made numerous balloon ascents in England, 1823–1853.

5348. **Jotti,** da Badia, polesine. Pagine di storia aeronautica: da Pierjacopo Martello ad Elisa Garnerin. illus. port. (Aeronautica, Milano, 1929, anno 3, p. 234–236.) **3–VDS**
Quotations from the poet Martello (18th century). Account of balloonist and parachutist, Elisa Garnerin.

5349. **Lady** Bailey's record solo flight. 18,000 miles on a "Cirrus-Moth." map. (Flight, London, 1929, v. 21, p. 29–30.) **VDS**
Flight in 1928.

5350. **La Quiante,** Emmy. Damen im Korb. illus. (In: Karl Bröckelmann, Wir Luftschiffer, Berlin, 1909, p. 144–154.) **VDW**
Brief accounts of women balloon pioneers and of some of their successors.

5351. **Laura** Ingalls ends solo flight of more than 16,000 miles. (U. S. air services, Washington, 1934, v. 19, May, p. 18.) **VDS**
Miss Ingalls was the first woman to fly over the Andes mountains.

5352. **Lillian** Gatlin, the first woman to cross the continent by airplane. port. (Aeronautical digest, New York, 1922, v. 1, p. 209.) **VDS**

5352a. **Lindbergh,** Anne (Morrow). North to the Orient; with maps by Charles A. Lindbergh. New York: Harcourt, Brace and Company, 1935. 255 p. front., illus. maps. 8°. **VDYC**
Story of the flight made by Col. and Mrs. Lindbergh in 1931 from Washington, D. C., to China and Japan.

5353. **Madame** Blanchard, the aeronaut. illus. port. (Scientific American supplement, New York, 1879, v. 7, p. 3099.) **VA**

5354. **Marvingt,** Marie. Traversée nocturne de la Mer du Nord. illus. (L'Aérophile, Paris, 1910, année 18, p. 17–20.) **3–VDS**
Flight by author, woman aviator, Nancy to England over the North Sea in 1909.

5355. **Miss** [Amy] Johnson's triumph. (Aeroplane, London, 1930, v. 38, p. 1045–1046.) **VDS**
Brief account of flight, England-Australia, in 1930.

5356. **Miss** Earhart makes woman's speed record. (Aviation, New York, 1930, v. 29, p. 28.) **VDS**
Made average speed of 174.9 miles per hour at Detroit, 1930.

5357. **Miss** Harriet Quimby meets death at Boston meet. illus. (Fly, Philadelphia, 1912, v. 4, Aug., p. 20–21.) **3–† VDS**

5358. **Miss** Quimby flies the Channel. port. (Flight, London, 1912, v. 4, p. 345.) **VDS**
Brief account of flight in 1912.

5359. **Miss** Ruth Law now holds American non-stop cross-country record, and world cross-country record for women. port. (Aerial age weekly, New York, 1916–17, v. 4, p. 277–278.) **VDS**
Flight, Chicago-New York, in 1916.

5360. **Mrs.** J. A. Mollison's great flight to the Cape and back in record time. illus. (Flight, London, 1932, v. 24, p. 1207–1208.) **VDS**

5361. **Mrs.** Julia Clark killed. port. (Fly, Philadelphia, 1912, v. 4, July, p. 10.) **3–† VDS**
First fatality to a woman flier in America, June 17, 1912.

5362. **Mrs.** Mollison's return. port. (Aeroplane, London, 1932, v. 43, p. 1194, 1196.) **VDS**
Record flight, England to Capetown and return, 1932.

Women in Aeronautics, continued

5363. Mrs. Winnie Buller. port. (Aeronautics, London, 1912, v. 5, p. 339.) **VDS**

Brief account of British woman aviator.

5364. Moedebeck, Hermann W. L. Die deutschen Frauen und die Luftschiffahrt. ports. (Illustrierte aëronautische Mitteilungen, Strassburg, 1904, Jahrg. 8, p. 381–390.) **VDS**

Brief accounts of German women balloonists.

5365. —— Nochmals die deutschen Frauen und die Luftschiffahrt. illus. (Illustrierte aëronautische Mitteilungen, Strassburg, 1905, Jahrg. 9, p. 105–108.) **VDS**

Accounts of ascents, 1900 to 1904.

5366. Now it's "birdgirls" — and young ones at that. (Literary digest, New York, 1928, v. 96, Feb. 18, p. 64–66.) *** DA**

Experiences of Miss Cozette Parker, 13 years old, and of other members of famous family.

5367. Omlie, Phoebe Fairgrave. Women in the air races. illus. (Aero digest, New York, 1930, v. 17, p. 40, 42.) **VDS**

Account of the Women's National Air Derby, 1930.

5368. Pease, Richmond. The career of Matilde Moisant. illus. ports. (Popular aviation, Chicago, 1929, v. 4, p. 46, 48, 50.) **VDS**

Experiences of American woman aviator.

5369. Queens of speed. ports. (Western flying, Los Angeles, 1933, v. 13, July, p. 47.) **VDS**

Portraits and brief accounts of Louise Thaden, Mae Haizlip, and Ruth Nichols.

5370. Quimby, Harriet. How I made my first big flight abroad. My flight across the English Channel. port. (Fly, Philadelphia, 1912, v. 4, June, p. 8–10.) **3–† VDS**

5371. Renard, Paul. Ceux qui disparaissent. Madame Poitevin. port. (L'Aérophile, Paris, 1908, année 16, p. 196–197.) **3–VDS**

Portrait and biography of woman aeronaut who made 571 ascensions during the middle years of the nineteenth century.

5372. Ruth Elder's ocean flight. illus. ports. (Literary digest, New York, 1927, v. 95, Oct. 22, p. 9.) *** DA**

Newspaper comments on her unsuccessful trans-Atlantic flight with George W. Haldeman, 1927.

5373. Ruth Law's record breaking flight. (Flying, New York, 1916, v. 5, p. 454–455.) **VDS**

Chicago-New York flight, 1916.

5374. Ruth Nichols sets altitude mark for women. (U. S. air services, Washington, 1931, v. 16, April, p. 45.) **VDS**

Record of 28,743 feet in 1931.

5374a. Semple, Elizabeth Anna. Harriet Quimby, America's first woman aviator. ports. (Overland monthly, San Francisco, 1911, series 2, v. 58, p. 525–532.) *** DA**

5375. Sibour, Violette (Selfridge), vicomtesse de. Flying gypsies; the chronicle of a 10,000 mile air vagabondage. Maps drawn by the Vicomte Jacques de Sibour. New York: G. P. Putnam's Sons, 1930. iv, 306 p. illus. maps, ports. 8°. **VDY**

Lively account of flight of author and her husband, London to Indo-China, and across America.

5376. Silberer, Victor. Käthchen Paulus. port. (Wiener Luftschiffer-Zeitung, Wien, 1904, Jahrg. 3, p. 195–197.) **VDS**

Biography and portrait of German woman balloonist and associate of Lattemann, the inventor of the double parachute.

5377. Smith, Erle H. From cold cream jars she conjured wings, and a half-pint blonde comes to realize their responsibilities. port. (U. S. air services, Washington, 1930, v. 15, April, p. 50–51.) **VDS**

Account of American woman flier, Mildred Kauffman.

5378. Spencer, G. K. Pioneer women of aviation. illus. ports. (Sportsman pilot, New York, 1930, v. 3, May, p. 16–17, 49.) **VDS**

5379. Spooner, Winifred. A cruise to the Near East. An account of an 8,500 miles pleasure cruise across Europe to Egypt and Transjordania by light aeroplane. illus. (Air and airways, London, 1932, v. 9, p. 299–302.) **VDS**

Author's flight, 1932.

5380. Stevens, Leo. First woman passenger in an airship. (American aeronaut, St. Louis, 1907, v. 1, Oct., p. 19–20.) **3–VDS**

Stated to have been Mrs. C. J. S. Miller, of Franklin, Pa.

5381. Stinson, Marjorie. The diary of a country girl at flying school. ports. illus. (Aero digest, New York, 1928, v. 12, p. 168–169, 296–297.) **VDS**

5382. Thaden, Louise McPhetridge. The women's air derby. Experiences on the flight from Santa Monica to Cleveland as related by the winner of the heavy plane class. illus. port. (Aero digest, New York, 1929, v. 15, Oct., p. 62, 299.) **VDS**

Event of 1929.

5383. Vaughan, Katherine. Women and flight. ports. (Aero digest, New York, 1925, v. 6, p. 24–25.) **VDS**

Brief accounts of some of the latter-day women aviators.

5384. Wolff, Maurice. Les femmes dans l'aéronautique. illus. (Revue aérienne, Paris, Aug. 10, 1912, p. 421–425.) **3–VDS**

5385. A **Woman** hops the Atlantic. illus. ports., map. (Literary digest, New York, 1928, v. 97, June 30, p. 8–9.) *** DA**

Flight of Amelia Earhart in airplane *Friendship,* Newfoundland to Wales, 1928.

Women in Aeronautics, continued

5386. A **Woman** who flew in 1911. Early days at Mourmelons and Brooklands. An interview with Hilda B. Hewlett. illus. port. (Airways, London, 1927, v. 4, p. 127–128, 150.) **VDS**

Mrs. Hewlett was both a pioneer aviator and a manufacturer.

5387. **Women** of the air. ports. (Popular mechanics, Chicago, 1929, v. 51, p. 42–46.) **VFA**

Brief accounts of Katherine Stinson, Harriet Quimby, Matilde Moisant, Bernetta Miller, Lillian Todd, Phoebe Fairgrave Omlie.

5388. The **Women's** air derby. illus. (Literary digest, New York, 1929, v. 102, Sept. 7, p. 9.) *** DA**

California to Cleveland, 1929. Death of Miss Marvel Crosson, holder of the women's altitude record.

5389. **Women's** Committee of the Air League of the British Empire. ports. (Air, London, 1929, v. 2, p. 268–271.) **VDS**

Brief accounts of women prominent in British flying.

5390. **Woodhouse,** Henry. Pioneer women of the air. ports. (Flying, New York, 1913, v. 2, June, p. 22–25.) **VDS**

Portraits and brief accounts of Marie Louise Driancourt, Bernetta Adams Miller, Mrs. Maurice Hewlett, Jeanne Pallier, Baroness Raymonde de Laroche, Mlle. Galanschikoff, Jane Hervue, Mlle. Marvingt, Matilda Moisant, Helene Dutrieu, and Nellie Beese.

MISCELLANEOUS

5391. **Abel** Hureau de Villeneuve. port. (L'Aéronaute, Paris, 1898, année 31, p. 127-136.) **3–VDS**

Obituary discourses on the founder of the periodical *L'Aéronaute.* Born 1833; died 1898.

5392. **Acland,** W. R. D. Deck flying. illus. (Royal Aeronautical Society, Journal, London, 1931, v. 35, p. 372–400.) **VDS**

History in use of flying decks on ships. Also in *Aeroplane,* London, 1931, v. 40, p. 157–162, *VDS.* Abstract in *Aviation engineering,* Washington, N. J., 1931, v. 4, June, p. 27–28, *VDS.*

5393. **Aerial** fish patrol. (Scientific American, New York, 1920, v. 123, p. 89.) **VA**

Experiments in California for locating schools of fish at sea.

5394. **Aerial** police unit starts operation in New Jersey. illus. (American city, New York, 1929, v. 41, July, p. 86.) **SERA**

"Believed to be the first of its kind, in the East if not in the entire country."

5394a. **Aërial** transport a factor in the Dominion's mining industry. illus. map. (Engineering and mining journal, New York, 1935, v. 136, p. 549–550.) **† VHA**

5395. **Aeronautic** radio on the Federal airways system. illus. (United States. — Commerce Department, Air commerce bulletin, Washington, 1933, v. 5, p. 127–133.) **VDS**

5396. The **Aeronautical** advisors of our nation. ports. (Flying, New York, 1915, v. 4, p. 511, 519–520, 546–547.) **VDS**

Portraits and biographies of members of the United States Advisory Committee: Charles D. Walcott, George P. Scriven, Samuel Reber, Charles F. Marvin, Holden C. Richardson, Byron R. Newton, John F. Hayford.

5396a. **Aircraft** in exploration. illus. (Aircraft yearbook, New York, 1926, p. 79–86.) **VDS**

"For the first time in the history of aviation, airplanes were used in 1925 as important means of transport in exploration." Dr. A. Hamilton Rice to northwestern Brazil; Macmillan and Byrd to the Arctic; and Amundsen-Ellsworth Arctic expedition.

5397. **Aircraft** radio development by the Bureau of Standards. (United States. — Commerce Department: Aeronautics Branch, Aircraft commerce bulletin, Washington, 1929, v. 1, Aug. 15, p. 4–11.) **VDS**

5397a. The **Airplane** in mining. map. (Engineering and mining journal, New York, 1935, v. 136, p. 541–543.) **† VHA**

5398. **Airplane** radio up to date. (Literary digest, New York, 1927, v. 94, Aug. 27, p. 62.) *** DA**

Describes equipment of Byrd's airplane *America.*

5399. **Amans,** P. C. Du rôle des formes animales dans les progrès de la navigation aérienne et aquatique. (Bulletin scientifique de la France et de la Belgique, Paris, 1906, v. 40, p. 207–228.) **PQA**

5400. **American** Society of Aeronautic Engineers select Messrs. Wood and Sperry for the Advisory Board. (Flying, New York, 1915, v. 4, p. 660–661, 673–674, 676.) **VDS**

Biographies of Henry Wise Wood and Elmer Ambrose Sperry.

5401. **America's** aeronautical industry. A record of labors engaged in and victories won. illus. maps. (Aviation, New York, 1933, v. 32, p. 257–266.) **VDS**

5402. **America's** first aviator senator. Gov. Hiram Bingham elected to United States Senate. port. (Aviation, Highland, N. Y., 1925, v. 18, p. 46.) **VDS**

5403. **Angerstein,** Wilhelm. Zur Erinnerung an den Freiherrn Friedrich [Josef] vom Hagen. (Zeitschrift für Luftschiffahrt, Berlin, 1885, Jahrg. 4, p. 353–356.) **3–VDS**

Vom Hagen (1827–1885) was a prominent German writer on the history of aeronautics.

5404. **Arnold,** Henry H. Aerial forest fire patrol, 1920. maps. (Aviation, Highland, N. Y., v. 10, p. 45–47.) **VDS**

Operations in California and Oregon.

Miscellaneous, continued

5405. **Bacon,** John Mackenzie. Seeing beneath the waves. illus. (Knowledge, London, 1905, v. 2, p. 98–100.) **OA**

Author was commissioned by the British Admiralty to endeavor to take photographs of the sea bottom from a balloon.

5406. **Beij,** Karl Hilding. Aircraft speed instruments. illus. (United States. — National Advisory Committee for Aeronautics, 18th annual report, 1932, Washington, 1933, p. 381–435.) **VDY (U. S.)**

Historical introduction and bibliography of 100 titles.

5407. [**Bibliographies** of aeronautical nomenclature.] (In: Paul Brockett, Bibliography of aeronautics, Washington, 1910, p. 818, under the heading, *Terminology*. Continued in United States. — National Advisory Committee for Aeronautics, *Bibliography of aeronautics,* 1909–1931, Washington, 1921–35, volumes for 1909/16, p. 960–961; 1917/19, p. 332–333; 1920/21, p. 303; 1922, p. 117; 1923, p. 101; 1925, p. 130; 1926, p. 107–108; 1928, p. 148; 1929, p. 169; 1930, p. 179; 1931, p. 212.) **VDT**

5408. [**Bibliographies** of carrier pigeons.] (In: Paul Brockett, Bibliography of aeronautics, Washington, 1910, p. 671; continued in United States. — National Advisory Committee for Aeronautics, Bibliography of aeronautics, 1909–1931, Washington, 1921–35, volumes for 1909/16, p. 1042; 1917/19, p. 354; 1920/21, p. 325; 1926, p. 114; 1929, p. 178.) **VDT**

5409. [**Bibliographies** of radio in aeronautics.] (In: Paul Brockett, Bibliography of aeronautics, Washington, 1910, p. 896; continued in United States. — National Advisory Committee for Aeronautics, Bibliography of aeronautics, 1909–1931, Washington, 1921–35, volumes for 1909/16, p. 1317–1318, 1438–1442; 1917/19, p. 372, 478–480; 1920/21, p. 343–344, 438–439; 1922, p. 130, 165; 1923, p. 111, 141; 1924, p. 89, 111; 1925, p. 144, 185; 1926, p. 119, 151; 1927, p. 144; 1928, p. 164; 1929, p. 185–186, 237–238; 1930, p. 198–199, 257; 1931, p. 231–232.) **VDT**

5410. [**Bibliography** of the aerial compass.] (In: United States. — National Advisory Committee for Aeronautics, Bibliography of aeronautics, 1909–1931, Washington, 1921–35, volumes for 1909/16, p. 289–290; 1917/19, p. 98–99; 1920/21, p. 93; 1922, p. 31; 1923, p. 28; 1924, p. 20; 1925, p. 40; 1926, p. 30; 1927, p. 36; 1928, p. 44; 1929, p. 55; 1930, p. 48–49; 1931, p. 58.) **VDT**

5411. [**Bibliography** of aeronautics at Harvard University.] (In: United States. — National Advisory Committee for Aeronautics, Bibliography of aeronautics, 1909–1931, Washington, 1921–35, volume for 1909/16, p. 618–619.) **VDT**

5412. [**Bibliography** of aeronautics at Yale University.] (In: United States. — National Advisory Committee for Aeronautics, Bibliography of aeronautics, 1909–1931, Washington, 1921–35, volumes for 1909/16, p. 1472; 1917/19, p. 490.) **VDT**

5413. [**Bibliography** of coast patrol.] (In: United States. — National Advisory Committee for Aeronautics, Bibliography of aeronautics, 1909–1931, Washington, 1921–35, volumes for 1909/16, p. 279, 1012; 1917/19, p. 91; 1920/21, p. 83; 1922, p. 29; 1926, p. 28; 1927, p. 33; 1929, p. 52.) **VDT**

5414. [**Bibliography** of expositions, exhibitions, and competitions.] (In: Paul Brockett, Bibliography of aeronautics, Washington, 1910, p. 294.) **VDT**

Cites numerous references down to July, 1909.

5415. [**Bibliography** of hangars.] (In: United States. — National Advisory Committee for Aeronautics, Bibliography of aeronautics, 1909–1931, Washington, 1921–35, volumes for 1909/16, p. 607–610; 1917/19, p. 208–209; 1920/21, p. 188; 1922, p. 63; 1923, p. 56; 1924, p. 43; 1925, p. 74; 1926, p. 59; 1927, p. 74; 1928, p. 85; 1929, p. 97–98; 1930, p. 99–100; 1931, p. 119.) **VDT**

5416. [**Bibliography** of instruments.] (In: United States. — National Advisory Committee for Aeronautics, Bibliography of aeronautics, 1909–1931, Washington, 1921–35, volumes for 1909/16, p. 672–675; 1917/19, p. 229–231; 1920/21, p. 210–212; 1922, p. 72–73; 1923, p. 62–63; 1924, p. 49; 1925, p. 82–83; 1926, p. 66–67; 1927, p. 84; 1928, p. 95–96; 1929, p. 107; 1930, p. 114; 1931, p. 135–136.) **VDT**

5417. [**Bibliography** of safety in aeronautics.] (In: United States. — National Advisory Committee for Aeronautics, Bibliography of aeronautics, 1909–1931, Washington, 1921–35, volumes for 1909/16, p. 1174–1177; 1917/19, p. 394; 1920/21, p. 361–362; 1922, p. 139; 1923, p. 119; 1924, p. 96; 1925, p. 153; 1926, p. 126; 1927, p. 152; 1928, p. 173–174; 1929, p. 195; 1930, p. 211; 1931, p. 244–245.) **VDT**

5418. [**Bibliography** of signals.] (In: United States. — National Advisory Committee for Aeronautics, Bibliography of aeronautics, 1909–1931, Washington, 1921–35, volumes for 1909/16, p. 1226–1228; 1917/19, p. 406; 1920/21, p. 375; 1922, p. 143; 1926, p. 132; 1927, p. 158; 1929, p. 203.) **VDT**

5419. **Blanchet,** Georges. L'inventeur Arthur William Loth. port. (L'Aérophile, Paris, 1922, année 30, p. 21–22.) **3–VDS**

Biography of inventor of magnetic aircraft guiding system. See also entry no. 5491.

5420. —— Lieutenant-colonel Georges Espitallier. port. (L'Aérophile, Paris, 1908, année 16, p. 285.) **3–VDS**

Portrait and brief biography of prolific French aeronautical writer.

Miscellaneous, continued

5421. **Boeve,** Richard de. Nouveau traité pratique du pigeon voyageur actuel, contenant les dernières perfections et les secrets de l'élevage colombophile applicables à l'art militaire et maritime... Lille: Liégeois-Six, 1898. 101 p. illus. 8°. **QMM**

5422. **Bourdin,** J. Sur un instrument analogue au compas aéronautique décrit par M. Janssen. (Académie des sciences, Comptes rendus, Paris, 1871, tome 72, p. 256–257.) *** EO**

5423. **Breckel,** Harry F. Aircraft radio. Its early history and more recent developments. illus. (Aero digest, New York, 1925, v. 6, p. 82–83, 102.) **VDS**

5423a. **Brenard,** R. Africa mineral areas united by air service. illus. (Engineering and mining journal, New York, 1935, v. 136, p. 567–568.) **† VHA**

5424. **Brief** histories of the development of companies exhibiting at the International Aeronautical Exposition. Chicago, Ill., Dec. 1–9, 1928. ports. (Aviation, New York, 1928, v. 25, p. 1727–1765.) **VDS**

Brief accounts of American companies.

5425. A **Brief** history of airplane forest patrol. (United States. — Commerce Department: Aeronautics Branch, Air commerce bulletin, Washington, 1929, v. 1, July 15, p. 12–14.) **VDS**

5426. **Brief** history of launching catapults. illus. (Aviation, Highland, N. Y., v. 10, p. 278.) **VDS**

5427. **Brion,** Maurice. Évolution des méthodes de l'instruction au pilotage. (L'Aéronautique, Paris, 1928, année 10, p. 56–58.) **VDS**

Development of military pilot instruction methods in France.

5428. **Burgess,** George Kimball. Aircraft radio beacon development. illus. (Aviation, Highland, N. Y., 1928, v. 24, p. 1764–1765, 1798–1803.) **VDS**

5429. **Cailletet,** Louis Paul. Sur l'emploi de l'oxygène dans les ascensions à grandes hauteurs. (Académie des sciences, Comptes rendus, Paris, 1901, tome 132, p. 1017–1020.) *** EO**

5430. **Cameron,** Allan J. Skywriting pilots have covered United States. illus. (Aeronautical digest, New York, 1923, v. 3, p. 406–407.) **VDS**

5431. **Capon,** R. S. The reduction of aircraft noise. diagrs. (Engineering, London, 1933, v. 136, p. 475–476.) **VDA**

Accounts of experimental work carried out by the British government.

5432. **Carmody,** Frank J. Aerial traffic cops in London and Chicago. (Literary digest, New York, 1928, v. 99, Nov. 24, p. 57–58.) *** DA**

5433. —— "Wildcat" aviation stocks. (Literary digest, New York, 1928, v. 99, Oct. 20, p. 86–87.) *** DA**

5434. **Catapault** launching for aeroplanes. illus. (Aeroplane, London, 1931, v. 40, p. 327–328, 330, 332.) **VDS**

Brief introduction on history of launching planes from decks of ships.

5435. **Chicanot,** E. L. Flying with the sealing fleet. illus. (Aero digest, New York, 1926, v. 8, p. 332–333.) **VDS**

Experiences of C. S. Caldwell in locating seal fields in Newfoundland.

5436. —— New "eyes" for the sealing fleet. illus. (Scientific American, New York, 1928, v. 138, p. 409–411.) **VA**

Experiences of C. S. Caldwell.

5437. **Christmann,** Léon. Zur Luftschiffersprache. Allerlei historische Ausgrabungen. (Deutsche Zeitschrift für Luftschiffahrt, Berlin, 1911, Jahrg. 15, March 22, p. 19–20, 22–23; April 19, p. 16–18.) **VDS**

Early use of the words Luftschiff, Luftball, and other aeronautical expressions.

5438. **Christopher,** Luke, and R. F. de Marolles. The life history of a world's record. illus. (Aviation, New York, 1930, v. 29, p. 158–161.) **VDS**

How a record is measured and recorded.

5439. **Church,** Glenn J. Battling aerial bootleggers. illus. (Popular aviation, Chicago, 1932, v. 11, p. 93, 126–127.) **VDS**

Story of U. S. government combat airplane on the Canadian border.

5440. **Clancy,** Carl Stearns. Aviation's patron saint. (Scientific American, New York, 1929, v. 141, p. 283–286.) **VA**

Flying experiences and comments of Will Rogers.

5441. **Cleaves,** Prentiss, jr. Joe Doakes, the miracle man of aviation. illus. (Popular aviation, Chicago, 1931, v. 8, Feb., p. 18–19, 63.) **VDS**

Biography of famous American "wing-walker."

5442. **Cooper,** Courtney Ryley. Frozen North. The story of a Canadian prospector as told to Courtney Ryley Cooper. illus. (Saturday evening post, Philadelphia, 1929, v. 202, Sept. 21, p. 56, 58, 60, 62, 64.) *** DA**

Airplane prospecting in the Hudson Bay region.

5443. **Crowdus,** Walter C. Aviation insurance. (Journal of air law, Chicago, 1931, v. 2, p. 176–192.) **VDS**

Has historical introduction.

5444. **Cuthill,** R. W., and H. S. Hinchman. History and commercial value of the airplane launching catapult. illus. (Aero digest, New York, 1931, v. 19, Dec., p. 40–44, 98–99.) **VDS**

Miscellaneous, continued

5445. Dacy, George H. The coast patrol. illus. (Popular aviation, Chicago, 1928, Feb., v. 2, p. 10–12, 82.) **VDS**

Account of work of U. S. Coast Guard.

5446. Daily sales of United Cigar flying store averaged $200. illus. (Aviation, Highland, N. Y., 1927, v. 23, p. 1113.) **VDS**

5447. David, Evan John. Flying the frozen North. illus. (St. Nicholas, New York, 1929, v. 56, p. 811–814, 845.) **(Children's Room)**

How the Canadian government transported men, surveying instruments, and railway materials into the Hudson Bay region.

5448. Dellinger, John Howard, and others. Development of the visual type radio-beacon system, by J. H. Dellinger, H. Diamond, and F. W. Dunmore. illus. (United States. — Standards Bureau, Journal of research, Washington, 1930, v. 4, p. 425–459.) **VBDA**

Includes early history of radio beacons. Bibliography.

5448a. Development of mining aviation. (Engineering and mining journal, New York, 1935, v. 136, p. 553–554.) **† VHA**

A chronological list of writings appearing in the *Engineering and mining journal,* 1919–1935.

5449. Diamond, H., and F. W. Dunmore. A radiobeacon and receiving system for blind landing of aircraft. illus. (United States. — Standards Bureau, Journal of research, Washington, 1930, v. 5, p. 897–931.) **VBDA**

Includes brief reference to pioneer work.

5450. Dienstbach, Carl. The man of the hour. port. (Fly, 1909, v. 1, April, p. 5–6.) **3–† VDS**

Account of the efforts of Cortlandt Field Bishop to foster aeronautics in the United States.

5451. Dieuaide, Emmanuel. L'Exposition aéronautique de Londres à Alexandra Palace en 1885. (L'Aéronaute, Paris, 1886, année 19, p. 50–56.) **3–VDS**

Brief account of exhibits.

5452. Doane, Robert R. Aeronautical finance in 1930. charts. (Aviation, New York, 1931, v. 30, p. 45–49.) **VDS**

History of capital changes, stock movement, earnings, etc.

5453. —— Finance; a curtailment of losses. charts. (Aviation, New York, 1932, v. 31, p. 116–117.) **VDS**

Statistical study of stock movements during 1931.

5454. —— The industry's income for 1929. charts. (Aviation, New York, 1930, v. 29, p. 75–80.) **VDS**

5455. Dusting plane development in 1926. (Aviation, Highland, N. Y., 1927, v. 22, p. 1174.) **VDS**

5456. E., M. B. A history of the development of aircraft wireless. illus. (Aeroplane, London, 1920, v. 19, p. 443–444, 446, 480, 482, 484, 515–516, 518, 549–550, 552.) **VDS**

5456a. Elford, Harold S. Australian mining and the airplane. illus. (Engineering and mining journal, New York, 1935, v. 136, p. 568–569.) **† VHA**

5457. Elias, H. H. W. L. Moedebeck. port. (Deutsche Zeitschrift für Luftschiffahrt, Berlin, 1910, Jahrg. 14, Heft 5, p. 7–8.) **VDS**

Has a full-page portrait of prominent German writer on aeronautics.

5458. Elmer Sperry is awarded Fritz medal. A description of the Sperry airway beacon to be used on the lighted airways. illus. port. (Aviation, Highland, N. Y., v. 21, p. 1036–1037.) **VDS**

Includes brief biography.

5459. Esnault-Pelterie, Robert Albert Charles. L'astronautique. Paris: A. Lahure, 1930. 248 p. diagrs. 8°. **VDYR**

Introductory historical chapter gives an excellent résumé of the various investigations into the possibility of rocket propulsion.

5460. —— Considérations sur les résultats d'un allégement indéfini des moteurs. (Journal de physique, Paris, 1913, série 5, tome 3, p. 218–230.) **PAA**

Author's calculations on rocket propulsion.

5461. Executive Board of the Aero Club of America passes resolution of appreciation of work of Henry Woodhouse. port. (Aerial age weekly, New York, 1917, v. 5, p. 7–8.) **VDS**

Woodhouse was one of the founders of the periodical *Flying,* and a prolific writer on aeronautics.

5462. Fatal [airplane] accidents of the year [1910 and 1911]. (Aeronautical journal, London, 1910, v. 14, p. 157–158; 1911, v. 15, p. 27, 125, 147.) **VDS**

Chronologies. Title varies.

5463. Fighting boll weevil by means of airplanes. Army Air Service and Department of Agriculture cooperate in aerial extermination campaign. illus. (Aviation, Highland, N. Y., 1923, v. 15, p. 210–212.) **VDS**

Account of various experiments.

5464. The **First** airplane radio broadcasting. illus. (Aviation, Highland, N. Y., 1924, v. 17, p. 961–962.) **VDS**

Experiments made August 14, 1924, by Station WJZ of New York City.

5465. The **First** British aero show. (Flight, London, 1909, v. 1, p. 155–156.) **VDS**

Description of exhibits at the Olympia exhibition, 1909.

5466. The **First** flight exhibition and tournament of the Aeronautic Society. illus. (Scientific American, New York, 1909, v. 101, p. 29.) **VA**

Brief account of exhibition at Society's grounds at Morris Park, N. Y., in which Glenn Curtiss took a prominent part.

5467. The **First** Paris aeronautical salon. illus. (Flight, London, 1909, new series, v. 1, p. 6–7.) **VDS**

Brief account of 1908 exhibition.

Miscellaneous, continued

5468. Flint, Howard R. The fire eagles. illus. (American forests, Washington, 1928, v. 34, p. 67–69.) **VQN**

Describes methods of detecting and mapping forest fires, as conducted by the United States War Department and the Forest Service. Also in *Literary digest,* New York, 1928, v. 96, Feb. 25, p. 25, * *DA.*

5469. Fog piercing beacon. (Popular aviation, Chicago, 1929, v. 4, Jan., p. 25, 107.) **VDS**

Quartz neon-gas filled lamp of the General Electric Company.

5470. Fonvielle, Wilfrid de. L'aéronautique à l'exposition de 1889 [Paris]. (L'Aéronautique, Paris, 1889, année 22, p. 121–124, 145–148.) **3–VDS**

5471. Foreign trade back to 1929 levels. diagrs. (Aviation, New York, 1934, v. 33, p. 88–89.) **VDS**

Statistics of U. S. exports for several years shown by charts.

5472. Les **400** premières victimes de l'aviation. (In: L'Aéronautique, par Roland Garros [and others], Paris: Lafitte & Cie., 1914, p. 216–222.) **VDW**

List covers accidents to civil and military aviators in the various countries, 1909–1913.

5473. Goddard, Robert Hutchings. A method of reaching extreme altitudes. City of Washington: Smithsonian Institution, 1921. 2 p.l., 69 p. illus. 8°. (Smithsonian miscellaneous collections. v. 71, no. 2.) *** EA**

Publication no. 2540.

5474. Goldstrom, John. Safety in aviation improving. (New York times, Current history, New York, 1927, v. 26, p. 37–40.) **BAA**

5474a. Gordon, W. A. Aircraft speed opening of Canada's mineral resources. illus. (Engineering and mining journal, New York, 1935, v. 136, p. 544–545.) **† VHA**

The author was formerly Minister of Mines of Canada.

5475. Grey, Charles Grey. Eric Clift. [By George G. Grey.] (Aeroplane, London, 1922, v. 23, p. 440.) **VDS**

Brief biography of pioneer British inventor of aeronautical instruments.

5476. —— On twenty years of the Aeroplane. (Aeroplane, London, 1931, v. 40, p. 1017–1024.) **VDS**

History of well-known British periodical.

5477. H.M.S. Courageous, the flagship of our aircraft carriers. illus. (Flight, London, 1933, v. 25, p. 27–34.) **VDS**

This famous ship was laid down in 1915.

5478. Hagen, Freiherr vom. Die Internationale aëronautische Ausstellung im Krystallpalast von Sydenham im Jahre 1868. (Zeitschrift für Luftschiffahrt, Berlin, 1885, Jahrg. 4, p. 33–42.) **3–VDS**

5479. —— Die Brieftaube. (Zeitschrift für Luftschiffahrt, Berlin, 1883, Jahrg. 2, p. 225–235.) **3–VDS**

Good history of carrier pigeons from ancient times.

5480. Haskelite history. (Aero digest, New York, 1926, v. 9, p. 146–147.) **VDS**

History of firm manufacturing waterproof plywood.

5481. Hathaway, Levy M. The beginning of aviation medicine. (Air Corps news letter, Washington, 1928, v. 12, p. 323–328.) **† VDS**

5482. The **Heinkel** catapault on the S. S. "Bremen." illus. (Aviation, New York, 1929, v. 27, p. 263–264.) **VDS**

5482a. Hoernes, Marianne. Todesopfer der Luftfahrt. illus. (In: Hermann Hoernes, Buch des Fluges, Wien, 1912, Bd. 3, p. 331–392.) **VDW**

History of aeronautical accidents with numerous portraits.

5483. Hofmann-Genf, J. Erfinder-Schicksale im Gebiete der Flugtechnik. illus. (In: Karl Bröckelmann, Wir Luftschiffer, Berlin, 1909, p. 312–325.) **VDW**

Misadventures of pioneer balloonists and aviators.

5484. Hopper, James Marie. A school of aviation at Mourmelon, France. illus. (Collier's, New York, 1910, v. 45, no. 11, p. 19–20, 24, 26.) *** DA**

Describes methods of instruction.

5485. Horton, Harry M. Story of the first communication by radio from a flying plane to earth. port. (U. S. air services, Washington, 1926, v. 11, March, p. 21–23.) **VDS**

Story of the author's experiments culminating in success in 1910.

5486. Howard, Edward P. An analysis of aircraft accidents during 1928. (Aviation, New York, 1929, v. 26, p. 1501–1503.) **VDS**

5487. Instrument to turn on airway lights at nightfall and on dark days developed by Aeronautics Branch. (United States. — Commerce Department: Aeronautics Branch, Air commerce bulletin, Washington, 1932, v. 3, p. 413–415.) **VDS**

5488. Instruments carried in the international balloon race at St. Louis. illus. (Scientific American, New York, 1907, v. 97, p. 290.) **VA**

5489. Jacobs, A. M. First use of liquid oxygen at McCook Field. (U. S. air services, Washington, 1924, v. 9, April, p. 34–35.) **VDS**

Used for altitude flight at McCook Field, June 1, 1923.

5490. —— The motion picture in aviation. illus. (Popular aviation, Chicago, 1928, v. 2, Feb., p. 13–15.) **VDS**

5491. James, P. Les câble-guides de M. Loth. La direction des aéronefs sans aucune visibilité. illus. (L'Aérophile, Paris, 1922, année 30, p. 18–21.) **3–VDS**

Proposal to guide aircraft cross country automatically by magnetic cables strung on poles. See also entry no. 5419.

Miscellaneous, continued

5492. Janssen, Pierre Jules César. Compas aéronautique. illus. (Académie des sciences, Comptes rendus, Paris, 1871, tome 72, p. 222–225, 291–292.) *** EO**

5493. Jones, Ernest Lester. Bill boards of the air. (National Aeronautic Association review, Washington, 1925, v. 3, p. 117–118.)
Brief history of sky-writing. **VDS**

5494. Kelvin, Bottomley & Baird, Ltd. Pioneers in design and construction of aircraft compasses. illus. (Air annual of the British Empire, London, 1930, v. 2, p. 628–635.) **VDS**

5495. Klemin, Alexander. The handwriting on the sky — the most recent and compelling method of advertising. illus. (Scientific American, New York, 1923, v. 128, p. 323.) **VA**
First conceived by Major J. C. Savage, of England, in 1911.

5496. —— Investing in aviation. The recent boom in the aviation industry has made it doubly necessary for the prospective investor to investigate the situation thoroughly. illus. (Scientific American, New York, 1929, v. 140, p. 148–153.) **VA**

5497. Kremser, Victor Felix Karl. Wilhelm Angerstein. (Zeitschrift für Luftschiffahrt, Berlin, 1893, Jahrg. 12, p. 129–130.) **3–VDS**
Angerstein (1835–1893) was the founder of the Deutsche Verein zur Förderung der Luftschiffahrt.

5498. Landauer, Bella Clara. Book-plates from the aeronautics collection of Bella C. Landauer. New York: Privately printed at the Harbor Press, 1930. 12 l. illus. 8°. *** KP (Harbor)**

5499. —— Some aeronautical music from the collection of Bella C. Landauer. Paris: Privately printed, 1933. [9] p., 1 l., 23 facsims. obl. 8°. *** MF**
Reproductions of title-pages of instrumental and vocal music going back well into the 19th century.

5500. Lasser, David. The conquest of space. With an introduction by H. H. Sheldon. New York: Penguin Press, 1931. 271 p. illus. 8°. **VDY**
Cites the literature of flight and traces the development of the modern rocket idea.

5501. The **Lexington** and Saratoga. New airplane carriers are 888 ft. long and 106 ft. wide and are rated as being the highest powered vessels in the world. illus. (Aviation, Highland, N. Y., 1928, v. 24, p. 94–95.) **VDS**

5502. Loeffler, H. C. Aerial anti-pest activities. illus. (World's health, Paris, 1926, v. 7, p. 64–69.) **SPA**
Results obtained in combating the cotton boll weevil, mosquitoes, and other agricultural pests.

5503. —— Aerial forestry. illus. (World's health, Paris, 1926, v. 7, p. 288–297.) **SPA**
What has been done in Canada, the United States, and other countries in forest patrol work.

5504. —— Aerial health and relief work. illus. (World's health, Paris, 1925, v. 6, p. 501–505.) **SPA**
Developments in various countries.

5505. —— Aerial medicine and teaching. illus. (World's health, Paris, 1926, v. 7, p. 162–164.) **PSA**
How Australia proposes to provide "flying doctors."

5506. Ludlow, Israel. Aeronautics at the Jamestown exposition. illus. (American magazine of aeronautics, New York, 1907, v. 1, July, p. 3–6.) **VDS**

5507. McConnell, Burt Morton. The story of sky-writing. illus. (St. Nicholas, New York, 1928, v. 55, p. 439–440, 491.) **Children's Room**

5508. Magruder, Thomas Pickett. Aircraft carriers. illus. (Saturday evening post, Philadelphia, 1928, v. 200, June 23, p. 25, 91, 93.)
Early history. *** DA**

5509. Maine the first state to actually take steps toward founding a system of aerial coast protection. (Aerial age weekly, New York, 1915, v. 2, p. 30, 104, 201; 1916, v. 3, p. 176.) **VDS**
Title varies.

5510. Mamer, N. B. Protecting our forests from fire. illus. (Aero digest, New York, 1927, v. 10, p. 173, 246–248.) **VDS**
Author's experiences along the Idaho-Washington border.

5511. Marchis, Lucien René Andre Edmond. L'enseignement de l'aéronautique à l'Université de Paris. (Revue du mois, Paris, 1910, v. 9, p. 257–269.) *** DM**

5512. Martin, E. Stockton. The Boy Scouts in aviation. How 800,000 boys are receiving the training aviators need. (Aviation, Highland, N. Y., 1926, v. 20, p. 216–217.) **VDS**

5513. —— Facts on aircraft insurance. (Aviation, Highland, N. Y., 1922, v. 12, p. 10.)
Brief history. **VDS**

5514. Mays, Brock. Aviation slang. illus. (Popular science monthly, New York, 1928, v. 112, May, p. 72, 141.) *** DA**
Partly reprinted in *Literary digest,* New York, 1928, v. 97, May 12, p. 73–74, ** DA.*

5515. Mealing, R. H. S. The development of aviation lighting. illus. (Illuminating engineer, London, 1933, v. 26, p. 296–304.)
With discussion. **3–† VEXA**

5516. Mercy takes wings. illus. (Popular aviation, Chicago, 1928, v. 2, June, p. 19–42.) **VDS**
Saving lives by airplane, notably during the Mississippi flood.

5517. Miller, J. Earle. Mystery of port of missing planes. illus. (Popular mechanics, Chicago, 1928, v. 49, p. 360–363.) **VFA**
Discoveries of Captain Ocker and Surgeon D. A. Myers, of the United States Army, concerning hallucinations caused by defective working of the semicircular ear canals of aviators.

Miscellaneous, continued

5518. Mr. Grindell Matthew's "death-ray." (Flight, London, 1924, v. 16, p. 375.) **VDS**

The inventor of the alleged engine-stopping ray refuses to accept British test conditions.

5519. Moedebeck, Hermann W. L. Ueber den Luftschiffer-Kompass des Kais. Russischen Ingenieur-Oberst Alexandrowitsch Kosloff zur Bestimmung von Fahrtrichtung, Geschwindigkeit und Drehungen des Ballons. (Zeitschrift für Luftschiffahrt, Berlin, 1890, Jahrg. 9, p. 232–234.) **3–VDS**

5520. Montagnes, James. Making an Indian treaty by air in Canada. Time and trouble saved in visiting the Indians of the North. illus. (Airway age, New York, 1930, v. 11, p. 1442–1444.) **VDS**

5521. —— The Royal Mounted Aerial Police. illus. (Popular aviation, Chicago, 1933, v. 12, p. 17, 66.) **VDS**

Experiences of Canadian organization.

5522. New neon light for aviation. illus. (Scientific American, New York, 1928, v. 138, p. 352, 369.) **VA**

General Electric Company's fog-piercing lamp for airports.

5523. Newman, J. O. The Frankfort aeronautical exhibition. illus. (Cassier's magazine, New York, 1909, v. 37, p. 99–115.) **VDA**

Illustrated account of exhibits, including airplanes, airship halls, aerial guns, motors, hydrogen apparatus, etc.

5524. Noyes, Charles E. New York shows of the past. Six exhibitions from 1912 to 1930, have had widely varying purposes and results. illus. (Aviation, New York, 1930, v. 28, p. 880–884.) **VDS**

5525. O'Malley, Henry. Bureau of Fisheries. (National Aeronautic Association review, Washington, 1925, v. 3, p. 38–39.) **VDS**

History of fish patrols in the United States and other countries.

5526. Operations of the Aerial Forest Patrol, 1922. Air Service report shows that aerial patrols offer unparalleled means for suppressing forest fires. illus. (Aviation, Highland, N. Y., 1923, v. 14, p. 241–244.) **VDS**

5527. Our airplane carrier "Saratoga" launched. Greatest and speediest ship of its class in the world. illus. (U. S. air services, Washington, 1925, v. 10, May, p. 31.) **VDS**

5528. Our first big plane carrier. illus. (Literary digest, New York, 1925, v. 85, April 18, p. 11.) *** DA**

Account of the *Saratoga.*

5529. P., C. La prima scuola militare di aeronautica. (Rivista aeronautica, Roma, 1928, anno 4, Sept., p. 663–665.) **VDS**

Established at Meudon, France, 1791.

5530. Paddock, Paul D. Flying electric billboards and talking airplanes introduce novelty in outdoor advertising. illus. (Popular mechanics, Chicago, 1929, v. 51, p. 723–727.) **VFA**

5531. Philp, Charles G. Stratosphere and rocket flight (astronautics); a popular handbook on space flight of the future, including a section on the problems of interplanetary navigation. London: Sir Isaac Pitman & Sons, Ltd., 1935. vii, 106 p. illus. 12°. **VDYR**

Historical survey, p. 5–9.

5532. Piersol, James V. Adapting the airplane to the newspaper. A report covering an experiment by the Detroit News. illus. (Aero digest, New York, 1931, v. 18, Jan., p. 35–39, 122.) **VDS**

William E. Scripps began experiments in 1912 as the basis of subsequent successful operations.

5532a. Pioneer aërial prospector: John E. Hammell. illus. port. (Engineering and mining journal, New York, 1935, v. 136, p. 551–552.) **† VHA**

His experiences in Canada.

5533. Polk, George W. Instruments used by the world flyers. illus. (U. S. air services, Washington, 1925, v. 10, Feb., p. 31–34.) **VDS**

Used by U. S. Army flyers.

5534. Progress of aviation insurance by years. (Popular aviation, Chicago, 1932, v. 10, Jan., p. 39.) **VDS**

Chronology, 1918–1931.

5535. Putnam, George Rockwell. Radiobeacons and radiobeacon navigation. Washington: U. S. Gov. Prtg. Office, 1931. 42 p. illus. 8°. (United States. Lighthouse Service. Publication.) **TTF (U. S. Lighthouse)**

For historical notes see p. 34–35.

5536. Radio range beacon service now available on entire transcontinental airway between San Francisco and New York. map. (United States. — Commerce Department: Aeronautics Branch, Air commerce bulletin, Washington, 1931, v. 2, p. 438–439.) **VDS**

5537. Redington, Paul G. Aerial fire patrol on the national forests. (Aviation, Highland, N. Y., 1925, v. 19, p. 248–249.) **VDS**

Reviews three years' work in California and Oregon.

5538. Review of recent research on airplane radio receiving equipment. (United States. — Commerce Department: Aeronautics Branch, Air commerce bulletin, Washington, 1932, v. 3, p. 357–359.) **VDS**

5539. Rigg, E. H. The launch of the airplane carrier U.S.S. Saratoga. illus. diagrs. (Society of Naval Architects and Marine Engineers, Transactions, New York, 1925, v. 33, p. 147–160.) **VXA**

With discussion.
Difficulties and problems of launching. Illustrations and diagrams at end of volume. Abridged in *Engineering,* London, 1926, v. 121, p. 56–60, *VDA.*

Miscellaneous, continued

5540. **Rowland,** May. An airman's hymn. (Literary digest, New York, 1928, v. 96, Feb. 11, p. 30.) *** DA**
Author awarded prize given by the Hymn Society of America.

5541. **Sagendorph,** Kent. The airplane finds oil. illus. (Popular aviation, Chicago, 1929, v. 5, July, p. 16–18, 90.) **VDS**
Aerial prospecting for petroleum. This number of *Popular aviation* has the title of *Aeronautics.*

5542. **Scaroni,** Silvio. What racing did for the Italian industry. illus. (Aviation, New York, 1929, v. 27, p. 400–403.) **VDS**

5543. **Schloeder,** Nicholas S. Chronology of model aviation. (Aircraft, New York, 1913, v. 4, p. 182–183.) **VDS**
Records, 1909–1913.

5544. **Schnurmacher.** An adventure in aerial sleuthing. The true story of the part played by the New York Aerial Police in trailing down a gang of kidnappers. illus. (Popular aviation, Chicago, 1931, v. 8, Feb., p. 31.) **VDS**

5545. **Scientific American** and its intimate connection with the progress of aviation. illus. (Scientific American, New York, 1933, v. 149, p. 166–167.) **VA**
Chiefly a history of the Scientific American trophy.

5546. The **Second** annual exhibition of the Aero Club of America. illus. (Scientific American, New York, 1906, v. 95, p. 447–449.) **VA**
Has pictures of some of the exhibits, including aeronautical engines.

5547. **Small,** R. D. Financial survey of aeronautical industry and securities. (Popular aviation, Chicago, 1928, v. 3, Oct., p. 29–30, 112–113.) **VDS**

5548. **Smith,** J. E. Evolution of aircraft and marine radio beacons. illus. (Aero digest, New York, 1929, v. 15, Nov., p. 76–77, 288, 290.) **VDS**

5549. The **Society's** exhibit at the International Aero Exhibition, Olympia, July 16th–27th, 1929. A summary. illus. (Royal Aeronautical Society, Journal, London, 1929, v. 33, p. 980–1015.) **VDS**
Descriptions of exhibits of the Royal Aeronautical Society.

5550. **Stanton,** Charles V. Seeding of waste lands by airplane. illus. (Aviation, New York, 1929, v. 26, p. 243–245.) **3–VDS**
Account of experiments in southern Oregon.

5551. A **Submarine** aircraft carrier. illus. (Flight, London, 1931, v. 23, p. 759–763.) **VDS**
Account of British submarine *M.2.*

5552. **Success** in aerial boll weevil campaign. (Aviation, Highland, N. Y., 1924, v. 16, p. 92.) **VDS**
Tests by the U. S. Department of Agriculture.

5553. **Tegetmeier,** William Bernhard. Pigeons for land and sea service, with examples from recent experiments. illus. map. (Royal United Service Institution, Journal, London, 1892, v. 36, p. 483–498.) **VWA**
With discussion.

5554. **Titterington,** Morris M. Earth inductor compass used by Lindbergh. illus. (U. S. air services, Washington, 1927, v. 12, July, p. 25–26, 52.) **VDS**

5555. **United States.** — Army: Air Corps. Statistical studies of aircraft accidents and forced landings. Washington: U. S. Gov. Prtg. Off., 1930. 30 p. diagrs. 4°. (Air Corps information circular. no. 652.) **VDY**

5556. **United States.** — Air Service: Medical Section. Statistics compiled from reports on crashes in the U. S. Army Air Service during the calendar years 1918–1921, inclusive, and results of physical examinations for flying during the calendar years 1920 and 1921. Washington: Gov. Prtg. Off., 1922. 22 p. 4°. (Air Service information circular, Aviation. v. 4. no. 340.)

5557. **Uses** of aircraft in agriculture, forestry, etc. illus. (Aircraft yearbook, New York, 1926, p. 71–77.) **VDS**
Significant developments during 1925.

5558. **Vallot,** Joseph, and others. Expériences de télegraphie sans fil en ballon libre, par J. Vallot, Jean et Louis Lecarme. (Académie des sciences, Comptes rendus, Paris, 1900, tome 130, p. 1305–1307.) *** EO**

5558a. **Van Dusen,** William. Mine freight by air in South Africa. illus. (Engineering and mining journal, New York, 1935, v. 136, p. 569–570.) **† VHA**

5559. **Vorreiter,** Ansbert. Die Internationale Luftschiffahrt Ausstellung in Frankfurt a.M. illus. (Verein deutscher Ingenieure, Zeitschrift, Berlin, 1909, Bd. 53, p. 1244, 1613–1623, 1721–1724.) **VDA**

5560. **Wakeman,** S. W. Launch of the airplane carrier U. S. S. Lexington. illus. diagrs. (Society of Naval Architects and Marine Engineers, Transactions, New York, 1926, v. 34, p. 139–161.) **VXA**
With discussion.
Problems and difficulties of launching, April 7, 1925. Illustrations and diagrams at back of volume. Abridged in *Engineering,* London, 1926, v. 122, p. 737–741, 771–772, *VDA.*

5561. The **War** in the air — on mosquitoes. Use of planes more efficient. Affords operators new source of revenue. illus. (Airway age, New York, 1930, v. 11, p. 931–933.) **VDS**

5562. **Wead,** Frank Wilbur. Flying movies. illus. (Popular aviation, Chicago, 1929, v. 5, Sept., p. 20–22, 96.) **VDS**
Author's experiences in taking motion pictures from airplanes. This number of *Popular aviation* has the title of *Aeronautics.*

Miscellaneous, continued

5563. Werner-Bleines. Sauerstoff und Atmungsvorrichtungen für Luftschiffahrt. illus. (Illustrierte aëronautische Mitteilungen, Berlin, 1909, Jahrg. 13, p. 900–903, 1029–1040.) **VDS**

Includes a history of breathing apparatus for use in aircraft.

5564. West, A. G. The air patrol of the [U. S.] Coast Guard. illus. (Aero digest, New York, 1929, v. 15, August, p. 57–58, 284, 286, 288.) **VDS**

Interesting history of early experiences.

5565. What balloon racers did with receiving sets. illus. (Radio broadcast, Garden City, N. Y., 1923, v. 3, p. 507–509.) **TTFA**

Experiences during the National balloon race, July 4, 1923.

5566. Wilson, J. A. Gentlemen adventurers of the air. Many regions of Canada's vast wilderness, long hidden even from fur trappers, are now revealed by exploring airmen. illus. (National geographic magazine, Washington, D. C., 1929, v. 56, p. 597–642.) **KAA**

5567. Wimperis, Harry Egerton. Some recent developments in aircraft instruments. illus. (Royal Aeronautical Society, Journal, London, 1924, v. 28, p. 3–34, 45–48.) **VDS**

With discussion.

5568. Winters, S. R. Crop survey by airplane. illus. (Popular aviation, Chicago, 1928, v. 3, Sept., p. 11–14.) **VDS**

Work of the U. S. Department of Agriculture. Also in *Literary digest,* New York, 1928, v. 99, Nov. 10, p. 22–23, ** DA.*

5569. —— Guiding airplanes by radio beacons. illus. (Popular aviation, Chicago, 1927, Nov., p. 23–26.) **VDS**

Apparatus at College Park, Maryland, designed at the Radio Laboratory of the U. S. Bureau of Standards.

5570. —— Plant protection by airplane. illus. (Popular aviation, Chicago, 1929, v. 4, Feb., p. 11–14.) **VDS**

Work of U. S. Department of Agriculture.

5571. With the ice-plane patrol. illus. (Popular mechanics, Chicago, 1929, v. 52, p. 731–733.) **VFA**

How German airplanes spotted ice-bound ships during the winter of 1928–1929.

5572. Woodhouse, Henry. Aeroplanes as life savers. They have preserved armies in war times and rescued many by land and sea. illus. (Flying, New York, 1914, v. 3, p. 10–13, 28–29.) **VDS**

5573. —— The first Pan-American Aeronautic Exhibition most far-reaching in developing Pan-Americanism. illus. (Flying, New York, 1917, v. 6, p. 126–129, 147.) **VDS**

5574. Words, meanings, and styles. (Nature, London, 1925, v. 115, p. 253–255, 290–291.) **OA**

Discussion of coined aeronautical expressions. For reply see entry no. 1534.

INDEX OF AUTHORS

(Numbers refer to entries, not to pages.)

A

Abbe, Cleveland, 2607–2609.
Abbot, Charles Greeley, 1423.
Abbott, Mary, 5308.
Abell, C. F., 2675, 2676.
Abercron, Hugo Wilhelm von, 397.
Abrams, Talbot, 5083a.
Acland, W. R. D., 5392.
Acosta, Bertram B., and others, 2907.
Adams, Arthur A., 3522.
Adams, Briggs Kilburn, 3809.
Adams, Darwin J., 115.
Adams, George, 402.
Adams, Heinrich, 1424.
Ader, Clément Agnès, 1425, 1426, 3524.
Adler, Milton, 853.
Aeligius, 37.
Aero Club of America, 117.
Aero-Club von Deutschland, 4586.
Aero-club de France, 1434.
Aerodynamische Versuchsanstalt, Göttingen, 4653.
Aeronautic Society, 4523.
Aeronautisches Observatorium, Lindenberg, 4655.
Agafonov, V. K., 4656.
Agen, Myer, 1440.
Agnoletti, V., 123.
Ahmad Zakī, 39, 39a.
Aimé, Emmanuel, 411–414, 860, 861, 4657, 4658.
Air Conference, London, 4525.
Airy, Hubert, 5178.
Aktiebolaget C. E. Fritzes kungliga hofbokhandel, 2.
Alace, 125.
Alban, and Vallet, 866.
Albert I, prince of Monaco, 2533, 4659.
Albret, André, 3819.
Alcanter de Brahm, 3.
Alcock, F. J., 5087a.
Alexander, A., 4660.
Alexander, P. Y., 4661. *See also* Brewer, Griffith, and P. Y. Alexander.
Alexander-Katz, Bruno, 126, 1450.
Alford, Leon Pratt, 1451.
Allen, J. W., 2679.
Amans, P. C., 1454, 2828, 5179, 5399.
American Bar Association, 4588.
American Society of Mechanical Engineers, 127.
Ames, Joseph Sweetman, 4662, 4663.
Amundsen, Roald Engelbregt Gravning, 4432, 4433.
Amundsen, Roald Engelbregt Gravning, and L. W. Ellsworth, 4434, 4435.
Ancelle, Paul, 416.
Anderson, John W., 4175.
Anderson, S. Herbert, 4664.
André, Henri, 869.
Andrée, Salomon August, 418.
Andrews, Solomon, 870, 3535.
Angerstein, Wilhelm, 2476, 5403.
Angle, Glenn Dale, 2680.
Apitz, 1459.
Aracena, Diègo, 2924.
Arago, Dominique François Jean, 4665, 4666.
Archdeacon, Ernest, 420, 421, 2307.
Archen, 3536.
Archibald, Douglas, 128, 422.
Arco, 872.
Argyll, George Douglas Campbell, 8th duke, 5180.
Arlandes, marquis d', 423.
Armani, Armando, 3821.
Armengaud, J., the younger, 129, 873, 1460, 1461.
Armstrong, Douglas B., 4355.
Armstrong, E. R., 2308.
Arnesen, Odd, 4437.
Arnesen, Odd, and E. O. Sem-Jacobsen, 130.
Arnold, Henry H., 2926, 3539a, 5404.
Arnold, Stuart Amos, 425.
Arson, 2829.
Artingstall, F. D., 1463.
Ashmore, Edward Bailey, 3822.
Ashwell-Cooke, J. R., 2365.
Assman, Richard, 4669, 4670.
Assman, Richard, and Arthur Berson, 4671.
Association française aérienne, 4528.
Aubé, 3540a.
Austin, A. O., 4672.
Austin, Warren Robinson, 4356.
Averly, A., 4673.
Aviaticus, pseud., 3824.
Ayers, W. O., 1475.

B

B., G., 877.
B., L., 1476.
B., N., 878.
Babinet, Jacques, 4674.
Babinet, Léon, 428, 429.
Bach, Carl von, 879.
Baciocchi, Alighiero, 1477.
Bacon, Francis, 4675.
Bacon, Gertrude, 136, 430, 431, 4676.
Bacon, John Mackenzie, 432–434, 880, 4677, 5405.
Bacon, Roger, 4678, 4679.
Baddeley, William, 881–883.
Baden-Powell, Baden Fletcher Smyth, 137–139, 435, 2610, 2611, 3541, 3542.
Baeder, F. de, and G. Dubouchet, 436.
Baer, Herbert Christian, 2683.
Baer, Joseph & Company, 4.
Bahier, Eugène, 4438.
Bailey, Ethel H., 5313.
Bairstow, Leonard, 1478, 3825, 4680, 4681.
Baker, George S., 2309.
Balaban, Karl, 2830.
Balbo, Italo, 2935a, 2936, 2936a, 2936b.
Balchen, Bernt, 4439.
Baldwin, N. C., 4357.
Baldwin, W. H., 1479.
Balichou, 437.
Baltimore, J. Mayne, 886.
Banet-Rivet, P., 140–143, 887.
Banks, Charles A., 4178.
Bannerman, Sir Alexander, 3545.
Bannerman-Phillips, H., 144, 3546.
Bans, Georges, 2938, 5314.
Barbet, A., 4683, 4684.
Baring, Maurice, 3826.
Barkeley, Armand Sylvain, 888.
Barker, Fowler W., 2939.
Barlow, Lester P., 3547.
Barnard, William N., 3827.
Barnes, G. M., 3548.
Barnes, Henry, 444.
Barnes, John K., 3828.
Barnett, Gilbert, 3829.
Baroni Cavalcabo, Clemente, 4685.
Barr, Guy, 445. *See also* Rosenhain, Walter, and Guy Barr, 5007.
Barral, Jean Augustin, and A. J. Bixio, 4686.
Barré, L., 446.
Barrenbrugge, Charles C., 1480.
Barrington-Kennett, B. H., 3549.
Barrows, H. R. *See* Payne, E. G., and H. R. Barrows.
Barry, Edward A., 889.
Barry, R. A., 4359.
Barton, F. A., 890, 2310.
Baschin, Otto, 4440.
Basté, J. F., 5181.

Bates, Lindon Wallace, 1481.
Batson, A. S., joint author. *See* Irving, H. B., and A. S. Batson.
Bauer, 1482.
Bauer, Louis Hopewell, 4687, 4688.
Bauer, P. P. von, 4179.
Bauer-Keiper, Karl, 894.
Baumann, A., 3830.
Baur, Fritz, 3831.
Bausset, A. de, 895.
Bayles, J. C., 3550.
Bazin, Albert, 5182, 5183.
Beach, A. E., 4689.
Beach, S. Y., joint author. *See* Jones, E. L., and S. Y. Beach.
Beamish, Richard Joseph, 2941.
Beatty, W. D., and A. D. de R. Martin, 1483.
Beaufoy, Henry, 447.
Beaumont, Frederick F. E., 3551, 3552.
Beazeley, George Adam, 5088.
Beck, James Montgomery, 1484.
Beecroft, David, 3832.
Beery, L. L., 2942.
Běhounek, František, 4441.
Beij, Karl Hilding, 5406.
Beinhorn, Elly, 5315.
Bejeuhr, Paul, 4690.
Bek, B. Roustam, 145.
Bell, Alexander Graham, 1485–1487, 2612.
Bellah, James Warner, 4180.
Bellanca, Giuseppe, joint author. *See* Acosta, Bertram B., and others.
Bellet, Daniel, 2613.
Bellonte, M., joint author. *See* Costes, Dieudonné, and M. Bellonte.
Benn, Wedgwood, 3833.
Bennett, Cora Lillian (Orkins), 4442.
Bennett, Floyd, 4443.
Benson, S. R., 2311.
Benson, Stuart, 2945.
Berchem, Beda von, 2366.
Berg, Alfred, 449.
Berger, 3553.
Berget, Alphonse, 146, 147, 4691.
Bergner, Hans Heinz, 2946.
Bergonzi, P. C., 1488.
Bergstrom, Florence O., 5316.
Berlepsch, Franz von, 3554.
Bernard, Marc, 2947.
Bernardi, Mario de, 4360.
Berry, William Hibbert, 148.
Berson, Arthur, 4692. *See also* Assman, Richard, and Arthur Berson.
Berson, Arthur, and R. Süring, 4693.
Bert, Paul, 4694.
Berthenson, Georg, 5184.
Bertin, E. L., 1490.
Bertrand, Charles, 3555.
Bertùccioli, Americo, 5.
Bertùccioli, Americo, and B. Pinchetti, 41.
Besançon, Georges, 149, 450, 451, 898–903, 1491, 1492, 2684, 2948, 3556, 4695, 4696. *See also* Fonvielle, Wilfrid de, and Georges Besançon, 1045; Hermite, Gustave Besançon, 4861.
Bescherelle, Louis Nicolas, aîné, 150.
Béthuys, G., 151, 904, 905, 3557.
Bewsher, Paul, 3834.
Biard, Henry Charles, 2949.
Bibesco, Marthe Lucie, princess, 3825.
Biddle, Charles John, 3840.
Biddlecomb, C. H., 4725.
Bielenberg, Johannes, 4590.
Bienaimé, Maurice, 464.
Bierbaum, Paul Willi, 2962.
Bigelow, A. A., 4184.
Bilancioni, Guglielmo, 4726.
Billwiller, C. F., 916.
Bingham, Hiram, 1520, 2963.
Binney, George, 4446.
Biot, Jean Baptiste, 4727.
Bishop, William A., 3841.
Bixio, A. J. *See* Barral, Jean Augustin, and A. J. Bixio.
Blachère, André, 4591.
Black, Archibald, 4185–4187.
Black, Joseph, 465.
Blackburn, T. W., 2964.
Blackden, L. S., 2368, 2369.
Blacker, Latham Valentine Stewart, 2965.
Blackford, William W., 1521.
Blair, William Richards, 2615.
Blake, A. E., 1522.
Blake, Wilfred Theodore, 2966, 2967, 3842–3844.
Blanchard, Jean Pierre, 466–473.
Blanchet, Georges, 158, 159, 917, 918, 1523, 2696, 2697, 3845, 4728, 5419, 5420.
Blanke, John H. D., 2313, 2370.
Blathwayt, R. *See* Delmard, Alaricus, and R. Blathwayt.
Blee, Harry H., 4188.
Blenk, H., and F. Liebers, 2314.
Blériot, L., 2931, 2968.
Blériot, L., and E. Ramond, 2969.
Blessing, Arthur Reed, 3562.
Blind, Karl, 3563.
Blon, John L. von, 5092, 5093.
Bludworth, T. Francis, 3846.
Bobrovsky, T. N. de, 2537.
Bock, 1527.
Bölcke, Oswald, 3847.
Börnstein, Richard, 4729.
Boeve, Richard de, 5421.
Boffito, Giuseppe, 6, 7, 160–163, 474–481, 920, 921, 1529.
Boffito, Giuseppe, and Adolfo Mabellini, 1530.
Bogert, George Gleason, 4593.
Boltzmann, Artur, 1531.
Bonaparte, Napoleone Luigi, 922.
Bonbright, John Miller, 1532.
Bonfort, Helene, 1533.
Bonforte, John, 4189.
Bonnefon, Paul, 2971.
Bonnet, G., 2972, 2973.
Boone, Andrew R., 3567.
Booth, R. S., 923.
Boothby, F. L. M., 4730.
Bordé, 4531.
Bordeaux, Henri, 3848.
Borelli, Giovanni Alfonso, 5191.
Bosis, Lauro de, 42.
Bostwick, A. E., 1534.
Bothezat, Georges de, 2538.
Bott, Alan John, 3849.
Botts, R. H., 1535.
Botved, Anders Peter, 2974.
Botzmann, Artur, 1531.
Bouchard, C., 2931.
Bouché, Henri, 4190.
Bouchot, Jean, 43.
Bourdin, J., 5422.
Bourne, John Frederick, 2539.
Bourne, Ray, 5094.
Bouttieaux, Victor Paul, 924, 925, 5095.
Bouvé, Clement Lincoln, 4594.
Bouvet, 483, 484.
Bowen, Robert Sidney, 164, 1536, 2975, 2976, 4532.
Bower, F. F., 1537.
Boyé, Pierre, 485.
Boyer, Jacques, 2371.
Boyer-Guillon, Alfred, 2833, 2834.
Boyton, Carroll A., 3568.
Brabazon. *See* Moore-Brabazon, John Theodore Cuthbert.

Bradley, W. F., 2698, 3850, 3851.
Bradshaw, Stanley Orton, 3852.
Brancker, Sir Sefton, 4191.
Brandes, Heinrich Wilhelm, 2616.
Brant, Gerald, 3569.
Brearey, Frederick W., 1538–1541.
Brearley, Norman, 4362.
Breckel, Harry F., 5423.
Bregeault, Henry, 2977.
Bréguet, Louis Charles, 2540.
Breithaupt, Joachim, 928.
Bremer, V. E., 2980.
Brenard, Robert, 5423a. *See also* Harper, Harry, and Robert Brenard.
Bretonnière, J., 5192–5195.
Brett, R. Dallas, 1543.
Brewer, Griffith, 1544–1549.
Brewer, Griffith, and P. Y. Alexander, 166.
Bridgman, Leonard, 3853.
Brie, Reginald A. C., 2541.
Briggs, Lyman J., 4731.
Brilioth, Borge, 4192.
Brillouin, Louis Marcel, 1552.
Brion, Maurice, 5427.
Briscoe, Walter Alwyn, and H. S. Stannard, 3854.
British Arctic Air Route Expedition to Greenland, 4448.
Brittain, Sir Harry Ernest, 167, 4193.
Broadwell, L. W., 1556.
Brock, Norman H., 5096.
Brockett, Paul, 8.
Brodetsky, Selig, 4732.
Bröckelmann, Karl, 168.
Broszus, J. E., 486, 929–931.
Brower, Edward M., 1557.
Brown, Alphonse, 932.
Brown, Sir Arthur Whitten, 2986.
Brown, Cecil Leonard Morley, 169.
Brown, Enos, 933.
Brown, Fay C., 4733.
Brown, Harold A., 1558.
Brown, Harold S., 2835.
Brown, Margery, 5317.
Brown, Robert J., jr., 2987–2989.
Brown, William G., 1559.
Browne, C. Orde, 487.
Bruce, Eric Henry Stuart, 488, 934, 1560, 3572–3575, 3856, 4449, 4533.
Bruce, Mildred Mary (Petre), 5318.
Bruel, François Louis, 170.
Bruignac, Duroy de, 935.
Brunel, Georges, 4363.
Brunet y Viadera, Gaspar, 1561.
Bruno, H. A., 4734, 5097.
Bryan, F. I., 4735.
Bryan, George Hartley, 171, 1562–1564, 2372, 2373, 4736–4738, 5196.
Bryan, John Randolph, 3576.
Buchanan, J. S., 2990.
Buchholtz, F., 936–939, 3577–3579.
Budau, Arthur Adalbert Joseph, 4739.
Buddecke, Hans Joachim, 3857.
Bülow, Hilmer, Freiherr von, 3580.
Buist, H. Massac, 2991.
Bull, Lucien, 5197.
Bumpus, F. A., 3581.
Bunnell, Charles Beecher, 4740.
Buranelli, Albert, 2992.
Burgess, Charles Paine, 940–942.
Burgess, George Kimball, 5428.
Burgoyne, Sir Allan Hugh, 2617.
Burke, Newton, 1566.
Burkett, E. F., 5098.
Burnaby, Frederick Gustavus, 489, 490.
Burrell, George Arthur, 490a.
Burrell, Thomas, 491.
Bursey, Jack C., 3859.
Burtnett, Gerald B., 5319.
Burton, Walter E., 492, 943, 2374.
Busk, Mary (Ackworth), 4741.
Busset, Maurice, 3860.
Butman, Carl H., 1567.
Buttenstedt, Karl, 5198–5200.
Bygott, John Ford, 3861.
Byrd, Richard Evelyn, 2993, 2994, 4450.

C

C., D. L., 2996.
Cable, Boyd, pseud. *See* Ewart, Ernest Andrew.
Cabot, Godfrey L., 944.
Cailletet, Louis Paul, 5429.
Cailletet, Louis Paul, and E. Colardeau, 4742, 4743.
Caldwell, C. S., 2997.
Caldwell, Cy, 1568, 2998.
Caldwell, Frank Walker, 2836.
Camac, Harriet Julia Metcalfe, 5320.
Cambridge, Richard Owen, 44, 45.
Cameron, Allan J., 5430.
Camm, Frederick James, 1569.
Campbell, G. L., 3862.
Camus, Gaston, 2837.
Canada. — Air Board: Library, 5099.
Canada. — Department of National Defence, 4195.
Canada. — Interior Department, 5100.
Candelaria, Luis C., 2999.
Canovetti, Cosimo, 4744, 4745.
Capon, R. S., 5431.
Capper, Sir John Edward, 3582.
Carelli, Jules, 946, 2479.
Carll, George S., 4197.
Carmody, Frank J., 5432, 5433.
Caron, Félix, 497, 947, 948, 1572, 2542.
Carpenter, F. G., 1573.
Carpenter, Ford Ashman, 4747.
Carr, C. F., joint author. *See* Howard-Flanders, R. L., and C. F. Carr.
Carr, G. W., 1574.
Carra, 949.
Carriel, Cruse, 5101.
Carter, W. G., 2315.
Carver, Leland M., 3863.
Caslant, Charles Eugène Éloi, 172.
Cassé, Émile, 950.
Cassinis, Gino, 5103.
Casson, Herbert Newton, 1576.
Castañeda, Vicente, 1577.
Castel, P., 2543.
Castelli, Gaetano, 951.
Castelnau, F. de, 3864.
Castle, Irene (Foote), 3865.
Castro, Giovanni de, 498.
Cathcart-Jones, Owen, 3002a.
Cavallo, Tiberius, 499, 500, 2617a.
Cavara, Otello, 3866.
Cave-Brown-Cave, H. M., 3003.
Cavendish, Henry, 501.
Cayley, Sir George, 952–954, 4748–4753.
Cesare, Oscar, 1579.
Chalmers, C. H., 2544.
Chamberlin, Clarence Duncan, 3005, 4754. *See also* Acosta, Bertram B., and others.
Chambers, W. Irving, 3584.
Chandler, Charles de Forest, 504, 3585.
Chanute, Octave, 173–176, 955, 1580–1587, 2316, 2375–2379, 2699, 5201.
Charles, Jacques Alexandre César, 505.
Charlton, Lionel Evelyn Oswald, 3586.
Charnley, Mitchell Vaughn, 1589.
Charnock, Richard Everett, 1590.
Charpentier, Paul, 956.
Chatley, Herbert, 1591–1593, 4755.
Chaudy, F., 4756, 4757.
Chenoweth, Opie, 2700.
Chicanot, E. L., 4198, 4199, 5435, 5436.

Chichester, Francis C., 3011.
Choinski, Tim, 957.
Chorlton, Alan Ernest Leofric, 2701.
Christmann, Léon, 5437.
Christopher, Luke, and R. F. de Marolles, 5438.
Chubb, I. William, 5321.
Chumakov, D., 506.
Chung, De Senn, 1595.
Church, Glenn J., 5439.
Cierva, Juan de la, 2545–2547.
Cierva, Juan de la, and D. F. Rose, 2548.
Clancy, Carl Stearns, 5440.
Clapp, Frederick Mortimer, 3867.
Clark, Virginius Evans, 1596, 3868.
Clarkson, Coker Fifield, 3587.
Claudy, Carl Harry, 1597–1600, 2618, 3014, 3015, 3869.
Claxton, William J., 1601.
Clayton, Henry Helm, 2619.
Cleaves, Prentiss, jr., 5441.
Cléry, A., 507, 508, 959, 1602, 2317, 3016–3019, 4759, 5322.
Cleveland, Reginald M., 178, 3020, 4367.
Clifford, George Reid, 3870.
Close, Sir Charles Frederick, 5104.
Cobham, Sir Alan John, 3021–3026.
Coburn, F. G., 3871.
Cochran-Patrick, Charles Kennedy, 5105.
Cochrane, Charles H., 180.
Cody, Samuel Franklin, 1604.
Coe, C. C., 4760.
Cogliolo, Pietro, 4595.
Cohn, Georg, 2838.
Colardeau, E., joint author. *See* Cailletet, Louis Paul, and E. Colardeau, 4742, 4743.
Colas, François, 1605.
Colegrove, Kenneth Wallace, 4596.
Coleman, Charles H., 3027–3028.
Colladon, Daniel, 2620.
Collins, Archie Frederick, 1606.
Collins, Gertrude L., 4368.
Collins, John, 3031.
Collins, Thomas Byard, 960, 961, 5202.
Collinson, Clifford Whiteley, and Francis McDermott, 3032.
Colvin, Fred Herbert, 2702.
Colwell, J. H., 181.
Condroyer, L., 4762.
Coney, William D., 3036.
Congrès international d'aéronautique, 4534–4539.
Congrès international de législation aérienne, 4540.
Congrès international de la sécurité aérienne, 4541.
Congresso giuridico internationale per il regolamento locomozione aerea, 4542.
Congreve, Sir William, 1608.
Connor, Harry P. McLean, 3037–3038.
Coontz, Robert Edward, 3588.
Cooper, Bertram G., 4544.
Cooper, Courtney Ryley, 5442.
Cooper, Mabel C., 3039.
Coppens, Willy, 3873.
Corbin, Paul, and Nicolas Oulianoff, 5106a.
Cordenons, Pascal, 962, 963.
Cordero di Montezemolo, Vittorio, 4763.
Cordier, Henri, 46.
Cornu, Paul, 2550, 2551.
Costello, Dudley, 510a.
Costes, Dieudonné, 3040, 3041.
Costes, Dieudonné, and J. M. Le Brix, 3042.
Costes, Dieudonné, and M. Bellonte, 3043.
Coues, Elliott, 5203.
Coughtrie, 5204.
Coupland, R. C., 3589.
Cousin, Joseph, 5205.
Cowdin, Elliot C., 3874.
Cowley, W. L., 2318.
Coxwell, Henry, 514, 966.
Coyler, D. B., 4370.
Crandell, Roy, 1610.
Cressy, Edward, 182.
Crestadoro, Andrea, 515.
Crispolti, C., 47.
Crocco, Gaetano Arturo, 183, 967, 4764.
Crocé-Spinelli, J., 2704–2706, 2839–2840, 4452, 4765.
Crocé-Spinelli, J., and Th. Sivel, 4766.
Crosara, Leonardo, 184, 3590.
Cross, B. C. H., 2319.
Crouvezier, Gustave, 3875.
Crowdus, Walter C., 5443.
Crowther, James Gerald, 185.
Cumberland, G., 516.
Curry, John F., 186.
Curtis, Frank L., 1611.
Curtiss, Glenn Hammond, 1612, 1613, 2320, 3046.
Curtiss Aeroplane and Motor Corporation, 3047.
Cuthell, Chester Welde, 4597.
Cuthill, R. W., and H. S. Hinchman, 5444.
Cutlack, Frederic Morley, 3876.
Cyrano de Bergerac, Savinien, 48.

D

Daçay, Jean, joint author. *See* Mortane, Jacques, and Jean Daçay.
Dacey, Norman F., 1614.
Dacy, George H., 2707, 4767, 4768, 5445.
Daedalus, pseud., 2481.
Dallwitz-Wegner, Richard von, 968, 1615, 2841, 4769.
Dam, H. J. W., 1616.
Dandrieux, 1617.
Daniel Guggenheim Fund for the Promotion of Aeronautics, Inc., 1618, 4770, 4771.
Daniel Guggenheim International Safe Aircraft Competition, 3050.
Danilevski, Konstantin, 969.
Dantin, Charles, 2708.
Darce, Léon, 517.
Darmon, J. E., 10.
Daubanton, C. Abel, 972.
Daus, Edgar, 4598.
David, Evan John, 187, 5447.
Davidson, George L. O., 2554a, 5206.
Davies, Alec, 188. *See also* Davis, Alec.
Davis, Alec, 4371. *See also* Davies, Alec.
Davis, Elmer Holmes, 3051.
Davis, John Woodbridge, 2621.
Davis, Warren Jefferson, 4599, 4600.
Davy, M. J. B., 1620.
Dayton, Wright Airplane Company, 1621.
De ***, vte, 973.
Deburaux, Édouard Léopold Joseph, 519, 520. *See also* Dibos, Maurice, and others.
Deburaux, Édouard Léopold Joseph, and Maurice Dibos, 521.
Decazes, vicomte, 522, 2842, 4774.
Decker, Horace M., 2623.
Decremps, Henri, 49.
Degérando, 523.
Degner, Glenn, joint author. *See* Winter, Lumen, and Glenn Degner.
Degoul, M., 2383, 3053, 3054.
Degouy, Marcel, 524, 4775.
Delacommune, Charles, 3877.
Delaney, Peter J., 3591.
Dellinger, John Howard, and others, 5448.
Delmard, Alaricus, and R. Blathwayt, 3592.
Delteil, L., joint author. *See* Grand-Carteret, John, and L. Delteil.
Deminne, Edm., 978.
Demuyter, Ernest, 525.
Denham, Athel F., 1625, 2710, 2711.
Denza, Francesco, 979.
Deparcieux, Antoine, 526.

Deprez, Marcel, 2384, 2843.
De Ritis, B., 4206.
Desaguliers, John Theophilus, 4776.
Deslandres, Henri, 981.
Desmarest, Léopold, 2385, 2386.
Dettmann, Ludwig J. Chr., 985.
Deussen, Ernest, 527.
Devèze, 1627.
Dew, Arthur W., 3057.
Dex, Leo, pseud. *See* Deburaux, E. L. J.
Diamond, H., joint author. *See* Dellinger, John Howard, and others, 5448.
Diamond, H., and F. W. Dunmore, 5449.
Diard, Ennemonde, 5324.
Díaz Arquer, Graciano, and P. Vindel, 11.
Dibos, Maurice, joint author. *See* Deburaux, Édouard Léopold Joseph, and Maurice Dibos.
Dibos, Maurice, and others, 528.
Dickman, Joseph Theodore, 3596.
Didiée, Joseph, 1628.
Dienstbach, Carl, 987, 988, 1629–1633, 2387, 3879, 3880, 5450. *See also* Kaempffert, Waldemar Bernhard, and Carl Dienstbach.
Dieuaide, Emmanuel, 5451.
Dijk, Evert van, 3059.
Dillon, Charles, 4210.
Dime, Eric Adolphus, 529, 1634.
Dines, William Henry, 2625, 2626.
Dinger, Henry Charles, 2716.
Dixon, Charles, 2484, 3060, 5325.
Doane, Robert R., 190, 1635, 5452–5454.
Dobias, Frank, 191.
Dodoret, Thomas, 992, 993.
Dollfus, Charles, 192, 531–533, 994, 1636, 1637, 2485, 3062, 3063, 3881, 5110.
Domarus, M., 534.
Donaldson, Washington Harrison, 535.
Donalies, Hans, 193, 4777.
Donegan, James Francis, 3882.
Donovan, Thomas B., 3064.
Dornier, Claude, 2325, 2326.
Doughty, John G., 5111.
Douglas, Norman, 50.
Dow, J. H., 2717.
Dowd, R. E., 2388–2391.
Drake, Vivian, 3883.
Draper, O. Griswold, 194.
Dreisch, Th., 1639.
Dresel, A., 4211.
Drew, George A., 3884.
Driggs, Laurence La Tourette, 3885.
Drzewiecki, Stepan, 1640, 2844, 4778, 5207, 5208.
DuBois, H. M., 3066.
Dubosc, André, 536.
Dubouchet, G., joint author. *See* Baeder, F. de, and G. Dubouchet.
Dubrunfaut, 537.
Duckworth, J. Herbert, 3067.
Duerr, L., 995.
Dufaux, Armand, joint author. *See* Dufaux, Henri, and Armand Dufaux.
Dufaux, Henri, and Armand Dufaux, 1641.
Dunbar, William, 51.
Dunmore, F. W., joint author. *See* Dellinger, John Howard, and others, 5448; Diamond, H., and F. W. Dunmore, 5449.
Dunn, H. H., 3068.
Du Plessis de Grenédan, Jean Joseph Anne Marie Julien, comte, 996.
Duponchel, A., 997, 998.
Dupont, Paul, 4779.
Dupuis-Delcourt, Jules François, 538.
Dupuy de Lome, Stanislas Charles Henri Laurent, 539, 540, 999–1003.
Durand, William Frederick, 1643, 4780, 4781.
Durand-Claye, Léon, joint author. *See* Mangon, Hervé, and Léon Durand-Claye.
Durant, Charles Ferson, 541, 542.
Duriez, 543.
Duroy de Bruignac, Albert, baron, 4782, 4783.
Dutczynski, Alfred R. von, 5209, 5210.
Duté-Poitevin, Adrien, 4784.
Duval, Albert B., 4212.
Duvernoy, von, 1004.

E

E., H., 544.
E., M., 1005.
E., M. B., 5456.
Eagleton, Clyde, 4602.
Earhart, Amelia, 5328, 5329.
Ebel, Martin, 4603.
Eberhardt, J. C., 1646.
Eberhardt, Walter von, 3886.
Ebert, Hermann, 4785.
Eckener, Hugo, 1007–1012.
Eddy, William A., 2628–2631, 5112.
Edgeworth, Richard Lovell, 1013.
Edmonds, Charles Humphrey Kingsman, 3887.
Edwards, R. S., 1014.
Ehrhardt, Gustav, 3073.
Eielson, C. B., 4375, 4376.
Eiffel, Alexandre Gustave, 1648, 4787, 4788.
Ekholm, Nils, 4455.
Elford, Harold S., 5456a.
Elias, H., 5457.
Ellicott, Eugene C., 1015.
Elliott, Shelden D., 4604.
Ellis, W. E., 4605.
Ellsworth, Lincoln W., 4456. *See also* Amundsen, Roald Engelbregt Gravning, and L. W. Ellsworth.
Elworthy, Reginald Thomas, 547a.
Emmons, H. H., 2720. *See also* Mixter, George Webber, and H. H. Emmons.
Encyclopaedia Britannica, 195.
Engel, Adolf, 3598.
Engineering index, 12.
Erdmann, Hugo, 548.
Erickson, H. A., 3080.
Erk, Fritz, 4545.
Esclangon, Ernest, 5211.
Esnault-Pelterie, Robert Albert Charles, 1649–1651, 5459, 5460.
Espitallier, Georges, 549–555, 1016–1025, 1652, 1653, 3600, 4457, 4790, 4791, 5331.
Essén, Axel, 4458.
Esterno, Henri Philippe Ferdinand, comte d', 5212.
Esteve, Rafael M., 3081.
Estificief, 3601.
Estournelles de Constant, d', 2931.
Étévé, Albert, 1654, 5213.
Etzdorf, Marga von, 5332.
Evans, Edwards S., 1026.
Evans, John, jr., 1027.
Evans, S. Maurice, 1028.
Evans, Stanley H., 1655.
Ewart, Ernest Andrew, 3889.

F

Faccioli, Aristide, 4793.
Fackeldey, Ferdinand, 4606.
Fage, Arthur, 2845.
Fair, Ernest W., 2560.
Fairbairn, A. R., 1029.
Fairchild, Sherman M., 5114.
Fairey, C. R., 200.
Fales, Elisha Noel, 13.
Fanciulli, Jerome S., 1030.
Fansler, P. E., 4214.
Faragher, R. G., 4215.
Faraud, Louis, 4794.
Farcot, Eugène, 1031, 3603.
Faria, Antonio de, vicomte, 1032.
Farina, Emanuele, 53.
Farman, Maurice, 561.

Farman, T. F., 1664, 1665, 3604, 3891.
Farré, Henry, 3892.
Faujas de Saint-Fond, Barthélemy, 563–565.
Faulkner, G. L., 3893.
Faunce, Cy Q, 4216.
Faurote, Fay Leone, 1668, 4217.
Favaro, Antonio, 1669.
Fawcett, E. D., 3093a.
Fechet, James Edmond, 201, 3605–3607.
Fedden, A. H. R., 2724.
Feldhaus, Franz Maria, 54–56, 202, 203, 566, 1033.
Felgate, Russell A., 1670.
Fell, L. F. R., 2725.
Ferber, Louis Ferdinand, 1671–1674, 2393, 2394, 2726, 2931, 4795.
Fergusson, Sterling Price, 2632–2637, 4796.
Ferrarin, Arturo, 3094.
Ferris, Richard, 204.
Ferro, Guido Gattegno, 1034.
Field, Francis J., 4379.
Figuier, Guillaume Louis, 567, 568.
Fijnje van Salverda, J. G. W., 205.
Fingerhuth, 3608.
Finsterwalder, Sebastian, 5115, 5116.
Fischer, 1681.
Fischer, Johannes, 3899.
Fischer von Poturzyn, F. A., 3112a.
Fischer-Treuenfeld, Richard Friedrich Eberhard von, 3611.
Fiske, Bradley Allen, 3612.
Fitzgerald, George Francis, 2395–2396.
Fitzgerald, Maurice F., 4797.
Fitzmaurice, James C., 3613. *See also* Köhl, Hermann.
Fixel, Rowland W., 4607.
Flammarion, Camille, 4798. *See also* Glaisher, James, and others, 4824.
"Flight commander," pseud., 3900.
Flint, Charles Ranlett, 1682, 1683.
Flint, Howard R., 5468.
Flores Alvarez, Enrique, 3614.
Florman, Carl, 4380.
Fokker, Anthony Herman Gerard, 1689, 4222.
Fokker, Anthony Herman Gerard, and B. Gould, 1690.
Fonck, René, 3901.
Fontaine, Charles, 3120.
Fonvielle, Wilfrid de, 208, 573–581, 1043, 1044, 3616–3621, 4461, 4546, 4547, 4799–4803, 5334, 5470. *See also* Glaisher, James, and others, 4824.
Fonvielle, Wilfrid de, and Georges Besançon, 1045.
Ford, W. B., 1692.
Forlanini, Enrico, 1046, 2564.
Forman, Harrison, 4226.
Forrest, Joanna, 581a.
Fortmüller, 1047.
Fournier, Lucien, 1695–1697, 2565, 2730, 4804.
Fox, Albert Whiting, 3123.
Fracchia, Umberto, 3902.
Fradiss, O., 5335.
Franck, François, 1699.
Franco, Ramón, and J. Ruiz de Alda, 3124.
Frankenburg und Ludwigsdorf, A. V. von, 2487.
Frankenfield, Harry Crawford, 2638.
Franklin, Benjamin, 582–584.
Franklin Institute, 2731.
Frantzen, L. P., 2638a, 3903.
Fraser, Chelsea Curtis, 210, 211.
Freeburg, Victor Oscar, 1700, 4462.
Freeman, Harry Joseph, 4608.
Freeman, Lewis Ransome, 3904.
Freeman, William Marshall, 4609.
Freeston, Charles Lincoln, 3905.
Freire de Carvalho, Francisco, 1049.
French, Joseph Lewis, 212, 213, 3125.
Freyer, Wilhelm, 214.
Fribourg, André, 587.
Frion, O, 215, 1053, 4806.
Fritsche, Carl B., 1054, 1055.
Fromage, Lucien, 1056.
Frossard, Benjamin Sigismond, 589.
Fürstenberg zu Fürstenberg, A., Graf, 1057.
Fuller, J. F., 1701.
Fuller, Moyra, 60, 591.
Fullerton, Frederick H., 4229.
Fullerton, J. D., 216, 217, 1702.
Fulton, Garland, 1058.
Funcke, H., 5214.
Fyfe, George, 1704.
Fyfe, Herbert C., 1059.

G

G., C. von, 3623.
G., H., 61.
G., L., 592.
G., P. de, 3906.
Gabel, S. L., and H. C. Knerr, 1705.
Gaede, R., 1060.
Gärtner, Jean Paul, 593.
Gale, Charles H., 4230, 4383, 4548.
Gall, J. F., 1706.
Gallarza, Eduardo G., and J. Loriga Taboada, 3127.
Galloupe, Francis Ellis, 14.
Galvez, comte de, 1061.
Gamble, Charles Frederick Snowden, 3624, 3907, 3908.
Gammeter, H. C., 1707, 4807.
Ganswindt, Hermann, 1062.
Garber, Paul Edward, 1708, 3128, 3129.
Gardner, Lester Durand, 3130, 3131, 4231, 4232.
Garner, H. M., 2330.
Garner, James Wilford, 4610.
Garros, Roland, 3132.
Gasnier, René, 1709–1712.
Gasterstädt, Johannes, 2732.
Gatke, Heinrich, 5215.
Gatty, Harold, joint author. *See* Post, Wiley, and Harold Gatty.
Gauchot, Paul, joint author. *See* Pénaud, Alphonse, and Paul Gauchot.
Gaudart, Louis, 1713.
Gauthier, Aimé, 1714.
Gay-Lussac, Joseph Louis, 4808.
Geiger, Fr., 1063.
Geisse, J. H., 2733.
Genet, Edmond Charles, 4809–4811.
Geoffrey of Monmouth, 62.
Geoffreys, Peter, 3909.
Georgii, Walter, 2398, 2399.
Gerlach, A., 596.
Gerlach, Edmund, 2847, 3627, 4812, 5216.
Gerli, Agostino, 597.
Gerstner, Ferdinand, 4813.
Gerville-Réache, Léo, 1065.
Giacomelli, R., 218, 598, 4814–4816.
Giacomelli, R., and E. Pistolesi, 4817.
Gibson, Ada, 5336.
Giffard, Henri, 599, 600, 1066, 1067.
Giglio, Vittorio, 3911.
Gile-Nicaud, Gilbert, 3137.
Gill, Donald, 5117.
Gill, Napier John, 3629.
Gillette, Leslie S., 1718.
Gillingham, James, 1719.
Gillmore, William E., 3630.
Gilmore, W. L., 1720.
Girard, E., and A. de Rouville, 1069.
Giroud de Villette, André, 601.
Giudici, Davide, 4464.
Giulini, Gino, 602.
Glaisher, James, 603, 4818–4823.
Glaisher, James, and others, 4824.
Glassford, William Alexander, 3631, 3632.
Glassman, Don, 2488.
Glazebrook, Sir Richard Tetley, 4825.
Glenville, I. E., 2400.
Glover, Warren Irving, 4384, 4385.
Godard, Louis, 604.

Goddard, Robert Hutchings, 5473.
Goebel, Johannes, 3912.
Gold, Ernest, 4826.
Gold, Ernest, and W. A. Harwood, 4827.
Golding, Harry, 219.
Goldsborough, Paul, 4234.
Goldsmith, Margaret Leland, 1070.
Goldstrom, John, 220, 5337, 5474.
Golikere, R. K., 221.
Goodman, Henry, 605.
Gordini, L., 4612.
Gordon, W. A., 5474a.
Gore, Howard Mason, 4828.
Gorrell, Edgar S., 3913, 3914.
Goualt, Émile, 3915.
Gouilly, Alexandre, 1072.
Gould, Bruce, 1724, 5338. *See also* Fokker, Anthony Herman Gerard, and B. Gould.
Goupil, Albert Louis, 1073, 2734, 4829–4833.
Gouzy, René, 4465. *See also* Mittelholzer, Walter, and others.
Govi, Gilberto, 1074, 2848.
Grace, Richard, 3916.
Gradenwitz, Alfred, 1076, 1726–1727a, 3633, 4834.
Graffigny, H. de, pseud. *See* Marquis, Raoul.
Graham, Lloyd S., 5339.
Graham, W. H., 4835.
Grahame-White, Claude, 222, 1728.
Grahame-White, Claude, and Harry Harper, 223, 1729, 3917–3919.
Grahame-White Company, Limited, London, 1730.
Gran, Tryggve, 3920.
Grand-Carteret, John, and L. Delteil, 224.
Gray, George A., 3140.
Grayson, King Hamilton, 4235.
Great Britain. — Air Ministry, 3921.
Great Britain. — Civil Aviation Directorate, 4236.
Great Britain. — Courts, 1077.
Great Britain. — Patent Office, 225.
Greely, Adolphus Washington, 3634.
Green, Charles, 608, 609.
Green, F. M., 1732, 2735, 3924.
Green, Fitzhugh, 3146.
Greene, Warwick, 3925.
Greenhill, G., 63.
Greenhill, Sir George, 1733.
Greenly, H. *See* Marshall, A. W., and H. Greenly.
Greer, John Fulton, 2331.
Greer, Rowan A., 4613.
Gregg, Willis Ray, 1078, 3147, 3148, 4836–4839.
Gregor, 4386, 4387.
Gregory, Austin, 610.
Gregory, George, 611.
Grey, Charles Grey, 226–236, 1079–1082, 1734–1747, 2332, 2333, 2489, 2736, 3148a–3172, 3635, 3636, 3926–3938, 4237, 5118, 5340, 5341, 5475, 5476.
Grey, Charles Grey, and Walter Kleffel, 3173.
Grierson, John, 3174.
Griffiths, J. C., joint author. *See* Jones, Bennett Melvill, and J. C. Griffiths, 5125.
Grilleau, B. de, 1083.
Grimaldi, Gabbriello, 612.
Gronau, Johann Friedrich Wilhelm, 4840.
Gronau, Wolfgang, 3175, 3176.
Gross, Hans, 1084, 1085, 3637, 3638, 4388.
Grosse, Margarete, 613, 5342.
Grosvenor, Gilbert Hovey, 2639.
Grover, G. E., 3639.
Grüel, C. A., 614.
Guarini, Emile, 1086, 5217.
Gubbins, M. N. T., 4239.
Günther, S., 237, 4841.
Guggenheim, Harry Frank, 238, 1749, 4842.
Guidoni, Alessandro, 1750, 2335, 3640.
Guillaume, Charles Edouard, 4844.
Gulf Refining Company, 239.
Gustoas, Corrado, 2336.

H

H., 1753, 1754.
H., E., 616, 3641.
Hackstetter, K., 617.
Haedicke, H., 4845.
Haenig, Alfred, 2737.
Hänisch, Viktor, 1755.
Haenlein, Paul, 1088, 1089, 2850.
Haferkorn, Henry Ernest, 5119.
Hagen, Freiherr vom, 15, 240, 618–621, 1091, 1756, 2490, 3642, 4846, 5478, 5479.
Hagen, Friedrich Heinrich von der, 64.
Haines, D. B., joint author. *See* Haines, Lynn, and D. B. Haines.
Haines, Lynn, and D. B. Haines, 3178.
Hale, Albert, 3179.
Haling, Elbert, and James Locklear, 3180.
Hall, Bert, 3939.
Hall, Bert, and J. J. Niles, 3940.
Hall, Donald A., 1757.
Hall, James Norman, 3941.
Hall, James Norman, and C. B. Nordhoff, 3942.
Hall, Norman Shannon, 3943.
Halliburton, Richard, 3181.
Hamlin, Percy G., 3643.
Hammer, W. J., joint author. *See* Hudson, Maxim, and W. J. Hammer.
Hammon, W. H., 2640, 4847.
Hammond, E. V., 2851.
Hanawalt, Wilbur R., 1092.
Handley-Page, Frederick, 1758, 1759, 4240.
Hansen, 2852.
Hanson, Earl, 4241–4244, 4466.
Hanst, Charles E., 3644.
Hardecker, John F., 4549.
Harding, Charles, 4848.
Hargrave, Lawrence, 1760–1768, 2403, 2404, 2641–2646, 2738, 2739, 5218.
Harlaut, 2740.
Harmel, Falk, 2491, 3645.
Harmon, Benjamin F., 3646.
Harney, Laura B., 241.
Harper, Harry, 242, 1093, 1772, 1773, 3182, 3183, 4245. *See also* Grahame-White, Claude, and Harry Harper.
Harper, Harry, and Robert Brenard, 4389.
Harriman, John Emery, jr., 1771.
Harrington, Mark Walrod, 4849.
Hart, Ivor Blashka, 4850, 4851, 5219.
Hartson, Joseph Tracy, 2741.
Hartwig, Georg Ludwig, 243.
Harwood, Ernest, joint author. *See* Gold, Ernest, and W. A. Harwood, 4827.
Haslett, Elmer, 3944.
Hastings, C. W., 2853.
Hastings, George E., 622.
Hathaway, Levy M., 5481.
Hattoom, Fred L., 5343.
Hauvel, Charles du, 2854, 5220.
Havill, C. H., 2855.
Hawker, Harry George, and K. MacKenzie-Grieve, 3184.
Hawker, Muriel, 3185.
Hawthorne, H. L., 3647.
Hay, James, jr., 623.
Hayward, Charles Brian, 1774.
Hazen, H. A., 624, 4852, 4853.
Hearne, R. P., 244, 1094, 3945.
Heath, John, 625.
Heath, Sophie Mary (Peirce-Evans) Elliott-Lynn, 5344–5346.
Hegener, Henri, and others, 1776.
Heiland, C. A., 5119a.
Heilig, Sterling, 1095, 1096.
Heim, Arnold, joint author. *See* Mittelholzer, Walter, and others.
Heinen, Anton, 1097.

Heinsen, Walter A., 5120.
Heintz, Henry, 1777.
Heinz, F., 2742.
Heinze, Edwin P. A., 1778–1780.
Heitman, Charles, 2405.
Heldt, Peter Martin, 2743.
Hellmann, Gustav, 4854.
Helmar, N., 1098.
Helmholtz, Hermann Ludwig Ferdinand von, 4855, 4856.
Hemming, H., 5121, 5122.
Henderson, John, 1099.
Henderson, Paul, 4246, 4390.
Hengler, F., 626.
Hengler, Lorenz, 2492.
Henkelburg, Hans, 3946.
Hennig, Richard, 65, 66.
Henrickson, H. B., 3187–3192.
Henrion, Francesco, 1101.
Henry, Edmond, 627. *See also* Soreau, Rodolphe, and Edmond Henry.
Henry, Louis Isidore, 1781.
Henry-Coüannier, André, 1782.
Herbert, Charles, 628.
Hergesell, Hugo, 629, 3648, 4857. *See also* Miethe, Adolf, and Hugo Hergesell.
Hermite, Gustave, 4858–4860.
Hermite, Gustave, and Georges Besançon, 4861.
Hermuth, Paul, 1787.
Herrick, Jack, joint author. *See* Hirth, Wolfram K. E., and others.
Herring, A. M., 1788, 2406.
Hersey, Henry B., 630.
Hervé, Henri, 631, 632, 1102, 1789, 4862.
Hervé, Henri, and Henri, comte de la Vaulx, 2857.
Hewlett, Hilda Beatrice (Herbert), 5347.
Heydemarck, Georg, 3947, 3948.
Higgins, John, 67.
Hild, Frederic C., 3949.
Hildebrandt, Alfred Louis Heinrich, 245–247, 1790.
Hildenbrand, 633.
Hildesheim, Erik, 3950.
Hill, E. W. *See* Woolley, James G., and E. W. Hill.
Hill, Frederick Thomas, 4864.
Hill, G. T. R., 1791.
Hill, Roderic, 4391.
Hill, Thomas A., 1792.
Hinchman, H. S., joint author. *See* Cuthill, R. W., and H. S. Hinchman.
Hinterstoisser, 3649.
Hinton, Walter, 3196.
Hirschauer, Auguste Édouard, 16.
Hirschauer, Ch., 634.
Hirschauer, Louis, 1793.
Hirth, Hellmuth, 3197.
Hirth, Wolfram K. E., and others, 2407.
Hoare, Sir Samuel John Gurney, 3205.
Hodges, Lawrence K., 1797.
Hodgins, Eric, and F. A. Magoun, 252. *See also* Magoun, Frederick Alexander, and Eric Hodgins.
Hodgson, J. E., 253, 254, 4550, 5347a.
Hoenerssen, G. A., pseud., 4467.
Hoeppner, Ernst von, 3951.
Hoernes, Hermann, 255, 1104, 2858, 4865.
Hoernes, Marianne, 5482a.
Hof, J. J., 256.
Hoffman, Bernard von, 636.
Hoffman, Frederick Ludwig, 3206.
Hofmann, A. W., 637.
Hofmann, Joseph, 1798, 1799.
Hofmann-Genf, J., 5483.
Hofwerberg, Johannes, 638.
Holland, John P., 1800, 1801.
Holland, Maurice, 4866.
Holland, Rupert Sargent, 257.
Hollands, Sidney H., 1802, 1803, 2746, 2860.
Hooke, Robert, 4867, 4868.
Hoop, Thomassen à T. van der. *See* Thomassen à Thuessink van der Hoop, A. N. J.
Hoover, Herbert Clark, 4249.
Hopkins, Albert Allis, 4869.
Hopkins, Manley, 639.
Hopper, James Marie, 5484.
Hopson, William L., 2493.
Horsfall, Jessie E., 1805, 3209.
Horton, Harry M., 5485.
Hotchkiss, Henry Greene, 4614.
Hotine, Martin, 5123.
Houston, George H., 2747.
Howard, Edward P., 5486.
Howard, W. H. L., 4392.
Howard-Flanders, Richard Leonard, 1807, 2408.
Hubbard, Thomas O'Brien, 2494.
Huber, François, 5221.
Hucks, B. C., 1809, 1810.
Hudson, Manley Ottmer, 4615.
Huenefeld, Ehrenfried Guenther, Freiherr von, 3212. *See also* Köhl, Hermann.
Huffaker, E. C., 5222–5224.
Huisman, Georges Maurice, 3955.
Humble, Richard, 3213.
Humphreys, F. E., 1812.
Hunsaker, Jerome Clarke, 258, 1105, 1106, 2338, 2339.
Hunt, Edmund, 1107.
Hunt, Frazier, 3956.
Hunter, H. Chadwick, 1813, 4870.
Hureau de Villeneuve, Abel, 259–261, 641, 1108, 1814, 1815, 2409, 2568–2570, 2748, 3652, 4871, 4872.
Hurry, H. C., 1816.
Hurst, Mont, 3214.
Hutchinson, W. N., 1109.
Huth, Fritz, 262.
Hutton, Frederick Wollaston, 5225.
Hyatt, Thaddeus, 1817, 2571.
Hydravion, pseud., 1110, 3215. *See also* Hegener, Henri, and others.
Hylander, Clarence John, 263.
Hynes, G. B., 2749.

I

Ide, John Jay, 3216, 3653, 3957, 4551, 4616.
Ijserman, J. M., 1820.
Immelmann, Max Franz, 3960.
Inarre, C. F. d', 645.
Indicator, pseud., 646.
Industrial arts index, 17.
Inglis, William, 1825.
Insley, Robert, 2750.
Institut égyptien, 5226.
Institut de France.—Académie royale des sciences, 647.
Institut de France.—Académie des sciences, Paris, 1117.
International Aeronautic Federation, 4552, 4553.
International Civil Aeronautics Conference, 4556, 4557.
International Commission for Air Navigation, 4558.
International Commission for the Investigation of the Upper Air, 4559.
International Conference on Aerial Navigation, 4561.
International Congress of Aerial Navigation, 4562.
Irish, W. E., 1826, 1827.
Irving, H. B., and A. S. Batson, 1828.
Irwin, R. Randall, 1829.
Isacco, V., 2572.
Italy. — Delegazione al Congresso internationale di aeronautica civile, 3218.
Italy. — Spedizione polare dell' aeronave "Italia," 4468.

J

Jackson, George Gibbard, 265, 1830.
Jackson, Joseph, 649.
Jacob, Emil, 4874, 5227.
Jacobi, Max, 68.
Jacobius, Helene, 69.
Jacobowsky, C. Vilhelm, 4469.

Jacobs, A. M., 1831, 5489, 5490.
Jacobsen, Sem. *See* Arensen, Odd, and Sem Jacobsen.
Jacques, P., 3655.
James, P., 5491.
Jamieson, Alexander, 650.
Jamin, J., 651.
Jane, Frederick Thomas, 3962.
Janet, 651a.
Janet, Charles, 5228.
Janssen, Pierre Jules César, 266, 652, 5492.
Japan advertiser, 3219.
Jaray, P., 1119.
Jarolimek, Anton, 2861, 2862, 4875–4877, 5229, 5230.
Jaubert, George F., 653.
Jauneaud, Marcel, 267.
Jeanjean, Marcel, 3963.
Jeffries, John, 654, 4878.
Jennings, W. Nicholson, 5124.
Jobert, Claude, 4879.
Joerg, Wolfgang Louis Gottfried, 4470, 4471.
Johns, William Earl, 268, 3964.
Johnson, George R., 1833, 3221, 4251.
Johnson, Harry Miles, 4880.
Johnson, John Berlin, 1834.
Johnson, Julius F., 1835.
Johnson, Robert, 1835a, 4252, 4253.
Johnson, Samuel, 70.
Jones, Bennett Melvill, 1836.
Jones, Bennett Melvill, and J. C. Griffiths, 5125.
Jones, Bradley, 655, 656, 3222.
Jones, Casey, 3223.
Jones, Ernest Lester, 71, 269, 3224, 3657, 3658, 5493.
Jones, Ernest Lester, and S. Y. Beach, 2751.
Jones, H. A., 3965. *See also* Raleigh, Sir Walter Alexander, and H. A. Jones.
Jones, H. B., 3659, 3660.
Jones, Thomas P., 4881.
Jordanoff, Assen, 3661.
Josselyn, Winsor, 2411.
Jotti, da Badia, polesine, 18, 270, 657–662, 1120, 1121, 1837, 2573, 2752, 5348.
Jouguet, Émile, 4882.
Joukovsky, N., 4883.
Joynson-Hicks, Sir William, 3966.
Judd, Neil Merton, 5126.
Jünger, Ernest, 271.
Jünke, Arnold, 3967.
Julien, Stanislas, 2647.
Julliot, H., 1122.
Junkers, Hugo, 1839.

K

K., 3662.
Kadarz, Theodor, 2863.
Kaemmerer, W., 1840.
Kaempffert, Waldemar Bernhard, 3969.
Kaempffert, Waldemar Bernhard, and Carl Dienstbach, 3968.
Kahle, 5127.
Kargl, Ludwig, 4885.
Karos, Willibald, 1841, 2574.
Kassner, Carl Julius Arminius, 4886.
Kay, John, 663.
Kean, 2341.
Kebra Nagast, 72.
Keller, H. R., 1123.
Kellett, Wallace, 1842.
Kempf, H., joint author. *See* Mittelholzer, Walter, and H. Kempf.
Kensche, Heinz, 2412.
Kérambrun, Paul, 1843.
Kerr, Mark Edward Frederic, 3970.
Kershaw, John Baker Cannington, 3971.
Keyhoe, Donald Edward, 3225.
Kimball, James Henry, 4887, 4888.
Kincaid, E. H., joint author. *See* Ramsey, Logan Carlisle, and E. H. Kincaid.
King, Ernest J., 3663.
King, W. A. de C., 3664.
Kingsford, A. R., 3972.
Kingsford-Smith, Charles Edward, and C. T. P. Ulm, 3227.
Kingston-McCloughry, Edward James, 3665.
Kircher, Athanasius, 4889.
Klassohn, Eduard, 2864.
Klaussman, Anton Oskar Eugen, 664.
Kleffel, Walter, joint author. *See* Grey, Charles Grey, and Walter Kleffel.
Klein, Felix, 4890.
Kleinenn, H., 665.
Klemin, Alexander, 272–275, 666, 1845–1847, 2575–2577, 2753–2755, 2865, 3230, 3231, 4255, 4891, 5128, 5129, 5495, 5496.
Klemmer, Harvey, 1848.
Klinckowstroem, Carl Ludwig Friedrich Otto, Graf von, 19, 73, 1124, 1125, 1849.
Kline, Gordon M., 1850.
Knappen, Theodore Macfarlane, 3973.
Knauss, Robert, 3232.
Knerr, H. C., joint author. *See* Gabel, S. L., and H. C. Knerr.
Knight, W., 4256, 4257.
Knolles, Richard, 74.
Knowles, James, 3974.
Koch, Gustav, 1851–1856.
Köhl, Hermann, 3233, 3234, 3975.
König von und zu Warthausen, Friedrich Karl, Freiherr von, 3235.
Köppen, Wladimir Peter, 2649, 4892.
Kohl-Larsen, Ludwig, 4472.
Kollman, Franz, 1126.
Konvalinka, Joseph George, 1857.
Koppe, Heinrich, 4893.
Koppen, G. A., 3236.
Koppen, Otto C., 4894.
Korf, Georg, 5231.
Kosch, Rudolph, 1858.
Kotzauer, Wenzel, 5232.
Kotzebue, August Friedrich Ferdinand von, 668.
Kousnetzow, V., 4895.
Kranzler, Wilhelm, 3976.
Kratzenstein, Christian Gottlieb, 1127.
Krauss, J. S., 276.
Krebs, A., joint author. *See* Renard, Charles, and A. Krebs.
Kreiss, Eugen, 2413, 2578.
Kremser, Victor Felix Karl, 669, 4896, 5497.
Kress, Wilhelm, 1859–1864, 2342, 2414, 2579, 2866, 4897, 5234.
Kretschmar, Julius, 1128.
Krogh, Chr. von, 277.
Kronfeld, Robert, 2415.
Kropp, H., 4395.
Krotki, Rudolf, 2416.
Krouchkoff, Michel, 4898.
Kruckman, Arnold, 670.
Krüger, Herbert, 75.
Krüger, Richard, 2756.
Krull, Fritz, 671.
Kuhn, Arthur Kline, 4617.

L

L., J. H., 1866.
L., L., 5234a.
L., W. J. S., 2417.
Labadié, Jean, 4899.
Lachambre, Henri, 3666.
Lachambre, Henri, and Alexis Machuron, 4473.
Lachmann, Gustav, 2418, 2419.
Lacroix, Pierre, 4618.
Ladrone, Francesco, 4619.
Lafargue, 4258.
Lafonte, Bernard, 3977.
Laforet, Auguste, 672.

La Frégeolière, Renaud de, 3978.
La Grange, Amaury de, 3979.
Lagrange, L., 1129, 3980.
Lahm, F. S., 673, 1868.
Lake, John, 1130.
La Landelle, Guillaume Joseph Gabriel de, 278, 1869, 2495, 2496.
Lallemand, Charles, 5130.
Lamarche, Paul E., jr., 4259–4266.
Lamb, Dean Ivan, 3237, 3981.
Lambert, comte de, 2931.
Lambrecht, Wolfgang, 1131.
Lamé, Maurice Luc Valère, 2580.
Lana Terzi, Francesco, 1132–1134.
Lancaster, I., 4900–4903, 5235.
Lanchester, Frederick William, 3667, 4904, 5236.
Landauer, Bella Clara, 5498, 5499.
Landauer, Bella Clara, and H. B. Weiss, 20.
Langley, Marcus, 1870, 4267.
Langley, Samuel Pierpont, 1871–1875, 4905–4909, 5237.
Langsdorf, Werner von, 2420.
Lansdowne, Zachary, 1135.
Laporte, Albert, 674.
La Quainte, Emmy, 5350.
Laredo de Mendoza, Saverio, 3982.
Larsen, Agnew E., 2581.
Lasser, David, 5500.
Laufer, Berthold, 76.
Launay, L. de, 675, 4911.
Laurain, Ernest, 1885.
Lauriol, P., 2422.
Laussedat, Aimé, 1137, 5132.
Lautier, Ferdinand, 1886, 1887.
La Vaulx, Henri, comte de, 280, 281, 676–680, 4912. *See also* Hervé, Henri, and Henri, comte de La Vaulx.
La Vaulx, Henri, comte de, and P. Tissandier, 681.
Lavergne, Gérard, 4474, 4475.
Lavisse, E., 2931.
Lavoisier, Antoine Laurent, 4913.
Lawrance, Charles Lanier, 1888, 1889, 2757–2759, 4396.
Lawrence, John, pseud., 4476.
Lawson, Alfred W., 3238.
Lawton, Hadold Walter, 77.
League, A. W., 3239.
Lebon, Émile, 4620.
Le Brix, J. M., joint author. *See* Costes, Dieudonné, and J. M. Le Brix.
Lecarme, Jean, joint author. *See* Vallot, Joseph, and others.
Lecarne, Louis, joint author. *See* Vallot, Joseph, and others.
Lecocq, Georges, 682.
Le Conte, Joseph, 1890, 1891.
Lecornu, Joseph, 2652, 4914.
Lecornu, Léon François Alfred, 1892.
Le Dantec, Louis Marie, 4915.
Ledeboer, John Henry, 282, 683, 1893, 3240, 4564, 4916.
Leechman, Douglas, 283.
Lees, T. Orde, 2487.
Lefebvre, Ed., 2867.
Leffman, Henry, 1894.
Lefranc, Jean Abel, 3984, 3985.
Léger, M., 2582, 2583.
Le Grand, G., 3241.
Leher, Max, 684–686.
Lehigh Portland Cement Company, 4269.
Lehmann, Ernst A., and Howard Mingos, 1139.
Leigh, Howard, 3986.
Leighton, B. G., 3242.
Leloup, Jules, 687.
Lendenfeld, Robert J., 5238.
Leo, Kurt, 4621.
Le Page, Wynn Laurence, 3245.
Lepère, G., 4917.
Lericolais, Eugène, 1141.
Lescarboura, Austin Celestin, 2760.
Lesueur, 688.
Letellier, R., 3247.
Létonné, A., 1142, 3668.
Levis, Howard Coppuck, 79.
Lewis, C. G., 5133.
Lewis, George M., 3248.
Lewis, H. Latane, 2d, 3249.
Lhoste, F., and J. Mangot, 690.
Liebers, F., joint author. *See* Blenk, H., and F. Liebers.
Liebmann, Louis, 284.
Liebmann, Louis, and Gustav Wahl, 21.
Liggett, Walter W., 3669, 3670, 4397.
Lilienthal, Gustav, 2423, 2424.
Lilienthal, Otto, 1897, 1898, 2425–2435, 5239, 5240.
Lin, Wo-chiang, 4622, 4623.
Lindbergh, Anne (Morrow), 3252, 5352a.
Lindbergh, Charles Augustus, 3253, 3254.
Lindsay, Orrin, 1900.
Lindstrom, Siegfried F., 3260.
Linke, Franz, 285, 692, 1144.
Lippert, P. W., 4918.
Lippish, A., 2438, 2439.
Liptrot, R. N., 2584.
Little, Ray, 3261.
Liversedge, A. J., 1145.
Livingstone, John H., 3262.
Lobo d'Avila Lima, J., 286.
Lochet, G., 5241.
Locklear, James, joint author. *See* Haling, Elbert, and James Locklear.
Loeffler, H. C., 5135, 5502–5505.
Loening, Grover Cleveland, 1901, 2343, 3263.
Lönnberg, Einer, joint author. *See* Sverdrup, Harald Ulrik, and Einar Lönnberg.
Loessl, F., Ritter von, 1146, 4919–4928.
Loewe, Karl F., 3264.
Logan, George Bryan, 4624.
Lomet, Antoine François, baron, 5136.
London.—Science Museum, 287, 288.
Loriga Taboada, J., joint author. *See* Gallarza, Eduardo G., and J. Loriga Taboada.
Loring, John, 694.
Lotti, Armeno, jr., 3270.
Lougheed, Victor, 289.
Louvrié, Charles de, 1903, 5242–5246.
Low, Archibald R., 1904.
Lowe, Thaddeus Sobieski Coulincourt, 3672.
Lucanus, Friedrich von, 5247.
Lucas, Frederic Augustus, 5248.
Lucas-Girardville, P., 2585.
Lucian, 80.
Lucy, de, 5249.
Ludlow, Israel, 1905–1909, 5506.
Lunardi, Vincenzo, 695, 696.
Lund, Fred, 3271.
Lundborg, Einar Paul Albert Muni, 4477, 4478.
Luntley, John, 1151.
Lupton, George W., jr., 4625.
Lutz, Earle, 3674.
Luyties, Otto G., 2586.
Lyle, Eugene P., 1152.
Lyman, Lauren D., 1153.
Lynch, Arthur, 3675.
Lyon, Albert B., 2654.
Lyons, Sir Henry George, 4930.

M

M., E., 290, 1910, 3676, 3677.
Maatschappij voor Landometing, 5137.
Mabellini, Adolfo, joint author. *See* Boffito, Giuseppe, and Adolfo Mabellini.
McAdie, Alexander George, 4931, 4932.
McAlery, C. M., 3678–3680.
McCauley, E. G., 2869.

McClean, Sir Francis Kennedy, 697.
McCluer, C. E., 1154.
McClure, Ivor H., 3272.
McConnell, Burt Morton, 5507.
McConnell, James Rogers, 3990.
MacCoull, Neil, 2763.
MacCracken, William Patterson, 4626.
McCudden, James Thomas Byford, 3991.
McCurdy, Edward, 4933.
McDarment, Corley, 2498, 2499, 2764, 3273.
McDermott, Francis, joint author. *See* Collinson, Clifford Whiteley, and Francis McDermott.
Macdonald, A. C., 4479.
McElroy, C. L., 3273a.
Maceroni, Francis, 1155, 1911, 2500.
Macgill, Ronald, 81.
MacGregor, John D., 4271.
Mach, Ludwig, 4934.
Machol, Morris R., 2765.
Machuron, Alexis, joint author. *See* Lachambre, Henri, and Alexis Machuron.
MacKaye, Percy Wallace, 1912.
Mackenzie, George, 2501.
MacKenzie-Grieve, K., joint author. *See* Hawker, Harry George, and K. MacKenzie-Grieve.
McKinley, Ashley Chadbourne, 5138.
Mackinnon, A., 1156.
Mackintosh, T. S., 1157.
McLaren, A. S. C. S., 3274.
McLaughlin, George F., 2766.
McLean, Robert Alexander, 3275.
McLean, William P., 1158.
MacLeod, Malcolm Neynoe, 5139.
MacMahon, John Robert, 1913, 1914.
McMahon, T. C., 2767.
Macmillan, Norman, 3992, 4272.
McMullin, C. R., 4273.
McNair, Arnold Duncan, 4627, 4628.
Macready, John A., 2768, 3276.
McReynolds, Charles F., 3276a, 4274.
MacSweeny, Joseph, 1159.
Maggs brothers, 22.
Magoun, Frederick Alexander, and Eric Hodgins, 292, 292a, 697a–697c, 1159a–1159d, 1914a–1914d, 2502a, 2586a, 3277, 3992a. *See also* Hodgins, Eric, and F. A. Magoun.
Magrini, Effren, 1915.
Magruder, Thomas Pickett, 5508.
Mahgni-Eltten, 3278.
Maigre, Étienne, 1916.
Maire, Albert, 23.
Maisoncourt, marquis de la, 698.
Maitland, Edward Maitland, 1160.
Maitland, Lester J., 1917.
Makower, Walter, 4935.
Maleire, E., 4936.
Mallet, Maurice, 2870.
Mamer, N. B., 5510.
Mance, H. Osborne, 699.
Mangin, Théodore, 3681.
Mangon, Hervé, and Léon Durand-Claye, 4937.
Mangot, J., joint author. *See* Lhoste, F., and J. Mangot.
Manly, Charles Matthews, 1918, 1919.
Mannesmann, O., 4938.
Manning, L. B., 4275.
Manning, William Oke, 2770.
Mansfield, Charles Blachford, 1162.
Marben, Rolf, 3994.
Marc, 3995.
Marchis, Lucien René André Edmond, 293, 294, 5511.
Marcillac, P., 4480.
Marey, Étienne Jules, 5250–5263.
Marey-Monge, Edmond, 700, 1163.
Marguet, Frédéric Phillipe, 4939.
Maria, A. de, 295.
Marie, Eugène Armand, 3282.
Marie, Félix, 3682.
Marion, Fulgence, 701, 702.
Markham, Christopher Alexander, 3996.
Marmonnier, L., 1921.
Marolles, Michel de, 82.
Marolles, R. F. de, joint author. *See* Christopher, Luke, and R. F. de Marolles.
Marquis, Raoul, 296–299, 703–708, 1922, 2771.
Marriott, William, 4940, 4941.
Marsh, William Lockwood, 24, 300, 301, 709, 1164–1166, 1923–1929, 2440, 2871, 3997, 3998, 4942.
Marshall, Alfred William, and H. Greenly, 302.
Marshall, Chapman Frederick Dendy, 1930.
Marshall, J., 1931.
Martell, P., 2503.
Martienssen, Oscar, 4943.
Martin, A. D. de R., joint author. *See* Beatty, W. D., and A. D. de R. Martin.
Martin, E. Stockton, 5512, 5513.
Martin, Glenn L., 1932.
Martin, James V., 1933.
Martin, Lajos, 1934.
Martin, Robert E., 3999.
Martin, Rudolf Emil, 303.
Martinet, Ludovic, 4481.
Martini, Tito, 1168.
Martinot-Lagarde, Pierre Léonard Camille, 2772.
Marvin, Charles Frederick, 710, 2656, 2657.
Marvingt, Marie, 5354.
Masfrand, Albert de, 1169–1171, 1936–1940, 2587, 5264.
Mason, Ethel I., 5141.
Mason, L. T., 3283.
Mason, Thomas Monck, 710a, 711.
Massau, J., 1172.
Massenet de Marancour, C., 4000.
Massip, 83.
Massy, H. S., 2345.
Masterman, Edward Alexander Dimsdale, 1173.
Mather, T. W., 1941.
Mathon de la Cour, 712.
Matthias, Joachim, editor, 1942.
Maxfield, Joseph E., 3683.
Maxim, Sir Hiram Stevens, 305, 1943–1954, 2872, 2873.
Maxim, Hudson, 2441, 3684.
Maxim, Hudson, and W. J. Hammer, 3284.
Maxson, Herburt W., 1175.
Mayo, Robert H., 2773.
Mays, Brock, 5514.
Mazet, Horace S., 84.
Mead, George J., 2774–2776, 4399.
Meade, F. N., 3685.
Mealing, R. H. S., 5515.
Means, James, 1956, 2442, 2443, 2658, 5265.
Mears, John Henry, 3285.
Mechanic, pseud., 1957.
Meek, Robert, 713.
Mehl, Adolf, 714.
Meili, Friedrich, 4629.
Meissel, F., 715.
Melikoff, 2588.
Meltzl, David, 1176.
Melville, George Wallace, 1961.
Memento, pseud., 1962.
Mendeleyeyv, D., 4945.
Merrill, A. A., 1963.
Mesa y Ramos, José, 1177.
Messimy, A., 3687.
Metcalfe, A. Wharton, 3688.
Mettler, C. G., 3689.
Mettler, Hans, 1965.
Meusnier, Jean Baptiste Marie, 1180.
Mewes, Rudolf, 85, 2504, 2505, 2874, 5266.
Meyer, Alex, 4568, 4630.
Meyer, E., 1966.
Meyer, Friedrich Albert, 4002.

Meyer, Willy, 1967, 4482.
Meylan, E., 716.
Micciollo, Alfred, 1181.
Middleton, Edgar Charles, 4003–4007.
Miethe, Adolf, and Hugo Hergesell, 4483.
Miles, Russell H., 1968.
Milla, Karl, 1182, 2506, 5267.
Miller, Charles M., 2659.
Miller, F., 1970.
Miller, Francis Trevelyan, 307, 4484.
Miller, H. B., 2507.
Miller, J. Earle, 5517.
Miller, William H., 4946.
Miller-Hauenfels, A. R. von, 5268.
Millet, J. B., 2660, 2661.
Millikan, Robert Andrews, 4947.
Mills, Manley, 1972.
Mills, S. P., 3695.
Mingos, Howard, 308, 1183, 1973, 1974, 4276. *See also* Lehmann, Ernst A., and Howard Mingos.
Minney, Rubeigh James, 3287.
Minor, Jacob, 25.
Mitchell, Harley W., 1185.
Mitchell, John W., 1975, 1976.
Mitchell, William, 309, 3696–3701, 4009, 4010.
Mittelholzer, Walter, 3289a, 3290–3292.
Mittelholzer, Walter, and H. Kempf, 3293.
Mittelholzer, Walter, and others, 3294.
Mix, Arthur, 3296.
Mixter, George Webber, and H. H. Emmons, 4011.
Mock, Richard M., 1977.
Mockler, Don, 3297, 3298.
Moedebeck, Hermann W. L., 86, 310, 719, 720, 1186–1195, 2508, 2778, 2876, 3704, 3705, 4569, 4570, 4949–4951, 5364, 5365, 5519.
Moffett, Cleveland, 2662.
Moffett, William Adger, 1196, 3299, 3706–3711, 4012.
Mohr, Adrian, 4485.
Moigno, F., 1197.
Mollison, James Alan, 3300a, 3301.
Molter, Bennett Arthur, 4013.
Monaghan, Frank, 721.
Monk, F. V., and H. T. Winter, 3303.
Montagnes, James, 4278–4280, 5142, 5520, 5521.
Montagu (2. baron), John Walter Edward Douglas-Scott-Montagu, 311.
Montgolfieri, Umbra, pseud., 725.
Montgomery, John J., 1981, 2444.
Moore, Alvin Edward, 2589.
Moore, R. F., 4953.
Moore, Richard Bishop, 726.
Moore, Samuel Taylor, 4401.
Moore, Willis Luther, 312.
Moore-Brabazon, John Theodore Cuthbert, 3304.
Mora, Alfred, 4954.
Morazzoni, Giuseppe, 727.
Morin, Arthur Jules, 4955.
Morris, Joseph, 4014.
Morris, Ralph, 87, 88.
Morris, Woodbridge E., 4402.
Morse, Daniel Parmelee, 4015.
Mortane, Jacques, 728, 1200, 1984, 1985, 3305–3307, 3712, 4017–4025.
Mortane, Jacques, and Jean Daçay, 4026.
Moses, Stanford E., 1201.
Mott, S. D., 2590, 4958, 5269.
Mouillard, Louis Pierre, 1986, 5270–5272.
Mourot, N. O., 4281.
Mowbray, Jay Henry, 314.
Moy, Thomas, 2509, 2877.
Müllenhoff, Karl, 729, 1202, 2446, 2447, 5273–5276.
Müller-Breslau, Heinrich Franz Bernard, 1203.
Munchausen, 89.
Munday, Albert H., 4027. *See also* Wemp, Bert, and A. H. Munday.
Munk, Max M., 1987.
Munro, Robert, 730, 731.
Munsey, Frank Andrew, 4028.
Murie, James, 5277.
Murillo, Abraham, jr., 2510.
Murphy, Charles J. V., 2511, 3308.
Murphy, Marvin, 2512.
Musschenbroek, Petrus van, 2663.
Myers, Carl E., 315, 732–735, 1204, 1205, 1988, 4959.
Myers, Denys P., 1989, 1990.
Mytton, R. G. V., 1991.

N

N., 736.
N., K., 737, 1206, 1207.
N., M., 1992.
Nadar, pseud. *See* Tournachon, Félix.
Nadaud, Marcel, 4029.
Nagel, Robert, 90.
Nansouty, Max Charles Emmanuel Champion de, 1993.
Narasimm, V. N., 91.
Narraway, Athos Maxwell, 5143.
National Geographic Society, 741.
Navier, Claude Louis Marie Henri, 4960.
Nayler, J. L., 4961.
Nayler, J. L., and E. Ower, 316.
Needham, Cecil Hugh Latimer, 2448.
Neely, Frederick R., 3314, 3315, 4486.
Nees, Chr., 5279.
Negro, Piero, 1996.
Neilson, Robert Morrison, 1997.
Némethy, Emil, 1998.
Neon, pseud., 4032.
Neue Augsburger Zeitung, 4962.
Neuhäuser, O., 743.
Neumann, 3713.
Neumann, Georg Paul, 4033, 4034.
Nevill, John T., 1999, 2000, 3316–3319, 4573.
Neville, Leslie E., 2001–2003.
New York Public Library, 27, 28, 4487.
New York Times, 3327.
Newcomb, J. Arthur, 3328.
Newcomb, Simon, 318, 2009, 2010.
Newell, Joseph Shipley, 2011.
Newman, J. O., 5523.
Newton, Byron Rufus, 2012–2016, 3329, 3330.
Nicholson, William, 748.
Nicolleau, A., 2017.
Niemeyer, Victor, 748a.
Nikel, Hugo L., 2667.
Niles, J. J., joint author. *See* Hall, Bert, and J. J. Niles.
Nimfuehr, Raimund, 319, 320, 2018, 2449, 4964.
Nippoldt, 1214.
Nobile, Umberto, 4488.
Noble, V. d'Oyley, 2021.
Noble, Walter, 4037.
Noel, Edmond Percy, 4285, 4574.
Nordhoff, Charles Bernard, 4038. *See also* Hall, James Norman, and C. B. Nordhoff.
Norman, William James, 5144.
North, H. C., 324.
North, J. D., 2022.
Norway, N. S., 1215.
Noyes, Charles E., 5524.
Nunn, Wilfrid, 4040.
Nutt, Arthur, 2782.

O

Oakes, Edward, 1218, 2450.
Obermayer, A. von, 325.
Ochoa, V. L., 2024.
Oelerich, Reinhold, 3342.
Oertel, Walter, 2783.
Offermann, Erich, 4287.
Ogilvie, A., 2025.
Ogilvie, Carl B., 4042–4046.
Oliver, Robert Shaw, 3715.
O'Malley, Frank Ward, 2026.

O'Malley, Henry, 5525.
Omlie, Phoebe Fairgrave, 5367.
Omri, 1219.
Oppermann, C. A., 4406.
Orcy, Ladislas d', 328, 1223, 1224, 2030, 2031, 2346, 3348, 3349, 3716, 3717, 4288–4290.
Orlebar, Augustus Henry, 3350.
Orthlieb, 4048, 4049.
Osborn, Earl D., 2036, 2592.
Osman, William H., 4050.
Ostrop, Wilhelm, 4631.
Oulianoff, Nicolas. *See* Corbin, Paul, and Nicolas Oulianoff.
Ovid, 92.
Ovington, Adelaide (Alexander), 3355, 4407.
Owen, Russell, 3356.
Ower, E., joint author. *See* Nayler, J. L., and E. Ower.

P

P., C., 5529.
P., E. L., 329.
P., F. X., 1227.
Paddock, Paul D., 5530.
Pagé, Victor Wilfred, 2038, 2452, 2784.
Page, William H., 5145.
Paine, Ralph Delahaye, 4053.
Painlevé, P., 2931.
Painter, K. M., 5146.
Pallier, T., joint author. *See* Sircos, A., and T. Pallier.
Palmer, Paul J., 3357.
Paltock, Robert, 93.
Pannell, John Robert, 4965.
Panzer, Friedrich, 94.
Pappafava, Vladimir, 760.
Parish, William Francis, 2039.
Parkes, Oscar, 3719.
Parkin, J. H., 4966.
Parmelin, Agenor, 3358.
Parrott, R. J., 2040.
Parseval, August von, 761, 762, 1228–1230, 2453, 5280.
Parsons, Bernard, 2041.
Parsons, Sir Charles Algernon, 2785.
Partridge, J. S., 1231.
Pascalis, Felix, 4967.
Patrick, Mason Mathews, 95, 3360, 4054, 4055.
Patterson, T. T., 1234.
Pattinson, Lawrence Arthur, 4056.
Paullin, Charles Oscar, 3721.
Paulus, Käthe, 2516.
Pawlowski, Felix Wladyslaw, 2042, 4968.
Payne, Enoch George, and H. R. Barrows, 330.
Peal, S. E., 5281.
Pearson, William, 1235.
Pease, Richmond, 5368.
Pégoud, Adolphe Célestin, 3361, 3362.
Pelletier Doisy, George E. C. M., 3364.
Pénaud, Alphonse, 2454, 2786, 4969, 5282–5284.
Pénaud, Alphonse, and Paul Gauchot, 2045.
Penny, R. E., 2347.
Peppler, Albert Gustav, 4970.
Percheron, Maurice, 2046, 4058.
Pérez de Guzmán, Juan, 4971.
Perfetti, R., 4059.
Perkins, Frank C., 1237, 2047.
Perrin, H. E., 4292.
Perrotin, Henri, 763.
Perry, H. W., joint author. *See* Bradley, W. F., and H. W. Perry.
Pesce, G. L., 2593, 4490.
Peterson, C. G., 2787.
Petit, Robert, 2048.
Petliakov, V. M., 4972.
Pettigrew, James Bell, 5285, 5286.
Peyrey, François, 2044.
Phélip, Gaston, 2049.
Phillips, Donald B., 2879.
Phillips, Harvey, joint author. *See* Randolph, Stella, and Harvey Phillips.
Phillips, John Scott, 2050, 2051.
Phillips, Stanley, 4410.
Philos, 3366, 4974, 4975.
Philp, Charles G., 5531.
Piazza, Lino, 331.
Picard, Émile, 2053.
Picca, Paolo, 1239.
Piccard, August, 4976.
Pichault, S., 4977.
Pierce, G. Wyndham, 3722.
Pierre, J., 1240.
Piersol, James V., 5532.
Pilâtre de Rozier, François, 764.
Pilcher, Percy Sinclair, 2455, 2456.
Pillet, Louis, 2880.
Pinchetti, B., joint author. *See* Bertùccioli, Americo, and B. Pinchetti.
Pinedo, Francesco de, 3367–3370.
Pistolesi, E., joint author. *See* Giacomelli, R., and E. Pistolesi.
Pitcairn, Harold F., 2594.
Pitter, Joseph, 1241, 1242.
Pitts, Nelson F., jr., 5148.
Platte, August, 332, 765, 1243, 2058, 2595, 4979–4981.
Plumb, C. M., 1244.
Pochhammer, Bruno, editor, 1245.
Poe, Edgar Allen, 96, 97.
Poeschel, Johannes, 766.
Poidebard, A., 5149.
Poincaré, L., 2347a.
Poirier, Jules, 4060.
Pole, William, 1246–1248, 4576.
Polk, George W., 5533.
Pollard, Alfred Oliver, 4061.
Pollner, P. B., 2059.
Pollock, Granville A., 3723.
Pollog, Carl Hanns, 2060, 4294.
Pomortzev, Mikhail Mikhailovich, 4982.
Pompeïn-Piraud, J. C., 2061, 2062.
Ponton d'Amécourt, Gustave, vicomte de, ed., 2063.
Poole, John, 767.
Popper, Josef, 333, 4983, 4984.
Porta, Giovanni Battista della, 4985.
Porter, Harold Everett, 4062.
Post, Augustus, 2064–2066, 3372, 3373, 4295.
Post, Wiley, and Harold Gatty, 3374.
Postnikov, F. A., 3724.
Potter, Leslie S., 4413.
Pouchet, Paul, 2067.
Poujoula, R., 2668.
Poulter, R. M., 4986.
Powell, O. A., 2788.
Prandtl, Ludwig, 4987, 4988.
Prechtl, Johann Joseph, Ritter von, 768, 5287.
Prepositi, Clemente, 2069, 4062a, 4063.
Preston, R. A. D., 770.
Priestley, Joseph, 771.
Prieur, C. A., 2517, 2518.
Primbault, H., 1249, 4989.
Pritchard, Henry Baden, 3725.
Pritchard, John Laurence, 2070, 3376.
Pritchard, Robert J., 4415.
Probert, Richard J., 2071.
Proctor, Richard Anthony, 1250, 2072.
Prohaska, Franz, 2519.
Protheroe, Ernest, 334.
Pulford, C. W. H., 3378.
Putnam, George Palmer, 4494.
Putnam, George Rockwell, 5535.
Pye, David Randall, 2789–2790.

Q

Quattrini, Antonio G., 4495.
Quimby, Harriet, 5370.

R

R., S., 2077.
Radford, A. W., 5150.
Rae, John, 775.
Raleigh, Walter, 3382.
Raleigh, Sir Walter Alexander, and H. A. Jones, 4065.
Ramakers, L., 2598.
Rameau, Louis, 335, 1259, 2078, 3727.
Ramond, E., joint author. *See* Blériot, Louis, and E. Ramond.
Rampont, 4416.
Ramsay, Sir William, 776.
Ramsey, DeWitt C., 3728.
Ramsey, Logan Carlisle, and E. H. Kincaid, 4993.
Randers-Pehrson, Nils Hendrik, 4994.
Randolph, Stella, and Harvey Phillips, 2079.
Ranney, H. F., 3729.
Ranwell, William, 1260.
Rappaport, C. E., 29.
Rateau, Auguste Camille Edmond, 2883.
Rathbun, John B., 1261, 2081, 2082.
Rawlinson, Alfred, 4066.
Rayleigh, John William Strutt, 3rd baron, 4995, 5288, 5289.
Raynham, F. P., 3383.
Read, A. C., joint author. *See* Westervelt, G. C., and others, 3499a.
Reber, Samuel, 3730.
Recouly, Raymond, 3386.
Redden, Charles F., 3387.
Redempt, 3732.
Redington, Paul G., 5537.
Reece, Robert Henry, 4069.
Reed, E. Stoy, 3388.
Reed, Henry Albert, 5151.
Reed, Thomas R., 4996.
Reichenbach, Oscar, 5290.
Reid, A. Cunningham, 4070.
Reid, Walter F., 3733.
Reis, P., 1263.
Reissner, H., 4997.
Renard, Charles, 778, 1264–1270, 2087–2089, 2599, 2884, 2885, 4998–5001.
Renard, Charles, and A. Krebs, 1271.
Renard, Paul, 337–340, 1272–1274, 2090, 2791, 3734, 4071, 5002–5004, 5371.
Renoir, 2600, 2601.
Rentschler, F. B., 341.
Renvoise, Ange, 4300.
Reuterskiold, 2092.
Rey, Robert, 98, 99, 5006.
Reyneker, F., joint author. *See* Hegener, Henri, and others.
Rhees, William Jones, 3735.
Riabouschinsky, Dmitri. *See* Ryabushinski, Dmitri.
Riat, Georges, 1278.
Ricardo, Harry Ralph, 2792–2793.
Richardson, Holden Chester, 2095, 2350–2352, 3395. *See also* Westervelt, G. C., and others, 3499a.
Richardson, James M., 4073–4075.
Richet, Charles, joint author. *See* Tatin, Victor, and Charles Richet.
Richmond, V. C., 1279.
Richthofen, Manfred Albrecht, Freiherr von, 4076.
Rickenbacker, Edward Vernon, 4077.
Riddell, M. R., 2096.
Riddle, Karl, 5152.
Rigg, E. H., 5539.
Riley, C. V., 2097.
Ritter, Hans, 3736.
Rivera y Sempere, Manuel, 1280.
Rivière, Pierre, 2353.
Robert, brothers, 779.
Roberts, E. M., 4078.
Roberts, E. W., 2098.
Robertson, Étienne Gaspard, 780, 781.
Robertson, Frederic Arthur de Vere, 2099, 2100, 3396, 3737, 3738.
Robertson, Hugh W., 4496.
Roch, Eugène, 782.
Roché, J. A., 2101.
Rodgers, John, 3397.
Rodnykh, Aleksandr A., 343, 344.
Roe, Sir Alliot Verdon, 2103, 2104.
Roer, Max, 4633.
Roeser, 1282.
Rogers, Bogart, 1283.
Rohlfing, Charles Carroll, 4634.
Rojas, Francisco de Paula, 345.
Rolland, Louis, 4635.
Rolls, Charles Stewart, 783.
Romas, Jacques de, 2669a.
Romocki, S. J. von, 784.
Roper, Albert, 4636.
Rose, D. F., joint author. *See* Cierva, Juan de la, and D. F. Rose.
Rose, Friedrich, 785.
Rosendahl, Charles Emery, 1285.
Rosenhain, Walter, and Guy Barr, 5007.
Rosette, Breese, 4302.
Ross, Malcolm, 2105.
Rotch, Abbott Lawrence, 346, 786, 787, 4577, 5008–5015. *See also* Teisserenc de Bort, Léon Philippe, and A. L. Rotch.
Rothschild, Zeta, 1286.
Rouland, 788.
Rousseau, Jean Jacques, 100.
Roussilhe, H., 5153.
Rouville, A. de, joint author. *See* Girard, E., and A. de Rouville.
Roux, Louis François, 2106.
Rowell, Ross E., 3740.
Rowland, John T., 789.
Rowland, May, 5540.
Roy, Jack Martin, 4079.
Roy, Maurice, 5016.
Royal Meteorological Society, 30.
Roze, Louis, 1288.
Rue, Larry, 3410.
Ruggles, C. L'H., 3740a.
Ruiz de Alda, J., joint author. *See* Franco, Ramón, and J. Ruiz de Alda.
Rummel, Richard M., 2108, 2109.
Rummel, Walter, Freiherr von, 1290.
Rumpler, Edmund, 1291, 2110, 2794.
Runkel, Peter, 4637.
Ryabushinski, Dmitri, 2886, 2887, 5017–5020.

S

S., W. H., 2111.
Saager, Adolf, 1292.
Sabatier, J., 1293.
Sacharow, 791.
Sachse, H., 2888.
Saco, A., 5291.
Sagendorph, Kent, 5154, 5155, 5541.
Saint-Fégor, L. de, 347.
Saint-Félix, T., 793.
Saint Loup, 5021.
St. Martin, Henri Reda, 2113.
Sala, George Augustus Henry, 2520.
Saladini, Girolamo, 794, 5022.
Salier, Cecil W., 2670.
Salvatores, P., 3741.
Samoilovich, Rudolf Lazarevich, 4497.
Sampaio, Carlos, 1295.
Samson, Charles Rumney, 2114, 2354, 3413, 4080.
Samuels, Frank E., 2119.
Samuelson, Arnold, 2120, 5023, 5024, 5292, 5293.
Sander, Alexander, 796, 797.
Sanderval, de, 2121, 2457, 2458.
Sandford, W. T., 1296.

Sanson, A. J., 5025.
Santos-Dumont, Alberto, 1297.
Sarigny, René de, 4419.
Sarnetzky, Heinrich, 5156.
Sartwell, Franklin G., 798.
Sassoon, Sir Philip A. G. D., 3rd bart., 3414.
Saunders, W. O., 2124.
Savage, E. W., 2125.
Savorgnan di Brazzà, Francesco, 3742.
Sayers, W. H., 348, 2126, 2127, 2355, 2459, 2795, 3415–3417, 4579, 5026.
Sayre, Daniel, 4304.
Sazerac de Forge, Léonide, 1304, 1305, 1306.
Scaroni, Silvio, 3418, 3743, 4081, 5542.
Schäffer, Ernst, 4082.
Schaerowitz, Ernst, 3419.
Schellbach, K. H., 5027.
Schempp, M. H., joint author. *See* Hirth, Wolfram K. E., and others.
Schiavone, Mario, 1307.
Schildhauer, Edward, 1308.
Schiller, Hans von, 1309, 4083.
Schio, Almerico, conte da, 1310, 1311.
Schleiffarth, W. L., 3744.
Schloeder, Nicholas S., 5543.
Schmeckebier, Laurence Frederick, 4306.
Schneider, von, 1312.
Schnepf, John, 2130.
Schnurmacher, 2131, 2132, 5544.
Schofield, Harry Methuen, 4084.
Scholle, Howard A., 3745.
Schott, Gaspar, 5028.
Schroeder, Rudolph William, 3423, 3424.
Schumann, Herbert, 101.
Schunck, Helen Kay, 2133, 2134.
Schwager, Otto, 2796–2797.
Scoble, Walter A., 2889.
Scott, A. J. L., 4085.
Scott, C. L., 3425.
Scott, Charles William Anderson, 3425a.
Scott, G. H., 1314.
Scott, W. H., 3426.
Scott, Wisner Gillette, 3427.
Scrutator Mechanicus, pseud., 1315.
Sée, Alexandre, 2138, 5294.
Seewald, F., 2356.
Selby, Francis James, 5032.
Sellers, M. B., 2602.
Sem-Jacobsen. *See* Arnesen, Odd, and Sem-Jacobsen.
Semple, Elizabeth Anna, 5374a.
Serph, André, 799.
Serrell, Edward Wellman, 3746.
Seux, Edmond, 2140, 2141.
Shackleton, Sir Ernest Henry, 4498.
Shannon, Homer H., 4308.
Shaughnessy, E. H., 4420.
Shaw, B. Russell, 800, 4309.
Shaw, Sir William Napier, 5033–5035.
Sheldon, William, 2798.
Shepherd, Edwin Colston, 4310.
Shepherd, William Gunn, 2142.
Sherbondy, E. H., 2799, 2800.
Sherbondy, E. H., and G. D. Wardrop, 2801.
Sherman, W. C., 4085a.
Shoemaker, James M., 2802, 3435.
Short, H. Oswald, 2357.
Short, Mac, 2143.
Sibbald, Andrew T., 349.
Sibley, Hi, 3436.
Sibour, Violette (Selfridge), vicomtesse de, 5375.
Silberer, Herbert, 801, 802.
Silberer, Victor, 1321, 5376.
Silberschlag, Essias, 5295.
Silva, Louis de, 3437.
Simmonds, O. E., 4311.
Simonin, L., 3747.
Sinclair, D., 4312.
Sinclair, Mrs. J. Herbert, 350.
Sircos, A., and T. Pallier, 803.
Sivel, Th., joint author. *See* Crocé-Spinelli, J., and Th. Sivel.
Slawson, H. H., 2460.
Slosson, Edwin Emery, 2148.
Small, Julia D. Sophronia, 805.
Small, R. D., 5547.
Smedal, George, 3438–3440, 4313.
Smith, A. M., 4499.
Smith, Earl Hamilton, 4086.
Smith, Erle H., 5377.
Smith, Ernie, 3441.
Smith, Floyd, 2521.
Smith, George Otis, 5157.
Smith, J. E., 5548.
Smith, James Walter, 1322.
Smith, Laurence Yard, 351.
Smith, Philip S., 5157a.
Smith, Robert A., 3442.
Smith, Sir Ross MacPherson, 3443, 3444.
Smith, William Harold, 2150.
Smyth, W., 5296.
Snodgrass, Robert Evans, 5297.
Snow, V. G., 2461.
Snowden Gamble, Charles Frederick. *See* Gamble, Charles Frederick Snowden.
Sohlman, J., 2603.
Somerville, Harry A., 4314, 5158.
Soreau, Rodolphe, 1323, 2155–2158, 5037–5040, 5298.
Soreau, Rodolphe, and Edmond Henry, 2890.
Soreau, Rodolphe, and A. Volmerange, 352.
Soucek, Apollo, 3446, 3447.
Souvestre, Pierre, and August Wimille, 3449.
Spaight, James Molony, 3749, 4638, 4639.
Spangler, Robert W., 811.
Spanner, Edward Frank, 1324–1327.
Speight, W. L., 2463.
Spencer, G. K., 2159, 5378.
Sperry, Elmer Ambrose, 2160, 2161.
Spooner, Winifred, 5379.
Sprat, Thomas, 5041.
Sprigg, Christopher St. John, 1328.
Sprigg, Theodore Stanhope, 1329, 3451.
Spuller, Eugène, 3750.
Squier, George Owen, 31, 2164, 2165, 3751, 5042.
Squires, J. Duane, 3751a.
Stack, T. Neville, 3452.
Stade, H., 353, 5043.
Stannard, H. S., joint author. *See* Briscoe, Walter A., and H. S. Stannard.
Stanton, Charles V., 5550.
Stauber, Josef, 3753.
Stearns, Myron Morris, 2167.
Steele, G. W., 1332.
Stefánsson, Vilhjálmur, 4500, 4501.
Steinmetz, Charles Proteus, 2464.
Stentzel, A., 2804.
Stern, Kenneth D., 2522.
Sterne, Carus, 812.
Stevens, Albert W., 3455.
Stevens, Leo, 5380.
Stevens, Truman, 2168.
Stevens, Walter Le Conte, 354, 355.
Stewart, Andrew, 1334.
Stewart, Charles John, 5044.
Stewart, N. C., 5159.
Stewart, Oliver, 3754.
Stinson, Marjorie, 5381.
Stockbridge, Frank Parker, 2169, 2359.
Stockhausen, von, 3755.
Stokes, Allen, 3456.
Stoll, Bert, 4315.
Stone, Francis Gleadowe, 3756.
Stout, R. Cheyne, 2806.
Stout, Wesley Winans, 4316.
Strahlmann, Fritz, 4088.

Strain, L. H., 4089.
Strait, H., 1335.
"Strasbourg," pseud., 2171.
Straus, Émile, 813.
Streett, St. Clair, 3460.
Strindberg, Nils, 4502.
Stuart, John, 4503.
Sturm, Johann Christoph, 1336.
Sturmey, Henry, 2172.
Suain, L. G., 357.
Süring, Reinhard Joachim, 5046–5048. *See also* Berson, Arthur, and R. Süring.
Sueter, Murray Fraser, 4090.
Sullivan, Alan, 4091.
Sumner, Percy James Hammond, 358.
Supf, Peter, 359.
Suplee, Henry Harrison, 360, 2173, 2174, 4504.
Surcouf, Édouard, 102, 814, 1338, 4505, 5049.
Sutner, von, 3758.
Sutton, B. E., 4092.
Sutton, George W., 3759.
Svehla, George, 2891, 4318.
Svenska sällskapet för antropologi och geografi, Stockholm, 4506–4508.
Sverdrup, Harald Ulrik, und Einar Lönnberg, 5050.
Swan, Andrew, 2807.
Swedenborg, Emanuel, 2175.
Sweetser, Arthur, 4093.
Swinehart, D. E., 2176.
Swinton, Alan Archibald, 2177.
Swoffer, Frank Arthur, 3760.
Sykes, Sir Frederick Hugh, 4094.
Sykes, H. M., 5300.
Sykes, William H., 5051.
Symons, William L., 2808.

T

Taffoureau, Edgar, 2605, 2892.
Talbot, Frederick Arthur Ambrose, 1339, 3760a.
Tarnowski, J., 2181.
Tastes, de, 2893.
Tatin, Victor, 2182–2187, 2894, 5301, 5302.
Tatin, Victor, and Charles Richet, 2188.
Taylor, Charles E., 2811, 3462.
Taylor, Charles Fayette, 2809–2810.
Taylor, H. W., 5160.
Taylor, John, 103.
Teale, Edwin Way, 2189.
Teffé, baron de, 1343.
Tegetmeier, William Bernhard, 5553.
Teisserenc de Bort, Léon Philippe, and A. L. Rotch, 5052.
Templer, D., 3761.
Templer, J., 3762.
Tennant, John Edward, 4096.
Tennant, John Kingston, 3463.
Terne, Carus, 812.
Teros, A. P., 361.
Tetens, Hans, 2193.
Teutenberg, A., 105.
Thaden, H. V., 3465.
Thaden, Louise McPhetridge, 5382.
Thayer, Russell, 1344, 1345, 3763.
Thébault, Louis, 4097.
Theilhaber, Felix A., 4098.
Thenault, Georges, 4099.
Theuermeister, Robert, 362.
Thibout, André, 4640.
Thiéry, Maurice, 4100.
Thomas, H. J., 2194.
Thomas, Hugh Hamshaw, 5161, 5162.
Thomas, Lowell Jackson, 3466, 3467.
Thomas, Pierre, 5053.
Thomas, W. E., 2812.
Thomassen à Thuessink van der Hoop, A. N. J., 3207, 3468.
Thomson, G. P., 2195.
Thomson, Jay Earle, 363.
Thornhill, Gertrude, 3469.
Thorpe, W. H., 106.
Thurston, Robert Henry, 364, 365.
Tidswell, Ella, 2465.
Tilgenkamp, Erich, 2360.
Tilley, N. N., 2813.
Tinker, Clifford A., 366.
Tissandier, Gaston, 32, 367, 816–826, 1346–1353, 2523, 2524, 2671, 2814, 2815, 3764–3771, 5054, 5055, 5163, 5164. *See also* Glaisher, James, and others.
Tissandier, P., joint author. *See* La Vaulx, Henri, comte de, and P. Tissandier.
Titterington, Morris M., 5554.
Todd, Frederick, 368, 2196.
Tolmachoff, I., 4582.
Tomaselli, Cesco, 4509.
Tomlinson, Daniel Webb, 4102.
Tomlinson, F. L., joint author. *See* Valentine, E. Seton, and F. L. Tomlinson.
Torres, L., 1355, 1356.
Touche, P. E., 5056.
Toulmin, Harry Aubrey, 2198, 4104.
Tournachon, Félix, 2199, 3773.
Tousley, Bennett E., 4105.
Towers, J. H., 4106.
Tracy, M. D., joint author. *See* Van Every, Dale, and M. D. Tracy.
Trenchard, H., 4107.
Treusch von Buttler Brandenfels, Freiherr, 4108.
Triaca, Albert C., 1360.
Tridon, J. Louis, 827.
Tronnier, Adolph, 1361.
Truby, Harry A., 4424.
Tschudi, Georg Julius Friedrich von, 1362.
Tucker, Ray A., 3479.
Turbiaux, E., 3774.
Turgan, L., 369.
Turnbull, W. R., 2203, 2204.
Turner, Charles Cyril, 370–372, 4109, 4583.
Turner, E. E., 1363.
Turner, George Kibbe, 2205.
Turnor, Christopher Hatton, 373.
Tuttle, Thomas Taft, 2361.
Tziolkovski, K., 374, 1366, 5057.

U

Ullendorff, Hans, 4322.
Ulm, C. T. P., joint author. *See* Kingsford-Smith, Charles Edward, and C. T. P. Ulm.
Unge, Eric, 831.
United States. — Aeronautics Bureau (Navy Department), 3776.
United States. — Air Service, 2209, 3777, 3778, 4111.
United States. — Air Service: Medical Section, 5556.
United States. — Aircraft Production Bureau, 4114.
United States. — Army: Chief Signal Officer, 3779.
United States. — Army: General Staff, 4115.
United States. — Army: Signal Office, 3780.
United States. — Army and Navy Board, 3781.
United States. — Army War College, 4116.
United States. — Census Bureau, 376.
United States. — Coast and Geodetic Survey, 5165.
United States. — Commerce Department: Aeronautics Branch, 4323–4326.
United States. — Director of Military Aeronautics, 4117.
United States. — Expenditures in the War Department Committee, 4118.
United States. — Forest Products Laboratory, 5059.
United States. — Interstate and Foreign Commerce Committee (House), 4642.
United States. — Justice Department, 4119.
United States. — Library of Congress, 33, 5166.

United States. — Military Affairs Committee (House, 70:1), 2210, 4121.
United States. — Military Affairs Committee (Senate, 65:2), 4120.
United States. — National Advisory Committee for Aeronautics, 34, 5058.
United States. — Naval Operations Bureau, 4122.
United States. — Navigation Bureau, 3482a.
United States. — President's Aircraft Board, 4123, 4124.
United States. — Spruce Production Corporation, 4125.
United States. — Standards Bureau, 5060.
United States. — State Department, 3483.
United States. — War Department: Air Corps, 2526.
Upshaw, H. C., 832.
Upson, Ralph Hazlett, 833, 1368, 1369.

V

Valentine, E. Seton, and F. L. Tomlinson, 377.
Valer, Paul, 2895.
Vallet, joint author. *See* Alban, and Vallet.
Vallot, Joseph, and others, 5558.
Van der Maas, H. J., 5062.
Van Dusen, L., 2466.
Van Dusen, W. I., 4330, 5558a.
Van Everdingen, E., 5061.
Van Every, Dale, and M. D. Tracy, 3485.
Vaniman, Melvin, 4510.
Van Muffling, Adrian, 378, 379, 2214.
Van Orman, Ward Tunte, 835.
Van Zandt, John Parker, 3486, 3487, 4331, 4332, 4428.
Vaughan, Katherine, 5383.
Vaughan-Fowler, R., 3488.
Veit, Sidney B., 4584.
Vélez, Miguel, 1370.
Ventou-Duclaux, L., 2215.
Venturini, Galileo, 35.
Veranzio, Fausto, 2527.
Verhaegen, P., 837.
Verne, Jules, 107.
Vernillat, 4126.
Vernon, 2467.
Very, Frank Washington, 2217.
Vest, J. R. W., 5168.
Veyrin, Émile, 2218, 2896.
"Vigilant," pseud., 4128.
Villiers du Terrage, Marc de, baron, 3786.
Vincent, J. G., 2818.
Vinci, Leonardo da, 5063.
Vindel, P., joint author. *See* Diaz Arquer, Graziano, and P. Vindel.
Violette, Marcel, 2220.
Vischer, Peter, 3787.
Vissering, Harry, 1371.
Vives y Vich, Pedro, 3788.
Vivian, Evelyn Charles H., 380.
Vivien, F. Louis, 381.
Voemel, Alexander, 1372, 1373.
Vogt, H. C., 2897, 2898.
Voisin, André Paul, 3789, 4129, 4130.
Voisin, Gabriel, 2222.
Volmerange, A., joint author. *See* Soreau, Rodolphe, and A. Volmerange.
Volumard, Maurice, 2606.
Vorreiter, Ansbert, 382, 1374, 1375, 2223–2225, 5559.
Voyer, 840, 841, 1376–1379, 5064, 5065.

W

W., I. H., 841a.
W., P. T., 841b.
W., R. C., 3490.
W., V. G., 3491.
Waack, Karl, 4511.
Wade, Walter Lewis, 1380, 1381, 2226, 4131.
Wäsch, Curt, 4645.
Wager-Smith, Curtis, 1382, 1383.
Wahl, Gustav, joint author. *See* Liebmann, Louis, and Gustav Wahl.
Wait, William, jr., 3492, 3493.
Wakeman, S. W., 5560.
Walcott, Charles Doolittle, 2227, 2228, 4585. *See also* Wright, Orville, and C. D. Walcott.
Walcott, Stuart, 4132.
Walford, Eric Weston, 2819.
Walker, Donald F., 2468, 2469.
Walker, Frederick, 383.
Walker, G. W., 5066.
Walker, George L., 5169.
Walker, John Bernard, 1384.
Walker, Thomas, 2229.
Walker, William George, 2899.
Wallin, B. H., 2230.
Walmsley, Leo, 3494.
Walton, Francis D., 1385, 3495.
Wampler, Rollin H., 2528, 2529.
Ward, William Edward, 1386.
Wardrop, G. D., joint author. *See* Sherbondy, E. H., and G. D. Wardrop.
Warner, Edward Pearson, 384, 1387, 2231, 2470, 3791, 4333, 4334, 4646.
Wasner, Adolf, 4136.
Waters, R. S., 3792.
Watkins, Henry George, 4512.
Watkins, John Elfreth, 385, 386, 3793.
Watson, Sir Charles Moore, 3794.
Watson, Geoffrey, 3795.
Watter, Michael, 2232, 5067.
Watts, Henry C., 2900.
Wead, Frank Wilbur, 1388, 2233, 3496, 3796, 5562.
Webb, L. D., 2234, 2820.
Webber, John E., 5170.
Weck, Hermann, 4647.
Weekes, W. H., 842.
Weems, Philip van Horn, 3498.
Weick, Fred Ernest, and D. H. Wood, 2901.
Weir, William Weir, 1st baron, 4137.
Weiss, Harry Bischoff, joint author. *See* Landauer, Bella Clara, and H. B. Weiss.
Weiss, Joseph Bernard, 2471.
Weisse, H., 5302a, 5303.
Weldon, 1389.
Wellman, Walter, 4513, 4514.
Wellman, William Augustus, 4138.
Wellner, Georg, 387, 388, 843, 1390, 1391, 2235–2238, 2902–2904, 5068–5070.
Wells, G. M., 3797.
Wells, Linton, 3499.
Wemp, Bert, and A. H. Munday, 4139.
Wenham, Francis Herbert, 389, 2239–2242.
Wenneman, Joseph H., 4335.
Wenz, Émile, 2672, 5171.
Wenz zu Niederlahnstein, Oskar von, 843a.
Werner, 5304.
Werner, A., 2905, 3798.
Werner-Bleines, 5563.
Werthner, William, 2244.
West, A. G., 5564.
West, Rupert E., 2245.
Westervelt, G. C., and others, 3499a.
Weyerbacher, Ralph D., 2246, 4336.
Weyher, Ch., 2247, 5305.
Whale, George, 1392.
Wheat, George Seay, 2821.
Whipple, F. J. W., 5072.
Whitcomb, Russell, 1394.
White, Bert, 2530.
White, Thomas D., 4337.
Whitehead, Gustave, 2251.
Whitehead, R. F., 5172.
Whitman, Roger Bradbury, 1395.
Whittaker, W. E. de B., 390, 3500, 3501.

Whymper, Frederick, 844.
Wicks, Z. W., 845, 845a.
Wien, Noel, 3502, 3503.
Wild, Horace B., 846.
Wild, Ralph Kirkby Bagnall, 2255.
Wilde, Henry, 2531.
Wilhelm, A., 4648.
Wilhelm, Balthasar, 108, 1398–1401.
Wilkins, Sir George Hubert, 1402, 4517–4519.
Wilkins, H. F., 5173.
Wilkins, John, bishop of Chester, 5075, 5076.
Willard, Charles F., 3505.
Willcox, H. Case, 4338, 4339, 5174.
Willgoos, A. V. D., 2822.
William of Malmesbury, 109.
Williams, Alford Joseph, 3506.
Williams, Archibald, 392.
Williams, Bertram W., 2259, 4141, 4142.
Williams, C. L., 5175.
Williams, W. E., 2260. *See also* Bryan, George Hartley, and W. E. Williams.
Williams, W. T. S., 4340, 4341.
Willows, E. T., 1403.
Willson, Robert W., 5077.
Wilson, E., 2261, 2261a.
Wilson, E. E., 2823.
Wilson, Edwin Bidwell, 5078.
Wilson, Elwood, 5176.
Wilson, J. A., 5566.
Wilson, N. W., 4143.
Wimille, August, joint author. *See* Souvestre, Pierre, and August Wimille.
Wimperis, Harry Egerton, 5079, 5567.
Winchester, Clarence Arthur Charles, 2262.
Windham, William, 847.
Wines, James P., 3507.
"Wings," pseud., 4144.
Winslow, Carroll Dana, 4146.
Winter, H. T., joint author. *See* Monk, F. V., and H. T. Winter.
Winter, Lumen, and Glenn Degner, 393.
Winter, Wilhelm, 5306.
Winters, S. R., 2264, 2265, 2532, 5568–5570.
Wise, Hugh D., 2673.
Wise, John, 848–850, 5080.
Wisting, Oscar, 4521.
Wittemann, A., 1406.
Witzleben, von, 3800.
Wolff, Maurice, 111, 112, 3801, 5384.
Wolff, Maximilian, 1408.
Wolfin, Hugh M., 4342.
Wood, D. H., joint author. *See* Weick, Fred Ernest, and D. H. Wood.
Wood, Eric, 4147.
Wood, George F. Campbell, 2266–2268, 3509, 3510.
Wood, H. C., 3511.
Woodhouse, Henry, 394, 2269–2271, 2362, 3512, 3513, 3802–3807, 4148–4155, 4429, 4430, 4522, 4649, 5390, 5572, 5573.
Woodman, Dorothy, 3807a.
Woodward, Calvin Milton, 2906.
Woolley, James G., and E. W. Hill, 4343.
Woolson, L. M., 2824.
Worcester (2. marquis), Edward Somerset, 113.
Wortley, Rothesay Stuart, 4160.
Wraxall, Sir Frederick Charles Lascelles, 3rd bart., 851.
Wright, Milton, 2274, 2275.
Wright, Orville, 2276, 2277.
Wright, Orville, and C. D. Walcott, 2278.
Wright, Orville, and Wilbur Wright, 2279, 2280.
Wright, Wilbur, 2281–2285, 2473, 2474. *See also* Wright, Orville, and Wilbur Wright.
Wright Company, 2291, 2292.
Wronsky, Martin, 4346, 4347.
Wurzer, Ferdinand, 1411.
Wyatt, B. H., 5081, 5177.
Wynne, Clarence P., 852.

Y

Young, Clarence Marshall, 2298, 4348–4350.
Young, Edward H., 2299, 2300.
Young, Kennedy, 2301.

Z

Z., O., 1412.
Zabel, Eugen, 1413.
Zacharia, August Wilhelm, 5307.
Zahm, Albert Francis, 36, 396, 2302, 2303, 2475, 5082, 5083.
Zamagna, Bernardo, 35.
Zarzar, V., 4351.
Zeigler, S. J., 4161.
Zeppelin, Ferdinand Adolf August Heinrich, Graf von, 1414–1418.
Zimmerman, Charles, 2304, 2674.
Zimmermann, Paul G., 2363.
Zollmann, Carl Friedrich Gustav, 4650.

INDEX OF SUBJECTS

(Numbers refer to entries, not to pages.)

A

Abate, Stanislao:
  rocket propulsion, 160.
Abbe, Cleveland:
  kites and meteorology, 2609.
Abercron, V. von:
  balloon ascensions, 397.
Abyssinia. *See* Italo-Abyssinian war.
Académie Française:
  medal, 4547.
Ackermann, Gebhard:
  wing regulation of balloons, 527.
Acosta, Bertram B.:
  biography, 3045.
  "Columbia" endurance airplane flight, 2907, 2943, 3095.
  portraits, 3045, 3079.
Acoustics:
  experiments of John M. Bacon, 4676.
  See also Noise prevention.
Adams, Briggs Kilburn:
  World war flying experiences, 3809.
"Adams-Farwell" engine, 2677, 2698, 2728.
Ader, Clément:
  bibliography, 1426.
  biography, 157, 1636, 1962.
  bird flight studies, 1905.
  claims priority of Wright brothers, 1923, 2284.
  military airplane, 3523.
  portrait, 1636, 1962.
  steam-driven airplane, 142, 175, 176, 380, 1425–1428, 1706, 1895, 1914b, 1962, 1992, 2044.
Advantages (relative) of airplanes and airships, 338.
Aerial derby, 2910.
Aerial Experimental Association, 1612, 1679, 1838, 3102, 4699.
Aerial race between Britain and German, 44, 45.
Aerial Survey Company, Ltd., 5109.
Aerial Transit Company, 1914b.
Aero Club of America:
  exhibition (1906), 118, 5546.
Aéro club de France, 4546.
Aerodynamics:
  bibliography, 4700.
  history, 253, 4780, 4781, 4817, 4968, 5078.
  progress, *1910–1935,* 4864; *1928,* 4891; *1929,* 273; *1931,* 127.
  similitude law, 4882.
  World war developments, 4137.
  See also Air resistance; Airplanes, stability and control; Bird flight; Insect flight; Laboratories; Stability. Also the following names: Eiffel, Genet, Gerlach, Giffard, Göttingen Experiment Station, Jones, T. P., Joukovsky, Kretschmer, Lander, Lanchester, Langley, Lössl, Meusnier, Phillips, H., Ryabushinski, Samuelson, Soreau, Witoszynski.
Aeronautical Association (1837), 4527.
Aeronautical Chamber of Commerce, 308.
"Aeronautical fish":
  early constructions, 994.
  See also Duponchel, Geiger, Pauly, Scott (Baron).
Aeronautical Research Committee, 5079.
Aeronautical Society of Great Britain. *See* Royal Aeronautical Society.
Aeronautics:
  bibliography, 1–36, 220, 253, 292, 359.
  general works, 115–396, 4674, 4736.
  industrial and social aspects, 303.
  slow progress, 185, 236, 309, 366, 2082, 3534.
"Aeroplane," periodical, 5476.
Afghanistan:
  civil aviation, 4341.
Africa:
  civil aviation, 4341.
  Ferguson's balloon journey across (1862), 851.
  flights around, Bailey, 5340; Cobham, 3026; Italian, 3403.
  flights to, Marie, 3282; Mittelholzer, 3290, 3292, 3294, 3295.
  Wilson Airway, Ltd., 4293.
  See also Air mail, Africa; Cape of Good Hope; South Africa; Sahara desert.
Agriculture:
  airplane as an aid, 4828, 5557.
  seeding experiments, 5550.
  See also Insects.
Ahlborn, Fr.:
  bibliography of works, 5187.
Air Conference, London:
  proceedings (1920–1923), 4525.
Air mail:
  Africa:
    British, 4357.
    South, 4419.
  airplanes:
    post-war developments, 4334.
  Alaska, 4375, 4376.
  Australia, 4362.
  Bagdad line, 4391, 4413.
  bibliography, 4361.
  developments various countries, *1920–1922,* 4387; *1922–1926,* 4386.
  first permanent service (Italy), 4360.
  France:
    claims first airplane test, 4414.
  Germany, 4386, 4387, 4395.
  Great Britain:
    England-India line, 4377.
    history, 4389, 4393.
    tests, 4245.
  Italy, 4360.
  night flying, 275, 4369, 4378, 4380, 4390.
  ocean liner, delivery, 4373.
  South America:
    Paris-Buenos Aires line, 3199, 4237.
    West coast, 4271.
  United States:
    Black-McKellar bill, 4409.
    Chicago-Cheyenne line, 4405.
    Chicago-Dallas line, 4282, 4365, 4422.
    chronology, 4366.
    contracts, 4353, 4356, 4367, 4427.
    fire losses, 4352.
    first experiments, 4364, 4374.
    first flight, 4407.
    Ford airmail, 4381.
    general, 4246, 4282, 4297, 4316, 4327, 4329, 4333, 4335, 4348, 4370, 4374, 4384, 4385, 4396, 4415, 4420–4423, 4426.
    Key West-Havana line, 4394.
    legislation, history, 4417.
    Little (Ray), experiences, 3261.
    Los Angeles-Salt Lake line, 4398.
    manufacturers, relation to, 308.
    monopoly, 4397.
    New York-Chicago line, 4403.
    New York-Cleveland line, 4402.
    New York-Washington line, 4404, 4429, 4430.
    night flying, 4378, 4380, 4390.
    Praeger (Otto) organizes service, 4400.
    progress (1920), 4412.
    route history, 4418.
    Shaughnessy (E. H.), biography, 4372.
    transcontinental route, 4425, 4428.

Air mail — United States, *continued*
Watres act, 4354, 4417.
See also Franco-Prussian war; Pigeons; Postage stamps.

Air resistance:
bibliography, 4697, 4745, 4938.
history, 4755, 4807, 4840.
See also Wind tunnels; also the following names: Bryan, Chaudy, Cailletet, Cannovetti, Castelli, Desagulier, Faccioli, Faraud, Genet, Jacob, Jobert, Jones (T. P.), Lancaster, Le Dantec, Lössl, Mannesmann, Mendeleyeyv, Renard (C.), St. Loup, Samuelson, Schellbach, Thomas (P.), Touche, Wellner, Zahm.

"Airco" airplanes, 1624.

Aircraft carriers:
Courageous, 5477.
history, 3806, 3807, 5508.
Japan, 3719.
Langley, 1123.
Lexington, 5501, 5560.
Saratoga, 1442, 5501, 5527, 5528, 5539.
submarines, 5551.

Aircraft year book, 2913a.

Airplanes:
American and British compared, 2028.
biplane, Wenham (1866), 1580.
chronology, *1861–1932,* 1526; *1879–1909,* 1651; *1890–1910,* 2086; *1905–1926,* 2037.
classification, 1591.
design drawings, 248.
design progress (1903–1920), 380.
enclosed, 2197.
flights:
aviators, early, 3214, 3300a, 3377.
chronology (1903–1909), 1586.
famous, *1903–1928,* 2976; *1903–1930,* 3335; *1910,* 1914a; *1919–1932,* 264; *1927,* 3079; *1927,* 3079; *1928,* 3144; *1929–1932,* 3198–3200a, *1933,* 3336a.
Great Britain, *1918–1932,* 3142; *1928,* 3144.
history, 1557, 3300a.
map, 239.
Pacific coast fliers, 2939.
photographs, early, 3377.
post-war, 3277.
pre-war, 280, 3284.
See also Alps, Altitude flights, Andes mountains, Desert flying, Disasters, English Channel, Exploration, Far East, Mediterranean Sea flights, Mount Everest, Mount Whitney, National Air Derby, National Air Races, National Air Tour, Polar flights, Prizes, Records, Refueling flights, Reliability tests, Rheims meeting, Royal Air Force, Sahara desert, Schneider Trophy Contest, Seaplanes, contests, Tour of Europe race, Trans-American flights, Transatlantic flights, Trans-Australian flights, United States Army, United States Navy, Women in aeronautics, World flights.
four-motored, 2729.
frames. *See* Frames, airplane.
general, 172, 174–176, 216, 222, 245, 253, 257, 264, 295, 297, 298, 386, 1424, 1430, 1460, 1461, 1472, 1488, 1550, 1560, 1561, 1569, 1583–1586, 1588, 1596, 1601, 1652, 1653, 1668, 1672, 1673, 1685, 1724, 1728, 1756, 1771, 1773, 1776, 1790, 1794, 1798, 1802, 1830, 1914a, 1914b, 1914c, 1915, 1922, 1930, 1974, 1985, 2039, 2044, 2054, 2069, 2070, 2081, 2164, 2173, 2182, 2185, 2187, 2220, 2223, 2233, 2252, 2969, 4334, 5038.
introduction into foreign countries, 2270.
light, 1972.

Airplanes, *continued*
military:
British, 3570.
classification, 4141.
development from racing types, 3605.
France, postwar, 3653, 4126.
history, 3594, 3600, 3802, 3811, 3981.
possibilities, 3598, 3687, 3796.
torpedo planes, 3612.
United States, first airplane, *1907,* 3739.
World war:
descriptions, 3964, 4005, 4137, 4149.
France, 4017, 4041, 4068, 4134.
general, 3717, 3924, 3986.
Germany, 3572, 3830, 3856, 3906, 3950, 3958, 4142.
tactics, 4000, 4042.
United States, 4011, 4041, 4068, 4134.
models, 1787.
names: lists, 292, 371.
See also Airco, America, Antoinette, Atlantic, Aviatik, Avro, Bluebird, Bremen, Bristol, Britannia, Cape Cod, Century of progress, Christmas bullet, City of Chicago, Columbia, Detroiter, Do. X, Flying carpet, Friendship, Gotha, Jesus del gran poder, Josephine Ford, June-bug, Justice for Hungary, Miss Veedol, Monomail, Mystery S, Question mark, Red-wing, St. Louis robin, Silver dart, Southern cross, Spirit of St. Louis, N. C., Nieuport, Old glory, Pathfinder, P. N., Tanager, Tingmissartoq, Winnie Mae, Woolaroc, Yellow bird.
patents. *See* Patents.
progress reports, *1892–1907,* 217; *1903–1918,* 148; *1905–1926,* 2037; *1907,* 312; *1908,* 1436, 1920, 1959; *1908–1935,* 1901; *1909,* 382, 1710, 2110, 2215, 2224; *1909–1910,* 1615; *1910,* 1711, 1914a; *1911,* 2269, 1712; *1911–1914,* 1810; *1914–1917,* 1809; *1916–1920,* 2046; *1919,* 1847; *1919–1923,* 2209; *1927,* 1902, 2214; *1928,* 2003, 2143; *1929,* 1977, 2001; *1930,* 2020; *1931,* 274; *1933,* 1437.
See also Airplanes, chronology.
racing type, 1720.
size limits to, 1899.
stability and control:
automatic:
Aveline, 1466, 1467.
Curtiss, 1916.
Doutre, 1984.
Hargrave, 2139.
history, 1446.
Macy, 1692.
Marçay, 1440.
Marmonnier, 1921.
Mettler, 1965.
progress (1931), 2101, 2234.
Sperry, 1752, 2160–2162.
Thayer, 1344.
Wright, Orville, 2034, 2166.
Wright brothers, 1465, 1916.
bibliography, 4721.
history, 1564, 1836, 1928, 2038, 2044.
See also Bertin, Brillouin, Bryan, Constantin, Cousin, Crocco, Duchêne, Etévé, Goupil, Henry (L. I.), Irving and Batson, Martin (J. V.), Miller, Reissner, Seux, Soreau, Turnbull, Wellner, Williams, Wright (Orville), Wright brothers, Zahm.
tailless:
Arnoux, 1440.
developments (1932), 189.
Hill, 1791, 2179.
history, 2127.
technical history, 380, 1732, 1793, 2011, 2022, 2101, 2155, 2190, 2246, 4680, 4795, 4825, 4832, 4864, 4894, 4917, 5026.
tests: pilotless collision, Dayton, Ohio, 1566.

Airports:
aerodrome, first private in U. S., 4221.
approaches, legislation, 4604.
bibliography, 4181.
John Rodgers airport, 4285.
Le Bourget, 4260, 4300.
Lehigh design competition, 4268, 4269.
lighting, 4176.
municipal, history, 4335.
Oregon, 4175.
"Port Aviation," 4218.
San Diego, Cal., 4194.
United States, 127, 4171, 4172, 4184, 4189.
Wright field, 4345.
See also Seadromes.
Airships:
aluminium. *See* Metal construction.
bibliography, 906, 1325, 1326.
bird propelled:
condors, 1156.
eagles, 101, 857, 1159d, 1260.
large birds in general, 1157, 1995.
chronology (1784–1912), 863.
constructors and designers. *See* Almerico da Schio, Baldwin, Barton, Bausett, Bechtel, Beedle, Bettoni, Billwiller, Blanchard, Boerner, Bonaparte, Bradsky, Carelli, Choinsky, Clément-Bayard, Comaschi, Compagnon, Danilevsky, Deminne, Deutsch de la Meurthe, Duponchel, Dupuy de Lôme, Genet, Giffard, Giradot, Greth, Gusmão, Haenlein, Heaton, Kircher, Lana, Lanzillo, La Vaulx, Lebaudy, Lennox, L'Hoste, Lieband, Lohmeier, Mager, Martyn, Mason, Merritt, Meudin, Myers, Olivier, Ossari, Pétin, Porter, Renard-Krebs, Ritchell, Sanson, Santos-Dumont, Severo, Schütte-Lanz, Schwarz, Tissandier brothers, Turbini, Unger, Vaniman, Vickers, Willows, Wolff, Wood, Zeppelin.
control at sea, 1338.
dangers, 131, 1221, 1273. *See also* Disasters.
details (1852–1903), 1390.
drawings, 248.
early types:
Italian, 161.
strange, 1120.
Europe (1914), 1106.
first American craft to carry three passengers, 1037.
first commercial route established by Zeppelin, 1159b.
France:
developments (1907), 924.
history, 1045, 1172.
See also Dupuy de Lôme, Lebaudy, Renard-Krebs, Santos-Dumont, and other constructors.
French and German craft compared, 140, 1374.
general histories, 147, 172, 174–175, 196, 216, 253, 255, 257, 263, 277, 292, 294, 297, 298, 549, 869, 880, 887, 925, 955, 968, 991, 994, 996, 1023, 1069, 1083, 1085, 1091, 1119, 1144, 1150, 1159a–1159d, 1188, 1191, 1192, 1200, 1272, 1304, 1306, 1328, 1351, 1360, 1381, 1388, 1403.
Germany, 692, 923, 1150, 1282, 1308, 3527, 3543, 3648, 3950. *See also* "Zeppelin" airships.
Great Britain, 1028, 1325, 1326, 1329, 1392.
horse propelled, 1159d, 4809, 4967.
impracticability, 1155, 1185, 1250, 1283, 1324–1326.
interior keel:
Torres, 1117, 1355.
international list (1917), 1224.
man propelled:
Dupuy de Lôme, 709.
Shaw, 1313.
metal construction. *See* Metal construction.
military:
Archen's book, 3536.
British, history, 3602, 3998.
difficulty of attack, 3545.
forecasts, *1906,* 3582; *1907,* 3597.
Germany, 3543, 3950, 4083.
Airships — Military, *continued*
inflation of, war period, 3952.
possibilities, 3598, 3647, 3685, 3687, 3713.
United States specifications (1907), 3537.
World war, 3815, 3881, 3897, 3909, 3950, 3998, 4005, 4083, 4088.
See also "Zeppelin" airships.
names:
lists, 371, 1224.
See also Aeron, Akron 1st, Akron 2d, America, American legion, Atlantic, Avitor, Bodensee, California arrow, California messenger, Celeste, Clément-Bayard, De Bradsky, Demoiselle, Deutschland, Dixmude, Eagle, L'Esperance, Ezekiel, La France, Graf Zeppelin, Italia, Langley, Lebaudy, Los Angeles, M–3, Mayfly, Norge, Nulli secundus, Patrie, Pax, R–31, R–34, R–101, République, Roma, Shenandoah, Spiess, Toledo, La Ville de Paris, ZMC–2, ZR–1.
patents:
British, 1325, 1326.
possibilities:
Eckner, 1010.
Liversedge, 1145.
Pole, 1246.
Proctor, 1250.
Renard (C.), 1266.
Upson, 1368.
progress:
general, *1884,* 935; *1907,* 312; *1909,* 382; *1909–1930,* 1166; *1914–1918,* 864; *1914–1919,* 1342; *1918–1928,* 1387; *1919–1931,* 942; *1927,* 1296; *1928,* 940; *1929,* 941; *1930,* 1337.
United States, *1916–1919,* 1105; *1930,* 1058; *1931,* 274.
propulsion, electric. *See* Electric propulsion.
publications:
before the year 1800, 866, 949, 973, 980, 992–993, 1061, 1101, 1127, 1180, 1227, 1252, 1336, 4889, 5028.
years 1800–1850, 857, 871, 881–884, 922, 953–954, 1013, 1039, 1049, 1099, 1107, 1130, 1155–1157, 1159, 1163, 1167, 1176, 1184, 1216, 1219, 1220, 1222, 1231, 1235, 1241, 1242, 1260, 1315, 1335, 4809–4811.
rigid type, 1164, 1196.
Russian prototype of the Zeppelin, 145.
sail propelled, 903.
speed:
critical, Renard (C.), 1270.
measurement, Renard (P.), 1274.
stability:
bibliography, 4721.
Beachy and Knabenshue, 896.
Crocco, 967.
Goupil, 1073, 4832.
Hervé, 1102.
Renard, 1265.
Soreau, 5039.
Torres, 1356.
steam driven. *See* Steam propulsion.
theory:
history (1877–1887), 333.
investigators, Baumgarten, 1047, Boltzmann, 1531, Daubanton, 972, Duroy de Bruignac, 4783, Goupil, 4832, Helmholtz, 4855, 4856, Hunsaker, 3775, 4866, Koch, 910, 1087, 1243, Lössl, 4981, Mansfield, 1162, Marey-Monge, 1163, Schiavone, 1307, Soreau, 1323, Voyer, 1377, Wölfert, 1047.
See also Theory of flight; Meusnier.
See also Balloons, directing of.
Airy, Hubert:
bird flight experiences, 5178.
Akron no. 1, airship, 982.
Akron no. 2, airship, 943, 1196, 1261.
Akroyd-Stuart, Herbert:
patents heavy-oil engine, 2790.

Aktiebolaget Aviatorer, 2230.
Alaska:
  adventures, 3388, 3502, 3503.
  aerial survey, 5092, 5093, 5098, 5113, 5141, 5150, 5157a, 5172.
  air mail, 4375, 4376.
  civil aeronautics, 4177.
  Fairbanks Airplane Corporation, 4281.
  U. S. Army flight to, *1920*, 2915, 3460; *1929*, 3198.
Alban and Vallet:
  dirigible balloon (1785), 866.
Albret, André:
  World war experiences, 3819.
Alcock, Sir John:
  biography and portrait, 3501.
  death (1919), 3151.
  transatlantic flight, 210, 257, 2896, 3057, 3060, 3088, 3147, 3340, 3472.
Alexander, A.:
  power in relation to weight, 4660.
Alexander, Patrick Y.:
  biography and portrait, 2056.
Alexander the Great:
  account by Feldhaus, 56.
  French manuscript of (1320), 55.
  Roman sculpture of 12th century, 63.
Allen, James:
  "Father of army aviation," 3585.
Allsop, Clifford:
  World war exploits, 4075.
Almerico da Schio, Count:
  Airship *Italia*, 918, 921, 1310, 1311.
Alps:
  flights:
    Bider, airplane, 3064.
    Chavez, airplane, 223, 3010, 3098, 3478.
    first balloon, Arban, 657, 661, 697a.
    geological survey, 5106a.
    history, 613.
    Mittelholzer, airplane, 3293.
    Mount Blanc, 2977, 3358.
    Spelterini, balloon, 143, 411, 438, 441, 628.
    Usuelli and Crispi, balloon, 607.
Altitude ascensions:
  balloons, lists, 604, 809.
  Berson and Spencer, balloon (1898), 749.
  Berson and Süring, balloon (1900), 4693, 4951.
  bibliography, 2950.
  Champion, airplane (1927), 3078.
  events by years, *1908–1914*, 3342; *1928*, 3187, 3192; *1929*, 3188, 3198; *1930*, 3189.
  Hawker, airplane (1916), 3517.
  historical chart (1876), 4784.
  history, 4931.
  Icare, balloon (1913), 464.
  Latham, airplane (1910), 3384.
  Macready, airplane flights, 3096, 3276, 3321.
  Oelerich, airplane (1914), 3342.
  Paulhan, airplane (1910), 3384.
  physiological effects on early balloonists, 729.
  Rohlf, airplane (1919), 3398.
  Schroeder, airplane flights, 3195, 3280, 3337, 3423, 3424, 4931.
  Soucek, airplane (1929), 3250, 3446, 3447.
  Unwins, airplane (1932), 2983, 4961.
  women's airplane flights, Ruth Nichols, 5374; Marvel Crosson, 5388.
  See also Ballooning, scientific, Glaisher, Stratosphere.
Aluminium:
  copper-plated aluminium for balloons, 743.
  kites, 2617.
  Schwarz's airship, 867, 1084, 1161, 1164, 1187, 1190, 1243, 1322, 1330.
Amans, P. C.:
  animal flight, 5179, 5399.
  propeller experiments, 2828.
Ambulances:
  airplanes, World war, 3882.
  history, 1626, 1714.
  Sweden, 2092.
America, airship, 1159c, 4503, 4504, 4510, 4513–4516.
America, seaplane. *See* Byrd, R. E., transatlantic flight.
American Bar Association:
  first report on air legislation, 4588.
American legion, airship, 3992a.
Amherst College:
  experimental hot-air balloon (1827), 616.
Amundsen, Roald:
  Polar flights, 210, 257, 4431–4436, 4456, 4460, 4467, 4489, 4495, 4521.
Andervolti, Leonardo:
  airship, 162, 871.
Andes mountains:
  flights:
    Aracena, airplane (1923), 2924.
    Candelaria, airplane (1918), 2999.
    early attempts, 3179.
    Ingalls (1934), 5351.
    Lamb, airplane (1921), 3237.
Andreani, Paolo:
  ascent with Gerli brothers (1784), 602.
  pioneer Italian balloonist, 498.
Andrée, Salomon August:
  bibliography, 4445, 4455.
  biography and portrait, 4469.
  Polar flight, 196, 697a, 748a, 4436, 4438, 4454, 4455, 4458, 4461, 4462, 4469, 4473–4475, 4479, 4485, 4494, 4502, 4506–4508, 5050.
Andrews, Solomon:
  airships, 1174, 1244, 1395, 3535.
  proposes aerial navigation company, 870, 1174, 3535.
Angerstein, Wilhelm:
  bibliography, 4722.
  biography, 5497.
  portrait, 4570.
Angius, Vittorio:
  airship, 161.
Animal propelled airships. *See* Airships, bird propelled; Airships, horse propelled; Airships, man propelled.
Animals:
  balloon ascents on horse back, 373, 537.
  carried in Montgolfier's balloon, 403, 688.
  forms in design, 37, 49, 812, 5399.
Annunzio, Gabriele d':
  exploits, 3911, 3959, 3982.
Anti-aircraft devices, 3548, 3553, 3593, 3646, 3647, 3689, 3723, 3733, 3740a, 3756, 3758, 3769, 3797, 3800, 4085a, 4135.
Antoine's propeller formula, 2895.
Antoinette airplanes, 2044, 2066, 2084, 2085, 2090.
"Antoinette" engine, 2698, 2708, 2718, 2726.
Anzani, Alexandre:
  invents "Anzani" engine, 2696, 2708, 2728.
  portrait, 2696.
Arabia:
  aerial survey, 5106.
Aracena, Diego:
  Andean airplane flight, 2924.
Arago, D. F. J., writings on scientific ballooning, 4665, 4666.
Arago, balloon, 420, 421.
Arban, Francesco:
  balloon ascensions, 657, 661, 697a.
Archaeology:
  use of aerial photography, 5113, 5126, 5135, 5149.
Archdeacon, Ernest:
  airplanes, 2307.
  bibliography, 1498.
  biography, 579.
  seaplanes, 2307, 2362.
Archibald, E. Douglas:
  captive kite balloon, 422.

Archytas:
  quotations regarding him, 373 (p. 23–24).
  wooden dove, 56, 292a.
Argand, Aimé:
  references concerning him, 253.
"Argus" engine, 2708.
Argyll, George Douglas Campbell, 8th duke:
  bird flight, 5180.
Ariosto:
  Astolpho's flight to the sources of the Nile, 373 (p. 13–19).
Armstrong seadrome, 4241.
Armstrong-Whitworth airplane, 1462.
Army-Navy standards conferences (1924–1927), 4549.
Arnold, Stuart Amos:
  balloon and parachute adventure (1785), 425, 806, 2502a.
  biography, 806.
Arnoux, René:
  tailless airplane, 1440.
Arnstein, Karl:
  airplane designer, 2274.
  portrait, 2274.
Arson:
  propeller studies, 2829.
Arsonval, Arsène d':
  hydrogen process, 547.
Art in aeronautics:
  balloon caricatures (1784), 1278.
  balloon in decorative art, 436.
  balloons on old French fabrics, 491.
  chinaware, 18th century, 682.
  Conte's career as artist, 549.
  See also Prints, engravings, etc.
Artingstall, F. D.:
  ornithopter, 1463, 1914d.
Ascanio, d':
  helicopter, 2554.
Assmann, Richard:
  bibliography, 4670.
  biography, 4690, 4970.
  Lindenberg observatory, 4655, 4834.
  pioneer work in aerology, 5046.
  portraits, 4690, 4970.
Association française aérienne, 4528.
Associations, societies, etc.:
  first American convention of aero clubs (1910), 4574.
  history, 4551.
  light airplane clubs in England, 4565.
  See also names of organizations, entries 4523–4585.
Assolant, Jean:
  transatlantic flight (1929), 3198, 3270, 3437a.
Astolpho's flight to the sources of the Nile (Ariosto), 373 (p. 13–19).
Astronomy:
  Dorothy Klumpke's experiences, 5334.
  relation to aeronautics, 4914.
Atlantic, airship, 179.
Atmosphere:
  bibliography, 4705.
  history, 5015.
  See also Balloons, sounding; Meteorology.
Atwood, Harry N.:
  experiences, 3067.
  St. Louis-New York airplane flight (1911), 2928.
Audemars, Edmond:
  experiences, 3068.
Australia:
  air mail, 4362.
  civil aeronautics:
    events (1934–1935), 4213.
    post-war period, 4341.
    transcontinental air service, 4238.
  flights:
    Houdini, 2930, 3099.
    Trans-Australian, 2908.
Australia, *continued*
  flights around:
    Brinsmead (1924), 3404.
    Goble and MacIntyre (1924), 3404.
  flights to or from:
    events of 1933, 3336a.
    See also Cathcart-Jones, Chichester, Cobham, Hinkler, Kingsford-Smith, Mollison (Amy J.), Mollison (J. A.), Pinedo, Robertson (Sir M.), Royal Air Force, Scott (C. W. A.), Smith (Ross).
  mapping, 5085.
  World war, Flying Corps, 3876.
Austria-Hungary:
  military aeronautics, 3649, 3714.
Autogyro:
  fatality, first, 2561.
  history, 189, 2541, 2581.
  police service, use in, 2567.
  See also Cierva, Kellett, Lamé, Pitcairn.
Automatic pilots. *See* Airplanes, stability and control, automatic.
Automobiles:
  progress compared with aeronautics, 341.
Aveline automatic pilot, 1466, 1467.
Averly, A.:
  flight theory, 4673.
Aviatik, airplane, 1468.
Aviators. *See* Airplanes, flights; Pilots; Women in aeronautics; World war, aviators.
Avitor, airship, 868.
Avro, airplane, 1969, 2032, 2040.
Ayres, W. O.:
  flying machine (1885), 1475, 2004.
Azores islands:
  Haldeman-Elder flight (1927), 3079.
  See also United States Navy, transatlantic flight (1919).

B

Babinet, Jacques:
  biography and portrait, 4871.
  flight theory, 4674.
Babylonia:
  brick sculpture, 38, 54, 56, 76.
Bacon, Francis:
  ideas on flight, 4675.
Bacon, Gertrude M.:
  biography and portrait, 2056.
Bacon, John M.:
  acoustical experiments, 4676.
  balloon experiences, 431, 433, 4948.
  bibliography, 434.
  biography and portrait, 431.
  scientific experiments, 4677.
  submarine observations, 5405.
  valveless balloon, 834.
Bacon, Roger:
  ideas on aeronautics, 56, 110, 292a, 867, 4678, 4679.
Bacqueville, Marquis de:
  attempted wing flight, eighteenth century, 46, 292a.
Bada y Leblich, Domingo:
  scientific ascension (1795), 4971.
Baddeley, William:
  comments on airships, 881, 882.
  direction of balloons, 883.
Baden-Powell, Baden Fletcher Smythe:
  ballooning as a sport, 435.
  bibliography, 139.
  biography and portrait, 2056.
  kite experiments, 2610, 2611.
Bagdad air mail, 4391, 4413.
Bailey, Lady:
  flights, 5340, 5349.
Balaschoff, Pierre de:
  Russian patron of aeronautics, 451.

Balbo, Italo:
  airplane flight, Italy to Aegean sea, 2936.
  transatlantic flight, *1931,* 2936b, 3200, 3400; *1933,* 2935a, 2937, 3336a.
  various flights, 2936a.
Balchen, Bernt:
  Antarctic experiences, 4439.
  biography, 3045, 4476.
  portraits, 3045, 3079.
Baldwin, Thomas Scott:
  airship, 131, 888, 1159c, 1367.
  bibliography, 914.
  biography, 157, 328, 610, 2992.
  builds first U. S. army airship, 328, 1030, 3731.
  California arrow, airship, 1237.
  contract for U. S. navy airship, 988.
  experiences with Bell and Curtiss, 1610.
  parachutes, 2486.
Baldwin, W. H.:
  ornithopter, 1479.
Balkan-Turkish war (1912):
  Adrianople, fall of, 3720.
  Jordanoff's airplane experiences, 3661.
Ball:
  biography and portraits, 3854.
Ballast. *See* Balloons, ballast.
Ballonets, interior:
  La Vaulx, 678.
  Leloup, 687.
  Meusnier, 549, 840, 905, 1159d, 1180.
  Prechtl, 768.
  Rojas, 345.
  Voyer, 840.
Ballons sondes. *See* Balloons, sounding.
Balloon farm:
  C. E. Myers, 753.
Balloon jumping, 804.
Ballooning, scientific:
  bibliography, 4698, 4719, 4729.
  general, 697a, 4827, 4846, 4849, 4912, 4949, 4950, 5029, 5030, 5033, 5035, 5047a.
  See also names: Arago, Assmann, Bacon (J. M.), Bada y Leblich, Barral and Bixio, Berson, Besançon, Biot, Börnstein, Coe, Coxwell, Crocé-Spinelli, Flammarion, Fonvielle, Gay-Lussac, Glaisher, Hazen, Hermite, Hervé, Jannsen, Jeffries, Jobert, Kremser, Pénaud, Pomortzeff, Rykatchew, Robertson (E. G.), Süring, Tissandier (G.), United States Signal Corps, Wise. *See also* Balloons, sounding; Eclipses, Meteorology; Stratosphere.
Balloons:
  altitudes attainable, 608.
  America, early, 622.
  ascensions:
    aeronauts, early, 292, 373, 592, 750, 774a.
    events, *1783–1784,* 596; *1803–1904,* 604.
    general, 253, 280, 292, 294, 372, 803, 821, 844.
    hot-air balloons. *See* Balloons, warm air.
    list (1783–1848), 538.
    long distance, 426, 519, 572, 708, 829, 831, 841.
    physics of, 4984.
    satires, 373 (p. 521).
    See also Altitude flights; Ballooning, scientific; Baltic sea; Mediterranean sea; Records; also names of balloonists as follows: Abercron, Andreani, Arban, Arnold, Bacon (J. M.), Barral and Bixio, Berson, Bienaimé, Biggin, Biot, Blanchard (J. P.), Blanchard (M. M. S. A.), Cagnola, Charles, Chartres, Coxwell, Crocé-Spinelli, Crosbie, Donaldson, Durant, Farman (M.), Flammarion, Forster, Gale, Gay Lussac, Glaisher, Godard (A.), Godard (L.), Graham (M.), Green, Hersey, Holland, Jeffries, La Mountain, Landriani, La Vaulx, L'Host, Lunardi, McCoy, Maistre, Marquis, Mason, Money, Mosment, Opitz, Pilâtre de Rozier, Poeschl, Potain, Regibus, Robert brothers, Robertson (E. G.), Robertson (Eugène), Rykatchew, Sadler, Schelcher, Silberer (H.), Silberer (V.), Sivel, Spelterini, Süring, Testu-Brissy, Tissandier (G.), Tournachon, Turner, Tytler, Veneziani, Wild, Wise, Zambecari.

Balloons — Ascensions, *continued*

  auxiliary:
    Carra, 949.
    Gale, 444.
  ballast:
    carbonic acid gas, 731.
    Dex and Castillon de St. Victor, 517.
  baskets, closed, 827.
  Belgium:
    first balloons, 837.
  bibliography, 453, 4719.
  bursting, 373 (p. 521), 646.
  Cambodia, early use, 102.
  captive:
    Abyssinian war (1888), 3557, 3741.
    accidents, 633.
    advantages, 3647.
    Archibald, 422.
    Austro-Franco war (1796), 3628.
    bibliography, 454.
    equilibrium, 524.
    Franco-Chinese war, 3528, 3764.
    Franco-Moroccan war (1906–1907), 3622.
    free and dirigeable balloons compared with, 3541, 3582, 3755.
    Giffard, 417, 495, 496, 599, 600, 758, 803, 819, 820.
    history, 633.
    Japan, 3601.
    Lachambre, 3637, 3666.
    Myers (C. E.), 774.
    Paris exposition (1889), 151, 817.
    signalling, use in, 3575.
    Surcouf, 574.
    Templer, 3592.
    various countries, 549.
    World war, 3903.
    Yon, 456, 3557, 3637.
  collodion spray, 614.
  copper:
    Crestadoro, 515.
    Marey-Monge, 510, 867.
    See also Lana.
  defects, *1832,* 626; *1839,* 842; *1863,* 1389.
  Deparcieux's designs (1783), 526.
  direction of:
    difficulties (1823), 1220.
    early references, 373 (p. 522).
    French opinion (1784), 544.
    history, 811a, 1137.
    New London magazine, 1227.
    pamphlet (1793), 973.
    See also Alban and Vallet, Andervolti, Baddeley, Blanchard (J. P.), Bowlder, Bruno, Carra, Castelli, Cayley, Costa, Coxwell, Curzio, Dodoret, Edgeworth, Evans (John, jr.), Franchot, Galvez, Green (C.), Grimaldi, Henderson (J.), Henrion, MacSweeny, Meltzl, Montgolfier, Morin, Muzzi, Omri, Soreau, Strait, Wise, Zambeccari.
    See also Airships.
  double, using different gases, 730.
  early types, details, 325.
  fire balloons, early, 56.
  form:
    Chumakov, 506.
    Henry (E.), 627.
    Marvin, 710.
    Myers (C. E.), 1205.
    Omri, 1219.

Balloons, *continued*
general, 196, 222, 245, 253, 257, 263, 292, 294, 297, 298, 372, 430, 442a, 499, 500, 503, 533, 545, 549, 567, 568, 611, 617, 639, 645, 651, 672, 701, 702, 707, 711, 714, 728, 743a, 755, 792, 800, 803, 821, 849, 3575, 4820.
geographical discovery, 715.
Germany, early, 310, 546, 685.
Hervé Mangon's improved balloon for meteorology, 4863.
hot air and gas combined balloon, 483, 484, 621.
iron shell proposed, 482.
Italian cities, 475–481, 598, 658–659.
kite type:
Archibald, 422.
bibliography, 461.
Hervé, 1194.
Parseval, 761, 762, 814, 3633.
Riedinger, 578.
large:
Coe, 419.
Ogal, 606.
lenticular type:
Capazza, 493.
Petersen, 494.
See also Airships.
life preservers, 691.
Lindenberg observatory, 4655.
manoeuvring, 733.
manufacture:
France, 442, 450, 451, 523, 549, 576.
history, 618, 758.
military:
Abyssinian war (1888), 3557, 3741.
Boer war, 3542, 3574, 3660, 3675, 3686, 3761.
Confederate army, 3576.
France, 373 (p. 524), 3668, 3692, 3725, 3745, 3765, 3790, 3792.
Germany, 3543, 3554, 3648.
Great Britain, 3592, 3617, 3659, 3744, 3794.
history, 245, 697a, 3541, 3596, 3634, 3639, 3674, 3691, 3721.
Italy, 373 (p. 524).
juvenile work, 212.
reconnaissance, early, 3551.
Russia, 3540.
Spanish experiments (1792), 3662.
United States, 373 (p. 525), 3779.
value, 3713, 3753.
various countries, 3544, 3611, 3659.
See also Civil war, America; Balloons, captive; Franco-Prussian war; Woolwich arsenal.
mooring devices, 644.
names:
list, 373 (p. 521).
See also Arago, Belgica, Bristol, Centaure, City of St. Louis, L'Entreprenant, Explorer, Le Géant, Goodyear, Icare, Leipzig, Mammoth, Minerva, Nassau, Ornen, St. Louis, Svenske, La Ville de Paris, Volta, Zenith.
parachute combination, 785.
pilot, 463, 4659. *See also* Balloons, sounding.
propeller for raising, 429, 522, 742, 841. *See also* Helicopters.
publications before 1800, 399–403, 406, 423, 425, 437, 439, 466–473, 499–501, 509, 526, 544, 557, 558, 560, 563–565, 589, 590, 594, 597, 601, 612, 635, 638, 645, 647, 689, 695, 696, 698, 712, 724, 736, 752, 754, 764, 769, 771, 773, 777, 779, 788, 806–808, 815, 838, 839.
publications, 1800–1850, 398, 404, 408, 409, 415, 429, 442a, 443a, 444, 447, 465, 510a, 581a, 584, 592, 606, 608, 609, 611, 614, 616, 644, 644a, 646, 650, 651a, 663, 668, 700, 710a, 711, 717, 718, 725, 730, 731, 738, 743a, 747, 748, 750, 756, 767, 774a, 775, 780–782, 791, 794, 811a, 811b, 836a, 841a, 841b, 842, 849, 5022.

Balloons, *continued*
regulation:
change of center of gravity, 992.
pressure belt, 625.
wings, 486, 527, 765, 843, 993.
rotary:
Lassie, 1159a.
Luntley, 253, 952, 1151, 1159a.
sea control:
chronology, 1338.
Surcouf, 1338.
sectional, 744.
sounding:
bibliography, 4706, 4712.
references, 396, 787, 4657, 4659, 4796, 4799, 4800, 4803, 4827, 4836, 4844, 4849, 4858–4860, 4895, 4929, 5010, 5011, 5012, 5052, 5054.
See also Balloons, pilot.
speed records, 708.
sport, 435.
studies. *See* Cavallo, Charles, Chumakov, Cortenovis, Crocé-Spinelli, Degouy, Euler (L.), Henry (E.), Hofwerberg, Mangon, Renard (C.), Saladini, Stella.
sun, effect, 775.
theory, *1786*, 638; *1791*, 815; *1794*, 402; *1796*, 773; *1825*, 465; *1827*, 650; *1838*, 711; *1881*, 778; *1901*, 506.
thunder storms, 432.
valveless, 834.
vertical movement, 629.
warm air:
Amherst college, 616.
ascensions, 621, 640, 743a, 756.
combined hot air and gas, 483, 484, 621.
history, 796.
Menier's, 723.
See also Montgolfier brothers.
Baltic sea:
balloon flights, 748a.
Baltimore navy seaplane races (1924), 3148a, 3313.
Bane, Thurman H.:
biography and portrait, 1607.
Banks, Sir Joseph:
references to, 253 (p. 421).
Baranowski:
steam driven airplane, 1431.
Barber, Horatio:
biography and portrait, 2056.
Barling bomber, 1717.
Barnes, B. F.:
airship, 1113.
Barnwell, F. S.:
biography and portrait, 2056.
Barnwell, Harold:
pioneer British airplane builder, 1734.
Barral, J. A., and A. J. Bixio:
scientific balloon voyage (1850), 4665, 4686, 4884, 5031.
Barrier, René:
airplane experiences, 3068.
Barton, F. A.:
airship, 890–893, 1212, 2310.
seaplane, 2310.
Basser, Gustav:
biography and portrait, 291.
Basté, J. F.:
bird flight, 5181.
Bath balloon centenary, 783.
Battey, S. B.:
airship propelled by explosives, 1210.
Baumann, Alexander:
biography and portrait, 291.
Baumgarten, G.:
airship investigations with Wölfert, 1047.
Bausset, A. de:
steel vacuum airship, 867, 895.

Bayard-Clément airship. *See* Clément-Bayard.
Bazin, Albert:
bibliography, 1496.
pioneer in glider study, 2383.
portrait, 2383.
Beach, Walter:
wins National Air Tour (1926), 2940, 3139.
Beachey, Hillery:
biography and portrait, 3464.
Beachey, J. Lincoln:
airplane flights, 3296.
airship adventures, 1382, 1474.
airship stabilization, 896.
biography, 3357.
death (1915), 3251.
Beacons. *See* Lighting of airways.
"Beardmore engine," 2701, 2795.
Beaufoy, Henri:
journal describing balloon voyage (1811), 447.
Beaumont, André:
wins Circuit of Britain race (1911), 3013.
Bechtel, Charles E.:
airship, 986.
Bedford, Duchess of:
Cape of Good Hope flight (1930), 5327.
flight, India-England (1929), 5326.
Beedle:
airship frame of bicycle tubing, 897.
Beinhorn, Elly:
airplane experiences, 5315.
mentioned in Halliburton's book (see entry no. 3181).
Belgica, balloon, 525.
Belgium:
civil aeronautics, 4256, 4259, 4291.
first balloons, 837.
See also Nélis.
Bell, Alexander Graham:
bibliography, 1487.
biography, 157.
experiences with Curtiss and Baldwin, 1610.
experiences with Langley, 1486.
tetrahedral kites, 1485, 2192, 2439a, 2612, 2618, 2619, 2639.
Bell, Hugh:
airship, 1159a.
Bellanca, Giuseppe Mario:
airplane endurance record (1927), 2907, 2943.
invents tractor airplane, 1568, 1721, 2274.
portraits, 1568, 2274.
transatlantic flight of Bellanca airplane. *See* Chamberlin.
Bellonte, M.:
transatlantic flight with D. Costes (1930), 3041, 3043.
Benn, Wedgwood:
World war experiences, 3833.
Bennet, John B.:
portrait, 4086.
prominent in work of U. S. Signal Corps, 4086.
Bennett, Floyd:
biography and portrait, 4442.
illness (1928), 3391.
Bennett, James Gordon:
biography and portrait, 157.
Benoist, Tom W.:
biography and portrait, 3464.
"Benz" engine, 2701.
Bergerac, Cyrano de, 68, 407, 894.
Berget, Alphonse:
bibliography, 4691.
Berliner, Emile:
biography and portrait, 2719.
helicopter, 2534, 2560, 2586a, 2596.
invents gyro motor, 2719.
rocket propelled airplane, 1489.
Berliner Verein für Luftschiffahrt:
history, 4570.
Bermuda:
flights to (1930), 3199.
Williams' non-stop flight (1930), 3038, 3114.
Bernard, Marc:
transafrican seaplane flight (1926), 2947.
Berson, Arthur:
altitude balloon ascensions, 749, 4693, 4951.
bibliography, 4692.
portrait, 4570.
Bertaud, Lloyd W.:
duration flight (1921), 3220.
fatal transatlantic flight (1927), 3121.
Bertelli, Achille:
airplane wing forms, 920.
biography, 1910.
bird flight models, 5217.
flying machines, 1910.
Berthenson, Georg:
bird flight studies, 5184.
Bertin, E. L.:
airplane gyration studies, 1490.
helicopter, 2208.
Besançon, Georges:
ascension with Hermite, 4861.
biography, 158, 413.
portrait, 413.
sounding balloons, 4657, 4799.
Besler, W. J. and G.:
steam driven airplane engine, 2754.
Besnier, 1885, 2023, 4867.
Betti, Nicolò:
flying machine (1810), 1530.
Bettoni, Carlo:
airship, 920.
Biard, Henry Charles:
World war career and Schneider Trophy races, 2949.
Bible:
Ezekiel, airship, 862.
references, 81, 110.
Bibliography:
general, 1–36, 220, 253, 292, 359, 5500.
mythological and mediaeval, 66.
See also as a sub-head under various subjects.
Biddle, Charles J.:
World war experiences, 3840.
Bider, Oscar:
Alpine flight (1913), 3064.
Bienaimé, Maurice:
French balloonist, 508.
Bienvenu and Launoy:
helicopter (1784), 2542.
Biggin, George:
balloon ascents, 253 (p. 422), 5312a.
Billwiller, C. F.:
airship, 916.
Bingham, Hiram:
biography and portrait, 5402.
Binns, Joseph:
claims to have navigated a balloon against the wind (1785), 1159d.
Biography:
Blue book of aviation (1932), 1526.
Who's who in American aeronautics (1922–1928), 391.
Biot, J. B.:
ascension with Gay-Lussac (1804), 4727, 4785.
Bird flight:
bibliography, 5185, 5186, 5200, 5201, 5235, 5236, 5263, 5276, 5284.
history, 5205, 5274, 5298.
See also Airy, Amans, Argyll, Basté, Bertelli, Berthenson, Borelli, Bretonnière, Bryan, Buisson, Buttenstedt, Chabrier, Chanute, Collins, Coues, Cousin, Davidson, Drzewiecki, Duroy de Bruignac, Esclangon, Esterno, d'Esterre, Éteve, Funcke, Gatke, Gerlach, Hargrave, Hauvel d'Audreville, Huber, Huffaker, Hutchinson (F. W. H.), Hut-

Bird flight, *continued*
ton, Jarolimek, Korf, Kotzauer, Kress, Lancaster, Lanchester, Lendenfeld, Lilienthal, Louvrie, Lucanus, Lucas, Lucy, Marey, Mewes, Milla, Miller-Hauenfels, Morènes, Mott, Mouillard, Müllenhoff, Navier, Nees, Parseval, Peal, Pénaud, Pettigrew, Pompéien-Piraud, Prechtl, Rayleigh, Reichenbach, Samuelson, Soreau, Sykes, Tatin, Vinci, Weisse, Werner, Weyher, Winter, Zacharia.
Bird propelled airships. *See* Airships, bird propelled.
Birkigt, Marc:
introduces "Hispano-Suiza" engine, 2744, 2747.
Bishop, Cortlandt Field:
biography, 157, 5450.
portrait, 5450.
Bishop, William A.:
World war aviator, 3841.
portrait, 3300a.
Bixio, A. J. *See* Barral, J. A., and A. J. Bixio.
Black, Joseph:
balloon discussion, 465.
references, 253 (p. 422).
small hydrogen balloons, 776.
Black, Van Lear:
Amsterdam-Batavia flight (1927), 2921.
Black-McKellar air mail bill, 4409.
Blackburn, Robert:
early airplane builder, 2056, 2297.
Blackden, L. S.:
gliding experiments, 2368, 2369.
Blackford, William W.:
airplane (1875), 1521.
Bladud, king of England:
mythological flight, 62, 67, 79, 103.
Blake, W. T.:
airplane flight, London-India (1922), 2966, 2967.
Blanchard, Jean Pierre:
airship, 803 (p. 120).
ascensions, 253 (p. 422), 407, 466–473, 485, 510a, 534, 560, 594, 697b, 719, 773, 803, 821, 4878.
bibliography, 473.
biography, 401.
centenary, 818.
direction of balloons, 1159d.
English Channel, 437, 654, 675, 697b, 759, 803, 818, 821.
nativity, 660.
parachute, 2502a.
portraits, 401, 803.
Blanchard, Jerome B.:
airplane, 180.
Blanchard, Marie Madeline Sophie Armant:
ascensions, 476, 533, 774a, 803, 5353, 5384.
portrait, 813.
tomb, 5384.
Blériot, Louis:
airplanes, 1524, 1525, 1671, 1697, 1914a, 1980, 2044, 2084, 2085, 3109, 3247.
bibliography, 1506.
biography, 157, 223, 2948.
English Channel flights, *1909,* 143, 380, 1525, 1914a, 2931, 2970, 2996, 3089, 3108, 3120, 3449; *1929,* 3480.
flights, 196, 223, 3097.
seaplanes, 2362.
Blimps, 789, 1159c.
Blue Hill Observatory:
bibliography, 4707.
kites, 2316, 2633, 2634, 2636, 2637, 2648, 2660.
See also Eddy, William A.; Rotch, A. Lawrence.
Bluebird, airplane, 5318.
Boardman, Russell N.:
flight, New York–Turkey (1931), 3200.
Bodensee, airship, 1097.
Böcklin, Arnold:
heavier-than-air experiments, 1413.
Boeing, William E.:
biography and portrait, 2274.
Boeing Aircraft Company:
airplanes, 1528, 2109.
history, 2108.
metal construction, history, 1574.
transcontinental service, 4205, 4252.
Boeing-Kirsten swivel propeller, 2888.
Boelcke, Oswald:
biography, 3824.
letters, 3847.
World war exploits, 3976, 4002, 4128.
Boer war:
balloons, 3542, 3574, 3660, 3675, 3686, 3761.
Boerner:
semi-rigid airship, 919.
Börnstein, R.:
writings on scientific ballooning, 4729.
Boglioni, Carlo Emanuele:
flying machine, 161.
Boltzmann, Artur:
airplane and airship theory, 1531.
Bombing apparatus:
balloon (1849), 3676.
Barling, 1717.
development (1919–31), 3540a.
explosives, 3733.
first American bomber planes, 3820, 3954.
first bomb dropped from airplane (1911), 3677.
Fiske's invention, 3612.
Guidoni launching apparatus, 3640.
history (1794–1910), 3733.
Rodeck's apparatus, 3772.
survey (1928), 3581.
United States Navy, 3606.
Ursinus bomber, 1476.
World war:
France, 4130.
lessons learned, 4085a.
night bombing, 3834, 4069.
United States, 4011.
See also London: air raids; Paris: air raids; Warship bombings.
Bonaparte, Napoleon Luigi:
airship, 922.
Book-plates:
Bella C. Landauer's collection, 5498.
Booth, H. P.:
ornithopter, 2006.
Borelli, Giovanni Alphonso:
theory of ornithopter, 1914d, 5191.
Borneo:
air survey, 3383.
Bothezat, George de:
helicopter, 2538, 2560, 2604.
portrait, 2538.
Botts, R. H.:
flying machine, 1535.
Botved, Anders Peter:
airplane flight, Copenhagen-Tokio (1926), 2974.
Bourdin, J.:
aerial compass, 5422.
Bourgeois, David:
inclined planes for balloon propulsion (1784), 1159d, 2881.
opinion of aeronautics (18th century), 587.
Bourne, John Frederick:
model helicopter (1843), 2539.
Bournonville, General:
parachute experience, 2490.
Boutet, Charles:
airship, 1240.
Bouttieaux, Commandant:
biography and portrait, 3556.
Bower, F. F.:
ornithopter, 1537.

Bowlder:
balloon direction, 3690.
Boy scouts, 5512.
Boyd, J. Errol:
transatlantic flight (1930), 3037, 3199.
Boyman, Richard:
steel cylindrical airship (1866), 1159a.
Brabazon. *See* Moore-Brabazon.
Bradley, Stuart:
Manufacturers' Aircraft Association, 308.
Bradsky-Laboun, Ottokar de:
airship, 877, 901, 939, 1020, 1206, 1207, 1312.
bibliography, 911.
propeller researches, 2874.
Brakes, 1637, 1639, 2101.
Brancker, Sir Sefton:
biography and portrait, 3935.
Brand, C. J. Q.:
London-Cape of Good Hope flight (1920), 3265.
Brandes, Heinrich Wilhelm:
kite experiments (1826), 2607, 2616.
Braun, Martin:
electric propelled airship (1887), 926.
Brearey, Frederick W.:
balloons for Arctic use, 4491.
bibliography, 1541.
biography, 1881, 1882.
experiments, 253 (p. 422), 1538–1542, 1802.
ornithopter, 1914d.
Bréguet, Louis:
bibliography, 1507, 2540.
biography, 1523, 1602.
carries 11 passengers in airplane, 3320.
helicopter, 1884, 2540, 2586a, 2606.
portrait, 1523, 1602.
Bremen, airplane, 212, 2978, 3060, 3134, 3135, 3171, 3212, 3233, 3234, 3391.
Bremer, V. E.:
flight, Finland–Cape of Good Hope (1932), 2980.
Bretonnière, J.:
bird flight, 5192–5195.
Brewer, Griffith:
biography and portrait, 2056.
visits Wright brothers, 1549.
See also author index.
Brewer, W. J.:
combined airplane and balloon, 856.
Bright, Henry:
helicopter (1859), 2586a.
Brillouin:
theory of airplane stability, 1564.
Brindejonc des Mouilinais, Marcel:
air trespass case, 4632.
biography, 3305.
Paris-Warsaw flight (1913), 3307.
portrait, 3307.
Brinsmead, H. C.:
around Australia flight (1924), 3404.
trans-Australian flight (1925), 2908.
Brissonnet, Anatole:
balloon manufacturer, 451.
Bristol, airplane, 1553.
Bristol, military balloon, 3660.
British Arctic Air Route Expedition (1930–1931), 4447, 4448, 4498.
Brock, W. L.:
wins Aerial Derby (1914), 2910.
Brock, William S.:
Brock–Schlee flight to Japan (1927), 3079.
portrait, 3079.
Brock surveying camera, 5096.
Brodie, Otto W.:
biography and portrait, 3464.
Bronte, Emory:
flight to Hawaii (1927), 3079, 3431, 3445.
portrait, 3079.
Brooklands airport, 371, 1807.
Brown, Sir Arthur Whitten:
portrait, 3087.
See also Alcock, Sir John.
Brown, Winifred:
wins King's Cup race (1930), 3226.
Brozus, J. E.:
bibliography, 931.
discusses steam engine for airships, 930.
See also author index.
Bruce, Eric Stuart:
bibliography, 455.
See also author index.
Bruce, Mildred Mary (Petre):
Bluebird's flight, 5318.
Brun, Stéphane:
"Father of French military aviation," 3805.
Bruno, J.:
attempts Mediterranean balloon flight, 476.
suggestions for directing balloons, 479.
Bryan, George Hartley:
air resistance study, 4738.
bibliography, 4738.
biography, 4732.
bird flight, 5196.
longitudinal stability, 1564, 4737.
theory of flight, 4732, 4736, 5078.
Buchholtz, Franz Hugo:
bibliography, 3579.
portrait, 4570.
See also author index.
Budau, A.:
flight theory, 4739.
Budd, Edward G., Co.:
builds first stainless steel airplane, 1678.
Buddecke, Hans Joachim:
experiences in Turkish service, World war, 3857.
Buisson, Pierre Lucien:
bird flight studies, 5324.
Bull, Lucien:
insect flight studies, 5197.
Buonaccorsi di Pistoja, Count:
aluminium airship, 1178.
Burattini, Tito Livio:
flying machine (1648), 68, 163, 1669.
Burgess, W. Sterling:
bibliography, 1518.
Burma:
Cobham's flight (1924–1925), London-Rangoon, 3022
mapping, 5084, 5105.
Burnham, Walter E. *See* Rawdon, Herbert.
Busk, Edward Teshmaker:
investigations, 4741.
Buttenstedt, Carl:
flight theory, 4964, 5198–5200, 5227.
portrait, 4570.
Byrd, Richard Evelyn:
biography, 2994, 3045, 3058, 3146.
dispute with Fokker, 2133.
North Pole flight (1926), 210, 257, 2994, 4442, 4443, 4451, 4453, 4496, 4500.
South Pole flight (1928), 212, 2994, 4439, 4450, 4471, 4487.
transatlantic flight (1927), 210, 1846, 2913, 2917, 2920, 2993–2995, 3060, 3079, 3476, 4725.

C

Caballero de los Olivos, Estanislao:
combined helicopter and airplane, 1432.
Cabot, Godfrey:
patron of aeronautics, 4008.
Cagnola, Luigi:
pioneer Italian balloonist, 498.
Cailletet, Louis Paul:
air resistance studies, 4742, 4743.
biography and portrait, 4802, 4929, 4956.
paper on the use of oxygen, 5429.
Cairo-Karachi route, 4220.

Calcina, Melano:
  airship, 161.
Caldwell, C. S.:
  airplane experiences in Canada, 2997.
  with sealing fleet, 5435, 5436.
California:
  aviation history, 3436.
  fire patrol, 5404, 5537.
California messenger, airship, 933.
Calthrop, E. R.:
  biography and portrait, 2056.
  obituary, 2489.
Camac, Harriet J. M.:
  India-England flight, 5320.
Cambridge University:
  air survey experiments, 5125.
Caminez, Harold:
  invents cam type engine, 2710, 2714.
Campbell, Peter:
  airship, 1147.
Canada:
  aviation history, 229, 4195, 4196, 4198, 4213, 4217, 4242, 4278, 4279, 4314.
  British Columbia, 4229.
  exploration, 5566.
  Hudson bay transportation, 5447.
  surveys, 5100, 5113, 5138, 5139, 5142, 5143, 5158, 5159, 5176, 5447.
  treaty making with Indians, 5520.
  See also World war.
Cannon, B.:
  airship *Ezekiel,* 862.
Canovetti, Cosimo:
  air resistance studies, 920, 4683, 4684, 4744–4746, 4936.
  bibliography, 4745.
Capazza, Louis:
  biography, 1121, 4975.
  Clément-Bayard, airship, 1129.
  combined balloon and parachute, 785, 1120, 2513, 2524.
  lenticular balloon, 493.
  portrait, 4975.
Cape Cod, airplane, 3200.
Cape of Good Hope:
  flights to or from:
    Bedford (1929), 5327.
    Bremer (1932), 2980.
    Chalmers-Mitchell (1920), 257.
    Cobham (1925–1926), 3024, 3266.
    Humble (1931), 3213.
    Junkers' airplanes (1934), 3112a.
    Mittelholzer, 3294, 3295.
    Mollison, J. A. (1932), 3301.
    Mollison, Mrs., 3200a, 3324, 5360, 5362.
    Royal Air Force (1926), 3378, 3380, 3381, 3407; (1927), 3413.
    Van Rynevelt and Brand (1920), 3265.
Carburetors:
  Stromberg, 2715.
Carelli, Jules:
  airship, 946.
  controlled parachute, 2479.
Carlingford, Viscount:
  airplane (1857), 2221.
Carlstrom, Victor:
  Chicago-New York airplane flight (1916), 3002.
Carra:
  balloon direction, 949.
Castel, P.:
  helicopter, 2543, 2586a.
Castelli, Gaetano:
  air resistance studies, 951.
  balloon direction, 951.
Castillon de St. Victor:
  balloon regulation, 517.
Castle, Vernon:
  World war letters, 3865.
Caterpillar Club, 2491, 2512, 2526, 2529, 3303.
Cathcart-Jones, Owen:
  flights, 3002a.
Caudron, Gaston:
  French airplane manufacturer, obituary, 1623.
Cavallo, Tiberius:
  experiments, 208, 240, 253, 499, 500.
  kites, 2607, 2617a.
  references to, 253 (p. 423).
Cave-Brown-Cave, H. M.:
  seaplane flight, England-Far East (1927–1928), 3003, 3158.
Cavendish, Henry:
  experiments with hydrogen, 501.
  various investigations, 253 (p. 423).
Cayley, Sir George:
  airship, 954.
  balloon direction, 953.
  biography, 253 (p. 423), 254.
  Cayley's top, 2586a.
  engine, internal combustion, 1914b.
  investigations, 253, 254, 380, 1159a, 1918, 2031, 2044, 2089, 2439a, 2471, 4748–4753, 4946, 5036.
  Luntley's rotary balloon, discussion, 952.
  portrait, 254.
Céleste, airship:
  imaginary adventures, 932.
Centaure, balloon, 677.
Central Aero-Hydro-Dynamic Institute, Russia, 4972.
Central America. *See* Latin America.
Century of Progress, airplane. *See* Mattern, James.
Chabrier:
  bird flight studies, 5234a.
Chalais-Meudon, Paris:
  Historical exhibit, 16.
Challe, Léon:
  transatlantic flight (1929), 3198.
Challenge internationale de tourisme, 3004, 3491.
Chalmers, C. H.:
  helicopter, 2544.
Chalmers-Mitchell, P.:
  airplane flight, Cairo-Cape, 257.
Chamberlin, Clarence Duncan:
  biography, 3005.
  endurance flight (1927), 2907, 2943, 3095.
  portrait, 3079.
  ship-to-shore flight (1927), 3434.
  transatlantic flight (1927), 210, 2944, 3005, 3006, 3060, 3079, 3325, 3430, 3482, 4754.
Champion, Carlton C.:
  airplane altitude flight (1927), 3078.
Chandler, Charles de F.:
  biography and portrait, 3872.
Chanute, Octave:
  appreciation by Hudson Maxim, 2441.
  bibliography, 1485, 1587.
  biography, 157, 2440, 2442, 2451, 2460.
  bird flight, 5201.
  gliding experiments, 380, 1802, 2282, 2285, 2375–2380, 2439a, 2471.
  ideas on engines, 2699.
  portrait, 2442.
  thirty-four uses for flying machines, 1587.
Charles, J. A. C.:
  ascensions, 697c, 499 (p. 95–107), 505, 752, 769, 788, 799, 803, 821, 839.
  balloon investigations, 697c.
  portrait, 803.
Chartres, duc de:
  balloon ascension (1784), 415.
Chatley, Herbert:
  bibliography, 1593.
  rocket propulsion, 1592.
  See also author index.

Chaudy, F.:
air resistance studies, 4756, 4757.
Chauvière:
airplane, 1594.
Chavez, Georg:
Alpine flight (1910), 223, 3010, 3098, 3478.
Chen, Wen Lin:
airplane flight, London-China (1929), 3488.
Chichester, Francis C.:
airplane flight, London–Sydney (1930), 3011.
Chile:
military aviation, 3614.
China:
airship of 1860, 373 (p. 271).
ancient, 76.
aviation history, 1595, 2144, 2176.
Bert Hall instructs aviators, 3260.
captive balloons, 3767.
Chen and Johannsen's flight to (1929), 3488.
Coste's flight to (1929), 3040.
first American airplane, 4337.
Knauss's flight to (1926), 3232.
manufacture and transport, 4226, 4273.
Shanghai bombarded, 3760.
Choinski, Tim:
airship (1881), 957.
Christmas, W. W.:
biography, 2167.
Christmas bullet, airplane, 2132, 2167.
Christofferson, Silas:
flight over Mt. Whitney (1914), 3427.
Chronology:
air laws, 4586.
airplane flights, *1873–1931,* 2927; *1903–1909,* 1586; *1910–1934,* 3032; *1919–1928,* 2946.
airplanes, *1879–1909,* 1651; *1890–1910,* 2086.
airplanes and seaplanes (1908–1929), 3012.
airships (1784–1912), 863.
Atlantic flights, 2946, 3060.
aviation (1905–1926), 2037.
United States (1861–1932), 1526.
balloon ascensions, *1783–1848,* 538; *1803–1904,* 604.
control of balloons at sea (1784–1902), 1338.
disasters, 377.
English Channel crossings (1784–1908), 667.
flights (11th century–1908), 1581.
general, 144, 149, 177, 184, 211, 239, 242, 262, 264, 269, 283, 292, 301, 311, 325, 357, 373 (p. 526).
hydrogen, 151.
insurance, 5534.
Italy, 125.
military aeronautics, 3530, 3654.
military balloons, Great Britain (1862–1901), 3794.
propellers, 2044.
speed of airplanes, 2231.
United States, air mail, 4366.
United States, air transport, 4323.
upper air investigations, 4886.
World war, 144, 301, 4047, 4133.
Wright brothers, 1551, 1775.
Chumakov, D.:
studies of balloon forms, 506.
Cierva, Juan de la:
autogiro, 2545–2548, 2559, 2560, 2575, 2589, 2594, 5079.
Cigar store, aerial, 5446.
Circuit of Britain race:
won (1911) by André Beaumont, 3013.
"Cirrus" engine, 2806.
City of Chicago, airplane, 3199, 3323.
City of New York, airship, 179.
City of Oakland, airplane. *See* Smith, Ernest L.
City of St. Louis, balloon, 636.
Civil aeronautics:
Afghanistan, 4341.
Africa:
post-war period, 4341.
Wilson Airway, Ltd., 4293.
See also Congo; Sahara desert; South Africa.
Alaska, 4177, 4281.
Australia, 4213, 4238, 4341.
Belgium:
progress, *1920–1930,* 4291; *1922,* 4256; *1928,* 4259.
See also Congo.
bibliography, 4183.
Canada:
British Columbia, 4229.
history, 4195, 4196, 4198, 4213, 4217, 4242, 4248, 4278, 4279.
progress, *1929,* 4314; *1930,* 4242; *post-war,* 4341.
China:
first American airplane, 4337.
history, 4226, 4273.
Congo, 4259, 4291.
Denmark, 4262.
Europe, 4256, 4287, 4291, 4294, 4296, 4331, 4340.
France:
Compagnie internationale de navigation aérienne, 4290.
Le Bourget airport, 4260, 4300.
"Port Aviation" aerodrome, 4218.
progress, *1922,* 4256; *1920–1930,* 429; *post-war,* 4341.
South American lines, 4230, 4237, 4289.
statistics, 4201, 4228, 4263, 4301.
world development, status in, 4227.
general, 200, 4170, 4193, 4239, 4240, 4245, 4246, 4249, 4272, 4294, 4304, 4320, 4331, 4333, 4343.
Germany:
Berlin-Moscow line, 4231.
Deutsche Luft-Reederei, 4347.
Deutsche Lufthansa, 4284.
history, 4186, 4264, 4291, 4326, 4346.
statistics, 4201.
Great Britain:
British possessions, 4213.
Civil Aviation Directorate, 4236.
Croydon aerodrome, 4209.
English Channel, 4215.
general strike (1926), 4292.
Imperial Airways, 4245, 4298, 4301.
Marine air lines, 4311.
progress, *1920–1930,* 4291; *1922,* 4256; *1928,* 4261; *1934–1935,* 4310.
statistics, 4201.
Hawaii, 4285.
Holland:
progress, *1920–1930,* 4291; *post-war period,* 4341.
Royal Air Traffic Co., 4319.
Royal Dutch Indian Airways, 4267.
India, 4213, 4341.
International Civil Aeronautics Conference, 4556, 4557.
Italy:
history, 4250, 4341.
Mussolini, 4265.
operations (1926), 4165.
progress (1922), 4256.
Latin America:
history of companies, 4338, 4339.
post-war developments, 4341.
Lawson's (A. W.) pioneer work, 4216.
mining, aids in, 4166, 4178, 4199, 4211, 4233, 4280, 4302, 4342.
night flying, 4212, 4239.
Pan-American Aeronautic Federation, 4295.
Persia, 4190, 4256.
railroad competition, 4210.

Civil aeronautics, *continued*
Russia:
air lines (1929), 4351.
Deruluft line, 4231.
five-year plan, 4322.
progress, 4231, 4235, 4303, 4341.
safety, 4191.
Sahara desert, proposed service, 4258.
South Africa, 4213.
South America:
airline to North America, 4180.
Colombia, 4179.
French lines to, 4230, 4237, 4270, 4289.
Pan American Airways, 4271, 4330.
Peru, 4251.
Peru-Brazil line, 4270.
post-war developments, 4341.
See also Latin America.
Spain, 4341.
statistics:
foreign (1930), 4224.
world, 4225, 4331, 4341.
See also names of countries as subheadings under Civil aeronautics; *also* Airways.
subsidies discontinued in Europe, 4222.
Sweden, 4192, 4313.
Switzerland, 4266.
United States:
Aeronautics Branch, 4163, 4188, 4306.
airports:
bibliography, 4181.
general, 127, 4171, 4172, 4175, 4184, 4189, 4194.
lighting, 4176.
bibliography, 4306.
Boeing Aircraft Co., 4205, 4252.
chronology, 4323.
Congressional action (1926), 4197.
duration flights, value of, 4309.
European and U. S. lines compared, 4234, 4288, 4348.
express service, 4219, 4243, 4245, 4308, 4333.
Ford, Henry, 4203, 4223, 4247.
Ford Air Transport, 4223.
Government sale (1921), 4283.
history, 272, 4246, 4316, 4318, 4324, 4336, 4344.
Key West-Havana line, 4254, 4394.
mileage statistics, 4173, 4174, 4323.
National Air Transport, Inc., 4282.
Northwest Airways, 4286.
problems, 4349.
progress, *1920–1930*, 4291; *1922*, 4257; *1926–1934*, 4297; *1928*, 4185, 4208, 4255; *1929*, 4187, 4207, 4255, 4350; *1930*, 4169, 4274, 4348, 4350; *1931*, 127, 4168; *1933*, 4321; *post-war*, 4341.
railroads adopt aviation, 4299.
reciprocal arrangements with other nations, 4328.
Saint Petersburg-Tampa line, 4164, 4214.
statistics, 4200, 4201, 4276, 4305, 4327, 4329.
Stout Air Lines, 4315.
United Air Lines, 4253.
See also United States Commerce Department.
Yugoslavia, 4167.
See also Routes.
Civil war, United States:
Andrews' airship, 3535.
balloons, 697a, 3631, 3632, 3634, 3672, 3721, 3735, 3751a, 3762.
Beaumont's experiences, 3551, 3552.
Confederate army, 3576.
helicopter experiment, 3746.
illustrations in *New York illustrated news*, 3672.
Lowe's experiences, 3568, 3631, 3643, 3658, 3672, 3721, 3735.
Metiviev offers an airship, 1052.
references, 373 (p. 525).
Clark, Julia:
death, 5361.
Clarke, T. W. K.:
biography and portrait, 2056.
Clayton, H. H.:
kites, 2619, 2665.
Clément, Adolph:
biography and portrait, 1129.
Clément-Bayard airships:
bibliography, 908.
descriptions, 1159c.
flights, 1171.
See also Capazza, Louis; Clément, Adolph.
"Clément-Bayard" engine, 2708, 2728, 2730.
Clerget, Pierre:
invents engine, 2697, 2728.
portrait, 2697.
Clifford, George Reid:
military aviator, 3870.
Clift, Eric:
instrument maker, 5475.
Clippings:
Alcock-Brown transatlantic flight (1919), 3340.
Andrews' airship *Aeron*, 1174.
Hoffman's scrapbook (1927–1928), 3206.
New York Times (1927–1928) on Lindbergh's flights, 3327.
notable flights (1931–1932), 3336.
years, *1830–1852*, 747; *1840–1890*, 642; *1858–1880*, 179.
Clout, George:
helicopter, 2153.
Coanda, Henry:
airplane, 1603.
Coast patrol:
bibliography, 5413.
United States, 2264, 5445, 5509, 5564.
Cobham, Sir Alan John:
around Africa flight (1928), 3026.
Australian flight (1926), 3021, 3023, 3076.
London-Cairo-Cape flight (1925–1926), 3024, 3266.
London-Rangoon flight (1924–1925), 3022.
Mt. Everest flight (1925), 2914.
various flights, 3025, 3516.
Cockburn, G. B.:
aviator, 2056, 3155.
Cocking, Robert:
parachutist, 253 (p. 423), 372, 2477, 2482, 2483, 2490, 2502a, 2505, 2507a.
Codos, Paul:
flight, Indo-China-Paris (1932), 3490.
flight, New York-Syria (1933), 3336a.
Cody, Samuel Franklin:
appreciations, 1638, 2114.
bibliography, 1604.
biography, 1866, 2112.
builds airship *Nulli secundus*, 1367, 1570.
experiences with power kite, 1604.
kites, 2666.
portrait, 2112, 3300a.
Coe, Carlos C.:
monster balloon (1860), 419.
observation of submarine objects, 4760.
Coey, C. A.:
balloonist, 811.
Colardeau, E.:
air resistance studies, 4742, 4743.
Cole, Moses S.:
airship, 1217.
Coli, François:
attempts transatlantic airplane flight (1927), 3339.
Colladon, Daniel:
kite experiments (1844), 2620.
Collier Trophy:
awards, 2480, 3029, 3030.
Packard Motor Car Co., 2781.
Pitcairn, H. F., 2549, 2563.
Colliex:
propeller calculations, 2048.

Collins, T. Byard:
bird flight studies, 5202.
Collodion:
Gruel proposes spray for balloons, 614.
Collyer, C. B. D.:
world flight (1928), 3285.
Columbia, airplane:
Connor and Boyd's transatlantic flight (1930), 3037, 3199.
Williams (R.), Bermuda flight (1930), 3038, 3114.
See also Acosta, Chamberlin.
Colvin, Charles H.:
instrument inventor, 2274.
Comaschi, Antonio:
airship, 161.
Comfort:
development during 1931, 274.
Comité juridique internationale de l'aviation:
Fifth congress (1922), 4600.
Compagnie générale aéropostale, 4237.
Compagnie internationale de navigation aérienne, 4290.
Compagnon, Armand de:
sail propelled airship, 903.
Compass:
bibliography, 4762, 5410.
Bourdin, 5422.
earth inductor type, 5554.
history, 4762.
Janssen, 5492.
Kelvin, Bottomley and Baird, Ltd., 5494.
Kosloff, 5519.
Competitions:
bibliography, 5414.
Coney, William D.:
trans-American flight (1921), 3036, 3106.
Congo:
commercial aviation, 4259, 4291.
Congrès international d'aéronautique, 4534–4539, 4564, 4575, 4579, 4581.
Congrès international de législation aérienne, 4540, 4568.
Congrès international de la sécurité aérienne, 4541.
Congrès de météorologie aéronautique de St. Pétersbourg, 4531.
Congresso giuridico internationale per il regolamento della locomozione, 4542.
Congresso giuridico internationale per il regolamento della navigazione aerea, 4612.
Congreve, Sir William:
possibility of mechanical flight (1828), 1608.
Conical balloons. *See* Balloons, lenticular.
Connor, Harry P. M.:
airplane flights, 3037, 3038, 3199.
Constantin, Louis:
airplane stability study, 1628.
Conté, Nicholas Jacques:
artist and aeronaut, 549.
balloon manufacturer, 523.
hydrogen, 523.
Coppens, Willy:
experiences in World war, 3873.
Copper balloon shell:
copper globe with ballonets, 768.
copper plated aluminium, 743.
Crestadoro, 515.
Marey-Monge, 510, 700.
See also Lana.
Cordenons, Pasquale:
airship studies, 921, 962, 963.
Cordero di Montezemolo, Vittorio:
flight theory, 4763.
Cornelius, William:
ornithopter, 1914d.
Cornell University:
U. S. Army School of Military Aeronautics, 3827.
Cornu, Paul:
bibliography, 2551.
helicopter, 2550–2553, 2565, 2586a.
Cortenovis, A. M.:
balloon studies, 162.
Cossus:
helicopter (1845), 2586a.
Costa, Marco Antonio:
balloon direction, 479.
Costes, Dieudonné:
four-continent airplane flight (1927), 2922, 3042, 3044, 3126.
Paris-China flight (1929), 3040.
Paris-New York flight (1930), 3041, 3043.
Coues, Elliott:
bird flight, 5203.
Coulomb:
ornithopter, 2208.
Cousin, Joseph:
airplane stability study, 1628.
bird flight, history, 5205.
Coutelle, J. M. J.:
early French balloonist, 549, 803, 821.
hydrogen manufacture, 554, 3770.
Coutinho, Gago:
flight, Spain-Brazil (1922), 3486.
Coxwell, Henry:
ascensions, 488, 514, 697a.
autobiography, 514.
balloon for Arctic exploration, 4493.
balloon direction, 966.
bibliography, 514.
biography, 159, 720.
portrait, 488, 514, 720.
See also Glaisher, James.
Cremorne Gardens, London:
balloon ascensions, 591.
Crespi. *See* Usuelli and Crespi.
Crestadoro, Andrea:
copper balloon, 515.
Cripe, Calvin:
airplane flight, Fairbanks, Alaska, to Cape North (1929), 4486.
Crocco, Arturo:
airplane stability theory, 1564, 4764.
bibliography, 4764.
wind tunnel, 4994.
Crocé-Spinelli, J.:
balloon stability studies, 4765.
bibliography, 4765.
biography, 4694.
engine studies, 2704–2705.
portrait, 803 (p. 448).
propeller studies, 2839.
scientific balloon ascensions, 816, 4667, 4668, 4766
tomb, 813.
Crop estimation:
U. S. Agriculture Department, 4828, 5568.
Crosbie, Richard:
balloon ascensions, 253 (p. 424), 697b.
Crosson, Marvel:
women aviator, death, 5388.
Croydon airport, 167, 4209.
Crystal Palace, London:
balloon ascensions of W. D. Dale, 253 (p. 424).
Exhibition (1868), 206, 2748, 2770, 5478.
Curtiss, Glenn H.:
Aeronautic Society exhibition, 5466.
airplanes, 1612, 1914a, 2275.
Albany–N. Y. flight (1910), 1448, 3046, 3048, 3138.
Baldwin and Bell, experiences with, 1610.
bibliography, 1613.
biography, 157, 1722, 2169.
first private aerodrome, 4221.
Hudson-Fulton celebration, 3519.
marine trophy race, 3049, 3056, 3083, 3249.
Newton, recollections, 2013.

Curtiss, Glenn H., *continued*
portrait, 3300a.
Scientific American trophy, 1838, 3115.
seaplanes, 2320, 2321, 2331, 2340, 2362.
U. S. army and navy, 2361.
"White wing" airplane, 3102.
wing tips, movable, 1838.
Wright brothers patent contest, 2291–2293.
Zahm, recollections, 2302.
Curtiss Aeroplane and Motor Corporation:
scientific research, 2136.
"Curtiss" engine, 2698.
Curzio, Vincenzo:
balloon direction, 479, 1034.
Czechoslovakia:
aviation, 2145.

D

Daedalus. *See* Icarus and Daedalus.
Dagron, Prudent René P.:
pigeon post, Franco-Prussian war, 3619, 3621.
Dahlberg, Karl Theodore:
airship (1784), 1252, 1361.
portrait, 1361.
Daily Mail gliding competition:
duration record of A. Maneyrol (1922), 2371a.
Daimler, Gottlieb Wilhelm:
biography and portrait, 2778.
engine, 2701.
Dale, W. D.:
balloon ascensions, 253 (p. 424).
Dallwitz-Wegner, Richard von:
aircraft calculations, 4769.
propeller studies, 2841.
Damblanc:
helicopter, 2558, 2560, 2586a.
Damian, John, abbott of Tungland:
flight legend, 51, 253 (p. 56–57).
Dandrieux:
ornithopter, 1617, 1914d.
Daniel Guggenheim Fund for the Promotion of Aeronautics, 4204, 4735, 4772, 4842, 4843, 5005.
Daniel Guggenheim International Safe Aircraft Competition, 1559, 1618, 2180, 3050.
Danilevsky, Konstantin:
airship, 180, 969–971, 1059.
bibliography, 969.
Dante, Giovanni Battista, of Perugia:
gliding experiments, 47, 58, 85.
Dante Alighieri, 53.
Dauthey, Dom.:
all metal airship (1783), 867.
Davidson, George L. O.:
airplane, 1978, 2007.
bird flight, 5206.
helicopter, 2554a, 2555.
Davis, Elmer:
New York-Texas airplane flight (1924), 3051.
Davis, William V., jr.:
Hawaiian flight (1927), 3061, 3079, 3514.
portrait, 3079.
Death ray, 5518.
Death Valley, California:
aerial survey, 5108.
"De Bradsky" airship, 877, 901, 1206, 1207.
Deburaux, E. L. J. (Leo Dex, pseud.):
ballast regulator, 517.
project for Sahara balloon voyage, 519–521, 528.
Decazes, Vicomte:
flight theory, 4774.
vertical propeller, 2842.
Decker, Horace M.:
kite mechanics, 2623.
Deeds, Edward A.:
vindication, 3828.
Degen, Jakob:
ornithopter, 58, 85, 1914d, 1482, 1756.
Degouy:
study of balloon dynamics, 4775.
"De Havilland" airplanes, 1624.
Delacommune, Charles:
World war experiences, 3877.
Delagrange, Léon:
airplanes, 223, 1914a, 2008, 2044, 3055.
bibliography, 2956.
biography, 2938, 3093, 3243, 3244.
death (1910), 3366.
engine, 2698.
portraits, 2044, 2938, 3300a.
Delaney, Peter J.:
experiences in Spanish-American war, 3591.
Del Prete, Carlo:
Rome-Brazil airplane flight (1928), 3094, 3218, 3399.
Deminne, Edmond:
airship, 978.
Demoiselle, airship, 1302.
Demuyter, E.:
wins Gordon Bennett balloon race, *1922,* 448; *1924,* 525.
Denmark:
civil aviation, 4262.
Ellehammer's airplane, 2059, 2208.
military and naval aviation, 3615.
See also Scandinavia.
Deparcieux, Antoine:
balloon design (1783), 526.
Deprez, Marcel:
bibliography, 5189.
gliding theory, 2384.
propeller calculations, 2843.
Deruluft air line, Berlin-Moscow, 4231.
Desaguliers, J. T.:
air resistance experiment (1717), 4776.
Desert flying:
Arabian desert, 3275.
dangers, 167.
surveying, 5106.
Syrian desert, 3081, 3386.
See also Sahara desert.
D'Esterre, C. R.:
bird flight, 5190.
"Deteuil-Chalmers" engine, 2708, 2728.
Detroit Arctic Expedition. *See* Wilkins, Sir George H.
Detroiter, airplane:
wins National Air Tour (1927), 3129.
Deutsch de la Meurthe, Henri:
airship, 336, 899.
biography, 157, 313.
Deutsch cup, 1198, 3157, 3511.
Deutsche Luft-Reederei:
history, 4347.
Deutsche Lufthansa:
organized (1926), 4284.
routes, 4232.
Deutsche Verein zur Förderung der Luftschiffahrt:
Angerstein's biography, 5497.
history, 4569.
sponsors balloon ascensions (1888–1895), 4670, 4671.
Deutschland, airship, 1035.
Devèze:
ornithopter, 1627.
Dex, Leo, pseud. *See* Deburaux, E. L. J.
Diard, Ennemonde:
experiences as aviator, 5324.
Dibos, M.:
projected balloon voyage across the Sahara desert with Deburaux, 521, 528.
Diemer, Franz Z.:
biography, 291.
Dienstbach, Carl:
bibliography, 154.
Diesel engine, 275, 2700, 2732, 2764, 2781, 2820, 2824.
See also Akroyd-Stuart.

Dieuaide, Emmanuel:
picture collection, 261.
steam driven helicopter, 2586a.

Dines, William Henry:
bibliography, 2626.
kite-flying from ship, 2625.

Dinort:
record gliding flight (1929), 2416.

Dinwiddie, J.:
references to balloon experiments, 253 (p. 424).

Disasters:
airplanes, *1896–1911*, 1728; *1908–1910*, 2266; *1908–1912*, 371; *1909–1910*, 1434.
airships (1909–1930), 1337.
analysis for 1929, 5486.
autogiro (1932), 2561.
captive balloons, 633.
chronology, 377, 5472.
collisions (1910–1912), 1989.
French army (1909–1913), 3752.
general, 168, 292, 386, 664, 674, 5482a, 5483.
parachutes, 2497.
records, *1783–1786*, 686; *1785–1884*, 619; *1785–1900*, 372 (p. 57); *1804–1902*, 294.
transatlantic attempts (1927), 3121.
United States:
airmail fire losses, 4352.
civil aviation (1927), 4325.
military and naval, 3532.
United States Army, 5555, 5556.
various countries (1909–1913), 5472.
World war:
Allied armies, 4016.
Germany, 3886, 3894, 4083.
Zeppelins, 3894.
See also *Akron*, airships, Alcock, *American legion* airship, Amundsen, Andrée, Arban, Baumgarten, Blanchard (Marie), Bradsky, Chavez, Clark (Julia), Cocking, Cody, Coli, Crocé-Spinelli, Delagrange, *Dixmude*, Erbslöh, Fonck, Groof, Helwin, *Italia*, L'Hoste, *M–3* airship, Mangot, Morrell, Mosment, National Airship Co., Nungesser, *Old Glory* airplane, Pilâtre de Rozier, Pilcher, Quimby (Harriet), *R. 38* airship, *R–101* airship, *République* airship, Rolls, *Roma* airship, Sadler, Severo, *Shenandoah* airship, Sivel, Vaniman, Willows, Wölfert, Zambeccari.
Zeppelins, 1340, 1410, 3894.

Disks:
Gerlach's studies, 4812.
Lössl's studies, 4919, 4920, 4923, 4983.
Ryabushinski's studies, 5018, 5019.
Wellner's studies, 5070.

Dixmude, airship, 915, 996, 1050, 1159b.

Dixon, Cromwell:
16-year-old airship constructor, 1383, 3082.

"Do. X" seaplanes. *See* Dornier seaplanes.

Doakes, Joe:
wing walking, 5441.

Donaldson, Washington Harrison:
balloon ascensions, 535.
biography, 535.
proposed Atlantic flight, 798.

Doolittle, James H.:
biography, 3439.
blind flight (1929), 2932.
portraits, 3439, 3457.
Schneider cup (1925), 3145.
speed record (1932), 3065.
trans-American records, 3200, 3457.

Dope for wings:
bibliography, 1850.
history, 1670, 1795, 2150.

Dorand:
airplanes, 2049.

Dornier seaplanes:
Gronau's Greenland-Chicago flight (1931), 3118a.
history, 2313, 2325, 2326, 2336, 2358a, 2360.
transcontinental flights, 3200, 3200a.
trials, 3198, 3435.

Doughty, John G.:
photographic experiences, 5111.

Doutre, Adolphe:
invents double automatic stabilizer, 1984.
portrait, 1984.

Dow's steam turbine, 2717.

Drag rope and anchor:
Andrée, 418.
Deburaux's use in Sahara desert, 519–521, 528, 737
first used by S. A. King (1857), 795.
history, 631, 632, 699.
L'Hoste's brake, 3575.

Dragon balloons:
early types, 566, 784.
See also Balloons, kite.

Drew, Andrew:
aviator, 3464.

Drouet:
parachute experiences (1794), 2490.

Dry ice:
use in airplane construction, 1835a.

Drzewiecki, Stephan:
airplane experiments, 1640.
bibliography, 4778.
bird flight, 5207–5208.
ideas on laboratories, 4778.
propeller calculations, 2048, 2844.

Duchêne:
studies on airplane stability, 2138.

Ducretet:
hydrogen process, 547.

Dufaux, H., and A.:
airplane, 1641, 1642, 2047.
engine, 317.
helicopter, 2585, 2587.

Dunne, J. W.:
biography and portrait, 2056.

Duponchel, A.:
"Aerial fish" airship, 997, 998.

Dupuis-Delcourt, Jules François:
all-metal airship (1824), 867.
bibliography, 538.
portrait, 803 (p. 328).

Dupuy de Lôme, C. H. L.:
airship, 253 (p. 36–37), 305, 549 (p. 153–164), 709, 874, 935, 936, 999–1003, 1159a, 1165, 1246, 1248, 1259, 3618.
hydrogen, 549 (p. 159), 554.
portrait, 549 (p. 155), 1259.

Durant, Charles F.:
Castle Garden ascension (1833), 542.
first American aeronaut, 529.
newspaper clippings, 747.
posters, 541.

Duroy de Bruignac, Albert:
airship form, 4783.
bird flight, 4782, 4783, 5207, 5208.
flight theory, 4782.
theories discussed by Soreau, 5040.

Dutczynski, Ladislaus R.:
biography, 1182.
insect flight, 5209, 5210.

Du Temple, Felix and Louis:
flying machine (1857), 2261.
steam driven airplane, 2031.

Dynamometers:
developments during 1931, 274.

E

Eagle, airship. *See* Lennox, Comte de.
Earhart, Amelia:
  Los Angeles-New York flight (1932), 3065.
  newspaper clippings (1931–1932), 3336.
  non-stop transatlantic flight (1932), 3200a, 5328.
  reminiscences, 5328.
  seaplane transatlantic flight (1928), 3107, 5329, 5385.
East Indies:
  flights to:
    Black (1927), 2921.
    Hewlett (1932), 5347.
    Koppen (1927), 3236.
    Lancaster (1927), 3001.
    statistics (1924–1927), 3072.
    Van der Hoop (1924), 3207.
Eastleigh, England:
  U. S. Navy station during World war, 3846.
Eckener, Hugo:
  bibliography, 1012.
  biography, 1153.
Eclipses:
  balloon observation (1927), 4730.
  U. S. Navy observation (1923), 5081.
Eddy, William A.:
  bibliography, 2631.
  flying and leaping machine, 2191.
  kites, 2628–2631, 2662, 5009.
  photography, 5112.
Edgeworth, Richard Lovell:
  balloon direction, 1013.
Edison, Thomas Alva:
  helicopter experiments (1880), 1814, 1815, 2189, 2586a.
Edwards, Charles J.:
  biography and portrait, 157.
Egg, Durs. *See* Pauly, Samuel John.
Egypt:
  carvings, 85, 86.
  Napoleon I, 373 (p. 123), 3786, 3801.
  winged animal forms, 37.
Ehrhardt, Gustav:
  airplane flight to Mediterranean countries (1928), 3073.
Eielson, Carl B.:
  flight, Point Barrow-Spitzbergen (1928) with Wilkins, 4459, 4500.
Eiffel, Alexandre Gustav:
  bibliography, 4788.
  biography, 4759, 4773.
  cable testing of airplanes, 1648.
  experiments, 4787–4789.
  portrait, 4759.
  wind tunnel, 4994.
Eiffel Tower experiments:
  bibliography, 4710.
Eighth Naval Squadron, Great Britain:
  history, 4031.
Elder, Ruth:
  Haldeman-Elder Azores flight (1927), 3079, 5372.
  portraits, 3079, 3300a, 5372.
Electric propulsion:
  advantages and disadvantages, 961.
  Braun, 926.
  history, 1191.
  opinion of William Pole, 1247.
  See also Renard-Krebs airship; Tissandier brothers.
Electricity:
  Antwerp balloon operated from ground, 1208.
  charges on balloons, 445, 4935, 5032.
  equipment for balloons, 705.
  Eskdalemuir Observatory experiments on charges, 5066.
  See also Balloons, thunder storms; Electric propulsion.
Ellehammer, J. H. C.:
  airplanes, 2059, 2153, 2208.
  engine, 2708.
Ellington, Sir Edward:
  Chief of English Air Staff, 3927.
Ellsworth, Lincoln. *See* Amundsen, Roald.
Elmer of Malmesbury. *See* Oliver of Malmesbury.
Elmira soaring competition, 2388–2391, 2407.
Ely, Eugene:
  death, 3082.
  originates ship-shore flying (1910), 3031, 3101, 3123.
Endres, Georg:
  New York-Hungary flight (1931), 3200.
Engines:
  air-cooled, 2676, 2701, 2716, 2750, 2757, 2758, 2774–2776, 2821.
  airships, 388, 1191, 1323, 2675.
  altitude performance, 2743.
  Belgium, 2708.
  bibliography, 2685, 2692.
  cam type, 2710, 2714.
  carbonic acid, Stenzel (1897), 2804.
  Cayley suggests internal combustion motor (1809), 1914b.
  Crocé-Spinelli, theories, 2704, 2705.
  Crystal Palace exhibition (1868), 2748, 2770.
  developments, *1892,* 305; *1892–1907,* 217; *1894,* 2771; *1909,* 360, 2215, 2708; *1918–1924,* 2683; *1919–1923,* 2209; *1926,* 2823; *1928,* 2712, 2813, 3607, 3630; *1929,* 2802; *1930,* 2700; *1931,* 127, 274, 2101, 2234; *1932,* 178, 189, 383; *1933,* 2810.
  ethylene glycal as cooling agent, 2753.
  first engine-driven flights, 222.
  France, 2708, 2772.
  fuel ignition, Pratt & Whitney, 2822.
  general, 148, 275, 380, 1728, 2164, 2701, 2723, 2733, 2735, 2737, 2759, 2763, 2766, 2767, 2782–2784, 2788, 2789, 2792, 2794, 2797, 2799, 2801, 2809, 2812.
  Gerardin's studies, 2063.
  Germany, 2708, 2819. *See also* Engines, World war, Germany.
  Goupil's studies, 2734.
  Great Britain, 1554, 1555, 2807.
  gyro motor of Berliner, 2719.
  Hargrave's experiments, 2738, 2739.
  hot air, Sheldon's (1850), 2798.
  International Exhibition (1909), 2678.
  inverted type, first British (1927), 2795.
  Italy, 4206.
  large types, 2680.
  Lawrance's writings and investigations, 2274, 2716, 2731, 2741, 2757, 2758.
  Levavasseur, 2726, 2736, 2761.
  lists, 371 (p. 367), 380.
  lubricants, bibliography, 2690.
  Manly (C. M.) builds engine for Langley, 1914a.
  Maxim (Sir Hiram), 305, 1835.
  Mort (G. F.), designer, 2709.
  multimotored airplanes, 2729, 2787.
  Otto gas engine, 2756.
  Paris Exposition (1868), 2478.
  power requirements:
    Heinz, 2742.
    Holland, 1800.
    Hollands, 2746.
    Pichault, 4977.
  pursuit type, American, 2713.
  Renard's rotary motor (1885), 2791.
  Ricardo on internal combustion, 2793.
  Royce (H.), biography, 2803.
  spark gap, 2765.
  spark plugs, bibliography, 2695.
  starters, 2679.
  supercharging, 2724, 2768.
  trans-American reliability tests, 1680.

Engines, *continued*
United States:
government surplus sold (1921), 4283.
progress, *1925–1931*, 1473; *1928*, 1536; *1930*, 2125.
specifications (1911), 2751.
United States Standards Bureau, 2707, 2743.
water cooling, 2725.
weight (1903–1917), 2722.
World war:
general, 3992a, 4036, 4058, 4137.
Germany, 2721, 2740, 2796, 2797, 3950, 3984.
United States, 2720, 4011, 4041, 4113.
See also Carburetors; Electric propulsion; Magnetos; Steam propulsion.
Also the following names: Adams-Farwell, Akroyd-Stuart, "Antoinette," "Anzani," "Argus," "Beardmore," "Benz," "Cirrus," "Clément-Bayard," Clerget, "Curtiss," Daimler, Delagrange, "Deteuil-Chalmers," Diesel, Dufaux, Ellehammer, Esnault-Pelterie, "Farcot," "Farman," Ferber, "Fiat," "Foillard," "Gnome," "Gobron-Brillié," "Green," "Gyro," "Hispano-Suiza," "Korwin-Rebikoff," Landur, "Liberty," Maybach, "Mercedes," "Napier," "Packard," "Panhard-Levassor," "Pipe," Ponton d'Amécourt, "Renault," Rumpler, Schwarz, "Sunbeam," "Vivinus," "Wolseley," "Wright."
England, E. C. Gordon:
biography and portrait, 2056.
English Channel crossings:
Blanchard and Jeffries, 437, 654, 675, 697b, 759, 803, 818, 821.
Blériot, 143, 380, 1525, 1914a, 2931, 2970, 2996, 3089, 3108, 3120, 3449, 3480.
Burnaby, balloon (1882), 490.
chronology (1785–1909), 3449.
commercial lines, 4215.
Latham, 143, 1914a, 3089.
L'Hoste and Mangot, 420–421, 573, 690.
lists, 667, 748a, 3215.
Quimby (Harriet), 5323, 5358, 5370.
Stott's hoax (1876), 1644.
Entreprenant, balloon:
used by French against Austrians (1794), 605, 803.
Erbslöh, Oscar:
German airship operator, 876.
"Escadrille 210," 4097.
Eschwege, von:
exploits in World war, 4128.
Esclangon, Ernest:
bird flight, 5211.
Eskdalemuir Observatory:
electrification experiments, 5066.
Esnault-Pelterie, Robert:
airplane, 1649, 1650, 1884.
biography, 157, 1492.
chronology of airplanes (1879–1909), 1651.
portrait, 1492.
rocket propulsion, 5459, 5460.
Espenlaub, Gottlob:
German aviator, 1942.
Espitallier, Georges:
bibliography, 155.
biography, 340, 4728, 5420.
portrait, 4728, 5420.
Esterno, Comte Ferdinand C. P. d':
bird flight, 5212.
ornithopter theory, 1914d, 2031.
Esterre, C. R. d':
bird flight studies, 5190.
Esteve, Rafael M.:
adventure in Syrian desert (1926), 3081.
Etana:
flight to Heaven, 54.
Étevé, A.:
bird flight studies, 5213.
Etrich, Igo:
airplane, 1656, 1939, 2272, 2305.
bibliography, 1502.
experiments with Franz Wels, 1936, 2216.
gliders, 2471.
Etzdorf, Marga von:
flights, 5332.
Euler, August:
biography and portrait, 291.
builds hydro-triplane (1914), 2329.
European Aeronautical Society, 881, 884, 1039, 1315.
Evans, Edward S.:
world flight (1926), 3499.
Evans, John, jr.:
balloon direction (1815), 1013, 1027.
Exhibitions:
Aero Club of America, 118, 5546.
Aeronautic Society (1909), 5466.
bibliography, 5414.
Crystal Palace, London (1868), 206, 2748, 2770, 5478.
International, Frankfurt (1909), 21, 371, 449, 5523, 5559.
International, Paris, *1878*, 819; *1909*, 2678.
Jamestown (1907), 5506.
London (1885), 5451.
New York (1912–1930), 5524.
Olympia, London, 251, 371, 1823, 5465, 5549.
Pan-American (1917), 5573.
Paris, Aeronautical Salon, 244, 371, 1439, 1677, 5467.
Paris Expositions, 151, 446, 817, 5470.
Science Museum, London, 287, 288, 1925.
Vienna (1888), 1454.
Expenditures:
European governments (1925–1929), 120.
foreign countries (1925–1930), 209.
United States:
appropriations, history, 3533, 3538.
military and naval (1921–1929), 3532.
World war, 4054, 4118.
years 1920–1925, 197.
Exploration:
aircraft, use of, 5396a.
Canada, 5566.
events of (1929), 5113.
Parima river, 4734.
See also Archaeology; Polar flights.
Explorer, balloon, 741.
Explosives as motive power:
Battey (1892), 1210.
Expositions. *See* Exhibitions.
Express service:
Canadian-Pacific (1929), 4243.
development (1924), 4333.
first (1910), 4219.
Kohler Aviation Corporation, 4308.
Ezekial, airship, 862.

F

Fabre, Henri:
biography and portrait, 2317.
seaplane, 2362.
Faccioli, Aristide:
air resistance experiments, 4793.
Fairbanks Airplane Corporation:
pioneer flights of Eielson and Wilkins, 4281.
Fairchild, Sherman Mills:
develops aerial photography and mapping, 2274.
portrait, 2274.
Fairchild Aerial Surveys, Inc.:
work in Florida, 5152.
work in Mexico, 5154.
Fairchild Camera Corporation, 5114.
Fairey, C. R.:
biography and portrait, 2056.

Far East:
Gallarza-Loriga, Madrid-Manila flight (1926), 3127.
Pinedo (1926), 3368, 3370.
Royal Air Force (1927–1928), 3003, 3090, 3158, 3408, 3409.
Sibor (1929), 5375.
statistics of flights (1924–1927), 3072.
See also Australia; Burma; China; East Indies; India; Japan.
Faraud, L.:
air resistance calculations, 4794.
Farcot, Eugène:
airship, 1031, 1044.
"Farcot" engine, 2728.
Farman, Henri:
airplanes, 1666, 1667, 1884, 1914a, 2044, 3097.
Belgium flights, 3091.
bibliography, 1493.
biography, 157, 223, 1938, 3513.
portraits, 1938, 2044, 3513.
wins Michelin prize, 3053.
Farman, Maurice:
balloon ascensions, 561.
bibliography, 1493.
biography and portrait, 580.
Farman brothers:
achievements, 1842, 2267, 2934.
"Farman" engine, 2698, 2729.
Farré, Henri:
war pictures, 4150.
World war experiences, 3892.
Faujas de Saint-Fond, B.:
references to, 253 (p. 425).
See also Author index.
Faure, Félix:
experiments with lifting propeller, 2868.
Faure, Jacques:
balloon manufacturer, 576.
Faustus, Doctor:
legends, 56, 57, 110.
Fawcett, E. D.:
airplane flights (1934), 3093a.
Feathers, 1827.
Fechet, James E.:
biography and portrait, 3625.
Fédération aéronautique internationale:
world records (1934), 3518.
Félipe, Julian:
helicopter, 2566.
Ferber, Louis Ferdinand:
airplane, 2083.
airplane flights, 2931.
airplane stability, 1564.
bibliography, 1674.
biography, 223, 1983.
engine (1908), 1929.
experiments, 1491, 1571, 1672, 1929.
gliding, 2393.
mathematics of flight, 4795.
portraits, 1491, 1983.
Ferguson, H. G.:
first airplane flight in Ireland (1909), 3100.
Ferguson, Samuel:
trans-African balloon voyage (1862), 851.
Fergusson, S. P.:
bibliography, 2637, 2665.
kites, 2632–2637.
sounding balloons, 5010.
Ferrarin, Arturo:
Rome-Brazil flight (1928), 3094, 3218, 3399.
Rome-Tokio flight (1928), 3094.
"Fiat" engine, 2728.
Fiction. *See* Literature and tradition.
50th Aero Squadron, U. S. Army:
World war record, 4015.
Finsterwalder, Sebastian:
principles of aerial surveying, 5115, 5116.
Fire hazard, 274.
Fire patrol. *See* Forest patrol.
Fire-resistant fabric:
bibliography, 1850.
First aerial bombardment (1849), 163, 803 (p. 409, 424), 3676.
First aerial propeller (1784), 2876.
First aerial radio broadcasting, New York City (1924), 5464.
First air mail flight in United States, Earle Ovington (1911), 4407.
First airplane crossing of the English Channel, Blériot (1909), 257, 1914a.
First airplane express carrier, Dayton-Columbus (1910), 4219.
First airplane flight over Andes mountains (1918), 2999.
First airplane flight in England:
claimed by A. V. Roe, 1924.
claimed by J. T. C. Moore-Brabazon, 3069, 3304.
First airplane flight in Ireland (1909), 3100.
First airplane flight over Mount Blanc (1914), 3358.
First airplane flight around the world. United States Army Air Service (1924), 210, 257, 3277.
First airplane night flight, Grahame-White (1910), 3278.
First airplane passenger carried:
claimed by Delagrange, 1914a.
claimed by Wright brothers, 1675.
First airship. Claimed for Hugh Newell (1841), 1015.
First airship flight around the world. *See* Graf Zeppelin.
First American aeronaut. *See* Durant, Charles F.
First American seaplane to rise from the water, Curtiss (1911), 2331.
First American woman aviator killed, Julia Clark, 5361.
First Atlantic air mail, Paris-Buenos Aires (1930), 3199.
First autogiro accident (1932), 2561.
First balloon ascension in America, Blanchard, 697a.
First balloon ascension of a human being. *See* Pilâtre de Rozier.
First balloon to cross the Alps, Arban (1846), 657, 661, 697a.
First balloon experiments in England. *See* Zambeccari.
First balloon voyage in England. *See* Lunardi.
First blind solo airplane flight, A. F. Hegenberger (1932), 3186.
First bomb dropped from an airplane (1911), 3677.
First British airship, *Nulli secundus*, 976.
First British inverted engine (1927), 2795.
First commercial airplane line in America, St. Petersburg-Tampa (1914), 4164, 4214.
First commercial airship route, Zeppelin (1910), 1159b.
First commercial seaplane service (1919), 2327.
First cylindrical balloon, claimed for Solomon Andrews (1863), 1395.
First East-West transatlantic airplane. *See* Bremen airplane.
First English "female aerial traveller," Mrs. L. A. Page, 253 (p. 125).
First English parachutist, John Hampton, 253 (p. 233).
First explosion of hydrogen in a balloon, 697b (p. 57).
First fighting airplane squadron, Balkan-Turkish war (1912), 3661.
First 500 aeronauts, 373 (p. 457–462).
First German aeronaut, Von Lütgendorf (1786), 684.
First helicopter to fly a circular kilometer (1923), 2562, 2597.
First important aviation meeting in the United States, Los Angeles (1910), 2919, 3351.
First inflation of a balloon with hydrogen, Charles, 536, 697a (p. 26).
First meteorological balloon in America, U. S. Signal Office (1885), 4682.
First military balloon, Spain (1792), 3662.

First non-stop airplane flight across America (1921), 3036, 3106.
First non-stop Atlantic flight. *See* Alcock, Sir John.
First non-stop Atlantic flight by a woman. *See* Earhart, Amelia.
First non-stop flight, Chicago-New York (1919), 3104.
First non-stop flight to India (1929), 3105.
First non-stop trans-Pacific flight, Herndon and Pangborn (1931), 3200.
First permanent air mail:
  Italy (1917), 4360.
  United States (1918), 4429.
First Polar flight. *See* Amundsen, Roald.
First power-driven airplane to rise from the water, Fabre (1910), 2362. *See also* Curtiss, entry no. 2331.
First power-driven airship model, Thomas Monck Mason (1842), 1159a.
First private aerodrome in America, Glenn Curtiss, 4221.
First radio message to earth, Harry M. Horton, 5485.
First rigid airship, Schwarz (1897), 867, 1084, 1159c.
First round trip night air mail letter, 4369.
First scientific balloon voyage, Jeffries (1785), 4854.
First seaplane flight to Europe, Stultz, Gordon, and Earhart (1928), 3107.
First skywriting suggested, J. C. Savage (1911), 5495.
First stainless steel airplane (1933), 1678.
First successful airship, Giffard (1852), 1159a.
First successful engine-driven airplane. *See* Wright brothers.
First successful helicopter, E. Berliner and son (1919), 2596.
First suit for damages (1785), 4592.
First 36 pioneer aviators, 2273.
First trans-American airplane flight, C. P. Rodgers (1911), 3092, 3110, 3462, 3463, 3479.
First transatlantic airplane flight. *See* United States Navy.
First transatlantic airship flight. *See* R-34, airship.
First transatlantic solo flight by a woman, Amelia Earhart (1932), 3200a, 5328.
First transpacific airplane flight, Kingsford-Smith (1928), 3112.
First United States Army airship, Baldwin (1908), 328, 1030.
First use of aircraft in war, battle of Fleurus (1794), 605.
First use of coal gas, J. P. Minckelers (1783), 797.
First use of the drag rope, S. A. King (1857), 795.
First war airplane for the United States Army (1914), 3539.
First woman American transcontinental aviator (1922), Lilian Gatlin, 5352.
First woman aviator, Thérèse Peltier, 5322.
First woman aviator in Great Britain, Mrs. Gavin, 5333.
First woman to ascend in a balloon, Madame Tible (1784), 5342.
First woman to fly over the Andes mountains, Laura Ingalls (1934), 5351.
Fischer, Johannes:
  experiences in World war, 3899.
Fish:
  fish movements:
    Samuelson, 5292.
    Zacharia, 5307.
  fish patrol:
    California, 5393.
    United States and abroad, 5525.
    See also "Aeronautical fish;" Sealing fleet.
Fisher, Mrs. Wayne H.:
  experiences as aviator, 5339.
Fiske, Bradley A.:
  torpedo launching apparatus (1917), 3612.
Fitzgerald, George F.:
  soaring experiments, 2395.
Fitzgerald, Maurice F.:
  theory of ornithopter, 4797.
Fitzmaurice, James. *See* Bremen, airplane.
Flammarion, Camille:
  scientific balloon ascensions, 4824.
Flight, theory. *See* Theory of flight.
Flights. *See* Airplanes, flights; Ballooning, scientific; Balloons, ascensions; Polar flights; Refueling flights; Stratosphere flights; Trans-American flights; Transatlantic flights; Trans-Australian flights.
Flint, Charles R.:
  relations with the Wright brothers, 1683.
Florez, Luis de:
  wins Scientific American award for airplane improvements (1933), 2135.
"Flying architect" of Persia, 76.
Flying carpet, airplane, 3181.
"Flying circus." *See* Richthofen.
"Flying monk." *See* St. Joseph of Copertino; Mohr, Caspar.
Fog penetration:
  Doolittle's blind flight (1929), 2932.
  early experiments, 4664.
  flying aids, 2101.
  General Electric Company, 5469, 5522.
  Guggenheim Fund experiments (1930), 4770, 4771.
"Foillard" engine, 2706.
Fokker, Anthony:
  airplanes, 1796, 1821, 2152, 3992a, 3999.
  biography, 1458, 1663, 1689, 1690, 2134, 2274.
  Byrd, Admiral, dispute, 2133.
  portraits, 1663, 1690, 2274.
  welded steel tube construction, 1689.
Fonck, René:
  biography and portrait, 3845, 3901, 4072.
  transatlantic attempt (1926), 2933.
Fontaine, J. A. A.:
  inclined balloon railway, 745.
Fonvielle, Wilfrid de:
  bibliography, 208.
  biographies and portraits, 4658, 4798, 5073, 5074.
  experiences in Franco-Prussian war, 3620.
  scientific balloon ascensions, 4824.
  See also Author index.
Ford, Henry:
  aeronautical accomplishments, 1718, 1999.
  air mail, 4381.
  builds "Liberty" engine, 2702.
  buys Stout Metal Airplane Co., 1783.
  cooperates with U. S. government in commercial aviation, 4203.
  Ford Air Transport, 4223, 4247.
Fordney, Chester L.:
  stratosphere flight with T. W. G. Settle (1933), 656.
Forest patrol, 5404, 5425, 5468, 5503, 5510, 5526, 5537.
Forlanini, Enrico:
  bibliography, 1046.
  comparison of helicopter and airship, 1046.
  helicopter, 290, 2557, 2564, 2570, 2573, 2586a, 2593.
  portrait, 2593.
  rocket propulsion, 290.
  seaplane, 2362.
  steam engine, 290, 1046.
Forster, T.:
  balloon ascension (1831), 398.
Foulois, Benjamin D.:
  commands air squadron to Mexico, 3609.
  portrait, 213.
  prominent in American military aviation, 213, 2274.
Fradiss, O.:
  woman aviator, 5335.
Frames, airplane, 1438.
France:
  aviation, civil and military, 1165, 4017, 4201, 4227, 4228, 4256, 4263, 4301, 4341.

France, *continued*
balloons, history, 373 (p. 519).
balloons, military, 373 (p. 524).
Compagnie internationale de navigation aérienne, 4290.
developments, *1872–1892,* 339; *post-war,* 309, 4126.
engines (1918–1930), 2772.
exhibitions, 371.
investigators, prominent, 129.
Joffre, General, 4017.
Le Bourget airport, 4260, 4300.
Lorraine, first balloons, 485.
Meudon, school, 825, 5529.
Mourmelon, school, 5484.
Polar expeditions proposed, 4452, 4463, 4505.
police regulations (1784), 641.
progress questioned, 375.
public opinion, 18th century, 587.
South America, lines to, 4230, 4237, 4289.
world record attempts (1924–1925), 2909.
See also Airships, France; Military aeronautics, France; Montgolfier brothers; World war.
Franchot, L.:
balloon direction (1850), 2063.
Francisco di Mendoza:
speculations, 292a.
Franco, Ramon:
attempts Spain-New York flight (1929), 3198.
flight, Palos-Buenos Aires (1926), 3124, 3128, 3450.
Franco-Chinese war (1883–1885):
captive balloon operations, 3528, 3764.
Franco-Moroccan war (1906–1907):
captive balloon operations, 3622.
Franco-Prussian war (1870–1871):
air laws, 4648.
Fonvielle's experiences, 3620.
general, 298, 549, 3618.
Mangin's experiences, 706.
Mangon's experiences, 3681, 4863.
pigeon post, 3619, 3621, 4358, 4388.
siege of Paris, 208, 294, 442, 549, 708, 803, 3563, 3577, 3652, 3681, 3727, 3747, 3750, 3762, 3764. 4358, 4388, 4408, 4416.
Tissandier's experiences, 3771.
Tournachon's experiences, 3773.
Turbiaux's experiences, 3774.
Franklin, Benjamin:
letters and works, 582–586, 786.
Freaks, 2259.
French monks of Siam, 58.
Friendship, airplane, 3060, 3107, 5329, 5385.
Fromage, Lucien:
airship dispensing with balloon, 1056.
Frost, Edward P.:
biography and portrait, 2056.
ornithopters, 1687, 2263, 5190.
Froude tank. *See* William Froude National Experiment Tank.
Fruscione, Vincenzo:
mechanical studies, 160.
Fučik:
makes wings from wood and tin (1760–1770), 2145.
Fuente la Peña, Antonio de:
discussion of his book *El ente dilucidado* (1677), 1577.
Fullerton, J. D.:
biography and portrait, 2056.
Funcheon, Daniel:
"Trolley flyer," 180.
Funcke, H.:
bird flight, 5214.

G

Gabel, S. L.:
metal tubing construction, 1705.
Gaede, R.:
airship (1873), 1060.
Gale, George Burcher:
auxiliary balloons, 444.
balloon ascensions, 253 (p. 426).
Galien, Joseph:
speculations, 68, 240, 253 (p. 426), 697c.
Gallarza, Edouard G., and J. L. Taboada:
Madrid-Manila flight (1926), 3127.
See also Esteve (entry 3081).
Galvez, Comte de:
airship (1784), 1005, 1061.
Gammeter, H. C.:
ornithopter, 1707.
Ganswindt, Hermann:
airship, 1062.
Gardner, Lester D.:
flights over commercial lines (1926), 3130.
Garnerin, André Jacques:
experience with Napoleon Bonaparte, 502.
parachute and balloon experiences, 253 (p. 426), 593, 774a, 783, 803, 821, 2490, 2502a, 2515.
Garnerin, Elisa:
parachute experiences, 2490, 5348, 5384.
portrait, 5384.
Garros, Roland:
bibliography, 3132.
biography, 3132, 3306.
experiences, 3068.
Mediterranean flight (1913), 3286.
Garvani:
fanciful airship flight Paris to Algiers, 855.
Gases:
bibliography, 457.
coal gas, 757, 796, 797.
early references, 373 (p. 522).
history of various gases, 796.
Montgolfier's gas, 697c (p. 25).
potato gas, 399.
steam for balloons, 414, 1331.
wood gas, siege of Paris, 3652.
See also Helium; Hydrogen.
"Gastambide-Mangin" airplane. *See* "Antoinette" airplane.
Gatke, Heinrich:
bird flight studies, 5215.
Gatlin, Lilian:
first woman transcontinental aviator (1922), 5352.
Gatty, Harold:
portrait, 3300a.
Post-Gatty world flight (1931), 213, 3200, 3222, 3331, 3336, 3374.
Gauchot, Paul:
assistant to A. Pénaud, 2045.
Gaudron, Auguste E.:
balloon ascensions, 830.
portrait, 830.
Gavin, Mrs.:
first woman aviator in Great Britain, 5333.
Gay-Lussac, Julius Louis:
scientific balloon ascensions, 697a (p. 65), 803, 4727, 4785, 4808.
Le Géant, balloon, 405, 697a, 3773.
Geiger, Fr.:
"Flying fish" airship, 1063.
General Electric Company:
automatic lighting of airports, 4176.
fog piercing beacon, 5469, 5522.
Genet, Edmond Charles:
horse-driven airship (1826), 4967.
memorial on air pressure (1825–1827), 4809–4811.
Geography:
aeronautical aids, 715, 4691, 5149.
See also Exploration.
Geology:
aeronautical aids, 5083a, 5087a, 5106a, 5117, 5119a, 5122, 5142, 5157, 5174, 5177.
Gerardin, A.:
engine studies (1867), 2063.

Gerlach, Edmond:
  bird flight experiments, 5216.
  flat disks, 4812.
  lifting propeller, 2847.
Gerli brothers:
  ascension with Andreani (1784), 602.
  balloon constructed by them (1785), 597.
German Rundflug (1925), 3159.
Germany:
  aerology, 5046.
  air laws:
    history, 4647.
    protest against restrictions (1925), 4611.
  air mail, 4386, 4387, 4395.
  airplanes:
    design trend (1919), 2200.
    general history, 1657, 1942, 1967, 2226.
    "Gotha" type, 3968.
    progress (1911–1912), 1527.
  airships. *See* Airships, Germany.
  aviation, 1657, 1942, 1967, 2193, 2226.
  bibliography, 19, 25.
  biographies, 291.
  civil aeronautics, 4186, 4232, 4264, 4291, 4326, 4346.
  Deruluft air line, 4231.
  Deutsche Luft-Hansa, 4232, 4284.
  Deutsche Luft-Reederei, 4347.
  early references, 373 (p. 519).
  engines:
    tests (1913), 2819.
    World war, 2721, 2740, 2796, 2797, 3950, 3984.
  exhibitions, history, 371.
  flights and competitions (1909–1912), 3419.
  general, 271, 1413, 4570.
  gliding, 2399, 2418.
  ice patrol, 5571.
  industry, *1908–1928,* 2193; *1911–1912,* 1527.
  Lütgendorf, first German aeronaut (1786), 684.
  military aeronautics. *See* Military aeronautics, Germany.
  modern developments, 285, 310.
  Mohr, the "Flying monk," 108.
  Schweikert the miller, 108.
  scientific ballooning, 5035.
  societies, 310.
  warships bombarded by United States. *See* Warship bombings.
  women balloonists, 5365.
  See also Military aeronautics, Germany; World war; Zeppelin.
Gibson, Hugh C.:
  biography, 157.
Gibson, William:
  man-operated ornithopter (1868), 1541.
Giffard, Henri:
  aerodynamics (1853), 2063.
  bibliography, 1067.
  biography, 821, 1072.
  captive balloons, 417, 495, 496, 599, 600, 803, 819, 820, 1072.
  hydrogen process, 554, 824.
  obituaries, 948.
  patents, 1072.
  portrait, 821.
  steam-driven airship, 305, 549, 931, 935–937, 950, 1066, 1067, 1072, 1159a, 1197, 1248.
Gilbert, Eugène:
  1000 kilometer flight (1913), 3307.
Gilbert, Octave:
  airplane, 2208.
Gill, Howard:
  airship designer, 1808, 3464.
Gillespie, G. Curtis:
  airplane, 1774.
Giotto's carving, 60.
Giradot, Léonce:
  airship, 917.
Giroud de Villette:
  experience with Pilâtre de Rozier (1783), 601.
Givaudan:
  airplane, 1695.
Glaisher, James:
  ascensions and investigations, 253 (p. 426), 651, 697a, 4818–4824, 4848, 4910, 4940, 4986, 5029, 5048.
  balloon instruments, 4669.
  bibliography, 4823.
  biography, 4941.
  portrait, 4940, 4941.
Gliders and gliding:
  bibliography, 2367, 2449.
  Elmira Soaring Competition (1930–1934), 2388–2391, 2407.
  first American club (1894), 2464.
  general history, 196, 255, 1488, 2044, 2365, 2370, 2374, 2394, 2401, 2408, 2415, 2420, 2439a, 2445, 2448, 2452, 2459, 2461, 2462, 2469, 2471.
  Germany, 2399, 2418.
  Hawks uses glider for transcontinental flight, 3199.
  progress, *1921–1931,* 2438; *1922,* 2470; *1930,* 2468; *1931,* 2101.
  Rhone Competition, 2398, 2412.
  See also Bazin, Blackden, Bryan, Chanute, Dante (G. B.), Deprez, Dinort, Ferber, Fitzgerald, Hargrave, Herring, Household, Koch, Kress, Le Bris, Lilienthal, Lindbergh, Lippisch, Maloney, Means, Montgomery, Mouillard, Nimfuhr, Parseval, Pénaud, Pilcher, Sanderval, Spencer (C.), Weiss (J.), Wenham.
"Gnome" engine, 2684, 2686, 2701, 2708, 2760.
Gnospelius, O. T.:
  biography and portrait, 2056.
"Gobron-Brillié" engine, 2708, 2728.
Godard, Eugène:
  balloon factory, 442.
  biography, 793.
Godard, Louis:
  long-distance balloon flight, 829.
  lifting propeller, 2868.
  picture of factory, 3656.
  tomb, 3656.
Goddard, Robert H.:
  rocket propulsion, 5473.
Goebel, Arthur:
  Hawaiian flight (1927), 3061, 3079, 3514.
  Los Angeles-New York flight (1928), 3341.
  portrait, 3079.
Goethe, Johann Wolfgang von:
  writings, 1413.
Göttingen Aerodynamic Experimental Station, 4651, 4653, 4987, 4988.
Goldbeater's skin, 555, 1234.
"Goodyear" balloon, 770, 835.
Goodyear Tire and Rubber Company:
  acquires Zeppelin rights (1923), 1071.
Gordon, Louis:
  transatlantic flight (1928), 3060, 3107.
Gordon Bennett races, 448, 513, 525, 571, 636, 655, 693, 694, 697, 833, 2945, 3510.
  bibliography, 452.
"Gotha" airplane, 3968.
Goupil, A.:
  airplane experiments, 176, 1725, 1914b, 4830, 4832.
  airship stability studies, 1073, 4833.
  bibliography, 4833.
  biography, 4974.
  engine studies, 2734, 4832.
  flight theory, 4829.
  launching formulas, 4831.
  portrait, 2734, 4974.
Grace, Cecil S.:
  biography, 223.
Graf Zeppelin, airship:
  comments by E. P. Warner, 1387.

Graf Zeppelin, airship, *continued*
flights of (1931), 3200.
four-continent flight (1930), 3199.
Polar flight (1931), 4440, 4472, 4483, 4511.
transatlantic flights, 927, 928, 985, 1007, 1131, 1159b, 1285, 1385.
world flight (1929), 1008, 1065, 1092, 1159b, 1285, 1402, 3198.
Graffigny, Henri de, pseud. *See* Marquis, Raoul.
Graham, George:
references to balloon ascensions, 253 (p. 427), 651a, 841a, 5347a.
Graham, Margaret:
references to her balloon ascensions, 253 (p. 427), 651a, 5347a.
Grahame-White, Claude:
bibliography, 2951.
biography, 223, 2056.
experiences, 1728, 3067.
Grahame-White Company, 1730.
London-Manchester flight, 380, 1914a, 3116, 3183, 3267, 3268.
portraits, 2056, 3300a.
Gran, Tryggve:
aviation experiences, 3920.
Grand Canyon:
flight of R. V. Thomas (1922), 3217.
Gray, George A.:
barnstorming experiences, 3140.
Gray, Hawthorne C.:
stratosphere flight (1927), 3281.
Great Britain:
air mail, 4389, 4393.
air policy, 3636, 3966.
aircraft industry, 1554, 1555, 2027, 2126, 2194.
airplane flights, notable (1918–1932), 3142.
airships. *See* Airships, Great Britain.
aviation (1908–1914), 1543.
Bagdad air mail, 4391, 4413.
balloon corps, early, 3592, 3617.
biographies of pioneers, 2056.
British Arctic Air Route Expedition, 4447, 4448, 4512.
civil aviation, 4256, 4261, 4291, 4310.
Civil Aviation Directorate, 4236.
dominions, 2100.
dominions in World war, 3572.
early references, 373 (p. 518, 519).
engines (1934), 2807.
England-India mail service, 4377.
exhibitions, early, 371.
flyers, first, 380.
general history, 253.
Government's relation to flying, 371.
Imperial Airways, 167, 4245, 4298.
Independent Air Force, 4107.
marine air lines, 4311.
post-war developments, 309, 4090.
propeller research (1923), 2900.
races, early airplane, 371.
Royal Aircraft Factory, 2107, 2749, 3837, 4741.
Royal Naval Air Force, 4157.
strike (1926), 4292.
Turner (C. C.), recollections, 371.
Wright brothers, English opinions of their airplane, 1682.
See also Military aeronautics, Great Britain; Royal Air Force; World war.
Greece:
legends, 110.
sculpture, 85.
Green, Charles:
balloon ascensions, 372, 572, 581a, 592, 710a, 767, 841b.
balloon direction, 644a, 811b.
biography, 253 (p. 241–260).
ideas on possible heights, 608.
Green, Charles, *continued*
model airship (1840), 1216.
Nassau balloon, 717, 738.
portrait, 711.
projected transatlantic flight, 409, 609, 697a, 718.
Vauxhall balloon, 836a.
"Green" engine:
bibliography, 2687.
Greene, Warwick:
letters, 3925.
Greene, William:
airplane, 3461.
Greenland:
Gronau's flights, 3118a, 3175, 3176, 3199.
See also Lindbergh, Charles A.
Greth, August:
airship, 886, 1294.
Grey, Charles G.:
bibliography, 236.
See also Author index.
Grey, Spenser D. A.:
biography and portrait, 2056.
Grierson, John:
airplane flight, England-Russia (1932), 3174.
Griffin, Bennett:
non-stop New York-Berlin flight (1932), 3200a, 3336.
Griffiths, J. C.:
aerial surveying contributions, 5162.
Grimaldi, Gabbriello:
balloon direction (1788), 612.
Gronau, Wolfgang von:
Greenland flights, 3118a, 3175, 3176, 3199.
world flight (1932), 3200a, 3406, 3489.
Groof, Vincent de:
ornithopter (1874), 1572, 1914d.
Gross, Hans:
airship, 3527.
bibliography, 458.
portrait, 4570.
"Guardian angel" parachute. *See* Calthrop, E. R.
Guericke, Otto von:
flight ideas, 68.
Guggenheim, Harry F.:
biography, 2274.
fog penetration, 4770, 4771.
portrait, 2274.
See also Daniel Guggenheim Fund for the Promotion of Aeronautics; *also* Daniel Guggenheim International Safe Aircraft Competition.
Guiche, duc de:
aerodynamic experiments, 4652.
Guiding of aircraft:
Loth's magnetic system, 5419, 5491.
See also Lighting of airways; Radio.
Guidoni, Alessandro:
pioneer seaplane designer, 2332.
torpedo launching device, 3640.
See also Author index.
Guillie, L. C.:
airship (1814) with inclined planes, 1159d.
Gusmão. *See* Lourenco de Gusmão.
Guynemer, George M. L.:
bibliography, 3836.
biography, 3848, 3851, 3980, 4029.
death, 4046.
portrait, 3848, 3851, 3980.
Guyton de Morveau:
hydrogen process, 554.
portrait, 803.
Gypson, Richard:
balloon ascensions, 253 (p. 427).
"Gyro" engine:
bibliography, 2688.
Gyroscope:
bibliography, 4711.

H

Haenlein, Paul:
  airships (1865–1872), 885, 931, 936, 1053, 1088, 1090, 1159c, 1193, 1214.
  bibliography, 1089.
  portrait, 1193.
  propeller investigations, 2850.
Hagen, Friedrich Josef, Freiherr vom:
  bibliography, 240, 5403.
  parachute theory, 2490.
  portrait, 4570.
  See also Author index.
Hague conferences:
  comments on law of aerial warfare (1914), 4605.
  conference of (1899), 4613.
Haiti. *See* Hispaniola.
Haizlip, James:
  Los Angeles-New York flight (1932), 3065.
Haizlip, Mae:
  biography and portrait, 5369.
Haldeman, George:
  Haldeman-Elder flight to Azores (1927), 3079, 5310.
  portrait, 3079.
Hall, Bert:
  instructs Chinese aviators, 3260.
  Lafayette Escadrille, 3939, 3940.
Halliburton, Richard:
  flight across Africa, Asia, and the United States (1932), 3181.
Hamel, Gustave:
  appreciation, 3240.
  bibliography, 2953.
  portrait, 3240.
Hamilton, Charles Keeney:
  biography, 157, 2992, 3007, 3052, 3067, 3329.
  flight, New York-Philadelphia (1910), 3372.
  portrait, 3007, 3329.
Hamilton Rice Scientific Expedition (1924), 3455, 5396a.
Hammell, John E.:
  aerial prospector, 5532a.
Hammon, W. H.:
  kite experiments, 2640.
Hammond, E. V.:
  biography and portrait, 2056.
  multiple propellers, 2851, 2878.
Hampton, John:
  parachute experiences, 253 (p. 427), 2490, 2502a, 2507b.
Handasyde, G. H.:
  biography and portrait, 2056.
Handley-Page, Frederick:
  airplane, 1634.
  bibliography, 1759.
  biography and portrait, 2056.
Hangars:
  bibliography, 5415.
  history, 1029.
  world's largest, 1409.
Hansen:
  study of wing propellers, 2852.
Hanson's seadrome, 4244.
Hargrave, Lawrence:
  bibliography, 2646.
  biography, 2651, 2670.
  bird flight, 5218.
  gliding experiments, 2403, 2404, 2439a.
  kites, 2641–2646, 2649, 2661, 2738.
  model airplanes, 176, 1659, 1760–1766, 1769, 1770, 1802, 1803, 1898, 1914b.
  motors, 2738, 2739.
  obituary, 2622.
  ornithopter, 1914d.
  portraits, 1802, 2669, 2670.
  stabilizer, airplane, 2139.
  wings, curved, 1767, 1768.
Harmon, Clifford B.:
  biography, 157.
Harper, Harry:
  early flying experiences, 3182.
Harriman, J. Emery, jr.:
  ornithopter, 1771.
Harris, H. R.:
  parachute experiences, 2499.
Hartness, James:
  patents sectional balloon (1874), 744.
Harvard University:
  bibliography, 5411.
Haskelite Manufacturing Corporation:
  history of plywood, 5480.
Haslett, Elmer:
  World war experiences, 3944.
Hastings, C. W.:
  biography, 2853.
  flight theory, 2853.
  propeller, 2853.
Hauvel d'Audreville, Ch. du:
  bird flight study, 5220.
  propeller study, 2854.
Havilland, Geoffrey de:
  biography and portrait, 2056.
Hawaiian islands:
  aerial survey, 5101.
  commercial aviation, 4285.
  Goebel-Davis flight (1927), 3061, 3079, 3514.
  Maitland and Hegenberger flight (1927), 2925, 3079, 3279, 3476.
  naval maneuvers (1925), 3569.
  Rodgers and "PN" seaplane flight (1925), 210, 3027, 3028, 3231, 3397, 4996.
  Smith and Bronte flight (1927), 3079, 3431, 3445.
Hawker, Harry G.:
  altitude record (1916), 3517.
  appreciation, 3153.
  biography, 3185, 3359.
  portrait, 3087, 3473.
  transatlantic flight (1919), 257, 3060, 3087, 3184, 3473.
Hawks, Frank Monroe:
  flights (1931), 3200, 3382.
  trans-American speed records, 3000, 3198, 3199, 3309, 3336a.
Hawley, Alan Ramsay:
  biography and portrait, 3818.
Hazen, H. A.:
  balloon voyages, 4853.
  first meteorological balloon in America (1885), 4682.
  See also Author index.
Hearst, William Randolph:
  biography, 157.
Heart's content, airplane:
  Ireland-America flight, J. A. Mollison (1932), 3119, 3200a, 3301, 3302.
Heath, Edward:
  pioneer American airplane builder, 1647.
Heath, Sophie Mary (Peirce-Evans) Elliot-Lynn, 3300a, 5344, 5345.
Heaton:
  California messenger, airship, 933.
Hegenberger, Albert F.:
  Hawaiian flight (1927), 2925, 3079, 3279, 3476.
  portrait, 3079.
  world's first blind solo flight (1932), 3186.
Heinkel catapault, 5482.
Heintz, Henry:
  airplane, 1777.
Helicopters:
  bibliography, 2535.
  British competition (1923–1924), 2586a, 2597.
  developments (1909), 382.
    American, 2586a.
    World war, 2586a.

Helicopters, *continued*
early machines, 253 (p. 428).
first flight, Berliner, 2596.
first to fly a circular kilometer (1923), 2562.
general, 295, 2044, 2537, 2560, 2573, 2577, 2584, 2586a, 2602.
gyroscopic action, bibliography, 2536.
power requirements, 2603, 2605.
See also Ascanio, Berliner, Bertin, Bienvenu and Launoy, Bothezat, Bourne, Bréguet, Bright, Caballero de los Olivos, Castel, Cayley, Chalmers, Clout, Cornu, Cossus, Damblanc, Davidson, Dieuaide, Dufaux, Edison, Felipe, Hyatt, Isacco, Karos, Kress, La Landelle, Leger, Liais, Lorenzon, Luyties, Mayer, Mélikoff, Mott, Oehmichen, Pénaud, Pescara, Petroczy-Karman, Phillips (W. H.), Pillet, Platte, Ponton d'Amécourt, Renaud (C.), Renoir, Santos-Dumont, Trouvé, Villard, Wenham.
Helium:
bibliography, 459.
general, 490a, 492, 547a, 726, 832, 845, 845a, 1293, 1334, 2164.
Helmar:
airship, 1098.
Helmholtz, Hermann F. L. von:
airship theory (1873), 4855, 4856.
bibliography, 4856.
investigations, 85.
Helwin, Maxwell:
fatal airplane accident (1902), 3014.
Henderson, David:
helps to organize Royal Air Force, 3928.
Henderson, John:
balloon direction, 1099.
Hengler, Lorenz:
parachute improvements (1832), 2492, 2504, 2505.
Henkelburg, Hans:
World war experiences, 3946.
Henrion, Francesco:
balloon direction (1789), 1101.
Henry, Edmond:
balloon forms, 627.
Henry, Joseph:
interest in American Civil war aeronautics, 3658.
letter to John Wise on proposed transatlantic voyage, 588.
Henry, Louis Isidore:
study of airplane stability, 1781.
Henson, William Samuel:
airplanes, 253 (p. 428), 380, 1481, 1541, 1620, 1645, 1784–1786, 1914b, 1918, 1926, 1941, 1957, 2021, 2031, 2044, 2105.
patent description (1842), 1450.
reconstruction of 1840 model, 1708.
Hergesell, Hugo:
bibliography, 4857.
biography, 291.
investigates vertical movement of balloon, 629.
Hermann, Kurt:
biography and portrait, 291.
Hermite, Gustave:
balloon ascensions, 4858, 4859, 4861.
bibliography, 4860.
biography and portrait, 615, 4657.
sounding balloons, 396, 4657, 4799, 4800, 4803, 4860.
Herndon, Hugh, jr.:
world and first non-stop Pacific flight (1931), 3200.
Herring, A. M.:
bibliography, 1788.
biography, 1825.
gliding and airplane experiments, 1629, 1630, 1788, 1825, 2378, 2406.
Hersey, Henry B.:
balloon experiences, 630.
Hervé, Henri:
bibliography, 632.
kite balloon, 1194.
propeller, 2857.
suggests improvements in meteorological balloons, 4862.
Hewlett, Hilda Beatrice (Herbert):
early experiences, 5386.
London-Batavia flight (1932), 5347.
Heydemarck, Georg:
World war experiences, 3947, 3948.
Hild, Frederick C.:
World war experiences, 3949.
Hildebrandt, Alfred L. H.:
German aeronaut and writer, 959.
Hill, G. T. R.:
tailless airplane, 1791, 2179.
Hill, James De Witt:
fatal transatlantic flight (1927), 3121.
Hillig, Otto:
transatlantic flight (1931), 3200.
Hinkler, Bert:
appreciation, 3344, 3345, 3396.
London-Australia flight (1928), 3141, 3177, 3289.
portraits, 3300a, 3344, 3345, 3396.
Hinton, Walter:
discovers source of Parima river (1925), 4734.
Hirth, Hellmuth:
experiences as aviator, 3197.
Hispaniola:
aerial survey, 5146.
"Hispano-Suiza" engine, 2701, 2744, 2747, 2800.
Hite, Charles E.:
airship, 1042.
Hoar, J. W.:
balloon ventures, 253 (p. 428).
Hoare, Sir Samuel J. G.:
England-India flight (1927), 3205.
Hoaxes:
Lindsay's magnetized alloy (1847), 1900.
Poe, balloon hoax, 96.
Rittenhouse and Hopkinson (1783), 649.
Stott's English Channel flight (1876), 1644.
Hoeppner, Ernst von:
biography and portrait, 291.
Hoernes, Hermann:
Austrian engineer and writer, 2856.
propeller studies, 2856.
Hoffman, E. L.:
parachute improvements, 2480, 2522.
Hofmann, Josef:
airplanes, 1681, 1726, 1798, 1799.
Hofwerberg, Johannes:
balloon calculations (1786), 638.
Hoiriis, Holger:
transatlantic flight (1931), 3200.
Holland, John P.:
engineering problems of airplanes, 1800, 1801.
Holland, Robert:
portrait, 711.
voyage in *Nassau* balloon (1836), 572.
Holland. *See* Netherlands.
Hollands, Sidney H.:
steam engine and propeller (1885), 2746.
Holman, Charles W.:
biography and portraits, 3440.
Holmes, John P.:
man-operated airplane, 1822.
Hooke, Robert:
biography, 4868.
experiments, 253 (p. 428).
possibility of flight, 4868.
theory of ornithopter, 1914d.
Hoop, Thomassen à T. van der:
Amsterdam-Batavia flight (1924), 3207.
Hopkinson, Bertram:
scientific contributions, 5079.

Hopson:
parachute experiences, 2493.
Horse-propelled airships. *See* Airships, horse-propelled.
Houdini, Harry:
airplane flight in Australia (1910), 2930, 3099.
House, Harry A.:
builds engine for Hiram Maxim, 1835.
Household, Goodman:
glider flight (1875), 2463.
Houston Mount Everest expedition. *See* Mount Everest.
Hoxsey, "Arch":
biography, 2026, 2992, 3067, 3513.
death (1910), 3471.
portraits, 3471, 3513.
Hoyt, Colgate:
biography, 157.
Hoyt, Richard F.:
portrait, 2274.
promotes civil aviation, 2274.
Hoyt, Ross G.:
New York-Nome flight (1929), 3198.
Hubbard, T. O'B.:
biography and portrait, 2056.
Huber, François:
bird flight study, 5221.
Huenefeld, Gunther von. *See Bremen,* airplane.
Huffaker, E. C.:
bird flight studies, 5222–5224.
Humble, Richard:
Cape of Good Hope flight (1931), 3213.
Hunsaker, Jerome Clarke:
bibliography, 258.
biography, 3775, 4866.
Massachusetts Institute of Technology, 4944.
portrait, 3775.
Hunt, Edmund:
airship (1848), 1107.
Hunter brothers:
refueling endurance flight (1930), 3199, 3323.
Hureau de Villeneuve, Abel:
bibliography, 261.
experiments, 176.
obituary, 5391.
portrait, 5391.
"steam bat," 1914d.
Hurry, H. C.:
airplane studies (1867), 1816.
Hutchinson, F. W. H.:
bird flight studies, 5190.
Hutchinson, W. N.:
airship, 1109.
Hutton, F. W.:
bird flight, 5225.
Hyatt, Thaddeus:
helicopter (1882), 2571.
offers prize for flying machine (1860), 1817.
Hydrogen:
airships, World war, 3952.
Andrée's polar balloon, 4502.
Arsonval process, 547.
bibliography, 460.
Cavendish, 501, 697c.
chronology, 151.
compression, 552, 1100, 1114, 1313.
condensation, Weeke's process, 842.
Conte's process, 523.
Coutelle's process, 554, 3770.
diffusion through fabric, 5007.
Ducretet process, 547.
Duponchel, heating and cooling, 997.
Dupuy de Lôme's process, 549, 554.
early references, 253 (p. 428).
explosion, first, 697b.
general history, 637, 796.
Giffard's process, 554, 824.
Hydrogen, *continued*
Great Britain, 3744.
Guyton de Morreau's process, 554.
hydrolith, 653, 671.
inflation, first, 536, 697c.
isolation, 501, 697c.
Lachambre's apparatus, 3637.
Latchinoff's process, 547, 716, 3744.
liquid hydrogen, 548.
Lorraine apparatus, 485.
Priestley's process, 565, 771.
Renard's process, 554.
Rindine's proposal to obtain from the upper air, 1443
Selle de Beauchamp's apparatus, 3770.
Serrati's process, 979.
Tissandier's process, 554.
Trouvé's hydrogen motor, 2202.
See also Yon, Gabriel.
Hydrolith. *See* Hydrogen.
Hymn, airman's. *See* Music.

I

Ibn Firnis:
reputed early heavier-than-air machine, 39, 39a, 56.
Icare, balloon, 464.
Icaromenippus (Lucian's satire), 80.
Icarus and Daedalus, legends, 42, 64, 82, 99, 104.
Ice-plane patrol:
German experiences, 5571.
Iglesais, Ignacio:
transatlantic flight (1929), 3198.
Ijserman, J. M.:
ornithopter, 1820.
Immelmann, Max:
death, 3859.
letters, 3960.
World war exploits, 3976, 4002, 4128.
Imperial Airways, 167, 4245, 4298.
Independent Air Force, Great Britain:
bombs German industrial centers, 4107.
India:
air mail to England, 4377.
Bedford, Duchess of, flight to England, 5326.
Blake's airplane flight from England (1922), 2966.
Camac's flight to (1929), 5320.
civil aviation, 4213, 4341.
flight statistics (1924–1927), 3072.
Hoare's flight to (1927), 3205.
Jones-Williams and Jenkins flight to (1929), 3105.
legends, 59, 76, 91.
Minney's trans-Indian flight (1920), 3287.
Sassoon's flight to (1928), 3414, 3425.
Stack's flight to (1926), 3452.
surveys, 3383, 5088, 5133, 5139, 5144.
India rubber:
Pénaud's use in model motors (1872), 2786.
Industrial arts index, 17.
Ingalls, David S.:
military aviator, 3898.
Ingalls, Laura:
Andes flight (1934), 5351.
Insect flight:
bibliography, 5188, 5265.
history, 5274.
See also Bull, Dutczynski, Janet, Jarolimek, Liais, Lucy, Marey, Müllenhoff, Pettigrew, Snodgrass.
Insects:
boll weevil elimination, 5463, 5552.
catching by airplane, 4761.
dusting planes, 5455.
mosquito elimination, 5561.
United States Agriculture Department, 4828, 5552, 5570.
Insignia:
French escadrilles, 4036.
United States, 3914.

Institut aérodynamique de Koutchino, 4804, 4895.
Instruments:
  anemometric balloon, 4790.
  bibliography, 5406, 5416.
  inventions of:
    Assmann, 4669.
    Cailletet, 4714, 4956.
    Clift, 5475.
    Colvin, 2274.
    Glaisher, 4669.
    Hermite, 615.
  modern developments, *1924*, 5567; *1928*, 5044.
  speed instruments, 5406.
  United States Army, world flight, 5533.
  World war, 4036.
  See also Compass.
Insurance:
  chronology, 5534.
  history, 5443, 5513.
Interior balloons. *See* Ballonets, interior.
International Aeronautic Federation, 4552, 4553, 4555.
International Aeronautical Conference, Strassbourg (1898), 4563, 4577.
International Aeronautical Congress, New York (1907), 4554.
International Air Conference, Gothenburg (1923), 4584.
International air meet:
  Belmont Park, N. Y. (1910), 3510.
  St. Louis (1923), 3352, 3487.
International Civil Aeronautics Conference (1928), 4556, 4557.
International Commission for Aeronautical Maps (1911), 4544.
International Commission for Air Navigation (1922–1932), 4558.
International Commission for the Investigation of the Upper Air (1898–1928), 4559.
International Commission for Scientific Aeronautics, 4560, 4857.
International Conference on Aerial Navigation, Chicago (1893), 4561, 4576.
International Congress of Aerial Navigation (1921–1930), 4562.
International Convention for Air Navigation (1919), 4646.
International Society for the Exploration of the Arctic Regions by Means of Aircraft, 4582.
International Technical Committee of Aerial Legal Experts, sessions, 4616.
Internationale aëronautische Kommission (1898), 4545.
Internationale Luftschiffahrts Ausstellung zu Frankfurt (1909), 21, 371, 449, 5523, 5559.
Ireland:
  first airplane flight (1909), 3100.
  Irish Air Force, 3613.
Irish, W. E.:
  experiments with feathered surfaces, 1827, 2400.
  jet-driven metal airplane, 1826.
Irminger, Johann:
  wind tunnel, 4994.
Iron proposed for balloon shell, 482.
Irving, H. B., and A. S. Batson:
  lateral control of airplanes, 1828.
Isacco, V.:
  helicopter and airplane, 2572, 2576.
Italia, airship:
  Almerico da Schio's, 918, 1310, 1311.
  Umberto Nobile's, 1159b, 4437, 4441, 4464, 4468, 4477, 4478, 4482, 4488, 4497, 4509.
Italo-Abyssinian war (1888):
  captive balloons, 3557, 3741.
Italo-Turkish war (1911), 3546, 3594, 3742.
Italy:
  anthology, 41.
  aviation, 1996.
  balloon ascensions, Turin, 161.
  balloonists, pioneer, 498, 979.
Italy, *continued*
  bibliography, 6, 7, 18.
  chronology, 125.
  civil aviation, 4165, 4250, 4256, 4341.
  early references, 373 (p. 520).
  engine and airplane companies (1929), 4206.
  first aerial bombardment (1849), 163, 803 (p. 409, 424), 3676.
  first balloon ascensions, 772.
  general history, 218, 331.
  Italian cities, 480, 598, 658, 659, 920, 921.
  military aeronautics, 3732, 3743, 4062a.
  post-war developments, 309.
  war balloons, early, 373 (p. 524).
  World war, 3911, 3961, 4059, 4063.
  World war aviators, 3821.
  World war, defense of Venice, 3902.

J

Jackson, Dale:
  refueling airplane flights, 2918, 3198, 3199, 3239, 3412.
Jacob, Emil:
  air resistance investigations, 4874.
  bibliography, 4874.
  criticizes Buttenstedt's bird flight theory, 5227.
Jamestown exposition, 5506.
Jamin, Jules:
  biography and portrait, 4806.
Janet, Charles:
  insect flight studies, 5228.
Janin, A. S.:
  suit against Glenn Curtiss, 2340.
Jannus, Anthony:
  biography and portrait, 3464.
Janssen, Pierre Jules César:
  bibliography, 4716.
  biography, 652, 4978.
  compass, 5422, 5492.
  portraits, 652, 4978.
Japan:
  aviation, history, 2301.
  flights to:
    Botved (1926), 2974.
    Brock and Schlee (1927), 3079.
    Ferrarin (1920), 3094.
    Herndon and Pangborn (1931), 3200.
    Lindbergh (1931), 3200, 3219.
    Pelletier-Doisy (1924), 3137.
  kites, 3704.
  See also Military aeronautics, Japan.
Jarolimek, Anton:
  bibliography, 4877.
  bird and insect flight, 5230.
  flight theory, 4875–4877.
Java:
  flying gods, 86.
Jeffries, John:
  balloon ascensions, 253 (p. 428), 407, 675, 4854, 4878.
  diary, 654.
  portrait, 803, 4854.
  See also English Channel.
Jenkins, N. H.:
  non-stop flight to India (1929), 3105.
Jesus del Gran Poder, airplane:
  transatlantic flight (1929), 3198.
Jet propulsion:
  Boyman (1866), 1159a.
  Irish (1900), 1826.
  Kinsella (1863), 1114.
  Lake (1837), 1130.
  Lorin, 2755.
  Miolan and Janinet (1784), 1159d.
  Reis (1859), 1263.
  Thayer (1885–1886), 1345, 3763.

Jewish aviators in World war, 4098.
Jezzi, Leo:
  biography and portrait, 2056.
Jimenez, Francisco:
  transatlantic flight (1929), 3198.
Jobert, Claude:
  air resistance study, 4879.
  chronology, 4952.
  ornithopter, 1914d.
  scientific balloon ascensions, 4667, 4668.
Joffre, Joseph Jacques Césaire:
  pioneer in French military aviation, 4017.
Johannsen, Christian:
  flight, London-China (1929), 3488.
"John Daniel" (fiction), 87, 88.
John Rodgers airport, 4285.
Johnson, Amy. *See* Mollison, Amy Johnson.
Johnson, George:
  aerial views of Peru, 3221.
Johnson, Walter E.:
  aviator, 3464.
Johnstone, Ralph:
  aviator, 2026, 3067, 3510.
Jones, Charles Sherman ("Casey"):
  commercial pilot, 2274.
  experiences, 3223.
  portrait, 2274.
Jones, James H.:
  airplane inventor, 1832.
Jones, Thomas P.:
  reply to Genet (1827), 4881.
Jones-Williams, A. G.:
  non-stop flight to India (1929), 3105.
Josephine Ford, airplane, 4443.
Joukovsky, N.:
  aerodynamical studies, 4883.
Jovis:
  proposed transatlantic flight (1887), 410.
Joy stick:
  invented by Wright brothers, 1914c.
Judd, Neil M.:
  archaeological work in Arizona, 5126.
Jullien, Pierre:
  power-driven airship model (1850), 1159a.
Julliott, Henri:
  describes Lebaudy airships, 1122.
  See also Lebaudy brothers.
Junebug, airplane, 1612, 1838, 3508.
Junkers, Hugo:
  biography, 291, 1811, 2060.
  Deutsche Luft-Hansa, 4284.
  German company difficulties (1926), 1575.
  metal airplanes, 1839.
  portraits, 291, 1811.
Jurisprudence:
  airport approaches, 4604.
  American Bar Association, 4588.
  bibliography, 4588–4591, 4593, 4595, 4598, 4599, 4603, 4606, 4614, 4618, 4621, 4628, 4629, 4631, 4633, 4634, 4637, 4639, 4640, 4644, 4645, 4647.
  bombing of cities, 4637a.
  China, 4623.
  Comité juridique internationale de l'aviation, 4600.
  Congrès internationale de législation aérienne, 4540.
  first air trespass, 4632.
  first suit for damages, 4592.
  Franco-Prussian war, 4648.
  general history, 4590, 4591, 4593, 4594, 4596, 4598, 4607, 4609, 4613–4615, 4617, 4618, 4624–4629, 4636, 4639, 4645.
  Germany, 4611, 4647.
  Hague conferences, 4605.
  international conferences, 4335, 4649.
  International Convention for Air Navigation, 4646.
  International Technical Committee of Legal Experts, 4616.
  laws previous to 1914, 4622.
Jurisprudence, *continued*
  laws of various countries, 4586, 4595.
  monopoly of air asked (1822), 4587.
  North Pole, 4630.
  Pan American Convention, 4646.
  Roman law, 4619.
  South Pole, 4602.
  United States:
    Air Commerce Act (1926), 4642.
    decisions (1822–1932), 4643.
    general, 4597, 4608, 4634, 4641.
    state laws, 4643, 4650.
  World war, effects on laws, 4618, 4620, 4635, 4638.
Justice for Hungary, airplane, 3200.
Juvenile works, 20, 191, 212, 219, 223, 241, 330, 362, 363, 394, 1589, 1599, 1831.

K

"K" factor, determination, 295.
Kadarz:
  propeller researches, 2863.
Kantner, Harold:
  aviator, 3464.
Kapferer, Henry:
  airplane, 2017.
Kargl, Ludwig:
  calculations for heavier-than-air engines (1870), 4885.
Karman helicopter. *See* Petroczy-Karman.
Karos, Willibald:
  helicopter and glider combination, 1841, 2574.
Kaufmann, J. M.:
  ornithopter, 116, 1914d, 1433.
Kaufmann, Mildred:
  biography and portrait, 5377.
  experience as aviator, 5339.
Kay's portraits (1837–1842), 663.
Keegan, Allen:
  pioneer balloon constructor, 253 (p. 428).
Kellett Autogiro Company, 2556.
Kelly, James:
  refueling endurance flight (1929), 3426.
Kelly, Oakley G.:
  non-stop trans-American flight (1923) with Macready, 210, 257, 2916, 3088, 3204.
  portrait, 3300a.
Kelly Air Mail Act, 4417.
Kelvin, Bottomley and Baird, Ltd.:
  compass makers, 5494.
Kepner, William E.:
  stratosphere flight (1934), 666, 5045.
Kerr, Mark:
  reminiscences as British admiral, 3970.
Kimball, Wilbur R.:
  multiple propeller, 2878.
King, Samuel Archer:
  biography, 852.
  first aeronaut to use drag-rope (1857), 795.
King's Cup races, *1922,* 3162; *1923,* 3163; *1924,* 3172; *1925,* 3417; *1926,* 3164; *1927,* 3165; *1928,* 3166; *1930,* 3226; *1933,* 3167.
Kingsford, A. R.:
  World war experiences, 3972.
Kingsford-Smith, Charles E.:
  Australia-London flight (1929), 3447a.
  portrait, 3300a.
  transatlantic flight (1930), 3059, 3060, 3199, 3228, 3229.
  transpacific flight (1928), 3112, 3141, 3143, 3227, 3495.
Kinsella, Arthur:
  airship driven by air jet (1863), 1114.
Kipfer, P.:
  biography, 4962.
  stratosphere flight with Piccard (1931), 4962.

Kircher, Athanasius:
flying machine (1671), 4889.
Schott's comments (1657–1659), 5028.
speculations, 68.
Kirsten-Boeing swivel propeller, 2888.
Kite balloons. *See* Balloons, kite.
Kites:
Abbe's writings, 2607–2609.
aluminium, 2617.
American development, 2608.
Baden-Powell, 2610, 2611.
Bell's tetrahedral, 1485, 2192, 2439a, 2612, 2618, 2619.
bibliography, 2614, 2624, 2659, 2672.
Brandes, 2607, 2616.
Cavallo, 2607, 2617a.
China, 2647, 2671.
Clayton's writings, 2619.
Cody, 2666.
Colladon's experiments, 2620.
construction (1909), 2659.
Davis's experiments, 2621.
Decker on mechanics, 2623.
Dines' writings, 2625–2626.
Eddy's kites, 2628–2631, 2662, 5009.
general, 76, 245, 247, 253 (p. 429), 1756, 2044, 2607, 2624, 2652, 5015.
Hargrave, 2641–2646, 2649, 2738.
international ascensions, 2635, 4826.
Japan, 3704.
Lecornu, 2653.
life-saving, 2668.
Lindenberg Aeronautical Observatory, 4655.
McAdie, 2640.
man-lifting kites, 2650, 2655, 2674, 3627.
Marvin on mechanics, 2657.
Means on value of kites, 2658.
military uses, 3582, 3627.
Mount Weather, Va., 2615.
Newton, Sir Isaac, 2609.
Nikel, 2667.
photography, use in, 5112, 5164, 5171.
Pocock's kite-driven carriage, 2627.
Porta, 4985.
progress (1914–1925), 2638a.
Romas, 2607, 2669a.
ships, kite flying from, 2625.
Thayer (D.), combination kite-balloon, 2664.
United States Weather Bureau, 2638, 2640, 2654, 2656.
Van Musschenbroek, 2607, 2663.
wire, early use, 2632.
Wise (H. D.), 2673.
Young, 2607.
See also Rotch, A. Lawrence.
Klassohn, Eduard:
airship propeller calculations (1865), 2864.
Klemin, Alexander:
biography, 4891.
See also Author index.
Klumpke, Dorothée:
woman aviator, 5334.
Knabenschue, Roy:
airship carries three passengers (1908), 1037.
New York City experiences, 1158.
stabilization of airships, 896.
Knauss, Robert:
Berlin-Peking flight (1926), 3232.
Knerr, H. C.:
metal tube construction, 1705.
Knoller, Richard:
biography and portrait, 291.
Koch, Gustave:
airship, 1087, 1243.
bibliography, 910.
glider, 1853.
steam-driven airplane, 1851, 1852, 1854–1856.
Koehl, Hermann:
autobiography, 3975.
See also Bremen, airplane.
König-Warthausen, F. K.:
world flight (1929), 3235.
Köppen, W.:
mechanics of flight, 4892.
Kohler Aviation Corporation:
breaks express-handling record, 4308.
Koninklijke-Indische-Luchtvaart Maatschappij. *See* Royal Dutch Indian Airways.
Konvalinka, J. G.:
man-power airplane, 1857.
Korf, George:
book on bird flight, 5231.
"Korwin-Rebikoff" engine, 2728.
Kosch, Rudolph:
ornithopter, 1858.
Kosloff, Alexandrowitsch:
aerial compass, 5519.
Kotzauer, Wenzel:
book on bird flight, 5232.
Kramp, Christian:
flight theories (c. 1800), 4841.
Krebs, A. C. *See* Renard and Krebs.
Kreiss, Eugen:
helicopter, 2578.
theory of flight, 2578, 4964.
Kremser, Victor:
biography and portrait, 5043.
Kress, Wilhelm:
airplane, 1454, 1753, 1754, 1859, 1861, 1863.
biography, 2225, 2256, 2257, 2472.
bird flight studies, 5234.
glider, 1859, 2414, 2472.
ornithopter, 1859.
portraits, 2225, 2257, 2472.
propeller for airship, 2866.
relation of speed to flight, paper, 4897.
seaplane, 2316, 2342, 2362.
Kretschmar, Julius:
aerodynamic calculations, 1128.
Krupp anti-aircraft gun (1876), 3769.

L

L–59, airship, 3813.
Laboratoire d'essais du Conservatoire national des arts et métiers:
propeller experiments, 2833, 2834.
Laboratories:
bibliography, 4713.
Central Aero-Hydro-Dynamic Institute, Russia, 4972.
Drzewiecki's ideas, 4778.
Göttingen University, 4651, 4890, 4987, 4988.
Institut aérodynamique de Koutchino, 4804.
Lindenberg Laboratory, 4655, 4834.
National Physical Laboratory, 4702, 4741, 4994, 5007, 5079.
Saint-Cyr, France, 4779.
United States Forest Products Laboratory, 5059.
See also Massachusetts Institute of Technology; Research; Rotch, A. Lawrence; United States Standards Bureau; United States Weather Bureau.
Lachambre, Henri:
biography, 450.
captive balloon, 3666.
hydrogen apparatus, 3637.
portrait, 450.
Lacour, Paul:
wind tunnel, 4994.
Lafayette Escadrille, 3840, 3874, 3939–3942, 4099, 4138.
Lafont, Bernard:
World war exploits, 3977.
La France, airship. *See* Renard and Krebs airship.

La Frégeolière, Renaud de:
 World war experiences, 3978.
Lahm, Frank S.:
 pioneer aviator, 3039.
Laird, E. M.:
 12 H. P. airplane, 1806.
Lake, John:
 air jet airship (1837), 1130.
La Landelle, Gabriel de:
 biography and portrait, 2568.
 helicopter, 2568.
 pioneer in heavier-than-air flight, 1941.
Lamb, Dean Ivan:
 Andes flight (1921), 3237.
Lambert, A. B.:
 biography and portrait, 157.
Lambert, Comte de:
 airplane flight (1909), 2931.
Lamberte, V.:
 balloon direction, 479.
La Mountain, John:
 American aeronaut, 648.
Lana Terzi, Francesco, 68, 110, 240, 253 (p. 429), 292a, 379, 867, 979, 1239, 1336, 1398, 1400.
 his *Prodromo,* 516, 1132, 1168.
 See also Author index.
Lancaster, I.:
 air pressure and gravity studies, 4900, 4901.
 bibliography, 5235.
 bird flight studies, 5235, 5236.
 flight theory, 4902, 4903.
Lanchester, Frederick W.:
 aerodynamics (book), 4904.
 bibliography, 5236.
 biography and portrait, 2056.
 bird flight studies, 5236.
 stability of airplanes, 1564.
Lanchester, W. N.:
 airplane flight, London-East Indies (1927), 3001.
Landing gear:
 bibliography, 1505.
Landriani, Marsilio:
 balloon ascensions (1783–1784), 662.
Landur, N.:
 engine calculations, 2063.
 heavier-than-air calculations, 2063.
Langley, Samuel Pierpont:
 aerodynamic experiments, 1478, 1894, 1914c, 2228, 2249, 4689, 4905, 4908, 4909.
 airplane, 175, 196, 205, 305, 380, 1487, 1544, 1567, 1573, 1598, 1662, 1703, 1871, 1872, 1875–1879, 1890, 1894, 1914c, 1918, 1919, 1975, 2033, 2044, 2075, 2087, 2089, 4870.
 Bell (A. G.), experiences with, 1486, 1872.
 bibliography, 1485, 4909.
 biography, 157, 1582, 1883, 1914c, 2033, 2118.
 importance of his work, 1976, 2116.
 portraits, 157, 2118.
 possibilities of airplanes, 1873, 1874.
 Tatin's airplane compared with Langley's, 1886.
 United States government, his relations with, 3785.
 See also Wright-Langley controversy; *also* Author index.
Langley, aircraft carrier, 1123.
Lanzillo, L. M.:
 airship, 161.
Laporte, Ferdinand:
 aviator, 3063.
Larre-Borges, Tadeo:
 transatlantic flight (1929), 3198.
Laskowitz, I. B.:
 sail-wheel flying machine, 2865.
Lassie, Jean Baptiste:
 rotating balloon, 1159a.
Latchinoff:
 hydrogen process, 547, 716, 3744.
Latecoere air line, France-South America, 4230.
Latham, Hubert:
 altitude record (1910), 3384.
 bibliography, 1500.
 biography, 223, 3500.
 death (1912), 3211.
 flights, 143, 1914a, 3089.
 portrait, 3510.
Latin America:
 civil aviation, post-war, 4341.
 commercial companies (1930), 4338, 4339.
 flights to (1930), 3199.
 Lindbergh's tour (1927–1928), 3079, 3255, 3258, 3259, 3322, 3356.
 temple hunting, R. A. Smith, 3442.
 United States Army flight, Panama to Guatemala (1924), 2942.
 See also names of countries.
Lattemann:
 invents double parachute, 2516, 2525.
 portrait, 2525.
Launching:
 airplanes from airships, 1036, 1365.
 airplanes from ships, 3031, 3101, 3123, 3434, 4383, 5392.
 formulas (Goupil's), 4831.
 Heinkel's catapault, 5482.
 history, 5426, 5434, 5444.
Laurus, Lauretus:
 speculations, 292a.
Laussedat, A.:
 balloon photography, 5132.
 biography and portrait, 5132.
La Vaulx, Henri de:
 airship, 977.
 bibliography, 680.
 biography, 532.
 Mediterranean balloon voyage, 511, 512, 553, 676, 679.
 propeller, 2857.
 Russian flight, 677.
Lavoisier, Antoine Laurent:
 memoirs, 4913.
Law, Ruth:
 Chicago-New York airplane flight (1916), 5359, 5373.
Lawrance, Charles L.:
 air-cooled engine development, 2757, 2758.
 engine design, 2274, 2759.
 portrait, 2274, 2741.
 "Wright" engine, 2716, 2731, 2741.
Laws and regulations. *See* Jurisprudence.
Lawson, Alfred W.:
 pioneer in commercial airplanes, 4216.
Lebaudy brothers:
 airships, 336, 898, 1018, 1019, 1022, 1086, 1122, 1138, 1159c, 1169, 1170, 1194, 1249, 1305, 1376, 1378.
 bibliography, 912.
 La Patrie, airship, loss (1907), 1148.
 Robert Lebaudy, portrait, 1281.
Leblanc, Alfred:
 aviator, 3513.
Le Bourget airport, 4260, 4300.
Le Bris:
 gliders, 253 (p. 49), 2031, 2439a, 2471.
Le Brix, Joseph. *See* Costes, Dieudonné.
Lecornu, Joseph:
 aeronautics as astronomical aid, 4914.
 cellular kite, 2653.
 theory of airplane stability, 1564.
Le Dantec, Abbé:
 air resistance study, 4684, 4915.
Lefèvre, René:
 transatlantic flight (1929), 3198, 3270, 3437a.
Legagneux, Georges:
 aviator, 3018.
Legends. *See* Literature and tradition.

Leger, M.:
helicopter, 2153, 2533, 2582, 2583, 2585.
Leggo and Company, New York. *See* Universal Aerial Navigation Company.
Lehigh Portland Cement Company:
airport competition, 4268, 4269.
Lehmann, Ernst:
experiences with Count Zeppelin, 1139.
Leinberger, L. A.:
steam-driven airship, 1033.
Leipzig, balloon, 426.
Lendenfeld, R. von:
bird flight studies, 5238.
Lennox, Comte de:
airships, 253 (p. 429), 881, 884, 1039, 1159a, 1235.
Lenormand, Sébastien:
parachute experiences (1783–1784), 803, 2490, 2502a, 2518.
Lepère-Weymann method of building standard wing sections, 2094.
Lepinte:
parachute-airplane, 2168.
Lestani, Lodovico:
flight theory, 162.
Letur:
ornithopter, 1914d.
parachute, 2490, 2502a.
Levavasseur, Léon:
biography, 2736.
engines, 2726, 2736, 2761.
portrait, 2736.
Levine, Charles A. *See* Chamberlin, Clarence D.
L'Hoste, Frédéric:
airship, 960.
ascensions, 412.
biography, 412, 420.
English Channel flight, 420, 421, 573, 690.
floating brake, 3575.
portrait, 412, 420.
Liais, Emmanuel:
bird flight, insect flight, and helicopter studies, 2063.
"Liberty" engine:
bibliography, 2689.
effect on World war operations, 2780.
first shipment to France, 2727.
Ford (Henry) builds engine, 2702.
general, 275, 2164, 2681, 2703, 2817, 2818, 3828.
Liberty engine builders' trophy race, 3083.
name copyrighted, 2808.
United States production (1918), 4041.
Liberty loan, United States:
assisted by aeronautics, 4105.
Library of Congress:
Aeronautical Division, 33, 36.
bibliography of aeronautical surveying, 5166.
Lieband, Theodore:
airship, 180.
Life saving:
kites, 2668.
life preserver for balloonists, 691.
Mississippi flood, 5516.
Van Paulsen's airplane, 2264.
See also Coast patrol.
Lighting of airways:
automatic, 5487.
Chicago-Cheyenne line, 4405.
history, 5515.
New York-Cleveland line, 4402.
Lightning:
Van Orman's investigation, 4672.
See also Balloons, thunder storms; Electricity.
Lilienthal, Gustav:
bibliography, 2435.
biography, 2366, 2402.
gliding experiments, 2423, 2424.
portrait, 2366.
Lilienthal, Otto:
appreciations:
Müllenhoff, 2446.
Wilbur Wright, 2473.
bibliography, 1485, 2436.
biography, 54, 223.
bird flight, 5239, 5240.
death, cause of, 2381, 2385.
gliders, 175, 253, 271, 372, 380, 1790, 1798, 1802, 2044, 2282, 2386, 2405, 2409, 2413, 2417, 2422, 2423, 2437, 2439a, 2456, 2467, 2471.
Platte's criticisms, 4979.
Lindbergh, Anne (Morrow), 3252, 5352a.
Lindbergh, Charles Augustus:
bibliography, 3254.
biography, 210, 2941, 2964, 2981, 3485.
carries serum to Floyd Bennett (1928), 3391.
glider license, 2466.
honors conferred on him, 3411.
inductor compass, 5554.
juvenile work about him, 212.
Langley medal, 2935.
Latin American tour (1927–1928), 3079, 3255, 3258, 3259, 3322, 3356.
New York-Paris flight, 210, 238, 257, 2969, 3033, 3060, 3079, 3150, 3160, 3209, 3253–3254, 3257, 3269, 3326, 3357, 3373, 3483, 5067.
North Atlantic flight (1933), 3252, 3256, 3336a, 3498.
parachute jumps, 2529.
Pueblo and Maya survey, 5113.
scrapbook (1927–1928), 3327.
Tokyo flight (1931), 3200, 3219, 5352a.
Transcontinental Air Transport Company, 3009.
United States tour (1927), 3079, 3225.
Lindbergh, Charles Augustus, sr.:
biography, 3178.
Lindenberg, Prussia, Aeronautical Observatory:
balloon and kite work, 4655, 4834.
Lindsay, Orrin:
magnetized alloy (1847), 1900.
Lippert, P. W.:
flight theory, 4918.
Literature and tradition:
bibliography:
Italy, 6.
Tissandier, 32.
Venturini, 35.
chapters by Cavallo, 499; Hodgson, 253; Magoun and Hodgkins, 292a; Turner, 372; Turnor, 373; Vivian, 380.
French fiction, 704, 707.
general, 110, 373, 379, 385, 1941.
Germany, 1413.
See also Alexander the Great; Animal forms; Archytas; Ariosto; Babylonia; Bergerac; Bible; Bladud; "Celeste"; China; Dante Alighieri; Egypt; Etana; Faustus; "Flying architect"; French monks of Siam; Garvani; Giotto's carving; Goethe; Greece; Hoaxes; Ibn Firnis; Icaromenippus; Icarus and Daedalus; India; Java; "John Daniel"; "Man in the moone"; Mohr; "Monck Mason's flight"; Moslems; Oliver of Malmesbury; Ovid; Persia; Pfall; Poetry; "Rasselas"; Rome; Rousseau; St. Joseph of Copertino; Saracen; Scandinavia; Schweikert the miller; Sheba, Queen of; Simon the magician; Tasso; Wagner, Christopher; Wieland's shirt of feathers.
Little, Ray:
early experiences in U. S. postal service, 3261.
Livingstone, John H.:
wins National Air Tour (1929), 3262.
Lockheed airplanes, 1829.
Locklear, James:
"stunt" aviator, 3180.

Loening, Grover:
bibliography, 2343.
experiences in American aviation, 1901.
seaplane, 2344.
Lössl, Friedrich, Ritter von:
air resistance measurement, 1454, 4922, 4924–4926, 4928.
airship lifting power, 4981.
bibliography, 4928.
biography and portrait, 4921.
researches, history, 5069.
study of thin plates, 4919, 4920, 4923, 4983.
theory of flight, 4927, 4964.
Löwenstein-Wertheim, Princess:
aviator, 3121.
Lohmeier, Philipp:
airship idea similar to Lana's, 240, 1411.
Lomet, Antoine F.:
biography and portrait, 5136.
use of balloons for surveying (1801), 5136.
London, air raids, World war, 3814, 3822, 3929, 4066, 4637a.
London-Manchester airplane race (1910), 380, 1728, 1914a, 3116, 3183, 3267, 3268.
London-Paris airlines:
history, 4215.
Loop-the-loop:
bibliography, 2957.
See also Pégoud.
Loraine, Robert:
biography and portrait, 2056.
Lorenzen, C.:
helicopter experiments, 2586a.
Loriga Taboada, J. *See* Gallarza, E. G.
Lorraine, France:
early experiences in, 485, 763.
Los Angeles, airship, 210, 1245, 1285, 1309, 1332, 1406, 1422.
Los Angeles aviation meeting (1910), 2919, 3351.
Lost battalion, World war, 4073.
Loth, Arthur William:
biography and portrait, 5419.
magnetic aircraft guide, 5491.
Lotti, Armeno:
transatlantic flight (1929), 3198, 3270, 3437a.
Loup, Michael:
ornithopter, 1914d.
Lourenco de Gusmão, Bartholomeo:
airship, 68, 110, 207, 240, 253 (p. 427), 292a, 379, 1032, 1049, 1110, 1124, 1125, 1141, 1186, 1239, 1398, 1399, 1401.
bibliography, 1401.
correspondence with Elisabeth von Braunschweig-Blankenburg, 1057.
satirical poems about him, 979a.
Louvrié, Charles de:
airplane experiments, 2078.
bibliography, 5246.
bird flight studies, 5242–5246.
Lowe, Thaddeus S. C.:
balloon experiences in American Civil war, 697a, 3568, 3643.
Lubricants:
bibliography, 2690.
developments, 189.
Lucanus, Friedrich von:
bird flight studies, 5247.
Lucas, F. A.:
bird flight studies, 5248.
Luckey, W. S.:
wins race around Manhattan island (1913), 3066, 3113.
Lucy, de:
flight theory, 1541, 2023, 5249.
Ludlow, Israel:
airplanes, 1906–1908.
biography, 157.
Lütgendorf, Joseph Maximilian:
first German aeronaut (1786), 684.
Luft-Hansa. *See* Deutsche Luft-Hansa.
Luke, Frank:
military aviator, 3943, 4043.
Lunardi, Vincent:
balloon ascensions, 253 (p. 429–430), 373, 406, 478, 479, 569, 697b, 727, 808, 821.
first aerial voyage in England (1784), 603, 695, 838.
first ascension in Italy (1789), 772.
five voyages in Scotland, 696.
oars for propelling balloons, 1159d.
portraits, 663, 727, 746.
Lund, Fred:
"stunt" aviator, 3271.
Luntley, John:
rotary balloon, 253 (p. 430), 952.
Luyties, Otto G.:
helicopter, 2586.

M

"M" wing sections, 1987.
M-3, airship, 987.
McAdie, Alexander:
kite experiments, 2640.
McClure, Ivor H.:
European flight (1927), 3272.
McConnell, James:
World war experiences, 3990.
McCoy, James Comly:
pioneer American balloonist, 504.
McCudden, James T. B.:
experiences in Royal Flying Corps, World war, 3991.
McElroy, C. L.:
experiences in Mexican jungle (1932), 3273a.
Maceroni, F.:
claims to have discovered airplane principle (1824), 1911.
parachutes, 2500.
Macfarlane, Irene:
parachute experience, 3303.
Macfie, Robert F.:
biography and portrait, 2056.
Mach, Ludwig:
photography of stream lines, 4934.
wind tunnel, 4994.
Machine guns, 3526, 3999, 4011.
MacKensie-Kennedy, J. H. C.:
biography and portrait, 2131.
Mackenthun, Walter:
biography and portrait, 291.
Mackenzie, George:
improved parachute (1837), 2501.
MacKenzie-Grieve, Kenneth:
portrait, 3300a, 3473.
transatlantic flight with H. Hawker (1919), 257, 3060, 3087, 3184, 3473.
McLaren, A. S. C. S.:
attempts world flight (1924), 3274.
McLennan, J. C.:
helium development, 547a.
Macmillan, Donald:
Polar airplane flight (1925), 210.
Macmillan, Norman:
World war experiences, 3992.
Macready, John A.:
altitude flights, *1921,* 3321; *1926,* 3096, 3276.
non-stop trans-American flight (1923), 210, 257, 2916, 3088, 3204.
parachute experiences, 2499, 2502.
portrait, 3300a.
supercharger experiences, 2768.
MacRobertson. *See* Robertson, Sir Macpherson.

MacSweeny, Joseph:
balloon direction (1844), 644a, 1159, 1159d.
Macy, Alfred J.:
airplane stabilizer patents, 1692.
Madelung, George H.:
builds world's record glider, 2397.
portrait, 2397.
Mager, Henri:
airship, 1100.
Magnetic observations, aerial:
history, 4785.
Magnetos:
"Scintilla," 2745.
World war, 3945.
Magyr, Alexander:
New York-Hungary flight (1931), 3200.
Maistre, Xavier de:
balloon ascension (1784), 416, 826.
Maitland, E. M.:
biography and portrait, 2056.
Maitland, Lester J.:
Hawaiian flight (1927), 2925, 3079, 3279, 3476.
portrait, 3079.
Maloney, Dan:
assistant to J. J. Montgomery, 2411.
Mammoth, balloon, 697a.
Man in the moone (Godwin), 77.
Mangin, Gabriel:
French aeronaut and inventor, 706.
Mangin, Théodore:
experiences in siege of Paris, 3681.
Mangon, Hervé:
balloon studies, 4937.
biography, 4863.
Mangot, Joseph:
biography, 420.
English Channel balloon flight, 420, 421, 573, 690.
portrait, 420.
Manly, Charles Matthews:
builds Langley's engine, 1632, 2172.
experiences with Langley, 1914c.
his history of aviation (1917), 1918.
portrait, 1632.
Mannesmann, O.:
air resistance studies, 4938.
Manning, W. O.:
biography and portrait, 2056.
Mansfield, Charles Blachford:
studies in airship construction (1851), 1162.
Manufacturers Aircraft Association, 308, 4566, 4567.
Map of famous flights, 239.
Maps. *See* Surveying and mapping.
Marçay:
automatic airplane, 1440.
Marey, Étienne Jules:
animal flight, 5204, 5291, 5250–5263.
bibliography, 5263.
wind tunnel, 4994.
Marey-Monge, Edmond:
airship construction (1847), 1163.
copper balloon, 510, 867.
Necklace, airship (1783), 1159d.
Marie, Eugène Armand:
Paris-Congo flight (1928), 3282.
Marquis, Raoul:
aeronautical tales, 707.
balloon ascensions, 707, 708.
See also Author index.
Marriott, Frederick:
airplane (c. 1842), 2021.
Marsh and Ranwell:
airship (1840), 1167.
Marshall, J.:
ornithopter experiments, 1931.
Martienssen, O.:
flight theory, 4943.
Martin, Glenn L.:
bibliography, 1499.
biography, 1723, 3464.
history of his factories, 1932.
portrait, 3464.
Martin, James V.:
airplane stabilizer, 1933.
Martin, Lajos:
ornithopter, 1934.
Martin and Handasyde. *See* "Martinsyde" airplanes.
"Martinsyde" airplanes:
history, 1935.
Martyn, Thomas:
references to his airships and parachutes, 253 (p. 430), 2502a.
Marvin, C. F.:
bibliography, 2657.
kites, 2656–2657.
Marvingt, Marie:
balloon experiences, 5314, 5354.
portrait, 5314.
Mason, Thomas Monck:
balloon ascensions and ellipsoidal balloon, 253 (p. 430).
model airship (1842), 1159a, 1184.
portrait, 711.
theory of aeronautics, 711.
voyage in *Nassau* balloon (1836), 572.
Massachusetts Institute of Technology:
bibliography, 4703.
work, 4654, 4944.
Mattern, James:
non-stop New York-Berlin flight (1932), 3200a, 3336.
Matthews, Grindell:
invents death ray, 5518.
Mattison, G. C.:
aerial survey of the Mississippi delta, 5165.
Maughan, Russell L.:
trans-American flight (1924), 210, 3433.
wins Pulitzer race (1922), 3349.
Maxim, Sir Hiram S.:
airplane, 176, 196, 207, 305, 380, 1605, 1616, 1733, 1789, 1835, 1914b, 1943–1955, 2097, 2098, 2158, 2201.
bibliography, 1954.
biography and portrait, 157.
power required for airplane, 1944.
propeller, 2872, 2873.
wind tunnel, 4994.
See also Author index.
Maxim, Hudson:
bibliography, 1501.
biography and portrait, 157.
Maybach, Wilhelm:
biography and portrait, 2827.
"Maybach" engine, 2701, 2827.
Mayer, David:
helicopter (1828), 2586a.
Mayfly, airship, 1159b.
Means, James:
bibliography, 1956.
glider models, 2443.
kites, 2658.
Mears, John H.:
world flight (1928), 3285.
Mechanics who have become famous, 372.
Medicine, aviation:
Australia's flying doctors, 5505.
bibliography, 4687, 4688.
health relief work, 5504.
history, 5481.
physiological effects, 729, 4715, 4726, 4880, 5517.
United States Army statistics, 5556.
See also Ambulances.

Mediterranean sea flights:
Bruno's balloon (1890), 476.
Comte de la Vaulx (1902), 511, 512, 553, 676, 679, 748a.
Garros (1913), 3286.
Mittelholzer (1928), 3073.
Samson (1921–1922), 2354.
Meerwein, Karl Friedrich:
ornithopter, 1684, 1914d, 2243.
quotations from paper (1784), 1960.
Meisinger, Clarence Le Roy:
portrait, 4758.
upper air investigations, 4758.
Meissel, E.:
balloon for geographical work, 715.
Mélikoff:
helicopter, 2586a, 2588.
Meltzl, David:
balloon direction (1824), 1176.
Mendeleyeyv, D.:
air and liquid resistance, 4945.
Mendell, Loren:
refueling flight (1929), 3338, 3481.
Menicali, Abdon:
biography and flying machines, 1529.
Menier:
warm air balloon, 723.
"Mercedes" engine, 2691, 2701, 2708.
Merrill, A. A.:
"fool proof" airplane, 1691.
wing curve experiments, 1963.
Merritt, F. R.:
airship, 180.
Metal construction:
airplanes, history, 1834, 1870.
airships, 1055, 1179, 1369.
bibliography, 1369.
Boeing Airplane Co., 1574.
Buonaccorsi di Pistoja's airship, 1178.
duralumin, 2328, 2879, 2891.
early projects, 620, 867, 1369.
Ford buys Stout Metal Airplane Co., 1783.
Hamilton monoplane, 1845.
Irish, airplane, 1826.
Junkers, airplanes, 1839.
Lepère-Weymann wing section, 2094.
magnesium-aluminium propeller, 2846.
metal tubes, 1705.
propellers, 2869, 2875, 2887a, 2879, 2891.
seaplanes, 2363.
Short brothers metal hull, 2328.
Torres, airship, 1177.
Tziolkovski, airship, 1366.
Unger, steel airship, 956.
"ZMC" all metal airship, 984, 1026, 1054.
See also Aluminium; Copper; Iron; Lana; Schwarz; Steel construction.
Meteorology:
Angerstein's work, 4722, 5497.
balloons:
equipment, Hervés, 4862.
history, 4950.
bibliography, 4698, 5047.
general history, 4837, 4838, 5033, 5034, 5047, 5061.
Germany, 5046.
International Commission for the Investigation of the Upper Air, 4559, 4857.
Kremser's investigations, 5043.
Lavoisier's memoirs, 4913.
Meisinger's studies, 4758.
post-World war developments, 4930.
Teisserenc de Bort's studies, 4712, 4932, 4957, 5004, 5052.
transatlantic flights, 4839, 4887, 4888, 4993.
United States Army, 2164, 4682, 4847, 4853, 4947.
United States Navy, Hawaiian flight (1925), 4996.
Meteorology, *continued*
upper air investigations, 4747, 4758, 4800, 4827, 4849, 4858, 4860, 4886, 4896, 4958, 4982, 5001, 5008, 5009, 5012–5015, 5033, 5034, 5052, 5072.
World war, 4137, 4930.
See also Ballooning, scientific; Balloons, sounding; Kites; Rotch, A. Lawrence; Stratosphere: United States Weather Bureau; Wind.
Metiviev:
offers airship during American Civil war, 1052.
Meudon:
airship, 872.
Meudon aeronautical school, 825, 5529.
Meusnier, J. B. M.:
account of writings and of famous atlas, 1142.
airship, 253 (p. 430), 549 (p. 139–147), 1159d, 1352, 1379.
ballonets, 549 (p. 144), 840.
biography, 549 (p. 140).
memoir on aerostatic equilibrium, 1074, 1180, 5064, 5065.
propeller, 2867, 2881.
Mewes, Rudolf:
bibliography, 4718.
bird flight studies, 5266.
Mexico:
aerial survey, 5154.
McElroy's jungle experiences, 3273a.
Villa's rebellion, 3609.
Micciollo, Alfred:
airship, 1181.
Michelin, André:
biography and portrait, 3062.
Michelin prize:
bibliography, 2958.
history, 3509.
won by Henri Farman (1909), 3053.
Miethe, Adolf:
biography and portrait, 291.
Military aeronautics:
air power policy, 3562, 3701, 3732.
airships, 1094, 1144.
See also Zeppelin.
Austro-Hungary, 3649, 3714.
bibliography, 3558, 3642.
chronology, 3530, 3654.
Denmark, 3615.
effect on aeronautical development, 185.
Europe:
maneuvers (1912), 3546.
post-World war, 3736.
See also World war.
France:
air policy (post-war), 3732.
airplanes, post-World war, 4126.
developments, *1909–1910*, 3682; *1910*, 3734; *1912*, 3549; *1913*, 3555.
general history, 549, 803, 821, 3712, 3716.
Meudon school, 825, 5529.
pilot instruction (1928), 5427.
revolutionary wars, 605, 3608, 3668.
See also Balloons, military; Franco-Chinese war; Franco-Moroccan war; Franco-Prussian war; Napoleon I; World war.
general history, 208, 209, 247, 255, 257, 294, 297, 568, 3580, 3586, 3590, 3604, 3629, 3642, 3648, 3742, 3749, 3803, 3806.
Germany:
air policy (post-war), 3732.
airships (1908–1909), 3527.
developments (1935), 3807a.
history, 3578, 3580, 3641, 3655.
Prussian army aeronautical branch established (1883), 3578.
See also World war.

Military aeronautics, *continued*
Great Britain:
air policy, 3636, 3732, 3966.
airship policy, 1325, 1326.
airships in World war, 3998.
developments (1932), 3722.
early balloon corps, 3592, 3617.
history, 3624, 3737, 3738.
military airships, 3602.
military ballooning, 3659, 3744, 3794.
rebellions, subjugation, 3665, 3808.
Royal Aircraft Factory, 2107, 2749, 3837, 4741.
Templer's balloon experiences, 3592, 3693, 3762.
Woolwich arsenal, 487, 3617, 3690, 3744.
See also Royal Air Force; World war.
Italy:
general, 3732, 3743, 4062a.
See also World war.
Japan:
aircraft carriers, 3719.
captive balloons, 3601.
history, 3599, 3704.
Russo-Japanese war, 3623.
Shanghai bombarded (1932), 3760.
Peru, 4251.
possibilities, 295, 3550, 3586, 3616, 3664, 3667, 3684, 3971, 4032.
progress, *1928,* 3607; *1930,* 272; *1932,* 178.
reconnaissance, 3789, 4062.
Russia, 133, 145, 3522, 3540, 3705, 3732.
status in various countries, *1887,* 3766; *1894,* 3638; *1908,* 3751; *1911,* 3635; *1913,* 3555; *1915,* 3730; *1929,* 3695; *1930,* 3694.
United States:
airplane guns, 3795.
appropriations, 3533, 3538, 3913, 4118.
backward condition, 3534, 3670, 3787, 4001.
Congressional inquiries, 4111, 4112, 4118, 4120, 4121.
contracts, 3669, 4110.
Coolidge air policy, 3696.
developments, *1918–1923,* 3777; *1931,* 127; *post-war,* 3791.
general history, 3645, 3806, 4093.
Mitchell (Gen.) attacks policies, 3531, 3697, 3698, 3701–3703, 3729, 4123.
Nicaragua operations (1927), 3567, 3740.
President's Aircraft Board, 4123a, 4124.
statistics of operations, expenditures, etc., 3532, 3533.
surplus engines and airplanes after World war, 4283.
See also Civil war, America; Spanish-American war; United States army; World war.
See also Airships, military; Balloons, military; Civil war, America; Machine guns; Naval aeronautics; World war.
Milla, Karl:
bird flight studies, 5267.
Miller, W.:
ornithopter, 1914d, 1971.
Miller von Hauenfels, Albert R.:
bibliography, 4704.
biography, 4980.
bird flight studies, 5268.
Minckelers, J. P.:
first to inflate balloon with coal gas (1783), 797.
Mineola Army Aviation School, 4051.
Minerva, balloon, 697a, 781.
Mining:
aerial transportation, 4166, 4178, 4199, 4211, 4233, 4280, 4302, 4342, 5394a, 5397a, 5423a, 5442, 5448a, 5456a, 5474a, 5532a, 5558a.
Minney, Rubeigh J.:
trans-India flight (1920), 3287.
Miolan and Janinet:
airship with jet propulsion (1784), 1159d.
Miss Veedol, airplane:
world flight (1931), 3200.
Mitchell, William:
attacks United States military policy (1925), 3531, 3697, 3698, 3701–3703, 3729, 4123a.
flies 224.38 miles an hour (1922), 3133.
his book *Winged defense,* 3701.
See also Warship bombings.
Mittelholzer, Walter, photographer:
flights:
Abyssinian flight (1934), 3289a.
Alpine, 3293.
Central Africa (1927), 3290.
Mediterranean (1928), 3073.
Persia (1924–1925), 3291.
Sahara desert (1930–1931), 3292.
Zürich-Cape Town (1926), 3294, 3295.
Models:
bibliography, 1510.
Moedebeck, Hermann:
balloons for scientific purposes, 4949, 4950.
bibliography, 310.
biography, 157, 279, 353, 5457.
portraits, 157, 4570, 5457.
Moffett, William Adger:
biography, 854, 3524.
portrait, 3524.
Mohr, Caspar:
German "flying monk," 108.
Moisant, John B.:
aviator, 2992, 3300, 3330.
bibliography, 3300.
death (1910), 3471.
portraits, 3300, 3330, 3471.
Moisant, Matilde:
aviator, 3464, 5368.
Mollison, Amy Johnson:
Australian flight (1930), 5325, 5355.
Cape of Good Hope flight (1932), 3200a, 3324, 5360.
portrait, 3300a.
See also Mollison, J. A.
Mollison, J. A.:
flights:
Australia-England (1931), 3301.
Cape of Good Hope (1932), 3301.
England-Rio de Janeiro (1933), 3077.
Ireland-America, with his wife (1932), 3119, 3200a, 3301, 3302.
"Monck Mason's flight" (Poe), 96.
Money, John:
balloon ascensions and his treatise on military balloons, 253 (p. 430).
Monge. *See* Marey-Monge.
Monospar:
developed by H. J. Stieger, 1979.
Montgolfier, Étienne:
biography, 539, 540, 681.
Montgolfier, Joseph:
biography, 539, 540, 681.
parachute, 2502a, 2517.
portrait, 821.
Montgolfier brothers:
Académie des sciences, report to, 647, 777.
balloons and ascensions, 208, 253 (p. 430–431), 400, 407, 539, 557, 558, 563, 589, 590, 634, 635, 645, 681, 688, 689, 697c, 712, 725, 752, 788, 803.
bibliography, 462.
chapters in Cavallo's work, 500.
monument, 497.
references to, 373 (p. 523).
See also Montgolfier, Étienne; Montgolfier, Joseph.
Montgolfier's gas, 292 (p. 25).

Montgomery, John J.:
bibliography, 2444.
biography, 2382, 2410.
gliders, 380, 2439a, 2444, 2471.
1000 meter altitude glider flight (1905), 2387.
wing warping, 1893, 1981, 2364, 2382.
See also Maloney, Dan.
Moore, Beecher:
rocket propelled flying machine, 180.
Moore, R. F.:
flight experiments, 4953.
Moore-Brabazon, J. T. C.:
biography, 2056.
officially made the first airplane flight in Great Britain, 3069, 3304.
portraits, 2056, 3304.
Mooring devices:
balloons (1837), 644.
evolution of airship moorings, 1173, 1314.
Mora, Alfred:
flight theory, 4954.
Morènes, Ramon de:
bird flight studies, 2063.
Moretti, Virgilio:
flight theory, 162.
Morin, Arthur:
balloon direction, 4955.
Morrell, John A.:
airship, 1068.
Mort, George Frederick:
engine designer, 2709.
Moscow-New York airplane flight (1929), 3198.
Moslems:
general, 76.
Ibn Firnis, 39, 39a, 56, 76.
Mosment:
fatal balloon ascension with horse and parachute (1806), 537, 2502a.
Mosquito pest. *See* Insects.
Motion pictures:
developing, 5128.
flying movies, 5562.
"Squadron of death," 3916.
use in aviation, 5490.
Mott, S. D.:
bird flight studies, 5269.
helicopter, 2590.
Mouillard, Louis Pierre:
bird flight investigations, 176, 5182, 5226, 5241, 5270–5272.
glider construction, 1986, 2471, 5183.
monument, 5241.
Wilbur Wright's opinion of him, 2285.
wing warping and vertical rudder, 1782, 4939.
Mount Blanc. *See* Alps.
Mount Carmel Aerial Navigation Co., 1136.
Mount Everest:
Cobham's flight (1925), 2914.
Halliburton's flight (1932), 3181.
Houston expedition (1933), 2965, 3070, 3084, 3118, 3210, 3230, 3303, 3336a, 3354, 3429.
photographic work of Houston expedition, 5147.
Mount Weather, Va.:
meteorological kites, 2615.
Mount Whitney:
Christofferson's flight (1914), 3427.
Mountain flights:
dangers, 167.
Mourmelon, France:
aviation school, 5484.
Moy, Thomas:
bibliography, 1516.
heavier-than-air experiments, 253 (p. 431), 1914b, 1941, 2031.
propeller, 2872, 2877.
Müllenhoff, Karl:
bibliography, 5276.
bird and insect studies, 5273–5276.
Munk, Max Michael:
bibliography, 1509.
theoretical studies, 4663.
Murillo, Abraham, jr.:
parachute experiences, 2510.
Murray:
parachute, 58, 253 (p. 325).
Muscular power:
Hooke's ideas, 4868.
Müllenhoff's studies, 5273.
Strasser's studies, 4898.
See also Bird flight; Insect flight.
Music:
airman's hymn, 5540.
Landauer's collection, 5499.
Musschenbroek. *See* van Musschenbroek.
Mussolini, Benito:
directs Italian commercial aviation, 4265.
Mustin, Henry C.:
"Father of naval aviation," 3583.
Muzzi, Muzio:
balloon direction, 1222.
Myers, Carl E.:
aerial torpedo, 3748.
airship, 180, 315, 1204.
balloon farm, 753.
balloon shapes, 1205.
bibliography, 735.
captive balloons, 774.
natural gas for balloons, 734.
opinion of the airplane, 735.
rain-making, 315.
Mystery S, airplane, 1480.
Mytton, R. G. V.:
gasoline driven airplane (1903), 1991.

N

"N C." seaplanes, 210, 257, 3060, 3087, 3148, 3196, 3395, 3474, 3482a, 3499a.
Nadar. *See* Tournachon, Félix.
"Napier" engine, 2769, 2816.
Napoleon I.:
balloons in Egypt, 373 (p. 123), 3786, 3801.
relations with Garnerin, 502.
Nassau, balloon, 697a, 717, 738.
National Aeronautic Association, 4571, 4580.
National Aeronautic Conference (1927), 4532.
National Air Derby, 3314.
National Air Races:
history, 2963, 3507.
results, *1924*, 3348; *1926*, 3310, 3454; *1927*, 3314, 3333, 3343; *1928*, 3311; *1929*, 3034, 3317; *1931*, 3297; *1932*, 3385, 3515; *1934*, 3334.
National Air Tour:
results, *1925*, 3390, 3392; *1926*, 2940, 3139; *1927*, 3129, 3453; *1928*, 3312, 3318; *1929*, 3262, 3316; *1930*, 3319; *1931*, 3298.
National Air Traffic Conference (1929), 4572, 4573.
National Air Transport, Inc.:
operation of Chicago-Dallas air mail, 4282, 4365, 4422.
National Airport Conference (1929), 4548.
National Airship Company of California, 945.
National balloon race:
results, *1913*, 770; *1922*, 559; *1923*, 424; *1924*, 835, 3465; *1925*, 740; *1926*, 739.
National Geographic Society:
contributions to aeronautics, 4526.
National Physical Laboratory:
balloon fabrics report, 5007.
bibliography, 4702.
Busk, E. T., work of, 4741.
contributions, 5079.
wind tunnel, 4994.

Natural gas for balloons:
experiences of C. E. Myers, 734.
Naval aeronautics:
bibliography, 3559.
Denmark, 3615.
France (post-war), 3653.
history, 3807.
Italy, 3961.
Liberty motor, effect on, 2780.
Mustin, Henry C., 3583.
progress in various countries, 3555, 3571, 3688, 3694, 3726, 4151.
Santo Domingo, 3644.
seaplanes (1911–1926), 2319.
United States, 3595, 3651, 3663, 4106.
appropriations, 3538, 3807.
See also Royal Naval Air Force; United States Navy; World war.
Navarre, Jean:
military aviator, 4021.
Navier, Claude L. M. H.:
bird flight study, 5242.
formulas, 4674, 4960.
Navigation:
bibliography, 4717.
Nees, Christian:
bird flight studies, 5279.
Nélis, Georges:
pioneer Belgian aviator, 3149.
Némethy, E.:
airplanes, 1727, 1727a, 1998.
Netherlands:
airplanes, history, 5062.
civil aviation, 4291, 4341.
research, history, 5062.
Royal Air Traffic Co., 4319.
Royal Dutch Indian Airways, 4267.
New England:
aviators in World war, 4035.
history, 194.
New York City:
aerial surveys, 5114, 5140.
Durant's ascension (1833) from Castle Garden, 542.
exhibitions, 5524.
Knabenshue's airship, 1158.
Robertson's ascensions, 408, 782.
sham attack (1929), 3757.
Newell, Hugh:
airship, 1015.
Newspapers:
delivery, 4162.
reporting, 5532.
Newton, Sir Isaac:
kite experiments, 2609.
Nicaragua:
aerial reconnoitering (1927), 3567, 3740.
Nicholl, Vincent:
military aviator, 3938.
Nichols, Ruth:
altitude record, 5374.
biography and portrait, 5369.
Niéport, Édouard de:
biography and portrait, 2067.
Nieuport airplane:
bibliography, 1511.
Night flying:
air mail, 4369, 4378, 4380, 4390.
bibliography, 2959.
commercial development, 4212, 4239, 4333.
development (1923), 275.
first flight (1910), 3278.
night photography, 5173.
surveying with search lights, 5145.
Nikel, Hugo L.:
registering kite, 2667.
Nimfuhr, Raimund:
ornithopter, 2018.
"99" Club (women pilots), 5317.
Nobile, Umberto:
Polar flight in airship *Italia* (1928), 4437, 4441, 4464, 4468, 4477, 4478, 4482, 4488, 4497, 4509, 4630.
Noble, Walter:
World war experiences, 4037.
Noise prevention:
British government experiments, 5431.
researches (1928), 2146.
"Romino" device (1926), 1843.
United States Standards Bureau, 2091, 2265.
Nomenclature:
bibliography, 5407.
derivation of the word aeroplane, 1534, 5574.
German words, 5437.
See also Slang.
Nordhoff, Charles Bernhard:
World war experiences, 4038.
Norge, airship. *See* Amundsen, Roald.
Northwest Airways, 4286.
Norway:
modern aeronautics, 130.
See also Scandinavia.
Noville, George O.:
aviator, 3045, 3079.
Nulli secundus, airship, 976, 1367.
Nungesser, Charles:
attempts transatlantic flight (1927), 3339.
portrait, 3300a.
Nye, James:
airship with rocket propulsion (1852), 1159a.

O

Oakes, E.:
airship, 1218.
O'Brine, Forest:
refueling flight (1929), 2918, 3198, 3199, 3239, 3412.
Oehmichen, Étienne:
helicopter, 2562, 2586a, 2591, 2597.
Oelerich, Reinhold:
altitude airplane flight (1914), 3342.
Ogal, Selim:
mammoth balloon (1825), 606.
Ogilvie, A.:
British aviation pioneer, 2056.
Old Glory, airplane:
lost at sea (1927), 3121.
Oliver of Malmesbury, 56, 58, 83, 109.
Olivier, Arsène:
wing regulated airship, 904.
Olmstead, R. S.:
wins National balloon race (1923), 424.
Olympia exhibitions, London, 251, 371, 1823, 5465, 5549.
O'Malley, Frank Ward:
recollections of early Wright flying days, 2026.
Omri:
balloon direction, 1219.
Opel, Fritz von:
rocket propulsion, 1779, 1780, 2029.
Opitz, Richard:
German balloonist, 669.
Oregon:
airports, 4175.
fire patrol, 5404, 5537.
seeding experiments, 5550.
Orlebar, Augustus H.:
experiences in Schneider Cup Races, 3350.
flies 357.7 miles an hour (1929), 3198.
Ornen, balloon, 697a.

Ornithopters:
advantages (1893), 1903.
bibliography, 1494.
Cambridge University experiments, 2263.
Fitzgerald's theory, 4797.
general, 295, 1914d, 2044.
Wenham's theory, 380, 1914d.
See also: Artingstall; Baldwin (W. H.); Besnier; Booth; Bower; Cornelius; Coulomb; Dandrieux; Degen; Devèze; Esterno; Étevé; Frost; Gammeter; Groof; Hargrave; Harriman; Hureau de Villeneuve; Ijserman; Jobert; Kaufmann (J. N.); Kosch; Kress; Letur; Loup; Martin (L.); Marshall; Meerwein; Miller (W.); Nimfuhr; Phillips (J. S.); Pompéien-Piraud; Quartermain; Quimby (W. F.); Roux; Samuelson; Spencer (C.); Stenzel; Struvé; Trouvé; Twining; Walker; Warshavsky; Wilkins (John); Zimmerman.
Ossari:
airship (1848), 161.
Otto gas-engine, 2756.
Ovid:
story of Phaeton, 373 (p. 9).
Ovington, Earl L.:
experiences, 3355, 4407.
makes first U. S. mail flight (1911), 4407.
Oxford University Arctic Expedition, 4446.
Oxygen:
use at high altitudes, 5429, 5489, 5563.

P

Pacific coast:
early aviators, 2939.
Pacific flights. *See* Australia; Hawaiian islands; World flights.
"Packard" engine:
bibliography, 2693.
Diesel type, 2824.
Paddle-wheel flying machines. *See* Laskowitz; Platt (H. H.); Wellner.
Pan American Aeronautic exhibition (1917), 5573.
Pan American Aeronautic Federation:
history, 4295.
Pan American Airways:
history, 4271, 4330.
Pan American Convention for Air Navigation (1928), 4646.
Pan American flight (1927), 3079.
Pangborn, Clyde:
world and first non-stop Pacific flight (1931), 3200.
"Panhard-Levassor" engine, 2728.
Parachutes:
airplane carried by parachute (1926), 2514.
backward development (1889), 2509.
bibliography, 2478.
disasters, 2497.
double parachute, 2516.
early references, 373 (p. 523).
French types (1929), 2485.
general, 221, 245, 247, 253 (p. 432), 639, 774a, 803, 849, 1756, 2044, 2476, 2484, 2487, 2488, 2490, 2494–2497, 2502a, 2503, 2507, 2508, 2511, 2521, 2523, 5324.
Hagen, theory, 2490.
kite-type, 2506.
Myers condemns parachute, 733.
progress (1931), 2101.
Prohaska, theory, 2519.
steering, 2479, 2481.
United States, 2498, 2499, 2528.
United States Army (1928), 2532.
women parachutists, 5339.
World war, 4036.
Parachutes, *continued*
See also Arnold; Baldwin (T. S.); Blanchard (J. P.); Bournonville; Calthorpe; Capazza; Carelli; Caterpillar Club; Cocking; Drouet; Garnerin (A. J.); Garnerin (Elise); Hampton; Harris (H. R.); Hengler; Hoffman (E. L.); Lenormand; Lepinte; Letur; Lindbergh; Maceroni; Macfarlane (Irene); Mackenzie (G.); Macready; Martyn; Mosment; Montgolfier (Joseph); Murillo; Sala; Veranzio; Vinci; White (Bert); Wilde; Wise (John).
Paris:
air raids, World war, 3814, 3864, 4060.
See also Franco-Prussian war.
Paris Aeronautical Salons, 244, 371, 1439, 1677, 5467.
Paris exhibitions, 151, 446, 817, 5470.
See also Paris Aeronautical Salons.
Parker, Cozette:
aviator, experiences, 5366.
Parmelee, Philip:
aviator, 3365, 3464.
Parmelin, Agenor:
first aviator to fly over Mt. Blanc (1914), 3358.
Parseval, A. von:
airships, 196, 1076, 1115, 1159c, 1228, 1291, 3527.
bibliography, 152.
biography, 291.
bird flight studies, 5280.
criticizes Wellner's airship, 1230.
gliders, 2453.
kite balloon, 761, 762, 814.
portraits, 291, 1228.
Parsons, Charles A.:
steam-driven airplane model (1893), 2177, 2785.
Partridge, J. S.:
steam-driven airship (1843), 1231.
Paschius, Georg:
reputed flight in 17th century, 68.
Patents:
airships, 1325, 1326.
American, 1997.
British, 166, 225, 1997.
French, 165, 215, 342.
German, 126.
See also Curtiss (Glenn H.); Wright brothers.
Pathfinder, airplane. *See* Williams, Roger Q.
Patrick, Mason M.:
biography and portrait, 3626.
Patrie, airship. *See* Lebaudy brothers.
Paulhan, Louis:
altitude record (1910), 3384.
appreciation, 3183.
bibliography, 1508.
biography, 223, 3054.
London-Manchester race (1910), 380, 1728, 1914a, 3116, 3183, 3267, 3268.
portrait, 3054.
Wright-Paulhan conflict (1910), 1909, 2268, 2294.
Paulus, Käthchen:
balloonist, 5376.
Pauly, Samuel John:
airships, 1159d.
Pax, airship. *See* Severo.
Payn, Lucien:
balloon inclined railway, 440.
Peal, S. E.:
bird flight studies, 5281.
Pegelius, Magnus:
his *Thesaurus rerum selectarum* (1604), 240.
Pégoud, Adolphe:
bibliography, 3361.
biography, 2971, 2972, 2992.
death (1915), 4057.
experiences, 2968.
loop-the-loop, 2911, 2929, 2969, 3085, 3361–3363.
portraits, 2971, 2972.

Pelletier Doisy, G. E. C. M.:
Paris-Tokio flight (1924), 3137, 3364.
Peltièr, Thérèse:
aviator, 5322.
Pénaud, Alphonse:
airplane patent (1876), 2045.
bibliography, 5284.
bird flight studies, 5282–5284.
gliders, 2454.
heavier-than-air investigations, 253 (p. 432), 4916, 4969.
helicopters, 2586a, 4969.
model motors, 2786.
scientific balloon ascensions, 4667.
Pennington:
airship, 1136, 1226, 1236.
Periodicals:
The *Aeroplane,* 5476.
pioneer English, 371.
Perkins' engine, 2777, 5307.
Persia:
civil aviation, 4190.
"Flying architect," 76.
Mittelholzer's flight to (1924–1925), 3291.
Peru:
aviation history, 1833, 4251, 4270.
Johnson's photographs, 4251.
Santiago de Cardenas (1762) and Pedro Ruiz (1878), Books, by, 1869.
Pesaro, Francesco:
rocket propelled balloon, 163.
Pescara helicopter (1924), 2560.
"Peter Wilkins" (fiction), 68, 93.
Petersen, Carl W.:
sectional airship, 1209.
Peterson, P. H.:
airship, 1238.
Pétin, Ernest:
airships, 803, 1108, 1159a.
Petroczy-Karman, helicopter, 2558, 2560.
Petroleum prospecting, 5174, 5541.
Pettigrew, James Bell:
animal flight studies, 5204, 5277, 5285, 5286.
bibliography, 5286.
Pfaal's (Hans) voyage to the moon, 97.
Phaeton (poem). *See* Ovid.
Phillips, Horatio F.:
air current studies, 4792, 4994.
biography, 1804.
experiments discussed by Sir Hiram Maxim, 1950.
portrait, 1804.
steam-driven airplane, 175, 1914b, 1958, 2052, 2111, 4973.
wing forms, 1914b.
Phillips, John Scott:
ornithopter, 2050, 2051.
Phillips, W. H.:
helicopter, 380, 396, 2586.
references to him, 253 (p. 432).
Photography:
Alps, geological survey, 5106a.
bibliography, 5089, 5099, 5134.
Doughty's experiences, 5111.
Eddy's experiences, 5112.
general, 245, 247, 255, 5087, 5110, 5135, 5161, 5163.
kites, 5112, 5164, 5171.
Massachusetts, view, 5129.
night work, 5173.
pigeons, 5102.
progress, *1932,* 189; *1934,* 167.
Schelcher's pioneer work, 507.
stream lines, 4934.
submarine, 5405.
telephotography, 5095.
United States government, 5087.
Woodbury's aerial camera, 3725.
Photography, *continued*
World war, 3831, 3984, 5120.
See also Fairchild; Johnson; Mittelholzer; Motion pictures; Surveying and mapping; Tournachon.
Physiology. *See* Medicine.
Piccard, Auguste:
stratosphere flights, 656, 3119, 3303, 4899, 4962, 4976, 4990.
Pichault:
flight theory, 4977.
Pigeons:
ancient use, 76.
bibliography, 5408.
Boeve's book, 5421.
Franco-Prussian war, 3619, 3621, 4358.
general, 5479.
Neubronner's camera, 5102.
progress, *1890,* 4382; *1892,* 5553.
World war, 4050.
Pilâtre de Rozier, François:
Arlandes, Marquis de, experiences, 423, 509.
balloon ascensions, 208, 253 (p. 432), 407, 428, 562, 564, 570, 697b–698, 712, 724, 788, 803.
biography, 803, 843a.
Cavallo's account of first ascension, 500.
death:
accounts of, 543, 821.
pictures, 803 (p. 193), 821 (p. 95).
English Channel balloon, 530.
Giroude de Villette, letter, 601.
King of France, letter to, 764.
portraits, 803, 821.
tomb, 803 (p. 181).
Pilcher, Percy Sinclair:
biography, 159, 2421.
death (1899), 2392.
gliders, 175, 372, 380, 2282, 2436, 2439a, 2455, 2456, 2465, 2471.
portrait, 2421.
Pillet, Louis:
helicopter, 2600.
propeller researches, 2880.
Pilot balloons. *See* Balloons, pilot.
Pilots:
American list (1913), 3402.
list (1911), 1914a.
old compared with new, 2998.
photographs (21 year period), 3377.
statistics U. S. Navy and Army (1921–1929), 3532.
Pinedo, Francesco de:
Australia and Orient flight (1925), 3368, 3370.
career, 3418.
portraits, 3369, 3371.
transatlantic flights, 3367, 3369, 3371.
"Pipe" engine, 2708, 2728.
Pitcairn, Harold E.:
autogiro, 2556, 2592.
receives Collier trophy (1930), 2549, 2563.
Pitter, Joseph:
Archimedean, airship (1847), 1241, 1242.
Pixton, C. Howard:
pioneer of British aviation, 2056.
Platt, H. H.:
sail-wheel flying machine, 2865.
Platte, August:
bibliography, 4981.
flight theory, 4964, 4981.
helicopter suggestion, 2595.
Plesman, A.:
Director Royal Air Traffic Co. of Holland, 4319.
Plüschow, Gunther:
World war aviator, 4128.
Plywood, 2174, 5480.
"PN" seaplanes, 210, 3231, 3470.
Pocock, George:
kite-driven road carriage (1854), 2627.

Poe, Edgar Allan:
balloon hoax, 96.
See also Author index.

Poeschel, Johannes:
balloon ascensions, 766.

Poetry, 35, 42, 44, 45, 51–53, 69, 82, 94, 99, 103, 330, 373 (p. 523, 524), 498, 538, 727, 979a, 1912.

Poitevin, Madame:
biography and portrait, 5371.

Polando, John:
flight, New York-Turkey (1931), 3200.

Polar flights:
balloon possibilities, *1875*, 4481; *1879*, 4491; *1904*, 4480.
balloons for signalling, 4449.
bibliography, 4444.
British Antarctic Expedition (1904), 4498.
Coxwell's proposal, 4493.
general, 372, 1159b, 4465, 4470, 4484, 4490, 4501, 4511, 4522.
International Society, 4582.
legal aspects, 4602, 4630.
proposed French expeditions, *1872*, 4452; *1897*, 4463, 4505.
suitability of aircraft, 4466.
See also Amundsen; Andrée; Balchen; British Arctic Air Route Expedition; Byrd; Cripe; Eielson; *Graf Zeppelin*, airship; Macmillan (Donald); Nobile; Oxford University Arctic Expedition; Wellman; Wilkins (Sir G. H.).

Police, aerial, 2567, 5394, 5432, 5439, 5521, 5544.

Police regulations:
France (1784), 641.

Polte, Willy:
German aviator, 1942.

Pomortzeff, Michael:
scientific balloon ascensions, 4692.
upper air investigations, 4982.

Pompéien-Piraud, J. C.:
bird flight studies, ornithopter, airship, and airplane, 2061, 2062.

Ponton d'Amécourt, Gustave:
biography, 2569.
engine, 2063.
helicopter, 2063, 2569, 2586a.

Popper, Josef:
discusses theories of Von Lössl, 4983.
physics of balloon ascensions, 4984.

"Port Aviation," aerodrome near Paris, 4218.

Porter, Rufus:
airship, 116, 991, 1154.

Porto Rico:
aerial survey, 5160.

Portraits:
Kay's portraits (1837–1842), 663.
pioneers and fliers, 239, 291, 2056.

Portugal:
contributions to aeronautics, 286.

Post, Wiley:
portrait, 3300a.
world flight (1931), 213, 3200, 3222, 3331, 3336, 3374.
world solo flight (1933), 3336a, 3504.

Postage stamps, 188, 4355, 4359, 4363, 4368, 4371, 4379, 4410, 4424.

Potain, Doctor:
balloon flight over Irish sea, 253, 697b.
portrait, 803.

Praeger, Otto:
organizes U. S. air mail, 4400.
portrait, 4400.

Prandtl, Ludwig:
bibliography, 4988.
contributions, 4987, 5079.
organizes Aerodynamical Experiment Station, Göttingen, 4651.
wind tunnel, 4994.

Pratt and Whitney, fuel injection, 2822.

Prechtl, Johann Josef:
bird flight studies, 5287.
proposes copper balloon, 768.

Prévost, Maurice:
biography and portrait, 2973.

Priestly, Joseph:
hydrogen process, 565, 771.

Prince, Norman:
World war experiences, 4039.

Prints, engravings, etc., 3, 4, 6, 9–11, 16, 18–19, 21, 22, 29, 82, 89, 104, 115, 119, 170, 191, 192, 224, 261, 281, 300, 307, 373, 381, 449, 596, 642, 663, 675, 681, 803, 805, 821, 3860, 3892, 4150.

Prohaska, Franz:
parachute theory, 2519.

Propellers:
anchored tests, 2889.
antiquity, 373, 2871, 2881.
balloons, 429, 487, 522, 742.
bibliography, 2831, 2832, 2845.
Camus's book on design, 2837.
chronology, 2044.
first used by Robert brothers (1784), 2876.
general history, 295, 2044, 2048, 2836, 2838, 2855, 2860–2862, 2897, 4781.
Great Britain, research, 2900.
lifting screws, history, 2830, 2868, 2892.
metal propellers, 2869, 2875.
duralumin, 2879, 2891.
magnesium-aluminium, 2846.
Reed, 2887a.
multiple:
Hammond, 2851, 2878.
Kimball, 2878.
See also Amans; Arson; Bradsky-Labrun; Crocé-Spinelli; Dallwitz-Wegner; Decazès; Deprez; Drzewiecki; Faure; Gerlach; Godard (Louis); Haenlein; Hansen; Hauvel; Hervé; Hoernes; Kadarz; Klassohn; Kress; Laboratoire d'essais de Conservatoire national; La Vaulx; Maxim (Sir Hiram); Meusnier; Moy; Pillet; Rateau; Renard (C.); Riabouschinsky; Sauvage; Soreau; Tastes; Tatin; Trouvé; Valer; Veyrin; Vinci; Vogt; Walker (W. G.); Wellner; Werner (A.); Woodward; Wright brothers.

Prophecies:
airplanes, early opinions, 1704, 2014.
anonymous, *1836*, 395; *1887*, 410.
Baden-Powell (1907), 137, 138.
balloon, doom of (1909), 1813.
balloon direction (1784), 544.
balloon possibilities (1884), 489.
Baroni Cavalcabo (1753), 4685.
Biblical, 81.
Bingham's reply to Melville (1901), 1520.
Chanute (1904), 173.
Esterno (1864), 5212.
forecast (1908), 121.
future service to science (1784), 754.
Krauss (1901), 276.
Langley (1892), 1873, 1874.
Leconte (1894), 1890, 1891.
Melville (1901), 1961, 2068.
Myers' reply to Melville (1902), 1988.
Newcomb (1901), 318, 2009, 2010, 2068, 2148.
romances, 95.
Thurston (1890), 364.
Zahm, 396.

Pulitzer Trophy Races, 275, 3056, 3083, 3201, 3349, 3352, 3379, 3496, 3506.

Q

Quartermain, W.:
ornithopter, 1914d.
Quentin:
flying machine, 1993.
Question mark, airplane, 3080, 3199, 3276a, 3497.
Quimby, Harriet:
biography, 5374a.
death, 5357.
English Channel flight, 5323, 5358, 5370.
portraits, 5323, 5374a.
Quimby, W. F.:
ornithopter, 1686.

R

R-31, airship, 1175.
R-33, airship, 1253, 1365.
R-34, airship, 210, 1078, 1159b, 1160, 1277, 1324, 1358, 1359, 1363, 3060.
R-38, airship, 1159b, 1254.
R-100, airship, 1215, 1255, 1256, 3060, 3199.
R-101, airship, 1077, 1116, 1149, 1159b, 1257, 1258, 1327, 1388.
Races, 371, 1914a, 3451.
See also Collier Trophy; Daily Mail Gliding Competition; Gordon Bennett Races; Hyatt (Thaddeus); International Aviation Meet; Michelin Prize; National Air Races; National Air Tour; National Balloon Race; Pulitzer Races; Schneider Cup Races; Tour of Europe; Women's National Air Derby.
Racing circuits, 3512.
Radio:
beacon development, 5428, 5448, 5449, 5458, 5535, 5536, 5548, 5569.
bibliography, 5409, 5448.
broadcasting, 5464.
Byrd's airplane *America,* 5398.
civil aviation (1925), 4312.
first message to ground, 5485.
history, 5423, 5456, 5538.
National balloon race (1923), 5565.
United States airways, 5395.
United States Standards Bureau, 5397.
Vallot's early experiments, 5558.
weather information, 4963, 4992.
World war, 4011.
See also Radio control.
Radio control:
Alexander (1901), 4661.
Marçay (1923), 1440.
Raffard:
flying machine, 1892.
Raiche, Bessica:
America's first flying sportswoman, 213, 5319.
Railway (inclined) with balloon as motive force:
Fontaine, 745.
Payn, 440.
Railways:
competition with aviation (1925), 4210.
railway-aviation services (U. S. 1929), 4299.
Rain-making, 315, 4959.
Rasselas (fiction), 70, 106.
Rateau, A. C. E.:
propeller theories, 2883.
wind tunnel, 4994.
Ravaud seaplane, 2349.
Rawdon, Herbert, and W. E. Burnham:
Mystery S, airplane, 1480.
Rayleigh, John Wm. Strutt, 3rd baron:
bird flight theory, 5288, 5289.
flight theory, 4995.
Read, Albert C. *See* "N. C." seaplanes
Records:
Aircraft yearbook (1919–1934), 2913a
balloons (1803–1904), 604.
bibliography, 2960.
chronology, 3012, 3224, 3284.
Fédération aéronautique (1906–1923), 3074.
measurement methods, 5438.
years, *1905–1930,* 200; *1906–1923,* 3074; *1906–1926,* 249; *1906–1927,* 2969; *1907–1912,* 3521; *1907–1914,* 3520; *1918–1935,* 358.
Recouly, Raymond:
Syrian desert flight (1922), 3386.
Red wing, airplane, 1612.
Reed, Sylvanus Albert:
invents metal propeller, 2274, 2887a.
Refueling flights:
events (1923–1930), 3389.
Hunter brothers (1930), 3199, 3323.
Jackson and O'Brine (1929), 2918, 3199, 3239, 3412.
Mendell and Reinhart (1929), 3338, 3481.
Robbins and Kelly (1929), 3426.
Smith and Richter (1925), 2926.
U. S. Army (1929), 3080, 3276a, 3497.
Regibus, Francesco Maria:
balloon ascensions (1783–1784), 662.
Reichenbach, Oscar:
bird flight studies, 5290.
Reichenbach's engine, 5307.
Reinhart, Roland:
refueling flight (1929), 3338, 3481.
Reissner, H.:
airplane stability theory, 1564, 4997.
Renard, Charles:
air resistance studies, 4994, 4998–5000.
airplane theory, 2087, 2088.
airship studies, 1265, 1266.
appreciation by Surcouf, 5049.
aviation conferences, 5002.
balloons:
sounding, 5001.
variable volumes, 778.
bibliography, 5001.
biography, 549, 1199, 4791, 5003.
critical speed, 1270.
fins for airships, 1268.
La France, airship, electrical equipment, 1267.
helicopter investigation, 2599.
hydrogen process, 554.
motor, rotary, 2791.
portraits, 4791, 5049.
propeller, 2884, 2885.
See also Renard and Krebs airship.
Renard, Paul:
biography and portrait, 4695.
Renard and Krebs airship *La France,* 176, 253 (p. 433), 396, 410, 549 (p. 265–283), 889, 931, 935, 936, 947, 1051, 1159c, 1189, 1247, 1250, 1264, 1267, 1269, 1271, 1275, 1347, 3575, 3592.
electric equipment, 1267.
Soreau's studies on, 1323, 5039.
"Renault" engine, 2694, 2698, 2708, 2728.
Renoir helicopter, 2586a, 2601.
République, airship, 1159c.
Research:
bibliography, 4701.
history, 200.
See also Laboratories.
Reynolds, Osborne:
scientific contributions, 5079.
Rheims meeting (1909), 1914a, 2912, 2931, 3035, 3332, 3393, 3394.
bibliography, 3394.
Rhone glider competition, 2398, 2412.
Ribeiro de Souza:
airship, 1343.

Ricardo, Harry R.:
research on internal combustion engine, 2793.
Rice, A. Hamilton. *See* Hamilton Rice Scientific Expedition.
Richards, George Tilghman:
biography and portrait, 2056.
Richardson, H. C.:
biography and portrait, 3775.
Richthofen, Manfred von:
memorial volume, 4136.
plan to rescue body from enemy, 4044.
portrait, 3300a.
World war exploits, 4076, 4128.
Rickenbacker, Edward V.:
biography and portrait, 3956.
exploits, 3832, 4077.
fighting the "Flying circus," 212.
Rieckert, Herman A. J.:
airship, 1112.
Riedinger, Auguste:
biography, 427.
invents kite balloon, 578.
portrait, 578.
Rindine, Professor:
proposes to obtain hydrogen from the upper air, 1443.
Rio de Janeiro:
air survey, 5159a.
Ritchell, Professor:
airship, 207, 1394.
Rittenhouse, David, and Francis Hopkinson:
reputed balloon ascension (1783), 649.
Rivera y Sempre, Manuel:
airship, 1280.
Robbins, Reginald L.:
refueling flight (1929), 3426.
Robert brothers:
balloon ascensions, 406, 499 (p. 95–107), 505, 556, 752, 779, 788, 803, 810, 821, 828, 839.
first to use aerial propeller (1784), 2876.
Roberts, E. M.:
World war experiences, 4078.
Robertson, Étienne Gaspard:
balloon ascensions, 697a, 748, 780, 4666.
Minerva, balloon, 697a, 781.
relations with Charles, 780.
Sacharoff's report to St. Petersburg Academy of Sciences, 791.
tomb, 813.
Robertson, Eugène:
balloon ascensions:
New York City, 408.
various, 782.
Robertson, Sir Macpherson:
sponsors London-Melbourne race (1934), 3075, 3194.
Robida, Henri:
flight, Indo-China-Paris (1932), 3490.
Rocket propulsion:
Abate, 160.
Berliner, 1489.
bibliography, 1592.
Chatley, 1592.
Esnault-Pelterie, 5459, 5460.
Forlanini, 290.
general history, 1837, 2102, 5459, 5460, 5500, 5531.
Goddard, 5473.
Moore, 180.
Nye, 878.
Opel, 1779, 1780, 2029.
Pesaro, 163.
Rodeck, George:
torpedo balloons, 3772.
Rodgers, Calbraith P.:
biography, 3479.
trans-American flight (1911), 3092, 3110, 3462, 3463, 3479.
Rodgers, John:
airplane flight to Hawaiian islands (1925), 210, 3027, 3028, 3231, 3397.
John Rodgers airport, 4285.
Roe, Sir Alliot Verdon:
bibliography, 2104.
biography, 1470, 1522, 2056.
first to fly in England (1907), 1925.
invents tractor airplane, 1924.
portraits, 1470, 1924, 2056.
reminiscences, 2103, 2104.
See also Avro airplane.
Rogers, Will:
flying experiences, 5440.
Rohlf, Roland:
altitude flight (1919), 3398.
Rolls, Charles S.:
airplane flights, 223, 3437.
bibliography, 1497.
biography, 157, 1772, 3008.
Roma, airship, 958, 1143, 1159c, 1276, 1284.
Romaine, M.:
accompanies Pilâtre de Rozier on last ascension, 698.
Romas, Jacques de:
review of his work on kites (1776), 2607, 2669a.
Rome:
history, 480.
legends, 110.
"Romino" silencer, 1843.
Roosevelt, Quentin:
military aviator, 3974.
Roosevelt, Theodore, 3739.
Roosevelt field:
runway, 3401.
Roshon, J. W.:
airplane, 1532.
Rossi, Maurice:
New York-Syria flight (1933), 3336a.
Rotch, A. Lawrence:
bibliography, 5015.
biography, 157, 5077.
Blue Hill Observatory, 2613, 2633, 2634, 2636, 2637, 2648, 2660, 2665, 5009.
meteorological observations, 346, 5008–5015, 5052.
portrait, 157.
Roussilhe, H.:
paper on surveying, 5153.
Routes:
atlas, 4202.
bibliography, 4182.
Cairo-Karachi, 4220.
Deruluft, 4231.
English Channel, 4215.
France-Buenos Aires, 4230.
Great Britain, marine lines, 4311.
history, 4202.
Imperial Airways, 167, 4245, 4298.
Key West-Havana line, 4254.
Luft-Hansa, 4232.
New York-Cleveland line, 4402.
St. Petersburg-Tampa line, 4164, 4214.
South America, 4179, 4180, 4270.
United States, 4255, 4306.
world routes, 167.
Roux, P.:
steam-driven ornithopter, 2106.
Rowland, May:
airman's hymn, 5540.
Royal Aero Club, 4583.
Royal Aeronautical Society:
annual reports, 4578.
exhibition (1929), 5549.
history, 4524, 4533, 4550.

Royal Air Force:
Arabian desert, survey, 5106.
Baring, Maurice, 3826.
bibliography, 3560, 3838.
Bott, Alan, 3849.
Brancker, Sir Sefton, 3935.
Cairo-Cape Town flight (1927), 3413.
Cape Town flight (1926), 3378, 3380, 3381, 3407, 3842–3844, 3852.
Clifford, G. R., 3870.
Ellington, Sir E., 3927.
Far East flight (1927–1928), 3003, 3090, 3158, 3408, 3409.
Henderson, D., 3928.
history, 309, 3678, 3680, 3737, 3738, 3842–3844, 3852, 3853, 3862, 3887, 3917, 3921, 3922, 3933, 3934, 4006, 4061, 4065, 4092, 4095, 4144, 4147, 4156, 4158.
Kerr, M., 3970.
McCudden, J. T. B., 3991.
pigeons, World war, 4050.
pioneer work, 167.
Roberts, E. M., 4078.
Salmond, Sir G., 3936.
Squadron 47, 3965.
Squadron 60, 4085.
Squadron 99, 4056.
Trenchard, Sir H., 3931.
World war, African experiences, 3494.
See also Royal Naval Air Force.
Royal Aircraft Factory, 2107, 2749, 3837, 4741.
Royal Condor Co., 1156.
Royal Dutch Indian Airways, 4267.
Royal Mounted Aerial Police, Canada, 5521.
Royal Naval Air Force:
bibliography, 3839.
history, 3932, 3970, 4157.
North Sea station, 3908.
Royal Society of London:
glass ball experiments (1667), 5041.
Royal Vauxhall balloon. *See Nassau,* balloon.
Royce, Sir Henry:
biography and portrait, 2803.
Roze:
airship, 1006, 1016, 1288, 1289, 4954.
Rudder, vertical:
Mouillard, 1782.
Rue, Monsieur de, pseud. *See* Ferber, Louis Ferdinand.
Ruiz, Pedro:
writings, 1869.
Ruiz de Alda, J.:
flight, Palos-Buenos Aires (1926), 3124, 3128.
Rumania:
civil aviation (1922), 4256.
Rumpler, E.:
airplane, 1778.
biography, 291.
engine, 2708.
Russell, Frank H.:
associate of Wright brothers and G. H. Curtiss, 1452.
Russia:
air policy (post-war), 3732.
airlines, 4351.
airplane deficiency in World war, 4152.
Central Aéro-Hydro-Dynamic Institute, 4972.
civil aviation, post-war, 4304, 4341.
Deruluft air line, 4231.
five year aviation program, 4322.
history, 343, 344, 373 (p. 520).
Institut aérodynamique de Koutchino, 4804, 4895.
military airplanes, post-war, 3522.
progress, *1884,* 132; *1917–1921,* 1471; *post-war,* 309; *1934,* 133.
See also Military aeronautics, Russia; Russo-Japanese war.
Russo-Japanese war, 3623, 3724.
Ryaboushinski, Dmitri:
aerodynamics, 5016–5020.
bibliography, 5020.
Institut aérodynamique de Koutchino, 4804, 4895.
propeller researches, 2886, 2887.
wind tunnel, 4994, 5017.
Rykatchew, Michael A.:
scientific balloon ascensions, 4801.

S

Saccadura, Captain:
Spain-Brazil flight (1922), 3486.
Sacharoff, Akademicus:
ascension in Russia with Robertson, 748, 791.
Sadler, James:
balloon ascensions, 253 (p. 434), 447, 697b, 807, 847.
Sadler, William Windham, 253 (p. 434), 697b.
Safety:
airplanes in life-saving, 5572.
airships, Paul Renard's suggestions, 1273.
bibliography, 5417.
Brancker's ideas, 4191.
engine-dropping airplane, 2057.
Guggenheim Safety Competition, 1559, 1618, 2180, 3050.
ice-plane patrol, 5571.
Lepinte's parachute airplane, 2168.
progress, *1920,* 2255; *1927,* 5474; *1931,* 2101; *1932,* 178.
See also Coast patrol.
Sage, Mrs. L. A.:
balloonist, 253 (p. 434).
Sahara desert:
ballast regulator for use there, 517.
bibliography, 2952.
Deburaux's proposed balloon voyage, 519–521, 528, 737.
Dibos' project, 521, 528.
Mittelholzer's flight (1930–1931), 3292.
proposed commercial service (1910), 4258.
St. Cyr, France, aeronautical laboratory:
history, 4779.
St. Joseph of Copertino, Italian "Flying monk," 50.
St. Louis, balloon, 680.
St. Louis robin, airplane, 2918, 3198, 3199, 3239, 3412.
Saint-Loup:
air resistance studies, 5021.
St. Martin, Henri Reda:
airplane model (1867), 2113.
St. Petersburg-Tampa air line, 4164, 4214.
Sala, George Augustus:
parachute experience (1851), 2520.
Saladini, Girolamo:
balloon calculations (c. 1800), 794, 5022.
Salmond, Sir Geoffrey:
chief of British Air Force, 3936.
Salmond, Sir John:
marshall of British Air Force, 3930, 3937.
Samson, Charles Rumney:
biography, 3926.
commands flight, Cairo-Cape Town (1927), 3413.
Mediterranean seaplane flights (1921–1922), 2354.
World war experiences, 4080.
Samuelson, Arnold:
aerodynamics, 5024.
air resistance studies, 5023.
bibliography, 5024.
bird flight studies, 5292, 5293.
fish movements, 5292.
ornithopter, 2120.
San Diego, California:
airport, 4194.

Sander. *See* Opel.
Sanderval, de:
gliding experiments, 2121, 2457, 2458.
Sanson, A. J.:
airship, 5025.
Santiago de Cardenas:
writings (1762), 1869.
Santo Domingo:
naval aeronautics, 3644.
See also Hispaniola.
Santos-Dumont, Alberto:
airplanes, 372, 1449, 1914a, 2008, 2044, 2085, 2122, 2123.
airships, 257, 263, 860, 873, 939, 990, 1024, 1095, 1096, 1103, 1152, 1159c, 1211, 1299, 1301–1303, 5039.
autobiography, 1297.
bibliography, 1297.
biography, 157, 223, 372, 1093.
death, 1232, 1298.
Deutsch prize, 1198, 1300.
helicopter, 2598.
portraits, 860, 1449.
Saracen, The unknown, 56, 58, 74–76.
Sassoon, Sir Philip A. G. D.:
flight to India (1928), 3414, 3425.
Sauvage, Frédéric:
propeller patent (1836), 2867.
Savage, J. C.:
skywriting invention, 5495.
Scadta Company, South America, 4179.
Scandinavia:
legends, 43, 64.
See also Denmark; Norway; Sweden.
Scaroni, Silvio:
World war experiences, 4081.
Schelcher, André:
balloonist and pioneer in aerial photography, 507.
Schellbach, K. H.:
air resistance study (1871), 5027.
Schio. *See* Almerico da Schio.
Schlee, Edward F.:
Brock-Schlee flight to Japan (1927), 3079.
Schlick, Hans C. von:
biography and portrait, 291.
Schneider, Franz:
biography and portrait, 291.
Schneider, Jacques:
altitude flight in balloon *Icare,* 464.
bibliography, 2954.
Schneider trophy contests:
Biard's recollections, 2949.
bibliography, 2954.
events, *1919,* 3168; *1923,* 3169, 3416, 3458, 3484; *1925,* 2990, 3145; *1927,* 2318, 3131, 3170, 3422, 4090; *1929,* 2343, 2984, 3198, 3263, 3350, 3420; *1931,* 3200, 3350.
history, 2990, 3161, 3202, 3203, 3299, 3448, 3459, 3496.
Schofield's recollections, 4084.
winners (1913–1929), 3421.
Schnepf, John:
flying machine with revolving sails, 2130.
Schofield, Harry Methuen:
World war aviator, 4084.
Schott, Gasparus, 68, 240, 292a.
Schroeder, R. W.:
altitude flights, 3195, 3280, 3337, 3423, 3424, 4931.
Schütte-Lanz airships:
bibliography, 913.
craft, 1228.
history of firm, 1413.
portrait of Johann Schütte, 1228.
Schwarz, C.:
aluminium airship, 867, 1084, 1161, 1164, 1187, 1190, 1243, 1322, 1330.
engine, 1187.
Schweikert the miller, 108.
Science and aeronautics:
astronomy, 4914.
early research, 372.
general, 208, 245, 247, 294, 297, 4662, 4671, 4677, 4729, 4825, 4846, 4852, 4864.
geography, 4691, 4734.
geology, 5117, 5157, 5174, 5177.
physics, 4873, 5079.
World war period, 4680.
See also Aerodynamics; Air resistance; Atmosphere; Ballooning, scientific; Balloons, sounding; Eclipses; Instruments; International Commission for Scientific Aeronautics; Laboratories; Medicine; Meteorology; Research; Surveying and mapping; Theory of flight; Wind; Wind tunnels.
Science Museum, London:
description of aeronautical exhibits, 287, 288.
Wright airplane, 1925.
Scientific American (periodical), 304, 1838, 2135, 3115, 3138, 5545.
Scotland, 373 (p. 520).
Scott, Baron:
airship (1789), 1159d.
Scott, Blanche Stewart:
aviator, 3464.
Scott, C. W. A.:
appreciation, 3288.
autobiography and portrait, 3425a.
flights to Australia, 3288, 3425a, 3428.
Scrapbooks. *See* Clippings.
Seadrome Ocean Dock Corporation, 4307.
Seadromes, 3020, 4241, 4244, 4307.
Seagrave, Henry:
aviator and engineer, 3193.
Sealing fleet, 5435, 5436.
Seaplanes:
bibliography, 2312.
British navy (1924), 3415.
Chilean navy (1924), 3415.
first all-metal hull (1923), 2363.
first commercial service (1919), 2327.
floats, early history, 2351.
French navy, 3915, 3957.
German contests (1926), 2314, 2356, 3136.
Great Britain, World war, 4030, 4087.
history, 2308, 2311, 2323, 2324, 2335, 2337, 2338, 2345, 2347, 2353, 2357, 2359, 2362, 4087, 4311.
Janin versus Curtiss patent suit, 2306, 2340.
military use (1911–1926), 2319.
Monaco race meet (1912), 3103.
racing type, history, 2315, 2341.
research, history, 2330.
technical development, post-war, 2347a.
testing. *See* Tanks, seaplane.
World war, 3992a.
See also Archdeacon; Barton; Curtiss; Dornier; Euler; Fabre; Farman (H.); Forlanini; Guidoni; Kress; Marine Experimental Aircraft Depot; Ravaud; Schneider trophy contests; Short brothers; Sikorsky; U. S. Navy; Voisin; Wright brothers.
Sée, Alexandre:
bird flight studies, 5294.
Seguin, Louis:
biography, 3016.
"Gnome" engine, 2684.
portraits, 2684, 3016.
Seguin, Marc:
biography and portrait, 5055.
Selfridge, Thomas E.:
pioneer American aviator, 3015.
Selle de Beauchamp, de:
hydrogen apparatus (circa 1795), 3770.
Serrati, Serafino:
airship proposed (1787), 979.
hydrogen manufacture (1787), 979.

Settle, T. G. W.:
stratosphere flight (1933), 656.
Seux, Edmond:
airplane, 2140.
airplane stability, 2141.
biography and portrait, 4786.
Severo, Auguste:
bibliography, 907.
biography, 900, 1321.
Pax, airship, 900, 934, 939, 983, 1017, 1025, 1295, 1316, 1317, 1321, 1357.
portrait, 983.
Sham air attacks:
London (1928), 3671.
New York City (1929), 3757.
United States (1921), 3564.
Shaughnessy, E. H.:
in charge of U. S. air mail, 4372.
Shaw, Thomas L.:
man-propelled airship (1863), 1313.
Sheba, Queen of, 72.
Sheldon, William:
hot-air engine (1850), 2798.
Shenandoah, airship, 210, 275, 944, 1135, 1201, 1225, 1285, 1318–1320, 1324, 1341, 1364, 1384, 1396, 1397.
Shill, R. E.: associate of Thomas Moy, 1914b.
Ship-shore flying. *See* Launching.
Short, Eustace:
British airplane manufacturer, 1622.
Short brothers:
seaplanes, 2056, 2328, 2333.
Shrigley, Alfred R.:
biography and portrait, 157.
Sibour, Violette de:
London-Far East flight (1929), 5375.
Siegert, D.:
biography and portrait, 291.
Signalling:
balloon signalling for Polar work, 4449.
bibliography, 5418.
captive balloons, 3575.
electric, 3573.
See also Lighting of airways; Radio.
Sikorsky, Igor Ivan:
bibliography, 1503.
biographies and portraits, 1614, 1818, 1819, 2274.
seaplane, 2358.
Silberer, Herbert:
balloon ascensions, 801.
Silberer, Victor:
bibliography, 156.
biography and portrait, 581.
Silberschlag, Essias:
bird flight studies, 5295.
Silver dart, airplane, 1612, 1698.
Simon, René:
aviation experiences, 3068.
Simon the magician, 56.
Sippe, S. V.:
pioneer in British aviation, 2056.
Sivel, Th.:
biography, 4694.
portrait, 803 (p. 441).
scientific balloon ascensions, 816, 4667, 4668, 4766.
tomb, 813.
Sky-writing, 5430, 5493, 5495, 5507, 5530.
Slang, 5514.
Slotted wings, 274, 1655, 1758.
Smeaton, John: wind tunnel, 4994.
Smith, Elinor:
endurance record (1929), 5338.
Smith, Ernest L.:
portrait, 3079.
Smith-Bronte flight to Hawaii (1927), 3079, 3431, 3445.
tribulations, 3441.
Smith, Robert A.:
temple hunting in Central America, 3442.
Smith, Ross:
airplane flight, London-Australia (1919), 257, 3246, 3443, 3444.
portrait, 3300a, 3443.
Smithsonian Institution:
aeronautical services, 4585.
bibliography of publications, 1485, 4720.
See also Wright-Langley controversy.
Smyth, W.:
bird flight studies, 5296.
Snodgrass, R. E.:
studies of insect flight, 5297.
Société française de navigation aérienne:
work of members, 129, 4546.
See also Jamin, Jules.
Somerville, William:
airplane builder, 3464.
Sopwith, T. O. M.:
airplanes, 2154.
bibliography, 1515.
biography, 223, 2056, 2178.
portraits, 2056, 2178.
Soreau, Rodolphe:
bibliography, 2158.
biography and portrait, 4696.
bird flight, 5298.
flight theory, 2156–2158, 5037, 5298.
history of aviation, 5038.
motors, 1323.
propeller theory, 2890.
resistance studies, 1323.
Soucek, Apollo:
world altitude record (1929), 3250, 3446, 3447.
South Africa:
air mail, 4419.
civil aviation (1934–1935), 4213.
South America. *See* Air mail, South America; Andes mountains; Civil aeronautics, South America; Hamilton Rice Scientific Expedition; Johnson (George), Peruvian views. Also the following flights: Balbo; Coutinho; Ferrarin; Franco; Jiminez; Larre-Borges; Mollison (J. A.); Pan American flight.
Souther, Henry:
standardization work for U. S. Army, 3587.
Southern Cross, airplane, 3059, 3060, 3112, 3141, 3143, 3199, 3227, 3229, 3447a, 3495.
Spain:
aviation progress (1924–1925), 1477.
bibliography, 11.
book by Fuente de la Peña (1677), 1577.
civil aviation, post-war, 4341.
first scientific ascension (1795), 4971.
military aeronautics, history, 3788.
military ballooning (1792), 3662.
post-war developments, 309.
Spanish-American war:
balloons, 3632, 3683.
Delaney's experiences, 3591.
Spark plugs:
bibliography, 2695.
Speed:
bibliography, 1512, 4893.
chronology, 2231.
determination, 981, 4893.
developments, *1932,* 189; *1933,* 1441.
high-speed racing, 3492, 3493.
Kress, relation of speed to flight, 4897.
limitations, 1464, 2231.
See also Records.
Speed indicators:
bibliography, 1513.
Spelterini, Edward:
balloon ascensions, 143, 411, 438, 441, 628.
biography and portrait, 411.

Spencer, Charles:
glider-ornithopter, 1541, 1914d.
Spencer, Stanley:
high balloon ascension (1898), 749.
Spencer brothers:
airship, 1339, 1386.
See also Spencer, Stanley.
Sperry, Elmer Ambrose:
airway beacon, 5458.
automatic pilot, 1752, 2160–2162.
bibliography, 4709.
biography, 5400.
Fritz medal, 5458.
"Spiess" airship, 1048.
Spirit of St. Louis, airplane, 210, 238, 257, 1757, 1968, 2163.
Spooner, Winifred:
English aviator, 5341.
Near-East flight, 5379.
Spruce production, 1797, 4114, 4119, 4125.
Squier, George O.:
chief of aviation, U. S. Army, 3828, 3987, 3988.
Stability:
bibliography, 4721.
See also Airships, stability; Airplanes, stability and control.
Stack, T. Neville:
light-airplane flight to India (1926), 3452.
Stainforth, G. H.:
flies 400 miles an hour (1931), 3122, 3303.
Stamps. *See* Postage stamps.
Stanley, Charles:
airship, 1014.
Stanton, T. E.:
wind tunnel, 4994.
Stark, Theodore A.:
man-driven airplane, 1896.
Starr, John:
experiences in U. S. Weather Bureau, 4835.
Starters. *See* Engines, starters.
Steam as a balloon gas, 414.
Steam propulsion:
history, 2752.
See also Ader; Besler; Brozus; Crocé-Spinelli; Dow; DuTemple; Forlanini; Giffard; Goupil; Hargrave; Henson; Hollands; Kargl; Koch; Langley; Leinberger; Maxim (Sir H.); Parsons; Partridge; Perkins; Phillips (H.); Reichenbach's engine; Roux; Stringfellow.
Steel, Tom:
proposes transatlantic flight (1843), 404.
Steel construction:
Bausett, airship, 867, 1369.
Boyman, airship, 1159a.
Budd, stainless steel airplane, 1678.
Fokker, welded steel tubes, 1689.
Unger, airship, 956.
Steinmetz, Charles P.:
organizes first American glider club (1894), 2464.
Stella, Fr.:
balloon studies, 162.
Stenzel, Arthur:
carbonic acid gas motor, 2804.
ornithopter, 180, 1887.
Stevens, A. Leo:
airship, 1140, 1333.
biography, 157, 670.
portrait, 670.
Stevens, Albert W.:
discovers source of the Parima river, 4734.
stratosphere flight, 666, 5045.
Stieger, H. J.:
develops "monospar," 1979.
Stinson, Edward A.:
duration flight (1921), 3220.
wins National Air Tour (1927), 3129.
Stinson, Marjorie:
experiences at flying school, 5381.
Stocks:
United States, 5452, 5453, 5496, 5547.
"wildcats," 5433.
Stott, Ralph:
flying hoax (1786), 1644.
Stout, William B.:
biography and portrait, 1973, 2274.
relations with Henry Ford, 1999.
Stout Air Lines, 4315.
Stout Metal Airplane Co.:
sold to Henry Ford (1925), 1783.
Strait, H.:
balloon direction (1834), 1335.
Strasser, Peter:
airship attacks on England, 3983, 4108.
portrait, 3907.
Stratosphere flights:
discovery of stratosphere, 4932.
Gray (1927), 3281.
Kepner and Stevens (1934), 666, 5045.
Piccard (1931–1932), 656, 3119, 4899, 4962, 4976, 4990.
Russia, 133, 656, 790.
Settle and Fordney (1933), 656.
U. S. Army balloon *Explorer* (1934), 741.
See also Teisserenc de Bort.
Stream lines:
bibliography, 1514.
photography, 4934.
Stringfellow, John:
airplane, 380, 1541, 1620, 1914b, 1927, 1941, 2031.
appreciation, 1719.
bibliography, 1504.
references, 253 (p. 435).
Struvé:
ornithopter, 1914d.
Studensky, Paul:
biography and portrait, 3464.
Stultz, Wilmer:
transatlantic flight (1928), 3060, 3107.
Submarine aircraft carriers. *See* Aircraft carriers.
Submarine observations, 5405.
Submarines versus aircraft in warfare, 3895, 3962, 4087, 4089, 4148, 4153.
Subsidies:
discontinuance in Europe, 4222.
effects (1924), 4333.
Europe (1931), 4332.
U. S. air mail contracts, 4332.
world chart (1921–1930), 200.
Süring, Reinhard:
balloon ascension with A. Berson (1900), 4693, 4951.
Sueter, Murray F.:
founder of Royal Navy Air Force, 3932.
Sun's effect on balloons, 775.
"Sunbeam" engine, 2701.
Supercharging. *See* Engines, supercharging.
Surcouf, Édouard:
biography, 574.
control of aircraft at sea, 1338.
Deutsch de la Meurthe airship, 336.
Surveying and mapping:
Aerial Survey Co., Ltd., 5109.
Africa, 5121, 5131, 5169.
Alaska, 5092, 5093, 5098, 5150, 5157a, 5172.
Alps, 5106a.
Arabia, 5106.
Australia, 5085.
bibliography, 5090, 5091, 5117, 5119, 5123, 5137, 5138, 5156, 5166.

Surveying and mapping, *continued*
Borneo, 3383.
British possessions, 167.
Brock process, 5096.
Burma, 5084, 5105.
Cambridge University, 5125.
Canada, 5100, 5113, 5139, 5142, 5143, 5158, 5159, 5176.
Colorado oil shales, 5177.
Death Valley, 5108.
economic development, use in, 5094.
Fairchild Aerial Surveys, Inc., 5152, 5154.
Fairchild Camera Corporation, 5114.
Finsterwalder's methods, 5115, 5116.
Florida, 5152.
general, 5122, 5123, 5127, 5132, 5137, 5138.
geology, uses in. *See* Geology.
Griffith's contributions, 5162.
Hawaii, 5101.
Hispaniola, 5146.
history, 5124, 5135, 5151, 5170.
India, 3383, 5088, 5133, 5139, 5144.
International Commission for Scientific Maps, 4544.
Lomet's paper, 5136.
marks, international, 5130.
Mexico, 5154.
Michigan, 5083a.
military maps, 5136.
Mississippi river delta, 5165.
New York City, 5114, 5140.
night work, 5145.
Porto Rico, 5160.
progress, *1924,* 5103, 5104; *1927,* 5118; *1929,* 5113.
Rhodesia, 5121.
Rio de Janeiro, 5159a.
Roussilhe's paper, 5153.
Syracuse, N. Y., 5148.
timber surveys, 5176.
United States Coast and Geodetic Survey, 5086, 5165.
United States government, 5087.
United States Hydrographic Office, 4768, 5167.
Svenske, balloon, 550, 551.
Sweden:
aviation history, 1700.
civil aviation, 4192, 4313.
See also Scandinavia.
Swedenborg, Emanuel:
flying machine, 1849, 2175.
Switzerland:
civil aviation, 4266.
Sykes, Sir F. H.:
pioneer in British aviation, 2056.
Sykes, H. M.:
bird flight studies, 5300.
Syracuse, N. Y.:
aerial survey, 5148.
Szalay, Emil:
New York-Hungary flight (1931), 3200.

T

Tabuteau, Maurice:
aviator, 3513.
Tanager, airplane, 1559, 2180, 3050.
Tanks, seaplane:
progress (1931), 274.
William Froude National Experiment Tank, 2309.
Tarbell, Gage E.:
biography and portrait, 157.
Tarnowski, J.:
flying machine, 2181.
Tasso:
verses, 373 (p. 19–22).
Tastes, de:
vibrating disk propeller (1871), 2893.
Tate, William J.:
recollections of the Wright brothers, 2245.
Tatin, Victor:
airplane compared with Langley's, 1886.
airplane models, 1696, 1699, 1914b, 2005, 2031, 2183, 2185, 2186, 2188.
bibliography, 2187.
bird flight, 5264, 5301, 5302.
experiments in Marey laboratory, 2184.
propeller experiments, 2894.
La Ville de Paris, airship, 1021.
"Taube" airplane. *See* Etrich, Igo.
Teisserenc de Bort, L.:
bibliography, 4712.
biography, 4957.
portrait, 4932.
stratosphere investigations, 4932, 5004.
Templer, J.:
military balloon experiences, 3592, 3693, 3762.
See also Woolwich Arsenal.
Tennant, J. E.:
experiences on Turkish front, World war, 4096.
Tests. *See* Airplanes, tests.
Testu-Brissy, Pierre:
ascensions, 415, 803.
ascent with horse, 373 (p. 91), 803.
Thaden, Louise:
biography and portrait, 5369.
Women's Air Derby, 5382.
Thayer, David:
combination kite-balloon, 2664.
Thayer, Russell:
airships, 1064, 1344, 1345.
jet propulsion, 1345, 3763.
Theory of flight:
history, 292, 333, 1904, 4977.
See also Aerodynamics; Air resistance; Alexander (A.); Boltzmann; Congreve; Deprez; Duroy de Bruignac; Ferber; Fitzgerald; Goupil; Hastings; Helmholtz; Holland; Hooke; Hurry; Kreiss; Lancaster; Lestani; Lucy; Maceroni; Martienssen; Mora; Moretti; Platte; Rayleigh; Veyrin, Weyher.
Thomas, G. Holt:
British pioneer in civil aviation, 2056, 4277.
Thomas, Herbert:
British aviation pioneer, 2056.
Thomas, Lowell Jackson:
European airplane tour (1927), 3466.
Thomas, Pierre:
air resistance studies (1870), 5053.
Thomas, Royal V.:
flight into Grand Canyon of the Colorado (1922), 3217.
Thomassen à Thuessink van der Hoop, A. N. J.:
Amsterdam-Batavia flight (1924), 3468.
Thomson, Lord, of Cardigan:
biography, 1081, 3835.
killed on airship *R-101,* 1149.
portrait, 3835.
Thunder storms. *See* Balloons, thunder storms.
Tible, Madame:
first woman to make a balloon ascension, 5342.
Timber surveys, Canada, 5176.
Times Aerial Derby around Manhattan (1913) won by W. S. Luckey, 3066, 3113.
Tingmissartoq, airplane, 3498.
Tire, largest (1923), 1880.
Tissandier, Albert:
biography, 703, 902, 1940.
Tissandier, Gaston:
biography, 722, 802.
portrait, 803 (p. 457).
scientific ascensions, 822, 4668, 4824.

Tissandier brothers:
airship, 305, 549 (p. 164–171), 935, 936, 1159c, 1247, 1346, 1348–1351, 1353.
bibliography, 367.
electric motors, 2814, 2815.
hydrogen process, 554, 751.
See also Tissandier, Albert; Tissandier, Gaston.
Toledo, airship, 1158.
Tombs of aeronauts:
Godard (Louis), 3656.
Père-Lachaise cemetery, 813.
Pilâtre de Rozier, 803.
Tomlinson, D. W.:
World war experiences, 4102.
Torpedoplane, 4103.
Torres, Leonardo:
airships, 1117, 1177, 1251, 1355.
stability studies, 1356.
Touche, P. E.:
air resistance studies, 5056.
Tour of Europe race (1930), 3173.
Tournachon, Félix (Nadar, pseud.):
aerial photography, 5095, 5107.
bibliography, 2199.
biography, 157.
comments on French Academy, 2199.
Le Géant, balloon, 697a, 3773.
obituary, 683.
portrait, 2044.
prints and manuscripts, 3, 9.
references, 253 (p. 231).
Toys, 451, 823, 1756.
Tradition. *See* Literature and tradition.
Training:
France, 4146, 5427.
United States, 2164, 4013, 4051.
Trans-American flights, airplane:
civilian races (1927), 3314.
Gatlin, first woman, 5352.
records (1932), 3200a.
records broken (1930), 3199.
reliability tests (1920), 1680, 3111.
See also Coney; Doolittle; Goebel; Hawks; Kelly; Macready; Maughan; Rogers (C. P.); Tucker; Turner (R.).
Trans-American flights, airships:
Shenandoah, 210.
Transatlantic flights:
bibliography, 2961.
history, 3060.
list (1873–1929), 3477.
See also Graf Zeppelin; Los Angeles; R–34; R–101; names of aviators.
Transatlantic flights, airplane:
bibliography, 2961.
chronology and history, 2946, 3032, 3060, 3300a.
early history, 3283.
flights of 1931, 3328.
list, *1927,* 3121; *to 1929,* 3477; *to 1931,* 2927.
meteorology, 4839, 4887, 4888, 4993.
tabular data (1919–1929), 3475.
See also Alcock; Balbo; Bellonte; *Bremen;* Brown; Byrd; Challe; Chamberlin; Coli; Connor; Costes; Coutinho-Saccadura; Del Prete; Earhart; Elder; Ferrarin; Fonck; Gordon; Haldeman; Hawker; Hill; Iglesais; Jiminez; Kerr; Kingsford-Smith; Larre-Borges; Le Brix; Lindbergh; Mackenzie-Grieve; Mollison; *NC* seaplanes; Nungesser; Pinedo; Stulz; Williams (R. Q.); Yancey; *Yellow bird.*
Transatlantic flights, balloon:
Green's proposal, 409, 609, 697a, 718.
history, with bibliography, 531.
Jovis's proposal, 410.
references, 373 (p. 521).
Steel's proposal, 404.
Transatlantic flights, balloon, *continued*
Wise's proposal, 588, 798.
See also Hoaxes.
Trans-Australian flights:
Brinsmead (1925), 2908.
Transcontinental Air Transport, Inc.:
Lindbergh's relation to, 3009.
Travel air, airplane, 2940.
Treaty-making by air:
Canadian Indians, 5520.
Trenchard, Sir Hugh:
chief of British Air Force, 3931.
Trouvé, Gustave:
biography, 2849.
helicopter, 2586a.
ornithopter, 176, 1914d, 2202.
propeller, 2849.
Tschudi, Georg J. F. von:
autobiography, 1362.
portrait, 4570.
Tubing. *See* Metal construction, metal tubes.
Tucker, Harry:
Los Angeles-New York flight (1928), 3341.
Turbiaux, E.:
experiences, Franco-Prussian war, 3774.
Turbini, Gaspare:
airship, 920.
Turnbull, W. R.:
airplane stability studies, 2203, 2204.
Turner, Charles Cyril:
biography, 282.
reminiscences, 371.
Turner, Roscoe:
Trans-American flight, 3336a.
Twining, H. La V.:
ornithopter, 1824, 2076.
Tytler, James:
balloon ascensions, 697b.
biography, 713.
portraits, 663, 713.
references, 713 (p. 435).
Tziolkovski, K.:
metal airship, 1366.

U

Udet, Ernst:
German aviator, 291, 1942.
Uncles:
eagle-drawn airship, 1159d.
Unge, Eric:
balloon construction, 831.
balloon *Svenske,* 550, 551.
Unger, Gustave:
steel airship, 956.
United Air Lines, 4253.
United States:
airships:
evolution, 1285.
manufacture during World war, 1223.
progress, *1916–1919,* 1105; *1933,* 923.
ballooning, early, 622.
Blanchard's ascensions, 595, 623, 721.
chronology, 1526.
early references, 373 (p. 520).
heavier-than-air craft, history, 1917, 2164.
industries:
exports, 1456, 1457, 5471.
general, 308, 376, 1445, 5402.
manufacturing companies, histories, 1455, 5424.
progress, *1909–1928,* 1444; *1909–1930,* 190; *1914–1925,* 2093; *1918–1919,* 3828; *1919–1929,* 1635; *1925–1931,* 1473; *1927,* 2019; *1928,* 1536, 2036; *1928–1931,* 2073; *1928–1932,* 2248; *1929,* 5454; *1930,* 2125; *1931,* 164.
progress, *1911,* 1429; *1922,* 2211; *1927,* 1902, 2214; *1930,* 1058.

United States, *continued*
Rittenhouse and Hopkinson's reputed ascension, 649.
Robertson (Eugene), ascensions, 783.
slow development, 309, 366.
See also Civil war, American; Military aeronautics, United States; Spanish American war; also the various departments under the heading United States.
United States. — Aeronautical Board:
reorganized (1924), 3525.
United States. — Agriculture Department:
airplane as an aid, 4828, 5570.
boll weevil campaign, 5552.
crop estimation, 4828, 5568.
United States. — Army:
accident statistics, 5555, 5556.
Air Corps, history, 3650, 4093.
Air Service, annual report, 3778.
airship specifications (1907), 3537, 3731.
Alaska expeditions, *1920,* 2915, 3460; *1929,* 3198.
Allen, James, career, 3585.
archaeological work, 5126.
aviation, early history, 3715.
Barling bomber, 1717.
Central American flight (1924), 2942.
five year expansion plan, 3610.
Gray's altitude flight. *See* Gray.
Hawaiian flight. *See* Hawaii.
medicine, 4688.
night surveying and photography, 5145, 5173.
parachute, large (1928), 2532.
Pursuit group flight to Canada, using skiis, 3315.
Schroeder's altitude flight. *See* Schroeder.
Souther, Henry, career, 3587.
Squier, G. O., career, 3987, 3988.
Stevens' stratosphere flight. *See* Stevens (A. W.).
World flight (1924), 210, 257, 2982, 2987–2989, 3086, 3360, 3405, 3467, 5533.
Wright brothers' contract (1907), 2165.
See also Military aeronautics, United States; Mitchell, William; *Question mark,* airplane; *R–31,* airship; *Roma,* airship; Warship bombings.
United States. — Coast and Geodetic Survey:
charts, 5086.
Mississippi river delta survey, 5165.
United States. — Commerce Department: Aeronautics Branch:
Aeronautic Development Service, 4188.
broadcasting of weather data, 4963.
civil aviation statistics, 4200.
history, 4306.
reorganization (1933), 4163.
United States. — Forest Products Laboratory:
record of achievements, 5059.
United States. — National Advisory Committee for Aeronautics:
annual reports, 5058.
bibliography of publications, 26.
portraits and biographies, 308, 5396.
United States. — Naval Aircraft Factory, 3871.
United States. — Navy:
Aeronautics Branch, annual reports, 3776.
Alaska air survey, 5150, 5172.
Baldwin (T. S.), contract for airship, 988.
eclipse observations (1923), 5081.
hangars, 1409.
history (1908–1928), 3711.
Hydrographic Office, 4768, 5167.
operations (1911–1919), 4122.
progress, *1913,* 3584; *1918–1923,* 3706; *1924,* 2350; *1925–1930,* 3707; *1928,* 3710; *1929,* 3708; *1930,* 3709; *1933–1934,* 3663; *post-war,* 3728, 4012.
R–38, airship disaster, 1159b, 1254.
Shenandoah, airship, 210, 275, 944, 1135, 1159b, 1318.
torpedo and bombing planes (1917–1933), 3606.
United States — Navy, *continued*
transatlantic flight (1919), 210, 257, 3047, 3060, 3087, 3148, 3196, 3395, 3474.
West Indian flights, *1920,* 3242; *1924,* 3470.
ZMC–2, metal-clad airship, 984, 1026, 1054.
ZR–1, airship, 1041, 1421.
See also Aircraft carriers; Hawaiian islands; Naval aeronautics, United States; World war.
United States. — Signal Office:
Allen, James, career, 3585.
annual reports, 3779, 3780.
Bennet, J. B., career, 4086.
bibliography, 3561.
Chandler, Charles de F., career, 3872.
Hazen, H. A., career, 4682.
meteorological ascensions, 4682, 4847, 4853.
See also Civil war, American; Spanish-American war.
United States. — Standards Bureau:
engine performance at high altitudes, 2743.
engine testing, 2707.
radio, 5397.
services, 2164, 4731, 4733, 4767, 5060.
United States. — Weather Bureau:
kite work, 2638, 2640, 2654, 2656.
Meisinger, C. L., 4758.
Starr, J., 4835.
Universal Aerial Navigation Co., 859.
Université de Paris:
aeronautical course, 5511.
catalogue of the library, 23.
University of Toronto, 4966.
Unwins, Cyril:
airplane altitude flight (1932), 2983, 4961.
Urlwin, J.:
claims airplane model the first to use gasoline motor (1910), 2213.
Ursinus, Hugo:
pioneer builder of bombing airplanes, 1476.
Usuelli and Crespi:
balloon flight over Alps (1906), 607.

V

Valer, Paul:
adapts Antoine's formula to aerial propellers, 2895.
Vallot, Joseph:
radio experiments, 5558.
Van Hecke:
vertical propellers for raising balloons, 429, 742.
Van Musschenbroek, Peter:
kite theory (1762), 2607, 2663.
Van Orman, W. T.:
lightning investigations, 4672.
wins National Balloon races, 739, 740, 835, 836.
Van Paulsen, C. C.:
life-saving airplane, 2264.
Van Reinveld, H. A.:
London-Cape of Good Hope flight (1920), 3265.
Vaniman, Melvin:
airship of constant displacement, 974.
biography, 982.
death (1912), 1159c.
Védrines, Jules:
early French aviator, 2955, 3019.
Velez, Miguel:
airship, 1370.
Veneziani, Giacomo:
balloon ascensions (1783–1784), 662.
Veranzio, Fausto:
parachute, 163, 2490, 2502a, 2527.
Verville, Alfred V.:
builder of fast airplanes, 1565, 1848.
Veyrin, E.:
propeller, 2896.
theory of flight, 2218.

Vickers airships and airplanes, 306, 1625.
Victoria cross, 3829.
Vidal, Eugene L.:
  heads U. S. Aeronautics Branch, 4163.
Vienna exhibition (1888), 1454.
Villard, Henri:
  helicopter, 2219.
La Ville de Paris, airship, 899, 1249.
La Ville de Paris, balloon, 803.
Vinci, Leonardo de:
  air-screw, 2848, 2881.
  bird flight, 5219, 5265.
  investigations, 56, 85, 196, 292a, 295, 379, 4740, 4777, 4814–4816, 4850, 4851, 4869, 4872, 4933, 5006, 5063, 5179.
  manuscripts, history, 4911.
  parachute, 2490, 2502a, 2848.
Viscosity in air support, 2217.
"Vivinus" engine, 2708, 2728.
Vogt, H. C.:
  propeller researches, 2897, 2898, 4994.
Voisin, Charles:
  biography and portrait, 1937.
Voisin, Gabriel:
  biography and portraits, 157, 1937, 2222.
Voisin airplane, 2084, 2362.
  bibliography, 1517.
Volta, balloon, 244, 652.
Voyer:
  ballonets, 840, 841.
  balloon theories, 551.
  long distance balloon equipment, 841.
Vuia:
  airplane, 2153.

W

Wagner, Christopher:
  reputed exploits in India and China, 110.
Wakefield, Sir Charles:
  English patron of aviation, 2147.
Walker, Thomas:
  ornithopter, 380, 1914d, 2229.
  references about him, 253 (p. 436).
Walker, William George:
  lifting propeller, 2899.
Walpole, Horace:
  references about him, 253 (p. 436).
Wanamaker, Rodman:
  American patron of aviation, 1469.
Warm air balloons. *See* Balloons, warm air.
Warner, Edward P.:
  American investigator and writer, 4708.
Warping. *See* Wings, warping.
Warshavsky, A. W. H.:
  ornithopter, 1606.
Warship bombings, experimental:
  United States, 3529, 3565, 3566, 3699, 3700, 3718, 3781–3784.
Warthausen. *See* König-Warthausen.
Washington, George:
  occasion of Blanchard's ascension in Philadelphia, 595, 721.
Water cooling. *See* Engines, water-cooling.
Watkins, H. E.:
  pioneer of English aviation, 2056.
Watres act, 4354.
Wechmar, Ernst von:
  man-operated flying machine, 1454.
Weczera, Anton:
  flying machine model, 1556.
Weed, Frank:
  motion pictures from airplane, 5562.
Weir, J. G.:
  British pioneer of aviation, 2056.
Weiss, José:
  biography and portrait, 2056.
  gliders, 2471.
Weisse, H.:
  bird flight studies, 5302a, 5303.
Wellman, Walter:
  airship *America,* 1159c, 4503, 4510, 4513–4516.
  bibliography, 4514.
  biography, 213, 4503.
  Polar expedition, 4457, 4504, 4510, 4513, 4514.
  portraits, 4503, 4513.
Wellner, Georg:
  air resistance studies, 5068, 5070.
  airplane stability, 2237.
  bibliography, 2238.
  biography, 2859, 2882.
  disk experiments, 5070.
  flight theories, 5068.
  propellers, 2858, 2861, 2862, 2902–2904.
  sail-wheel flying machine, 486, 843, 1230, 1391, 1533, 1993, 2235, 2236, 2238.
Wells, D. D.:
  claims to have warped wings in 1897, 2128.
Wells, Franz. *See* Etrich, Igo.
Wells, Linton:
  world flight (1926), 3499.
Wenham, Francis Herbert:
  airplane theories, 2089, 2239.
  bibliography, 2242.
  biography, 2239.
  biplane, 1580.
  gliders, 2439a.
  helicopter, 2586a.
  ornithopter, 380, 1914d.
  references about him, 253 (p. 436).
  wind tunnel, 4994.
Werner:
  bird flight, 5304.
Werner, A.:
  vane propeller (1882), 2905.
Werthner, W.:
  recollections of the Wright brothers, 2244.
Weyher, Ch.:
  bird and human flight, 2247, 5305.
Weyman, Charles:
  aviator, 3017.
"Whirlwind" engine. *See* "Wright" engine.
White, Bert:
  parachute experiences, 2530.
White, E. F.:
  non-stop flight, Chicago-New York (1919), 3104.
White wing, airplane, 1612, 3102.
Whitehead, Gustave:
  airplane, 1660, 1751, 2079, 2251.
Wichita, Kansas:
  leader in airplane manufacture, 2000.
Wieland's shirt of feathers, 64.
Wien, Noel:
  airplane flight Fairbanks, Alaska, to Cape North (1929), 4486.
Wild, Horace B.:
  airship operator, 157, 846, 3464.
Wilde, Henry:
  vibratory parachute, 2531.
Wilkins, Sir George Hubert:
  Polar expeditions, 212, 4459, 4492, 4499, 4500, 4517–4520.
Wilkins, John, bishop:
  flying chariot, 5075, 5076.
  writings, 110, 253 (p. 436), 499, 1701.
Willard, Charles F.:
  first aviator taught by Glenn Curtiss, 3505.
William Froude National Experiment Tank. *See* Tanks, seaplane.

Williams, Alford J.:
  "speed king," 3438.
Williams, Roger Q.:
  non-stop flight, Bermuda and return (1930), 3038, 3114.
  transatlantic flight (1929), 3198, 3308.
Willows, Ernest Thompson:
  airships, 1082, 1404.
  bibliography, 909.
  biography, 2056.
  death, 1080.
  portrait, 2056.
Wilson, Alexander:
  kite experiments (1749), 5009.
Wilson, E.:
  airplane experiments, 2261a.
Wilson Airway, Ltd.:
  operations in East Africa, 4293.
Wind:
  bibliography, 4723.
  Langley on internal work, 4906.
  Soreau, 5038, 5039.
  Zahm, 5083.
Wind tunnels:
  bibliography, 4724.
  developments (1931), 274.
  history, 4942, 4965, 4994.
  Marey, 4994.
  Maxim, 4994.
  Ryabushinski, 5017.
  U. S. National Advisory Committee, 2901.
  Wright brothers, 4994.
Windham, William:
  references about him, 253 (p. 436).
Wings:
  Bertelli, forms, 920.
  bibliography, 1495.
  cantilever type, with bibliography, 1966.
  Curtiss, movable tips, 1838.
  curves, history, 2299.
  Hargrave, forms, 1767, 1768.
  Lepère-Weymann metal sections, 2094.
  "M" sections, 1987.
  Merrill, forms, 1963.
  monospar, 1979.
  Oakes, curves, 2450.
  Phillips, forms, 1914b.
  trussing, history, 2042.
  warping:
    bibliography, 1519.
    Montgomery, 1893, 1981, 2382.
    Mouillard, 1782, 4939.
    Wells, 2128.
  See also Wright brothers.
Wings, flapping. *See* Ornithopters.
Wings for balloons:
  Ackermann, 527.
  bibliography, 765.
  Platte, 765.
  Wellner, 486.
Winnie Mae, airplane, 213, 3200, 3222, 3331, 3336, 3374.
Winter, Wilhelm:
  bird flight studies, 5306.
Wise, Hugh D.:
  kite experiments, 2673.
Wise, John:
  American civil war, 3721.
  balloon ascensions, 179, 263, 437a, 518, 639, 697a, 849, 850, 4852.
  balloon direction, 848.
  biography, 673.
  parachute experiences, 2490.
  portraits, 386, 673.
  proposes transatlantic voyage, 588, 798.
Witoszynski, C.:
  aerodynamics, 5016.
Wölfert:
  airships, 1047, 1202, 1213, 1407.
Wolff, Maximilian:
  airship, 929, 1089, 1408.
"Wolseley" engine, 2701.
Women in aeronautics:
  adaptability, 5335.
  aviators:
    American, 5387.
    British, 5344, 5363, 5389.
    lists, 5336, 5337.
  balloonists, early, 581a, 607a, 697b, 5330, 5331, 5350.
  biographies of noted pilots, 1526.
  British factories, World war, 5321.
  German aeronauts, 5364, 5365.
  history, 5308, 5345, 5350, 5378, 5390.
  "99" Club, 5317.
  records (1931), 3200.
  Woman's National Air Derby, 5367, 5382, 5388.
  See also Bailey; Bedford; Beinhorn; Blanchard (M.); Bruce (M.); Camac; Clark (J.); Crosson; Diard; Earhart; Elder; Etzdorf; Fisher; Fradiss; Garnerin (Elisa); Gatlin; Gavin; Graham (M.); Haizlip (M.); Heath; Hewlett; Ingalls (L.); Kaufmann (M.); Klumpke; Law; Löwenstein-Wertheim; Marvingt; Moissant (M.); Mollison (Amy); Nichols; Parker; Paulus; Peltier; Poitevin; Quimby (H.); Raiche; Sage; Scott (B.); Sibour; Smith (E.); Spooner; Stinson (M.); Thaden; Tible; Wright (K.).
Women's National Air Derby, 5367, 5382, 5388.
Wood, Alanson:
  airship, 865, 1354.
Wood, Henry A. Wise:
  engineer, 5400.
Wood, Walter B.:
  military aviator, 3861.
Woodbury, Walter:
  aerial camera (1878), 3725.
Woodhouse, Henry: writer, 5461.
Woodward, Calvin Milton:
  propeller discussion (1907), 2906.
Woolaroc, airplane, 3079, 3514.
Woolwich Arsenal:
  military balloon experiments, 487, 3617, 3690, 3744.
  See also Templer, J.
World flights:
  Blake's attempt (1922), 2967.
  Graf Zeppelin, airship, 1008, 1065, 1092, 1159b, 1285, 1402, 3198.
  Gronau (1932), 3200a, 3406, 3489.
  Herndon-Pangborn (1931), 3200.
  König-Warthausen (1929), 3235.
  McLaren's attempt (1924), 3274.
  Mears and Collyer (1928), 3285.
  Post (1933), 3336a, 3504.
  Post-Gatty (1931), 213, 3200, 3222, 3331, 3336, 3374.
  U. S. Army Air Service (1924), 210, 257, 2982, 2987–2989, 3086, 3273, 3360, 3405, 3467, 5533.
  Wells-Evans (1926), 3499.
World war:
  aeronautical development, 196, 222.
  aeronautics, effect of war on, 1543, 1668, 2030, 2171, 3905, 4058, 4079.
  airplane tactics, 4000, 4003, 4045, 4070, 4085a, 4160.
  airplanes, 2046, 2195, 3760a, 3815, 3825, 3828, 3830, 3850, 3856, 3868, 3890, 3891, 3906, 3924, 3950, 3964, 3969, 3981, 3986, 4005, 4028, 4058, 4131, 4134, 4149.
  airships, 3760a, 3815, 3881, 3897, 3909, 3950, 3998, 4083, 4088.
  Australian Flying Corps, 3876.

World war, *continued*
aviators: general lists, 3964, 4159.
Allies, 3152, 3885, 4018–4020, 4023–4026.
American, 1526, 4093, 4127.
Canadian, 3884, 4027.
French, 3892, 4036.
German, 3152, 3885, 3976, 3993, 4034, 4128.
Italian, 3821.
Jews, 4098.
New England, 4035.
See also Adams; Albret; Allsop; Annunzio; Ball; Benn; Biard; Biddle; Bishop, W. A.; Boelcke; Bott; Buddecke; Castle; Clifford; Coppens de Houthulst; Delacommune; Escadrille "210"; Eschwege; Farré; Fischer; Fonck; Gran; Greene, W.; Guynemer; Hall, B.; Haslett; Henkelburg; Heydemarck; Hild; Immelmann; Ingalls; Kingsford; Koehl; Lafayette Escadrille; Lafonte; La Frégeolière; Luke; McConnell; McCudden; Macmillan; Navarre; Nicholl; Noble; Nordhoff; Pégoud; Plüschow; Prince; Richthofen; Rickenbacker; Roberts; Roosevelt; Samson; Scaroni; Schofield; Strasser; Tennant; Tomlinson; Wood.
balloons, captive, 3903.
bibliography, 5.
Canada, 229, 2096, 3884, 4091, 4139.
chronology, 144, 301, 4133.
engines, 4058.
equipment of various nations, 3875, 3953.
France:
air raids, Paris, 3814, 3864, 4060, 4100.
caricatures, 3963.
"Escadrille 210," 4097.
general, 3888, 3955, 4036, 4049, 4129, 4146.
manufacturing firms, 4036.
seaplanes, 3915, 3957.
training, 4146.
Germany:
aeronautical developments, 3880, 3886, 3904, 3906, 3908, 3910, 3912, 3950, 3951, 3984, 3989, 4003, 4033, 4083, 4088.
airplanes, 3807a, 3830, 3856, 3906, 3984, 4142.
airplanes, identification, 3958.
armament, 3985.
bombardment of German cities, 4637a.
"Den of the Zeppelins," 3904.
engines, 2721, 2740, 2796, 2797, 3950, 3984.
Gotha airplanes, 3986.
See also World war, aviators, Germany.
Great Britain:
air raids, 3814, 3822, 3929, 3983, 3996, 4014, 4108, 4637a.
airships, 3998.
chronology, 4047.
dominions, 3572.
effect of war on aviation, 1543.
Eighth Naval Squadron, 4031.
general, 3624, 3816, 3853, 3855, 3869, 3921, 3922, 4090, 4108.
naval operations, 3887.
operations in general, 4095.
seaplanes, 4030.
Tigris operations, 4040.
training, 3883, 3893.
women in factories, 5321.
hydrogen, 3952.
Italy, 3902, 3911, 3961, 4059, 4062a, 4063.
land forces, cooperation, 3889.
laws, aerial, effect of war on, 4618, 4620, 4635, 4638.
magnetos, 3945.
medical services, 3882.
observations, 4062.
operations, 148, 280, 3911, 4007, 4036, 4048, 4049, 4095.

World war, *continued*
photography, 5120.
Russia, 145, 4152.
seaplanes, 2319, 3961, 3992a, 4087, 4154.
submarines, 3895, 3962, 4087, 4089, 4148, 4153.
torpedoplane, 4103.
United States:
expenses, 3913, 4054, 4118.
first airplanes shipped, 3896.
first coastal airship patrol, 3897.
general, 3969, 4054, 4055, 4068.
Liberty loan helped by aeronautics, 4105.
Lost battalion, 4073.
manufactures, difficulties, 308.
naval aeronautics, 3878.
operations, 1526, 3923, 4010, 4055, 4115.
preparations, 3858, 4001.
supply system, 3850, 3969, 4114, 4123.
technical developments, 2164, 3973, 4052, 4068, 4101.
U. S. Army School, Cornell University, 3827.
U. S. Naval Aircraft Factory, 3871.
value of aeronautics, 4032, 4064, 4085a.
Wright, Howard T.:
English aviation pioneer, 2056.
Wright, Katherine:
assists the Wright brothers, 5313.
Wright, Orville:
biography, 157, 1914, 2142.
stability, paper on, 2277.
stabilizer, 2034, 2166.
Wright, Wilbur:
appreciations, 1547, 2254.
biography, 1545, 1914, 2142, 2271.
death (1912), 1646, 2253.
flights in France, 1548, 1868, 1964.
Hudson river flight (1909), 1484, 1912, 3519.
interview, 1579.
Wright brothers:
Ader claims priority, 2284.
assistants, early, 2026.
bibliography, 2280.
Brewer, Griffith, visit (1910), 1549.
celebrations, 1447, 1578, 1844, 2043.
chronology, 1551, 1775.
Curtiss airplane compared, 2065.
English criticism, 1682, 1693.
European flights, 1914a.
Fort Myers tests (1908), 1558, 1597, 1600, 1676, 2035, 2212, 2300.
general, 131, 140, 196, 257, 372, 380, 1424, 1451, 1474, 1545, 1576, 1599, 1609, 1621, 1633, 1658, 1709, 1715, 1759, 1790, 1805, 1812, 2002, 2044, 2170, 2205, 2250, 2276, 2279–2283, 2286–2289, 2295, 3657, 5071.
gliding experiments, 2378, 2439a.
Herring-Curtiss Co., lawsuit, 2291–2293.
juvenile works about them, 212, 223, 1589, 1599.
Langley medal (1910), 2296.
Martin stabilizer lawsuit, 1933.
Newton, B. R., recollections, 2012, 2013, 2015, 2016.
O'Malley, F. W., recollections, 2026.
patents, 1459, 1548, 1613, 1674, 1716, 1731, 1792, 1914c, 1990, 2024, 2198, 2232, 2291.
Paulhan-Wright lawsuit (1910), 2268, 2294.
predecessors, debt to, 2283.
propellers (1908), 2835.
Saunders, W. O., recollections, 2124.
seaplane, 2362.
Science Museum, London, exhibit, 1925.
scientific methods, 1643, 1888, 2025.
stabilization, 1465, 1916.
Tate, W. J., recollections, 2245.
U. S. Army contract, first, 2165, 3657, 3739.
war, use of airplane, 1812.
Werthner, W., recollections, 2244.
Westcott, R. L., recollections, 1913.

Wright brothers, *continued*
wind tunnel, 4994.
wing warping, 292.
Wright Company a menace, 1909.
See also Wright-Langley controversy.
"Wright" engine:
air cooling, 2716.
"E4" endurance tests, 2805.
early types, 2698, 2701, 2811.
"J" type, 2711.
"Superfighter," 2713.
"Whirlwind," 2731, 2741, 2762, 2825, 2826, 2907, 2943, 3006.
Wright field, 4345.
Wright-Langley controversy, 1423, 1546, 1631, 1994, 2129, 2207, 2210, 2227, 2278, 2290.

Y

Yale University:
bibliography, 5412.
first war unit, 4053.
Yancey, Lewis A.:
transatlantic airplane flight (1929), 3198, 3308.
Yellow bird, airplane:
transatlantic flight (1929), 3198, 3270, 3437a.
Yon, Gabriel:
bibliography, 456.
biography and portrait, 575.
captive balloons and hydrogen, 3557, 3637, 3768.
Young, Thomas:
kite theory (1807), 2607.
Yugoslavia:
air transport development, 4167.

Z

ZMC–2, airship, 984, 1026, 1054.
ZR–1, airship, 1041, 1421.
ZR–2, airship. *See R–38,* airship.
ZR–3, airship. *See Los Angeles,* airship.
Zacharia, August Wilhelm:
bird and fish movements, 5307.
Zahm, Albert Francis:
air friction experiments, 4991, 4994, 5082, 5083.
bibliography, 5083.
Zambeccari, Francesco:
balloon, 474, 481.
biography, 760.
control of balloon, 665.
first experiments in England, 499, 758.
flights, 668, 697a, 760.
interview, 668.
references, 253 (p. 436).
Zeise, H.:
airship (1850), 1412.
Zenith, balloon, 803, 4668.
Zens brothers:
airplane (1908), 2083.
Zeppelin, Ferdinand, Graf von:
American visit during Civil war, 213, 1286.
bibliography, 1070, 1418.
biography, 157, 263, 291, 372, 853, 1011, 1070, 1079, 1111, 1290, 1292, 1371–1373.
encouragement by German engineers, 879.
personality, 1009.
portraits, 853, 1009, 1228, 1371–1373.
"Zeppelin" airships:
comparison with former types, 936.
disasters, 1340, 1410, 3894.
general, 143, 148, 168, 257, 263, 336, 964–965, 995, 1004, 1009, 1012, 1035, 1038, 1040, 1104, 1119, 1126, 1139, 1146, 1159b, 1183, 1195, 1203, 1228, 1249, 1262, 1375, 1393, 1406, 1414, 1419, 3527.
Goodyear Company buys American rights, 1071.
World war, 1159b, 1228, 3813, 3907, 3908, 3910, 3912, 3919, 3967, 3976, 3994, 4108.
See also Graf Zeppelin, airship.
Zimmerman, Charles:
ornithopter, 2304.

CORRECTIONS (1971)

page 2, item 21: *change classmark to* VDT
page 3, item 39: *change classmark to* VDT
page 28, item 545, third line: *insert following pagination:* [these pages follow page 160]
page 182, item 4032: *change author to* Acworth, M. W. (Neon, pseud.)
page 182, item 4036: *change classmark to* VDYD
+
Mil
page 255: *insert* Acworth, M. W., 4032 *in index following* Acosta, Bertram B.